Rising Contagious Diseases

Rising Contagious Diseases

Basics, Management, and Treatments

Edited by

Seth Kwabena Amponsah
Department of Medical Pharmacology
University of Ghana Medical School
Accra, Ghana

Ranjita Shegokar
Chief Scientific Officer (CSO)
Capnomed GmbH, Germany

and

Yashwant V. Pathak
USF Health Taneja College of Pharmacy
University of South Florida
Tampa FL, USA
and
Faculty of Pharmacy
Airlangga University
Surabaya, Indonesia

WILEY

Published by John Wiley & Sons, Inc., Hoboken, New Jersey.
Published simultaneously in Canada.

Library of Congress Cataloging-in-Publication Data
Names: Amponsah, Seth Kwabena, editor. | Shegokar, Ranjita, editor. | Pathak, Yashwant, editor.
Title: Rising contagious diseases : basics, management, and treatments / edited by Seth Kwabena Amponsah, Ranjita Shegokar and Yashwant V. Pathak.
Description: Hoboken, New Jersey : Wiley, [2024] | Includes bibliographical references and index.
Identifiers: LCCN 2023039506 (print) | LCCN 2023039507 (ebook) | ISBN 9781394188710 (cloth) | ISBN 9781394188727 (adobe pdf) | ISBN 9781394188734 (epub)
Subjects: MESH: Communicable Diseases, Emerging–prevention & control | Communicable Disease Control–trends
Classification: LCC RA566 (print) | LCC RA566 (ebook) | NLM WA 110 | DDC 616.9/8–dc23/eng/20231211
LC record available at https://lccn.loc.gov/2023039506
LC ebook record available at https://lccn.loc.gov/2023039507

Cover Design: Wiley
Cover Image: © Baac3nes/Getty Images

Set in 9.5/12.5pt STIXTwoText by Straive, Pondicherry, India

Dedication

This book is dedicated to all health workers who continuously work hard to save the lives of many people with infectious diseases. Thank you for your dedication, commitment, and courage. Deepest gratitude and admiration to all scientists who have made contributions toward getting a better understanding of infectious pathogens and also to researchers who work tirelessly to develop new drugs and vaccines.

Additionally, this book is dedicated to all the rishis, sages, shamans, and medicine men and women of ancient traditions and cultures who have contributed to the development of drugs and nutraceuticals. They have kept health science alive for the past several millennia.

Contents

List of Contributors

Abdullah Abdelkawi
Taneja College of Pharmacy
University of South Florida
Tampa, FL
United States

Ismaila Adams
Department of Medical Pharmacology
University of Ghana Medical School
Accra
Ghana

Ofosua Adi-Dako
Department of Pharmaceutics and Microbiology
School of Pharmacy
University of Ghana
Accra
Ghana

Selorme Adukpo
Department of Pharmaceutics and Microbiology
School of Pharmacy
University of Ghana
Accra
Ghana

Alka
Amity Institute of Pharmacy, Lucknow
Amity University Uttar Pradesh
Noida, Uttar Pradesh
India

Seth K. Amponsah
Department of Medical Pharmacology
University of Ghana Medical School
Accra
Ghana

Abigail Aning
Department of Clinical Pathology
Noguchi Memorial Institute for Medical Research
University of Ghana
Accra
Ghana

Orhan E. Arslan
Department of Cellular Biology and Pharmacology
Florida International University Herbert Wertheim
College of Medicine
Miami, FL
United States

Arti Bagada
Department of Pharmaceutical Sciences
Faculty of Health Sciences
Saurashtra University
Rajkot, Gujarat
India

Shwetlana Bandyopadhyay
MGM College of Pharmacy
Chiraura, Patna
India

Praful D. Bharadia
Department of Pharmaceutics
L. M. College of Pharmacy
Ahmedabad, Gujarat
India

Sachin P. Bhatt
Gyanmanjari Pharmacy College
Gujarat Technological University
Bhavnagar, Gujarat
India

Eugene Boafo
Department of Medical Pharmacology
University of Ghana Medical School
Accra
Ghana

Chinmoyee Borah
Department of Pharmaceutical Sciences, Faculty of
Science and Engineering
Dibrugarh University
Dibrugarh, Assam
India

Kwasi A. Bugyei
Department of Medical Pharmacology
University of Ghana Medical School
Accra
Ghana

Ananya Chanda
Department of Pharmaceutical Technology
School of Medical Sciences
ADAMAS University
Kolkata, West Bengal
India

Chandrakanta
Department of Pharmaceutical Sciences
Faculty of Technology
Sir J. C. Bose Technical
Campus Bhimtal, Kumaun University Nainital
Nainital, Uttarakhand
India

Moitreyee Chattopadhyay
Department of Pharmaceutical Technology
Maulana Abul Kalam Azad University of Technology
Kolkata, West Bengal
India

Neha Chintapally
University of South Florida Morsani College of Medicine
Tampa, FL
United States

Mehul Chorawala
Department of Pharmacology and Pharmacy Practice
L. M. College of Pharmacy, Gujarat University
Ahmedabad, Gujarat
India

Ankit Choudhary
Department of Pharmaceutical Sciences
Faculty of Science and Engineering
Dibrugarh University
Dibrugarh, Assam
India

Carina Copley
College of Agriculture and Life Sciences
The University of Florida
Gainesville, FL
United States

Jeannez Daniel
University of South Florida Morsani College of Medicine
Tampa, FL
United States

Aparoop Das
Department of Pharmaceutical Sciences
Dibrugarh University
Dibrugarh, Assam
India

Debankini Dasgupta
Department of Pharmacology
MGM College of Pharmacy
Patna, Bihar
India

Sandipan Dasgupta
Department of Pharmaceutical Technology
Maulana Abul Kalam Azad University of Technology
Kolkata, West Bengal
India

Akua O. Forson
Department of Medical Laboratory Sciences, School of Biomedical and Allied Health Sciences
University of Ghana
Accra
Ghana

Mainak Ghosh
Department of Pharmaceutical Sciences
Faculty of Science and Engineering
Dibrugarh University
Dibrugarh, Assam
India

Urvashee Gogoi
Department of Pharmaceutical Sciences
Dibrugarh University
Dibrugarh, Assam
India

Bismarck A. Hottor
Department of Anatomy
University of Ghana Medical School
Accra
Ghana

Avani N. Joshi
Gyanmanjari Pharmacy College, Gujarat Technological University
Bhavnagar, Gujarat
India

Bibhuti B. Kakoti
Department of Pharmaceutical Sciences, Faculty of Science and Engineering
Dibrugarh University
Dibrugarh, Assam
India

Avinash D. Khadela
Department of Pharmacology
L. M. College of Pharmacy
Ahmedabad, Gujarat
India

Nirjari Kothari
Department of Pharmacology and Pharmacy Practice
L. M. College of Pharmacy, Gujarat University
Ahmedabad, Gujarat
India

Awo A. Kwapong
Department of Pharmaceutics and Microbiology, School of Pharmacy
University of Ghana
Accra
Ghana

Monali Lahiri
Department of Pharmaceutical Sciences, Faculty of Science and Engineering
Dibrugarh University
Dibrugarh, Assam
India

Anjali Mahapatra
University of South Florida Morsani College of Medicine
Tampa, FL
United States

Subhasundar Maji
Department of Pharmaceutical Technology
Maulana Abul Kalam Azad University of Technology
Kolkata, West Bengal
India

Rahul K. Maurya
Amity Institute of Pharmacy, Lucknow
Amity University Uttar Pradesh
Noida, Uttar Pradesh
India

Deepak Mishra
Amity Institute of Pharmacy, Lucknow
Amity University Uttar Pradesh
Noida, Uttar Pradesh
India

Badrud D. Mohammad
Department of Pharmaceutical Chemistry
GRT Institute of Pharmaceutical Education and Research
Tirutanni, Tamil Nadu
India

Lucy Mohapatra
Amity Institute of Pharmacy, Lucknow
Amity University Uttar Pradesh
Noida, Uttar Pradesh
India

Muhasina K. Muhamadkazim
Department of Pharmacognosy
JSS College of Pharmacy
JSS Academy of Higher Education & Research
Ooty, The Nilgiris, Tamil Nadu
India

Vishwas H. Nagendra
Department of Pharmacy Practice
JSS College of Pharmacy
JSS Academy of Higher Education & Research
Ooty, Tamil Nadu
India

Meera Nagpal
University of South Florida Morsani College of Medicine
Tampa, FL
United States

Lawandashisha Nongrang
Department of Pharmaceutical Sciences, Faculty of Science and Engineering
Dibrugarh University
Dibrugarh, Assam
India

Lynn Nguyen
University of South Florida Morsani College of Medicine
Tampa, FL
United States

Eric N. Y. Nyarko
Department of Chemical Pathology
University of Ghana Medical School
Accra
Ghana

Emmanuel K. Ofori
Department of Chemical Pathology
University of Ghana Medical School
Accra
Ghana

Verner N. Orish
Department of Microbiology and Immunology
School of Medicine
University of Health and Allied Sciences
Ho
Ghana

Vaishnavi M. Oza
Department of Pharmaceutics
L. M. College of Pharmacy
Ahmedabad, Gujarat
India

Aanshi Pandya
Department of Pharmacology and Pharmacy Practice
L. M. College of Pharmacy, Gujarat University
Ahmedabad, Gujarat
India

Sambit K. Parida
Seth Vishambhar Nath Institute of Pharmacy
Barabanki, Uttar Pradesh
India

Rihana B. Patnool
Department of Pharmacy Practice
JSS College of Pharmacy
JSS Academy of Higher Education & Research
Ooty, Tamil Nadu
India

Komal Parmar
Department of Pharmacy, ROFEL
Shri G.M. Bilakhia College of Pharmacy
Vapi, Gujarat
India

Anita Patel
R & D Department
Samrajya Aromatics Pvt. Ltd.
Gandhinagar, Gujarat
India

Dhara Patel
Pioneer Pharmacy College
Vadodara, Gujarat
India

Geeta Patel
Shree S K Patel College of Pharmaceutical
Education and Research
Ganpat University
Kherva, Gujarat
India

Grishma Patel
Pioneer Pharmacy College
Vadodara, Gujarat
India

Jayvadan K. Patel
Formulation and Development Aavis Pharmaceuticals
Hoschton, GA
United States
Aavis Pharmaceuticals, Hoschton, Georgia, USA
and
Faculty of Pharmacy, Sankalchand Patel University
Visnagar, Gujarat, India

Kshama Patel
Judy Genshaft Honors College
The University of South Florida
Tampa, FL
United States

Manish P. Patel
Department of Pharmaceutics
L. M. College of Pharmacy
Ahmedabad, Gujarat
India

Nisarg Patel
Department of Pharmacognosy
APMC College of Pharmaceutical
Education and Research
Himatnagar, Gujarat
India

Priya Patel
Department of Pharmaceutical Sciences
Faculty of Health Sciences
Saurashtra University
Rajkot, Gujarat
India

Vivek Patel
Sun Pharmaceutical Industries Ltd.
Vadodara, Gujarat
India

Kalyani Pathak
Department of Pharmaceutical Sciences
Dibrugarh University
Dibrugarh, Assam
India

Sivasankaran Ponnusankar
Department of Pharmacy Practice
JSS College of Pharmacy, JSS Academy of Higher Education & Research
Ooty, Tamil Nadu
India

Manash P. Pathak
Faculty of Pharmaceutical Sciences
Assam Down Town University
Guwahati, Assam
India

Yashwant V. Pathak
USF Health Taneja College of Pharmacy
University of South Florida
Tampa, FL
United States
Faculty of Pharmacy
Airlangga University
Surabaya
Indonesia

Bhupendra G. Prajapati
Department of Pharmaceutics and Pharmaceutical Technology
Shree S K Patel College of Pharmaceutical Education and Research, Ganpat University
Mehsana, Gujarat
India

Charles Preuss
Department of Molecular Pharmacology & Physiology
University of South Florida Morsani College of Medicine
Tampa, FL
United States

Jasmine Primus
Judy Genshaft Honors College
The University of South Florida
Tampa, FL
United States

Archana N. Sah
Department of Pharmaceutical Sciences
Faculty of Technology
Sir J. C. Bose Technical
Campus Bhimtal, Kumaun University Nainital
Nainital, Uttarakhand
India

Riya Saikia
Department of Pharmaceutical Sciences
Dibrugarh University
Dibrugarh, Assam
India

Aayushi Shah
Department of Pharmacology and Pharmacy Practice
L. M. College of Pharmacy, Gujarat University
Ahmedabad, Gujarat
India

Ishika Shah
Department of Pharmacology and Pharmacy Practice
L. M. College of Pharmacy, Gujarat University
Ahmedabad, Gujarat
India

Ranjita Shegokar
Capnomed GmbH
Tübingen
Germany

Aliyah Slim
Taneja College of Pharmacy
University of South Florida
Tampa, FL
United States

Hemangi B. Tanna
Department of Pharmaceutics
L. M. College of Pharmacy
Ahmedabad, Gujarat
India

Alok S. Tripathi
Department of Pharmacology
ERA College of Pharmacy, ERA University
Lucknow, Uttar Pradesh
India

Elizabeth Twum
Department of Pharmaceutical Sciences
College of Pharmacy, University of Tennessee Health Science Center
Memphis, TN
United States

Nasir Vadia
Department of Pharmaceutical Sciences
Faculty of Health Sciences
Marwadi University
Rajkot, Gujarat
India

Vernon Volante
University of South Florida Morsani College of Medicine
Tampa, FL
United States

Julia Wang
University of South Florida Morsani College of Medicine
Tampa, FL
United States

Mohammad Yasir
Amity Institute of Pharmacy, Lucknow
Amity University Uttar Pradesh
Noida, Uttar Pradesh
India

Zaineb Zinoune
Taneja College of Pharmacy
University of South Florida
Tampa, FL
United States

Foreword

Infectious diseases are known to cause morbidity and mortality in several populations. Despite significant progress made over the years, infectious diseases are still a threat to global health. A number of infectious agents have emerged and re-emerged in recent decades due to the dynamic host–pathogen interplay, niche adaptation, and increase in human mobility and density. Some of the infectious agents include Severe Acute Respiratory Syndrome (SARS) viruses, Zika virus, Middle East Respiratory Syndrome (MERS) viruses, and Ebola virus. Furthermore, in the last three years, the world has had to deal with the novel coronavirus disease 2019 (COVID-19), which has caused many deaths across the world. In all these, active and passive surveillance systems with prompt reporting and analysis of data are important in the early detection of emerging and re-emerging infectious diseases. The quest to promote ONE HEALTH by the World Health Organization is proving a lead and guide to global efforts in containing emerging and re-emerging infectious diseases.

After carefully going through the literature, I found that there are few books available on the market that compile the various emerging and re-emerging infectious diseases. This edited book has 30 chapters that highlight relevant aspects of infectious diseases the world has faced over the last few decades. Some of the emerging and re-emerging infectious diseases in the book include Ebola virus disease, bird flu, swine flu, COVID-19, mpox, Chikungunya fever, Chagas disease, among others. Renowned scientists and researchers in infectious diseases have made contributions to this book, with each chapter well written, concise, and easy-to-understand. I believe that this book will be a great resource for clinicians, scientists, researchers, and students all over the world.

I am very delighted to write the foreword for this book, *Rising Contagious Diseases: Basics, Management, and Treatments*, edited by Seth K. Amponsah, Ranjita Shegokar, and Yashwant V. Pathak. I want to congratulate the editors of this book for a great job done. To all contributing authors, I say kudos. I look forward to seeing this book on the market for the use of all relevant stakeholders and cannot wait to get a personal copy.

July 8, 2023

Prof. Edwin Alfred Yawson
Public Health Consultant and Dean, University of Ghana Medical School
Accra, Ghana

Preface

Infectious diseases such as COVID-19, Ebola virus disease, swine flu, and mpox have caused morbidity and mortality worldwide. Many of these infectious diseases have emerged and/or re-emerged in recent decades due to dynamic host–pathogen interactions, anthropogenic selection, and climate change. The aforementioned factors present major challenges to public health; hence, there is a need for appropriate preventive measures, cost-effective diagnostic procedures (assays), and effective therapeutic strategies. Additionally, real-time epidemiological surveillance can improve the detection of infectious disease outbreaks or new infections of public health importance.

Over the years, there has been growing interest in the field of emerging and re-emerging infectious diseases; however, there appear to be few books that cover recent trends. The current book gives a compilation of information on several emerging and re-emerging infectious diseases. Due to the fact that medicine is a fast-evolving field, it is important that information on current trends be documented. This book offers thorough, but succinct, details into many aspects of infectious diseases (transmission dynamics, epidemiology, clinical manifestation, advances in diagnostics, and management), such as COVID-19, mpox, Ebola virus disease, bird flu, swine flu, Zika virus, Chikungunya fever, Chagas disease, toxoplasmosis, and neurocysticercosis, among others.

Due to high priority for prevention, diagnosis, and treatment of infectious diseases, most of the primary audience (health care professionals, academic institutions, especially academicians who are working in the fields of microbiology, immunology, infectious diseases, pathology, pharmacology, and public health) and secondary audience (undergraduate and postgraduate students) will be interested in this book. The book contains 30 chapters with rich content, presenting fundamental facts, as well as practical and clinically related data.

Renowned scientists and researchers in the field of infectious diseases are the contributors to this book which highlights molecular surveillance and epidemiology (including newly characterized zoonotic pathogens or their variants), innovative strategies for pathogen detection, drug and vaccine development, and disease prevention and control. The editors believe that this book is timely and will meet the needs of both primary and secondary audience.

The editors appreciate the efforts of all contributors who shared their knowledge through chapters. The editors are also grateful to John Wiley & Sons, Inc. for facilitating all processes involved in getting this book published.

July 8, 2023

Seth Kwabena Amponsah
Accra, Ghana

Ranjita Shegokar
Tübingen, Germany

Yashwant V. Pathak
Tampa, FL, United States

Biographies

Seth Kwabena Amponsah is currently a senior lecturer and head of the Department of Medical Pharmacology, University of Ghana Medical School. He has an MPhil and Ph.D. in pharmacology. He has had post-doctoral fellowships under the BANGA-Africa Project and BSU III (DANIDA – Denmark). He has over 12 years' experience in teaching and research. He teaches students in the medical school, school of pharmacy, school of nursing and midwifery, and school of biomedical and allied health sciences. His research focus includes clinical pharmacology (infectious disease and antimicrobial stewardship): prudent use of antimicrobials, antimicrobial level monitoring, and efficacy of antimicrobials in patients. He also has experience in population pharmacokinetic modeling, non-compartment pharmacokinetic estimation, and pharmacokinetic evaluation of new drug formulations. He has supervised several undergraduate and postgraduate students. He has published over 50 research articles, 1 book, 10 book chapters, and several conference abstracts. He is an academic editor for *PLOS One* and an associate editor for *Pan African Medical Journal*.

Ranjita Shegokar holds a PhD degree in Pharmaceutical Technology from the SNDT University, India, and has been a postdoctoral researcher in the Department of Pharmaceutics, Biopharmaceutics and NutriCosmetics at the Free University of Berlin, Germany. For the last many years, she has been working with various multinational pharmaceutical companies in technical/R&D leadership roles. Currently, she serves as Chief Scientific Officer (CSO) at Capnopharm GmbH, Germany. She has authored several research articles and book chapters and presented her research at many national/international conferences. She has filed multiple patent applications in the areas of drug delivery and targeting. Besides that, she has edited many trending books in the area of pharmaceutical nanotechnology and drug delivery aspects. For her research, she has received many prestigious national and international awards, among them include recently received prestigious German Innovation Award 2022. Her areas of interest include polymeric nanoparticles, nanocrystals, lipid nanoparticles (SLNs/NLCs), nanoemulsions, cancer drug targeting, and the role of excipients in delivery systems.

Yashwant V. Pathak has over 15 years of versatile administrative experience in an institution of higher education as dean (and over 30 years as faculty and as a researcher in higher education after his PhD). Presently holds the position of associate dean for Faculty Affairs and Tenured Professor of Pharmaceutical Sciences. He is an internationally recognized scholar, researcher, and educator in the areas of health care education, nanotechnology, drug delivery systems, and nutraceuticals. He has received many international and national awards including four Fulbright Fellowships, Endeavour Executive Fellowship by the Australian Government, four outstanding faculty awards, and he was selected as Fellow of the American Association for Advancement of Science (AAAS) in 2021. He has published over 350 research publications, reviews, and chapters in various books. He has edited over 60 books in various fields including nanotechnology, nutraceuticals, conflict management, and cultural studies. He is also actively involved in many non-profit organizations, to mention a few, Hindu Swayamsevak Sangh, USA; Sewa International USA; International Accreditation Council for Dharma Schools and Colleges; International Commission for Human Rights and Religious Freedom; and Uberoi Foundation for Religious Studies, among others.

1

Emerging and Re-Emerging Infectious Diseases of the Decade: An Overview

Ranjita Shegokar[1], Seth K. Amponsah[2], and Yashwant V. Pathak[3]

[1] Capnomed GmbH, Tübingen, Germany
[2] Department of Medical Pharmacology, University of Ghana Medical School, Accra, Ghana
[3] USF Health Taneja College of Pharmacy, University of South Florida, Tampa, FL, United States

1.1 Introduction

Infectious diseases are caused by microorganisms such as bacteria, viruses, fungi, or parasites. There are a number of factors that can spread infections, some of which are summarized in Table 1.1. Infections can be transmitted by human contact, environmental means, animal contact, or insect bites [1]. Typical infectious diseases have symptoms like fever, diarrhea, fatigue, and muscle aches or can have other symptoms. Many infectious diseases are treated with antibiotics (specifically bacteria-caused infections), antiviral medications (virus-mediated infections), and antifungals (fungi-related infections). Over the years, however, a number of pathogens responsible for these infectious diseases have become resistant to drugs due to mutations [2]. For some infectious diseases vaccines are currently available. For e.g. the Centers for Disease Control and Prevention (CDC) recommends two doses of chickenpox vaccine for individuals who have never had chickenpox. Also, for coronavirus disease 2019 (COVID-19) up to two to four shots of messenger ribonucleic acid (mRNA)-based vaccine are recommended [3].

Some common emerging infectious diseases (EIDs) and re-emerging infectious diseases (REIDs) include– chickenpox, Lyme disease, diphtheria, COVID-19, influenza, West Nile virus, among others. This chapter gives an overview of EIDs and REIDs and their impact on human lives and social structure.

1.2 Infectious Diseases and the Economy

Data suggests that there is some association between infectious disease burden and the economic state of a country–per capita income or gross domestic product (GDP), as illustrated in Figure 1.1. COVID-19 is the best example to reference for economic impacts [4]. A number of developing countries have struggled to deal with the impact of infectious diseases and running their economies. This very much depends on the financial capacity of each country. The question here is, "Do these developing countries put in place policies that mimic developed countries without understanding affordability, impacts, or the massive investment needed in dealing with infectious diseases?" It is also questionable how the challenges of mental and stress impact, woman/child abuse, and suicidal attempts associated with countrywide restrictions are helpful to developing countries [4]. Indeed, a country's economy is run on a "healthy population" which serves as the main workforce. The repercussions of a diseased population (the workforce) due to a lack of medical care and national and internal travel bans imposed on people cannot be overemphasized [5]. There are many more factors on all fronts that affect not only the spread but also the prevention of such communicable diseases. Any EID and REID are a threat to humanity and needs to be tackled with pre-planning and good medical access to the population by each country.

Rising Contagious Diseases: Basics, Management, and Treatments, First Edition. Edited by Seth Kwabena Amponsah, Ranjita Shegokar, and Yashwant V. Pathak.

Table 1.1 Factors that aid the spread of infections.

Human	Pathogens	Vectors and reservoirs
Age	Type of pathogen	Genetic plasticity
Lifestyle	Mutation rate	Vector capacity
Food habits	Adaptive emergence	Geographical mutations
Immunity	Contamination frequency	Transfer frequency
Bioterrorism	Resistance to drugs	Organ reservoirs

Access to healthcare, food and water quality, animal-human consumption and contact, air pollution, availability of detection methods, cost of therapy, and societal-medical layers.

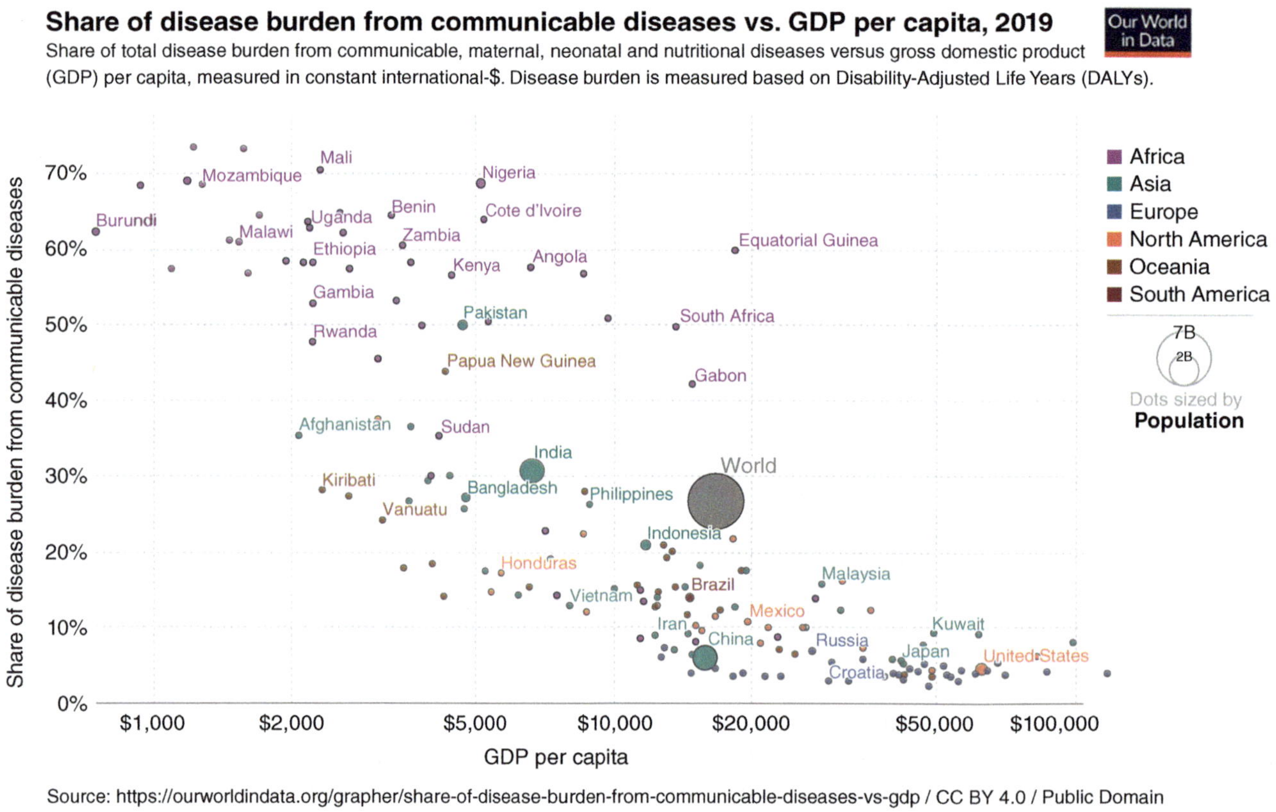

Figure 1.1 Impact of diseases on GDP.

1.3 The Main Factors Involved in EIDs and REIDs

The world is dramatically changing on all fronts and requires changing attitudes, policies, and science to tackle infectious diseases [6]. Factors like sudden changing environmental factors, demographic influences, lifestyle, and technologically connected factors not only affect the risk of pathogen emergence, and alterations, but also disturbs the whole economies of countries (Figures 1.2 and 1.3). More futuristic, faster research and related processes, regulation of both wild and domestic animal populations, and inter country collaborations are important aspects of the prevention and management of EIDs and REIDs [7].

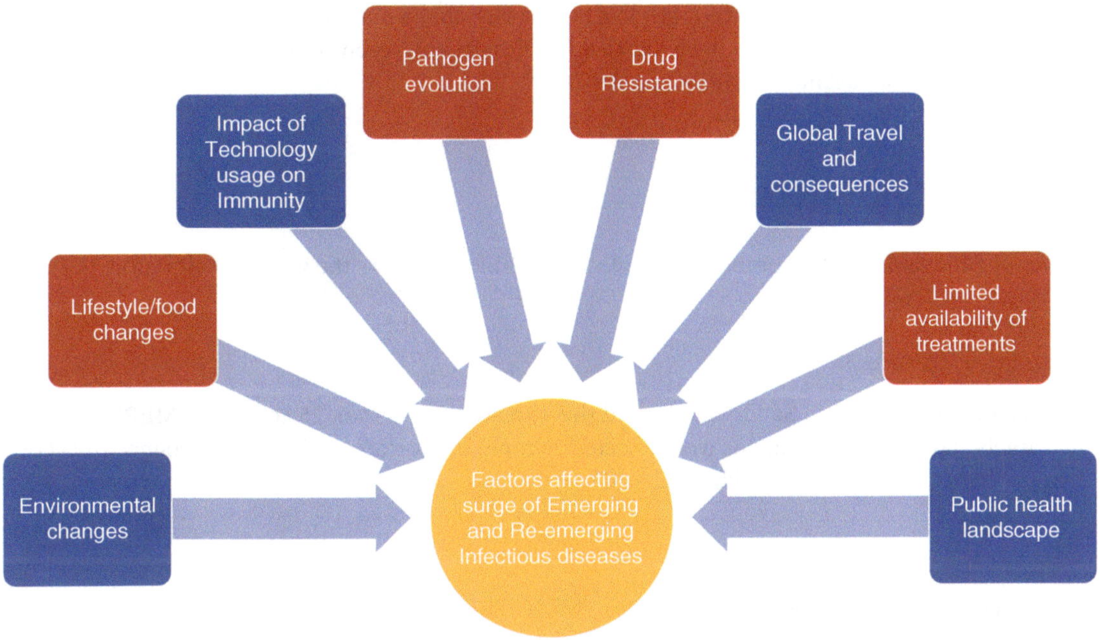

Figure 1.2 Factors affecting the surge of emerging and re-emerging diseases.

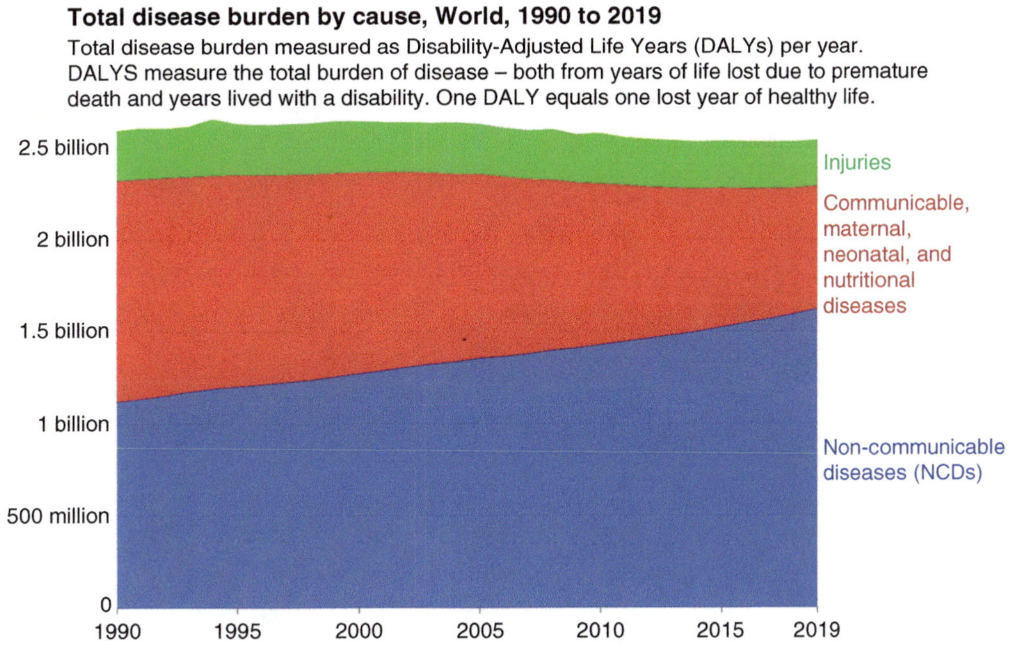

Figure 1.3 Burden of disease spread into communicable, noncommunicable, and injuries. *Source:* Our World in Data / https://ourworldindata.org/grapher/total-disease-burden-by-cause / CC BY 4.0 / Public Domain

1.4 Prevention of EIDs and REIDs

Increased capital investment in infectious disease outbreak countries is key in preventing the spread. For e.g. World Health Organization (WHO) recently formed a hub for pandemic and epidemic intelligence. Further, efforts are being made to develop universal vaccines that could provide a monumental leap to treat infectious diseases. Each country has learned its lessons from COVID-19 and understood the importance of preparedness in fighting infectious diseases at the detection, treatment, hospitalization, and vaccine development front [4]. It is noteworthy that a majority of infectious diseases are of zoonotic origin, thus, this aspect also needs to be effectively tackled [7]. Besides this, each country has to secure in-house production of medicines for effective management of EIDs and REIDs.

1.5 Infectious Diseases of the Last Decades

There have been a number of EIDs and REIDs over the last few decades. Table 1.2 shows the year of their outbreak.

1.5.1 Middle East Respiratory Syndrome (MERS)

The first case of MERS was reported in 2012 in Saudi Arabia, and this spread to more than 25 countries. MERS is known to have originated from camels and eventually humans were infected. Symptoms typically include fever, cough, and shortness of breath, and often progress to pneumonia, similar to COVID-19 [8].

Table 1.2 The spread of infectious diseases both emerging new diseases (blue) and re-merging (gray) diseases.

Year	EIDS/REIDS
1981	HIV/AIDS
1982	*E. coli*
1989	Hepatitis C virus
1993	Hantavirus pulmonary infection
1997	H5N1 influenza
1998	Nipah virus
1999	West Nile virus
2000	Dengue
2001	Anthrax infection
2002	West Nile
2002	SARS-CoV
2004	Marburg/MARV
2004	Dengue
2005	Chikungunya
2006	XDR tuberculosis
2006	Cholera
2009	Swine flu/H1N1
2010	Measles
2012	MERS
2013	Ebola
2013	Chikungunya – H7N9 influenza
2015	Zika
2016	Yellow Fever
2018	Lassa disease
2018	Ebola
2019	Covid-19

1.5.2 Severe Acute Respiratory Syndrome (SARS)

SARS originated from small mammals and emerged to infect humans sometime in 2002. Infection with the SARS virus causes acute respiratory distress (severe breathing difficulty) and is known to have a mortality rate of about 10% [9].

1.5.3 COVID-19

A number of chapters in this book will give relevant information about COVID-19. COVID-19 affected the mental, physical, and economic aspects of not only individuals but also many societies. Currently, the spread of this disease is on the decline, but COVID-19 has taught many nations the need to prepare for possible pandemics in the future.

1.5.4 Dengue

Dengue viruses are spread through the bite of an infected mosquito (*Ae. aegypti* or *Ae. albopictus*). Mild dengue symptoms last two to seven days while severe dengue can be life-threatening within a few hours. Dengue is caused by one of four related viruses (known as dengue virus 1, 2, 3, and 4). Up to 400 million people have been infected with dengue [10].

1.5.5 Ebola

Ebola is a rare and deadly disease caused by infection with a virus of the family *Filoviridae*, genus *Ebolavirus*. Ebola can cause disease in humans and other primates (monkeys, gorillas, and chimpanzees). Ebola was first discovered in 1976 near the Ebola River. Symptoms of the disease appears around 2–21 days after infection, but the average length of time is 8–10 days. Ebola vaccine is recommended by Center for Disease Control and Prevention (CDC) and Food and Drugs Authority (FDA).

1.5.6 Influenza

Influenza or flu is a contagious viral infection that can cause mild to severe symptoms and life-threatening complications, including death, even in healthy children and adults. Annual flu vaccination is the best way to prevent influenza. Treatment of flu with antiviral drugs can reduce influenza symptoms.

1.5.7 Viral Hepatitis

The most common types of viral hepatitis are hepatitis A, hepatitis B, and hepatitis C. Hepatitis A is usually a short-term infection and does not become chronic. Hepatitis B and hepatitis C can begin as short-term, acute infections, but they also have the potential to lead to chronic disease and long-term liver problems [11]. There are effective vaccines to prevent hepatitis A and hepatitis B. Although there is currently no vaccine for hepatitis C, there are effective treatments.

1.5.8 Mpox

Mpox virus belongs to the *Orthopoxvirus* genus. Mpox was first discovered in 1958 when two outbreaks of a pox-like disease occurred in colonies of monkeys kept for research. The first human case of mpox was reported in 1970. The United States declared a public health emergency for mpox in August 2022, as there was a surge in the number of cases. Symptoms of the disease usually start within three weeks of exposure.

1.5.9 Zika Virus

Zika virus disease is caused by a virus transmitted primarily by *Aedes* mosquitoes. The virus was isolated for the first time in 1947 in the Zika forest. Symptoms are similar to those of dengue or chikungunya, which are transmitted by the same type of mosquito. There is no vaccine or specific medicine currently available for its management.

1.5.10 Swine Flu

Swine flu or H1N1 flu is caused by the influenza A virus. In 2009, the WHO declared the H1N1 flu as a pandemic. Many antiviral drugs can be used to treat influenza infections including typically oseltamivir, baloxavir, peramivir, and zanamivir. Like typical flu, swine flu can lead to more serious problems including pneumonia, lung infection, and other breathing challenges.

1.6 Conclusion

The emergence and re-emergence of infectious diseases will continue, based on several factors. The question is how we tackle them. At a global level, we do not only need good medicines and hospital care, but rather the availability of health personnel, rapid and cost-effective diagnostic tools, information technology (IT) infrastructure, surveillance tools and less sophisticated regulatory systems to avoid delays in drug and/or vaccine approvals. Availability of well furbished medical facilities in both rural and urban cities are required. The financial capabilities of citizens to afford such treatments need to be taken into account when designing national policies. Collaboration between grant agencies, researchers, and manufacturers is key in all these. Further, the impact of new trends like artificial intelligence (AI) or ChatGPT on predictability and containment of EIDs and REIDS remain unclear; time will tell. In any case, each country needs to be ready to tackle EIDs and REIDS in an effective way, using current available tools and keeping futuristic tools and supporting fighting tools in prospective policies.

References

1 Tabish, S.A. (2009). Recent trends in emerging infectious diseases. *Int. J. Health Sci.* 3: V–VIII.

2 De Oliveira, D.M.P., Forde, B.M., Kidd, T.J. et al. (2020). Antimicrobial resistance in ESKAPE pathogens. *Clin. Microbiol. Rev.* 33: 00181.

3 Chavda, V.P., Soni, S., Vora, L.K. et al. (2022). mRNA-based vaccines and therapeutics for COVID-19 and future pandemics. *Vaccines (Basel)* 10 (12): 2150.

4 Ogunleye, O.O., Basu, D., Mueller, D. et al. (2020). Response to the novel corona virus (COVID-19) pandemic across Africa: successes, challenges and implications for the future. *Front. Pharmacol.* 11: 1205.

5 Afriyie, D.K., Asare, G.A., Amponsah, S.K., and Godman, B. (2020). COVID-19 pandemic in resource-poor countries: challenges, experiences and opportunities in Ghana. *J. Infect. Dev. Ctries.* 14 (8): 838–843.

6 Zumla, A. and Hui, D.S.C. (2019). Emerging and reemerging infectious diseases: global overview. *Infect. Dis. Clin. N. Am.* 33: 13–19.

7 Hill-Cawthorne, G.A. and Sorrell, T.C. (2016). Future directions for public health research in emerging infectious diseases. *Public Health Res. Pract.* 26: e2651655.

8 Memish, Z.A., Perlman, S., Van Kerkhove, M.D., and Zumla, A. (2020). Middle East respiratory syndrome. *Lancet* 395 (10229): 1063–1077.

9 Parry, J. (2003). WHO warns that death rate from SARS could reach 10%. *BMJ* 326 (7397): 999.

10 Sabir, M.J., Al-Saud, N.B.S., and Hassan, S.M. (2021). Dengue and human health: a global scenario of its occurrence, diagnosis and therapeutics. *Saudi J. Biol. Sci.* 28 (9): 5074–5080.

11 Kulkarni, A.V. and Duvvuru, N.R. (2021). Management of hepatitis B and C in special population. *World J. Gastroenterol.* 27 (40): 6861–6873.

2

Recent Trends and Possible Future Trajectory of COVID-19

Ismaila Adams[1], Ofosua Adi-Dako[2], Eugene Boafo[1], Emmanuel K. Ofori[3], and Seth K. Amponsah[1]

[1] *Department of Medical Pharmacology, University of Ghana Medical School, Accra, Ghana*
[2] *Department of Pharmaceutics and Microbiology, School of Pharmacy, University of Ghana, Accra, Ghana*
[3] *Department of Chemical Pathology, University of Ghana Medical School, Accra, Ghana*

2.1 Introduction

Coronavirus disease 2019 (COVID-19) is caused by the novel coronavirus, severe acute respiratory syndrome coronavirus 2 (SARS-CoV-2). COVID-19 has negatively affected global health, societies, and economies. COVID-19 was declared a Public Health Emergency of International Concern (PHEIC) by the World Health Organization (WHO) in January 2020, and a pandemic in March 2020 [1]. The evolving nature of the virus necessitated a comprehensive re-evaluation beyond the emergency phase. After more than three years (May 2023), the WHO stated that COVID-19 was no longer a PHEIC.

The global response to the pandemic has been significantly shaped by the remarkable advancements in therapeutics and vaccines, achieved through international collaboration [2]. Nonetheless, the journey toward controlling the SARS-CoV-2 spread was fraught with many challenges. A study on achieving herd immunity against COVID-19 in Africa highlighted vaccine hesitancy and logistical issues as significant barriers [3]. Furthermore, a paper on the global surplus of COVID-19 vaccines pointed to the difficulties in aligning vaccine manufacturing and delivery, suggesting the need to decelerate production, enhance community trust, and optimize logistics. Furthermore, the importance of preparedness and availability of favorable respiratory protective equipment (RPE) were highlighted during the COVID-19 pandemic [4].

Despite the fact that WHO has declared that COVID-19 is no longer a PHEIC, there is still a need to broaden our perspective, examine the virus's trajectory, and impact beyond the acute phase [5]. This re-evaluation encompasses a comprehensive assessment of evolving epidemiological patterns, advancements in therapeutics and vaccines, long-term health implications, socioeconomic and ethical considerations, and global preparedness for future pandemics [6]. Understanding recent global epidemiological trends is paramount to anticipating the potential trajectory of the pandemic. By critically assessing regional variations and disparities, key factors that shape the future course of COVID-19 can be identified, providing insights into potential hotspots, challenges, and effective mitigation strategies [7, 8].

Beyond the acute phase, understanding the long-term health implications of COVID-19 is crucial. There is emerging evidence of post-acute sequelae, commonly referred to as "Long COVID," encompassing a range of persistent symptoms and potential organ-specific complications [9]. Indeed, COVID-19 disrupted healthcare systems, delayed diagnoses, and worsened the burden of non-COVID-19 diseases. These aforementioned negative impacts of the COVID-19 pandemic reemphasize the need for holistic healthcare approaches and long-term management strategies.

The COVID-19 pandemic also unmasked socioeconomic disparities and posed ethical dilemmas; some of which included its disproportionate effects on marginalized populations, the strain on global economies, data privacy, and ethical considerations in vaccination strategies [10, 11]. As we transition to a post-PHEIC era, it is relevant that we reflect on lessons learned from COVID-19, highlighting the significance of robust surveillance systems, early warning mechanisms, resilient public health infrastructure, and international collaboration. By proactively strengthening these arms of public health, we can enhance our capacity to detect, respond to, and mitigate the impact of future outbreaks, fostering a more resilient global health landscape.

Rising Contagious Diseases: Basics, Management, and Treatments, First Edition. Edited by Seth Kwabena Amponsah, Ranjita Shegokar, and Yashwant V. Pathak.

In conclusion, this chapter provides a comprehensive exploration of current trends and potential future trajectories of COVID-19 by critically examining evolving epidemiological patterns, advancements in therapeutics and vaccines, long-term health implications, socioeconomic and ethical considerations, and global preparedness for future pandemics.

2.2 Overview of COVID-19

2.2.1 Summary of Timelines and Major Events

In late December 2019, reports emerged from Wuhan, China, of an unknown respiratory illness causing severe pneumonia. The causative agent was soon identified as a novel coronavirus, named SARS-CoV-2 due to its genetic similarities to the virus responsible for the 2002–2003 SARS outbreak [12]. By January 2020, the virus rapidly spread beyond China's borders, leading WHO to declare a PHEIC on 30 January 2020. This designation gave an indication of the urgency of the situation, and the need for international collaboration and coordinated efforts to reduce the spread of SARS-CoV-2 [13].

Throughout the first half of 2020, the pandemic spread with alarming speed. Many countries implemented stringent public health measures, including travel restrictions, lockdowns, and social distancing guidelines, to slow the transmission and "flatten the curve" of new cases. However, the virus continued its relentless spread across the globe [14]. As the pandemic evolved, scientists raced to understand the characteristics of the virus and developed novel diagnostic tools. By March 2020, widespread testing became crucial for detecting cases, isolating infected individuals, and initiating contact tracing efforts [15]. In the following months, the world witnessed the devastating effect of COVID-19 on healthcare systems. In many hospitals, intensive care units reached capacity, and healthcare workers were confronted to deal with many patients who were critically ill from COVID-19 [16].

Further, significant efforts were directed toward vaccine development. Multiple candidate vaccines were taken through clinical trials. The year 2020 ended with the approval and distribution of the first COVID-19 vaccine (Pfizer), marking a monumental achievement in the fight against the pandemic [17].

Throughout 2021 and 2022, vaccination campaigns gained momentum, providing hope for a path toward controlling the virus. However, the emergence of new variants of SARS-CoV-2, such as the Alpha, Beta, Gamma, and Delta, posed challenges to global eradication efforts. These variants exhibited increased transmissibility, higher disease severity, and, in some cases, reduced vaccine effectiveness [18, 19]. There were also instances of COVID-19 waves in different regions of the world. Countries faced complex decisions regarding the implementation of public health measures.

In May 2023, WHO announced the reclassification of COVID-19, as no longer being a PHEIC. This decision was due to the ongoing nature of the pandemic and the necessity to adapt strategies for long-term management [20]. Understanding these timelines and major events are essential in shaping effective strategies for possible future outbreaks of COVID-19.

2.2.2 Key Characteristics of SARS-CoV-2

SARS-CoV-2 belongs to the family of coronaviruses that have crown-like spike proteins (S1/S2) on their surface [21, 22]. Knowledge of the structure (Figure 2.1) and characteristics of the virus has been crucial in developing effective strategies to mitigate its spread.

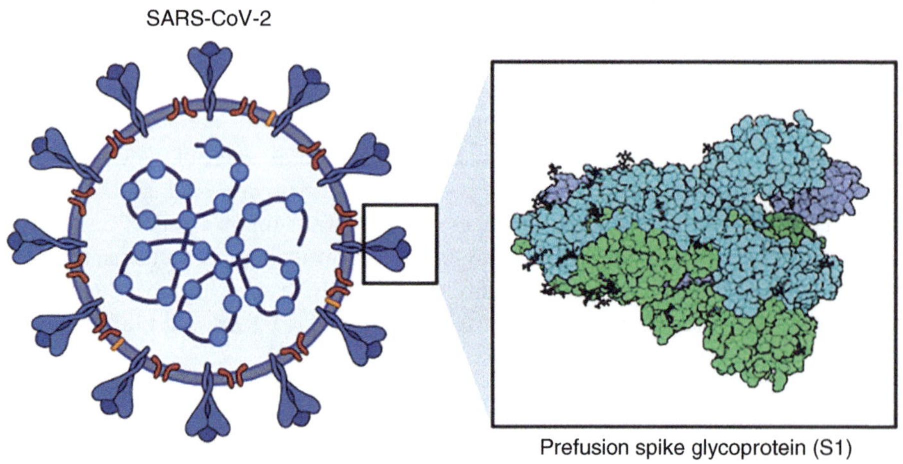

Figure 2.1 Structure of SARS-CoV-2.

SARS-CoV-2 primarily spreads through respiratory droplets generated when an infected person coughs, sneezes, talks, or breathes heavily. These respiratory droplets can contain the virus and infect individuals close by, typically at a distance of about 6 feet. One significant characteristic of SARS-CoV-2 is its potential for transmission by individuals who are asymptomatic or pre-symptomatic (infected but not yet showing symptoms). This feature poses challenges in controlling the spread, as infected individuals can unknowingly transmit the virus to others [23, 24].

Infected individuals can shed the virus through respiratory secretions, such as saliva, nasal discharge, and respiratory droplets. Viral shedding occurs during both symptomatic and asymptomatic stages of infection, contributing to the transmission dynamics of the virus. The incubation period of SARS-CoV-2 typically ranges from 2 to 14 days, with the average being around 5–6 days. However, it is important to note that individuals can transmit the virus even during the incubation period, heightening the challenges of early detection and control [25].

2.2.3 Transmission Modes of SARS-CoV-2

Understanding the modes of transmission of SARS-CoV-2 has been crucial in implementing effective preventive measures and devising strategies to reduce its spread. While respiratory transmission is the primary mode, other routes of transmission have been identified (Figure 2.2).

2.2.3.1 Close Contact Transmission

Close contact with an infected individual, particularly within about 6 ft, poses a significant risk of SARS-CoV-2 transmission. This can occur through direct physical contact, such as shaking hands, hugging, or kissing, or through respiratory droplets released during breathing, talking, coughing, or sneezing.

2.2.3.2 Airborne Transmission

In addition to respiratory droplets, recent evidence suggests that SARS-CoV-2 can also spread through small, aerosolized particles that remain suspended in the air for extended periods. This airborne transmission can occur in enclosed spaces with poor ventilation, particularly during activities that generate aerosols, such as singing, shouting, or exercising [14, 26].

2.2.3.3 Fomite Transmission

SARS-CoV-2 can survive on surfaces for varying periods, and transmission can occur when individuals touch contaminated surfaces (objects) and then touch their face, particularly the eyes, nose, or mouth. However, while fomite transmission is possible, it is considered less common compared to respiratory transmission [3, 27].

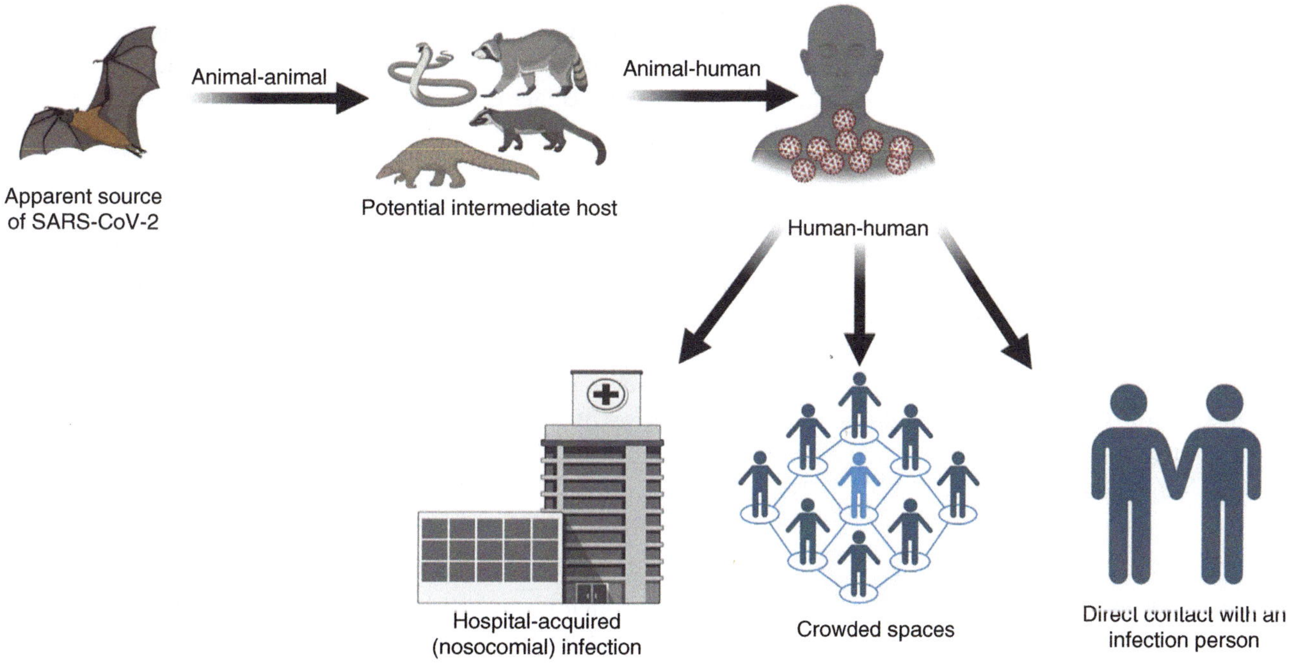

Figure 2.2 Modes of transmission of SARS-CoV-2.

2.2.3.4 Other Possible Transmission Routes

While less common, there have been suggestions of potential alternative transmission routes, such as fecal-oral transmission (viral particles present in fecal matter) and transmission through ocular surfaces (conjunctiva of the eyes). However, further research is needed to better understand the significance of these routes in overall transmission dynamics [28].

It is important to note that the understanding of characteristics of SARS-CoV-2 and transmission modes continues to evolve as new research emerges. Ongoing research is critical to refine our knowledge and guide public health interventions to effectively control the spread of the virus. Public health measures, including vaccination, mask-wearing, physical distancing, and improved ventilation, remain essential in mitigating transmission and preventing the spread of SARS-CoV-2.

2.3 Current State of the Pandemic

The COVID-19 pandemic continues to have a significant impact on a global scale, with diverse patterns observed in different regions. Understanding the current state of the pandemic is crucial for assessing the effectiveness of interventions.

2.3.1 Global and Regional Trends in COVID-19 Cases, Hospitalizations, and Deaths

The global impact of COVID-19 has been substantial, with millions of confirmed cases and deaths worldwide. Regional trends vary, reflecting variations in testing capacity, public health measures, healthcare infrastructure, and population demographics. Some regions have experienced multiple waves of infections, with fluctuations in case of numbers and severity. Ongoing surveillance and monitoring of cases, hospitalizations, and deaths provide critical insights into the geographic distribution and intensity of the pandemic, helping guide targeted interventions and resource allocation [15, 29].

2.3.2 Variants of Concern and Their Impact on Transmission and Severity

The evolution of SARS-CoV-2 has led to the emergence of several variants of concern, each introducing new complexities to the pandemic. Initially, variants such as Alpha, Beta, Gamma, and Delta gained attention due to their increased transmissibility. These variants were also associated with increased severity of illness and increased hospitalization.

However, the landscape of the pandemic took another turn with the emergence of the Omicron variant (B.1.1.529) toward the end of 2021. This variant, first identified in South Africa, raised global concern due to its large number of mutations, particularly in the spike protein, which is the target of most COVID-19 vaccines. Preliminary reports suggested that Omicron may be more transmissible than previous variants and could potentially have an impact on vaccine effectiveness, although this might not have been the case [30, 31].

In addition to transmissibility and disease severity, another concern with these variants was their potential resistance to certain treatments and their ability to evade immunity acquired from previous infections or vaccinations. Thus, monitoring of these variants and their potential impact on the effectiveness of vaccines is essential [30].

2.3.3 Pharmacotherapy of COVID-19

Pharmacotherapy for COVID-19 has evolved since the onset of the pandemic (Table 2.1). Antiviral drugs like remdesivir have been used to directly target the SARS-CoV-2 virus, while other medications such as dexamethasone have been employed to manage severe inflammatory response in COVID-19 patients [32]. Antithrombotic drugs have also been critical in managing COVID-19 complications, given the increased risk of blood clotting disorders in some patients [31]. Furthermore, the use of monoclonal antibodies, such as bamlanivimab and casirivimab/imdevimab, have shown promise in treating individuals with mild to moderate disease [35]. However, the treatment strategy is often tailored to the severity of the disease, the patient's overall health status, and the presence of any comorbid condition. It is important to note that while these treatments can help manage the disease, vaccination remains the most effective strategy for controlling the spread of COVID-19 and reducing the severity of the disease.

2.3.4 Vaccination

Vaccination has played a pivotal role in mitigating COVID-19 pandemic. As of August 2021, over 20 COVID-19 vaccines, utilizing different platforms such as messenger ribonucleic acid (mRNA), viral vectors, and protein subunits, had been approved for emergency use and deployed globally [39]. Vaccination rates vary across countries and regions, influenced by

Table 2.1 Pharmacotherapy in COVID-19.

Drug class	Some main drugs	Indications/comments	References
RNA-dependent RNA polymerase inhibitors	Remdesivir, molnupiravir	Early treatment of mild–moderate COVID-19 to reduce risk of progression	[32, 33]
Main protease (M^{pro}) inhibitors	Nirmatrelvir-ritonavir, ensitrelvir	Early treatment of mild–moderate COVID-19 to reduce risk of progression	[34]
Anti-spike monoclonal antibodies	Bebtelovimab, bamlanivimab, casirivimab/imdevimab	Early treatment of mild–moderate COVID-19; many no longer recommended due to resistance	[35]
Glucocorticoids	Dexamethasone, hydrocortisone	For hospitalized patients on oxygen, may harm early treatment	[32]
Janus kinase (JAK) inhibitors	Baricitinib, tofacitinib	For hospitalized patients on oxygen	[36]
Anti-IL-6R monocloncal antibodies	Tocilizumab, sarilumab	For hospitalized patients on oxygen who received glucocorticoids	[37]
Anticoagulants	Heparin, apixaban	For certain hospitalized non-ICU patients to reduce thrombosis	[38]

factors such as vaccine availability, distribution strategies, public acceptance, and logistical challenges [40]. Monitoring and evaluating vaccination rates are essential for assessing the progress of immunization campaigns and identifying populations that may require targeted interventions.

Furthermore, ongoing studies assess the effectiveness of vaccines against emerging variants, duration of protection, and the potential need for booster doses. Understanding vaccine effectiveness in real-world conditions is crucial for guiding vaccination strategies and public health policies [41]. The leading COVID-19 vaccines have shown efficacy of 70–95% against symptomatic infection, with generally lower efficacy against emerging variants like Delta. Common adverse reactions are usually mild and transient. However, unequal global distribution has been a major challenge, with high-income countries receiving the vast majority of doses while low-income countries lag behind [39].

Monitoring the current state involves analyzing multiple indicators, including case numbers, hospitalizations, deaths, variant surveillance, and vaccination coverage. These data will provide a comprehensive picture of the trajectory of the pandemic, enabling policymakers and public health authorities to implement evidence-based interventions, adapt strategies, and allocate resources effectively.

Although WHO has stated that COVID-19 is no longer a PHEIC, continuous surveillance, data sharing, and international collaboration remain crucial in comprehending global and regional trends, identifying emerging challenges, and implementing focused responses. Timely and accurate information continues to play an important role in adapting public health measures, optimizing vaccine distribution, and mitigating the impact of COVID-19 [42].

2.4 Epidemiological Projections and Modeling

Epidemiological projections and modeling are essential in understanding and predicting the future trajectory of COVID-19. By utilizing mathematical and statistical models, researchers can project the spread of the virus, assess potential impact of interventions, and inform public health decision-making stakeholders. Nonetheless, it is important to appreciate the complexities, uncertainties, and limitations associated with these modeling approaches.

2.4.1 Different Models Used to Predict Future Trajectory of COVID-19

Numerous modeling techniques have been used to simulate and project the course of the COVID-19 pandemic. Examples include compartmental models such as susceptible-exposed-infectious-recovered (SEIR) model developed by the Institute for Health Metrics and Evaluation. This has been used to forecast infections and deaths in the United States (US) and other countries [43]. Agent-based models like Covasim have simulated the spread of COVID-19 while accounting for factors such as social distancing [44]. Statistical approaches include a Bayesian model from Los Alamos National Laboratory that projected the spread of the virus across the US with uncertainty estimates, and machine learning models using random forests and neural networks to forecast cases and mortality [45]. Additional examples include a metapopulation SEIR model that analyzed the role of international travel [46], and an age structured SEIR model that estimated the impact of interventions across different age groups [47]. Each model is selected based on the specific research question and the data available [48, 49].

2.4.2 Factors Influencing the Projections

Projections generated by epidemiological models are influenced by various factors. Vaccination rates, for example, can have a significant impact on the future trajectory of outbreaks. High vaccination coverage has reduced transmission, severity of illness, and mortalities from COVID-19, leading to a decline in cases globally [50]. Public health measures, such as mask-wearing, physical distancing, and travel restrictions, also influenced projections by mitigating the spread earlier in the pandemic [8]. Other factors, including population density, age distribution, healthcare capacity, and adherence to public health guidelines, further shaped the trajectory as COVID-19 transitioned from a pandemic to an endemic disease [51]. Understanding the interplay between these factors was crucial for accurate modeling and effective decision-making during the pandemic.

2.4.3 Uncertainties and Limitations Associated with Modeling Approaches

Epidemiological modeling has always faced uncertainties and limitations, even when applied to diseases that are well-understood. Models rely on assumptions based on available data, and uncertainties arise due to incomplete information and variations in data quality across different regions. For example, early models of the spread of influenza in 2009 suffered from underestimating key parameters like the transmission rate [52]. Additionally, unforeseen events and changes in human behavior can introduce further uncertainties. A model of the Ebola outbreak in West Africa in 2014 initially projected a much smaller scale before the disease spread to densely populated urban areas [53].

It is essential to interpret model results with caution, considering their inherent limitations, and recognize that projections are not absolute predictions but rather scenario-based estimations. To enhance accuracy and reliability, ongoing data collection, validation, and refinement of models are essential, as demonstrated with malaria models refined over decades of data [54]. Collaboration between epidemiologists, mathematicians, statisticians, and policymakers also contributes to more robust models, as occurred in the global effort to model SARS-CoV-2. Overall, transparency in assumptions, sensitivity analyses, and regular reassessment based on emerging data are key to producing useful models.

2.5 Potential Scenarios for the Future

As WHO has now declared COVID-19 as being no longer a PHEIC, it is important to explore potential scenarios that may shape the trajectory of the disease in the coming years. These scenarios are based on different assumptions and variables, allowing the understanding of a range of possibilities and their associated implications.

2.5.1 Presentation of Multiple Scenarios Based on Different Assumptions and Variables

Epidemiological models can generate multiple scenarios by manipulating various factors, such as vaccination rates, effectiveness of public health interventions, and population behavior. These scenarios provide a spectrum of potential outcomes, helping policymakers and public health officials prepare for different situations and assess the impact of various strategies. Scenarios may consider factors like the emergence of new variants, changes in population behavior, and the duration of immunity conferred by vaccination or previous infection [47].

2.5.2 Examination of Best-Case, Worst-Case, and Most Likely Scenarios

Within these potential scenarios, three broad categories can be identified: best-case, worst-case, and most likely scenarios. The best-case scenario represents an optimistic outcome, where effective control measures and high vaccination rates lead to a substantial reduction in cases, hospitalizations, and deaths. The worst-case scenario reflects a more pessimistic outcome, where the virus continues to spread rapidly again, overwhelming healthcare systems, and causing significant morbidity and mortality. The most likely scenario lies between these extremes, taking into account prevailing factors, such as vaccination coverage, ongoing transmission dynamics, and the effectiveness of interventions [55].

2.5.3 Factors That Could Influence the Trajectory

Several key factors can influence the trajectory of the pandemic within each scenario. Public health interventions, including vaccination campaigns, mask-wearing, testing, contact tracing, and physical distancing, play a crucial role in mitigating transmission and reducing the burden on healthcare systems [15]. The level of public adherence to these measures and the consistency of their implementation can significantly impact the trajectory. Additionally, factors like vaccine hesitancy, the effectiveness of vaccines against emerging variants, and the potential need for booster doses may influence the course of the pandemic. Socioeconomic factors, population mobility, and global collaboration in response efforts also shape potential scenarios for the future [41].

It is crucial to emphasize that potential scenarios should not be treated as fixed predictions, but rather as valuable tools for planning and preparing purposes. The actual course of the pandemic will be influenced by a dynamic interplay of various factors, including unforeseen events and emerging scientific knowledge. To effectively navigate the complexities of the situation, it is essential to continuously monitor real-time data, adapt strategies based on the evolving circumstances, and remain flexible in response efforts [56].

By considering a range of potential scenarios, policymakers can develop adaptive strategies that account for uncertainties and respond effectively to changing circumstances. These scenarios inform decision-making, resource allocation, and communication strategies, facilitating a proactive approach to pandemic management.

Examining potential scenarios for the future of the COVID-19 pandemic helps us understand a range of possibilities and their associated implications. By exploring different assumptions and variables, including best-case, worst-case, and most likely scenarios, and considering factors such as public health interventions and vaccine hesitancy, we can better prepare for different trajectories and adjust strategies accordingly [40].

2.6 Impact of COVID-19 on Public Health and Healthcare Systems

The COVID-19 pandemic placed immense strain on public health infrastructure and healthcare systems worldwide. Analyzing the potential burden on healthcare systems in different scenarios allows us to assess the challenges and develop strategies for addressing them effectively.

2.6.1 Analysis of the Potential Burden of COVID-19 on Healthcare Systems in Different Scenarios

The potential burden of COVID-19 on healthcare systems can vary depending on the trajectory of the pandemic and the severity of cases. In worst-case scenarios (high transmission rates and limited control measures), healthcare systems may face overwhelming demands, exceeding their capacity to provide adequate care [57]. Hospitalizations, intensive care unit (ICU) admissions, and the need for ventilator support can place significant stress on critical resources and healthcare personnel. Conversely, in more optimistic scenarios as we have seen with effective control measures and high vaccination rates, the burden on healthcare systems has significantly reduced, allowing for better management of COVID-19 cases and the provision of essential non-COVID healthcare services [58].

2.6.2 Addressing Potential Challenges of COVID-19 on Healthcare

Dealing with the challenges posed by the COVID-19 pandemic requires a multifaceted approach. Adequate staffing, sufficient personal protective equipment (PPE), and well-equipped healthcare facilities are essential for managing COVID-19 patients. Maintaining surge capacity through flexible hospital bed management, including the creation of field hospitals or repurposing existing facilities, can help accommodate the influx of patients during periods of high transmission [57]. Additionally, enhancing testing capacity, contact tracing efforts, and the availability of diagnostic tools is crucial for early detection and containment of the virus. Collaborative and coordinated responses among healthcare facilities, public health agencies, and policymakers are necessary for resource allocation, ensuring equitable access to care, and implementing effective preventive measures [59].

2.6.3 Impact of COVID-19 on Non-COVID Healthcare Services and Long-Term Healthcare Planning

The impact of COVID-19 extends beyond COVID-19 patients. The pandemic has also affected non-COVID healthcare services and long-term healthcare planning. Disruptions in routine medical care, such as preventive screenings, elective surgeries, and management of chronic conditions, have been observed due to overwhelmed healthcare systems and redirected resources [60]. Delayed diagnoses and treatments may lead to adverse health outcomes for non-COVID patients. Ensuring the continuity of essential healthcare services while managing the demands of the pandemic requires innovative approaches, such as telemedicine, remote monitoring, and risk-stratified prioritization of care. Long-term healthcare planning must address the need for robust healthcare infrastructure, resilient supply chains, and sustainable financing models that can adapt to future public health emergencies [61].

The impacts of the pandemic on public health and healthcare systems highlight the importance of a holistic approach that considers both COVID-19 and non-COVID healthcare needs. Balancing the provision of optimal care for COVID-19 patients with the maintenance of essential services requires ongoing monitoring, data-driven decision-making, and the implementation of adaptive strategies [62]. Learning from the experiences of the pandemic, investing in public health infrastructure, strengthening healthcare systems, and fostering interdisciplinary collaborations are essential for mitigating the impacts of future public health crises.

The COVID-19 pandemic has had significant impacts on public health and healthcare systems. Analyzing the burden on healthcare systems in different scenarios, addressing potential challenges through effective strategies, and considering the impact on non-COVID healthcare services, and long-term planning are crucial for navigating the complexities of the pandemic [3]. By adopting a comprehensive approach, we can develop resilient healthcare systems that can effectively manage the demands of future pandemics while ensuring the provision of quality care for all patients.

2.7 Socioeconomic and Behavioral Considerations

The COVID-19 pandemic has had profound socioeconomic implications, reshaping societies and economies worldwide. Analyzing the socioeconomic impact of the pandemic and potential future outcomes allows us to understand the challenges and opportunities that lie ahead.

2.7.1 Analysis of the Socioeconomic Impact of the Pandemic and Potential Future Outcomes

The pandemic has led to significant disruptions in economic activities, including widespread job losses, business closures, and supply chain disruptions. The impact has been felt across various sectors, with vulnerable populations, low-income individuals, and marginalized communities being disproportionately affected. Economic recovery trajectories can vary depending on the speed and effectiveness of vaccination campaigns, the resilience of local economies, and the implementation of targeted support measures. Potential future outcomes include scenarios of robust recovery, uneven economic rebound with persistent inequalities, or prolonged economic downturns, depending on factors such as vaccination rates, policy responses, and global cooperation [10, 63].

2.7.2 Behavioral Changes and Their Long-Term Implications

The pandemic has triggered substantial behavioral changes, such as remote work, online education, and increased utilization of telemedicine services. These changes have had both immediate and long-term implications. Remote work, for instance, has transformed traditional work models and highlighted the importance of flexible work arrangements [64]. The accelerated adoption of telemedicine has demonstrated the potential for expanding access to healthcare services, particularly in underserved areas. Understanding the long-term implications of these behavioral changes, including their impact on productivity, job markets, and healthcare delivery, can inform future planning and decision-making [65].

2.7.3 Exploration of Societal Responses and Adaptive Measures

Societal responses to the pandemic have varied, encompassing a wide range of measures and approaches. These include the implementation of public health interventions, such as mask mandates, travel restrictions, and physical distancing guidelines, as well as the development and distribution of vaccines [14]. Adaptive measures, such as the expansion of digital infrastructure, investment in research and development, have played critical roles in mitigating the pandemic. Effective responses have often required strong leadership, transparent communication, and collaboration across sectors, highlighting the importance of resilient governance structures and community engagement.

Societal responses have also brought to the forefront issues of social equity, including access to healthcare, digital resources, and education. The pandemic has underscored existing inequalities, amplifying disparities in healthcare outcomes, educational attainment, and economic well-being. Addressing these disparities and promoting social cohesion are essential for building a more inclusive and resilient society [64].

The COVID-19 pandemic has had wide-ranging socioeconomic impacts and triggered behavioral changes with long-term implications. Analyzing the socioeconomic impact, understanding potential future outcomes, and exploring adaptive measures are essential for informing policies, supporting recovery efforts, and fostering resilient and inclusive societies. By addressing inequalities, leveraging technological advancements, and promoting adaptive strategies, societies can mitigate the long-term effects of the pandemic and build a more sustainable and equitable future.

2.8 Global Preparedness and Response

The COVID-19 pandemic has underscored the importance of global preparedness and coordinated response efforts. Evaluating the lessons learned from the pandemic, discussing global cooperation, and exploring the role of technology and innovation are important for building resilience in the face of future outbreaks.

2.8.1 Lessons Learned from the Pandemic and Their Application to Future Outbreaks

The pandemic has provided valuable lessons that can inform future outbreak preparedness and response. These lessons include the need for early detection and rapid response systems, robust surveillance networks, effective communication strategies, and agile decision-making processes. Understanding the strengths and weaknesses of global and national responses to pandemics can guide the development of strategies that are adaptive, evidence-based, and culturally sensitive. Evaluating lessons learned from the COVID-19 pandemic also involves examining the role of scientific research, data sharing, and cross-sectoral collaborations [8].

2.8.2 Global Cooperation and Preparedness Efforts

The pandemic has showed that global cooperation is decisive in addressing public health emergencies. Collaborative efforts among countries, international organizations, and public health agencies have been instrumental in sharing data and resources. Strengthening global preparedness requires ongoing investments in building resilient health systems, enhancing surveillance capacities, and developing rapid response mechanisms. Additionally, equitable access to vaccines, therapeutics, and diagnostics is essential to ensure that no population is left behind. Strengthening global health governance structures and fostering solidarity and partnerships are pivotal in effective preparedness and response efforts.

2.8.3 Exploration of the Role of Technology and Innovation in Pandemic Response

Technology and innovation have played a key role in the pandemic response. Telemedicine, remote monitoring, and digital contact tracing applications have enabled the provision of healthcare services and facilitated early detection of cases. Rapid development and deployment of vaccines, aided by innovative vaccine technologies, have been important milestones in the pandemic response. Advances in data analytics, artificial intelligence, and genomics have supported surveillance, modeling, and identification of emerging variants. Embracing technological innovations, leveraging data-driven approaches, and promoting research are essential in enhancing preparedness, response, and recovery in future outbreaks [66].

Addressing global preparedness and response requires sustained political will, resource allocation, and long-term investments. Strengthening health systems, training healthcare workers, and prioritizing research and development are significant components. Furthermore, promoting cross-border collaborations, sharing best practices, and facilitating technology transfer can enhance global cooperation and response capabilities.

2.9 Conclusion and Recommendations

This chapter has examined various aspects of the COVID-19 pandemic, including epidemiological trends, the impact on public health and healthcare systems, socioeconomic considerations, and global preparedness. Key findings and insights include–

1) Understanding the evolving epidemiological patterns of COVID-19 is essential in anticipating the potential trajectory of the pandemic. Regional variations, vaccination rates, and the emergence of new variants significantly influence the spread and severity of COVID-19.
2) The pandemic has placed a significant burden on healthcare systems, necessitating adaptive strategies, surge capacity planning, and prioritization of essential healthcare services. Addressing the impact on non-COVID healthcare and long-term healthcare planning is essential for maintaining overall population health.
3) Socioeconomic implications, including job losses, economic disparities, and behavioral changes, have reshaped societies and economies. Leveraging the opportunities presented by remote work, telemedicine, and digital innovation can shape future developments and improve access to healthcare and education.
4) Global preparedness and response efforts are vital for effective pandemic management. Evaluating lessons learned, fostering global cooperation, and harnessing technology and innovation can enhance response capabilities and strengthen health systems.

2.9.1 Recommendations for Policymakers, Public Health Officials, and Researchers

Based on the insights gathered, the following recommendations can guide policymakers, public health officials, and researchers for future pandemics–

1) Prioritize equitable access to vaccines, ensuring widespread coverage and addressing vaccine hesitancy. Emphasize public health campaigns that promote vaccine confidence, targeting communities at higher risk, and considering the unique challenges of different populations.
2) Maintain and strengthen public health measures, such as mask-wearing, testing, contact tracing, and physical distancing, particularly in areas experiencing surges or facing the risk of new variants. These measures should be tailored to local contexts and backed by clear communication strategies.
3) Invest in healthcare system resilience and capacity building, considering long-term healthcare planning and the need for flexible responses to future pandemics. Strengthen healthcare infrastructure, enhance telemedicine capabilities, and prioritize the provision of essential non-COVID healthcare services.
4) Foster international collaboration and recognize the importance of global cooperation in addressing public health emergencies. Promote the equitable distribution of resources, knowledge, and technology to enhance preparedness and response capacities across nations.
5) Support research and development efforts to advance the understanding of pandemics, including the dynamics of viral transmission, long-term health implications, and the effectiveness of interventions. Prioritize studies that address disparities, health equity, and the social determinants of health to ensure inclusive and equitable outcomes.

2.9.2 Identification of Areas for Further Research and Study

While significant progress has been made in understanding COVID-19, several areas warrant further research–

1) Long-term health consequences and post-acute sequelae of COVID-19 require continued investigation, particularly the impact of the disease on different organ systems, risk factors, and management strategies.
2) The effectiveness of vaccination against emerging variants, the duration of immunity conferred by vaccines, and the potential need for booster doses.
3) The social, psychological, and economic impacts of the pandemic on individuals and communities warrant in-depth exploration, including the assessment of mental health outcomes, socioeconomic inequalities, and the interplay between social determinants and health disparities.
4) Enhancing pandemic modeling and forecasting capabilities, incorporating real-time data, and refining assumptions can contribute to more accurate projections and aid in decision-making at both global and local levels.

By embracing these recommendations and building upon the key insights gained, we can navigate the complexities of the pandemic more effectively, mitigate its impact, and foster resilience in the face of future public health challenges.

References

1 WHO. (2020). COVID-19 Public Health Emergency of International Concern (PHEIC) Global Research and Innovation Forum. https://www.who.int/publications/m/item/covid-19-public-health-emergency-of-international-concern-(pheic)-global-research-and-innovation-forum

2 Farlow, A., Torreele, E., Gray, G. et al. (2023). The future of epidemic and pandemic vaccines to serve global public health needs. *Vaccines* 11 (3): 690. https://doi.org/10.3390/vaccines11030690.

3 Ogunleye, O.O., Basu, D., Mueller, D. et al. (2020). Response to the novel corona virus (COVID-19) pandemic across Africa: successes, challenges, and implications for the future. *Front. Pharmacol.* 11: 1205. https://doi.org/10.3389/fphar.2020.01205.

4 Quan, N.K., Anh, N.L.M., and Taylor-Robinson, A.W. (2023). The global COVID-19 vaccine surplus: tackling expiring stockpiles. *Infect. Dis. Poverty* 12 (1): 21. https://doi.org/10.1186/s40249-023-01070-7.

5 Torner, N. (2023). The end of COVID-19 public health emergency of international concern (PHEIC): and now what? *Vacunas* https://doi.org/10.1016/j.vacun.2023.05.002.

6 WHO, R&D Blueprint. (2022). How Global Research Can End This Pandemic and Tackle Future Ones. https://www.who.int/publications/m/item/how-global-research-can-end-this-pandemic-and-tackle-future-ones

7 Dyson, L., Hill, E.M., Moore, S. et al. (2021). Possible future waves of SARS-CoV-2 infection generated by variants of concern with a range of characteristics. *Nat. Commun.* 12: 5730. https://doi.org/10.1038/s41467-021-25915-7.

8 Telenti, A., Arvin, A., Corey, L. et al. (2021). After the pandemic: perspectives on the future trajectory of COVID-19. *Nature* 596 (7873): 495–504. https://doi.org/10.1038/s41586-021-03792-w.

9 Proal, A.D. and VanElzakker, M.B. (2021). Long COVID or post-acute sequelae of COVID-19 (PASC): an overview of biological factors that may contribute to persistent symptoms. *Front. Microbiol.* 12: 698169. https://doi.org/10.3389/fmicb.2021.698169.

10 Kantamneni, N. (2020). The impact of the COVID-19 pandemic on marginalized populations in the United States: a research agenda. *J. Vocat. Behav.* 119: 103439. https://doi.org/10.1016/j.jvb.2020.103439.

11 Kooli, C. (2021). COVID-19: public health issues and ethical dilemmas. *Ethics Med. Public Health* 17: 100635. https://doi.org/10.1016/j.jemep.2021.100635.

12 Pollard, C.A., Morran, M.P., and Nestor-Kalinoski, A.L. (2020). The COVID-19 pandemic: a global health crisis. *Physiol. Genomics* 52 (11): 549–557. https://doi.org/10.1152/physiolgenomics.00089.2020.

13 Singh, S., McNab, C., Olson, R.M. et al. (2021). How an outbreak became a pandemic: a chronological analysis of crucial junctures and international obligations in the early months of the COVID-19 pandemic. *Lancet (London, England)* 398 (10316): 2109–2124. https://doi.org/10.1016/S0140-6736(21)01897-3.

14 Afriyie, D.K., Asare, G.A., Amponsah, S.K., and Godman, B. (2020). COVID-19 pandemic in resource-poor countries: challenges, experiences and opportunities in Ghana. *J. Infect. Dev. Ctries.* 14 (8): 838–843. https://doi.org/10.3855/jidc.12909.

15 Wang, X., Du, Z., James, E. et al. (2022). The effectiveness of COVID-19 testing and contact tracing in a US city. *Proc. Natl. Acad. Sci. U. S. A.* 119 (34): e2200652119. https://doi.org/10.1073/pnas.2200652119.

16 Arabi, Y.M., Azoulay, E., Al-Dorzi, H.M. et al. (2021). How the COVID-19 pandemic will change the future of critical care. *Intensive Care Med.* 47 (3): 282–291. https://doi.org/10.1007/s00134-021-06352-y.

17 Wouters, O.J., Shadlen, K.C., Salcher-Konrad, M. et al. (2021). Challenges in ensuring global access to COVID-19 vaccines: production, affordability, allocation, and deployment. *Lancet (London, England)* 397 (10278): 1023–1034. https://doi.org/10.1016/S0140-6736(21)00306-8.

18 Duong, D. (2021). Alpha, beta, delta, gamma: what's important to know about SARS-CoV-2 variants of concern? *CMAJ* 193 (27): E1059–E1060. https://doi.org/10.1503/cmaj.1095949.

19 Islam, S., Islam, T., and Islam, M.R. (2022). New coronavirus variants are creating more challenges to global healthcare system: a brief report on the current knowledge. *Clin. Pathol.* 15: 2632010X221075584. https://doi.org/10.1177/2632010X221075584.

20 WHO. (2023). Statement on the Fifteenth Meeting of the IHR (2005) Emergency Committee on the COVID-19 Pandemic. https://www.who.int/news/item/05-05-2023-statement-on-the-fifteenth-meeting-of-the-international-health-regulations-(2005)-emergency-committee-regarding-the-coronavirus-disease-(covid-19)-pandemic

21 Abdelrahman, Z., Li, M., and Wang, X. (2020). Comparative review of SARS-CoV-2, SARS-CoV, MERS-CoV, and influenza A respiratory viruses. *Front. Immunol.* 11: 552909. https://www.frontiersin.org/articles/10.3389/fimmu.2020.552909.

22 Pal, M., Berhanu, G., Desalegn, C., and Kandi, V. (2020). Severe acute respiratory syndrome coronavirus-2 (SARS-CoV-2): an update. *Cureus* 12 (3): e7423. https://doi.org/10.7759/cureus.7423.

23 Dhand, R. and Li, J. (2020). Coughs and sneezes: their role in transmission of respiratory viral infections, including SARS-CoV-2. *Am. J. Respir. Crit. Care Med.* 202 (5): 651–659. https://doi.org/10.1164/rccm.202004-1263PP.

24 Gao, W., Lv, J., Pang, Y., and Li, L.-M. (2021). Role of asymptomatic and pre-symptomatic infections in covid-19 pandemic. *BMJ* 375: n2342. https://doi.org/10.1136/bmj.n2342.

25 Puhach, O., Meyer, B., and Eckerle, I. (2023). SARS-CoV-2 viral load and shedding kinetics. *Nat. Rev. Microbiol.* 21 (3): 147–161. Article 3. https://doi.org/10.1038/s41579-022-00822-w.

26 Güner, R., Hasanoğlu, İ., and Aktaş, F. (2020). COVID-19: prevention and control measures in community. *Turk. J. Med. Sci.* 50 (3): 571–577. https://doi.org/10.3906/sag-2004-146.

27 CDC. (2020). Coronavirus Disease 2019 (COVID-19). Centers for Disease Control and Prevention. https://www.cdc.gov/coronavirus/2019-ncov/more/science-and-research/surface-transmission.html

28 Meyerowitz, E.A., Richterman, A., Gandhi, R.T., and Sax, P.E. (2021). Transmission of SARS-CoV-2: a review of viral, host, and environmental factors. *Ann. Intern. Med.* 174 (1): 69–79. https://doi.org/10.7326/M20-5008.

29 Sawicka, B., Aslan, I., Della Corte, V. et al. (2022). The coronavirus global pandemic and its impacts on society. In: *Coronavirus Drug Discovery* (ed. C. Egbuna), 267–311. Elsevier 10.1016/B978-0-323-85156-5.00037-7.

30 Aleem, A., Akbar Samad, A.B., and Vaqar, S. (2023). Emerging variants of SARS-CoV-2 and novel therapeutics against coronavirus (COVID-19). In: *StatPearls*. StatPearls Publishing http://www.ncbi.nlm.nih.gov/books/NBK570580.

31 Sheikh, A., Kerr, S., Woolhouse, M. et al. (2022). Severity of omicron variant of concern and effectiveness of vaccine boosters against symptomatic disease in Scotland (EAVE II): a national cohort study with nested test-negative design. *Lancet Infect. Dis.* 22 (7): 959–966. https://doi.org/10.1016/S1473-3099(22)00141-4.

32 Amponsah, S.K., Tagoe, B., Adams, I., and Bugyei, K.A. (2022). Efficacy and safety profile of corticosteroids and non-steroidal anti-inflammatory drugs in COVID-19 management: a narrative review. *Front. Pharmacol.* 13: 1063246. https://doi.org/10.3389/fphar.2022.1063246.

33 Malin, J.J., Suárez, I., Priesner, V. et al. (2020). Remdesivir against COVID-19 and other viral diseases. *Clin. Microbiol. Rev.* 34 (1): e00162–e00120. https://doi.org/10.1128/CMR.00162-20.

34 Hashemian, S.M.R., Sheida, A., Taghizadieh, M. et al. (2023). Paxlovid (nirmatrelvir/ritonavir): a new approach to Covid-19 therapy? *Biomed. Pharmacother.* 162: 114367. https://doi.org/10.1016/j.biopha.2023.114367.

35 Aleem, A. and Vaqar, S. (2023). Monoclonal antibody therapy for high-risk coronavirus (COVID 19) patients with mild to moderate disease presentations. In: *StatPearls*. StatPearls Publishing http://www.ncbi.nlm.nih.gov/books/NBK570603.

36 Dupuis, D., Fritz, K., Ike, E. et al. (2022). Current use of baricitinib in COVID-19 treatment and its future: an updated literature review. *Cureus* 14 (9): e28680. https://doi.org/10.7759/cureus.28680.

37 Zeraatkar, D., Cusano, E., Martínez, J.P.D. et al. (2022). Use of tocilizumab and sarilumab alone or in combination with corticosteroids for covid-19: systematic review and network meta-analysis. *BMJ Med.* 1 (1): e000036. https://doi.org/10.1136/bmjmed-2021-000036.

38 Farkouh, M.E., Stone, G.W., Lala, A. et al. (2022). Anticoagulation in patients with COVID-19. *J. Am. Coll. Cardiol.* 79 (9): 917–928. https://doi.org/10.1016/j.jacc.2021.12.023.

39 Rahman, S., Rahman, M.M., Miah, M. et al. (2022). COVID-19 reinfections among naturally infected and vaccinated individuals. *Sci. Rep.* 12 (1): 1438. https://doi.org/10.1038/s41598-022-05325-5.

40 Lazarus, J.V., Wyka, K., White, T.M. et al. (2023). A survey of COVID-19 vaccine acceptance across 23 countries in 2022. *Nat. Med.* 29 (2): 366–375. https://doi.org/10.1038/s41591-022-02185-4.

41 Chi, W.-Y., Li, Y.-D., Huang, H.-C. et al. (2022). COVID-19 vaccine update: vaccine effectiveness, SARS-CoV-2 variants, boosters, adverse effects, and immune correlates of protection. *J. Biomed. Sci.* 29 (1): 82. https://doi.org/10.1186/s12929-022-00853-8.

42 Allan, M., Lièvre, M., Laurenson-Schafer, H. et al. (2022). The World Health Organization COVID-19 surveillance database. *Int. J. Equity Health* 21 (3): 167. https://doi.org/10.1186/s12939-022-01767-5.

43 IHME (2020). *IHME | COVID-19 Projections*. Institute for Health Metrics and Evaluation https://covid19.healthdata.org.

44 Kerr, C.C., Stuart, R.M., Mistry, D. et al. (2021). Covasim: an agent-based model of COVID-19 dynamics and interventions. *PLoS Comput. Biol.* 17 (7): e1009149. https://doi.org/10.1371/journal.pcbi.1009149.

45 Chimmula, V.K.R. and Zhang, L. (2020). Time series forecasting of COVID-19 transmission in Canada using LSTM networks. *Chaos, Solitons Fractals* 135: 109864. https://doi.org/10.1016/j.chaos.2020.109864.

46 Chinazzi, M., Davis, J.T., Ajelli, M. et al. (2020). The effect of travel restrictions on the spread of the 2019 novel coronavirus (COVID-19) outbreak. *Science (New York, N.Y.)* 368 (6489): 395–400. https://doi.org/10.1126/science.aba9757.

47 Tuite, A.R., Fisman, D.N., and Greer, A.L. (2020). Mathematical modelling of COVID-19 transmission and mitigation strategies in the population of Ontario, Canada. *CMAJ* 192 (19): E497–E505. https://doi.org/10.1503/cmaj.200476.

48 Adiga, A., Dubhashi, D., Lewis, B. et al. (2020). Mathematical models for COVID-19 pandemic: a comparative analysis. *J. Indian Inst. Sci.* 100: 793–807.

49 Shamil, M.S., Farheen, F., Ibtehaz, N. et al. (2021). An agent-based modeling of COVID-19: validation, analysis, and recommendations. *Cogn. Comput.* 1–12. https://doi.org/10.1007/s12559-020-09801-w.

50 Damijan, J.P., Damijan, S., and Kostevc, Č. (2022). Vaccination is reasonably effective in limiting the spread of COVID-19 infections, hospitalizations and deaths with COVID-19. *Vaccines* 10 (5): 678. https://doi.org/10.3390/vaccines10050678.

51 Amponsah, S.K., Tagoe, B., and Afriyie, D.K. (2021). Possible future trajectory of COVID-19: emphasis on Africa. *Pan Afr. Med. J.* 40: 157. https://doi.org/10.11604/pamj.2021.40.157.31905.

52 Coburn, B.J., Wagner, B.G., and Blower, S. (2009). Modeling influenza epidemics and pandemics: insights into the future of swine flu (H1N1). *BMC Med.* 7 (1): 30. https://doi.org/10.1186/1741-7015-7-30.

53 Meltzer, M.I., Atkins, C.Y., Santibanez, S. et al. (2014). Estimating the future number of cases in the Ebola epidemic – Liberia and Sierra Leone, 2014–2015. *MMWR Suppl.* 63 (3): 1–14.

54 Huynh, J., Li, S., Yount, B. et al. (2012). Evidence supporting a zoonotic origin of human coronavirus strain NL63. *J. Virol.* 86 (23): 12816–12825. https://doi.org/10.1128/jvi.00906-12.

55 Schwarze, M.L., Zelenski, A., Baggett, N.D. et al. (2020). Best case/worst case: ICU (COVID-19) – a tool to communicate with families of critically ill patients with COVID-19. *Palliat. Med. Rep.* 1 (1): 3–4. https://doi.org/10.1089/pmr.2020.0038.

56 Massaro, E., Ganin, A., Perra, N. et al. (2018). Resilience management during large-scale epidemic outbreaks. *Sci. Rep.* 8 (1): 1859. Article 1. https://doi.org/10.1038/s41598-018-19706-2.

57 Filip, R., Gheorghita Puscaselu, R., Anchidin-Norocel, L. et al. (2022). Global challenges to public health care systems during the COVID-19 pandemic: a review of pandemic measures and problems. *J. Pers. Med.* 12 (8): 1295. https://doi.org/10.3390/jpm12081295.

58 Liu, Y., Procter, S.R., Pearson, C.A.B. et al. (2023). Assessing the impacts of COVID-19 vaccination programme's timing and speed on health benefits, cost-effectiveness, and relative affordability in 27 African countries. *BMC Med.* 21 (1): 85. https://doi.org/10.1186/s12916-023-02784-z.

59 Kim, Y.J. and Koo, P.-H. (2021). Effectiveness of testing and contact-tracing to counter COVID-19 pandemic: designed experiments of agent-based simulation. *Healthcare* 9 (6): 625. https://doi.org/10.3390/healthcare9060625.

60 Pujolar, G., Oliver-Anglès, A., Vargas, I., and Vázquez, M.-L. (2022). Changes in access to health services during the COVID-19 pandemic: a scoping review. *Int. J. Environ. Res. Public Health* 19 (3): 1749. https://doi.org/10.3390/ijerph19031749.

61 Vazquez, J., Islam, T., Gursky, J. et al. (2021). Access to care matters: remote health care needs during COVID-19. *Telemed. J. E Health* 27 (4): 468–471. https://doi.org/10.1089/tmj.2020.0371.

62 Kabwama, S.N., Wanyenze, R.K., Kiwanuka, S.N. et al. (2022). Interventions for maintenance of essential health service delivery during the COVID-19 response in Uganda, between March 2020 and April 2021. *Int. J. Environ. Res. Public Health* 19 (19): 12522. https://doi.org/10.3390/ijerph191912522.

63 Nicola, M., Alsafi, Z., Sohrabi, C. et al. (2020). The socio-economic implications of the coronavirus pandemic (COVID-19): a review. *Int. J. Surg. (London, England)* 78: 185–193. https://doi.org/10.1016/j.ijsu.2020.04.018.

64 Mouratidis, K. and Papagiannakis, A. (2021). COVID-19, internet, and mobility: the rise of telework, telehealth, e-learning, and e-shopping. *Sustain. Cities Soc.* 74: 103182. https://doi.org/10.1016/j.scs.2021.103182.

65 Garfan, S., Alamoodi, A.H., Zaidan, B.B. et al. (2021). Telehealth utilization during the Covid-19 pandemic: a systematic review. *Comput. Biol. Med.* 138: 104878. https://doi.org/10.1016/j.compbiomed.2021.104878.

66 Sharma, A., Virmani, T., Pathak, V. et al. (2022). Artificial intelligence-based data-driven strategy to accelerate research, development, and clinical trials of COVID vaccine. *Biomed. Res. Int.* 2022: 7205241. https://doi.org/10.1155/2022/7205241.

3

Mpox: New Challenges with the Disease

Julia Wang[1], Lynn Nguyen[1], Vernon Volante[1], Jeannez Daniel[1], and Charles Preuss[2]

[1] *University of South Florida Morsani College of Medicine, Tampa, FL, United States*
[2] *Department of Molecular Pharmacology & Physiology, University of South Florida Morsani College of Medicine, Tampa, FL, United States*

3.1 History of *Mpox virus*

Mpox virus (MPXV) is a disease caused by a zoonotic virus from the *Orthopoxvirus* genus in the family *Poxviridae*. MPXV was first discovered during an outbreak in monkeys by Danish scientists in 1958 [1–3]. The monkeys had been imported from Africa to a laboratory in Denmark for research, and MPXV was found in the vesicles of infected primates. Due to these primate hosts, the name "mpox" was determined. MPXV has since been noted in several other animal species. The first case of MPXV in humans was reported in 1970, when the virus infected a child in the Democratic Republic of the Congo (DRC) [1–4]. Since then, MPXV has been considered primarily animal-to-human transmission in Central and West Africa [1, 2, 5] before becoming human-to-human in the primary outbreak in 2022. MPXV cases were reported in non-endemic nations outside of Africa beginning in 2003. However, a significant increase in reported MPXV cases with no previous history of travel from endemic areas occurred in multiple countries during 2022, raising global concerns about MPXV.

3.2 Characteristics

MPXV is a large, brick-shaped, double-stranded, DNA-enveloped virus [3, 5]. It measures 197 kb linearly and 200–250 nm in size. It is part of the *Chordopoxvirinae* subfamily, *Poxviridae* family, and *Orthopoxvirus* genus. It has virions contained in a lipoprotein outer membrane. Infection is initiated by either the intracellular mature virion or the extracellular enveloped virion, which express different surface glycoproteins from each other. The MPXV virions bind to the glycosaminoglycans in the membranes of mammalian cells and complete their replication cycle in the cytoplasm of those host cells. MPXV has a wide range of potential hosts. However, the exact animal reservoir of MPXV is unknown. The specific method of transmission to humans is also unknown, but the literature suggests that the virus can be transmitted to humans from animals, other humans, and the environment [1–3, 5].

As a member of the *Orthopoxvirus* genus, MPXV is related to viruses like Variola, which causes smallpox. Due to shared antigenic and genetic features, orthopoxviruses have immunological cross-reactivity and cross-protection. This means that infection by any species of the *Orthopoxvirus* genus provides some protection between *Orthopoxvirus* species. As a result, it is suspected that vaccines for another orthopoxvirus could confer some immunity to MPXV as well.

3.3 Epidemiology

Since the 1970 discovery of MPXV in humans, it has been endemic to animals in the tropical rainforest areas of Central and West Africa. Cases of human MPXV occur sporadically or during local outbreaks when MPXV is transmitted from wild animals to humans. As of now, human MPXV cases have been reported in the DRC, the Republic of the Congo, Cameroon,

Rising Contagious Diseases: Basics, Management, and Treatments, First Edition. Edited by Seth Kwabena Amponsah, Ranjita Shegokar, and Yashwant V. Pathak.

the Central African Republic, Nigeria, Ivory Coast, Liberia, Sierra Leone, Gabon, and South Sudan [3–5]. The rising incidence of human MPXV cases is considered a consequence of the decrease in cross-protective and herd immunity among the population after universal smallpox vaccination ended in the early 1980s after the eradication of smallpox. Subsequently, the incidence of MPXV is highest in younger age groups that never received the smallpox vaccine [1]. Another factor contributing to the incidence of MPXV is increasing contact between humans and small mammals potentially carrying MPXV. Social factors such as civil wars, refugee displacement, farming, deforestation, and climate change cause humans to enter the natural environments of species known to be reservoirs for MPXV. As a result, the incidence of MPXV is also highest in forested regions [1, 4, 5].

Cases of MPXV reported outside of African countries are typically associated with international travel or the importation of animals infected with MPXV. The first cases of human MPXV outside Africa were detected in the United States in 2003 [1, 3, 4]. Exposure to infected pet prairie dogs transmitted MPXV to several people in Illinois, Indiana, Kansas, Missouri, Ohio, and Wisconsin. Molecular investigation and epidemiologic studies led to the discovery that the virus had been imported into the United States via a shipment of small mammals, including African rodents. The animals had been shipped from Ghana to Texas. Some of the infected animals were sheltered with the prairie dogs that would later be sold as pets. Other cases of international travel-associated MPXV have since been reported in countries including Israel, the United Kingdom, and Singapore [1, 3, 5]. These cases were linked to exposures during the 2018 outbreak in Nigeria [1].

Between 2010 and 2019, the median age of MPX cases increased from young children (4 years old) in the 1970s to young adults (21 years old) [2]. This was due to the Nigerian outbreak in 2018, in which the median age of infected cases was around 30 years. During this outbreak, men were infected at a higher rate than women. The majority of cases in Nigeria were from urban and peri-urban areas in the country's southern regions during the first three decades of the outbreak. In contrast, the majority of cases in DRC during that same period were from rural villages surrounded by tropical forests. Human MPXV cases in non-endemic countries were reported by people who had recently traveled to Nigeria or had contact with someone who had a confirmed MPXV infection outside of Africa following the 2018 Nigerian outbreak.

In 2022, several non-endemic countries reported human MPXV cases despite no direct animal or travel-related exposure risks, and the source of these infections is unknown. This led to concerns about the virus's evolutionary capability. These cases are now considered to be part of the ongoing 2022 multi-country outbreak. During this outbreak, it has been determined that MPXV primarily affects young men with a median age of 36 years [3]. Based on case reports, the outbreak primarily affects men who have sex with men (MSM), who are homosexual, bisexual, and other males who have MSM. These men reported recent sex with one or more partners. Over 95% of cases with sexual orientation were reported to be identified as members of the gay, bisexual, and other men who have MSM community [2, 3]. Sexual encounters were the most frequently reported type of transmission, accounting for 91.3% of all transmission events [2]. The vast majority of patients do not know their HIV status, with over 41% of those who do know their HIV status reporting being HIV-positive [2, 3]. However, it must be noted that MSM is not the only patient population affected by the outbreak, with 1.9% of case reports describing female patients [2].

Additionally, data on the race and ethnicity of reported cases during the 2022 MPXV outbreak are provided by the United Kingdom Health Security Agency (UKHSA) and the Centers for Disease Control and Prevention (CDC) for Europe and the Americas [2]. According to the CDC, race or ethnicity information is available for 59.2% of cases, with 33.1% being non-Hispanic White, 31.1% being Hispanic of any race, 31.1% being non-Hispanic Black, and 3.7% being non-Hispanic Asian. Less than 1% of cases with an available race/ethnicity are American Indian or Alaska Native, Native Hawaiian or other Pacific Islander, or multiple races. According to the UKHSA, 76.4% of respondents were White, 9.1% were mixed, 7.1% were Asian, 4.3% were Black, and 0.1% were other [2].

3.4 Transmission

The exact animal reservoir of MPXV and the method of transmission of MPXV to humans are unknown. However, direct (touches, scratches, or bites) or indirect contact with live or dead animals is assumed to cause zoonotic, or animal-to-human, transmission [1–4]. This contact may take the form of encountering the bodily fluids of infected animals, handling infected animals, and consuming inadequately cooked meat, otherwise known as bushmeat, from infected animals. Poverty and civil unrest force people to hunt small mammals for food. This increases exposure to wild animals that have the potential to carry MPXV. Besides monkeys and other nonhuman primates, animals that have been found capable of MPXV infection include Gambian giant squirrels, tree squirrels, terrestrial rodents, rats, rabbits, dormice, porcupines, other rodents, antelopes, and gazelles [1, 2].

The transmission of MPXV between humans can occur through direct contact with an infected person's skin, mucous membranes, or mucocutaneous lesions. The virus is thought to enter the body through broken skin, mucosal surfaces (such as those of the eyes, nose, or mouth), or the respiratory tract. Droplets from oral or respiratory secretions may transmit MPXV if an infected individual maintains prolonged close contact with an uninfected person. Consequently, MPXV has also been linked to nosocomial infections, with hospital staff, caretakers, and family members at risk. The prospect of sexual transmission was considered due to the significant prevalence of infection during the 2022 outbreak. This was supported by the discovery of high viral loads of MPXV in the saliva, rectal swab, sperm, urine, and fecal samples of infected individuals. Finally, MPXV can undergo mother-to-fetus, or vertical, transmission [2]. This leads to congenital MPXV, a condition that results in developmental anomalies in the newborn after birth.

Humans may also contract MPXV through direct contact with fomites contaminated by an infected individual's lesion fluid, bodily fluids, crust, or scab [2, 5]. Consequently, an infected person's sheets, clothing, or towels may transmit the virus. Orthopoxviruses also tend to be relatively resistant to environmental stress and have a relatively high level of environmental stability. Depending on room conditions, some pox viruses can also survive in the environment and on contaminated surfaces for up to 56 days [2]. This hardiness may be reflected in the environment-to-human transmission of MPXV, but there is currently a shortage of information on such environmental transmission as well as the presence of MPXV in human wastewater. The various modes of MPXV transmission are shown in Figure 3.1.

3.5 Pathogenicity

There are two phylogenetic clades of MPXV viruses. This means that there are two different groups of MPXV classified based on their genetic sequences. The two clades are: (i) Clade 1, otherwise known as the Central African or Congo Basin clade, and (ii) Clade 2, the West African clade [1, 2, 4, 5]. Disease severity varies between the clades, with the West African clade demonstrating a more self-limiting, less severe illness in humans [4, 5]. The case-fatality ratio for West African viral strains is approximately 3–6% [2]. On the other hand, the Central African viral strains are associated with higher transmissibility and have a case-fatality ratio of as high as 10%. In total, more suspected cases of the Central African clade have been reported than cases caused by the West African clade.

Genetic variation may be the source of the antigenic differences between the two clades. For example, Central African MPXV may produce an agent that prevents T-cell receptormediated T-cell activation [4]. This suppression of host T-cell responses promotes immune evasion by prohibiting the production of inflammatory cytokines based on previous MPXV infections. Additionally, the MPXV virus inhibitor of complement enzymes protects virally infected host cells from getting

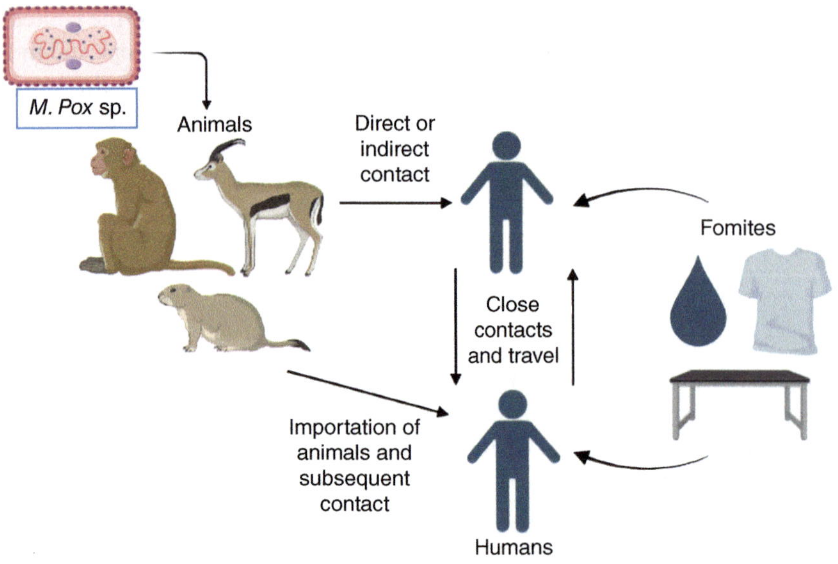

Figure 3.1 Pathogenesis of MPXV.

phagocytized by innate immune cells and lysed. This gene is absent in West African strains and may be a factor in the increased virulence of Central African strains. Furthermore, Central African MPXV strains downregulate host responses, specifically the apoptosis of infected host cells. They may even silence the transcription of genes involved in host immunity during infections.

Phylogenomic analysis of MPXV viral sequences associated with the 2022 multi-country outbreak of MPXV revealed the development of a third clade [2, 3]. Clade 3 is composed of the human MPXV virus-1A (hMPXV-1A) clade and four newly discovered lineages: A.1, A.1.1, A.2, and B.1 [2]. Lineage B.1 consists of all MPXV genomes from the 2022 outbreak. Compared to the other clades, the 2022 MPXV virus has coding regions that differ by an average of 50 single-nucleotide polymorphisms (SNPs). It is also associated with an increased mutational signature compared to its predecessors [3]. This may contribute to the 2022 MPXV virus's higher transmission rates compared to those of its ancestors as well as the rising number of cases in the ongoing MPXV outbreak. In addition, this altered mutational signature may indicate a newly diverging branch of the virus that presents with a more rapid evolution and hence mark MPXV as a disease that requires further surveillance in preparation for possible future outbreaks [2].

3.6 Risk Factors

Populations that have a higher risk of severe disease or complications include children, pregnant women, men who have MSM, people living with HIV (PWLH) and taking HIV pre-exposure prophylaxis (PreP), and workers with high exposure to infectious materials, such as healthcare workers and diagnostic laboratory staff [6], as shown in Table 3.1. Vertical transmission of MPXV has been documented, with the passing of the infection from mother to child in a small study that suggests that MPXV infection of a pregnant individual could lead to adverse outcomes for the fetus, either infection, spontaneous abortion, or stillborn delivery [8]. With the 2022 global outbreak, risk groups are considered for vaccinations with smallpox and an MPXV-specific vaccine as both pre- and post-exposure prophylaxis [6].

For PWLH, particularly those with poorly controlled disease, they are at risk of more severe sequelae of the disease, such as genital ulcers, secondary bacterial infection, and a longer duration of illness [8]. The WHO suggests that PWLH continue antiretroviral therapy with careful consideration of drug–drug interactions with antivirals for MPXV.

For MSM, MPXV has been seen with atypical manifestations as well as high co-infection with other sexually transmitted diseases, thus having a higher risk of transmission in this population. The WHO recommends that all patients abstain from sex until all MPXV lesions have crusted, fallen off, and resolved with the formation of a new layer of skin. Additionally, the consistent use of condoms for 12 weeks after recovery can help prevent the potential transmission of MPXV, as there is still uncertainty about whether MPXV is able to remain infectious within semen, saliva, and other bodily fluids [8].

Table 3.1 High-risk populations as defined by the CDC who are recommended for pre-exposure MPXV vaccination.

High-risk populations
Occupations exposed to orthopoxviruses
Research laboratory personnel
Clinical laboratory personnel handling diagnostic testing
Response teams designated by appropriate public health and antiterror authorities
Gay, bisexual, and other men who have MSM, and transgender or nonbinary people (including adolescents) who in the past six months have had:
New diagnosis of one or more sexually transmitted diseases
More than one sex partner
Sexual partners of people with above risks
People with HIV infection or other causes of immunosuppression with recent or anticipated potential MPXV exposure

Source: Data from Ref. [7].

3.7 Diagnostic Testing

To diagnose MPXV, a healthcare provider will typically perform a physical exam and review the patient's symptoms and medical history. Laboratory testing is also commonly used to confirm an MPXV diagnosis. The most common diagnostic test for MPXV is a polymerase chain reaction (PCR) test, which detects the genetic material of the virus in a patient's blood, skin lesions, or other body fluids. PCR testing is highly sensitive and specific, meaning it can accurately detect the virus in infected individuals. PCR testing involves amplifying the virus's genetic material in a laboratory to detect its presence. PCR testing is quicker than viral culture and can detect smaller amounts of the virus, but it may produce false-negative results in the early stages of the infection [9].

Another test that can be used to diagnose MPXV is serology, which looks for antibodies produced by the body in response to the virus. However, this test is less commonly used and may not be as accurate as PCR testing. Lastly, viral culture is another diagnostic test that can be used to detect MPXV. Viral culture involves taking a sample of fluid from a skin lesion or respiratory secretions and growing the virus in a laboratory. This method takes longer than PCR but is more sensitive and can confirm the presence of the virus [9].

It is important to note that MPXV is a rare disease, and the symptoms can be similar to those of other viral infections, such as chickenpox or measles. Therefore, it's important to consult a healthcare provider if you develop symptoms and have been in an area where MPXV is known to occur (Figure 3.2).

3.8 Symptoms

MPXV has an incubation period of 5–21 days, followed by the onset of clinical symptoms for up to 21 days. During the incubation period, MPXV is not considered contagious. MPXV can be clinically present in three stages: the febrile stage, the prodrome stage, and the skin eruption phase. The infectious period of MPXV is considered from the onset of prodromal symptoms to the complete resolution of symptoms, which can last for nearly two to five weeks [10]. Between 0 and 5 days after infection, the first stage is characterized by fever, intense headache, muscle pains (myalgia), generalized weakness, lymphadenopathy, or the swelling of lymph nodes (lymphadenopathy) in response to infection. Myalgia typically presents as back pain. Lymphadenopathy is a crucial feature that distinguishes MPXV from other illnesses that present similarly, such as chickenpox and measles. After one to three days after the fever, lesions begin to appear on the face, oral mucosa, limbs, soles, and palms. The lesions progress slowly from a macule (a flat lesion) to a papule (a raised lesion), to a vesicle (a raised lesion with fluid inside), and to a pustule (a large pus-filled lesion) throughout the next two to four weeks. The pustules eventually progress to a dry crust that will fall off during the following one to two weeks [10]. The lesions range in size

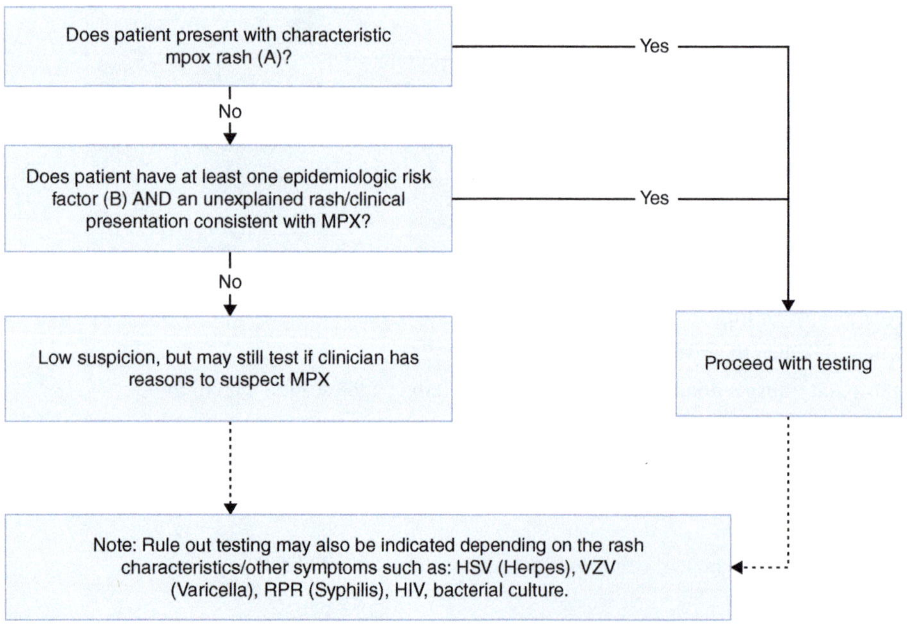

Figure 3.2 Algorithm for MPXV diagnostic testing.

Table 3.2 Differential diagnosis for MPXV.

Virus	MPXV	Smallpox	Measles	Chickenpox
Rash distribution	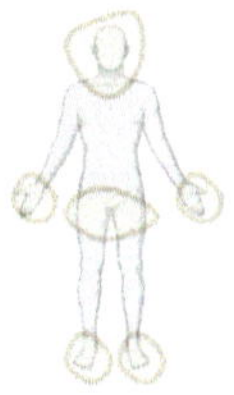		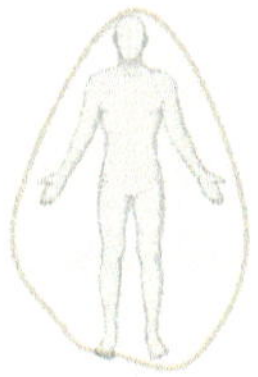	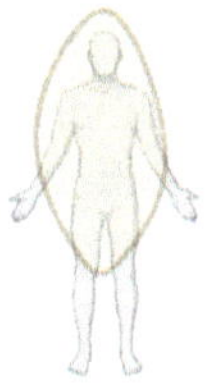
Length	*Incubation:* 5–21 days *Symptoms:* 2–4 weeks	*Incubation:* 7–19 days *Symptoms:* 2–5 weeks	*Incubation:* 7–21 days *Symptoms:* 1–2 weeks	*Incubation:* 10–21 days *Symptoms:* 1–2 weeks
Associated symptoms	• **Backache** • **Lymphadenopathy** • Fever, chills • Headache • Fatigue, muscle aches	• Fever, chills • Virus is since eradicated	• Fever, chills • Cough • Conjunctivitis • Runny nose	• Fever, chills • Headache • Fatigue, muscle aches
Appearance of rash	Raised spots that become small, fluid-filled blisters that become crust before falling off	Small red spots on tongue and pus-filled boils on face, arms, and legs	Non-itchy spots that form a red/brown blotchy patch **Associated with small white spots in mouth**	Red, **itchy** vesicles that ooze fluid. Drying into a crust

between 2 and 10 mm, are firm to the touch, and develop slowly with a distribution concentrated on the face (95%), rather than the "centrifugal distribution" on the trunk. The lesions range from a few to thousands. Oral lesions may affect the individual's ability to eat and drink, potentially leading to dehydration and malnutrition [8]. The illness is self-limiting and will clear itself in three to four weeks; after the crust falls off of the patient, the patient is no longer considered infectious. After healing, the lesions may leave pale marks before turning into dark scars [11].

In the recent 2022 outbreak, the most common clinical symptoms were the classic rash (affecting 99.5% of patients), general feeling of illness or malaise (85.2%), sore throat (78.2%), and lymphadenopathy (98/6%) most commonly in the cervical region, inguinal region, and lesions in the mouth and throat [8]. Many cases have reported genital and peri-anal lesions along with oropharyngeal symptoms [11]. The involvement of the anorectal mucosa was associated with anorectal pain, inflammation of the lining of the rectum, diarrhea, or tenesmus (the feeling of needing to pass a stool even with an empty rectum). The involvement of the oropharynx was accompanied by inflammation of the pharynx and epiglottis, pain upon swallowing, and lesions within the oral cavity, including tonsils [11].

MPXV skin lesions can resemble other infectious diseases, such as *Varicella Zoster* Virus (VZV), colloquially called "chickenpox," smallpox, herpes simplex virus (HSV), measles, and *Rickettsia* [11]. While Smallpox also presents with fever, generalized malaise, headache, and hard, well-circumscribed lesions similar to MPXV, smallpox typically does not present with lymphadenopathy [11]. VSV presents with an incubation period of one to three weeks, fever, malaise, and headache, but the lesion will appear as a vesicular, "dew drop" appearance with fluid within lesions with irregular borders [11]. The vesicular rash caused by VSV also progresses quicker, tends to be centrally located on the trunk rather than having a centrifugal distribution similar to MPXV, and will be seen as multiple lesions in different stages of development. MPXV will show lesions in the same stage of development and lesions on the palms and soles of hands and feet (Table 3.2).

3.9 Long-Term Effects and Complications

Unvaccinated patients had more severe symptoms than vaccinated patients, with an average case fatality rate of up to 11% [11]. Vomiting and diarrhea can occur in the second week of the illness, leading to further dehydration, electrolyte abnormalities, and a potential risk of shock. The skin lesions can lead to hyper- or hypo-pigmentation of the skin or skin exfoliation [10]. If the classic MPXV lesions merged, they were susceptible to further infection of the skin and soft tissue, such as cellulitis, abscesses, and necrotic soft tissue infections [10]. During the crusting phase of MPXV, the accumulation of fluid under the skin can lead to intravascular fluid depletion and shock [8].

Serious complications include one patient with encephalitis (swelling of the brain tissue), another with sepsis (overactive immune response to viral infection), and pneumonia [11]. Severe pneumonia can lead to respiratory distress; bronchopneumonia is a frequent late manifestation of the disease from a secondary lung infection. The lymphadenopathy of cervical lymph nodes could lead to a retropharyngeal abscess; signs include pain in swallowing, poor oral intake, and respiratory distress. Infections of the eye are also a possibility, leading to keratitis, conjunctival and corneal scarring, and possibly permanent loss of vision [11].

During infection, laboratory abnormalities in patients with MPXV can show elevated white blood cells (leukocytosis), elevated transaminases, low blood urea nitrogen, and low levels of albumin. In one-third of the patients evaluated, the elevation of lymphocytes and decreased levels of platelets were also seen [8].

While >90% of individuals with MPXV report complete resolution of infection regardless of smallpox vaccination status, some individuals still develop long-term complications. The most widespread, long-term complication of infection is pitted scarring. The pitted scarring of these skin lesions can lead to pockmarks on the skin. Blindness from ocular infections was also a long-term complication, albeit less commonly than pitted scarring. There is some data suggesting that patients may be at risk of developing mental health complications [8].

3.10 Vaccinations

Currently, there is no specific vaccine available to treat MPXV, but previous studies have suggested that vaccination against smallpox can protect against it. A 1980s study examined MPXV secondary attack rates (the rate of spread within households) and found that households that had received smallpox vaccinations had a significantly lower secondary attack rate than unvaccinated households [12]. Therefore, the vaccines currently being considered for MPXV are those that were originally developed to treat smallpox, including ACAM2000® and JYNNEOSTM, both of which are available in the US Strategic National Stockpile (SNS). Additionally, an experimental vaccine called Aventis Pasteur Smallpox Vaccine (APSV) is available for use under an investigational new drug (IND) protocol [13].

ACAM2000 was approved by the FDA in 2007 for active immunization against smallpox in individuals deemed to be at high risk for infection. This vaccine is derived from a live, replication-competent vaccinia virus, a poxvirus related to smallpox that causes a milder disease. Due to its ability to replicate, there is a higher risk of adverse effects compared to a replication-deficient virus. These adverse effects can include eczema vaccinatum [14], progressive vaccinia [15], and myopericarditis [16], particularly in certain individuals. Eczema vaccinatum and progressive vaccinia are both caused by the systemic spread of the vaccinia virus, with eczema vaccinatum being more common in individuals with atopic dermatitis or eczema, while progressive vaccinia is more commonly seen in immunocompromised individuals. Furthermore, vaccinia can be transmitted through close contact with the vaccination site or even vertically. Vertical transmission can result in fetal vaccinia, which can be fatal.

JYNNEOS is the more recently developed smallpox vaccine that was approved by the FDA for smallpox and MPXV disease prevention in 2019. It is manufactured by Bavarian Nordic (BN) and contains Modified Vaccinia Ankara (MVA), a replication-deficient vaccinia virus strain [17]. Due to the strain's inability to replicate, it is deemed safe to use in immunocompromised individuals and lacks the adverse effects seen in ACAM2000. However, although it is safe for use in immunocompromised populations, its effectiveness may be lower (CDC, 2023). Both JYNNEOS and ACAM2000 are approved for use in individuals 18 years of age or older who are deemed at high risk for infection.

APSV is a live, replication-competent vaccinia virus. Like ACAM2000, it can cause serious complications such as encephalitis, progressive vaccinia, and eczema vaccinatum [17]. It is an investigational product stored in the US SNS and can be used under IND protocol or via emergency use authorization (EUA) when the other two vaccines are unavailable.

Currently, the CDC suggests the use of smallpox vaccines to reduce transmission, prevent disease, or reduce disease severity. Vaccination prior to exposure is recommended for people in occupations with increased risk for exposure: research laboratory personnel working with orthopoxviruses, clinical laboratory personnel performing diagnostic testing for orthopoxviruses, and orthopoxvirus and healthcare worker response teams designated by appropriate public health and antiterror authorities [7]. Along with occupational exposure risk, vaccination against MPXV is recommended for the following populations, as shown in Figure 3.3.

- Gay, bisexual, and other men who have MSM, and transgender or nonbinary people with more than one sex partner or a new diagnosis of one or more sexually transmitted diseases in the past six months

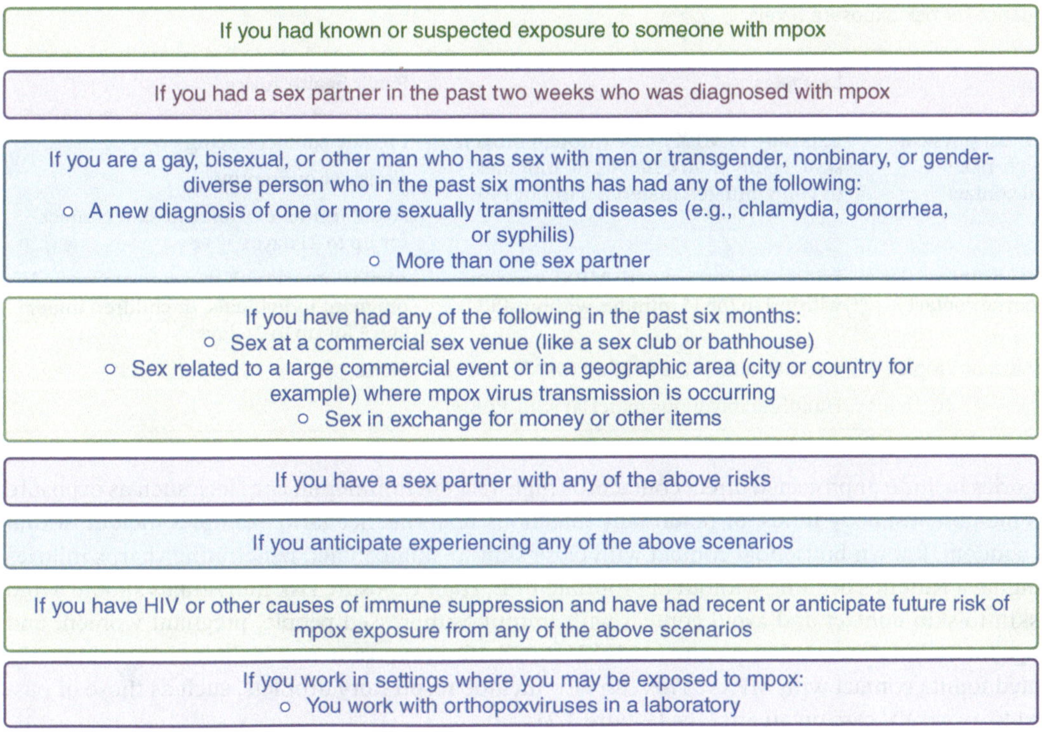

Figure 3.3 CDC recommendations for MPXV vaccination.

- People who, in the past six months, have had sex at a commercial sex venue or in association with a large public event in a geographic area where MPXV transmission is occurring
- Sex partners of people with the risks listed above
- People with HIV infection or other causes of immunosuppression with recent or anticipated MPXV exposure

Use of the vaccines for post-exposure prophylaxis is reserved for the following populations [7]:

- Known contacts of someone with MPXV who are identified by public health authorities (such as through contact tracing)
- People who are aware that a sex partner within the past 14 days was diagnosed with MPXV
- Gay, bisexual, MSM, and transgender or nonbinary people who, in the past 14 days, have had sex with multiple partners, sex at a commercial sex venue, or sex in association with an event, venue, or geographic area where MPXV transmission is occurring

The CDC recommends PEP be initiated as soon as possible after MPXV exposure. Ideally, PEP is done within 4 days of exposure, but 4–14 days after exposure are still shown to be effective. After 14 days, administration of the vaccine is recommended on a case-by-case basis.

3.11 Epidemic Management

MPXV prevention is recommended as contact avoidance with infected persons, safe sex practices with the use of protection, avoidance of multiple sexual partners, hand hygiene practice, disinfection of contaminated surfaces, use of masks and gloves in general practice, and use of personal protective equipment (PPE) if caring for an infected person [11].

In the event of an epidemic, clinical care requires early recognition of individuals with infection, rapid implementation of appropriate infection prevention and control, diagnostic testing and confirmation, symptomatic management, and monitoring for complications such as severe dehydration, pneumonia, and sepsis. Transmission occurs through fluid secretion through skin-to-skin contact or respiratory tract droplets. Thus, important prevention and control measures include case isolation, hand hygiene, the use of PPE, and standard contact and droplet precautions. Contact tracing is also highly encouraged for every individual to submit and control the spread of MPXV [8].

Table 3.3 Public health guidance for risk exposure levels.

	Exposure	Examples	Public health advice
High risk	Unprotected direct physical exposure or high-risk environmental contact	Exposure to MPXV case through broken skin, bodily fluids, mucus membranes, or contaminated material without PPE	Provide contact tracing Consider quarantine Avoid intimate or skin-to-skin contact for up to 21 days
Medium risk	Unprotected exposure to droplet or airborne contact	Intact skin contact with MPXV case, or within 1 m for 15 minutes without PPE	Avoid contact with immunosuppressed, pregnant individuals, or children under age 5 for up to 21 days
Low risk	Protected physical or droplet exposure	Contact within 1–3 m of an MPXV case Handling infectious material with PPE	None

High exposure risk categories include unprotected direct contacts or high-risk environmental contacts, such as exposure to broken skin or mucous membranes, body fluids, or potentially infectious material. Scenario examples include sexual contact with or without a condom, known household contact with close skin-to-skin contact, penetrating sharps injuries from used needles, or changing a patient's bedding without appropriate PPE. High exposure risk individuals should avoid both sexual contact and skin-to-skin contact and avoid contact with immunosuppressed people, pregnant women, and children under 5 years of age for 21 days from their exposure to MPXV [8]. Medium exposure risk includes intact skin only, bodily fluid, or contaminated fomite contact with MPXV. This can also include respiratory droplets, such as those of passengers seated directly next to an MPXV case on an airplane (Figure 3.2).

Public health advice recommends passive monitoring; the avoidance of sexual or skin-to-skin contact for 21 days from the last exposure, and international travel as inadvisable [8]. Low exposure risks do not require any further action and can include protected physical or droplet exposure, including contact within 3 m of an individual infected with MPXV (Table 3.3).

3.12 Pharmacology

While it is on the horizon, there is no specific antiviral therapy or vaccine available for MPXV. However, supportive care and symptom management are essential to managing the disease. People who develop a severe disease or are immunocompromised may be prescribed an antiviral agent known as tecovirimat. This medication was originally developed for smallpox and acts by interfering with a protein on the surface of orthopoxviruses, preventing virion release and, in turn, preventing viruses from reproducing normally, which slows the spread of the infection [18].

Other antiviral medications, such as acyclovir and cidofovir, have been used to treat other poxvirus infections and may be used off-label to treat severe cases of MPXV. However, their efficacy in treating MPXV specifically has not been well-established [19].

Acyclovir is an antiviral medication that is typically used to treat viral infections. It is FDA approved for the treatment of genital herpes and herpes simplex virus (HSV) encephalitis. Some non-FDA approved uses are for mucocutaneous herpes simplex virus (HSV), varicella-zoster virus (VZV), both shingles and chickenpox, and Epstein–Barr virus (EBV) [20]. The mechanism of action of acyclovir involves its conversion to an active form by the virus-encoded thymidine kinase enzyme. The active form of acyclovir inhibits viral DNA polymerase, an enzyme that is necessary for the replication of the virus. The active form of acyclovir competes with the natural nucleoside triphosphates that the virus needs to replicate its DNA, thereby causing chain termination during DNA synthesis and preventing the virus from replicating. Due to acyclovir being selectively activated by the virus-infected cells, it has a low toxicity profile and minimal side effects.

Cidofovir is an antiviral medication traditionally used to treat viral infections, particularly those caused by cytomegalovirus (CMV). In addition, cidofovir is mostly used off-label for the treatment of infections caused by several DNA viruses, such as papilloma- and polyomaviruses, and most recently, mpox, which do not encode their own DNA polymerases [21]. The mechanism of action of cidofovir is through inhibition of viral DNA synthesis. Cidofovir is a nucleotide analog, meaning it is similar in structure to the nucleotides that make up DNA. When cidofovir enters the infected cell, it is converted into an active form. The active form can then be incorporated into the growing viral DNA chain, but it lacks the necessary

3′ hydroxyl group to allow for further DNA chain elongation. This results in chain termination and ultimately inhibits viral DNA synthesis. Additionally, cidofovir also inhibits viral DNA polymerase, which is the enzyme responsible for adding new nucleotides to the growing viral DNA chain. By inhibiting both viral DNA synthesis and viral DNA polymerase, cidofovir is able to inhibit the replication of CMV and other viruses that rely on DNA synthesis for replication.

In addition to antiviral medications, treatment for MPXV includes the management of symptoms such as fever, headache, body aches, and rash. This may involve the use of pain relievers, anti-inflammatory medications, and antihistamines.

3.13 Future Implications

As the 2022 outbreak of MPXV was primarily concentrated among the population of MSM, this poses new challenges with prevention and treatment. In a population that experienced discrimination and health inequity during the heavily stigmatized AIDS epidemic, MPXV is an opportunity to provide equitable care for gay, bisexual, and other men who have MSM [22]. At the beginning of the outbreak in the United States, there was a dearth of vaccine doses available, and not every sexually active gay, bisexual, or MSM was able to receive the JYNNEOS vaccine if desired. Health equity comes into question when there is a limited amount of healthcare resources that are oftentimes most available and accessible to those who are connected and privileged, especially in a very racially diverse population. An important challenge with the prevention and treatment of MPX is to ensure equity is prioritized early in the outbreak management and to continue to ensure that every individual who desires a vaccine is able to access one [20].

Future implications for MPXV can include the potential for the emergence of new strains of the virus with increased pathogenicity or different antigenic profiles. Outbreaks of MPXV can be challenging to manage due to the lack of specific treatments and the potential for nosocomial transmission. Outbreak management will require close collaboration between public health authorities and medical professionals.

Prevention is a key to reduce the spread of MPXV. Public health surveillance, early detection, and vaccination are crucial to controlling the outbreak. Measures such as avoiding contact with infected animals, washing hands frequently, and using PPE can help reduce the risk of infection. Vaccines are currently being developed and tested, but none are yet available for use in humans.

3.14 Conclusion

MPXV, a rare zoonotic disease caused by the mpox virus, a member of the *Orthopoxvirus* genus. The virus is transmitted to humans through direct contact with infected animals or their bodily fluids, as well as through human-to-human transmission via respiratory droplets, contact with skin lesions, or fomites. MPXV is thought to be pathogenic in part due to its ability to evade host immune responses and replicate in a variety of cell types.

MPXV shares many similarities with smallpox but is less severe and typically results in a self-limiting illness characterized by fever, rash, and respiratory symptoms. Rarely, complications include secondary bacterial infections, encephalitis, and lesions on the eyes. Risk factors for severe disease include immunosuppression, pregnancy, and young age. In the recent 2022 outbreak, the outbreak centered around a population of gay, bisexual, and other MSM, leading to implications for health equity in the distribution of vaccines and limited antiviral treatment. There is currently no specific treatment for MPXV, and prevention relies on early recognition, strict infection control measures, and vaccination with the smallpox vaccine. Further research is needed to better understand the pathogenesis of the disease and to develop more effective treatments and vaccines.

References

1 Petersen, E., Kantele, A., Koopmans, M. et al. (2019). Human monkeypox: epidemiologic and clinical characteristics, diagnosis, and prevention. *Infect. Dis. Clin. N. Am.* 33 (4): 1027–1043. https://doi.org/10.1016/j.idc.2019.03.001. https://pubmed.ncbi.nlm.nih.gov/30981594/.

2 Ferdous, J., Barek, M.A., Hossen, M.S. et al. (2022). A review on monkeypox virus outbreak: new challenge for world. *Health Sci. Rep.* 6: e1007. https://doi.org/10.1002/hsr2.1007.

3 Patauner, F., Gallo, R., and Durante-Mangoni, E. (2022). Monkeypox infection: an update for the practicing physician. *Eur. J. Intern. Med.* 104: 1–6. https://doi.org/10.1016/j.ejim.2022.08.022.

4 McCollum, A.M. and Damon, I.K. (2014). Human monkeypox [published correction appears in Clin Infect Dis. 2014 Jun; 58(12): 1792]. *Clin. Infect. Dis.* 58 (2): 260–267. https://doi.org/10.1093/cid/cit703. https://academic.oup.com/cid/article/58/2/260/335791?login=true.

5 Titanji, B.K., Tegomoh, B., Nematollahi, S. et al. (2022). Monkeypox: a contemporary review for healthcare professionals. *Open Forum Infect. Dis.* 9 (7): ofac310. https://doi.org/10.1093/ofid/ofac310.

6 Bosworth, A., Wakerley, D., Houlihan, C.F., and Atabani, S.F. (2022). Monkeypox: an old foe, with new challenges. *Infect. Prev. Pract.* 4 (3): 100229. Published 2022 Jun 30. doi:https://doi.org/10.1016/j.infpip.2022.100229.

7 2023 United States Monkeypox Case. https://www.cdc.gov/poxvirus/monkeypox/outbreak/current.html. Accessed 28 March 2023.

8 World Health Organization. 2022. Clinical Management and Infection Prevention and Control for Monkeypox: Interim Rapid Response Guidance. https://www.who.int/publications/i/item/WHO-MPX-Clinical-and-IPC-2022.1WHO/MPX/Clinical_and_IPC/2022.

9 Martín-Delgado, M.C., Martín Sánchez, F.J., Martínez-Sellés, M. et al. (2022). Monkeypox in humans: a new outbreak. *Rev. Esp. Quimioter.* 35 (6): 509–518. https://doi.org/10.37201/req/059.2022. Epub 2022 Jul 6. PMID: 35785957; PMCID: PMC9728594.

10 Lahariya, C., Thakur, A., and Dudeja, N. (2022). Monkeypox disease outbreak (2022): epidemiology, challenges, and the way forward. *Indian Pediatr.* 59 (8): 636–642. https://doi.org/10.1007/s13312-022-2578-2.

11 Sapkal, A. and Agrawal, S. (2022). Monkeypox: the re-emerging terror. *Cureus* 14 (8): e28597. Published 2022 Aug 30. https://doi.org/10.7759/cureus.28597.

12 Fine, P.E., Jezek, Z., Grab, B. et al. (1988). The transmission potential of monkeypox virus in human populations. *Int. J. Epidemiol.* 17: 643–650. https://doi.org/10.1093/ije/17.3.643.

13 Rizk, J.G., Lippi, G., Henry, B.M. et al. (2022). Prevention and treatment of monkeypox. *Drugs* 82 (9): 957–963. https://doi.org/10.1007/s40265-022-01742-y. Epub 2022 Jun 28. Erratum in: Drugs. 2022 Aug; 82(12): 1343. PMID: 35763248; PMCID: PMC9244487. https://www.ncbi.nlm.nih.gov/pmc/articles/PMC9244487.

14 Reed, J.L., Scott, D.E., and Bray, M. (2012). Eczema vaccinatum. *Clin. Infect. Dis.* 54: 832–840. https://doi.org/10.1093/cid/cir952.

15 Bray, M. and Wright, M.E. (2003). Progressive vaccinia. *Clin. Infect. Dis.* 36: 766–774. https://doi.org/10.1086/374244.

16 Halsell, J.S., Riddle, J.R., Atwood, J.E. et al. (2003). Myopericarditis following smallpox vaccination among vaccinia-naive US military personnel. *JAMA* 289: 3283–3289. https://doi.org/10.1001/jama.289.24.3283.

17 Petersen, B.W., Damon, I.K., Pertowski, C.A. et al. (2015). Clinical guidance for smallpox vaccine use in a postevent vaccination program. *MMWR Recomm. Rep.* 64 (RR-02): 1–26.

18 Food and Drug Administration. Accessed 1 June 2022. https://www.fda.gov/news-events/press-announcements/fda-approves-first-drug-indication-treatment-smallpox

19 Zovi, A., Ferrara, F., Langella, R., and Vitiello, A. (2022). Pharmacological agents with antiviral activity against monkeypox infection. *Int. J. Mol. Sci.* 23 (24): 15941. https://doi.org/10.3390/ijms232415941. PMID: 36555584; PMCID: PMC9784635.

20 Taylor, M. and Gerriets, V. (2023). Acyclovir. [Updated 2022 Jun 18]. In: *StatPearls*. Treasure Island, FL: StatPearls Publishing https://www.ncbi.nlm.nih.gov/books/NBK542180.

21 Andrei, G., Topalis, D., De Schutter, T., and Snoeck, R. (2015). Insights into the mechanism of action of cidofovir and other acyclic nucleoside phosphonates against polyoma- and papillomaviruses and non-viral induced neoplasia. *Antivir. Res.* 114: 21–46. https://doi.org/10.1016/j.antiviral.2014.10.012. Epub 2014 Oct 30. PMID: 25446403.

22 Freckelton, I. and Wolf, G. (2022). Responses to monkeypox: learning from previous public health emergencies. *J. Law Med.* 29 (4): 967–986. PMID: 36763012.

4

Ebola Virus Disease: Transmission Dynamics and Management

Kshama Patel[1], Jasmine Primus[1], Carina Copley[2], and Yashwant V. Pathak[3]

[1] *Judy Genshaft Honors College, The University of South Florida, Tampa, FL, United States*
[2] *College of Agriculture and Life Sciences, The University of Florida, Gainesville, FL, United States*
[3] *College of Pharmacy, University of South Florida Health, Tampa, FL, United States*

4.1 Introduction

Ebola virus disease (EVD) is a hemorrhagic fever disease that occurs as a result of an infection of a virus belonging to the species Filoviridae and genus Ebola virus. EVD, also known as Ebola hemorrhagic fever (EHF), has been associated with an approximate 50% mortality rate [1]. EVD was first discovered in 1976 by Dr. Peter Piot, who was investigating what was believed to be a yellow fever case in Zaire, Africa (now known as the Democratic Republic of Congo) [2]. Ebola virus was named after it was first discovered near the Ebola River in the Congo. Ebola virus has six known species:

- *Zaire ebolavirus*,
- *Bundibugyo ebolavirus*,
- *Reston ebolavirus*,
- *Bombali ebolavirus*,
- *Sudan ebolavirus*,
- *Tai Forest ebolavirus*.

Z. ebolavirus, *B. ebolavirus*, and *S. ebolavirus* have caused the larger outbreaks affecting Africa. *Z. ebolavirus* was found to be the most lethal of all species. Two of the species, *R. ebolavirus* and *B. ebolavirus*, have not been found to cause disease in humans and are only known to have infected nonhuman primates and pigs [3]. Ebola case mortality rates have ranged from 25% to 90% since 1976. Due to modern treatments, patients are much more likely to survive infection following early treatment and care [4].

4.1.1 Origins of Ebola Virus

The main sources of the EVD outbreaks in Africa result from human interactions with infected animals, such as bats, chimpanzees, and gorillas [5]. Though it has not been possible to formally link Ebola virus with a specific animal reservoir, two main modes of transmission have been hypothesized:

1) Direct contact with a reservoir,
2) Contact with another animal that has contracted Ebola virus from the reservoir.

Nonhuman primates are unlikely to be the reservoir, as high pathogenicity has been found in wild populations of these primates. It is more likely that bats, such as the *Chaerephon pumilus* and *Mops condylurus*, which have been shown to survive experimental infection with the Ebola virus, represent the reservoir. This is further evidenced by the discovery of

Rising Contagious Diseases: Basics, Management, and Treatments, First Edition. Edited by Seth Kwabena Amponsah, Ranjita Shegokar, and Yashwant V. Pathak.

Ebola virus RNA in three different fruit bat species: *Epomops franqueti*, *Hypsignathus monstrosus,* and *Myonycteris torquata* [6]. Additionally, as fruit bats are frequently hunted and used for consumption as bushmeat, direct infection by bats is much more plausible [7].

4.1.2 Symptoms

The incubation period of EVD in humans ranges from 2 to 21 days. Symptoms typically present within 8 to 10 days following exposure to the Ebola virus [8]. Infected people are only contagious upon the onset of symptoms [9]. Early symptoms of the disease are similar to manifestations of other viral diseases and include constitutional symptoms such as fever, headache, myalgia, vomiting, and diarrhea. Later-onset symptoms include hemorrhagic rash as well as internal and external bleeding. The distinguishing symptoms of Ebola virus are abnormal and peculiar, occurring in late stages of EVD [2]. These symptoms consist of unexplained bleeding from the mouth or various bodily orifices [10]. Additional symptoms may persist following recovery from EVD. Such symptoms can include:

- Fatigue [9]
- Headache [9]
- Vision problems and eye pain [9]
- Hair loss and skin issues [9]
- Myalgia and arthralgia [9]
- Loss of appetite [9]
- Weight gain [9]
- Abdominal pain [9]
- Difficulty sleeping [9]
- Memory loss [9]
- Hearing loss [9]
- Anxiety and depression [9]

4.1.3 Risk Factors

Various behavior risk factors that increase the likelihood of being infected with Ebola virus exist and should be avoided as preventive measures. These risk factors include attending a funeral within one month of symptom onset and touching the body at the funeral. When combined, attending a funeral of or contacting a suspected infected patient was reported in 49.2% of case patients during the 2014–2016 West Africa Ebola epidemic. Healthcare workers are also at an increased risk of being infected while treating confirmed and suspected EVD patients. Risk factors for EVD did not change remarkably during the West African Ebola virus outbreak, except for the proportion of patient cases that reported attending a funeral prior to infection. This percentage decreased toward the end of 2014. Additionally, the percentage of healthcare workers reduced from 9.3% of all cases in August 2014 to 1.2% of cases in January 2015 [11]. Family members who do not use proper caution and infection control procedures when caring for patients with confirmed or suspected EVD are at the highest risk of contracting the virus. Travelers and the general public who have not been in contact with an Ebola patient are at low risk [12].

4.1.4 Diagnosis

Due to initial symptoms mimicking those of other infections, such as typhoid fever or malaria, the diagnosis of EVD sometimes presents a challenge. To receive a diagnosis of Ebola disease, a person must have a combination of symptoms suggestive of EVD as well as a possible exposure to an Ebola virus within 21 days of symptom onset. Possible methods of exposure include contact with any of the following:

- Blood or other bodily fluids from someone who was infected with or died from EVD,
- Objects that have been contaminated with bodily fluids of someone sick with or who died from EVD,
- Infected apes, monkeys, and fruit bats,
- Semen from a man who has been infected with and recovered from EVD.

EVD can be confirmed through positive laboratory testing. Ebola virus can be detected in the blood within three days after the onset of symptoms. The most common diagnostic technique is polymerase chain reaction (PCR), a method capable of detecting low levels of an Ebola virus in the blood. Over the course of EVD, PCR may become an ineffective diagnostic method if the virus reaches low enough blood levels. In this case, the detection of antibodies is another diagnostic technique that can be used to confirm a patient's infection with an Ebola virus [13].

4.2 Transmission Dynamics

Transmission of the Ebola virus is believed to have resulted from a spillover event, during which a virus becomes feasible in another species. After a human contracts the disease following contact with an infected animal, the virus spreads from human to human [12].

4.2.1 Human to Human Transmission

Transmission occurs primarily through close bodily contact with an infected person or their bodily fluids or through contact with contaminated surfaces or objects, such as clothing from infected patients. Unsafe burial practices with poor infection control often lead to high rates of transmission [2]. Ebola virus may sometimes persist in certain areas and organs of the body following recovery from the disease. Such areas include the testes, placenta, central nervous system, and interior of the eyes. There have been documented cases of transmission resulting from sexual contact with patients who have recovered or are recovering from EVD. Additionally, pregnant women who contract Ebola virus and recover from EVD may continue to carry the virus in their breastmilk [9]. There is no evidence of airborne transmission, such as through coughing or sneezing [14].

4.2.2 Different Outbreaks

In 1976, two concurrent outbreaks of Ebola virus occurred in Sudan and the Democratic Republic of Congo, with a reported 602 cases in total, 431 of which resulted in mortality. Since then, Ebola viruses have emerged intermittently, infecting people throughout various African countries [1].

The largest outbreak to date, the West African Ebola virus outbreak, occurred from 2014 to 2016. The initial cases were reported by WHO on 23 March 2014, in a small village in southeastern Guinea that bordered both Sierra Leone and Liberia. Due to their emergence in various major cities, such as Freetown (Sierra Leone) and Lagos (Nigeria), the increased potential sources of local and international disease propagation lead to rapid transmission [15]. Further, lack of infrastructure and surveillance systems reduced the capability of affected nations to detect and moderate the rapidly spreading virus. The outbreak spread exponentially, quickly overwhelming the limited treatment and testing resources of the affected countries [8]. The total reported number of cases reached 28,610 with 11,308 (39% of cases) becoming fatal [16].

The Center for Disease Control and Prevention's (CDC) response to the West African Ebola epidemic was the largest emergency response in the history of the CDC. Using experience from prior Ebola outbreaks, the CDC worked with African governments and partners in order to detect Ebola virus cases and terminate any lines of transmission, establishing and improving infrastructure for surveillance, laboratory testing, and systems information management and communication. The CDC also helped to support the creation of local emergency operations centers and set up an incident management system with the objective of accelerating and optimizing responses to cases. Techniques utilized included identification and isolation of infected patients, contact tracing, social mobilization, infection control, and sanitary burials [8].

Genomic technologies and massively parallel viral sequencing were used to determine when the spillover occurred that would lead to the 2014 West African outbreak. Genetic sequencing of 2014 samples suggested that the West African variant probably separated from the central African lineages of the virus in approximately 20c4, moved from Guinea to Sierra Leone in May of 2014, and has since shown continued transmission between humans, with no signs of other zoonotic origins. Studies have shown that there is a history of mutations in the virus that alter protein sequences, which may indicate the promising nature of continued genetic sequencing of Ebola viruses for the creation of novel therapies, diagnostics, and vaccines [15].

The most recent known cases of Ebola virus were of the species *S. ebolavirus* and occurred in Uganda in November 2022. This outbreak was the 6th outbreak in Uganda, five of which have been caused by *S. ebolavirus*. The outbreak was declared finished in January 2023, with 142 confirmed cases, 55 of which were fatal [16].

To date, Ebola virus outbreaks have occurred in the Democratic Republic of Congo, Guinea, Gabon, Liberia, Sierra Leone, Nigeria, South Sudan, and Uganda. Travelers from these countries have enabled the dissemination of EVD in other countries, such as Italy, Spain, Mali, Senegal, the United States, and the United Kingdom [3].

4.3 Disease Management

Since Ebola is an infectious disease that can spread rapidly from person to person, measures should be taken to help manage the disease. Taking preventative measures from the beginning is one of the best ways to help manage an infectious disease such as Ebola. However, if the preventative measures fail and an individual contracts the disease, there are some treatment methods to assist in managing the disease, as there is no cure at this moment [9].

4.3.1 Prevention

4.3.1.1 Proper Hygiene and Avoiding Contact

Like many other infectious diseases, Ebola is no different when it comes to preventative methods. The main goal is to keep yourself safe, follow proper hygiene to prevent being infected with the virus, and keep the virus from passing onto another person who may come into contact with an infected individual. Proper hygiene includes properly washing and sanitizing one's hands [9]. Furthermore, an individual living or traveling to another country where the Ebola virus is prevalent should be very cautious and take preventative measures to protect themselves from the virus. Most preventative measures consist of avoiding contact with certain people or animals that have or are suspected to have the virus. The following activities should be avoided when protecting yourself from the virus:

- Contact with body fluids, including blood, urine, feces, saliva, sweat, vomiting, breast milk, amniotic fluid, semen, and vaginal fluids, of infected individuals
- Contact with anything that an infected individual may have come into contact with, such as clothing, bedsheets, or medical equipment like needles
- Contact with animals like bats, forest antelopes, monkeys, and chimpanzees
- Attending funerals and burials where touching of an infected individual may be performed [17].

4.3.1.2 Medicinal Prevention Methods

While the main form of prevention is proper hygiene and avoiding contact with infected items and infected individuals, there is also a vaccine for Ebola that individuals can take prior to being exposed to the virus. The vaccine is called Ervebo, which helps in the prevention of the *Z. ebolavirus* strain. Ervebo was approved in the United States by the U.S. Food and Drug Administration (FDA) in 2019 for Merck & Co., Inc. after the outbreaks in West Africa between 2014 and 2016, which caused thousands of deaths. Ervebo is a single-dose injection administered to individuals eighteen or above as it is a live, attenuated vaccine that contains a protein from the *Z. ebolavirus* [18]. People who could be exposed to the virus, such as researchers, health care workers (HCW), and even travelers, are advised to take the vaccine [19]. Common side effects of the vaccine include pain, headaches, fever, joint and muscle aches, fatigue, and swelling and redness at the site of injection [18].

4.3.2 Medical Treatments

When patients are infected with Ebola, the EVD symptoms are usually managed rather than treated. This is known as supportive care, which is when the patient is provided with care that will help fight the symptoms of the disease. Supportive care for Ebola consists of the following:

- Providing an IV with fluids to keep the patient hydrated
- Providing oxygen or ventilation to circulate oxygen throughout the body
- Providing dialysis to eliminate any waste products from the blood
- Providing vasopressor drugs to help raise low blood pressure back to normal levels
- Providing medications to help with blood clotting [19].

However, climate control is also a critical factor when treating patients with EVD. Hence, HCWs have to be extremely careful when treating patients. For example, in a case study performed to explore the clinical case management of patients, researchers found that HCWs should be in a climate-controlled hospital where temperature and humidity control are provided to properly care for the patients. This is usually controlled in hospitals with individual patient rooms with biosecure cubes to separate the HCW from having direct patient interaction. This set-up allows for the patients to have transparent and air-conditioned rooms in which the HCW can monitor and have patient access through specialized ports with minimal personal protective equipment (PPE) [20].

4.3.2.1 Inmazeb

On 14 October 2020, the first treatment for the *Z. ebolavirus* to be used in both adults and pediatric patients was approved by the U.S. FDA to Regeneron Pharmaceuticals [21]. This treatment is commonly referred to as Inmazeb, which is a combination of three different fully-human monoclonal antibodies, including atoltivimab (REGN3470), maftivimab (REGN3479), and odesivimab (REGN3471). This treatment attacks the Ebola virus glycoprotein and is the treatment for the *Z. ebolavirus* [22]. The three antibodies combined allow the Ebola virus to be neutralized. This is done by blocking the virus from entering the host cells through the glycoprotein. Additionally, the antibodies enable the antibody-dependent effector function, in which the cells of the immune system will attack the infected cells. Both of these methods, when combined, help eliminate the infected cells from the individual. Inmazeb should be administered intravenously through a single infusion [23].

Infected patients who were given Inmazeb experienced different symptoms, varying from mild symptoms to severe, life-threatening symptoms. These included, but were not limited to:

- Fevers [21]
- Chills [21]
- Tachycardia [21]
- Tachypnea [21]
- Vomiting [21]
- Hypersensitivity [21]
- Hypotension [23]
- Diarrhea [23]
- Hypoxia [23].

There seemed to be a similarity in the symptoms of EVD and Inmazeb. Hence, it was difficult to determine if the symptoms were actually caused by the disease itself or if the treatment was causing the symptoms. However, one symptom that was definitely caused by the treatment was hypersensitivity at the infusion site [21]. Therefore, Regeneron Pharmaceuticals advises healthcare providers to slow down the treatment for mild infusion-associated reactions but stop the treatment if a severe reaction occurs [23]. Additionally, they advise patients not to combine any other live vaccination with Inmazeb, as this will hinder the efficacy of Inmazeb [21].

4.3.2.2 Ebanga

Shortly after the U.S. FDA approved Inmazeb as a treatment for EVD, another treatment was approved by Ridgeback Biotherapeutics, LP. On 21 December 2020, the U.S. FDA approved Ebanga as another treatment for EVD. This treatment was also approved to be used in infected adults and children. However, while Inmazeb consists of three different monoclonal antibodies, Ebanga is manufactured with one human monoclonal antibody. Ebanga helps treat EVD by blocking the virus from entering the cell by blocking the virus from binding to the cell receptor [24]. Ebanga is also administered via intravenous infusion with the assistance of a HCW. Ridgeback Biotherapeutics, LP reports the symptoms of Ebanga and the course of action for hypersensitivity reactions to be the same as those of Inmazeb [25].

4.3.2.3 Limitations to Treatments

Both Inmazeb and Ebanga have been approved to treat the *Z. ebolavirus*. However, there are multiple strains of the Ebola virus. Neither Inmazeb nor Ebanga have yet been tested to determine the efficacy of other species of Ebola virus and *Marburgvirus genera* [23, 25]. Additionally, like any other virus, the *Z. ebolavirus* can also evolve as time goes on. This means that the treatments against the virus would also have to change. Hence, both Regeneron Pharmaceuticals and

Ridgeback Biotherapeutics, LP have declared that the effects of the treatment can dwindle if the virus becomes resistant to the treatment or if the virus virulence increases [23, 25]. Furthermore, additional research is still required to determine the long term effects of both treatments since they are both fairly new treatments.

4.3.3 Experimental Therapies

Technically, there are only two treatments that are approved in the United States to be used for EVD. However, there are many experimental therapies and ongoing clinical trials that are being researched by scientists to see if they could possibly find other treatments or even a cure for EVD. Many experimental therapies focus on glycoprotein inhibitors or polymerase inhibitors [26].

4.3.3.1 ZMapp and Remdesivir

Both ZMapp and Redesivir are possible treatments that are still in the clinical trial phase stage and need more research to be done to see if they can be used to actually treat EVD. ZMapp is manufactured by Mapp Biopharmaceutical, Inc. [27], with three different monoclonal antibodies, which include 2G4, 4G7, and 13C6 [28]. This treatment works by basically attacking the main surface protein and preventing the progression of EVD [27]. While the preclinical models have shown that ZMapp helps protect against the Ebola virus, very little is known about the safety and the pharmacokinetics of the treatment. Hence, more research has to be completed before human clinical trials can even begin [29]. Even the World Health Organization has recommended against using ZMapp in patients who are confirmed to have EVD because they are not certain of how it will help or harm patients. Additionally, ZMapp requires a total of three doses administered by intravenous infusions by a healthcare professional, compared to Inmazeb and Ebanga, which require a single dose of an intravenous infusion. Hence, some difficulty can arise when treating the patients; however, the advantage of ZMapp is that it seems to be acceptable and feasible to most, if not all, patients [28].

While Remdesivir is similar to ZMapp in certain aspects, there are some vital differences. Remdesivir is a broad-spectrum antiviral that is an RNA-directed RNA polymerase inhibiting nucleoside. This means that with Remdesivir, replication of EVD will be prevented through RNA chain termination. Remdesivir is also known as Veklury. There is not enough research that has been conducted on Remdesivir. Therefore, the World Health Organization suggests not using it as a treatment for patients with confirmed EVD. The main difference between ZMapp and Remdesivir is that Remdesivir is a difficult treatment to effectively administer to patients. This is because it requires the intravenous infusion to occur over multiple days. These include the 200 mg initial loading dose in adults or the adjusted dose according to weight in pediatric patients. Then, a daily 100 mg dose of the infusion has to be administered as a maintenance dose in adults starting on day 2. The maintenance dose is administered for 9–13 continuous days, depending on the viral load. Hence, it is very inefficient and may be infeasible for some patients. Overall, neither ZMapp nor Remdesivir should be administered to patients as a treatment for EVD until further research has been completed and possibly a simpler method of administering the Remdesivir infusions [28].

4.4 Morbidity

EVD is an infectious disease characterized by its devastating impact and diverse range of clinical symptoms, such as bleeding, diarrhea, abdominal pain, vomiting, cough, and sore throat. These signs are thought to be associated with higher mortality rates in patients, highlighting the need for a thorough understanding of the disease to better contribute to effective diagnosis, treatment, and prevention [30]. This section aims to provide a comprehensive analysis of both the short-term and long-term physiological manifestations of EVD.

4.4.1 Short-Term Symptoms

The Ebola virus may infect a person through open wounds, cuts, and abrasions on the skin's surface, mucous membrane, or parenteral routes that bypass the digestive system. The initial phase of EVD infection is characterized by dry symptoms, which include fever, chills, muscle aches, or general fatigue, occurring 8–10 days from exposure [31]. However, distinguishing EVD from other infectious diseases based solely on dry symptoms can be challenging due to similarities in syndromes and overlapping geographical ranges. For instance, like EVD, malaria is prevalent in sub-Saharan Africa and surrounding tropical regions and a common cause of fever and malaise [31].

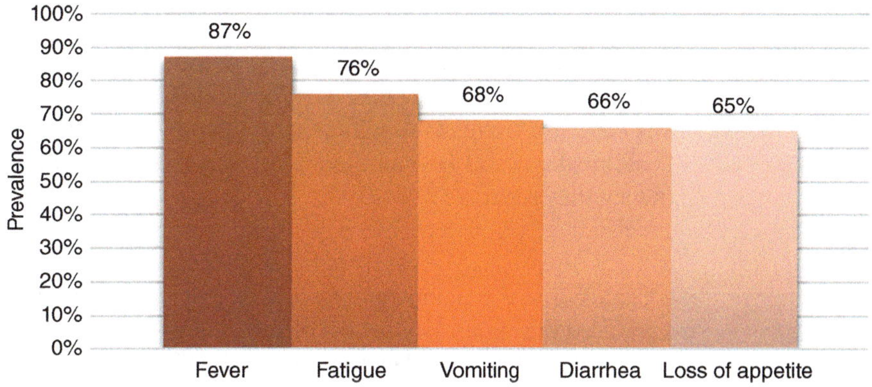

Figure 4.1 The most common symptoms reported during the West Africa outbreak from 2014 to 2016 [31]/U.S. Department of Health & Human Services /Public Domain

Approximately a week after the onset of dry symptoms, patients may present with wet symptoms, including gastrointestinal conditions like diarrhea, vomiting, or nausea. Other symptoms like conjunctival infection, chest and abdominal pain, shortness of breath, or headache may also develop [30]. More severe symptoms like bleeding develop in approximately one-fifth of patients and can manifest as bruising, blood in stool or vomit, or hemorrhages [31]. Between one-fourth and half of EVD patients have exhibited the development of a maculopapular rash that extends to the neck, torso, or arms. However, during the 2014 outbreak, this symptom was observed in only approximately 5% of those affected (Figure 4.1) [32].

Patients who experience more severe conditions, such as organ failure or septic shock, early in the course of the infection often succumb to the illness within 6–16 days from the onset of symptoms. The average time to fatality for these cases during the 2014–2016 West Africa outbreak was 7.5 days. However, in nonfatal cases, patients have commonly reported a persistent fever that subsides after the initial week of infection [31]. Survivors of EVD often experience a prolonged period of rehabilitation, referred to as convalescence. This critical phase is characterized by a gradual return to health and necessitates careful monitoring to address the lingering impacts of the disease.

4.4.2 Convalescence

Post-recovery symptoms experienced by EVD survivors are referred to as Ebola sequelae and can range from minor ailments like headaches to more significant conditions such as vision-altering uveitis and arthritis, resulting from inflammation of unknown pathogenic mechanisms [33]. Among survivors of acute EVD, approximately two-thirds have reported the persistence of two or more symptoms for duration of one to fouryears following their recovery. The most frequently reported long-lasting symptoms include fatigue and myalgia. Predictors of the onset of convalescent symptoms include age and gender; however, ongoing research on EVD survivors is crucial to understanding the natural progression of these conditions [34].

Additionally, the Ebola virus has been found to persist in various areas of the body, chiefly in the semen of male survivors, for several months after recovery from the disease. This persistence poses an added challenge as it increases the risk of transmission, further complicating the process of EVD survivors returning to normalcy [33]. Dedicated investigations aimed at identifying the underlying pathogenic mechanisms, including cellular indicators of inflammation, are essential for mitigating the impact of these symptoms on health and overall quality of life [34]. Furthermore, early recognition of the abrupt onset of fever, myalgia, malaise, bleeding, and gastrointestinal issues, among other conditions, is crucial for timely diagnosis and disease management. Comprehensive care must be established to address the complex needs of EVD survivors and reduce the long-lasting repercussions of this formidable disease.

4.5 Intersections

The emergence and spread of the EVD have demonstrated the intricate relationship between biological dynamics and the social, behavioral, and environmental factors that contribute to the transformation of localized incidents into devastating epidemics. While understanding the biological factors that contribute to the emergence of EVD is crucial, it is equally important to recognize the profound influence of these sociocultural intersections on the management of the disease.

4.5.1 Socioeconomic Status

EVD is commonly referred to as a social disease, primarily transmitted through interactions that involve the provision of care to the sick and the handling of the deceased during burials [35]. The disease thrives in settings where close contact between individuals is common, particularly within communal facilities. However, the factors that influence the extent and severity of EVD outbreaks extend beyond mere physical proximity and touch upon the complex web of social and cultural contexts in which they occur.

While socioeconomic status (SES) is recognized as a critical determinant of health in non-communicable diseases, its role in emerging infectious diseases can be observed in several geopolitical divisions. For instance, economically disadvantaged countries like Uganda and the Democratic Republic of the Congo, which rank among the top 20 poorest nations worldwide, have experienced the onset of outbreaks before their wealthier neighboring nations [36].

Particularly impoverished communities lacking access to adequate healthcare, resources, and education often face greater challenges in effectively responding to outbreaks. Limited healthcare workers and facilities impede the timely diagnosis, isolation, and treatment of infected individuals, amplifying the likelihood of disease transmission [35]. The vulnerability of populations with lower SES is compounded by factors such as overcrowded living conditions, a lack of clean water, and poor sanitation practices, all of which facilitate the spread of the virus [36].

Not only do lower SES communities serve as hotspots for EVD, but they also act as catalysts for epidemics, facilitating the transmission of the disease to other communities more readily than those from higher SES communities. A notable example is observed in Monrovia, where cases of individuals commuting in search of employment led to the migration of the disease [37]. Considering the pivotal role of SES in EVD outbreaks, it becomes evident that these communities should be prioritized in disease prevention efforts. However, the lack of public health infrastructure presents a formidable barrier. Allocation of resources, improving healthcare accessibility, and educating communities about the disease are just a few of the crucial structural adaptations necessary to control transmission.

4.5.2 Social Stigma

Fear and discrimination associated with EVD often result in the marginalization of affected individuals and communities, presenting significant challenges to outbreak control efforts. Social ostracization can discourage individuals from seeking timely medical attention or reporting their condition to others, leading to undetected and uncontrolled transmission.

The EVD-related stigma surrounding survivors is influenced by the social perception that they still pose a risk of transmitting the disease. During an active Ebola infection, the virus can be present in various bodily fluids, such as blood, urine, semen, and sweat. After a person recovers from Ebola, they are generally considered not infectious; however, ongoing research is being conducted to determine the duration of viral persistence in these bodily fluids [38]. Still, survivors may experience lingering effects known as post-Ebola sequelae, which can influence this perception. EVD-related stigma has been reported to be experienced by 35% of EVD survivors in the Democratic Republic of the Congo, 33% in Sierra Leone, and 26% in Guinea [38].

Discrimination regarding the Ebola virus is more commonly experienced and reported by female survivors than males. However, additional factors that can predict social apprehension toward the affected include age, educational attainment, and utilization of conventional medical services [38]. The consequences of EVD-related stigma are distressing, as survivors often endure alienation and discrimination from their communities, facing eviction by property owners, unemployment, and marital dissolutions. Other survivors have encountered limitations on utilizing public facilities, which have led to challenges in engaging in commerce at markets and accessing essential resources [39].

Furthermore, negative perceptions surrounding health facilities, including ambulances, treatment units, and hospitals, within EVD-exposed communities result in reduced utilization of conventional treatment among the affected populations. Particularly, the disruption of normal interactions between healthcare workers and patients during Ebola outbreaks results in varied levels of trust in medical services. Namely, during the peak of the epidemic, healthcare workers were shown to express more fear toward physical interaction with patients. To minimize the risk of infection, healthcare workers distanced themselves, inadvertently contributing to a lack of proper treatment for those affected [40].

Addressing social stigma, improving healthcare provider-patient interactions, ensuring timely and efficient transportation services, and fostering accountability within healthcare systems are crucial components for effective response and control of EVD outbreaks. By incorporating these findings into policies and interventions, we can work toward minimizing the impact of social stigma and enhancing community resilience in the face of infectious disease outbreaks.

4.5.3 Cultural and Traditional Practices

Culture remains a significant intersecting factor in the prevention and management of EVD. Traditional burial customs, often characterized by interaction with previously-infected deceased, such as washing and dressing the corpses for funerals, can contribute to the spread of the virus. Respecting cultural practices while implementing measures to reduce transmission risk is essential in order to effectively manage outbreaks and maintain the trust and cooperation of affected communities.

Over the course of Ebola outbreaks, there has been notable progress in knowledge about the disease dynamics and the adoption of preventive measures among individuals and communities. In tandem with this newfound comprehension, there is concurrent compliance with government policy regarding traditional burials, including widespread acceptance of government-provided Ebola burial teams [40]. However, despite the positive change in attitudes and the improvement in the promptness and quality of services provided by burial teams to affected families, there continues to be a tendency toward tradition. As a result of this cultural inclination, there is a heightened prevalence of prohibited practices, such as secret burials and traditional healings [40].

The transmission of EVD is heavily influenced by the suboptimal adherence to conventional treatments and the reliance on informal or non-integrated healthcare approaches, such as traditional and complementary medicine. Traditional healthcare is often preferred due to its affordability, accessibility, alignment with local values, and perceived effectiveness. In sub-Saharan Africa, approximately 58% of the general population is estimated to utilize traditional medicinal products, while 29% seek consultation from traditional healthcare practitioners [38].

Understanding this complex interplay between culture and epidemiology is paramount for successful outbreak control and prevention. Addressing these intersecting factors requires a holistic approach that integrates biomedical knowledge in a social and behavioral context. By acknowledging the multifaceted nature of EVD epidemics, we can develop comprehensive strategies to mitigate the impact of the disease, protect vulnerable communities, and foster resilience in the face of future incidences.

4.6 Conclusion

Ebola is a highly contagious infectious disease that can be fatal if supportive care to maintain the symptoms is not properly provided. Additionally, EVD occurs in many people due to a lack of proper hygiene and sanitation. Hence, if proper hygiene and sanitation were followed, the risk of EVD spreading would lessen severely. While there is no cure for EVD, thankfully, there are methods that will help lower the percentage of transmission or it will help in preventing getting the disease altogether. These include a vaccine, Ervebo, as well as avoiding contact with infected individuals, infected items, and animals that can transmit the disease. Furthermore, there are two treatments that have been approved for EVD in the United States, but the long-term effects of the treatments have not yet been researched in depth. Hence, more research is being conducted to help manufacture more treatments and possible cures for EVD, but it is proving to be difficult since the efficacy of the drugs is not as successful. There are still years of research to be performed to determine if the benefits of the treatments outweigh the adverse effects.

References

1 Kishore, S. and Singh, R. (2014). Ebola virus disease – an update. *Natl. J. Indian Assoc. Prev. Soc. Med.* 26 (4): 443–445.

2 Shamimul, H., Syed, A.A., Masood, R. et al. (2019). Ebola virus: a global public health menace: a narrative review. *J. Family Med. Prim. Care* 8 (7): 2189–2201.

3 Viral Hemorrhagic Fever Consortium: Ebola. https://vhfc.org/diseases/ebola/#:~:text=There%20are%20six%20known%20Ebolavirus,Ta%C3%AF%20Forest%2C%20and%20Zaire)

4 World Health Organization: Ebola Disease. https://www.afro.who.int/health-topics/ebola-disease#:~:text=Ebola%20case%20fatality%20rates%20have,early%20and%20given%20supportive%20care.

5 Mahwish, R., Khan, A., Fatima, M. et al. (2015). Knowledge and awareness of Ebola virus disease among medical students. *PJMHS* 9 (3): 852–855.

6 Saéz, A.M., Weiss, S., Nowak, K. et al. (2015). Investigating the zoonotic origin of the West African Ebola epidemic. *EMBO Mol. Med.* 7: 17–23.

7 Mickleburgh, S., Waylen, K., and Racey, P. (2009). Bats as bushmeat: a global review. *Oryx* 43 (2): 217–234.
8 Bell, B.P., Damon, I.K., Jernigan, D.B. et al. (2016). Overview, control strategies, and lessons learned in the CDC response to the 2014–2016 Ebola epidemic. *MMWR Suppl.* 65 (3): 4–11.
9 Ebola Virus Disease. https://www.who.int/news-room/fact-sheets/detail/ebola-virus-disease
10 Samaranayake, L., Scully, C., Nair, R.G. et al. (2014). Viral haemorrhagic fevers with emphasis on Ebola virus disease and oro-dental healthcare. *Oral Dis.* 21 (1): 1–6.
11 Dietz, P.M., Jambai, A., Paweska, J.T. et al. (2015). Epidemiology and risk factors for Ebola virus disease in Sierra Leone—23 May 2014 to 31 January 2015. *Clin. Infect. Dis.* 61 (11): 1648–1654.
12 Center for Disease Control and Prevention: Ebola Transmission. https://www.cdc.gov/vhf/ebola/transmission/index.html#:~:text=Risk,infected%20blood%20or%20body%20fluids
13 Center for Disease Control and Prevention: Ebola Diagnosis. https://www.cdc.gov/vhf/ebola/diagnosis/index.html
14 GOV.UK: Ebola: Overview, History, Origins and Transmission. https://www.gov.uk/government/publications/ebola-origins-reservoirs-transmission-and-guidelines/ebola-overview-history-origins-and-transmission#:~:text=Ebola%20virus%20can%20be%20transmitted,been%20contaminated%20with%20infectious%20secretions
15 Gire, S.K., Goba, A., Andersen, K.G. et al. (2014). Genomic surveillance elucidates Ebola virus origin and transmission during the 2014 outbreak. *Science* 345 (6202): 1369–1372.
16 Center for Disease Control and Prevention: History of Ebola Outbreaks. https://www.cdc.gov/vhf/ebola/history/chronology.html
17 Prevention and Vaccine. https://www.cdc.gov/vhf/ebola/prevention/index.html
18 First FDA-Approved Vaccine for the Prevention of Ebola Virus Disease, Marking a Critical Milestone in Public Health Preparedness and Response. https://www.fda.gov/news-events/press-announcements/first-fda-approved-vaccine-prevention-ebola-virus-disease-marking-critical-milestone-public-health
19 Ebola. https://www.hopkinsmedicine.org/health/conditions-and-diseases/ebola
20 Kiiza, P., Mullin, S., Teo, K. et al. (2020). Treatment of Ebola-related critical illness. *Intensive Care Med.* 46 (2): 285–297.
21 FDA Approves First Treatment for Ebola Virus. https://www.fda.gov/news-events/press-announcements/fda-approves-first-treatment-ebola-virus
22 Markham, A. (2021). REGN-EB3: first approval. *Drugs* 81 (1): 175–178.
23 Inmazeb: The First FDA-Approved Treatment for Zaire Ebolavirus. https://www.inmazeb.com
24 FDA Approves Treatment for Ebola Virus. https://www.fda.gov/drugs/news-events-human-drugs/fda-approves-treatment-ebola-virus
25 Ebanga. https://www.ebanga.com
26 Liu, C.H., Hu, Y.T., Wong, S.H., and Lin, L.T. (2022). Therapeutic strategies against Ebola virus infection. *Viruses* 14 (3): 579.
27 Study Finds Ebola Treatment ZMapp Holds Promise, Although Results Not Definitive. https://www.nih.gov/news-events/news-releases/study-finds-ebola-treatment-zmapp-holds-promise-although-results-not-definitive
28 Therapeutics for Ebola Virus Disease. https://apps.who.int/iris/rest/bitstreams/1459615/retrieve
29 Partnering to Rapidly Advance ZMapp for Ebola Treatment. https://www.sri.com/case-study/partnering-to-rapidly-advance-zmapptm-for-ebola-treatment
30 Kadanali, A. and Karagoz, G. (2015). An overview of Ebola virus disease. *North. Clin. Istanb.* 2 (1): 81–86.
31 Ebola Disease: Signs and Symptoms. https://www.cdc.gov/vhf/ebola/symptoms/index.html
32 Beeching, N.J., Fenech, M., and Houlihan, C.F. (2014). Ebola virus disease. *BMJ* 349: g7348.
33 Vetter, P., Kaiser, L., Schibler, M. et al. (2016). Sequelae of Ebola virus disease: the emergency within the emergency. *Lancet Infect. Dis.* 16 (6): e82–e91.
34 Tozay, S., Fischer, W.A., Wohl, D.A. et al. (2020). Long-term complications of Ebola virus disease: prevalence and predictors of major symptoms and the role of inflammation. *Clin. Infect. Dis.* 71 (7): 1749–1755.
35 Richards, P., Amara, J., Ferme, M.C. et al. (2015). Social pathways for Ebola virus disease in rural Sierra Leone, and some implications for containment. *PLoS Negl. Trop. Dis.* 9 (4): e0003567.
36 Grépin, K.A., Poirier, M.J.P., and Fox, A.M. (2019). The socio-economic distribution of exposure to Ebola: survey evidence from Liberia and Sierra Leone. *SSM Popul. Health* 10: 10c472.
37 Fallah, M.P., Skrip, L.A., Gertler, S. et al. (2015). Quantifying poverty as a driver of Ebola transmission. *PLoS Negl. Trop. Dis.* 9 (12): e00c4260.

38 James, P.B., Wardle, J., Steel, A. et al. (2020). An assessment of Ebola-related stigma and its association with informal healthcare utilization among Ebola survivors in Sierra Leone: a cross-sectional study. *BMC Public Health* 20: 182.

39 Rabelo, I., Lee, V., Fallah, M.P. et al. (2016). Psychological distress among Ebola survivors discharged from an Ebola treatment unit in Monrovia, Liberia – a qualitative study. *Front. Public Health* 4: 142.

40 Nuriddin, A., Jalloh, M.F., Meyer, E. et al. (2018). Trust, fear, stigma and disruptions: community perceptions and experiences during periods of low but ongoing transmission of Ebola virus disease in Sierra Leone, 2015. *BMJ Glob. Health* 3 (2): e00c410.

5

Avian Influenza Outbreaks over the Last Decade: An Analytical Review and Containment Strategies

Abdullah Abdelkawi[1], Zaineb Zinoune[1], Aliyah Slim[1], and Yashwant V. Pathak[1,2]

[1] *Taneja College of Pharmacy, University of South Florida, Tampa, FL, United States*
[2] *Faculty of Pharmacy, Airlangga University, Surabaya, Indonesia*

5.1 Background

5.1.1 Emergence and Spread of Avian Influenza (2013–2023)

The H5N1 variant of avian influenza A has been a persistent worry in the realms of public health and animal husbandry for the past decade [1]. This concern primarily stems from the virus's capacity to reassort, a process where two disparate viral strains infecting a single host exchange genetic material [2]. Such genetic exchange fosters novel strains, thereby amplifying the potential for pandemics [2]. H5N1 strains were initially restricted to birds, both domestic and wild [3]. However, the previous decade has witnessed an unsettling trend of mutations in the virus, primarily driven by genetic exchange between domestic poultry and wild birds [2]. This genetic interchange has been promoted by the dense coexistence of various bird species, especially on poultry farms, and has led to the emergence of a more virulent H5N1 strain [2]. This newly evolved strain displays enhanced transmissibility and a broader host range, implying that it can infect more species than before, augmenting its potential for widespread diffusion [4, 5]. The evolution and dispersion of the H5N1 virus can be charted through comprehensive global surveillance data [6]. Evidence suggests a noteworthy progression of the virus in terms of both host species and geographical locales [6]. Initial detection of the H5N1 virus in wild birds was reported across three continents – Europe, Asia, and Africa [5]. Notably, these detections were not restricted to a single species of wild birds but encompassed multiple species, indicating a broad host range and the virus's adaptability [5]. Following these initial detections, cases in domestic poultry populations soon emerged [3]. Poultry farms rapidly became primary epicenters for the virus, largely owing to the high density of potential hosts and conducive conditions for reassortment and mutation [3]. Regrettably, the dense living conditions and the regular transport of poultry for trade purposes further facilitated the virus's diffusion [3]. Moreover, the virus demonstrated the ability to transcend species boundaries, with H5N1 cases reported in various mammalian species, including pigs, cats, and dogs [4]. Such cross-species transmission is alarming as it paves a potential route for the virus to adapt to mammalian hosts, thereby escalating the risk of eventual transmission to humans [4]. Although these cross-species transmissions have been occasional, their presence underlines the adaptability of the H5N1 virus and its potential for broad-ranging impacts [4]. In terms of geography, H5N1 has shown the capability to transcend continents [6]. Initially localized in East Asia, the virus has steadily spread westward [2]. The transmission vectors for this geographic spread have been both wild migratory birds, capable of carrying the virus over lengthy distances, and human activity, particularly through the international trade of poultry [2]. As a result, H5N1 has been identified in diverse regions, including Southeast Asia, Europe, and Africa, instigating significant outbreaks in each of these areas [2]. The worldwide distribution of H5N1, along with its established ability to infect a broad range of species, presents a considerable obstacle for public health and animal husbandry [3]. Despite strenuous efforts to curb its spread, the virus has shown resilience and adaptability that continue to make it a persistent concern for future pandemics.

Rising Contagious Diseases: Basics, Management, and Treatments, First Edition. Edited by Seth Kwabena Amponsah, Ranjita Shegokar, and Yashwant V. Pathak.

5.2 Causative Microorganisms

5.2.1 H7N9 and H5N8 (2013–2023)

Parallel to H5N1, the H7N9 and H5N8 variants of AI have surfaced and spread significantly over the past decade. The H7N9 strain was initially reported in China in 2013 and has since triggered multiple outbreaks, predominantly in East and Southeast Asia [7]. Conversely, the H5N8 virus has been responsible for major outbreaks in Europe and Asia since it was first detected in South Korea in 2014 [8]. A key factor propelling the emergence and spread of both H7N9 and H5N8 is their capacity to undergo genetic reassortment. Similar to H5N1, these strains have reassorted with other circulating AI viruses, leading to the generation of novel genotypes [9]. However, the reassortment patterns have shown differences. For H7N9, reassortment has often occurred in live poultry markets, where multiple AI viruses co-circulate [9]. H5N8, on the other hand, seems to have reassorted mainly in wild waterfowl before spreading to domestic poultry [10]. In terms of species affected, both H7N9 and H5N8 have exhibited a broad host range, similar to H5N1. They have infected multiple species of birds, both wild and domestic, and, in some cases, mammalian species [11]. Notably, H7N9 has raised significant concerns due to its ability to infect humans, causing severe and frequently fatal disease [7].

Geographically, the distribution of H7N9 and H5N8 deviates from that of H5N1. While H5N1 has been primarily reported in East Asia, Europe, and Africa, H7N9 has been largely restricted to East and Southeast Asia, particularly China [7]. H5N8, meanwhile, has experienced major outbreaks in Europe and Asia, with some sporadic detections in North America [8].

The emergence and spread of H7N9 and H5N8 bear profound implications for public health and animal husbandry. The capacity of H7N9 to infect humans, coupled with its continued circulation in live poultry markets, presents a significant public health threat [7]. It also emphasizes the necessity for stronger surveillance in these markets to prevent human infections. H5N8, while not yet proven to cause human disease, has triggered massive outbreaks on poultry farms [8]. This has had substantial economic implications and underscores the need for enhanced biosecurity measures in these settings [8]. Both viruses, like H5N1, continue to present ongoing challenges in controlling their spread and minimizing their impact on public health and animal husbandry [11].

5.2.2 H5N1: Impact on Humans

The previous decade has observed H5N1, a variant of the AI virus, inflict significant harm on human health [12]. This virulent subtype, noted for its high mortality rate, has caused extensive illness and deaths, primarily in Southeast Asia and Egypt [13]. The World Health Organization (WHO) reported a disturbing increase in human H5N1 infections since 2013, with a case fatality rate exceeding 50% [13]. Direct contact with diseased poultry or exposure to environments harboring the virus has been the principal route for human infection [14]. These situations are predominantly seen in rural and farming communities where regular interaction with domestic poultry is common, highlighting the importance of managing the human-animal interface for disease control.

The repercussions of the H5N1 virus stretch beyond causing severe disease in humans, as it represents a formidable risk to public health due to its potential to ignite a global pandemic [12]. This forms a pressing concern for organizations like the Centers for Disease Control and Prevention (CDC) and equivalent global public health entities [14]. While the H5N1 virus is currently not proficient at human-to-human transmission, if it were to acquire this capability through genetic reassortments or mutations, it could instigate a global pandemic [12].

The public health risk posed by H5N1 is amplified by several factors. Firstly, its high pathogenicity leads to severe symptoms and potentially fatalities in a significant portion of the infected population [14]. Secondly, the absence of pre-existing immunity within humans leaves us vulnerable to a widespread infection if the virus develops efficient human-to-human transmission capabilities. Vaccination remains a key preventive measure; however, the virus's constant evolution poses a challenge to the development of effective and long-lasting vaccines.

Moreover, the socio-economic consequences of a human H5N1 pandemic could be massive [12]. Given the high fatality rate, a large-scale outbreak could lead to significant loss of life, overwhelm healthcare systems, and result in considerable disruptions in social order [14]. The economic impact, including direct medical costs and indirect expenses arising from societal disruption, could be devastating [12]. The 2020 COVID-19 pandemic serves as a stark warning of these potential impacts, underlining the necessity for continuous surveillance and prevention initiatives for H5N1.

5.2.3 H7N9: Impact on Humans

First observed in humans in China in 2013, the H7N9 variant of avian influenza A has posed a considerable public health challenge. Similar to H5N1, the virus has exhibited a high mortality rate, with severe cases frequently progressing to critical conditions such as pneumonia and acute respiratory distress syndrome (ARDS) [15]. Furthermore, sporadic but consistent human infections since its emergence have emphasized the ongoing risk associated with this virus [16]. The H7N9 AI virus has led to over 1500 human infections with a mortality rate of roughly 40% since the first human case in 2013 [17]. Limited human-to-human spread of the virus has been reported, though without sustained transmission [18]. The virus has mutated into highly pathogenic variants and has instigated five outbreak waves in humans [19]. The virus has been circulating silently in poultry, and understanding its activity in humans in China is essential [20]. The virus has developed resistance to neuraminidase inhibitors without losing its in vivo virulence or transmissibility [21]. The consequences of changes in viral protein on the adaptation and pathogenicity of H7N9 have been investigated to devise early warning or intervention strategies to prevent the circulation of highly pathogenic IAVs in live poultry and transmission to humans [22]. The emergence of new infectious diseases, such as H7N9, represents a serious threat to human health and life. H7N9 infections have been most common among individuals in direct contact with live poultry, such as those employed in live poultry markets or involved in the poultry industry. Notably, many reported cases were not associated with severe illness in birds, suggesting a "silent" circulation of the virus in poultry populations, making it challenging to control its spread [23]. The potential for H7N9 to instigate a pandemic is considerable. While the virus currently does not exhibit efficient human-to-human transmission, its high mutation rate presents the risk of the virus acquiring this ability. Moreover, as with H5N1, humans have no pre-existing immunity to H7N9, which means a possible pandemic, could result in widespread illness and severe outcomes [22]. The threat of H7N9 has galvanized global efforts to develop an effective vaccine, but challenges persist due to the virus's continuous evolution. This makes surveillance and rapid response measures vital to mitigating its public health impact. Socio-economic consequences associated with a possible H7N9 pandemic could be severe, reminiscent of the devastation caused by COVID-19. Such a pandemic would not only entail significant health costs due to increased morbidity and mortality but also exert a heavy toll on the global economy due to disruptions to trade, travel, and daily life. The potential for such a scenario underscores the importance of ongoing efforts to control the spread of H7N9.

5.2.4 H5N8: Impact on Humans

The H5N8 strain of avian influenza A, while less discussed in relation to human health, has been a substantial concern in the veterinary and wildlife health fields due to its high pathogenicity in birds [24]. In the year 2021, – the H5N8 virus, initially identified in birds, was detected in humans for the first time, with the incident being reported in Russia [25]. Cases of human infection with the H5N8 virus have largely been associated with direct interaction with birds that have been infected by the virus, a pattern reminiscent of what we have seen with H5N1 and H7N9 [26]. Despite the relatively limited instances of human cases at this juncture, the ability of H5N8 to transition across species underscores the potential for the virus to acclimate to new hosts. Even though H5N8's ability to transmit effectively among humans remains deficient at present, it possesses the potential to attain such capacity through genetic reconfiguration or mutation – a phenomenon we have seen occur with H5N1 and H7N9 [27]. In the unfortunate event that H5N8 develops this capability, the human population, devoid of pre-existing immunity, could be looking at a potential pandemic situation [26]. The scientific community is currently mobilizing to create a vaccine against H5N8, but the virus's continuous evolutionary nature adds a layer of complexity to these efforts [28]. For now, rigorous monitoring of H5N8 and prompt action in response to any fresh instances of human infection remain our best strategies to preempt a potential pandemic. If H5N8 were to spread widely, the socio-economic repercussions could be profound. Although the current risk remains on the lower side, a large-scale outbreak could potentially result in substantial loss of human life, impose immense pressure on healthcare systems, and disrupt socio-economic activities on a global scale [29]. In light of these potential consequences, the need for continued alertness and proactive steps to stifle the spread of H5N8 is of the utmost importance. Our experiences dealing with H5N1, H7N9, and COVID-19 serve as stark illustrations of the destructive impacts global pandemics can inflict.

5.2.5 H5N1: Impact on Animals

The H5N1 subtype of the AI virus, commonly known as bird flu, has dramatically influenced various animal species globally [11]. The primary victims have been avian species, including domestic poultry and wild birds, although sporadic infections in mammals have been documented, with varying disease severity [30]. Within domestic poultry, H5N1 infection often results in acute disease, leading to 100% mortality within 48 hours, thereby inflicting significant economic damage to global poultry industries [11]. The virus has ravaged poultry farms, necessitating widespread culling of birds to hinder further spread. The potential implications of this scenario for food security are grave, particularly in regions where the primary protein source is poultry. The carriers of AI viruses, including H5N1, are often migratory waterfowl, a subset of wild birds [31]. These birds exhibit no symptoms, yet they are known to serve as a conduit for introducing the virus to new geographical locales along their migratory paths.

The instances where H5N1 infections have been reported in mammals – including humans – have been relatively rare, yet their significance is undeniable [30]. The virus has been detected in various species, such as cats, dogs, pigs, and ferrets, which introduces a new dimension of concern regarding possible novel pathways of transmission and virus reservoirs. The transmission of H5N1 across species is an alarming prospect because it creates additional opportunities for the virus to mutate or reassort. Such changes could potentially enhance its virulence or its transmissibility [32].

Efforts to control H5N1 in animals largely involve swift detection and response. Typical farm biosecurity measures include implementing disinfection protocols and restricting access to poultry areas. During outbreaks, large-scale culling of infected or potentially exposed birds is usually carried out to curb the virus's spread. Poultry vaccination is another strategy utilized in some regions, although its efficacy can vary due to the virus's evolving nature [33].

Despite these attempts, the persistent circulation of H5N1 within animal populations and its continued threat to animal and human health demand ongoing vigilance, surveillance, and response capabilities. Addressing this issue also requires a One Health approach, acknowledging the interconnectedness of human, animal, and environmental health [34].

5.2.6 H7N9 and H5N8: Impact on Animals

In similarity to H5N1, the AI viruses H7N9 and H5N8 have had considerable effects on various animal species worldwide, predominantly affecting avian species, yet demonstrating the capability to infect mammals [31]. H7N9 primarily impacts domestic poultry such as chickens and ducks, and its ramifications on global animal health are significant. The infection often progresses into severe disease, resulting in substantial mortality rates and thereby inflicting considerable economic damage within the poultry industry [32]. It becomes a daunting task to contain H7N9 due to its ability to remain undetected in certain poultry, resulting in an unnoticed spread and extensive culling once an outbreak is eventually discovered. This, analogous to H5N1, poses threat to food security in regions that are heavily dependent on poultry.

Conversely, H5N8 proves especially fatal to wild birds. There have been records of outbreaks in diverse wild bird populations, from waterfowl to birds of prey [33]. The implications of these outbreaks are extensive, potentially disrupting ecological balances and facilitating the geographic spread of the virus. As asymptomatic carriers, these wild birds can distribute the virus along their migratory paths, effectively serving as a significant element in H5N8's global propagation.

Instances of H7N9 and H5N8 infections in mammalian species, including cats, dogs, and pigs, have been infrequent but nonetheless pivotal. Their capacity to cross the species barrier and infect mammals can lead to additional reservoirs and transmission routes, thus enhancing the complexity of control measures [11]. This cross-species transmission also provokes the fear of the viruses reassorting or mutating to increase their virulence or transmissibility, akin to the anxieties surrounding H5N1.

To address these two influenza subtypes, approaches similar to those employed against H5N1 are invoked. Swift detection and immediate response, along with improved biosecurity measures at the farm level, are critical. In case of an outbreak, culling of infected or potentially exposed birds is commonly practiced to prevent further spread. However, compared to H5N1, vaccination strategies for H7N9 and H5N8 are less developed due to their relatively recent emergence and ongoing evolution [34].

Much like with H5N1, the sustained circulation of H7N9 and H5N8 within animal populations and their continual threat to animal and human health necessitate constant vigilance, surveillance, and response capabilities. This highlights the importance of a One Health approach, recognizing the integral relationship between human, animal, and environmental health.

5.3 Strategies for Containment

AI, specifically the H5N1 strain, poses a significant threat to both human and animal health, and its containment is a global concern. The fundamental strategies for containment include biosecurity measures, vaccination, public education, surveillance, and host-adaptive strategies. Biosecurity measures play a crucial role in preventing the spread of the virus among poultry populations. These measures encompass controlling the movement of poultry, disinfecting poultry premises, and ensuring poultry workers utilize personal protective equipment [35]. By executing these measures, the transmission of AI viruses such as H7N9, H5N1, and H5N8 can be significantly reduced. Vaccination is another fundamental strategy that has been extensively employed, especially in countries where the virus is endemic to poultry. Vaccines aim to decrease the quantity of viruses circulating in poultry, thereby lowering the risk of human exposure [36]. In Egypt, for instance, where H5N1 is endemic, the control strategy has shifted toward mass vaccination of backyard birds using inactivated H5 vaccines, provided by the government free of charge [37]. Public education is another essential component of containment strategies. It is paramount to educate the public regarding the risks associated with exposure to infected poultry and to advocate behaviors that minimize the risk of exposure. This includes avoiding direct contact with live poultry, maintaining proper hand hygiene, and practicing safe food handling [38]. By increasing awareness and promoting these preventive measures, the spread of AI viruses can be further curtailed. Surveillance constitutes the fourth pillar of containment strategies. It entails monitoring poultry populations for signs of disease and conducting laboratory tests to confirm the presence of the virus. Surveillance data aids in tracking the spread of the virus, identifying areas where control measures need to be enacted, and assessing the effectiveness of these measures [39]. Techniques such as glycan-functionalized gold nanoparticles can also be harnessed for a rapid assessment of the potential threat of AI viruses to humans [40]. Furthermore, host-adaptive strategies, like the E627K substitution in the PB2 protein, which is critical for the replication of avian influenza A viruses in mammalian hosts, can be developed to contain AI viruses [41]. Understanding the genetic adaptations that permit AI viruses to infect mammals enables the development of targeted interventions to disrupt their replication and transmission. It's vital to acknowledge that emerging infectious diseases like avian influenza can have a significant impact on health, trade, and biodiversity. The circulation of highly pathogenic avian influenza viruses (HPAIVs) of various subtypes (such as H5N1, H5N6, H5N8, and H7N9) in poultry remains a global concern for animal and public health, and measures to control the spread of these viruses in poultry must be implemented [40]. Outbreaks stemming from the wildlife trade have highlighted the need for stricter regulations and enforcement to prevent the introduction and spread of AI viruses. In conclusion, a comprehensive approach that encompasses biosecurity measures, vaccination, public education, surveillance, and host-adaptive strategies is indispensable for effectively containing AI viruses and mitigating the risks they pose to both human and animal health. Collaboration between governments, healthcare organizations, and scientific communities is vital to combating this global threat and averting potential pandemic situations.

5.4 Ongoing Investigations and Prospective Look into Avian Influenza

Investigating the biology and epidemiology of the H5N1, H7N9, and H5N8 AI viruses is crucial [42]. This requires multidimensional research, ranging from surveillance and diagnostic tool creation to therapeutic measures and vaccine strategy examination [42]. International surveillance networks stand at the vanguard of this approach, providing indispensable tools for monitoring these viruses' dynamics within human and animal populations [42]. Utilizing a mix of active and passive surveillance methodologies, these networks can discern shifts in virus prevalence, distribution, and genomic makeup globally [43].

Real-time data from these networks serves as an essential resource for swift interventions during outbreaks and molding research agendas in vaccine and therapeutic development [44]. Advancements in diagnostics, including portable real-time polymerase chain reaction (PCR) systems, represent a paradigm shift in virus identification [42]. These technologies offer rapid, precise detection at the patient's side or an animal outbreak site, thus reducing diagnostic delays that could inadvertently propagate viral spread [43].

Despite the array of antiviral medications and vaccines available, the rapid mutation capacity of these AI viruses presents a significant challenge [44]. This highlights the need for ongoing research into resilient and enduring therapeutic interventions, including the exploration of monoclonal antibody treatments, antiviral drugs, gene therapies, and innovative vaccine technologies [42, 43].

The effectiveness of current vaccines for H5N1, H7N9, and H5N8 continues to be scrutinized [44]. Innovative strategies are being rigorously pursued, such as using adjuvants to amplify immunological responses, developing broader-spectrum vaccines offering protection against various strains, and pursuing universal vaccines [42]. AI remains a concern for animal and human health, necessitating large-scale surveillance efforts and focused research studies [45]. These include evaluating the effectiveness of poultry vaccination programs for AI control and eradication at the national and regional levels [46].

Prospective data from a cohort in Nigeria is being gathered to gain insights into the occurrence of zoonotic infections due to AI viruses [47]. In parallel, studies are investigating virus inactivation using disinfectants like glutaraldehyde [48]. The threat of an AI (H5N1) pandemic poses risks to human and animal health, emphasizing the importance of identifying both highly pathogenic avian influenza (HPAI) and low pathogenic avian influenza (LPAI) during outbreaks [48]. Laboratory systems and services play an instrumental role in providing diagnostic support for AI outbreak investigations.

Adopting a "One Health" approach is essential given the interconnected impact of these viruses on human health, animal health, and environmental conservation [43]. This approach promotes cohesive and cooperative research efforts across various sectors and geographical boundaries [44]. The future objective remains enhancing global collaborations and building capacities in regions most vulnerable to these AI outbreaks. The comprehensive and coordinated efforts strive to manage the global threat posed by H5N1, H7N9, and H5N8 effectively.

References

1 Imai, M., Watanabe, T., Hatta, M. et al. (2012). Experimental adaptation of an influenza H5 HA confers respiratory droplet transmission to a reassortant H5 HA/H1N1 virus in ferrets. *Nature* 7403 (486): 420–428. https://doi.org/10.1038/nature10831.

2 Kandeil, A., Patton, C., Jones, J.C. et al. (2022). Rapid evolution of A(H5N1) influenza viruses after intercontinental spread to North America. *Nat. Commun.* 14: 3082.

3 Horwood, P.F. (2021). Avian influenza H5N1: still a pandemic threat? *Microbiol. Aust.* 42 (4): 152–155. https://doi.org/10.1071/ma21044.

4 Tajudeen, Y.A., Ajide-Bamigboye, N.T., and Oladunjoye, I.O. (2021). Emerging strain (H5N8) of highly pathogenic avian influenza virus: an impending pandemic threat. *J. Infect. Dis. Epidemiol.* 7 (7): 217. https://doi.org/10.23937/2474-3658/1510217.

5 Juariah, S., Irawan, M.P., Surya, A., and Mahardika, D.P. (2020). Molecular epidemiology of avian influenza virus (H5N1), Sumatera Indonesia. *IOP Conf. Ser. Earth Environ. Sci.* 430 (1): 012019. https://doi.org/10.1088/1755-1315/430/1/012019.

6 Njoto, E.N., Scotch, M., Bui, C.M. et al. (2018). Phylogeography of H5N1 avian influenza virus in Indonesia. *Transbound. Emerg. Dis.* 65 (5): 1339–1347. https://doi.org/10.1111/tbed.12883.

7 Haider, N., Kock, R., Zumla, A., and Lee, S.S. (2023). Consequences and global risks of highly pathogenic avian influenza outbreaks in poultry in the United Kingdom. *Int. J. Infect. Dis.* 129: P162–P164.

8 Vigeveno, R.M., Poen, M.J., Parker, E. et al. (2020). Outbreak severity of highly pathogenic Avian Influenza A(H5N8) viruses is inversely correlated to polymerase complex activity and interferon induction. *J. Virol.* 94 (11): e00375-20. https://doi.org/10.1128/JVI.00375-20. PMID: 32238581; PMCID: PMC7269427.

9 Bi, Y. et al. (2016). Genesis, evolution and prevalence of H5N6 Avian Influenza Viruses in China. *Cell Host Microbe* 20: 810–821.

10 Lycett, S.J., Duchatel, F., and Digard, P. (2019). A brief history of bird flu. *Philos. Trans. R. Soc. B* 374: 20180257. https://doi.org/10.1098/rstb.2018.0257.

11 Liu, J., Xiao, H., Lei, F. et al. (2005). Highly pathogenic H5N1 influenza virus infection in migratory birds. *Science* 309 (5738): 1206. https://doi.org/10.1126/science.1115273. PMID: 16000410.

12 Belser, J.A., Johnson, A.R., Pulit-Penaloza, J.A. et al. (2017). Pathogenicity testing of influenza candidate vaccine viruses in the ferret model. *Virology* 511: 135–141.

13 de Jong, M.D. (2008). H5N1 transmission and disease: observations from the frontlines. *Pediatr. Infect. Dis. J.* 27 (10): S54–S56. https://doi.org/10.1097/INF.0b013e3181684d2d.

14 Oh, M., El-Shesheny, R., Kutkat, M.A. et al. (2011). Continuing threat of influenza (H5N1) virus circulation in Egypt. *Emerg. Infect. Dis.* 17 (12): 2306–2308. https://doi.org/10.3201/eid1712.110683.

15 Belser, J.A., Gustin, K.M., Pearce, M.B. et al. (2013). Pathogenesis and transmission of avian influenza A (H7N9) virus in ferrets and mice. *Nature* 501 (7468): 556–559. https://doi.org/10.1038/nature12391.

16 Mulligan, M.J., Bernstein, D.I., Winokur, P. et al. (2014). Serological responses to an avian influenza A/H7N9 vaccine mixed at the point-of-use with MF59 adjuvant. *JAMA* 312 (14): 1409. https://doi.org/10.1001/jama.2014.12854.

17 Chang, P., Sadeyen, J.R., Bhat, S. et al. (2023). Risk assessment of the newly emerged H7N9 avian influenza viruses. *Emerg. Microbes Infect.* 12 (1): 2172965. https://doi.org/10.1080/22221751.2023.2172965.

18 Hai, R., Schmolke, M., Leyva-Grado, V.H. et al. (2013). Influenza A(H7N9) virus gains neuraminidase inhibitor resistance without loss of in vivo virulence or transmissibility. *Nat. Commun.* 4 (1): 2854. https://doi.org/10.1038/ncomms3854.

19 Wang, D., Yang, L., Zhang, Y. et al. (2016). Two outbreak sources of influenza A (H7N9) viruses have been established in China. *J. Virol.* 90 (12): 5561–5573. https://doi.org/10.1128/jvi.03173-15.

20 Wang, D., Yang, L., Zhu, W. et al. (2016). Two outbreak sources of influenza A (H7N9) viruses have been established in China. *J. Virol.* 90 (12): 5561–5573. https://doi.org/10.1128/JVI.03173-15. PMID: 27030268; PMCID: PMC4886776.

21 Nishiura, H., Mizumoto, K., and Ejima, K. (2013). How to interpret the transmissibility of novel influenza A(H7N9): an analysis of initial epidemiological data of human cases from China. *Theor. Biol. Med. Model.* 10: 30. https://doi.org/10.1186/1742-4682-10-30.

22 World Health Organization (WHO) (2013). *Human Infection with Influenza A(H7N9) Virus in China – Update*. Geneva, Switzerland: WHO http://www.who.int/csr/don/2013_04_09/en/index.html. Accessed 18 April 2013.

23 World Health Organization (WHO). 2013. Background and Summary of Human Infection with Influenza A(H7N9) Virus – as of 5 April 2013. http://www.who.int/influenza/human_animal_interface/update_20130405/en/index.html. Accessed 18 April 2013.

24 Liu, D., Shi, W., and Gao, G.F. (2018). Poultry carrying H5N1 in live bird markets in China as a major threat to public health. *Emerg. Microbes Infect.* 2 (1): 1–4.

25 Liu, D., Shi, W., and Gao, G.F. (2023). First cases of human H5N8 infection in Russia. *Emerg. Microbes Infect.* 2 (1): 1–4.

26 Liu, D., Shi, W., and Gao, G.F. (2014). Potential for H5N8 to adapt to human hosts. *Emerg. Microbes Infect.* 3 (1): 1–4.

27 Liu, D., Shi, W., and Gao, G.F. (2018). Efforts to develop a vaccine against H5N8. *Emerg. Microbes Infect.* 4 (1): 1–4.

28 Liu, D., Shi, W., and Gao, G.F. (2019). Potential socio-economic impacts of an H5N8 pandemic. *Emerg. Microbes Infect.* 5 (1): 1–4.

29 Liu, D., Shi, W., and Gao, G.F. (2020). Experiences with H5N1, H7N9, and COVID-19 as reminders of global pandemics. *Emerg. Microbes Infect.* 6 (1): 1–4.

30 Kilpatrick, A.M., Chmura, A.A., Gibbons, D.W. et al. (2006). Predicting the global spread of H5N1 avian influenza. *Proc. Natl. Acad. Sci. U. S. A.* 103 (51): 19368–19373.

31 Chen, H., Smith, G.J., Zhang, S.Y. et al. (2005). Avian flu: H5N1 virus outbreak in migratory waterfowl. *Nature* 436 (7048): 191–192.

32 Reperant, L.A., Kuiken, T., and Osterhaus, A.D. (2012). Adaptive pathways of zoonotic influenza viruses: from exposure to establishment in humans. *Vaccine* 30 (30): 4419–4434.

33 Swayne, D.E., Pavade, G., Hamilton, K. et al. (2011). Assessment of national strategies for control of high-pathogenicity avian influenza and low-pathogenicity notifiable avian influenza in poultry, with emphasis on vaccines and vaccination. *Rev. Sci. Tech.* 30 (3): 839.

34 Jones, B.A., Grace, D., Kock, R. et al. (2013). Zoonosis emergence linked to agricultural intensification and environmental change. *Proc. Natl. Acad. Sci. U. S. A.* 110 (21): 8399–8404.

35 Abdelwhab, E. and Hafez, H. (2011). An overview of the epidemic of highly pathogenic H5N1 avian influenza virus in Egypt: epidemiology and control challenges. *Epidemiol. Infect.* 139 (5): 647–657. https://doi.org/10.1017/S0950268810003122.

36 Song, W., Wang, P., Mok, B.W. et al. (2014). The K526R substitution in viral protein PB2 enhances the effects of E627K on influenza virus replication. *Nat. Commun.* 5 (1): 5509. https://doi.org/10.1038/ncomms6509.

37 Wei, J., Zheng, L., Lv, X. et al. (2014). Analysis of influenza virus receptor specificity using glycan-functionalized gold nanoparticles. *ACS Nano* 8 (5): 4600–4607. https://doi.org/10.1021/nn5002485.

38 Lee, J., Kim, S., and Yoon, T. (2016). Treatment of various avian influenza virus based on comparison using decision tree algorithm. *MATEC Web Conf.* 69: 01004. https://doi.org/10.1051/matecconf/20166901004.

39 Zhao, H., Xu, K., Jiang, Z. et al. (2018). A neuraminidase activity-based microneutralization assay for evaluating antibody responses to influenza H5 and H7 vaccines. *PLoS One* 13 (11): e0207431. https://doi.org/10.1371/journal.pone.0207431.

40 Ulaankhuu, A., Bazarragchaa, E., Okamatsu, M. et al. (2020). Genetic and antigenic characterization of H5 and H7 avian influenza viruses isolated from migratory waterfowl in Mongolia from 2017 to 2019. *Virus Genes* 56 (4): 472–479. https://doi.org/10.1007/s11262-020-01764-2.

41 Ma, H., Dong, J.P., Zhou, N., and Pu, W. (2016). Military-civilian cooperative emergency response to infectious disease prevention and control in China. *Mil. Med. Res.* 3 (1): 39. https://doi.org/10.1186/s40779-016-0109-y.

42 Dudley, J.P. (2008). Public health and epidemiological considerations for avian influenza risk mapping and risk assessment. *Ecol. Soc.* 13 (2): 21. https://doi.org/10.5751/es-02548-130221.

43 Okoye, J., Eze, D., Krueger, W.S. et al. (2013). Serologic evidence of avian influenza virus infections among Nigerian agricultural workers. *J. Med. Virol.* 85 (4): 670–676. https://doi.org/10.1002/jmv.23520.

44 Bevins, S.N., Pedersen, K., Lutman, M.W. et al. (2014). Large-scale avian influenza surveillance in wild birds throughout the United States. *PLoS One* 9 (8): e104360. https://doi.org/10.1371/journal.pone.0104360.

45 Sakoda, Y., Endo, M., Sato, Y. et al. (2012). Effects of disinfectant containing glutaraldehyde against avian influenza virus. *J. Jpn. Vet. Med. Assoc.* 65 (4): 303–305. https://doi.org/10.12935/jvma.65.303.

46 Apisarnthanarak, A., Puthavathana, P., Kitphati, R. et al. (2006). Avian influenza H5N1 screening of intensive care unit patients with community-acquired pneumonia. *Emerg. Infect. Dis.* 12 (11): 1766–1769. https://doi.org/10.3201/eid1211.060443.

47 Parsons, L.M., Somoskovi, A., Lee, E. et al. (2011). Global health: integrating national laboratory health systems and services in resource-limited settings. *Afr. J. Lab. Med.* 1 (1): 11. https://doi.org/10.4102/ajlm.v1i1.11.

48 Payungporn, S., Chutinimitkul, S., Chaisingh, A. et al. (2006). Discrimination between highly pathogenic and low pathogenic H5 avian influenza A viruses. *Emerg. Infect. Dis.* 12 (4): 700–701. https://doi.org/10.3201/eid1204.051427.

6

Swine Flu: Current Status and Challenges

Lucy Mohapatra[1], Geeta Patel[2], Alok S. Tripathi[3], Alka[1], Deepak Mishra[1], Sambit K. Parida[4], Mohammad Yasir[1], and Rahul K. Maurya[1]

[1] Amity Institute of Pharmacy, Lucknow, Amity University Uttar Pradesh, Noida, Uttar Pradesh, India
[2] Shree S K Patel College of Pharmaceutical Education and Research, Ganpat University, Kherva, Gujarat, India
[3] Department of Pharmacology, ERA College of Pharmacy, ERA University, Lucknow, Uttar Pradesh, India
[4] Seth Vishambhar Nath Institute of Pharmacy, Barabanki, Uttar Pradesh, India

6.1 Introduction

Flu viruses infect several animal species, especially horses, pigs, birds, marine animals, and humans. Historically, pandemics induced by flu viruses often resulted in the deaths of millions worldwide [1]. The influenza A virus (IAV) of Orthomyxoviridae family causes swine flu, a serious respiratory condition that affects pigs and has been a pandemic across most parts of the world [2]. SIV often leads to significant fatalities of about 100% and moderate deaths in pigs suffering from the infection; however, it may also create a fatality rate of 10–15% in pigs that are not immune. Swine have been regarded as a promising animal model for influenza experimentation because infectious pigs exhibit clinical symptoms that are comparable to those seen in people [3]. The typical clinical indicators of an acute respiratory illness include fever, tiredness, coughing, sneezing, trouble breathing, and change in appetite [4]. Respiratory infections are frequent illnesses caused by germs (mainly viruses), which are more common in our regions in winter. Influenza viruses (IV) have a big effect, and becoming infected with one of them causes true influenza, which must be distinguished from the common cold, which is spread by rhinoviruses or other viruses [1]. Even though IV infections are often more harmful and can occasionally be fatal, they can be challenging to discern between diverse clinical indications in many situations [5]. The primary reason why swine flu costs pig farmers a lot of money is that sick pigs lose weight and do not gain as much weight. Nevertheless, certain instances can be substantially worse if other infections are also present. Together with other swine pathogens like the porcine reproductive and respiratory syndrome virus, *Ap-pneumoniae*, *B. bronchiseptica*, *M. hyopneumoniae,* and porcine circovirus 2, the SIV can lead to porcine respiratory disease complex, which causes significantly increased mortality rates and financial losses for the swine industry each year. The highly concentrated rearing model used in the swine business has made it unlikely that SIV is a periodic illness like human seasonal influenza, which mostly occurs during winter and spring [6]. The phrase "seasonal influenza" is often used to refer to illnesses caused by IV that are limited to the winter season. It seems like they happen annually. The word "pandemic influenza" refers to the spread of influenza infections that, in contrast to this seasonal virus, spread all over the world [5, 7]. Scientific knowledge is important to comprehend complicated policy challenges and help with decision-making. In that environment, scientific competence is progressively institutionalized in protocols as well as in organizations, governments, and international entities. However, there are drawbacks to modern science due to its unpredictability, which is a distinguishing aspect of complicated policy situations [7]. In this chapter, we examine how alternatives as well as complementary therapies have been successfully utilized by humans for many years to treat a variety of illnesses and how they can be used to focus host response during influenza epidemics [8]. The foundation of medical technology is a medical paradigm that primarily gives symptomatic treatment and places a greater emphasis on therapeutic interventions [8]. It focuses on the use of medications, hardness

Rising Contagious Diseases: Basics, Management, and Treatments, First Edition. Edited by Seth Kwabena Amponsah, Ranjita Shegokar, and Yashwant V. Pathak.

testing, intrusive therapies including surgery, and a passive attitude toward the patient. In medical advances, there are now two kinds of antiviral medications that can be used to cure seasonal human influenza: neuraminidase inhibitors like adamantanes (rimantadine and amantadine) and zanamivir (Relenza) and oseltamivir (Tamiflu) [9].

6.2 Current Status – Swine Flu

There are now three major forms of IAVs that are moving in swine herds all over the world: H1N1, H3N2, and H1N2. The 1918-like cH1N1 virus was discovered and spread to pigs globally just after the 1918 human pandemic (Figure 6.1) [10]. The cH1N1 has already been spreading in porcine species over the world for several years, but in certain regions, like Europe, it has vanished and been substituted by newly emerging SIV [11]. As SIV monitoring has been conducted in these regions for so many years and there is a great deal of information available, this review will emphasize on the present situation of SIVs in Asia, Europe, and North America.

6.2.1 Swine Influenza Viruses in Asia

The cH1N1 SV was initially identified in Asia in 1974, although it has long been believed to have been prevalent in bovines in China during the 1918–1919 human pandemic [12]. During early 1980s, influenza monitoring discovered widespread pig infection with the cH1N1 virus in numerous Asian nations and areas [13, 14]. The virus was later discovered to be prevalent in numerous Asian nations, including India, China, Republic of Korea, and Thailand [15–19]. Moreover, major analyses demonstrated that few cH1N1 samples in Japan and Hong Kong had imported pigs from North America, which is how they were first introduced to those countries [20].

A significant epidemic of the H1N2 virus containing the N2 component from the earlier human-like H3N2 virus as well as the remaining seven components from the cH1N1 viruses occurred in pigs with the usual influenza sickness from 1989 winter to 1990 spring in Japan [20]. Similar recombinant viruses from porcine brought over from China between the years 1999 and 2004 were already found in Hong Kong, proving the virus's distribution throughout Asian nations [21]. Soon after the Hong Kong epidemic, Taiwanese pigs became the first source of the H3N2 human-like influenza viruses in Asia [22]. Additionally, the human-like H3N2 variants have always been found in pigs across several Asian nations, such as mainland China, Korea, and Japan; moreover, these viruses were reassorted with prevalent viruses like cH1N1 and afterwards foreign European avian-like H1N1 viruses, failing to be retained and set up in swine herds [19]. Although AIV, like the H9N2 and H5N1 viruses, has been documented to infect people and was regularly found in swine in Asian nations, this virus did not have serious effects on affected animals and did not become established in pigs. It is feasible to create a new virus that may completely adapt to pigs, comparable to the Eurasian avian-like H1N1 virus, because of the increased entry of AIV into swine; alternatively, the avian viruses may reassort with indigenous viruses as shown in Figure 6.2 [23, 24].

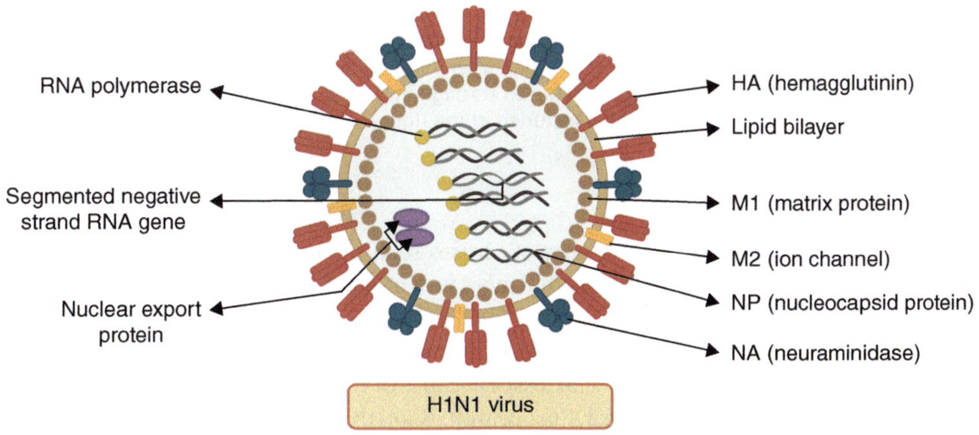

Figure 6.1 Schematic diagram of the H1N1 virus, which is responsible for the transmission of Swine flu. Influenza resonates with the Orthomyxoviridae family of viruses containing SS-RNA, also referred to as A viruses that have eight-segmented DNA. The viral proteins NA and HA surround membrane proteins and serve as the main targets for which humoral and cellular immune responses are directed.

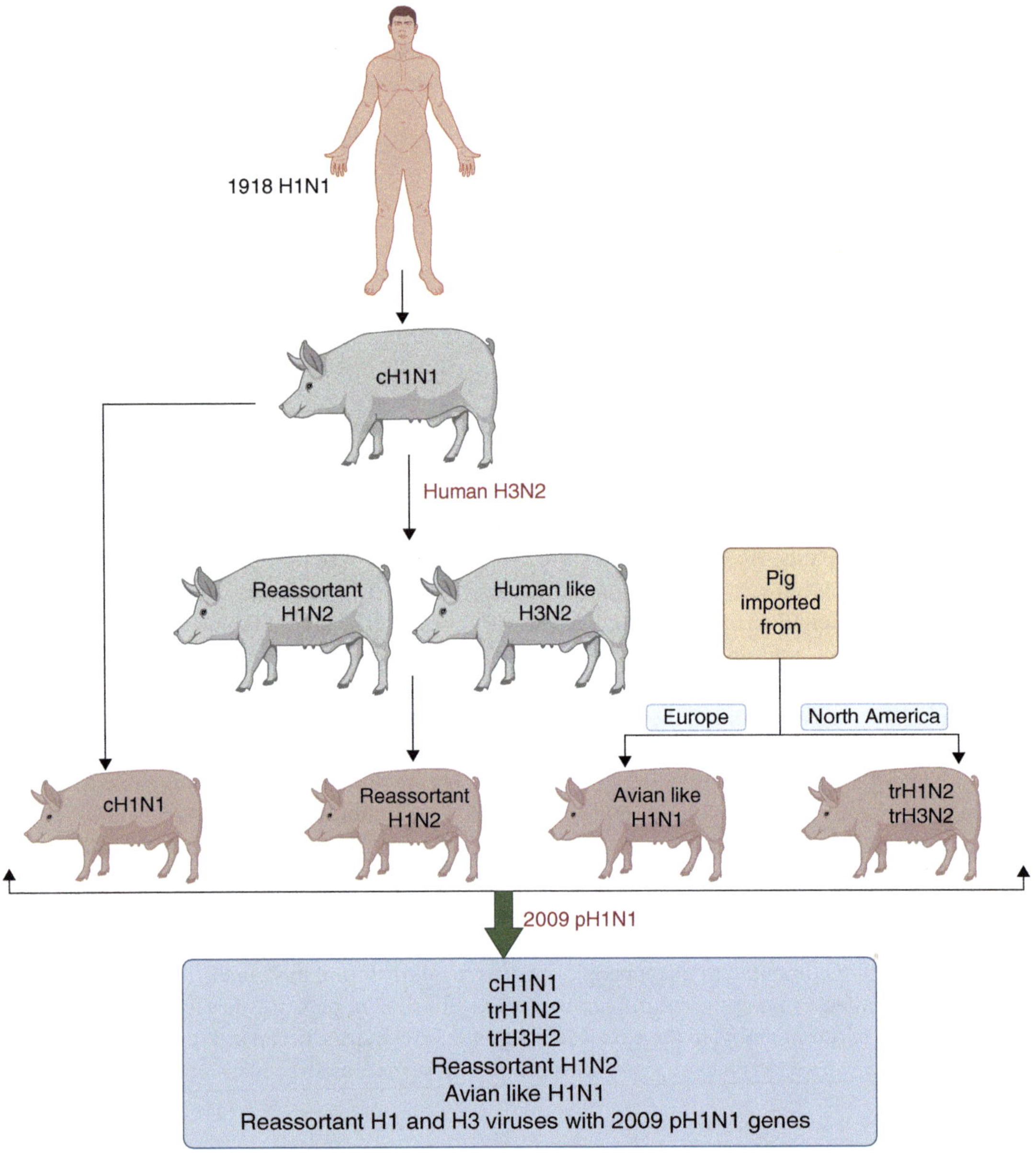

Figure 6.2 Swine influenza virus major subtypes and genotypes in Asia. Major subtypes of influenza virus that are found in Asia and nearby countries are depicted in the diagram. H1N1, which was transferred to swine after 1918, transformed into other major forms of SV like H1N2 and H3N2, and some other strains like trH1N1 and trH3N2 introduced through imported pigs from Europe and North America. Likewise, other reassortant strains of pH1N1 during 2009 are also included in the figure.

6.2.2 Swine Influenza Virus in Other Continents

Prior to the 1990s, the cH1N1 SIV was one of the significant sub-types of viruses seen in North American pig herds for about 70 years. It first appeared during the 1918 pandemic. The swine business was probably not severely affected by this virus [25]. A distinct triple-reassortant H3N2 virus with the HA and NA genes from the normal migratory influenza, the PA and PB2 genes from the AIV, and the three major genes from the cH1N1 SIV arose in the swine population in North America in the late 1990s [26]. Pig farmers have suffered large financial losses because of the extremely adaptable and communicable avian-like H1N1 viruses, which have spread to all major European nations that breed pigs [27]. As a result, a human-origin H3N2 virus with six key genomes and the HA and NA from the prevalent H1N1 virus, which mimics the 1968 H3N2-like pandemic virus, has been identified [28, 29] and started propagating in European pigs [30, 31]. Subsequently, a triple-reassortant H3N2 virus containing internal genes from 2009 pH1N1 and a human-origin HA from a seasonal virus from 2004 to 2005 has propagated in swine farms in Denmark [32]. The genetic diversity of migrating SIV has grown due to the reassortment of the 2009 pH1N1 virus and modern SV in European pig herds, which poses a serious threat of illness. On the other hand, the pig herds no longer include the cluster II H3 virus. Cross-protection across H1 or H3 viral clades (clusters) is either nonexistent or extremely restricted [31]. The extremely genetic and antigenic variety of SIV now present in North American pigs makes it more difficult for the pharmaceutical sector to develop timely and efficient vaccinations to control swine influenza and pig productivity [33–35].

6.2.3 Human to Swine IAV Transfer – Two-Way Transmission

IAVs have been detected in a two-way propagation between swine and humans. As an official for the USBAI in 1919, Koen observed that diseases either originated in pigs or humans, but then quickly expanded between the two of them at the same time [36]. Thereafter, human infections caused by the H1N1 or H3N2 strains of the SIV have been documented everywhere because of direct contact with pigs in many cases, and they have also proved deadly in certain instances [37]. Further evidence of two-way transmission of IAVs between people and pigs has been shown by the 2009 pH1N1 virus [38]. The 2009 influenza pandemic was brought on by this swine-to-human viral transmission [39]. Nelson and Vincent (2015) found the 2009 pH1N1 virus in swine populations across the world and it has been demonstrated that it can invade and spread disease in swine lacking any resistance in Australia and Norway [40, 41]. Even though there were no confirmed cases of human infection in Korea, it is significant to note that the related H3N2 variant SIV had been found there in pigs [42]. It seems most probably that this virus entered Korea through the import and export of swine from the United States. The dual-way transfer of IAVs has accelerated IAV development in both species and may increase the severity of IV infections since most people lack SV protection.

6.3 Pathogenesis Related to Swine Flu

6.3.1 Symptoms Related to Swine Flu

The H1N1 SF is an acute illness that affects the upper respiratory tract (URT). It can inflame these passageways, the trachea, and potentially even the lower respiratory tract (LRT). The usual time for most people to become ill with H1N1SF is two days; however, for certain people, it might take as long as seven days. The incubation period is reported to last from one to four days. Adults are infectious for approximately one day after symptoms appear and continue for approximately five to seven days following the onset of symptoms. In youngsters and others with compromised immune systems, the infectious period may last longer (e.g. 10–14 days) [43]. The viral syndrome, which comprises high fever, coryza, and myalgia, is brought on by the body's immunological response to the virus and the interferon response. People with long-term lung diseases and heart conditions and women who are presently pregnant are more likely to experience serious consequences such as hemorrhagic bronchitis, superimposed bacterial pneumonia, viral pneumonia, and even death. Within 48 hours of the commencement of symptoms, these problems might potentially materialize. From the moment of injection, the virus replicates largely in the URT and LRT passages, peaking in most patients' bodies 48 hours later. A five-day quarantine period is advised for the infected patient [44]. Most SF symptoms are found in the URT and LRT. The RT typically exhibits just a few pathologic alterations in mild instances, but pneumonia can manifest itself pathologically in severe cases. The pseudo-columnar and columnar epithelial cells may have been multifocally destroyed and desquamated. There may also have been submucosal hyperemia and edema that was apparent [37]. At the bronchiolar level, there could potentially be thrombus development. Occasionally, the AI might be severe and be accompanied by desquamative bronchiolitis and hemorrhagic trachea bronchitis, which can lead to bronchiolar wall necrosing. Polymorphs and mononuclear cells invade the afflicted region when necrosis develops. The fibro proliferative phase of ARDS and diffuse alveolar damage are characterized by these alterations. Moreover, in certain autopsy instances, bacterial co-infections were found [37]. The most frequently found bacteria were *H. influenzae*, *S. aureus*, *S. pneumoniae*, *S. pyogenes*, and community-acquired, methicillin-resistant *Staphylococcus aureus* [45].

6.3.2 Pathological Changes During Swine Flu

The preceding investigations, as well as other publications on interesting isolates, comparative studies, and vaccination trials, have all documented varying levels of emphasis on the clinical manifestations, infectious dissemination in body tissue, and virus-shedding processes in addition to macroscopic and microscopic abnormalities. Two of the first investigations, one of which used immune fluorescence and histology [46]. Although influenza has already been observed in circulation as well as other regions including the brain in a few investigations in porcine, [47] prolific IV replication with following tissue injury is really only found in the RT in this breed. The most common macroscopic sign of IVI, which affects a varied proportion of the lung since the virus attacks the lung and leads to cranio-ventral bronchopneumonia. In less severe conditions, small clusters of medium to gloomy lobules can occasionally be observed in the intermediate lung regions, the cranio-ventral region of the caudal lobule, and auxiliary lobes. These foci of consolidation can be found in the hilar area, where the lung lobes meet, in some pigs [48]. They are often linear and confluent in shape. The majority or even all the cranial and middle lung lobes, as well as the cranio-ventral sections of the caudal lobes, may congest in more severe or widespread infections. A diffusely engorged lung, substantial interlobular edema, and significant foam in the bronchi and trachea are less common signs of serious AI [48]. Such instances most likely reflect an excessive cytokine response, and in such lungs, the more widespread interstitial pneumonia and edema mask the distinctive cranio-ventral lobular consolidation. Influenza virus predominantly affects the epithelial cell [EC] lining of the RS in pigs, including the nasal mucosa and alveoli, yet it has been found in the glandular EC connected to the larger airways. Obliterating bronchitis and bronchiolitis are the defining microscopic lesions [49]. Prior to the appearance of epithelial damage, neutrophils start to build up in the vasculature next to bronchioles within the initial 24 hours post-injury (PI), and light, loose lymphocytic cuffs start to form around the airways. At 24–48 hours post-infusion, neutrophils migrate into the airway lumens, and airway EC slough and die [49]. All H1N1 viruses in HLE reproduced solely in type 2 pneumocytes, but only the HPAI H5N1 and seasonal viruses did so with high titers. When compared to seasonal viruses and the 2009 HPAI and LPAI elicited strong cytokine replies, the original SV had inadequate cytokine response and poor replication as seen in Figure 6.3 [50]. While swine are not easily infected by AV and HV, reports of entire AVI and HV in pigs exist. The triple reassortants that make up the

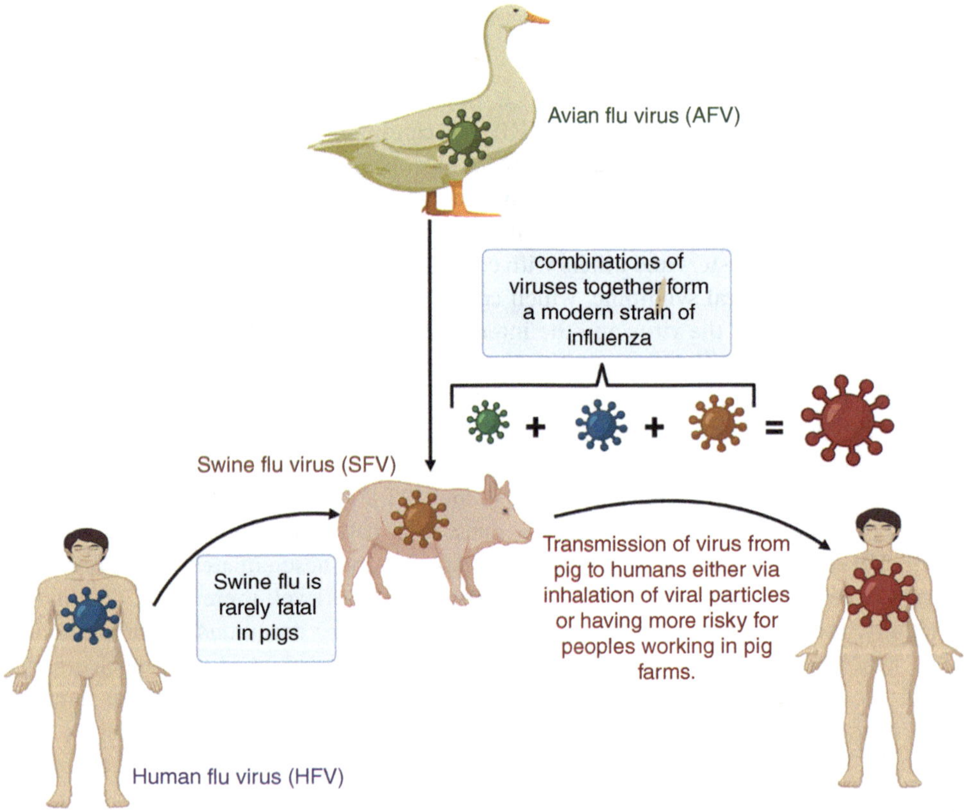

Figure 6.3 Transmission of virus from avian to pig and humans leading to the transmission of swine influenza virus. Although avian and human viruses need not readily invade porcine, it has been recorded that complete bird and human viral have been inoculated with swine. The triple reassortments, which also constitute the majority of the flu strains currently prevalent in swine, carry genes of bird and human descent, providing additional evidence of such illnesses. These triple reassortments viral character with a sequence of intracellular protein sequences from vectors with human, avian, and pig roots appear to have a larger ability for integrating genomes from infections that are presently prevalent in human populations.

majority of the IV which are now spreading in pigs also give evidence for such infections since they contain genes of avian and human origin [48]. These triple-reassortant viruses seem to have a greater capacity for incorporating genes from viruses already in circulation in human populations since they include an array of human, avian, and pig viral genesis intracellular families [51].

6.3.3 Pathogenesis and Host Response in Swine

Swine are used as models for the investigation of the pathophysiology of HI since an infection of influenza in the porcine is often described as an intense, reduced fatality condition. In addition to research on swine IAV inside the native host, the clinical signs and sores brought by the avian-like H1N1 pathogens adjusted to bovine-like numerous H1N1 [52, 53] H1N2, H3N2, and H2N387 reassortants (including the latest 2009 human pandemic H1N1), as well as other H1N1, H1N2, H3N2, and H2N387 pathogens are largely the same, despite the difference in the genetic structure. The preceding investigations, as well as other publications on interesting isolates, comparative studies, and vaccination trials, have all documented the gross and microscopic lesions, clinical symptoms, viral redistribution in tissues, and virus-shedding patterns with a varying emphasis [54].

IAV attacks the EC that line the pigs' URT and LRT, along with the tonsils, trachea, and bronchi, Pathogenesis and Vaccination of IAV in Pig bronchioles, and alveoli [55]. Nevertheless, swine IAV normally favors the lungs. Only through the respiratory pathway through nasal or oral secretions do viruses excrete and spread [56]. During 1–3 days of infection, the virus can be found in nasal secretions, and viral excretion usually lasts for 5–7 days [57]. According to the size and immunological condition of the susceptible population, swine influenza used to be a seasonal illness with varying kinetics of a herd epidemic that may be stretched out over many weeks as it spread through the herd. In certain big pig production facilities, where pigs of all ages are constantly coming in and going out of the barn, EII may lead to the year-round prevalence of IAV in the community. Despite the fact that contemporary disease outbreaks tend to peak in seasonal fluctuations, IAV has demonstrated that it can be reliably discovered yearly and in populations with and without apparent illness symptoms [58]. The generation of chemotactic and pro-inflammatory cytokines, such as TNF-a, IL-1b, IL-6, and type I IFN, helps promote the flow of migrants of neutrophils, macrophages, immune cells, and lymphocytes to the infection site. It is also frequently affiliated with viral antibody titers inside the airways and the clinical and pathological impacts of swine flu in pigs [59]. The innate response prompts cell migration and, in turn, triggers the immune system's adaptable component. Furthermore, when live attenuated influenza vaccines were prescribed to swine, for instance, multiple studies showed substantial immunity to infection and precisely defined it despite the utter lack of discovered HI antibody titer, proving that the exclusion of HI antibody titer does not always suggest an absence of security [60]. After a primary infection, the CD4+ helper cell and CD8+ cytotoxic T-cell response, known as the CMI response, is crucial for the clearance of the IAV virus and healing [61, 62]. T-cells initiating CMI toward influenza can attack both the highly conserved inner proteins like the NP and the exterior glycoproteins, potentially offering wider heterologous defense [63].

6.4 Diagnosis

Over time, methods for diagnosing IVI have become more accurate because of the accessibility of new technology and the need to stay up-to-date with viral evolution. The many assays used to diagnose IVI in pigs are described in a recent well-cited review paper (Table 6.1), along with the advantages and justifications for influenza monitoring programs [64].

Table 6.1 H1N1 and influenza A detection tests and their possible results.

H1N1 testing	Influenza A detection test	Possible result
Negative	Negative	No influenza A (and hence also no H1N1)
Negative	Positive	Influenza A but not swine flu
Positive	Positive	Influenza A H1N1 swine flu
Positive	Negative	No interpretation

Source: Adapted from Detmer et al. [64].

6.4.1 Histopathology

The detection of influenza infection in pigs may still be made via histopathologic analysis of lung tissue, despite the availability of a few tests, even though resolving lesions are far less reliable. There is no other frequently occurring respiratory infection in pigs that causes extensive necrotizing bronchitis and bronchiolitis [65]. The airways are not similarly harmed by even the most severe cases of purulent bronchopneumonia. Further testing may be necessary to distinguish between these possibilities since lesions from resolving influenza infections mirror those of *Mycoplasma hyopneumoniae* infections. A conspicuous lymphocytic cuff surrounds the closely packed, rather atypical, hyperplastic CEC that line the airways [65].

6.4.2 Pig Selection

It is more efficient to sample several pigs than one for diagnosing IVI in swine. This strategy could be crucial since not every infected pig in a group will be experiencing the illness at the same stage at any given moment [64]. The importance of picking the proper pig for an individual sample cannot be overstated given the brief duration of infection by this virus. These pigs had high fevers (105 °F), fatigue, aversion to mobility, and breathlessness, along with transparent nasal discharge. Cough is a clinical manifestation of influenza virus infection (IVI), and yet porcine at this stage has no symptoms of the illness that are going to have passed the phase where the viral load is greatest in the lungs [64].

6.4.3 Serology

The HI estimation has been the go-to serologic technique for detecting influenza virus-specific antibodies in pigs. ELISAs, a few of which are accessible as commercial applications, have indeed been created in order to eliminate the uncertainty of the assay system, to make it easier to test a lot of samples. A blocking ELISA has been created to identify antibodies against the nucleoprotein (NP), an intracellular protein found in all type A influenza viruses that are highly conserved. Further HI serologic assays can be performed using a panel of common virus strains if the blocking ELISA test results indicate a recent influenza infection. Moreover, an ELISA has been created to find antibodies to the NS1 protein. Only infected cells can create this protein, which is absent from intact virions in vaccinations that have been destroyed. Hence, swine that are infectious but not vaccinated will have their defenses against NS1 triggered by a naturally occurring DIVA vaccine that separates the infected from the vaccinated animals [53, 66].

6.4.4 Virus Characterization

Several veterinary diagnostic laboratories routinely determine the H and N subtypes; however, A subtype description by itself does not provide enough information about the background of the genes that make up a flu virus or the potential source of infection. Having the full or partial nucleic acid sequence of the HA1 section of the hemagglutinin peptide enables the creation of dendrograms. Such dendrograms contrast the present viral strain with other characterized pathogens that are common in swine populations and with earlier isolates from the same experiment. Although there are methods for characterizing and comparing all genes, routine situations often just involve the HA-type gene. The creation of M gene-specific PCR tests were utilized for surveillance to spot probable H1N1 virus entry into swine herds in 2009. Several veterinary diagnostic laboratories conduct research investigations that concentrate on the characterization of additionally identifying new reassortants or homologous recombination trends by examining genomes in field isolates [67].

6.4.5 Oral Fluids

The novel way of diagnosing includes collecting swine oral fluids to screen for primary infection in oral secretions in clusters. This method is especially useful for finding viruses that only infect an individual pig for a brief time [68]. Just 20 % of the swine in the group are shedders of the virus, but this method has a very high likelihood of identifying the virus; therefore, it may be used to infer the source of a disease that is now occurring. By suspending thick, absorbent cotton strings dangling from the marker's edges, oral fluids are collected. Pigs are typically inquisitive animals, so they will begin chewing on the rope to explore the new thing in their habitat. Pigs will examine things with their mates because they are sociable animals [69]. The rope will quickly become saturated with saliva. PCR methods are used to detect viruses, although certain samples may have less effectiveness in identifying the subtype. It has become increasingly challenging to isolate viruses.

Oral fluids include substances that might obstruct PCR detection, be hazardous to cell lines, or contain live pathogens [69]. Research is being done to eliminate or neutralize these chemicals and boost the effectiveness of tests carried out on oral fluids. As early as five days PI, antibodies can be seen in saliva. Moreover, it is possible to find vaccine-induced antibodies. For the best findings, certain modifications to the protocol may be required since the antibody titers in oral secretions are lower than those in serum [70].

6.5 Treatment and Management of Swine Flu

The choices for treating H1N1 swine flu are extremely restricted and only successful when given in the initial phases of transmission. In such a perfect scenario, the drugs would effectively relieve all symptoms and reduce the duration of symptomatology [71]. Those who are ineffective if administered too late also could be as little as 32–34 hours after the start of the illness. The creation of an effective vaccine is the best course of action in the event of a viral epidemic for controlling the illness and ultimately curing it. To meet the demands of the public, health authorities, and policymakers as well as to service the market, the vaccine manufacturers also launched a significant effort in this instance [71]. There are different approaches to combat the SIV as discussed in this chapter.

6.5.1 Vaccination Against Influenza in Swine

Although there are vaccinations present, they have not yet been shown to be 100% successful at preventing SI in pigs. Pigs can be protected against either the original or another kind of IV by the recombinant equine herpes virus-1 (EHV-1) [72].

6.5.1.1 Currently Available Vaccine for Swine Flu

US swine breeders and veterinarians frequently use the influenza vaccine to avoid disease and spread, but also to reduce clinical disease. Adjuvanted o/w, either multivalent or bivalent vaccinations made of entirely inactive viruses constitute the sole available commercial SI vaccines with the approval of the United States Two intramuscular shots separated by two to four weeks usually constitute a vaccination [25]. High blood antibody levels to the IAV HA can be produced following intramuscular injection of killed vaccines, but only modest cutaneous antibody production can be produced [73]. Protection from the WIV vaccine relies on the challenge virus and antigen initiation having similar antigens or matching antigens. When pigs are exposed to viruses that are closely related to WIV, injection of commercial WIV vaccinations has demonstrated to partly guard as opposed to clinical symptoms and decrease viral discharge in the nose [74]. Because of this, herd-specific autogenous killed vaccines have become more popular recently and account for half of the dosages of pig vaccines made in the United States [75]. An even more significant problem related to the application of killed adjuvanted IAV in the challenging element swine IAV epidemiologists is the phenomenon called vaccine-associated enhanced respiratory disease (VAERD), which is characterized by serious breathing disorder in heterologous contested pigs after vaccination with misaligned WIV [76].

6.5.1.2 Experimental Vaccine for Swine Flu

As a substitute for the subpar defense seen with the currently accessible inactive SIV, new versions using investigational vaccine technologies have been developed. Live-attenuated IAV (LAIV) vaccines have continuously been demonstrated to be safe as well as provide better protection against heterologous infections in experimental studies in pigs [77]. Later studies demonstrated that the intranasal (i.n.) route primed a mucosal antibody response. Moreover, immunization with H3N2 NS-1-truncated LAIV resulted in a virus-specific T-cell response and provided a limited level of cross-protection versus a provocation with a heterosubtypic H1N1 [78]. Pigs infected with a replication-defective HAd5 expressing IAV, NP, and HA proteins exhibited full resistance to a very similar challenging variant, and there was also substantial resistance when HA alone was expressed [79]. One i.n. dosage of an Ad5-vectored HA was observed to induce mucosal IgA and provoke an IAV-specific, cross-reactive IFN-c response in pigs, providing resistance toward a similar virus and some resistance against the divergent virus [80].

6.5.2 Modern Therapeutic Approach

The foundation of modern medicine is a medicinal approach that primarily offers symptomatic relief and sets a stronger emphasis on therapeutic interventions. It concentrates on the application of medications, mechanical evaluation, intrusive therapies such as surgery, and a passive attitude for the patient. Currently, there are two classes of antiviral drugs that may be employed to treat temporary human influenza in modern medicine. Neuraminidase inhibitors such as adamantanes, zanamivir, and oseltamivir also known as Tamiflu, rimantadine, and amantadine are currently available for therapy. It is indeed noteworthy that an herb is employed as the primary component in the creation of Tamiflu which is extracted from the plant *Illicium verum* [9]. According to molecular and phenotypic research, H1N1 is robust to adamantanes but vulnerable to oseltamivir and zanamivir. HA, NA, and the M-2 ion channel peptide are the prime targets of these drugs. A recent study emphasized the immediate need for augmentation of oseltamivir stockpiles with extra drug treatments, such as Zanamivir, depending on an assessment of the efficacy of these substances in the situation that the 2009 H1N1 "Swine Flu" NA were to obtain the Tamiflu-resistance (His274Tyr) mutant, which is presently broad in seasonal H1N1 varieties [5, 6]. Therapy with amantadine may result in drug prices becoming extremely expensive. Countless deaths due to virus types that have experienced mutations or have grown a more deadly resistance to drug immunization supply are uncertain [81]. Likewise, laboratory research has shown that it is possible that even the prophylactic use of inadequate amounts of these drugs may contribute to the development of drug sensitivity [82].

6.5.3 Traditional Approaches for the Therapeutic Interventions of Swine Flu

Different areas of the world have been using alternative and complementary therapies for so many decades to treat serious diseases. It is impossible to estimate the prevention-focused value of such complementary and alternative treatment strategies against conventional medicines in specific. In completing the process of healing, "complementary" suggests that it works in collaboration with conventional medicine. Replace conventional medicine with alternative native therapies. It functions as a substitute rather than aiding or improving conventional medicine [83]. It has rarely been combined with traditional medicine to provide synergistic benefits. Alternative and complementary medicine has served to improve the general public's well-being, especially when traditional contemporary medicine has not been successful [84]. The conventional and complementary medicine group includes TISM, which includes Ayurveda [85], TCM [85, 86], JTM, Unani, Siddha, and so on. Although substantial efforts have been made to establish an effective evidence-based quality control of Ayurveda, Siddha, Unani, TCM Therapy [87], and other CAM so it can properly fit into the modern medical structure, natural therapeutic systems have, for one purpose or the other, yet to be able to enter conventional medicine [88]. Herbs that improve the body's overall immunity by stimulating unique or broad immune system elements and improve a person's ability to fight off pathogens have been listed in Table 6.2. These medicinal plants primarily fight H1N1 transmission by boosting the person's overall immunity or by acting directly against the virus by inhibiting viral signaling pathways or by preventing viral replication. The various medicinal plants from around the world as listed in Table 6.2 have been evaluated for their efficacy against the virus [88].

6.6 Challenges During Combating Swine Flu

There have been various challenges during the Swine flu era and the condition turned out to be difficult for each and every one. Challenges related to controlling the swine influenza were related to the variation of the strains of the virus that created difficulty in the production of an efficient drug regime for the infected patients. This chapter also discusses the challenges for live stalks and human health [100].

6.6.1 Challenge to Control Swine Influenza

Worldwide, there are many extremely varied SIVs that are in circulation among swine groups, and pig infestations with bird and/or human IAV could always result in the development of novel genomes and subgroups of infections. As an illustration, take the unusual H2N3 influenza that has been identified in pigs found in the United States around 2006, which was contagious and extremely communicable in swine and ferrets [101]. Additionally, the movement of live pigs internationally has confounded the worldwide swine flu scenario and hastened virus mutation. As a consequence, new SIV, such as the triple-reassortant avian-like H1N1 SIV that has been imported into China from other Western countries, has been brought into misinformed swine populations. Creating efficient vaccines and managing swine influenza for the livestock business is a tremendous task. The benefit of live-virus vaccines is that they can offer excellent partial and heterosubtypic protection while not worsening illness [102]. The possibility for LAIV vaccines to reassort with currently prevalent endemic viruses is a drawback [103]. Due to reassortment with indigenous viruses, novel mutant viruses have been found in pig farms using the LAIV vaccine [93]. In contrast to IAVs in people, pig IAV development displays a distinct pattern that is

Table 6.2 List of medicinal plants found across the world that are believed to have anti-influenza properties and might be beneficial in the fight against Swine flu.

Sr. no.	Plant name	Family	Native country	Chemical constituent	Mechanism of action	Reference
1.	*C. indicum*	Asteraceae	China along with India	Camphor, thujone, alpha-terpineol, and cis-sabinol	Rich in vitamin, C. indicum tea is said to have a number of health benefits, including helping people recover from influenza	[89]
2.	Withaniasomnifera	Solanaceae	India	Beta-sisterol, anaferine, anahygrine, withaferin, chlorogenic acid, and with anolides	Immune system stimulant and highly effective modulator	[90]
3.	*C. siamensis*	Acanthaceae	Thailand	2-Propenol, strans-3-methylsulfinyl, and trans-3-methylsulfonyl	A greater amount of antagonist of IAV IgA antibody was generated when there was a higher generation of IgG(1) antibody	[91]
4.	*Narcissus tazetta*	Amaryllidaceae	China	A peptide that attaches to fetuin	Significantly reduce the cytopathic impact brought on by the influenza A (H1N1) virus and the plaque development brought on by the respiratory syncytial virus (RSV)	[92]
5.	*Cistus incanus*	Cistaceae	Mediterranean	Components of polymeric flavonoid	A topical administration at the entrance points of the pathogen may be a potential strategy to inhibit the HA capacity to adhere to receptor molecules and avoid the spread of IAV spread	[93]
6.	*Bergenia ligulata*	Saxifragaceae	Nepal	Tannins in condensed form	Suppresses viral peptide production and prevents the creation of viral RNA	[94]
7.	*C. verum*	Lauraceae	Sri Lanka	Mannitol and cinnamaldehyde	An absorber of radicals and antioxidative property. Pathogen entrance is prevented	[95]
8.	Junipers	Cupressaceae	Africa and Central America	Sabinene, myrcene, α-phellandrene, α-terpinene	It has high macrophage immunomodulatory efficacy	[96]
9.	*Echinacea purpurea*	Compositae	North America	Alcohol, ethyl acetate, and citric acid	Echinacea raises the body's amounts of the chemical properdin, which triggers the immune system component in charge of boosting the body's defenses against bacteria and viruses. It also works well as a photosensitizer for viruses	[97]
10.	*S. flavescens*	Leguminosae	Korea	Pterocarpans and flavanones	NA which is an enzyme essential for spreading the IAV, can be inhibited	[98]
11.	*Justicia pectoralis*	Acanthaceae	Latin America	Umbelliferone, coumarin	Shown a relaxant effect along with anti-inflammatory action and helps in recovery from influenza infection	[99]

influenced by regional areas, including the nation, region, and even individual farm levels [104]. Because of this, unlike the yearly human influenza vaccine, producers of pig influenza vaccines are unable to create a vaccine that is "universal" for all farms in various regions. Instead, they must independently choose the strains that will be included in their products. This shows that the potential pandemic virus can emerge anywhere and highlights the importance of pig surveillance in those other regions. Likewise, other swine pathogens, including the North American triple-reassortant H3N2 viral infection in the United States and the Eurasian avian-like H1N1 virus in Asia, have procured a growing human infection rate via homologous recombination, likely to result in the H3N2 modified viral infection and reassortant Eurasian H1N1-like virus in line with a freestanding that caused human infections [105]. The ecology of IAVs has to be best understood through an increase in global surveillance of IAVs in pigs, and better swine immunizations should be created to safeguard both animal and human health [106].

6.6.2 Challenge for Livestock and Human Health

A contagious lung illness known as swine flu frequently affects French pig farms. Although it generally does not cause any harm, it can become worse or come back on a farm, leading to serious health issues and financial damages. These pathogens are endemic risk factors (they can be transmitted to humans). The agency's efforts to combat these viruses are divided into various degrees of action and knowledge, including reference work, epidemiological surveillance, and study, which provide information for expert evaluations [107]. As a result, numerous distinct strains have been identified over time, thanks to monitoring efforts on swine farms. In the pig community, there are three circulating subtypes of influenza viruses (H1N1, H3N2, and H1N2), but each of these subtypes has numerous genetic lines (or genotypes) based on the place of origin of each viral gene. Pigs with flu infections typically exhibit only mild clinical symptoms, and only a portion of the piglets on the same farm are affected. Nevertheless, based on the virulence of the strain in question and the agricultural techniques used, infections can also worsen and almost become generalized. Other unfavorable variables that are still inadequately characterized may also have an impact on the severity of the illness [104], as well as concurrent infections with other pathogens that affect the respiratory system. Additionally, whereas influenza exposure typically results in an isolated (episodic) flu syndrome, a recurrent version of the illness has become more prevalent in fields in recent years. Repeated flu outbreaks cause the infected fields to become unstable and may encourage co-infections with various influenza virus types and consequent viral reassortments. To learn more about the factors that influence virus survival in pig farms, both on-farm (observational epidemiological studies) and in silico (epidemiological modeling) studies are being conducted [108].

6.7 Conclusion

Although the situation is largely under control now, the pandemic flu has presented a danger to the entire world. Despite years of warning and planning to handle such fatalities, society is still not adequately prepared to handle an epidemic of resistant strains if they ever materialize. There is presently no immunization in mainstream medicine. When the vaccine becomes accessible, antiviral medications will also be stocked up in case of emergency. However, directly following its debut, it would be hard to find and useless for treating the general populace in emerging and underdeveloped countries in the event of a pandemic. In addition to the development of drug resistance, other challenging issues include the emergence of mutant viral strains, the emergence of a more virulent strain, the prohibitive prices of existing medications, the delay in the creation of vaccines, and mass fatalities. Complementary and alternative medicine provides a wide range of intriguing opportunities to assist patients considering this. Herbs can be used to successfully manage the pandemic flu because they show a wide range of biological activities. Dietary and botanical strategies work together to create very effective tools for battling a variety of viral illnesses. This chapter presents an eclectic overview of the current situation and difficulties associated with the swine flu, as well as its treatment. It also discusses some of the herbs that are most likely to be helpful in managing the current pandemic situation as well as dealing with the potential for the next pandemic soon. Our strong conviction is that the plants described in the article would be helpful in treating patients with severe influenza in non-pandemic circumstances as well.

References

1 Chen, Q., Madson, D., Miller, C.L., and Harris, D.H. (2012). Vaccine development for protecting swine against influenza virus. *Anim. Health Res. Rev.* 13 (2): 181–195.

2 Alexander, D.J. and Brown, I.H. (2000). Recent zoonoses caused by influenza A viruses. *Rev. Sci. Tech.* 19 (1): 197–225.

3 McQueen, J.L., Steele, J.H., and Robinson, R.Q. (1968). Influenza in animals. *Adv. Vet. Sci.* 12: 285–336.

4 Ma, W., Lager, K.M., Vincent, A.L. et al. (2009). The role of swine in the generation of novel influenza viruses. *Zoonoses Public Health* 56: 326–337.

5 Boltz, D.A., Aldridge, J.R., Webster, R.G., and Govorkova, E.A. (2010). Drugs in development for influenza. *Drugs* 70: 1349–1362.

6 Schultz-Cherry, S., Olsen, C.W., and Easterday, B.C. (2013). History of swine influenza. *Curr. Top. Microbiol. Immunol.* 370: 21–28.

7 Thomann, E., Trein, P., and Maggetti, M. (2019). What's the problem? Multilevel governance and problem solving. *Eur. Policy Anal.* 5 (1): 37–57.

8 Alleva, L.M., Cai, C., and Clark, I.A. (2010). Using complementary and alternative medicines to target the host response during severe influenza. *Evid. Based Complement. Alternat. Med.* 7 (4): 501–510.

9 Lynch, J.P. and Walsh, E.E. (2007). Influenza: evolving strategies in treatment and prevention. *Semin. Respir. Crit. Care Med.* 28 (02): 144–158. Copyright© 2007 by Thieme Medical Publishers, Inc., 333 Seventh Avenue, New York, NY 10001, USA.

10 Brown, I.H. (2000). The epidemiology and evolution of influenza viruses in pigs. *Vet. Microbiol.* 74 (1–2): 29–46.

11 Zhu, H., Webby, R., Lam, T.T. et al. (2013). History of swine influenza viruses in Asia. In: *Swine Influenza* (ed. J.A. Richt and R.J. Webby), 57–68. Springer.

12 Arikawa, J., Yamane, N., Odagiri, T., and Ishida, N. (1979). Serological evidence of H1 influenza virus infection among Japanese hogs. *Acta Virol.* 23 (6): 508–511.

13 Nerome, K., Ishida, M., Oya, A., and Oda, K. (1982). The possible origin of H1N1 (Hsw 1N1) virus in the swine population of Japan and antigenic analysis of the isolates. *J. Gen. Virol.* 62 (1): 171–175.

14 Chatterjee, S., Mukherjee, K.K., Mondal, M.C. et al. (1995). A serological survey of influenza a antibody in human and pig sera in Calcutta. *Folia Microbiol.* 40: 345–348.

15 Guan, Y., Shortridge, K.F., Krauss, S. et al. (1996). Emergence of avian H1N1 influenza viruses in pigs in China. *J. Virol.* 70 (11): 8041–8046.

16 Kupradinun, S., Peanpijit, P., Bhodhikosoom, C. et al. (1991). The first isolation of swine H1N1 influenza viruses from pigs in Thailand. *Arch. Virol.* 118: 289–297.

17 Lee, C.S., Kang, B.K., Kim, H.K. et al. (2008). Phylogenetic analysis of swine influenza viruses recently isolated in Korea. *Virus Genes* 37: 168–176.

18 Song, D.S., Lee, C.S., Jung, K. et al. (2007). Isolation and phylogenetic analysis of H1N1 swine influenza virus isolated in Korea. *Virus Res.* 125 (1): 98–103.

19 Vijaykrishna, D., Smith, G.J., Pybus, O.G. et al. (2011). Long-term evolution and transmission dynamics of swine influenza A virus. *Nature* 473 (7348): 519–522.

20 Sugimura, T., Yonemochi, H., Ogawa, T. et al. (1980). Isolation of a recombinant influenza virus (Hsw1N2) from swine in Japan. *Arch. Virol.* 66: 271–274.

21 Yu, H., Zhou, Y.J., Li, G.X. et al. (2011). Genetic diversity of H9N2 influenza viruses from pigs in China: a potential threat to human health? *Vet. Microbiol.* 149 (1–2): 254–261.

22 Zhou, P., Hong, M., Merrill, M.M. et al. (2014). Serological report of influenza A (H7N9) infections among pigs in Southern China. *BMC Vet. Res.* 10: 1–5.

23 Butt, K.M., Smith, G.J., Chen, H. et al. (2005). Human infection with an avian H9N2 influenza A virus in Hong Kong in 2003. *J. Clin. Microbiol.* 43 (11): 5760–5767.

24 Dong, X., Xiong, J., Huang, C. et al. (2021). RETRACTED ARTICLE: human H9N2 avian influenza infection: epidemiological and clinical characterization of 16 cases in China. *Virol. Sin.* 36: 564.

25 Vincent, A.L., Ma, W., Lager, K.M. et al. (2008). Swine influenza viruses: a North American perspective. *Adv. Virus Res.* 72: 127–154.

26 Olsen, C.W. (2002). The emergence of novel swine influenza viruses in North America. *Virus Res.* 85 (2): 199–210.

27 Zell, R., Scholtissek, C., and Ludwig, S. (2013). Genetics, evolution, and the zoonotic capacity of European swine influenza viruses. In: *Swine Influenza* (ed. J.A. Richt and R.J. Webby), 29–55. Springer.

28 Castrucci, M.R., Campitelli, L., Ruggieri, A. et al. (1994). I. Antigenic and sequence analysis of H3 influenza virus haemagglutinins from pigs in Italy. *J. Gen. Virol.* 75 (2): 371–379.

29 Haesebrouck, F., Biront, P., Pensaert, M.B., and Leunen, J. (1985). Epizootics of respiratory tract disease in swine in Belgium due to H3N2 influenza virus and experimental reproduction of disease. *Am. J. Vet. Res.* 46 (9): 1926–1928.

30 De Jong, J.C., Van Nieuwstadt, A.P., Kimman, T.G. et al. (1999). Antigenic drift in swine influenza H3 haemagglutinins with implications for vaccination policy. *Vaccine* 17 (11–12): 1321–1328.

31 Simon, G., Larsen, L.E., Dürrwald, R. et al. (2014). European surveillance network for influenza in pigs: surveillance programs, diagnostic tools and swine influenza virus subtypes identified in 14 European countries from 2010 to 2013. *PLoS One* 9 (12): e115815.

32 Krog, J.S., Hjulsager, C.K., Larsen, M.A., and Larsen, L.E. (2017). Triple reassortant influenza A virus with H3 of human seasonal origin, NA of swine origin, and internal A (H1N1) pandemic 2009 genes is established in Danish pigs. *Influenza Other Respir. Viruses* 11 (3): 298–303.

33 Beato, M.S., Tassoni, L., Milani, A. et al. (2016). Circulation of multiple genotypes of H1N2 viruses in a swine farm in Italy over a two-month period. *Vet. Microbiol.* 195: 25–29.

34 Pippig, J., Ritzmann, M., Büttner, M., and Neubauer, J.A. (2016). Influenza a viruses detected in swine in southern Germany after the H1N1 pandemic in 2009. *Zoonoses Public Health* 63 (7): 555–568.

35 Watson, S.J., Langat, P., Reid, S.M. et al. Molecular epidemiology and evolution of influenza viruses circulating within European swine between 2009 and 2013. *J. Virol.* 89: 9920–9931.

36 Koen, J.S. (1919). A practical method for field diagnosis of swine disease. *Am. J. Vet. Med.* 14: 468–470.

37 Myers, K.P., Olsen, C.W., and Gray, G.C. (2007). Cases of swine influenza in humans: a review of the literature. *Clin. Infect. Dis.* 44 (8): 1084–1088.

38 Chastagner, A., Enouf, V., Peroz, D. et al. (2019). Bidirectional human–swine transmission of seasonal influenza A (H1N1) Pdm09 virus in pig herd, France, 2018. *Emerg. Infect. Dis.* 25 (10): 1940.

39 Mena, I., Nelson, M.I., Quezada-Monroy, F. et al. (2016). Origins of the 2009 H1N1 influenza pandemic in swine in Mexico. *elife* 5: e16777.

40 Hofshagen, M., Gjerset, B., Er, C. et al. (2009). Pandemic influenza A (H1N1) v: human to pig transmission in Norway? *Euro Surveill.* 14 (45): 19406.

41 Holyoake, P.K., Kirkland, P.D., Davis, R.J. et al. (2011). The first identified case of pandemic H1N1 influenza in pigs in Australia. *Aust. Vet. J.* 89 (11): 427–431.

42 Kim, S.H., Kim, H.J., Jin, Y.H. et al. (2013). Isolation of influenza A (H3N2) v virus from pigs and characterization of its biological properties in pigs and mice. *Arch. Virol.* 158: 2351–2357.

43 Calore, E.E., Uip, D.E., and Perez, N.M. (2011). Pathology of the swine-origin influenza A (H1N1) flu. *Pathol. Res. Pract.* 207 (2): 86–90.

44 Jilani, T.N., Jamil, R.T., and Siddiqui, A.H. (2023). H1N1 influenza. In: *StatPearls*. Treasure Island, FL: Stat Pearls Publishing.

45 Littauer, E.Q., Esser, E.S., Antao, O.Q. et al. (2017). H1N1 influenza virus infection results in adverse pregnancy outcomes by disrupting tissue-specific hormonal regulation. *PLoS Pathog.* 13 (11): e1006757.

46 Nayak, D.P., Twiehaus, M.J., Kelley, G.W., and Underdahl, N.R. (1965). Immunocytologic and histopathologic development of experimental swine influenza infection in pigs. *Am. J. Vet. Res.* 26 (115): 1271–1283.

47 Landolt, G.A. and Olsen, C.W. (2007). Up to new tricks – a review of cross-species transmission of influenza A viruses. *Anim. Health Res. Rev.* 8 (1): 1–21.

48 Landolt, G.A., Karasin, A.I., Phillips, L., and Olsen, C.W. (2003). Comparison of the pathogenesis of two genetically different H3N2 influenza A viruses in pigs. *J. Clin. Microbiol.* 41 (5): 1936–1941.

49 Weinheimer, V.K., Becher, A., Tönnies, M. et al. (2012). Influenza A viruses target type II pneumocytes in the human lung. *J. Infect. Dis.* 206 (11): 1685–1694.

50 Kuntz Simon, G. and Madec, F. (2009). Genetic and antigenic evolution of swine influenza viruses in Europe and evaluation of their zoonotic potential. *Zoonoses Public Health* 56 (6–7): 310–325.

51 Khatri, M., Dwivedi, V., Krakowka, S. et al. (2010). Swine influenza H1N1 virus induces acute inflammatory immune responses in pig lungs: a potential animal model for human H1N1 influenza virus. *J. Virol.* 84 (21): 11210–11218.

52 Vincent, A.L., Ma, W., Lager, K.M. et al. (2009). Characterization of a newly emerged genetic cluster of H1N1 and H1N2 swine influenza virus in the United States. *Virus Genes* 39: 176–185.

53 Vincent, A.L., Lager, K.M., Ma, W. et al. (2006). Evaluation of hemagglutinin subtype 1 swine influenza viruses from the United States. *Vet. Microbiol.* 118 (3–4): 212–222.

54 Vincent, A.L., Swenson, S.L., Lager, K.M. et al. (2009). Characterization of an influenza A virus isolated from pigs during an outbreak of respiratory disease in swine and people during a county fair in the United States. *Vet. Microbiol.* 137 (1–2): 51–59.

55 Nelli, R.K., Kuchipudi, S.V., White, G.A. et al. (2010). Comparative distribution of human and avian type sialic acid influenza receptors in the pig. *BMC Vet. Res.* 6 (1): 1–9.

56 Van Reeth, K. and Vincent, A.L. (2019). Influenza viruses. In: *Diseases of Swine* (ed. J.J. Zimmerman, L.A. Karriker, A. Ramirez, et al.), 480–504. Ames, IA: John Willey and Sons Inc.

57 Janke, B.H. (2013). Clinicopathological features of swine influenza. *Curr. Top. Microbiol. Immunol.* 370: 69–83.

58 Jo, S.K., Kim, H.S., Cho, S.W., and Seo, S.H. (2007). Pathogenesis and inflammatory responses of swine H1N2 influenza viruses in pigs. *Virus Res.* 129 (1): 64–70.

59 Loving, C.L., Vincent, A.L., Pena, L., and Perez, D.R. (2012). Heightened adaptive immune responses following vaccination with a temperature-sensitive, live-attenuated influenza virus compared to adjuvanted, whole-inactivated virus in pigs. *Vaccine* 30 (40): 5830–5838.

60 Vincent, A.L., Ma, W., Lager, K.M. et al. (2012). Live attenuated influenza vaccine provides superior protection from heterologous infection in pigs with maternal antibodies without inducing vaccine-associated enhanced respiratory disease. *J. Virol.* 86 (19): 10597–10605.

61 Larsen, D.L., Karasin, A., Zuckermann, F., and Olsen, C.W. (2000). Systemic and mucosal immune responses to H1N1 influenza virus infection in pigs. *Vet. Microbiol.* 74 (1–2): 117–131.

62 Platt, R., Vincent, A.L., Gauger, P.C. et al. (2011). Comparison of humoral and cellular immune responses to inactivated swine influenza virus vaccine in weaned pigs. *Vet. Immunol. Immunopathol.* 142 (3–4): 252–257.

63 Thomas, P.G., Keating, R., Hulse-Post, D.J., and Doherty, P.C. (2006). Cell-mediated protection in influenza infection. *Emerg. Infect. Dis.* 12 (1): 48.

64 Detmer, S., Gramer, M., Goyal, S. et al. (2013). Diagnostics and surveillance for swine influenza. *Curr. Top. Microbiol. Immunol.* 370: 85–112.

65 Vincent, L.L., Janke, B.H., Paul, P.S., and Halbur, P.G. (1997). A monoclonal-antibody-based immunohistochemical method for the detection of swine influenza virus in formalin-fixed, paraffin-embedded tissues. *J. Vet. Diagn. Investig.* 9 (2): 191–195.

66 Jung, T., Choi, C., and Chae, C. (2002). Localization of swine influenza virus in naturally infected pigs. *Vet. Pathol.* 39 (1): 10–16.

67 Thompson, C.I., Barclay, W.S., Zambon, M.C., and Pickles, R.J. (2006). Infection of human airway epithelium by human and avian strains of influenza a virus. *J. Virol.* 80 (16): 8060–8068.

68 Prickett, J.R. and Zimmerman, J.J. (2010). The development of oral fluid-based diagnostics and applications in veterinary medicine. *Anim. Health Res. Rev.* 11 (2): 207–216.

69 Romagosa, A., Gramer, M., Joo, H.S., and Torremorell, M. (2012). Sensitivity of oral fluids for detecting influenza A virus in populations of vaccinated and non vaccinated pigs. *Influenza Other Respir. Viruses* 6 (2): 110–118.

70 Panyasing, Y., Goodell, C.K., Wang, C. et al. (2014). Detection of influenza A virus nucleoprotein antibodies in oral fluid specimens from pigs infected under experimental conditions using a blocking ELISA. *Transbound. Emerg. Dis.* 61 (2): 177–184.

71 Meijer, A., Lackenby, A., Hungnes, O. et al. (2009). Oseltamivir-resistant influenza virus A (H1N1), Europe, 2007–08 season. *Emerg. Infect. Dis.* 15 (4): 552.

72 Klingbeil, K., Lange, E., Teifke, J.P. et al. (2014). Immunization of pigs with an attenuated pseudorabies virus recombinant expressing the haemagglutinin of pandemic swine origin H1N1 influenza A virus. *J. Gen. Virol.* 95 (4): 948–959.

73 Heinen, P.P., Van Nieuwstadt, A.P., de Boer-Luijtze, E.A., and Bianchi, A.T. (2001). Analysis of the quality of protection induced by a porcine influenza A vaccine to challenge with an H3N2 virus. *Vet. Immunol. Immunopathol.* 82 (1–2): 39–56.

74 Kitikoon, P., Nilubol, D., Erickson, B.J. et al. (2006). The immune response and maternal antibody interference to a heterologous H1N1 swine influenza virus infection following vaccination. *Vet. Immunol. Immunopathol.* 112 (3–4): 117–128.

75 Lee, J.H., Gramer, M.R., and Joo, H.S. (2007). Efficacy of swine influenza A virus vaccines against an H3N2 virus variant. *Can. J. Vet. Res.* 71 (3): 207.

76 Macklin, M.D., McCabe, D., McGregor, M.W. et al. (1998). Immunization of pigs with a particle-mediated DNA vaccine to influenza A virus protects against challenge with homologous virus. *J. Virol.* 72 (2): 1491–1496.

77 Kappes, M.A., Sandbulte, M.R., Platt, R. et al. (2012). Vaccination with NS1-truncated H3N2 swine influenza virus primes T cells and confers cross-protection against an H1N1 heterosubtypic challenge in pigs. *Vaccine* 30 (2): 280–288.

78 Vincent, A.L., Ma, W., Lager, K.M. et al. (2007). Efficacy of intranasal administration of a truncated NS1 modified live influenza virus vaccine in swine. *Vaccine* 25 (47): 7999–8009.

79 Wesley, R.D., Tang, M., and Lager, K.M. (2004). Protection of weaned pigs by vaccination with human adenovirus 5 recombinant viruses expressing the hemagglutinin and the nucleoprotein of H3N2 swine influenza virus. *Vaccine* 22 (25–26): 3427–3434.

80 Braucher, D.R., Henningson, J.N., Loving, C.L. et al. (2012). Intranasal vaccination with replication-defective adenovirus type 5 encoding influenza virus hemagglutinin elicits protective immunity to homologous challenge and partial protection to heterologous challenge in pigs. *Clin. Vaccine Immunol.* 19 (11): 1722–1729.

81 Bright, R.A., Medina, M.J., Xu, X. et al. (2005). Incidence of adamantane resistance among influenza A (H3N2) viruses isolated worldwide from 1994 to 2005: a cause for concern. *Lancet* 366 (9492): 1175–1181.

82 World Health Organization (2009). Swine influenza: frequently asked questions. *Wkly Epidemiol. Rec.* 84 (18): 149–151.

83 Patwardhan, B., Warude, D., Pushpangadan, P., and Bhatt, N. (2005). Ayurveda and traditional Chinese medicine: a comparative overview. *Evid. Based Complement. Alternat. Med.* 2 (4): 465–473.

84 Kaphle, K., Wu, L.S., Yang, N.Y., and Lin, J.H. (2006). Herbal medicine research in Taiwan appropriate person among the authors to be in contact for any further information on the status of CHM research in Taiwan and for opportunities to study and conduct research in the field of TCM in Taiwan. *Evid. Based Complement. Alternat. Med.* 3 (1): 149–155.

85 Kiyohara, H., Nonaka, K., Sekiya, M. et al. (2011). Polysaccharide-containing macromolecules in a Kampo (traditional Japanese herbal) medicine, Hochuekkito: dual active ingredients for modulation of immune functions on intestinal Peyer's patches and epithelial cells. *Evid. Based Complement. Alternat. Med.* 2011: 492691.

86 Azaizeh, H., Saad, B., Cooper, E., and Said, O. (2010). Traditional Arabic and Islamic medicine, a re-emerging health aid. *Evid. Based Complement. Alternat. Med.* 7: 419–424.

87 Firenzuoli, F. and Gori, L. (2007). Herbal medicine today: clinical and research issues. *Evid. Based Complement. Alternat. Med.* 4 (S1): 37–40.

88 Chang, I.M. Initiative for developing evidence-based standardization of traditional Chinese medical therapy in the western pacific region of the World Health Organization. *Evid. Based Complement. Alternat. Med.* 1: 337–341.

89 Nagai, T., Moriguchi, R., Suzuki, Y. et al. (1995). Mode of action of the anti-influenza virus activity of plant flavonoid, 5, 7, 4′-trihydroxy-8-methoxyflavone, from the roots of *Scutellaria baicalensis*. *Antivir. Res.* 26 (1): 11–25.

90 Shah, A. and Krishnamurthy, R. (2013). Swine flu and its herbal remedies. *Int. J. Eng. Sci.* 2 (5): 68–78.

91 Yale, S.H. and Glurich, I. (2005). Analysis of the inhibitory potential of *Ginkgo biloba*, *Echinacea purpurea*, and *Serenoa repens* on the metabolic activity of cytochrome P450 3A4, 2D6, and 2C9. *J. Altern. Complement. Med.* 11 (3): 433–439.

92 Mochida, K. (2008). Anti-influenza virus activity of *Myrica rubra* leaf ethanol extract evaluated using Madino-Darby canine kidney (MDCK) cells. *Biosci. Biotechnol. Biochem.* 72 (11): 3018–3020.

93 Mitani, T., Ota, K., Inaba, N. et al. (2018). Antimicrobial activity of the phenolic compounds of *Prunus mume* against enterobacteria. *Biol. Pharm. Bull.* 41 (2): 208–212.

94 Li, Y., Jiang, R., Ooi, L.S. et al. (2007). Antiviral triterpenoids from the medicinal plant *Schefflera heptaphylla*. *Phytother. Res.* 21 (5): 466–470.

95 Schepetkin, I.A., Faulkner, C.L., Nelson-Overton, L.K. et al. (2005). Macrophage immunomodulatory activity of polysaccharides isolated from *Juniperus scopolorum*. *Int. Immunopharmacol.* 5 (13–14): 1783–1799.

96 Lumaret, R. and Ouazzani, N. (2001). Ancient wild olives in Mediterranean forests. *Nature* 413 (6857): 700.

97 Yale, S.H. and Liu, K. (2004). Echinacea purpurea therapy for the treatment of the common cold: a randomized, double-blind, placebo-controlled clinical trial. *Arch. Intern. Med.* 164 (11): 1237–1241.

98 Ooi, L.S., Sun, S.S., and Ooi, V.E. (2004). Purification and characterization of a new antiviral protein from the leaves of *Pandanus amaryllifolius* (Pandanaceae). *Int. J. Biochem. Cell Biol.* 36 (8): 1440–1446.

99 Kappagoda, C., Isaacs, D., Mellis, C. et al. (2000). Critical influenza virus infection. *J. Paediatr. Child Health* 36 (4): 318–321.

100 Ma, W., Vincent, A.L., Gramer, M.R. et al. (2007). Identification of H2N3 influenza A viruses from swine in the United States. *Proc. Natl. Acad. Sci.* 104 (52): 20949–20954.

101 Richt, J.A., Lager, K.M., Janke, B.H. et al. (2003). Pathogenic and antigenic properties of phylogenetically distinct reassortant H3N2 swine influenza viruses cocirculating in the United States. *J. Clin. Microbiol.* 41 (7): 3198–3205.

102 Vincent, A.L., Perez, D.R., Rajao, D. et al. (2017). Influenza A virus vaccines for swine. *Vet. Microbiol.* 206: 35–44.

103 Sharma, A., Zeller, M.A., Li, G. et al. (2020). Detection of live attenuated influenza vaccine virus and evidence of reassortment in the US swine population. *J. Vet. Diagn. Investig.* 32 (2): 301–311.

104 Rajao, D.S., Anderson, T.K., Kitikoon, P. et al. (2018). Antigenic and genetic evolution of contemporary swine H1 influenza viruses in the United States. *Virology* (518): 45–54.

105 Lauterbach, S.E., Nelson, S.W., Robinson, M.E. et al. (2019). Assessing exhibition swine as potential disseminators of infectious disease through the detection of five respiratory pathogens at agricultural exhibitions. *Vet. Res.* 50: 1–5.

106 Sun, H., Xiao, Y., Liu, J. et al. (2020). Prevalent Eurasian avian-like H1N1 swine influenza virus with 2009 pandemic viral genes facilitating human infection. *Proc. Natl. Acad. Sci.* 117 (29): 17204–17210.

107 Opriessnig, T., Meng, X.J., and Halbur, P.G. (2007). Porcine circovirus type 2–associated disease: update on current terminology, clinical manifestations, pathogenesis, diagnosis, and intervention strategies. *J. Vet. Diagn. Investig.* 19 (6): 591–615.

108 Vander Veen, R.L., Loynachan, A.T., Mogler, M.A. et al. (2012). Safety, immunogenicity, and efficacy of an alphavirus replicon-based swine influenza virus hemagglutinin vaccine. *Vaccine* 30 (11): 1944–1950.

7

Zika Virus Disease: An Emerging Global Threat

Ofosua Adi-Dako, Awo A. Kwapong, and Selorme Adukpo

Department of Pharmaceutics and Microbiology, School of Pharmacy, University of Ghana, Accra, Ghana

7.1 Introduction

Zika virus (ZIKV) is ribonucleic acid (RNA) virus that causes the ZIKV disease. It is a mosquito-borne disease that has, over the years, become a public health concern. The ZIKV was first identified in 1947 in the Zika Forest of Uganda in East Africa. It is mainly transmitted by the Aedes mosquito, which also spreads other diseases like dengue fever and chikungunya. Until recently, ZIKV outbreaks were only sporadic and mainly confined to Africa and Asia. In 2015, however, the virus spread rapidly throughout the Americas, leading to a global public health emergency.

The ZIKV infection is associated with a range of symptoms, including fever, rash, joint pain, and conjunctivitis. While the symptoms are generally mild and self-limiting, the virus has been associated with severe birth defects in babies born to mothers who were infected during pregnancy. The most well-known of these birth defects is microcephaly, a condition in which a baby's head is smaller than expected and can lead to developmental delays, seizures, and other complications.

The rapid spread of ZIKV in the Americas has prompted significant public health concerns. In addition to the risk of birth defects, the virus has also been linked to an ascendancy in cases of Guillain-Barre syndrome, a rare neurological disorder. To combat the spread of ZIKV, public health officials have implemented a range of measures, including mosquito control programs, public education campaigns, and travel advisories for pregnant women.

Despite these efforts, ZIKV remains a major public health concern. Currently, there is no specific treatment or vaccine for the virus, and prevention efforts heavily rely on mosquito control measures and public education. In addition, research is ongoing to better understand the virus and develop new treatments and prevention strategies.

In all, the speedy spread of ZIKV in recent years emphasizes the need for ongoing investment in public health infrastructure and research. By continuing to work together to better understand and fight the virus, we can help protect the health and well-being of people around the world.

7.2 Epidemiology of the Zika Virus Infection

ZIKV is a human arbovirus, belonging to the Flaviviridae family. It is phylogenetically close to Japanese encephalitis, West Nile, dengue, yellow fever viruses and several others. It was first isolated from the blood of sentinel rhesus macaque monkey in the Zika forest of Uganda near Kampala, East Africa in April 1947 [1], followed by evidence of infection and disease in humans in several other African countries and parts of Asia in the 1950s. Currently, African and Asian-American lineages of the virus are recognized in circulation [2].

It is vectorized by *Ades africanus* mosquitoes from several of which the virus was isolated in January 1948 [3]. Though its infectiousness in humans was first described in 1952 and confirmed in 1954 when the virus was isolated from a Nigerian girl during an outbreak of jaundice, not much attention had been paid to it by the scientific community till recently, mainly

Rising Contagious Diseases: Basics, Management, and Treatments, First Edition. Edited by Seth Kwabena Amponsah, Ranjita Shegokar, and Yashwant V. Pathak.

due to the occurrence of only a few and sporadic outbreaks [4]. The ZIKV evoked only mild clinical symptoms then in areas limited to the African continent. In 2007, however, there was a major epidemic in the Federated State of Micronesia, first of its kind outside Africa, where more than 70% of the 7391 inhabitants were infected with ZIKV which then brought the infection back in the spotlight. Though there were no cases of fatality in general nor birth defect microcephaly among babies born to women who became infected with ZIKV while pregnant, the people exhibited clinically mild symptoms. Additional major outbreaks occurred in French Polynesia in 2013, and Brazil in 2015 where an estimated 1.5×10^6 people were infected [1]. The Brazilian or the Pacific outbreak spread rapidly possibly due to the abundance of an effective and efficient vector, *Ae. Aegypti*, which prompted the World Health Organization to declare ZIKV infection and disease as a public health emergency in February 2016. Analyses of circulating ZIKV have revealed the existence of two lineages, an African lineage, and an Asian-American lineage [5].

Associated with the virus's emergence, new modes of transmission have been described with several hypotheses put forward to explain the extent and rapid expansions of the most recent outbreaks. Among these are a switch from African and sylvatic Aedine mosquitoes such as *Aedes* (*Stegomyia*) *luteocephalus, Aedes* (*Diceromyia*) *furcifer, Aedes* (*Aedimorphus*) *dalzieli, Ae. aegypti, Ae. africanus, Aedes* (*Fredwardsius*) *vittatus, and Aedes* (*Diceromyia*) *taylori* to more urban and anthropophilic mosquito vectors, including *Ae. aegypti, Ae. albopictus, Ae. vittatus, Ae. hensilli, Ae. polynesiensis*, and *Ae. unilineatus* [6, 7]. Besides being horizontally transmitted through sexual intercourse, they can also be vertically transmitted from mother to fetus during pregnancy [8].

7.3 Infection and Pathophysiology

ZIKV transmission occurs through infective arthropod bite and non-arthropod routes. The non-arthropod routes include, nosocomial, mother to child, blood transfusion, bone marrow or organ transplantation, and coitus [9]. Primarily, the transmission occurs via viremic blood meal and injection of infectious saliva during blood feeding by mosquitoes of Aedes (Stegomyia) genus. This genus contains more than 700 species that are found all over the world except in Antarctica. They bite mostly during the day with two biting peaks, early in the morning and in the evening before dusk. Out of the lot, *Aedes aegypti* or the yellow fever mosquito which are black mosquitoes with white stripes on their back and legs is the main vector in tropical and subtropical regions. Other notable vectors include *Ae. Hensilii, Ae. Polynesiensis*, and *Ae. Albopictus* [9].

Following an infective bite, the virus invades the cells through receptor-ligand interactions, employing several cell-specific receptors as the virus is ubiquitous in receptor preference. Infection is established in several cell types of humans, including human dermal fibroblasts, epidermal keratinocytes, immature dendritic cells, and innate immune cells are permissible to it [10]. It starts with the viral proteins interacting with specific receptors on the surface of host cells. The primary receptor for ZIKV entry into a cell is a protein called AXL, which belongs to the TAM (Tyro3, AXL, and Mer) family of receptor tyrosine kinases. AXL is expressed on various cell types, including skin cells, immune cells, cells of the nervous system, and myeloid-lineage cells to phagocytose apoptotic cells and debris [10]. AXL interacts with the viral envelope protein (E protein) which contains three distinct domains, an N-terminal domain, a long finger-like structure that is responsible for the E protein dimerization, and a hydrophobic fusion loop involved in membrane fusion and immunoglobulin-like domain, and a ligand-binding domain on the AXL receptor play a key role during virus entry to host cell [11].

Apart from AXL, other receptors have been implicated in ZIKV infection, although their roles may vary depending on the cell type and context. For instance, the dendritic cell-specific intercellular adhesion molecule 3, also known as cluster of differentiation 50 grabbing non-integrin receptor expressed on dendritic cells, monocytes among others can facilitate viral attachment and entry. Additionally, T cell immunoglobulin and mucin domain-containing protein (TIM) receptors, particularly TIM-1 and TIM-4, heparan sulfate proteoglycans (HSPGs) have all been shown to mediate ZIKV infection in certain cell types [11].

It is important to note that while AXL is the primary receptor for ZIKV entry, other receptors or co-receptors or attachment factors may also be involved in the process. For example, heparan sulfate proteoglycans (HSPGs) on the cell surface can facilitate initial attachment of the virus particles to the host cells [11].

Understanding the specific receptors and ligands involved in ZIKV infection is important for developing targeted therapeutics and preventive strategies. By identifying and studying these interactions, researchers can gain insights into the mechanisms of viral entry and potential targets for intervention.

Generally, most of the people infected with ZIKV do not develop symptoms, but for those who do, the symptoms typically start 3–14 days post infection. These symptoms are mostly associated with other arboviral infections like dengue fever and include rash, fever, conjunctivitis, muscle, and joint pain, malaise, and headache, and are normally mild. The infections are, however, self-limiting, and the symptoms generally disappear within two to seven days. The virus is also transmitted vertically from mother to child during pregnancy as well as through sexual contact, transfusion of infected blood and blood products. There is a possibility of organ transplantation transmission as well [12].

Both symptomatic and asymptomatic ZIKV infections during pregnancy is linked to birth defects, microcephaly, and other congenital malformations in infants or congenital Zika syndrome which include high muscle tone, limb contractures, eye abnormalities, and hearing loss. Other associated clinical complication includes fetal loss, stillbirth, and preterm birth. The risk of the development of the syndrome is unknown, but approximately 5–15% of expectant mothers who contract ZIKV experience the syndrome. ZIKV infection can also lead to Guillain-Barré syndrome, neuropathy, and myelitis, particularly in older children and adults [12].

7.4 Immune Response and Vaccine

Following infection, the body becomes exposed to viral particles, and the immune system mounts a response to fight off the virus. The innate immune system recognizes viral components, including ZIKV envelope or NS1 protein, triggers a rapid response to limit viral replication, and becomes activated first to provide an initial defense against the virus. The involvement of innate immunity in the control of ZIKV stems from the observation of the fact that mice lacking type 1 interferon activity demonstrated pronounced viral pathology [13], while IFN-stimulated genes have restricted viral replication in vitro [14]. Additionally, innate immune cells such as dendritic cells, macrophages, and natural killer (NK) cells play important roles in the early immune response to ZIKV. These cells detect the presence of the virus and release antiviral cytokines and chemokines to draw other immune cells to the site of infection. Macrophages and NK cells are also involved in killing infected cells directly.

As the infection progresses, B and T cell responses kick in and work together with the innate immune system to eliminate the virus and provide long-term immunity. The B lymphocytes, therefore, become activated to produce antibodies, which bind to the ZIKV and neutralize with the neutralizing antibodies associating positively with clearance of the virus from the blood [15]. Some of these antibodies can also opsonize the virus, marking it for destruction by other components of the immune system, including phagocytes and complement molecules [16]. Some of these Zika-specific antibodies persist in the bloodstream for several months to years after infection [17], which is suggestive of the development of memory B cells and long-term immunity following initial encounter with the virus.

The infection also leads to the development of ZIKV-specific T cells which contribute to viral clearance as $CD4^+$ helper T cells and $CD8^+$ cytotoxic T cells are critical for eliminating infected cells and controlling the viral infection. $CD4^+$ T cells help coordinate the immune response by releasing cytokines, including interferons, and activating other immune cells, including B lymphocytes to produce antibodies while $CD8^+$ T cells directly recognizes and kills infected cells [18].

It is important to note that while immune response plays a crucial role in controlling ZIKV infection, some evidence suggests that the virus can evade immune detection and establish persistent infections in certain tissues, such as the central nervous system and testes. These hinge on observed lack of the presence of ZIKV-specific antibodies produced in the periphery in central cerebrospinal fluid due to failure to cross the blood-cerebrospinal fluid barrier [15]. In addition, and because of reduced immune response to accommodate the developing fetus, ZIKV infection during pregnancy poses significant risks to the developing fetus, as it could lead to congenital Zika syndrome punctuated by damage to eyes and/or part of the brain, which may affect their visual development.

This immune response involves the production of antibodies that can recognize and neutralize the virus. Human beings produce neutralizing antibodies which are detectable for at least 18-months in blood [17]. There is evidence from animal models that these antibody responses can confer long-lasting protection from reinfection. This has led to enormous effort to understand the biology of the virus and develop vaccines to contain the disease. Sadly, however, there is no efficacious vaccine currently to prevent the disease in humans [19]. The development of a safe and effective vaccine against the ZIKV should therefore be a priority for global health authorities. A number of vaccine candidates are in development, using different vaccine platforms, and are at various stages of testing and evaluation. The successful development of a ZIKV vaccine would provide a necessary tool to prevent the spread of the virus and protect vulnerable populations soon.

7.5 Zika Virus Infection Diagnosis and Management

ZIKV fever is diagnosed through a combination of clinical symptoms, travel history, and laboratory testing. The medical doctor may ask about your symptoms, including any recent travel to areas where ZIKV is prevalent. A physical examination for characteristic signs and symptoms of ZIKV infection, including fever, rash, muscle pain, joint pain, headache, conjunctivitis (red eyes), and fatigue is also conducted.

Additionally, body fluids may be used to conclusively diagnose ZIKV infection through Reverse Transcription-Polymerase Chain Reaction (RT-PCR) which detects the virus RNA. This molecular technique is very effective during the acute phase of infection when the virus is replicating actively.

Antigen–antibody test, directed to detecting ZIKV-specific antibodies raised against the virus by the immune system, is also used. The detection of a recent or an ongoing infection is achieved by testing for ZIKV-specific IgM while IgG antibody testing can indicate past exposure to the virus.

Once tested positive, follow the advice and instruction of your care giver. Getting a lot of rest while taking in plenty of fluid, including oral rehydration solutions or electrolyte-rich drinks to rehydrate are essential for recovery. ZIKV-associated fever is managed with acetaminophen (paracetamol) by following the recommended dosage instructions, while aspirin or non-steroidal anti-inflammatory drugs should be avoided until dengue fever is completely ruled out since these medications may increase the risk of bleeding. In case of itching and rashes, scratching the affected areas should be avoided to prevent secondary infections; a health-care provider may recommend antihistamines or calamine lotion for relief when necessary. Depending on the severity of symptoms and individual circumstances, however, a health-care provider may recommend regular monitoring of vital signs, blood tests, and follow-up appointments to ensure proper recovery and to detect any complications. If you have ZIKV infection, it is recommended to abstain from sexual activity for that period, as is advised by health-care professionals, particularly if you have a pregnant partner or are planning to conceive. In case of existing pregnancy or those planning to become pregnant and being diagnosed with ZIKV infection, close monitoring is essential to provide appropriate prenatal care, including ultrasounds to monitor fetal development and screening for potential birth defects.

Generally, it is crucial to consult a healthcare professional for personalized advice and management, more especially if you have specific medical conditions or concerns. Nevertheless, once infected, it is essential to take measures to prevent further mosquito bites and transmission by wearing protective clothing, as well as ensuring proper mosquito control measures in your environment.

7.6 Development of Antiviral Therapeutic Drug against Zika Virus

The development of antiviral therapeutic drugs against ZIKV has been an active area of research since the ZIKV outbreak in 2015–2016. Quite a few approaches have been explored to discover potential antiviral candidates. The strategies and developments in the development of antiviral drugs against ZIKV include among others.

Drugs repurposing: Drugs that are approved for other viral and non-viral infections hold potential for activity against ZIKV. Remdesivir, which was originally developed for Ebola, and favipiravir, which was developed for influenza, as well as doramectin, an approved veterinary antiparasitic drug, have been identified to have varying degrees of antiviral activity against ZIKA and inhibitory effects against ZIKV in laboratory studies [20, 21].

With the use of computational biology and high-throughput screening methods, small molecules which target specific viral proteins or enzymes important for viral replication, such as protease inhibitors, polymerase inhibitors, or entry inhibitors which may be inhibitory to viral replication, have also been explored. A couple of molecules based on the sequence of Claudin, more specifically from the N-terminal sequences of claudin-7 and claudin-1, designated as CL7.1 and CL1.1, respectively, have been identified to inhibit ZIKV infection through their ability to block the ZIKV E-mediated membrane fusion. Other molecules, including peptides from Japanese encephalitis virus E protein, have been assessed for their ability to block Japanese encephalitis virus and ZIKV infection [22]. Toll-like receptor agonist, R848 has been also identified to blocks ZIKV replication by inducing viperin [23], an important host restriction factor in the control of ZIKV infection [24].

ZIKV-specific antibodies neutralize the virus and have opened the door for the development and assessment of monoclonal antibodies like therapeutic agents. Hundreds of these have been investigated with several of these antibody preparations, including monoclonal and cross-reactive antibodies that bind to E proteins that have been characterized and observed to protect against lethal ZIKV challenge. They markedly reduce tissue pathology, decrease vertical transmission, and prevent ZIKV-induced microcephaly, which highlight the therapeutic potential of antibodies in preventing ZIKV-related damage [25].

RNA interference (RNAi): RNAi-based approaches aimed at using small RNA molecules to interfere with viral genome and prevent viral replication. RNAs (siRNAs) or short hairpin RNAs (shRNAs) targeting specific ZIKV genes to inhibit viral replication in humans [26] or strain transcending inhibition of the virus in the mosquitoes [27] have all been explored.

CRISPR-Cas9 ([clustered regularly interspaced short palindromic repeats]/Cas-9) gene editing: The CRISPR-Cas9 system directly targets and edits the viral genome of ZIKV, thereby disrupting viral replication [28], an approach that is still in the early stages of development and requires a lot of optimization and testing.

It is important to note that while significant progress has been made in identifying potential antiviral candidates against ZIKV, the current state of no ZIKV infections have negatively impacted the urgency and advancement of the drug development process, besides its complexity and time-consuming nature. However, the research and development efforts for ZIKV-specific drugs have bestowed on us valuable insights and tools that can be utilized for future outbreaks or emerging viral diseases.

7.7 The Way Forward

Currently, there is no specific antiviral treatment for ZIKV infection, nor a widely available approved ZIKV vaccine, so continued research is needed to ensure that vaccine candidates that have shown promise in preclinical, and early-stage clinical trials are developed to help control the disease. In the meantime, the supportive care provided to relieve symptoms, such as rest, hydration, and pain relief should be enhanced while those with ZIKV should avoid nonsteroidal anti-inflammatory drugs and aspirin until dengue is ruled out to prevent complications.

More research is, therefore, needed to understand the ZIKV, including its transmission dynamics, long-term health effects, and the potential link between ZIKV and other health conditions. Mosquito control programs to reduce Aedes mosquito populations need to be strengthened to minimize the risk of ZIKV transmission. There is the need to foster collaboration among global health organizations, governments, researchers, and health-care professionals to share knowledge, resources, and best practices in ZIKV prevention, control, and research.

Raising awareness about ZIKV prevention and control measures, particularly in areas where the virus is endemic or where outbreaks have occurred, taking an integrated approach that addresses multiple mosquito-borne diseases, as Aedes mosquitoes also transmit other viruses like dengue and chikungunya, need to be encouraged and embraced.

Practice the use of barrier methods during sexual intercourse if you or your partner have traveled to or lived in a ZIKV-affected area. Abstain or delay pregnancy if you have been exposed to ZIKV. Screening measures for blood donations should be implemented to prevent the transfusion transmission of ZIKV and protect life.

Meanwhile, it is important to seek the latest guidelines and information from reliable health authorities, such as the World Health Organization, for the most current and accurate information on ZIKV infection and other emerging or re-emerging viral infections like Human Immunodeficiency Virus, severe acute respiratory syndrome, Lyme disease, Hantavirus, Dengue fever, and West Nile virus, and practice them, as the field of virology is undergoing rapid evolution.

References

1 Campos, G.S., Bandeira, A.C., and Sardi, S.I. (2015). Zika virus outbreak, Bahia, Brazil. *Emerg. Infect. Dis.* 21 (10): 1885–1886. https://doi.org/10.3201/eid2110.150847.

2 Faye, O., Freire, C.C., Iamarino, A. et al. (2014). Molecular evolution of Zika virus during its emergence in the 20(th) century. *PLoS Negl. Trop. Dis.* 8 (1): e2636. https://doi.org/10.1371/journal.pntd.0002636.

3 Boyer, S., Calvez, E., Chouin-Carneiro, T. et al. (2018). An overview of mosquito vectors of Zika virus. *Microbes Infect.* 20 (11–12): 646–660. https://doi.org/10.1016/j.micinf.2018.01.006.

4 Smithburn, K.C. (1952). Neutralizing antibodies against certain recently isolated viruses in the sera of human beings residing in East Africa. *J. Immunol.* 69 (2): 223–234.

5 Ye, Q., Liu, Z.Y., Han, J.F. et al. (2016). Genomic characterization and phylogenetic analysis of Zika virus circulating in the Americas. *Infect. Genet. Evol.* 43: 43–49. https://doi.org/10.1016/j.meegid.2016.05.004.

6 Gutiérrez-Bugallo, G., Piedra, L.A., Rodriguez, M. et al. (2019). Vector-borne transmission and evolution of Zika virus. *Nat. Ecol. Evol.* 3 (4): 561–569. https://doi.org/10.1038/s41559-019-0836-z.

7 McKenzie, B.A., Wilson, A.E., and Zohdy, S. (2019). Aedes albopictus is a competent vector of Zika virus: a meta-analysis. *PLoS One* 14 (5): e0216794. https://doi.org/10.1371/journal.pone.0216794.

8 Chan, J.F., Choi, G.K., Yip, C.C. et al. (2016). Zika fever and congenital Zika syndrome: an unexpected emerging arboviral disease. *J. Infect.* 72 (5): 507–524. https://doi.org/10.1016/j.jinf.2016.02.011.

9 Musso, D. and Gubler, D.J. (2016). Zika virus. *Clin. Microbiol. Rev.* 29 (3): 487–524. https://doi.org/10.1128/CMR.00072-15.

10 Agrelli, A., de Moura, R.R., Crovella, S., and Brandão, L.A.C. (2019). ZIKA virus entry mechanisms in human cells. *Infect. Genet. Evol.* 69: 22–29. https://doi.org/10.1016/j.meegid.2019.01.018.

11 Khongwichit, S., Sornjai, W., Jitobaom, K. et al. (2021). A functional interaction between GRP78 and Zika virus E protein. *Sci. Rep.* 11 (1): 393. https://doi.org/10.1038/s41598-020-79803-z.

12 Rawal, G., Yadav, S., and Kumar, R. (2016). Zika virus: an overview. *Fam. Med. Prim. Care Rev.* 5 (3): 523–527. https://doi.org/10.4103/2249-4863.197256.

13 Lazear, H.M., Govero, J., Smith, A.M. et al. (2016). A mouse model of Zika virus pathogenesis. *Cell Host Microbe* 19 (5): 720–730. https://doi.org/10.1016/j.chom.2016.03.010.

14 Savidis, G., Perreira, J.M., Portmann, J.M. et al. (2016). The IFITMs inhibit Zika virus replication. *Cell Rep.* 15 (11): 2323–2330. https://doi.org/10.1016/j.celrep.2016.05.074.

15 Aid, M., Abbink, P., Larocca, R.A. et al. (2017). Zika virus persistence in the central nervous system and lymph nodes of Rhesus monkeys. *Cell* 169 (4): 610–620.e14. https://doi.org/10.1016/j.cell.2017.04.008.

16 Enlow, W., Bordeleau, M., Piret, J. et al. (2021). Microglia are involved in phagocytosis and extracellular digestion during Zika virus encephalitis in young adult immunodeficient mice. *J. Neuroinflammation* 18 (1): 178. https://doi.org/10.1186/s12974-021-02221-z.

17 Magalhães, I.C.L., Souza, P.F.N., Marques, L.E.C. et al. (2022). New insights into the recombinant proteins and monoclonal antibodies employed to immunodiagnosis and control of Zika virus infection: a review. *Int. J. Biol. Macromol.* 200: 139–150. https://doi.org/10.1016/j.ijbiomac.2021.12.196.

18 Tappe, D., Pérez-Girón, J.V., Zammarchi, L. et al. (2016). Cytokine kinetics of Zika virus-infected patients from acute to reconvalescent phase. *Med. Microbiol. Immunol.* 205 (3): 269–273. https://doi.org/10.1007/s00430-015-0445-7.

19 Turpin, J., El Safadi, D., Lebeau, G. et al. (2022). Apoptosis during Zika virus infection: too soon or too late? *Int. J. Mol. Sci.* 23 (3): 1287. https://doi.org/10.3390/ijms23031287.

20 Kim, J.A., Seong, R.K., Kumar, M., and Shin, O.S. (2018). Favipiravir and ribavirin inhibit replication of Asian and African strains of Zika virus in different cell models. *Viruses* 10 (2): 72. https://doi.org/10.3390/v10020072.

21 Zhu, Y., Liang, M., Yu, J. et al. (2023). Repurposing of doramectin as a new anti-Zika virus agent. *Viruses* 15 (5): 1068. https://doi.org/10.3390/v15051068.

22 Chen, L., Liu, Y., Wang, S. et al. (2017). Antiviral activity of peptide inhibitors derived from the protein E stem against Japanese encephalitis and Zika viruses. *Antivir. Res.* 141: 140–149. https://doi.org/10.1016/j.antiviral.2017.02.009.

23 Vanwalscappel, B., Tada, T., and Landau, N.R. (2018). Toll-like receptor agonist R848 blocks Zika virus replication by inducing the antiviral protein viperin. *Virology* 522: 199–208. https://doi.org/10.1016/j.virol.2018.07.014.

24 Van der Hoek, K.H., Eyre, N.S., Shue, B. et al. (2017). Viperin is an important host restriction factor in control of Zika virus infection. *Sci. Rep.* 7 (1): 4475. https://doi.org/10.1038/s41598-017-04138-1.

25 Wang, Q., Yan, J., and Gao, G.F. (2017). Monoclonal antibodies against Zika virus: therapeutics and their implications for vaccine design. *J. Virol.* 91 (20): e01049-17. https://doi.org/10.1128/JVI.01049-17.

26 Giulietti, M., Righetti, A., Cianfruglia, L. et al. (2018). To accelerate the Zika beat: candidate design for RNA interference-based therapy. *Virus Res.* 255: 133–140. https://doi.org/10.1016/j.virusres.2018.07.010.

27 Magalhaes, T., Bergren, N.A., Bennett, S.L. et al. (2019). Induction of RNA interference to block Zika virus replication and transmission in the mosquito Aedes aegypti. *Insect Biochem. Mol. Biol.* 111: 103169. https://doi.org/10.1016/j.ibmb.2019.05.004.

28 Li, Y., Muffat, J., Omer Javed, A. et al. (2019). Genome-wide CRISPR screen for Zika virus resistance in human neural cells. *Proc. Natl. Acad. Sci. U. S. A.* 116 (19): 9527–9532. https://doi.org/10.1073/pnas.1900867116.

8

Current Perspectives in Dengue Hemorrhagic Fever

Manish P. Patel[1], Vaishnavi M. Oza[1], Hemangi B. Tanna[1], Avinash D. Khadela[2], Praful D. Bharadia[1], and Jayvadan K. Patel[3]

[1] Department of Pharmaceutics, L. M. College of Pharmacy, Ahmedabad, Gujarat, India
[2] Department of Pharmacology, L.M. College of Pharmacy, Ahmedabad, Gujarat, India
[3] Formulation and Development, Aavis Pharmaceuticals, Hoschton, GA, United States

8.1 Introduction

Dengue hemorrhagic fever (DHF), which is caused by the dengue virus (DENV), affects millions of individuals each year. The presence of both *Aedes aegypti* and *Aedes albopictus* mosquitoes, along with the circulation of multiple types of the DENV in a specific region, tends to be linked with outbreaks of DHF and dengue shock syndrome (DSS).

Humans who are afflicted with DHF act as the primary hosts for the virus and are capable of spreading it during the initial three days of the illness, known as the viremic stage [1]. Mosquitoes that feed on an infected individual during the daytime typically become carriers of the virus after about two weeks and retain this infectivity throughout their lifespan. Dengue is typically prevalent in certain regions: Americas, the Middle East, Africa, and the Pacific Islands.

DENVs, originating from mosquitoes, underwent an evolutionary process that led to their adaptation first in non-human primates and later in humans. In humans, the viremia reaches high levels two days before the onset of fever (non-febrile phase) and continues for five to seven days after the fever begins (febrile phase). It is during these specific periods that vector species become infected. Subsequently, humans become dead-end hosts for transmission, as the movement of vectors is highly limited compared to the movement of the human host [2]. *A. aegypti* is the most common mosquito involved in the transmission of DENV, and it breeds in stagnant water (e.g., collections of water in containers, water-based air coolers, and tire dumps) [3].

Dengue fever and its severe manifestations, DHF and DSS (Figure 8.1), have emerged as significant global public health issues. In the last 30 years, there has been a substantial rise in the occurrence of dengue fever, DHF, and DSS, along with their epidemics, on a global scale. These diseases are prevalent in tropical and sub-tropical regions worldwide, primarily in urban and semi-urban areas, and are increasingly spreading to rural regions.

Approximately 3.9 billion individuals residing in 128 countries face the risk of contracting DENVs. According to recent estimates, there are around 390 million dengue infections annually, with about 96 million cases displaying clinical manifestations of varying severity. The reported cases have witnessed a significant rise, escalating from less than 0.5 million in 2010 to 4.2 million in 2019. Not only is the number of cases increasing, but the disease is also spreading to new regions, leading to sudden and widespread outbreaks.

Globally, around 500,000 individuals diagnosed with DHF need hospitalization each year. Among them, approximately 90% are children under the age of five years, and about 2.5% of those affected unfortunately succumb to the illness. In epidemic situations, the infection rate among individuals who have not previously been exposed to the virus often ranges from 40% to 50% but can occasionally escalate to 80–90%. It is noteworthy that multiple serotypes/genotypes of the virus are simultaneously circulating during these outbreaks.

Dengue fever and DHF have reached epidemic proportions in over 100 countries across various World Health Organization (WHO) regions, including Africa, the Americas, the Eastern Mediterranean, South-East Asia, and the Western Pacific. The South-East Asia and Western Pacific regions are particularly hard-hit by the diseases. The detection of all four serotypes of

Rising Contagious Diseases: Basics, Management, and Treatments, First Edition. Edited by Seth Kwabena Amponsah, Ranjita Shegokar, and Yashwant V. Pathak.

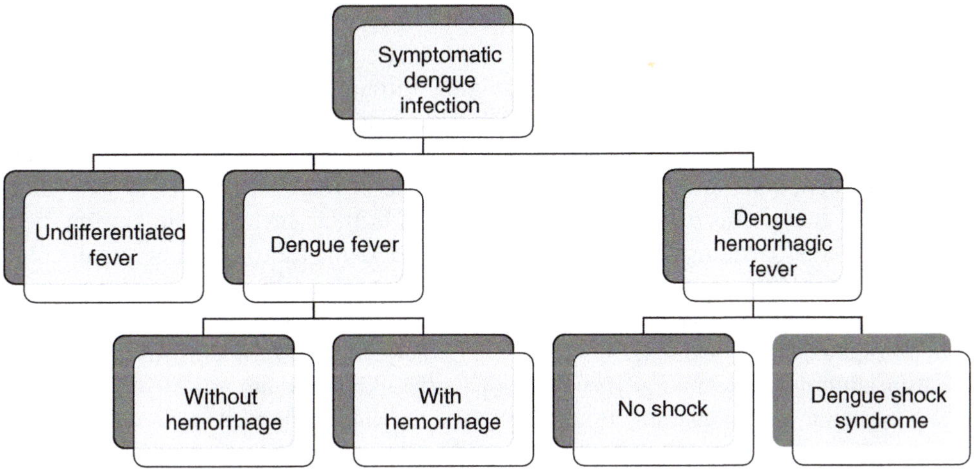

Figure 8.1 Sub-classification of dengue disease.

the DENV has led to these countries becoming hyperendemic. In the South-East Asia region, countries are categorized into three distinct groups: Category A (Bangladesh, India, Indonesia, Maldives, Myanmar, Sri Lanka, Thailand, and Timor-Leste): major public health problem, leading cause of hospitalization and death among children, hyperendemicity with all four serotypes circulating in urban areas, and spreading to rural areas; Category B (Bhutan, Nepal): Endemicity uncertain; Bhutan reported the first outbreak in 2004 and Nepal reported the first indigenous case in 2004. Category C (DPR Korea): no evidence of endemicity.

Severe dengue fever is a life-threatening medical emergency. Seek immediate medical attention if you have recently visited an area, in which dengue fever is known to occur, you have had a fever, and you develop any of the warning signs. Warning signs include severe stomach pain, vomiting, difficulty breathing, or blood in your nose, gums, vomit, or stools.

If you have recently traveled and developed a fever accompanied by mild dengue fever symptoms, it is advisable to contact your doctor. The following host factors contribute to more severe disease and its complications: infants and elderly, obesity, pregnancy, peptic ulcer disease, thalassemia and other hemoglobinopathies, congenital heart disease, chronic diseases such as diabetes mellitus, hypertension, asthma, ischemic heart disease, chronic renal failure, and liver cirrhosis [4].

The transmission of the DENV to humans occurs when an infected mosquito bites. Only specific mosquito species act as vectors for the DENV. Among the different types of vectors, arthropods are the most common. Arthropods are invertebrate animals characterized by an external skeleton, known as an exoskeleton. Examples of arthropod vectors include mosquitoes, ticks, lice, flies, and fleas. For instance, ticks can carry Lyme disease, while certain species of mosquitoes can transmit yellow fever, malaria, and dengue fever.

When a mosquito bites an individual with the DENV in their bloodstream, the mosquito becomes infected with the virus. Later, this infected mosquito can transmit the virus to other healthy individuals by biting them. It is important to note that dengue cannot spread directly from one person to another, and mosquitoes are necessary for the transmission of the DENV.

Dengue fever is a viral illness caused by the DENV, which does not stem from microorganisms like bacteria or fungi. The DENV belongs to the Flaviviridae family and is transmitted to humans primarily through the bites of infected female mosquitoes, particularly the *A. aegypti* mosquito.

There are four distinct serotypes of the DENV: DENV-1, DENV-2, DENV-3, and DENV-4. All four serotypes have the potential to cause dengue fever, a flu-like illness characterized by symptoms such as high fever, severe headache, muscle and joint pain, rash, and, in some cases, hemorrhagic manifestations.

When an Aedes mosquito infected with the DENV bites a person, the virus enters the bloodstream and infects various cells, including immune cells. It then reproduces within these cells and spreads throughout the body, resulting in the characteristic symptoms of dengue fever.

It is important to understand that dengue fever is not directly transmitted from person to person. Mosquitoes play a vital role as the primary vectors for transmitting the DENV. Preventing dengue involves efforts to control mosquito populations and implementing personal protective measures such as using mosquito repellents, wearing protective clothing, and eliminating mosquito breeding sites.

8.2 Life Cycle of Dengue Mosquitoes

The life stages of mosquitoes involve a complex transformation as they progress through different forms and habitats. Female mosquitoes typically deposit their eggs above the water level inside various water-containing containers, such as tires, buckets, birdbaths, water storage jars, and flower pots (Figure 8.2). Once the containers become filled with water, often after the rainfall, the mosquito eggs hatch, developing into mosquito larvae. These larvae are adapted to an aquatic lifestyle and dwell in the water, nourishing themselves on microorganisms present in their environment. During their growth, the larvae undergo three molting events, shedding their outer skin, which marks the transition between different larval stages known as the first to fourth instars [5].

Once the larva reaches the fourth instar and has fully developed, it undergoes a metamorphosis process into a pupa, resembling a "cocoon" stage for the mosquito. This pupal stage also occurs in an aquatic environment. After a span of approximately two days, the fully mature adult mosquito emerges from the pupa by breaking through its skin. At this point, the mosquito has completed its transformation and transitions to an adult form capable of flight. Unlike the previous stages, the adult mosquito is no longer reliant on an aquatic habitat and can be found in terrestrial environments.

8.3 Life Cycle of Dengue Virus

The life cycle of the DENV involves two primary hosts: humans and mosquitoes. During transmission, the female Aedes mosquitoes, particularly *A. aegypti*, acquire the DENV when they take a blood meal from an infected human. The virus enters the mosquito's bloodstream and infects its midgut [5]. Within the mosquito's midgut epithelial cells, the DENV undergoes replication, leading to an increase in viral particles (replication). The virus then spreads from the mosquito's midgut to other tissues, including the salivary glands, which are crucial for transmission (dissemination). When the

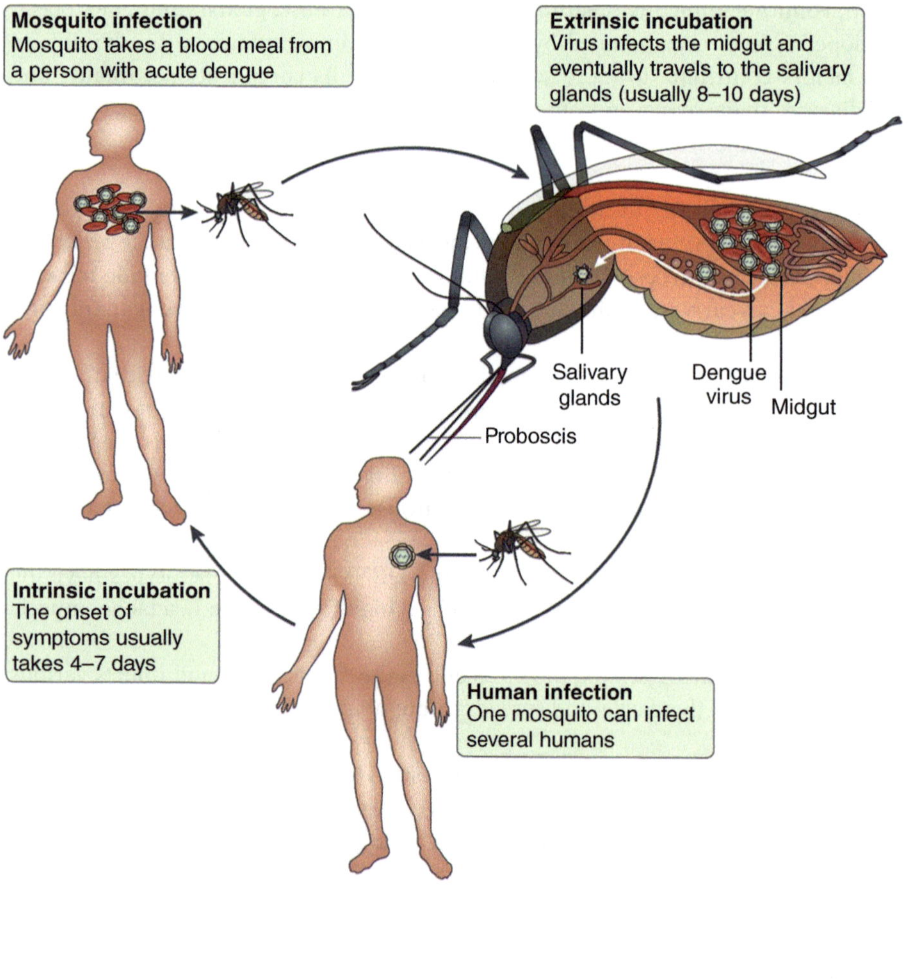

Figure 8.2 Life cycle of Aedes mosquito.

infected mosquito feeds on another human, the DENV can be injected into the new host along with the mosquito's saliva, entering the human bloodstream.

Within humans, the DENV replicates in various cell types, including immune cells. This can result in a range of clinical manifestations, from mild dengue fever to more severe forms such as DHF or DSS [6].

During the acute phase of infection, if a mosquito bites an infected human, it can acquire the virus and become infected itself. The virus can then continue its life cycle within the mosquito, replicating and enabling it to transmit the virus to other human hosts. The adult mosquito stage is distinct from the DENV life cycle and involves the mosquito transitioning into a fully developed form capable of flight. This adult mosquito is no longer aquatic and inhabits terrestrial environments.

8.4 Risk Factors Responsible for Dengue Fever

Mosquito bites: The primary cause of dengue fever is the transmission of the DENV through the bites of infected Aedes mosquitoes. These mosquitoes become infected with the virus when they bite an individual who is already infected, and they can subsequently transmit the virus to others through subsequent bites [7].

Travel to regions with endemic dengue: Dengue fever is more widespread in tropical and subtropical regions, particularly in urban and semi-urban areas. Traveling to these regions heightens the likelihood of exposure to the DENV.

Inadequate mosquito control measures: Insufficient efforts to control mosquitoes, along with poor sanitation and inadequate waste management, contribute to the proliferation of Aedes mosquitoes, thereby increasing the risk of dengue transmission.

Influence of climate conditions: Mosquito populations thrive in warm and humid environments. Climate factors such as increased rainfall, temperature, and humidity create favorable conditions for mosquito breeding and the subsequent transmission of dengue.

Weakened immune system: Individuals with weakened immune systems, including infants, young children, and older adults, may be more vulnerable to severe dengue infection.

Prior dengue infection: Having experienced dengue fever in the past does not guarantee immunity to other serotypes of the DENV. In fact, subsequent infections with different serotypes can heighten the risk of severe dengue, also known as DHF [1].

8.5 Pathophysiology of Dengue Fever

8.5.1 The Viral Genome

DENVs are members of the Flavivirus genus within the Flaviviridae family. These viruses exist in four serotypes and have an enveloped, spherical structure with an approximate diameter of 500 Å20. Each serotype's genome consists of around 11 kb of positive-sense, single-stranded RNA, which encodes 10 proteins [8].

The genome encodes three structural proteins: the membrane (M) protein, envelope (E) protein, and capsid (C) protein. Additionally, there are non-structural (NS) proteins, namely NS1, NS2A, NS2B, NS3, NS4A, NS4B, and NS5. The E protein and M protein play important roles in the structure and function of the virus. The delivery of the DENV genome into the host cytoplasm involves several steps, starting with the fusion of the viral membrane and the host plasma membrane. Subsequently, the virus is endocytosed into an endosome, followed by pH-dependent fusion of the viral and endosomal membranes [5].

The interior of the virus particle consists of RNA complexed with capsid proteins and is surrounded by a lipid bilayer membrane. This membrane contains externally anchored M protein and E protein, which work together to facilitate interactions between the virus and the host during entry.

The E protein is arranged as 90 tightly packed monomers on the surface of DENVs. It lies flat against the membrane and plays a crucial role in viral entry by binding to cellular receptors and mediating fusion between the viral and cellular membranes. Among its three domains (DI–DIII), DIII is responsible for receptor binding, and mutations in this domain have been found to affect receptor interaction.

The hinge connecting DI to DII is highly flexible and allows DII to maneuver within the low pH environment of the endosome. This movement exposes the fusion loop of DII, which then interacts with the endosomal membrane, facilitating fusion between the virus and the endosomal membrane. This fusion process leads to the release of viral RNA into the host cell.

After being released from infected cells, newly formed DENV particles can either have processed surface M protein, making them infective and considered "mature," or retain the Uncleaved precursor form of the M protein (prM) on their surface, remaining in an "immature" form. Some viral particles may have a combination of M protein and prM protein on their surface, and whether they are infective or not can vary [6].

8.5.2 Structure and Function of the NS Proteins

The NS proteins play a vital role in viral replication and packaging, both processes are closely associated with the functioning of the host's endoplasmic reticulum (ER) and secretory pathway. NS1, a glycoprotein weighing 46 kDa, has three forms: an ER-resident form, a membrane-anchored form, and a secreted form. Initially, NS1 is produced as a soluble monomer and undergoes dimerization in the ER lumen, leading to its association with the membrane. Recent determination of the crystal structure of NS1 has shown exposed hydrophobic domains in the dimer, which likely facilitates its interaction with the membrane.

Intracellularly, NS1 is involved in early viral RNA replication and localizes to vesicular compartments induced by the virus, which serve as sites for viral replication complexes. Additionally, NS1 is transported to the cell surface, where it remains associated with the cell membrane or is released as a soluble form known as sNS1. sNS1 exists as a lipid-associated hexametric species and can be detected in the blood of infected individuals from the early onset of symptoms. During the acute phase of infection, sNS1 circulates at varying levels, ranging from nanograms per milliliter to milligrams per milliliter. The blood levels of sNS1 correlate with the peak viremia and severity of the disease in secondary DENV infection.

Several studies have proposed that sNS1 plays a crucial role in dengue pathogenesis. For example highly purified recombinant NS1 (rNS1), free of bacterial endotoxin activity, directly activates mouse macrophages and human peripheral blood mononuclear cells through Toll-like receptor 4 (TLR4). This activation leads to the production and release of pro-inflammatory cytokines and chemokines. Moreover, exposure to NS1 has been shown to disrupt the integrity of endothelial cell monolayers in both in vitro and in vivo models of vascular leakage. These findings highlight the potential of NS1 as a key mediator in dengue pathogenesis [6].

8.5.3 Dengue Symptoms

The symptoms associated with dengue vary as days elapse after infection. They include flu-like symptoms, fever, and joint pain, among others (Table 8.1) [9–12].

8.6 Clinical Manifestation

DHF is an intense manifestation of dengue fever. The progression of the dengue illness can be categorized into three distinct phases: the febrile phase, the critical phase, and the recovery/convalescent phase [13, 14].

8.6.1 Febrile Phase

Clinical indicators include high fever accompanied by a positive Tourniquet test and leukopenia (white blood cell count $\leq$5000 cells/mm^3), with a positive predictive value of 70–83%. NS1 antigen test conducted during the febrile phase (first five days of fever) with a sensitivity of 60–70% and specificity above 99% PCR, which exhibits good sensitivity and specificity, but can be expensive and unavailable in many locations. ELISA-IgM and IgG tests are not suitable for early diagnosis as antibody levels significantly increase after the fifth day of fever. Management of a high fever: Administer paracetamol for fever reduction, and consider tepid sponge baths to help lower body temperature. Encourage oral intake: provide a soft diet, including milk, fruit juice, and oral rehydration solution (ORS). If there is no vomiting and moderate/severe dehydration, avoid administering intravenous (IV) fluids. Conduct a complete blood count (CBC) daily to track the patient's progress. Advise patients to return to the hospital as soon as possible if there is no clinical improvement despite the absence of fever, if they experience severe abdominal pain, vomiting, bleeding, restlessness, irritability, drowsiness, refusal to eat or drink (some patients may still feel thirsty), or if they have not passed urine for four to six hours [13].

Table 8.1 Symptoms associated with dengue fever.

Duration	Symptoms	Category
Two to seven days	"Flu-like" syndrome Retro-orbital pain Fever Rash Intense headache Joint pain Intense joint and muscle pain Nausea Swollen gland Pain behind the eyes	Dengue fever (DF)
After three to five days of fever	Plasma leakage Pleural effusion, bleeding Thrombocytopenia with Raise in hematocrit levels Restlessness Abdominal pain Vomiting Sudden drop in temperature	DHF
After three to five days of fever	Temperature reaches 37.5–38 °C After three to five days of fever Hypotension Decrease in platelet count leads to leakage of plasma subsequent shock Fluid accumulation with respiratory distress Critical bleeding Organ impairment Cardiorespiratory failure and cardiac arrest	DSS
The warning signs usually begin the first day or two after your fever goes away	Severe stomach pain Persistent vomiting Bleeding from your gums or nose Blood in your urine, stools, or vomit Bleeding under the skin, which might look like bruising Difficult or rapid breathing Fatigue Irritability or restlessness	Warning signs of severe dengue fever – which is a life-threatening emergency can develop quickly

8.6.2 Critical/Leakage Phase (Early Detection of Plasma Leakage/Shock)

Thrombocytopenia, indicated by a platelet count of ≤100,000 cells/mm^3, is the most reliable indicator of plasma leakage. When the platelet count ranges between 50,000 and 100,000 cells/mm^3, it signals the beginning of plasma leakage, as approximately half of the patients with dengue fever (DF) experience thrombocytopenia at this level. If the platelet count drops below 50,000 cells/mm^3, it strongly suggests DHF and typically indicates the occurrence of plasma leakage, often for duration of around 24 hours. For patients exhibiting thrombocytopenia along with a poor appetite or overall poor clinical conditions, admission to the hospital is recommended. High-risk individuals, such as infants, obese patients, those with prolonged shock (grade IV), bleeding, encephalopathy, underlying diseases, or pregnancy, should also be considered for admission. Identifying pleural effusion and ascites during the early phase of plasma leakage or even at the time of shock can be challenging through physical examination alone. However, alternative methods such as a right lateral decubitus chest film, ultrasonography, or a serum albumin level of ≤3.5 g% can be employed to detect plasma leakage. During the critical period, isotonic salt solutions such as 5% dextrose in normal saline solution (NSS), 5% Ringer Acetate, or

5% Ringer-Lactate are used. 5% dextrose in NSS is the preferred choice for severe cases requiring admission, especially those experiencing poor appetite, nausea/vomiting, and abdominal pain. The total fluid requirement during the critical period of 24–48 hours is estimated as maintenance plus 5% deficit (M + 5%D), including both oral and IV fluids. In patients with dengue shock syndrome (DSS), IV fluid duration may range from 24 to 36 hours, while in non-shock cases of DHF, it may extend to 48–60 hours. The rate of IV fluid should be adjusted based on clinical vital signs (blood pressure, pulse, respiratory rate, and temperature), hematocrit (Hct), and urine output (0.5 ml/kg/h). The recommended IV fluid rate for shock patients (DHF grade III) is 10 ml/kg/h, which is lower than the rate recommended for other types of shock (not exceeding 20 ml/kg/h). A larger amount of IV fluid may be required for DHF grade IV, but the rate should be reduced to 10 ml/kg/h as soon as blood pressure is restored.

If the clinical response is inadequate (e.g., re-shock, unstable vital signs, inability to reduce the IV fluid rate), the following laboratory data should be investigated and corrected:

Acidosis: Check blood gas (capillary or venous). If acidosis is present (blood pH < 7.35 and $HCO_3 < 15$ mEq/l), assess liver and renal functions and correct acidosis accordingly.

Bleeding: Evaluate hematocrit (Hct). If Hct is high, consider dextran administration. If Hct is low or not rising, contemplate blood transfusion and intravenous administration of vitamin K1.

Calcium and other electrolytes: Monitor sodium (Na) and potassium (K) levels. Administer calcium gluconate at a dose of 1 ml/kg, diluted twice with IV fluid, and administer slowly via IV push. The maximum dose per dose is 10 ml. Monitor blood sugar levels.

For colloidal solutions, the recommended plasma expander with higher osmolarity than plasma is 10% Dextran-40 in NSS. A bolus dose of 10 ml/kg/h in children or 500 ml/h in adults is recommended, and this generally reduces Hct by 10 points in cases with signs of fluid overload or persistently high Hct. In cases of significant bleeding, defined as >6–8 ml/kg of ideal body weight in children or 300 ml/min in adults, immediate blood transfusion is recommended. The transfusion amount should be equal to the estimated loss. If estimation is not possible (e.g., concealed internal bleeding), transfuse 10 ml/kg of fresh whole blood (FWB) or 5 ml/kg of packed red cells (PRC) in children to raise Hct by 5 points. In adults, transfuse 1 unit of FWB or PRC.

Platelets are indicated in cases with significant bleeding. However, if the patient exhibits signs of fluid overload, platelet transfusion should be avoided to prevent fluid overload and possible acute pulmonary edema. Platelet transfusion serves as adjunct therapy and is not a specific treatment. Platelet prophylaxis is not recommended for children, regardless of platelet count. Clinicians may consider prophylactic platelet transfusion in adults with underlying hypertension or heart disease.

8.6.3 Recovery/Convalescent Phase

Discontinue IV fluid administration when there are indications of recovery, such as convalescence rash, itching, increased appetite, or after a minimum of 30 hours following shock and 60 hours after plasma leakage. It's important to note that some patients may experience sinus bradycardia during this phase [13].

Patients with significant ascites and pleural effusions may require diuretics during the reabsorption period, which facilitates the return of extravasated plasma back into circulation.

It is possible that some patients may not experience a return of appetite during this phase. This could be attributed to diuresis and potassium loss through urine. In such cases, potassium supplementation may be necessary. Most patients prefer potassium-rich fruits (e.g., bananas, oranges) and fruit juices as a source of supplementation.

In adults, the convalescence period can last for two to four weeks, during which fatigue may persist.

8.7 Diagnosis of Dengue Fever

Medical practitioners can diagnose dengue infection with a blood test to check for the virus or antibodies to it. If you become sick after traveling or living near people who already have dengue infection, let your doctor know. This will allow your doctor to evaluate the possibility that your symptoms were caused by a dengue infection [7].

Medical practitioners may suggest a combination of blood tests and imaging tests to diagnose dengue fever infection because the body's immune response to the virus is complex and dynamic. The dengue infection is difficult to diagnose without laboratory and radiology tests because initial symptoms may be the same as those of other diseases, such as malaria. Nonetheless, a CBC, dengue serology test, among others (Figure 8.3) can be beneficial.

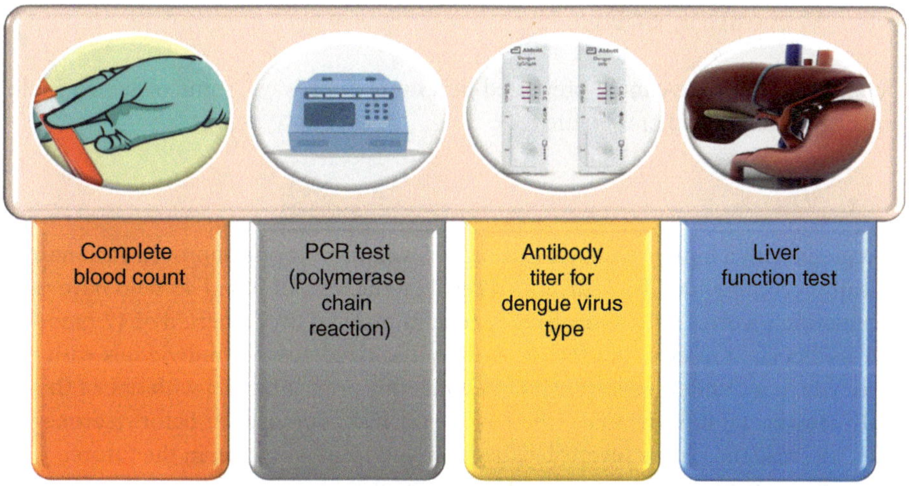

Figure 8.3 Diagnosis tests for dengue disease.

Complete blood count (CBC or CBP) – to check the platelet count typical of the later stages of the illness and to detect the decrease in hematocrit, hemoglobin, and red blood cell (RBC) count (evidence of anemia) that would occur with blood loss associated with severe dengue fever.

Dengue Serology Test (Dengue IgG and IgM) – to detect antibodies produced by the immune system when a person has been exposed to the virus; these tests are most effective when performed at least four days after exposure in both primary and secondary infections.

Dengue Virus Antigen Detection (NS1) – to confirm Dengue viral infection. This test is useful to diagnose early dengue infection and can be conducted within one to two days following Dengue infection.

In cases of severe symptoms, medical practitioners may suggest other blood tests and radiology imaging tests to determine the spread of dengue infection to other organs.

Liver function Tests (LFT) – to detect mild elevations in serum bilirubin, elevated transaminases, and derangements in serum albumin caused by the Aedes mosquito's hepatotoxic effects that can also lead to acute liver failure, with fatal outcomes.

Renal Function Test (RFT) – to check the elevation of the serum creatinine level that may cause a variety of renal diseases such as acute renal failure, acute tubular necrosis, hemolytic uremic syndrome, hypotension, rhabdomyolysis, proteinuria, glomerulopathy, nephrotic syndrome, or hemolysis.

Electrocardiogram (ECG) – to check the heart electrical disturbances. In many patients, ECG abnormalities were observed mainly sinus bradyarrhythmia's, ventricular asystole, sinus tachyarrhythmia's, supraventricular tachycardia (SVT), and ST- and T-wave changes due to electrolyte abnormalities of potassium, calcium, and magnesium caused by dengue infection.

Ultrasound abdomen (USG) – to check the conditions like serositis, fluid in the abdomen, edema of the gallbladder, pericholecystic fluid, and ascites (the buildup of fluid in spaces within your abdomen) mainly caused by dengue fever infection.

Chest X-ray – to check the pleural effusion (the build-up of excess fluid between the layers of the pleura [the tissues that line the lungs and the chest]), pericardial effusion (the build-up of the fluid in the saclike double-layered structure around the heart called as pericardium) caused by dengue fever infection.

D-dimer – to check the D-dimer value in the blood, D-dimer is a protein fragment made when a blood clot dissolves in your body. Dengue fever infection can increase the D-dimer levels in the blood, which can result in pain in the body, sharp chest pain, high fever, trouble breathing, and changes in the skin color of your arm or leg.

2D echocardiography (2D Echo) – to check the damage to heart muscles. Severe dengue fever affects the heart structurally and functionally. Cardiac complications secondary to DENV infection vary from self-limiting arrhythmias to severe myocardial infarction, leading to hypotension, pulmonary edema, and cardiogenic shock.

A fibrinogen test – to check fibrinogen levels. Fibrinogen is a blood protein made in the liver that helps blood clot. Low fibrinogen may make it difficult for blood to clot. In cases where complicated DHF patients have a tendency toward excessive bleeding, their doctor may go for this test to check their fibrinogen levels [12].

8.8 Prevention of Disease

Dengue fever disease can be prevented in a number of ways, some of which include the use of vaccines, mosquito bite prevention, among others (Figure 8.4).

8.8.1 Vaccine

In regions with a high prevalence of dengue fever, there is one approved vaccine known as Dengvaxia. This vaccine is specifically recommended for individuals between the ages of 9 and 45 who have previously experienced dengue fever. The vaccination involves a three-dose regimen administered over a period of 12 months.

It is important to note that the vaccine is authorized only for individuals with a documented history of dengue fever or those who have undergone a blood test confirming prior infection with one of the DENVs, which is referred to as seropositivity. However, for individuals who have never had dengue fever before (seronegative), receiving the vaccine may potentially increase the risk of severe dengue fever and hospitalization in the future.

Dengvaxia is not currently accessible for travelers or individuals residing in the continental United States. However, in 2019, the U.S. Food and Drug Administration granted approval for the vaccine's usage among individuals aged 9–16 who have previously contracted dengue fever and reside in the U.S. territories of American Samoa, Guam, Puerto Rico, and the U.S. Virgin Islands, where dengue fever is prevalent [11, 15, 16].

8.8.2 Prevent Mosquito Bites

According to the World Health Organization, it is important to understand that the dengue fever vaccine alone is not sufficient to effectively reduce the prevalence of the disease in areas where it is widespread. The primary measures for preventing the transmission of dengue fever continue to be avoiding mosquito bites and implementing mosquito control methods.

If you reside in or plan to visit an area where dengue fever is prevalent, the following recommendations may assist in reducing your exposure to mosquito bites: When you go into mosquito-infested areas, wear a long-sleeved shirt, long pants,

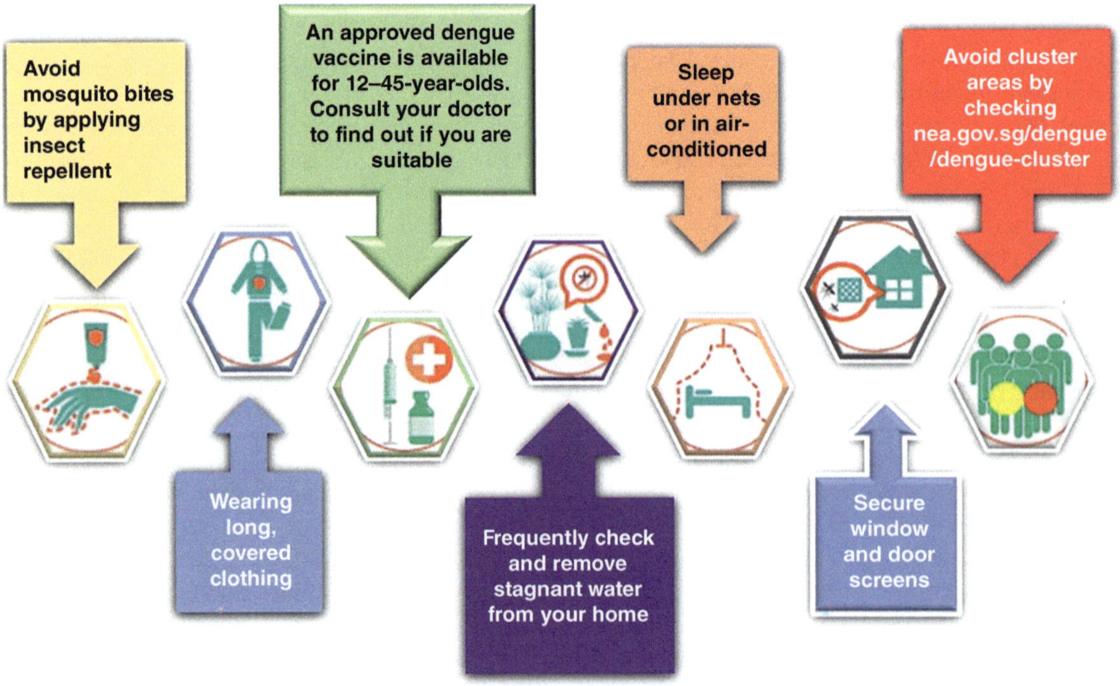

Figure 8.4 Preventive approaches for dengue fever.

socks, and shoes. Permethrin can be applied to your clothing, shoes, camping gear, and bed netting. You can also buy clothing made with permethrin already in it. For your skin, use a repellent containing at least a 10% concentration of N,N-diethyl-meta-toluamide (DEET).

8.8.3 Reduce Mosquito Habitat

The mosquitoes known for transmitting the DENV are commonly found in residential areas, particularly in and around houses. These mosquitoes breed in stagnant water, often accumulating in items like discarded tires. To reduce mosquito populations, it is important to remove their breeding grounds. Regularly empty and clean the containers that may collect standing water, such as plant pots, pet bowls, and vases, at least once a week. Additionally, keep these containers covered when not in use to prevent water accumulation.

8.8.4 Case Surveillance and Vector Surveillance

Improved outbreak prediction and detection strategies, along with continuous monitoring of case status and vector distribution, can contribute to reducing dengue morbidity. These efforts should align with integrated vector management principles, leading to enhanced vector control measures [9].

An accurate estimation of the dengue burden is crucial for assessing the effectiveness of prevention measures. To achieve this, more efficient surveillance systems and dedicated studies are necessary as part of dengue prevention and control plans.

Dengue has been reported in various regions worldwide, including the Eastern Mediterranean, the Americas, Africa, the Western Pacific, and South-East Asia. The Americas alone reported 2.35 million cases of dengue in 2015, with 10,200 cases diagnosed as DHF and resulting in 1,181 deaths. By 2016, the total cases across the Americas, Western Pacific, and South-East Asia exceeded 3.34 million.

The potential risk of dengue outbreaks now extends to Europe, with recorded cases of local transmission in France and Croatia in 2010, along with imported cases detected in three other European countries. Increased cases have also been observed in the West Pacific, including China, Cambodia, Malaysia, Singapore, the Philippines, and Vietnam.

Taiwan witnessed a large dengue outbreak in 2014, with 15,732 DF cases reported, followed by an even larger outbreak of 43,748 DF reported cases in 2015.

In 2016, there were significant dengue outbreaks worldwide, with the Western Pacific Region reporting over 375,000 suspected cases. Among these, the Philippines reported 176,411 cases, and Malaysia reported 100,028 cases.

Although several large dengue outbreaks were reported worldwide in 2016, there was a significant reduction in the number of cases in 2017. However, since 2019, there has been a sharp increase in cases.

The dominant strains responsible for dengue outbreaks are DENV-1 and DENV-2. As of July 2019, an increase in dengue cases has been observed in countries in South East Asia, such as the Philippines (130,463), Lao People's Democratic Republic (15,657), Vietnam (115,186), Malaysia (75,913), Singapore (8,020), and Cambodia (4,532).

In the Americas region, up until June 2019, a total of 1,191,815 dengue cases were reported, of which 546,589 (46%) were laboratory-confirmed and 5,599 (0.47%) were classified as DHF. The reported case-fatality rate was 0.02%. The majority of reported cases (93%) were from Brazil, Colombia, and Honduras, with a rising trend compared to the previous year and the circulation of all four DENV serotypes.

GIS mapping is currently suggested for surveillance of dengue foci to provide real-time information about the dengue status. By identifying and localizing positive dengue cases within a specific area, transmission of dengue can be controlled through the implementation of preventive strategies.

GIS mapping offers enhanced surveillance and community-based intervention programs to control dengue, facilitating the measurement of prevention success rates in the mapped areas.

The rapid spread of DENV can be attributed to global climate change, inadequate vector eradication efforts, and population expansion. The primary vectors responsible for DENV transmission are Aedes mosquitoes, particularly *A. aegypti* and *A. albopictus*, due to their susceptibility to the virus and efficiency in transmitting it to humans.

8.8.5 Community-Based Control Programs

Efforts to control and eradicate breeding sites for disease-carrying vectors have been a key focus of community-based programs. These programs aim to engage the community by dividing them into different groups based on their education level and knowledge about vectors and the disease. Implementing mosquito elimination programs has shown positive outcomes in certain countries where the population has demonstrated a high level of awareness regarding vectors.

However, the success of community-based control largely depends on the knowledge, education, and behavior of individuals, along with practical strategies. Providing training and disseminating information plays a crucial role in enhancing education levels and equipping individuals with the skills to identify and implement preventive measures for vector control and elimination. Furthermore, raising awareness among the population is vital. A study conducted in Thailand demonstrated the effectiveness of utilizing the media to promote awareness and knowledge.

8.8.6 Biological Control

The control of dengue disease primarily relies on reducing or eliminating the population of disease-carrying vectors through the implementation of biological vector control plans. Traditional methods of vector control have proven ineffective against this daytime biting mosquito, necessitating the development of new strategies.

One major strategy for biological protection against dengue is the use of genetically modified (GM) mosquitoes. The sterile insect technique (SIT), a conventional control measure, has been successfully employed in numerous countries. SIT involves compromising the reproductive capacity of male insects, thereby reducing their ability to compete with wild-type female insects for breeding.

The first control program focused on transgenic sterility has obtained approval for *A. aegypti* mosquitoes. These mosquitoes are genetically engineered to express a fluorescent marker and a late-acting conditional lethality trait, preventing most of their offspring from reaching adulthood.

By repeatedly releasing millions of genetically modified mosquitoes carrying lethal genes, the aim is to surpass the population of wild *A. aegypti* mosquitoes and reduce the overall adult mosquito population. Experimental releases of transgenic mosquitoes were conducted in 2008 in the Cayman Islands, Malaysia, Panama, and Brazil, resulting in significant reductions in the *A. aegypti* population. Separation of male and female pupae was performed based on size differences to eliminate female mosquitoes during the release of male *A. aegypti* into the environment.

However, this process is labor-intensive, time-consuming, expensive, and not entirely fool proof, with a contamination rate of 0.02% for females. Furthermore, releasing genetically modified male Aedes mosquitoes requires continuous and long-term releases of large quantities of transgenic males to suppress mosquito populations.

Another approach involves the release of mosquitoes transfected with the intracellular bacterium Wolbachia. Wolbachia naturally occurs in up to 60% of insect species, including butterflies, moths, and certain types of mosquitoes. Wolbachia can be utilized in two different strategies for mosquito population control: reducing the vector's reproductive capacity and suppressing RNA virus replication. However, *A. aegypti*, the primary mosquito species involved in dengue transmission, is not usually infected with Wolbachia.

Wolbachia functions by competing for iron and cholesterol, which are essential for virus replication. When Wolbachia is present, it consumes these molecules, making it more challenging for the viruses to grow.

Additionally, scientists propose utilizing the natural immune system of mosquitoes to enhance their resistance to virus infection through the induction of several signaling pathways of the innate immune system, including Toll, immune deficiency (Imd), and JAK–STAT. Activation of these signaling pathways leads to increased resistance against various arboviruses.

While Wolbachia shows promise as a viable vector control tool, it is necessary to assess whether the introduction of Wolbachia-infected mosquitoes into the environment can displace wild-type populations under field conditions. Field releases are crucial to understanding invasion dynamics and optimizing release strategies involving Wolbachia. Studies have reported stable and persistent Wolbachia infection, which contributes to the reduction of vector competence. Field releases of Wolbachia-infected *A. aegypti* were conducted near Australia (Cairns) and Vietnam in January 2011.

These releases successfully led to the invasion of Wolbachia-infected *A. aegypti* in all targeted areas. However, these studies primarily focused on establishing the sustainability of Wolbachia in wild *A. aegypti* populations. To evaluate its viability as a dengue control measure, ongoing releases are now taking place in Indonesia, Brazil, Australia, and Colombia.

8.8.7 Chemical Control

For several decades, control programs aimed at reducing vector populations have relied on the application of insecticides derived from chemicals or plants. Although these strategies have been convenient for vector control, long-term use of insecticides has been associated with certain side effects, including the development of insecticide resistance in the targeted vectors and negative impacts on the environment [7].

To address these issues, scientists have developed various alternative approaches, such as the utilization of plant-based insecticides derived from plant sources. These alternative agents have demonstrated sustained effectiveness against mosquitoes while posing less toxicity to both mosquitoes and the environment.

Plant derivatives are not limited to insecticides alone; they have also shown potential as repellents against *A. aegypti*. Additionally, another category of chemical compounds known as insect growth regulators (IGRs) has been effective in inhibiting the growth and development of insects. IGRs induce changes in the early stages of insect development, ultimately leading to the prevention of adult stage formation and the subsequent elimination of the insect. Among the various IGRs available, diflubenzuron, Cyromazine, pyriproxyfen, Spinosad, and Methoprene have been reported for their use in larval control, effectively reducing the populations of *A. aegypti*.

8.9 Management of Dengue Disease

8.9.1 Hydration

The conventional approach to managing DF/DHF involves rehydration therapy, with a preference for oral or intravenous fluids in hospitalized individuals (Figure 8.5). Fever control is achieved through sponging and the administration of paracetamol, while non-steroidal anti-inflammatory drugs are strictly contraindicated due to the risk of bleeding.

Initiating oral rehydration therapy (ORT) on the first day of illness in DF is crucial, as it prevents the development of DHF and reduces the likelihood of hospitalization [15].

It is important to highlight that platelet transfusion is not a universal solution for managing dengue. While thrombocytopenia is theoretically a risk factor for bleeding, the platelet count alone is not indicative of bleeding risk. The duration of

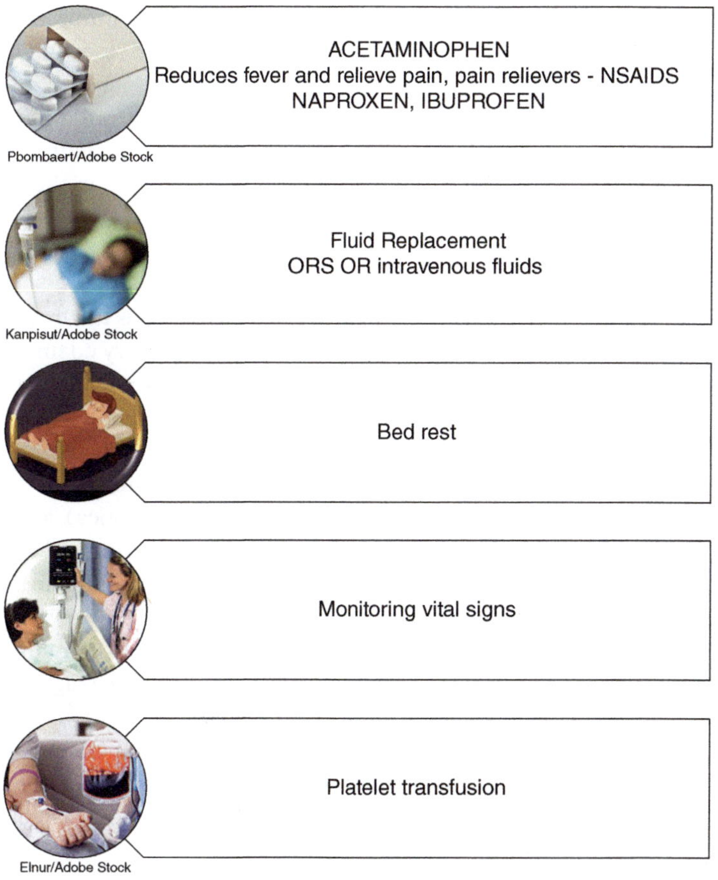

Figure 8.5 Management of dengue disease.

shock and the hematocrit levels during shock are more relevant risk factors for severe hemorrhage and subsequent mortality.

Although the threshold for prophylactic platelet transfusion in non-dengue patients is 10,000/mm^3, there is no specific threshold for dengue patients. Empirical platelet transfusion may be considered for patients with bleeding tendencies or a platelet count below 20,000–25,000/mm^3 since there are no other specific therapies available for DHF/DSS.

Multiple transfusions in non-dengue patients can lead to immunization against various HLA- and platelet-specific antigens, resulting in no increase in the post-transfusion platelet count. In dengue patients, this effect may be more pronounced due to additional factors contributing to thrombocytopenia. Therefore, it is preferable for dengue patients to receive single donor apheresis platelets (SDAP) instead of random donor platelets (RDP) to minimize the risk of alloimmunization.

Unfortunately, both patients and healthcare workers often excessively focus on platelet counts. However, maintaining adequate hydration, monitoring for bleeding symptoms, and avoiding panic are essential aspects of dengue treatment. Platelet counts between 50,000/mm^3 and 150,000/mm^3 in otherwise asymptomatic and improving patients (who are afebrile, have normal appetite, and feel well) should not cause alarm. This phenomenon of panicking over platelet counts in dengue patients who are otherwise clinically stable can be referred to as "Dengue panic syndrome."

The 2003 dengue outbreak in the northern states of India clearly demonstrated the prevalence of this panic syndrome. Therefore, the role of media and healthcare workers in promoting the importance of hydration in the early stages of illness cannot be overstated. Adequate hydration is the primary therapeutic intervention that can determine the difference between life and death for dengue patients.

On the downside, the "dengue panic syndrome" further burdens emergency services in tertiary care hospitals with patients who may not require urgent attention. As a result, genuine DHF/DSS patients in need of immediate care may face challenges in receiving prompt and appropriate treatment, despite the best efforts of hospital staff [15].

8.9.2 Ancillary Therapeutic Modalities

Additional approaches, both supportive and pharmacological, are being explored to reduce the mortality associated with DHF/DSS. Patients with DSS are at risk of developing acute respiratory distress syndrome (ARDS).

This phenomenon was also observed during a recent outbreak in northern India. In the past, these patients were managed by providing oxygen through a nasal cannula or resorting to tracheal intubation and mechanical ventilation when necessary.

A recent study has proposed the use of non-invasive ventilation as a potential strategy to improve the outcomes of ARDS in children with DHF/DSS in areas where dengue is prevalent. Furthermore, researchers are investigating the potential of mycophenolic acid (MPA), an immunosuppressive agent, in the treatment of dengue [15].

8.9.3 Management of Volume Overload

One of the primary challenges in managing DHF/DSS is the frequent occurrence of fluid overload, which can have serious consequences such as heart failure and acute pulmonary edema and, if left unaddressed, potentially even result in fatalities. Therefore, it is crucial to effectively and promptly manage fluid overload to prevent these complications [16]. Early indications of fluid overload include swollen eyelids, rapid breathing (tachypnea), and a distended abdomen accompanied by abdominal discomfort.

Late indications of fluid overload, which signify the need for a diuretic such as furosemide (1 mg/kg/dose), include symptoms like coughing, respiratory distress (dyspnea/orthopnea), a very tense abdomen, a wide pulse pressure (some individuals may experience a narrowing of pulse pressure), a strong and forceful pulse, hypertension during the reabsorption phase, and abnormal lung signs such as crepitation, rhonchi, and wheezing. In patients exhibiting late signs of fluid overload, it is recommended to insert a urinary catheter.

When assessing a patient, it is important to consider their current status, whether they are in the post-shock phase or post-plasma leakage phase. If a patient is still in the leakage phase, it is recommended to administer a dextran bolus 15–30 minutes before administering furosemide.

Furthermore, the patient's status, whether in a state of shock or non-shock, should be taken into account. For patients in a state of shock, a dextran bolus is suggested 15–30 minutes prior to furosemide administration.

Additionally, it is crucial to assess and address any associated complications, such as bleeding, electrolyte/metabolic/acid–base imbalances, and liver/renal failure.

Clinicians are advised to monitor vital signs at 15-minute intervals for a total of four times following furosemide administration, as the effects of furosemide typically last less than one hour 5. Management of DHF with Encephalopathy:

Most instances of encephalopathy in DHF patients are typically observed during or after the critical phase, although they can also occur early in the febrile phase. Encephalopathy is less common in dengue fever (DF) patients. Symptoms of encephalopathy can manifest as behavioral changes (aggression, violence, use of vulgar language), alterations in consciousness (irritation, agitation, confusion, hallucinations, coma), and seizures.

The majority of DHF patients with encephalopathy are those who experience prolonged shock and develop liver or renal failure. Other common causes include hyponatremia, hypoglycemia, and hypocalcaemia. Intracranial bleeding, while rare, is typically found in DSS patients who have experienced prolonged shock and multi-organ failure. It usually occurs very late, more than three days after the onset of shock.

The management approach for dengue encephalopathy is generally similar to that of hepatic encephalopathy and involves the following steps: ensuring proper airway and oxygenation through oxygen therapy. Intubation may be necessary for patients with respiratory failure or those who are semi-comatose or comatose.

Administer the least amount of intravenous (IV) fluid necessary to maintain appropriate intravascular volume, aiming to keep the total IV fluid intake below 80% of the maintenance requirement.

If the hematocrit (Hct) continues to rise or there is a significant need for a large volume of IV fluid in cases with severe plasma leakage, consider transitioning to a colloidal solution at an earlier stage.

If signs and symptoms of fluid overload are present, administer diuretics as appropriate. In cases of increased intracranial pressure (ICP), consider the use of steroids. It is recommended to administer Dexamethasone at a dosage of 0.5 mg/kg/day intravenously every six to eight hours.

8.9.4 Decrease Ammonia Production

Administer lactulose at a dosage of 5–10 ml every six hours to induce osmotic diarrhea. Local antibiotics may be prescribed to eliminate the intestinal flora. However, this may not be necessary if systemic antibiotics are already being administered. Maintain a blood sugar level above 60 mg% and recommend a glucose infusion rate ranging from 4 to 6 mg/kg/h. Correct imbalances in acid–base and electrolyte levels, such as by addressing hypo/hypernatremia, hypo/hyperkalemia, hypocalcaemia, and acidosis.

Administer intravenous vitamin K1: 3 mg for children under one year old, 5 mg for children under five years old, and 10 mg for individuals five years old and above, including adults. Anti-convulsants should be given to control seizures. Examples include phenobarbital, Dilantin, and diazepam, administered intravenously as indicated. Blood transfusion may be necessary, preferably with freshly packed red cells, based on indications. Platelets and fresh frozen plasma should be avoided to prevent fluid overload and increased intracranial pressure (ICP).

Empiric antibiotic therapy may be considered if there is suspicion of superimposed bacterial infections. H2-blockers or proton pump inhibitors may be administered to alleviate gastrointestinal bleeding. Minimize the use of unnecessary drugs, as most drugs require liver metabolism. In cases of clinical deterioration, plasmapheresis, hemodialysis, or renal replacement therapy may be considered.

8.10 Conclusion

Dengue is a viral illness that is transmitted through the bites of mosquitoes, particularly the *A. aegypti* species. It poses a significant public health concern in numerous tropical and subtropical regions across the globe. Over time, dengue has evolved into a global health burden, with an estimated annual occurrence of 390 million infections. The challenges associated with dengue are multifaceted, encompassing various aspects such as prevention, diagnosis, treatment, and vaccine development.

Addressing the impact of dengue requires a comprehensive approach that includes effective vector control measures, improved diagnostic capabilities, and the development of vaccines and therapeutics. It is crucial to enhance public awareness about dengue, strengthen surveillance systems, and foster international collaboration to effectively mitigate the spread of the disease and minimize its global transmission. By adopting these measures, we can work toward reducing the overall impact of dengue on public health worldwide.

References

1 Alhaeli, A., Bahkali, S., Ali, A. et al. (2016). The epidemiology of dengue fever in Saudi Arabia: a systematic review. *J. Infect. Public Health* 9: 117–124.

2 Dengue Guidelines for Diagnosis, Treatment, Prevention and Control Treatment, Prevention and Control Treatment, Prevention and Control. http://www.who.int/tdr.

3 Dengue, Dengue Haemorrhagic Fever and Dengue Shock Syndrome in the Context of the Integrated Management of Childhood Illness Department of Child and Adolescent Health and Development. (2005).

4 World Health Organization. Regional Office for South-East Asia (2011). *Comprehensive Guidelines for Prevention and Control of Dengue and Dengue Haemorrhagic Fever*. World Health Organization Regional Office for South-East Asia.

5 Gubler, D.J. (1998). Dengue and dengue hemorrhagic fever. *Clin. Microbiol. Rev.* 11: 480–496.

6 Guzman, M.G., Gubler, D.J., Izquierdo, A. et al. (2016). Dengue infection. *Nat. Rev. Dis. Primers* 2: 16055.

7 Schaefer, T.J., Panda, P.K., and Wolford, R.W. (2023). Dengue fever. In: *StatPearls*. Treasure Island, FL: StatPearls Publishing https://www.ncbi.nlm.nih.gov/books/NBK430732/.

8 Srinivas, V. and Srinivas, V.R. (2015). Dengue fever: a review article. *J. Evol. Med. Dent. Sci.* 4: 5048–5058.

9 Wang, W.-H. et al. (2020). Dengue hemorrhagic fever – a systemic literature review of current perspectives on pathogenesis, prevention and control. *J. Microbiol. Immunol. Infect.* 53: 963–978.

10 Sanyaolu, A. (2017). Global epidemiology of dengue hemorrhagic fever: an update. *J. Hum. Virol. Retrovirol.* 5: 00179.

11 Cleveland clinic. Dengue Fever. https://my.clevelandclinic.org/health/diseases/17753-dengue-feve.

12 Pace Hospital. Dengue Fever – Symptoms, Causes, Diagnosis and Treatment. https://www.pacehospital.com/dengue-fever-causes-symptoms-and-treatments.

13 Kalayanarooj, S. (2011). Clinical manifestations and management of dengue/DHF/DSS. *Trop. Med. Health* 39: S83–S87.

14 Umakanth, M. and Suganthan, N. (2020). Unusual manifestations of dengue fever: a review on expanded dengue syndrome. *Cureus* 12: e10678.

15 Ahluwalia, G. and Sharma, S.K. Dengue: current trends and challenges-an Indian perspective. *J. Assoc. Physicians India* 52: 561–563. http://www.japi.org561.

16 Simmons, C.P., Farrar, J.J., van Nguyen, V.C., and Wills, B. (2012). Dengue. *N. Engl. J. Med.* 366: 1423–1432.

9

West Nile Virus: Evolutionary Dynamics, Advances in Diagnostics, and Therapeutic Interventions

Bibhuti B. Kakoti, Lawandashisha Nongrang, Chinmoyee Borah, Monali Lahiri, Mainak Ghosh, and Ankit Choudhary

Department of Pharmaceutical Sciences, Faculty of Science and Engineering, Dibrugarh University, Dibrugarh, Assam, India

9.1 Introduction

In 1999, New York City experienced a dramatic outbreak of disease, which caused a serious outbreak of meningoencephalitis in a few people, which even led to death [1]. The disease extended its range to most of the eastern region of the United States and the western hemisphere [2]. It was then identified to be a flavivirus from the family of *Flaviviridae,* which is known as West Nile Virus (WNV) and is mainly transmitted by mosquito bites [3]. Ever since this outbreak, 20,000 symptomatic cases of WNV have been reported in the United States, and the virus infection is estimated to have affected more than a million people [4]. Over the next few years, the virus spread widely throughout the United States, Canada, and Central and South America. In all states except Hawaii, Alaska, and Oregon, WNV activity has been reported in humans, birds, animals, or mosquitoes, making it the flavivirus with the greatest number of cases being reported [3].

Initially, in the year 1937, the blood sample of a febrile patient from the West Nile Province of Uganda was collected and WNV was isolated and identified from it [5]. Due to the majority of cases being asymptomatic, and because WNV is not a significant pathogen among humans or animals, it is not considered to be of any significance. There was a close antigenic relationship between the WNV, Systemic lupus erythematosus (SLE) virus, and Japanese encephalitis (JE) during the 1940s [6]. It has been demonstrated that mosquitoes transmit the WNV under laboratory conditions, and it was also reported that the residents of eastern central Africa possess significant levels of antibodies neutralizing the WNV and related flaviviruses [7, 8]. Aspects of WNV infection have been linked to encephalitis and death in Israel in the following decades (1950s), France from 1962 to 1963, then in 1974 in South Africa, and 1980/1981 in India [9–11]. In Europe and the Mediterranean basin, WNV infection is a major problem for both public health and veterinary medicine after a severe outbreak in Romania in 1996 [12]. There was no major outbreak of WN viral disease documented from 1975 through 1993. However, there was an alarming rate of WN meningoencephalitis outbreaks in Europe, North Africa, the Middle East, and North America between 1994 and 2000 [2]. Horses have been linked to a number of outbreaks, including widespread urban epidemics like the Israeli pandemic of 2000, which resulted in more than 400 illnesses and 35 fatalities [9].

Humans can get different types of WN fever including asymptomatic infections, mild acute febrile illnesses, and neurological diseases, such as meningitis, encephalitis, and acute flaccid paralysis (AFP) [13–15]. The WNV host range is extensive, with more than 65 mosquito and tick species, 225 bird species, and 29 different animals having been detected. Currently, there is no vaccine or specific antiviral treatment for WN fever [16, 17]. Based on the genetic makeup of several viruses, different strains of WNV can cause limited outbreaks or wide epidemics. There have been many outbreaks of human encephalitis associated with strains of WNV that can be classified as lineage 1, demonstrating a correlation between viral genetic makeup and disease outcomes. Competent vectors (ornithophilic mosquito species) and avian hosts are necessary to sustain adequate viremia for the subsequent infection of successive mosquitoes for WNV to be transmitted and maintained.

Rising Contagious Diseases: Basics, Management, and Treatments, First Edition. Edited by Seth Kwabena Amponsah, Ranjita Shegokar, and Yashwant V. Pathak.

In addition, other animals can also produce viremia high enough for transmission by mosquitoes. However, it is not yet apparent how these hosts affect the dynamics of WNV transmission. In general, humans and horses are regarded as "dead-end" hosts because they are unable to produce sufficiently high viremic titers for the infection of mosquito hosts [3].

9.2 Geographical Distribution

The neurotropic flavivirus known as WNV is ubiquitous in most of the world. After its first discovery in 1937 [18], it became widespread across the globe and is currently the main mosquito-borne and epidemic encephalitis cause in the United States. The number of people infected with WNV has increased in Europe due to two major outbreaks of specific strains of the virus, and the appearance of a strain of the virus, which causes the illness has increased as well. In the United States, Europe, the Middle East, and Africa, WNV continues to represent a serious threat to public health [19]. The actual infection rates in the United States have been sought to be estimated in a variety of research. WNV IgG antibody seroprevalences have been observed to range from 19% to 14% in surveys of seroprevalence, done to establish cumulative infection incidence, although these investigations were constrained by the relatively few patients and narrow geographic areas they surveyed. The fraction of overall infections estimated as per West Nile Neuroinvasive Disorder (WNND) case has been multiplied by the total number of reported WNND cases to get the cumulative infection incidence. Each recorded case of WNND, according to the findings of a seroprevalence survey conducted in New York following the 1999 outbreak [20], was thought to be the consequence of 140 infections on average [21]. This ratio was then used to calculate the cumulative infection incidence in the United States, which was estimated to be 1,655,080 up to 2006. However, given that the 140 : 1 ratio of infection to WNND was based on a seroprevalence estimate (2%) with confidence ranges that ranged more than thrice (12–4%), this cumulative incidence estimate was prone to significant uncertainty [22]. Based on blood donor screening statistics, it is likely that for every documented case of WNND, there are up to 256–353 infections. A recent survey of blood donors and WNND reports from a state indicate that WNND cases require between 21 and 286 infections. The last study had enough data to estimate the infection-to-WNND ratios for each age group and gender [23], but it was unable to determine the ratio for children since the infection data were collected from blood donors. The age- and gender-stratified number of WNND cases was multiplied with onset up to the end of 2010 by the recently derived, corresponding estimated infection-to-WNND ratios, which aim to estimate to what extent there were adult WNV infections in each state in the United States by the end of 2010. This was done to account for the significant variation in risk across age and gender groups [24]. These were then divided by the population of the appropriate state to determine cumulative incidence (percent of the population already infected). As of the end of 2010, there were 12,823 WNND cases being documented and 79 of these cases had age and/or gender information that was missing from ArboNET. Those cases were proportionately divided into the various age and gender strata according to the state's share of WNND patients in each stratum. By taking into consideration the stratification by age and gender, estimates for each state's total number of previously infected individuals were also made. Based on state-specific totals (variances), Wald 95% confidence intervals (CIs) were calculated to estimate the total number of infections in the United States. At creative testing solution's donor screening facility in Tempe, Arizona, United States, plasma from 6999 consecutive voluntary blood donations received from November to December 2008 at North Dakota donation centers was preserved. Based on postal codes, samples were chosen to exclusively include those from donors who resided in North Dakota. The resultant 4514 samples were examined at Focus Laboratories USA using commercially available ELISAs for WNV-specific IgG and IgM if the IgG test findings were reactive [25].

By multiplying the corresponding stratum-specific WNV-specific IgG seroprevalence of blood donors by the North Dakota population, researchers were able to estimate the cumulative number of infections by age and sex strata in the state's population. The seroprevalence among those under 16 years of age could not be precisely measured since blood donors are only those who are >16 years old. As a result, only estimates for the adult population were computed. For this report, teenagers between the ages of 16 and 17 are treated as adults. By dividing the expected number of infections by the number of WNND cases reported in North Dakota with onset from the first reported case between 2002 and 2008, it was possible to derive the corresponding ratio of WNND cases to infections for each age and sex stratum [26]. Despite the fact that a few of the cell-specific pathways engaged in immunological prevention of infection have been identified by this body of work, it has also come to light that the transmission process is largely to blame for the complexity of WNV emergence, epidemiology, and pathogenesis. The evaluation of the mechanisms regulating immunological control, neuroinvasion, and viral tissue tropism, in addition to the description of the molecular characteristics regarding WNV adaptability to mammalian hosts and vectors. These molecular characteristics underlie transmission and pathogenesis [27].

9.3 The Virus and its Genome

Being a part of the large family of Flaviviridae, the WNV is serologically one of the members of the JE virus antigenic complex alongside St. Louis encephalitis, Murray Valley encephalitis, and Kunjin virus [1]. These viruses belonging to the JE complex share a close antigenic relationship, resulting in the cross-reactions in the diagnostic labs [28]. With the help of a cryo-electron microscope (Cryo-EM), Mukhopadhyay and his colleagues were able to determine the structure of this virus that led to the outbreak in New York City. The virion has an envelope enclosing an icosahedral capsid measuring approximately 500 Å in diameter and unlike other envelope-containing viruses, no projections or spikes are present on the WNV [29]. This virus has a 35–40 Å lipid bilayer membrane, which has mainly been derived from the host cell, this surrounds a nucleocapsid core that contains approximately 11 kb of positive-sense single-stranded RNA (ssRNA) [2], and here, translation and replication are facilitated by secondary structures within untranslated 5′ and 3′ regions [30]. Host-derived lipid bilayers contain 180 copies of the envelope (E) embedded in the genome along with the viral capsid protein (C) [31]. The surface protein E can induce a neutralizing antibody response by binding to receptors and fusing with the host membranes. The RNA genome lacks a polyadenylated tail at its 3′ end and has a 5′ cap at its 5 ends. The genome is translated into a single polyprotein, which is then cleaved into seven nonstructural (NS1, NS2A, NS2B, NS3, NS4A, NS4B, and NS5) and three structural (C, prM, and E) proteins by viral and host proteases (Figure 9.1) [32].

Moreover, around 100 nucleotides (nt) of 5′ and 3′ coding regions (NCRs) and 400–700 nt surround the ORF [30]. Viral replication takes place within the cytoplasm, in close proximity to the rough endoplasmic reticulum (ER), followed by virus assembly and release from the cell through the cell secretory pathway [33]. Additionally, when the E protein undergoes a structural reorganization, nucleocapsids, and viral RNA are delivered to the cytoplasm to be translated [34]. Around 3000 amino acids of viral proteins from the single polyprotein are processed by cellular and viral proteases. Progeny virions assemble in rough ER through intracellular membrane budding after RNA replication within cytoplasmic reticulum replication complexes. By exocytosis, mature virions are released into the cytoplasm after passing through the host secretory pathway, becoming fused with the plasmatic membrane, and finally released into the extracellular space [30, 34]. An ATP-dependent helicase is found in the NS3 protein, and a serine protease is found in the NS2B protein, which is necessary for the processing of virus polypeptides. In addition to being a methyltransferase, NS5 is an RNA-dependent RNA polymerase. A number of other proteins are small, generally hydrophobic, and serve a variety of purposes. Chung et al. demonstrated that the WNV NS1 protein has immunomodulatory functions. It was found to be a secreted glycoprotein, hence having immune invasion activity [35]. The role of NS2 protein is inhibition of interferon (IFN)-β promoter invasion [36, 37]. The ER is rapidly expanded and modified by NS4A to establish replication domains [36–40]. The IFN response is blocked by the NS4B protein [41–44]. For efficient replication, all NS proteins appear to be essential [45].

9.3.1 Phylogeny

Researchers found that WNV isolated from different geographical regions falls into two distinct genetic lineages with nucleotide differences of 25%–30% and several subclades or clusters based on nucleic acid sequence data corresponding to a 255-bp

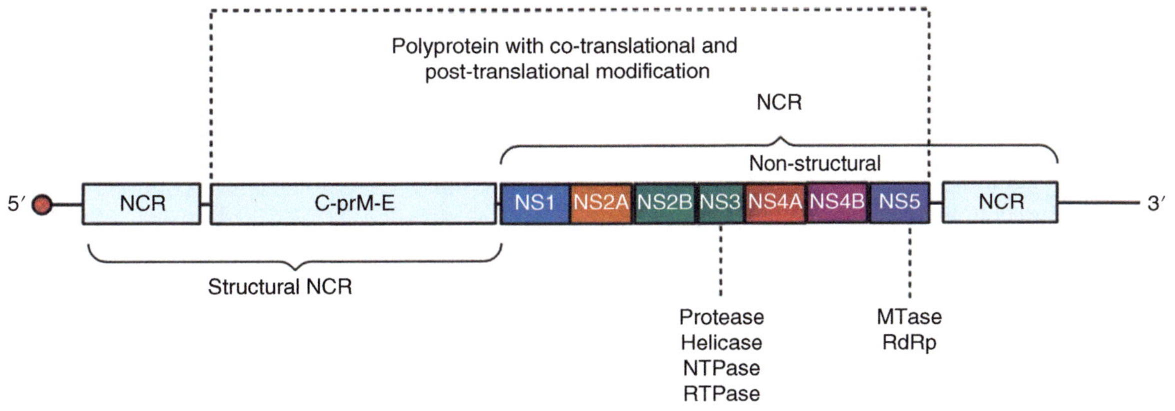

Figure 9.1 An RNA genome from the WN virus contains 5′ non-coding regions (NCRs; 100 nucleotides), a single open reading frame containing three structural proteins (capsid [C], membrane [M], envelope [E]), and 3′ nonstructural proteins.

region of the E glycoprotein gene [46–51]. Additionally, this lineage is also constituted of strains that are found in Eastern Europe, Australia, and the Middle East. This lineage is further subdivided into three subclades, i.e. 1a, 1b, and 1c. Moreover, Clade 1b, also known as the Kunjin virus, is only found in Oceania. Lineage 1 strains cause the majority of neurological disease outbreaks in Europe, Africa, and the Americas, with the exception of clade 1b, which rarely causes neurological disease [48, 52, 53].

In lineage 2, there are strains isolated exclusively in Africa and Madagascar, including the B 956 prototype strain [46, 48, 50]. As of 2004, lineage 2 was observed exclusively in Africa before being isolated from humans and birds in Hungary, Greece, and Italy [54–56]. The second lineage was also assumed to be less pathogenic than the first until it caused severe disease in South Africa and encephalitis in Europe [23, 57]. A variety of strains from both lineages are neuroinvasive in humans [58].

In addition to lineages 1 and 2, there are other less widespread lineages. There have been repeated isolations of lineage 3, otherwise known as Rabensburg virus, in the Czech Republic [58–60]. It has been reported that lineage 4 was isolated and reported from Russia [61]. Often believed to be a distinct clade of lineage 1 (clade 1c), the 5th lineage was isolated from India [47]. Using a small gene fragment, a 6th lineage has been proposed for Spain [62].

Recently, severe outbreaks of human and avian diseases have been linked to strains from lineage 1. Contrary to lineage 1, strains of lineage 2 appear less virulent for humans, having been isolated from asymptomatic or mild cases or even during a search for other pathogens [47].

9.4 Etiology and Pathogenesis

The earliest transmission and infection (the initial stage), secondary viral expansion (the vital organ infected stage), and after that involvement of neuroinvasion (CNS) are the three distinct phases of WNV pathogenesis that have been identified via research studies using animal experimental models of infection. Each of these stages is believed to summarize the events that occur in humans after a mosquito bite (Figure 9.2) [63].

The initial control of WNV is performed by the innate immune response, consisting of cell-intrinsic antiviral barriers, the type I interferon (IFN) reaction, and innate cell-mediated responses (which include neutrophils, natural killer [NK] cells, and T cells) [64]. By contrast, the response of the adaptive immune system, which offers humoral and adaptive immune cell-mediated reactions (CD4+, CD8+, and regulatory T cells) [65], controls WNV subsequently is crucial for the elimination of WNV and minimizing possible immune response-mediated harm in the final stages of infection [66].

9.4.1 The Transmission of the West Nile Virus Within the Mosquito Host

When female species of Culex mosquitoes bite on viremic birds, they carry up WNV. The mosquito's gastrointestinal tract epithelial cells replicate the WNV, which then flows to the salivary glands and other organs using the hemolymph [66, 67]. The middle gastrointestinal barrier, which serves as a physical and immune barrier by creating antimicrobial peptides and a peritrophic matrix (which is made up of chitin, proteins, glycoproteins, and proteoglycans) that collectively restrict the replication of viruses and propagation within the insect, is an essential phase in WNV spread and vector ability [68, 69].

According to a new investigation, C-type lectins assist in the transmission of WNV in mosquitoes14. MosPTP-1, a mosquito surface protein that is a homolog for human CD45, combines with mosGCTL-1 [70], a soluble C-type lectin protein, to promote viral binding and transmission. In the hemolymph [71], WNV attaches to secreted mosGCTL-1, enabling viral entryway and infiltration into multiple tissues of mosquito. Invertebrate natural immune systems are stimulated by WNV infection, which may mitigate proliferation. These consist of antimicrobial peptides, RNAi, immunological deficiency (IMD), Toll [72], and JAK–STAT (Janus kinase-signal transduction and accelerator of transcription) proteins, in addition to basic immunological signaling pathways. Furthermore, mosquitoes are transporters of Wolbachia spp., symbiotic microbes that block the insect from multiplying WNV [73].

In response to WNV infection, Wolbachia spp. promote oxidative damage and the production of reactive oxygen species, which trigger the Toll cascade [74] and promote the formation of peptides with antimicrobial properties, such as defensins and cecropins, which limit flavivirus.

9.4.2 Proliferation, Viral Multiplication, Dissemination, and the Initial Infection

For the purpose to obtain blood to feed, mosquitoes explore the skin for veins or extravasated blood. As mosquito introduces its vomit along with the virus particles, it harbors as a component of this method. Up to 106 plaque-forming cells of

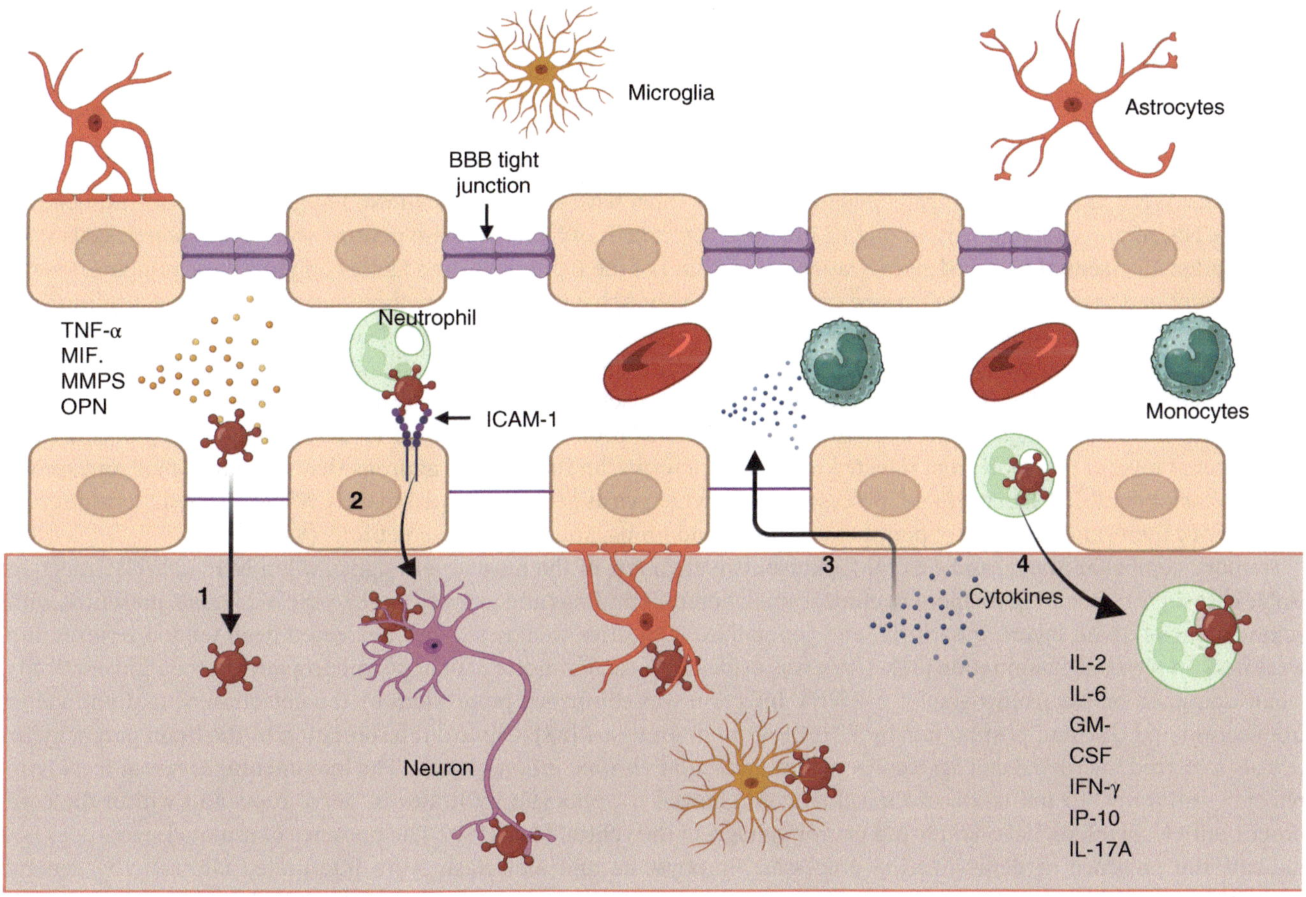

Figure 9.2 The possible mechanisms by which WNV enters the CNS include: (1) WNV infection induces the expression of TNF-α, MIF, MMP9, ICAM-1, and Opn, which directly or indirectly increase the permeability of the BBB allowing the virus to penetrate to the CNS; (2) WNV may infect endothelial cells in the cerebral microvasculature, from which progeny viruses may be released into the CNS; (3) WNV may enter the CNS from infected olfactory bulbs via olfactory neurons; (4) WNV-infected leukocytes, such as neutrophils via a "Trojan horse" transport of WNV to the CNS.

a transmissible virus can be transported into an individual per mosquito bite, according to the species [75]. Mosquito saliva carries compounds that suppress hemorrhage, diminish inflammation, and modulate host immunity, in addition to viral proteins that inhibit the host immune system [76]. When Culex or Aedes spp. mosquitoes, which are loaded with WNV, bites on mice exhibited quicker infection dynamics, raised viremia, and more rapid neuroinvasion than the mice that did not get any bite from a mosquito. Furthermore, mosquito saliva modifies the amount of cytokines, leading to local immunosuppressive and impairment in the movement of neutrophils, dendritic cells (DCs), and T cells to the main point of attack [77]. Although the human pathophysiology of WNV is not widely recognized, effective models of animals have assisted and shine a light on the molecular processes that drive WNV disease [78]. While several groups have used intraperitoneal and intravenous methods of administration, which prevent the virus-host connections during the dermis and exhausting lymphatic system node, WNV is usually given in animal infection models via a beneath-the-skin footpad infection to imitate mosquito-delivered virus [79]. WNV proliferation of the virus occurs in the superficial Keratinocytes and skin-resident dendritic cells, which also includes a dermal layer of dendritic cells along with Langerhans cells, describing the early phase subsequent subcutaneous infection. Following this, the draining lymph node undergoes viral amplification, resulting in viremia [80] and extends to the internal body parts, particularly the liver, which acts as a primary place for the replication of viruses in the surrounding tissues. In the liver, spleen, as well as remaining surrounding tissues, the exact target cellular environment for transmission with WNV is unknown. It is believed to consist of subsets of DCs, macrophages, and neutrophils. WNV is neurotropic and neuroinvasive [58]. The final stage in laboratory mice transmission is the migration of WNV into neural tissues, such as the vertebral column and brain; this process is anticipated to mimic every phase of pathology following human mosquito-borne infection [58, 81].

9.4.3 Neuroinvasion

In order for WNV to trigger neuropathogenesis, the virus must have the capacity allowed to get into the central nervous system (CNS) to propagate quickly among its intended target cells, which comprise neuron and hematopoietic cells [58]. In particular, an essential N-linked glycosylation on domain I of the E protein30 and factors on virus basic proteins29 serve an important part to regulate neuroinvasive capacity [82]. Although it is unclear how these variables impact neuroinvasion, it is expected that they promote adherence to and entry of capillary cells, which could lead to viral spread and penetration of the CNS. A different view suggests that WNV penetrates the CNS by crossing the blood–brain barrier (BBB). Moreover, increase in endothelial cell flexibility, which is regulated by vasoactive substances such as tumor necrosis factor (TNF) [83], may contribute to the collapse of the BBB. Additional predicted routes for the entry of viruses into the CNS [84] consist of direct axonal reverse transport from infected peripheral neurons, an infection of neurons that sense the smell and spread via the olfactory bulb, a "Trojan Horse" process through which virus gets transmitted to the CNS by infected immune cells, which move there, and infection of peripheral neurons and propagation from these neurons [85].

Tremors, cerebellar ataxia, and general Parkinsonism are a few of the most severe signs [86] in human WNV infection single case report, including premortem clinical and laboratory findings and autopsy. An 83-year-old female presented with acute confusion, high fevers, dysarthria, and generalized subjective weakness, with decreased deep tendon reflexes and weakness on physical examination [87]. There was evidence of sensorimotor axonal polyneuropathy in the right extremity when examined on electromyography. ++WNV infection was confirmed premortem by the detection of IgM antibodies from serum and CSF and postmortem by RT-PCR from brain tissue [88]. Clinical interpretation of the brain parenchyma reveals scattered microglial cell aggregates with perivascular chronic inflammation. The leptomeninges reveal focal lymphocytic infiltrates. Examination of the spinal cord showed lymphocytic infiltrates in nerve roots and within the cord proper, with focal microglial nodules and neuronophagia in the ventral horns [89]. The presence of mutated stains does not indicate the presence of demyelinating processes. Emphysema and atelectasis were interpreted clinically by general autopsy. The clinical manifestation in this type of case suggests direct viral infection at the spinal cord along with nerve roots, which suggests the probable mechanism of the flaccid paralysis that is often found in patients infected with WNV. Findings are reviewed in comparison with other reports of neuropathologic findings in human WNV [29] and symptoms linked to WNE. A polio-like paralysis with flaccidity triggered by WNP has been associated with invasion of the cells of the anterior horn (lower motor neurons) of the spinal cord, which can, at its most serious, lead to quadriplegia and difficulty breathing [33]. WNP differs from a more unusual Guillain-Barre syndrome that has been correlated with the transmission of WNV. Although the differences are essential, clinical signs of infection with WNV may occur in the form of any one or more of these illnesses. West Nile neuroinvasive disease has general histological characteristics that are similar among various viral encephalitis, involving perivascular lymphocytic infiltrates, microglial nodules, neuronal loss, and neuronophagia. Additionally, necrosis can be found in severe circumstances. The brainstem (medulla and pons), deep gray matter nuclei (substantia nigra of the basal ganglia and thalamus), and cerebellum are among the most regularly affected regions of the brain's CNS with extrapyramidal (movement-related) function, with gray matter getting the most seriously impacted [90].

The anterior horns (ventral horns) and anterior spinal nerves frequently experience damage in the spinal cord, and consequently, lower motor neuron depletion creates a weakness in the muscles [91].

9.5 Relationship Between Host-Pathogen in West Nile Virus Infection

WNV infection recently became a major public health concern in the western hemisphere. WNV is transmitted to humans primarily through the bite of infected mosquitoes [92] and is maintained in a bird-mosquito-bird transmission cycle. Although WNV has been detected in 65 different mosquito species and 326 bird species in the United States, only a few *Culex* mosquito species drive transmission of the virus in nature and subsequent spread to humans [93]. The WNV was first isolated in 1937 [94]. Is such a zoonotic illness, which mostly affects birds and is transmitted by mosquitoes, infects individuals. On the other hand, mosquito vectors are also capable of transmitting WNV to mammals, especially human beings and horses, which may result in severe neurologic conditions along with elevated fever. Even so, data on WNV occurrences have grown over the past 15 years, and as of today, the virus has spread to practically most tropical and subtropical areas of the planet. The formation of new, more virulent strains is thought to be related to the expansion of WNV. In humans, the majority of WNV infections are asymptomatic, although 20% of patients experience flu-like symptoms and a high fever. The first symptoms of the disease appear 2–14 days after infection. Encephalitis, meningitis, or flaccid paralysis are the three severe neuroinvasive medical conditions that WNV can cause in roughly 1% of patients. In particular, older and immune-suppressed persons are at risk [95]. WNV infections are often short-lived but leave behind a lifetime immunity. However, persistent infections have been demonstrated occasionally [96]. Horses have a higher risk of developing severe symptoms than mortality rates for people can approach 40%. WNV has a far wider spectrum of hosts and vectors than the majority of mosquito-borne viruses. More than 300 avian species are susceptible and many of these develop high viral serum titers during the acute phase of infection, most prominently passerine species, such as house sparrows and crows [97]. These WNV titers are sufficiently high to spread the virus to and infect blood-sucking mosquitoes during a window of three to seven days of infection. With only a limited number of exceptions (alligators and specific frogs), such serum quantities have only been seen in birds, hence mammalian hosts are a dead end for WNV. However, in rare cases, WNV transmission has been documented among humans via blood transfusion, organ transplantation, and the intrauterine route [14]. As a result, donations of blood from regions affected are checked in various nations to detect the detection of the WNV nucleic acids. Despite WNV now eradicated from more than 60 different mosquito species, the Culex genus of bird-feeding mosquitoes is the most significant insect carrier [98].

Structurally, the WN virus is a single-stranded RNA genome that is spherical in shape. Mainly protein C determinants that participate in RNA and protein interactions during nucleocapsid assembly have not been defined [99].

The most common route of WNV transmission to vertebrates is through the bite of an infected mosquito, and a broad range of mammals and non-mammals species like reptiles and amphibians are susceptible to natural or experimental infection with WNV. Some mammals (rodents, rabbits, squirrels) and reptiles (alligators) have been found to develop sufficient viremia to allow transmission to feeding mosquitoes [10]. Other routes of WNV transmission among humans can be direct blood product contact, organ transplant, transplacentally, or via milk ingestion by the newborn [11–13]. Although the precise mechanisms and locations of WNV spread after a mosquito bite are unknown, early replication is believed to take place in the skin and local lymph nodes and to result in an initial viral infection that seeds the retinal endothelial system (RES) [20]. Viruses can subsequently seed the CNS, according to the extent of additional viremia put on by RES replication. Infected individuals who are healthy may typically isolate the virus from their blood during the peak viremia period, which lasts until approximately two days before to approximately four days after the start of the illness but during the first day of illness, the rate of virus isolation significantly declines; this observation is most likely caused by an increase in macrophage clearance and the production of IgM antibodies. WN virus was recovered from the blood of an immune-compromised patient up to 28 days postinoculation, and some terminally ill persons intentionally infected with the WN virus developed high-titer viremia [14]. WNV infections in people typically have no symptoms. Most symptomatic patients present a mild flu-like illness (West Nile fever (WNF)), described in early reports as a self-limited febrile illness characterized by fever, headache, back pain, myalgia, and anorexia, sometimes associated with other symptoms like nausea vomiting, and diarrhea, and rarely with neurological symptoms [15]. The same patient may exhibit many clinical symptoms associated with one of these disorders at once. Additional neurological syndromes include hepatitis [17], myocarditis [18], pancreatitis [19], and hemorrhagic fever. The most significant risk factor underlying the emergence of encephalitic disorders is advanced age.

9.6 Clinical Presentation of West Nile Virus

Approximately 80% of WNV-infected individuals are thought to be asymptomatic [100]. The average incubation period of symptoms for WNV infection in humans is difficult to predict with any degree of accuracy, however, it is said that symptoms typically range from 2 to 15 days [101, 102]. Symptomatic infections typically present as a nonlife-threatening, mild febrile sickness. Most symptomatic individuals have a moderate sickness with fever, which is occasionally accompanied by headache, chills, myalgias, nausea, vomiting, and lymphadenopathy [2, 15, 103]. Additionally, some patients transiently exhibit widespread roseolar or maculopapular rashes on the torso, arms, or legs [5, 7]. With an average illness duration of less than seven days, these symptoms often follow a predictable pattern. Additional symptoms, such as rhabdomyolysis, chorioretinitis, myositis, and autonomic nerve involvement, have also been mentioned [104–106]. Other neurological disorders, such as hepatitis, pancreatitis, myocarditis, orchitis, uveitis, and vitreous, may also manifest despite their rarity [15]. Optic neuritis, chorioretinal inflammation, retinal hemorrhages, and vitreous inflammation are among the ocular side effects of WNV infection [104, 107]. Patients with ocular problems initially displayed blurred vision or "floaters" as well as systemic symptoms such as fever, exhaustion, myalgia, headaches, weakness, ataxia, or confusion [11–13].

The main clinical presentations associated with WNV infection are discussed below:

9.6.1 West Nile Fever (WNF)

The most common clinical presentation among WNV-infected people is WNF. People of all ages may be impacted; however, evidence points to younger people having a higher proportion of WNF [100].

Uncomplicated WN fever frequently manifests as an acute onset of fever (generally >39° C), headache, and myalgia, frequently associated with gastrointestinal disturbances, including nausea and vomiting, which may result from dehydration. Although the acute ailment often lasts no longer than a week, chronic weariness is prevalent [2].

WNF may occasionally be accompanied by a rash that predominates throughout the torso and extremities, excluding the palms and soles, and tends to be morbilliform, maculopapular, and non-pruritic [100, 108, 109]. In some people, the rash may be temporary, lasting not more than 24 hours. Additionally, younger people are more likely than older people to get rash [100, 110].

Even while older people with WNF are more likely to die than younger people with similar symptoms [111, 112] the majority of patients make a full recovery. However, some healthy people may continue to have chronic fatigue, headaches, and attention problems for days or weeks after infection. Cardiopulmonary problems are frequently to blame for the deaths of WNF patients, which mostly affect the elderly and those who are immunocompromised [100, 113].

9.6.2 West Nile Neuroinvasive Disease

The neuroinvasive disorders in which the virus enters the intrathecal space and infects parts of the CNS, affect less than 1% of infected people, can be divided into three clinical syndromes: West Nile Meningitis (WNM), West Nile Encephalitis (WNE), and West Nile poliomyelitis (WNP). Clinical characteristics of these syndromes may coexist in the same patient. It is unclear if these disorders are different facets of a continuous clinical spectrum or separate entities. The majority of patients with neuroinvasive illnesses can be classified as either having WNM or WNE, and individuals with the latter had a greater mortality rate and more serious consequences [15, 114].

9.6.2.1 West Nile Meningitis (WNM)

WNM and other viral meningitides (also known as "aseptic meningitis") are nearly identical. Fever and meningeal irritation symptoms, such as headache, stiff neck, nuchal stiffness, Kernig's and/or Brudzinski's signs of photophobia and phonophobia, as well as a quick onset of fever and headache, are the typical symptoms of WNM [15, 100]. The gastrointestinal symptoms that come along with them, such as nausea, diarrhea, and vomiting, might make you thirsty, which can make headaches and other systemic symptoms worse. If you have concomitant headaches, you might need to go to the hospital for pain relief. Although, like WNF, WNM is typically linked with a positive outcome, some patients do have persistent headaches, fatigue, and myalgias [100, 115, 116].

9.6.2.2 West Nile Encephalitis (WNE)

In clinical terms, WNE is typically an arboviral encephalitis disease. A prodrome of fever, headache, and other vague symptoms lasting for one to few days is seen by some patients (i.e. the typical WN fever). Others have a more sudden start of fever coupled with encephalitis symptoms and indications, particularly mental state changes and vomiting. The progression of cerebral dysfunction to coma occurs in roughly 15% of cases. Depressed deep tendon reflexes, generalized muscle weakness (typically accompanied by severe proximal muscle weakness), flaccid paralysis, and respiratory failure can accompany abnormalities [2, 117]. Movement disturbances (or dyskinesias) are a prominent component in reports of WNV encephalitis defined as parkinsonism or having elements of parkinsonism [118]. The abnormalities include bradykinesia, cogwheel rigidity, postural instability, and masked facies, which are characteristics of Parkinson's disease. There have also been reports of myoclonus, purposeful tremor, and bruxism (teeth grinding). In individuals with WNV encephalitis, tremor is frequently reported (up to 80–100% in one group of patients), while myoclonus is less frequent (approximately a third of cases) [118]. The majority of the time, these movement disturbances get better on their own with time; nonetheless, tremors and parkinsonism might linger in people who are recovering from severe encephalitis [118]. The cerebellum, substantia nigra of the basal ganglia, the thalami, the brainstem (especially the medulla and pons), and the deep gray matter nuclei are affected, notably the substantia nigra of the basal ganglia and the thalami, and the movement problems in WNE are usually impacted. Among patients recovering from WNE, neuropsychiatric symptoms such as depression, anxiety, and apathy have been documented. WNE mortality rates have ranged from 10% to 30%, with mortality rates being greater in the elderly and those with impaired immune systems [100, 119, 120].

9.6.2.3 West Nile Poliomyelitis (WNP) and Acute Flaccid Paralysis (AFP)

WNP typically appears within the first 24–48 hours after an illness's beginning. Limb weakness typically appears suddenly and progresses quickly. Another symptom is a central facial weakness that is typically bilateral. Sensory loss or numbness is typically absent; however, some patients report excruciating limb pain that lasts for days or weeks before the onset of weakening. The involvement of respiratory muscle innervation, which results in paralysis of the diaphragmatic and intercostal muscles and neuromuscular respiratory failure, necessitating immediate endotracheal intubation, is the most severe form of WNP [121, 122]. Similar to poliovirus infection, it appears that the lower brainstem, namely the motor nuclei of the vagus and glossopharyngeal nerves, is the primary pathophysiologic basis for this manifestation.

WNV infection has also been linked to other types of AFP, such as radiculopathy and the acute demyelinating polyradiculoneuropathy variant of Guillain–Barré syndrome (GBS) [123, 124].

These syndromes can be distinguished based on their clinical and electrophysiologic characteristics; however, they seem to be far less frequent than poliomyelitis. The weakness connected to GBS is typically symmetric, ascending (starting in the legs and moving on to the arms and muscles innervated by cranial nerves), and linked to sensory and autonomic abnormalities [100].

9.7 Management of West Nile Virus

The clinical care of WNV disease is supportive; there is no specific treatment available. Patients with severe meningeal symptoms frequently require rehydration, antiemetic medication, and pain management for headaches and related nausea and vomiting.

In order to treat the WNV infection, no antiviral or adjuvant medicines are permitted or advised; clinical management is supportive. Many case reports and case series have described the use of different products (like polyclonal immune globulins, polyclonal immune globulins that are hyperimmune, monoclonal immune globulins, interferon, ribavirin, and corticosteroids) for WNV disease. In controlled clinical studies, a number of these products have been investigated for infections caused by WNV or flaviviruses that are closely related to it (such as the St. Louis encephalitis and JE viruses). None have demonstrated a definite advantage. The studies frequently used tiny sample sizes, and several clinical trials' findings have not yet been made public. Since they are readily available, interferon alfa and polyclonal immune globulin have been used by certain doctors to treat [20].

The sections that follow provide summaries of pertinent articles that discuss the use of various items to treat WNV diseases as well as those with other closely associated flaviviruses.

9.7.1 Polyclonal Immune Globulin Intravenous (IGIV)

Interferon alfa-2b as well as polyclonal IGIV were administered to two prior organ transplant patients, according to a case research. In Arizona in 2015, a total of three patients (one kidney, one heart, and one kidney and heart) with meningoencephalitis triggered by the St. Louis encephalitis virus. Whenever invasive meningoencephalitis became suspected, the immune system suppression in both organ recipients decreased. Interferon alfa-2b (300,000 units each day) was administered to one patient for 14 days and the other for 10 days. Additionally, both patients got IGIV for five days (0.4 g/kg daily). Both patients finally made a full recovery with little to no neurologic aftereffects. An additional previous organ-transplant recipient (kidney) with St. Louis encephalitis virus meningoencephalitis died early in their hospital course before receiving any interferon or IGIV [21].

9.7.2 Polyclonal IGIV with High Titers of WNV Antibodies Derived from Blood Donors (Omr-IgG-Am)

Omr-IgG-am is an Israeli polyclonal IGIV product. Based on the seroprevalence of donor antibodies, it is believed to have a lot of WNV-neutralizing antibodies. The United States no longer carries Omr-IgG-am.

In a clinical experiment conducted in the United States between 2003 and 2006, 64 patients with AFP or WNV encephalitis were assigned at random to receive Omr-IgG-am, polyclonal IGIV, or a placebo. Only limited results have been published but they showed no differences in outcomes between the study groups [22].

9.7.3 WNV Recombinant Humanized Monoclonal Antibody (MGAWN1)

Forty healthy people participated in a phase 1 safety and dose-ranging pharmacokinetics research to assess MGAWN1; while 10 received a placebo, 30 underwent single infusion of the trial medication. Eleven drug-related side effects (diarrhea, chest pain, oral herpes, rhinitis, neutropenia, leukopenia, dizziness, headache, and somnolence) were experienced by six trial participants. One participant had schizophrenia diagnosed just 50 days after taking the research medication. The placebo group experienced no adverse events that were recorded. The highest dose of MGAWN1 had a half-life of 27 days and exceeded serum target levels by 28-fold [23].

9.8 Recent Advancements in Diagnostics, Clinical Assessment, and Treatment

9.8.1 Diagnosis and Clinical Assessment

Clinical indicators such as severe fever, anorexia, nausea, vomiting, eye pain, and other symptoms are observed during the clinical assessment of WNV infection in people. Arthralgia, lymphadenopathy, rash, headache, myalgia, and multifocal chorioretinitis have been shown to be possible indicators of severe WNV infection in people [28, 29].

The diagnosis of WNV infection in the laboratory is still difficult, and it frequently needs to be confirmed by secondary tests run in reference labs. In order to diagnose WNV infection, there are several barriers. They include low or absent viremia at the time of symptom onset, genetic diversity of WNV, cross-reactivity of antibodies among flaviviruses that must be verified by neutralization assays, and persistence of IgM antibodies for several months after the infection has been recovered. WNV RNA in blood or CSF, WNV isolation in cell culture, WNV-specific IgM antibodies in serum or CSF, and evidence of a fourfold increase in WNV IgG antibody titer between an acute and a convalescent serum sample are the main components of the laboratory diagnosis [125].

IgM and IgG antibodies against WNV are used as the foundation for the serological diagnosis. IgM antibodies alone or IgG seroconversion can indicate a WNV acute infection because (IgG) antibodies can typically be found by days 4 and 8, respectively, after the onset of symptoms. IgM in particular can be persistently found for up to two years, especially in horses, which limits its value in a diagnostic context [29–31].

In a laboratory context, blocking ELISA has been proven to be a trustworthy, affordable, and simple diagnostic technique. It entails assessing the patient's serum antibodies' capacity to prevent monoclonal antibodies from adhering to the NS1 and E protein epitopes [126]. The NS1-targeted ELISA is frequently utilized to distinguish between postinfection antibodies and postvaccinal antibodies, despite the fact that the majority of vaccine candidates are produced based on prM and E proteins [95, 127]. It would be beneficial to develop NS1 antibody-based ELISA assays for WNV and dengue virus in order to lessen cross-reactivity and produce more accurate diagnostic tests for flaviviruses. This assay has the benefit of being flexible, very simple to perform in a lab setting, having results that are generally repeatable, and having a portion of their process that can be automated. Because WNV IgM can linger in serum for weeks, months, or even years after infection, and because all flaviviruses exhibit significant cross-reactivity, it can be challenging to make a diagnosis of acute WNV infection using only these commercial kits [95, 126, 128].

Molecular detection of viral genomes, which can be accomplished in hours, has replaced conventional viral isolation in cell culture, which can take days to weeks, in the laboratory identification of viral infections since the development of molecular biology techniques in the 1980s [127]. Due to its sensitivity and specificity, industry-standard format, high levels of repeatability and reproducibility, quick turnaround time, and ease of use, quantitative Real-Time Polymerase Chain Reaction (qRT-PCR) has recently become the standard diagnosis method for many viral infections [129, 130].

Reverse transcription polymerase chain reaction (RT-PCR), and in situ hybridization are also the most frequently utilized molecular diagnostic procedures. The ability to quantify the viral genome gives qRT-PCR an edge over conventional

RT-PCR. SYBR® Green or other DNA-intercalating fluorescent dyes are used to monitor the buildup of double-strand DNA in order to quantify it [131]. The most preserved genome region in almost all flaviviruses, NS5, should be the focus of PCR primers. Due to RT-PCR's high sensitivity, it may be possible to identify viral RNA in animals that have received vaccinations with the killed WNV vaccine. As a result, when screening people who have had such vaccinations, PCR should be combined with additional diagnostic techniques, such as virus isolation. As an alternative, the WNV-NS5 should be the target of the RT-PCR to separate viral RNA from the vaccination from infecting the virus that is replicating. Samples of serum or cerebrospinal fluid (CSF) should be used for the testing [95].

Unfortunately, due to the low level and short-lived viremia generated by most flaviviruses, including WNF, molecular diagnosis by qRT-PCR of serum, plasma, and CSF samples is of little use for routine diagnosis [128, 132, 133].

There are a few commercial nucleic acid amplification tests (NATs) available, which are done in order to assess and lower the risk of WNV infection in organ transplants and blood recipients. These tests are fully automated, allow testing of hundreds of plasma samples per day, and are approved for use with blood from healthy individuals [133–135]. These commercial tests must detect WNV with extremely high sensitivity and specificity due to the modest quantities of WNV RNA that may be present in healthy, asymptomatic blood donors [126].

However, the development of next-generation sequencing (NGS) techniques in recent years – also known as deep or high-throughput sequencing – has increased the use of NGS to support the diagnosis and monitoring of infectious diseases brought on by both bacteria and viruses [37, 40, 41]. The objective detection of pathogens in clinical samples is perhaps the most significant diagnostic use for NGS. Unlike conventional viral detection techniques that depend on prior viral sequence or antigen information, NGS shotgun metagenomics analysis of clinical samples enables the detection of novel or unexpected infections [136]. When the presence of a new WNV strain is suspected but cannot be multiplied and sequenced using the current primer sets, this method may be beneficial to identify WNV RNA and sequence its entire genome.

Recently, there has been a breakthrough in the detection of flavivirus RNAs, such as those from Zika, dengue, and WNV, in bodily fluids. Fluorescent proteins were employed to monitor and detect flavivirus RNAs in patient samples using CRISPR-cas13 technology, creating a field-deployable viral diagnostic platform with high performance and little equipment or sample processing requirements [137]. This platform is quick (within two hours), comparable in sensitivity to qRT-PCR tests, and may be employed as a diagnostic platform in resource-constrained locations or even as a competitor to currently available molecular diagnostic equipment [125, 126].

Electrospray ionization mass spectrometry (PCR-ESI/MS), is a recent innovation that enables the identification of the base compositions of amplicons produced by multiplex PCR by determining the mass/charge ratio of single-stranded DNA, can also be used to characterize amplicons produced by PCR [138].

The following new technologies are anticipated to emerge in the field of WNV diagnostics in the future, in line with advancements in laboratory technology.

- Rapid point-of-care (POC) testing for the quick identification of WNV infection. Rapid point-of-care tests that are sensitive and specific for detecting dengue virus NS1 antigen, IgM and IgG antibodies, and maybe WNV antibodies, are now available [125].
- Increased understanding of the genetic diversity of WNV, the availability of numerous WNV genome sequences, the antigenic characteristics of WNV and other flaviviruses, and the kinetics and distribution of WNV in blood compartments and tissues will all contribute to the achievement of this objective [125].
- Syndromic methods for the diagnosis of neuroinvasive illnesses and/or imported viral infections spread by vectors. This can be done using multiplexed assays that target different pathogens, such as multiplex real-time PCR, DNA and protein microarrays, and Luminex arrays, for serology and molecular tests. For example, pan-flavivirus PCR followed by sequencing, array identification, or mass spectrometry [125].

9.8.2 Treatment

For the time being, there is no permanent cure for WNV infection. Therefore, preventing infection through mosquito bite prevention is crucial and the most significant public health action [100]. Supportive care is the main course of action. Despite the work of numerous laboratories and organizations as well as the vaccinations available for use in horses, there is currently no FDA-licensed vaccine to treat WN illness in people [139]. The fact that human viremia is transient, and that WNV is typically gone from the body by the time of clinical presentation is a significant theoretical barrier to the development of targeted antiviral medicines [100]. Patients with straightforward WNF typically do not need any special care, though headache control and rehydration may occasionally be necessary. The development of a more severe neuroinvasive illness should, however, be monitored in those with WN viremia and those with other risk factors, like advanced age and immunosuppression. Patients with severe WNM may also need pain medication for a severe headache, and they may need hospitalization for rehydration if they get dehydrated from accompanying nausea and vomiting. It is crucial to pay attention to the patient's degree of consciousness and airway protection in WNE patients. Although seizures and elevated intracranial pressure have very seldom been associated with WNE, they should be properly handled if they do occur [100, 121].

Testing medications that have been shown to be successful against other flaviviruses, such as dengue and hepatitis C, may be one of the more promising strategies. In addition, favipiravir (T-705, FujiFilm Pharmaceuticals) has generated interest as a possible treatment for WNV infection. The pyrazine-carboxamide derivative favipiravir mimics a purine nucleotide in its active triphosphate form, which prevents the RNA-dependent RNA polymerases of multiple families of RNA viruses, including various species of alphavirus and flavivirus, from working [103, 140]. As soon as WNV binds to the receptors on host cells, it enters the cells through an endocytosis process that is clathrin-independent. This is followed by a low-pH-dependent viral uncoating in the endosome to release the viral DNA into the cytoplasm for replication [141]. Inhibitors that prevent membrane fusion or stop the E protein's interaction with cell receptors are possible WNV antivirals. This antiviral tactic is supported by two effective HIV medications, Maraviroc (a CCR5 antagonist) and Enfuvirtide (a peptide inhibitor), which prevent membrane fusion and viral entry, respectively [142, 143]. Inhibitors may all have their sights set on the hydrophobic pocket, stem domain, receptor-binding domain, and other functional areas of the influenza virus E glycoprotein. One such target for the creation of small chemical inhibitors against flaviviruses is the hydrophobic ligand-binding pocket in a hinge region between domains I and II of E protein, which plays a significant role in the low-pH-mediated membrane fusion process [144, 145]. The majority of these inhibitors were created and tested against DENV, though some were reported to have antiviral activities against WNV [146]. DENV and other flaviviruses have been targeted with various screening techniques to find inhibitors that bind to this hydrophobic pocket and prevent structural changes to the E protein. These compounds, however, failed to be developed for future medications due to their unfavorable characteristics, such as their low solubility and cytotoxicity. The area of the E glycoprotein that mediates receptor binding may be targeted by the development of inhibitors that prevent viral attachment to host cell receptors. The effectiveness of this tactic has been demonstrated by neutralizing antibodies against the E protein. Additionally, it has been demonstrated that a number of substances prevent flu antivirus receptors from attaching to host cell receptors [147–149]. However, the effectiveness of this strategy has been limited by a paucity of knowledge on WNV cell receptors. The discovery of WNV cellular receptors and knowledge of virus–receptor interaction may open up new possibilities for the discovery of small chemical inhibitors that prevent WNV from attaching to host cell receptors [150].

9.9 Conclusion and Future Prediction

The knowledge of the clinical picture of illness, along with the short- and long-term effects linked to human WNV infection, has greatly expanded during the past 10 years. However, there are still certain clinical issues that need more clarification. There is a lack of evidence on the chronic neuroprotective effects on individuals surviving infection, and further research is required to determine whether there will be persistent cognitive damage following encephalitis from WNV. However, recurrent- or early-onset parkinsonism in such individuals due to the aging of dopaminergic neurons remains a theoretical potential. The parkinsonian characteristics linked to acute WNV infection are responsible to be temporary in the majority of instances and resolve with time. Innate immunity defensing mechanisms, type I IFN antiviral mechanisms, viral and mammalian host complement activation and regulation processes, humoral immunity has features of protection, and immune cell-regulated infection control mechanisms are just a few of the immunological correlates of protection that have been discovered over the past 10 years of research to knowing the processes of WNV transmission and immunity. Despite tremendous improvements in our comprehension of WNV pathogenesis, there are still several unanswered concerns. The discovery of genes that prevent WNV infection has come a long way, but further research will be required to determine the context-dependent roles of cell- and tissue-specific gene expression as well as the progression of the specified proteins. The mechanisms behind these evasion tactics still need to be completely understood, even though several investigations have discovered viral components that facilitate immune evasion. In order to better understand how the host immunity, including nonself-recognition response pathways and pro-inflammatory cytokine signaling response pathways, interacts with one another and eventually affects the establishment of protective immunity against WNV infection, more research is required. Researchers will probably be able to characterize these interactions using systems biology-based techniques, including RNA sequencing, bioinformatics tool, genomics, proteomics, lipidomics, metabolomics, and computational modeling. Though protective immunity against WNV infection depends on cell-mediated and humoral responses, it is unclear how these responses eliminate the WNV virus from infected cells, including non-cytolytic clearing of the virus from infected nerve cells. Furthermore, using mice models, tremendous progress has been achieved in understanding the host immune response to WNV virus infection, but more work is required to conduct similar investigations in relevant human cells. The opportunity to create WNV vaccines for humans is significant, and various businesses are now working on potential inactivated subunit or live-attenuated vaccines. The expense of creating a candidate vaccine and doing pricey clinical trials is a problem that any firm that produces vaccines must deal with. Despite the fact that clinical trials will be used to characterize the immunological responses (both humoral immunity and cell-mediated immunity) by vaccination, the protective effectiveness may be challenging to demonstrate because viral infection fluctuates annually by geographic location and occurrence. Government, academic, and industry partners will need to work together to address these concerns. There is not yet any WNV vaccine or licensed medicine that can be used to stop or treat WNV infection in people. However, a number of vaccine candidates are undergoing or have already finished early clinical studies. Several veterinary vaccinations have been approved for use in horses, birds, and other mammals that are susceptible to WNV infection in the United States. A few clinical studies utilizing IFN, humanized WNV-specific monoclonal antibodies, nucleic acid analogs, small-molecule inhibitors, and passive immunoglobulin treatment have been started in humans, which are yet to be unfolded.

References

1 Petersen, L.R., Marfin, A.A., and Gubler, D.J. (2003). West Nile virus. *JAMA* 290 (4): 524–528.

2 Campbell, G.L., Marfin, A.A., Lanciotti, R.S., and Gubler, D.J. (2002). West Nile virus. *Lancet Infect. Dis.* 2 (9): 519–529.

3 Monini, M., Falcone, E., Busani, L. et al. (2010). West Nile virus: characteristics of an African virus adapting to the third millennium world. *Open Virol J.* 4: 42–51.

4 Davis, L.E., DeBiasi, R., Goade, D.E. et al. (2006). West Nile virus neuroinvasive disease. *Ann. Neurol.* 60 (3): 286–300.

5 A Neurotropic Virus Isolated from the Blood of a Native of Uganda. https://www.cabdirect.org/cabdirect/abstract/19412700112?freeview=true (accessed 7 June 2023).

6 Smithburn, K.C. (1942). Differentiation of the West Nile virus from the viruses of St. Louis and Japanese B encephalitis. *J. Immunol.* 44: 25–31. https://journals.aai.org/jimmunol/article-abstract/44/1/25/100316.

7 Philip, C.B. and Smadel, J.E. (1943). Transmission of West Nile virus by infected *Aedes albopictus*. *Proc. Soc. Exp. Biol. Med.* 53 (1): 49–50.

8 Smithburn, K.C. and Jacobs, H.R. (1942). Neutralization-tests against neurotropic viruses with sera collected in Central Africa. *J. Immunol.* 44 (1): 9–23.

9 Weinberger, M., Pitlik, S.D., Gandacu, D. et al. (2001). West Nile fever outbreak, Israel, 2000: epidemiologic aspects. *Emerg. Infect. Dis.* 7: 686–691. https://www.ncbi.nlm.nih.gov/pmc/articles/PMC2631774.

10 Murgue, B., Murri, S., Triki, H. et al. West Nile in the Mediterranean Basin: 1950–2000. *Ann. N. Y. Acad. Sci.* 951: 117–126. https://nyaspubs.onlinelibrary.wiley.com/doi/full/10.1111/j.1749-6632.2001.tb02690.x?casa_token=EODjidyEfIIAAAAA%3A-Dgb3OJp2qS4hk5aGooz2nzpNqKrxr3Zp3Ejjb9WGTYj0ks2fJfieiIYQsD922HnYTUbgz3Z0KU8wqc.

11 George, S., Gourie-Devi, M., Rao, J.A. et al. (1984). Isolation of West Nile virus from the brains of children who had died of encephalitis. *Bull. World Health Organ.* 62 (6): 879–882.

12 Hubálek, Z. and Halouzka, J. (1999). West Nile fever – a reemerging mosquito-borne viral disease in Europe. *Emerg. Infect. Dis.* 5: 643–650. https://www.ncbi.nlm.nih.gov/pmc/articles/PMC2627720.

13 Murgue, B., Zeller, H., and Deubel, V. (2002). The ecology and epidemiology of West Nile virus in Africa, Europe and Asia. In: *Japanese Encephalitis and West Nile Viruses* [Internet], Current Topics in Microbiology and Immunology (ed. J.S. Mackenzie, B.A.D.T., and V. Deubel), 195–221. Berlin, Heidelberg: Springer doi:10.1007/978-3-642-59403-8_10.

14 Hayes, E.B., Sejvar, J.J., Zaki, S.R. et al. (2005). Virology, pathology, and clinical manifestations of West Nile virus disease. *Emerg. Infect. Dis.* 11: 1174–1179. https://www.ncbi.nlm.nih.gov/pmc/articles/PMC3320472.

15 Kramer, L.D., Li, J., and Shi, P.Y. (2007). West Nile virus. *Lancet Neurol.* 6 (2): 171–181.

16 Marra, P.P., Griffing, S.M., and McLean, R.G. (2003). West Nile virus and wildlife health. *Emerg. Infect. Dis.* 9 (7): 898–899.

17 McLean, R.G. (2003). West Nile virus: emerging threat to public health and animal health. *J. Vet. Med. Educ.* 30 (2): 143–144.

18 Centers for Disease Control and Prevention (CDC) (2011). West Nile virus disease and other arboviral diseases – United States, 2010. *MMWR Morb. Mortal. Wkly Rep.* 60 (30): 1009–1013.

19 Kramer, L.D., Styer, L.M., and Ebel, G.D. (2008). A global perspective on the epidemiology of West Nile virus. *Annu. Rev. Entomol.* 53 (1): 61–81.

20 Anderson, J.F., Andreadis, T.G., Vossbrinck, C.R. et al. (1999). Isolation of West Nile virus from mosquitoes, crows, and a Cooper's Hawk in Connecticut. *Science* 286 (5448): 2331–2333.

21 May, F.J., Davis, C.T., Tesh, R.B., and Barrett, A.D.T. (2011). Phylogeography of West Nile virus: from the cradle of evolution in Africa to Eurasia, Australia, and the Americas. *J. Virol.* 85 (6): 2964–2974.

22 Charrel, R.N., Brault, A.C., Gallian, P. et al. (2003). Evolutionary relationship between Old World West Nile virus strains. *Virology* 315 (2): 381–388.

23 Bakonyi, T., Ivanics, É., Erdélyi, K. et al. (2006). Lineage 1 and 2 strains of encephalitic West Nile virus, Central Europe. *Emerg. Infect. Dis.* 12 (4): 618–623.

24 Shahhosseini, N., Chinikar, S., Moosa-Kazemi, S.H. et al. (2017). West Nile virus lineage-2 in *Culex* specimens from Iran. *Tropical Med. Int. Health* 22 (10): 1343–1349.

25 Kolodziejek, J., Seidel, B., Jungbauer, C. et al. (2015). West Nile virus positive blood donation and subsequent entomological investigation, Austria, 2014. *PLoS One* 10 (5): e0126381.

26 Dawson, J.R., Stone, W.B., Ebel, G.D. et al. (2007). Crow deaths caused by West Nile virus during winter. *Emerg. Infect. Dis.* 13 (12): 1912–1914.

27 Banet-Noach, C., Simanov, L., and Malkinson, M. (2003). Direct (non-vector) transmission of West Nile virus in geese. *Avian Pathol.* 32 (5): 489–494.

28 Martin, D.A., Biggerstaff, B.J., Allen, B. et al. (2002). Use of immunoglobulin M cross-reactions in differential diagnosis of human flaviviral encephalitis infections in the United States. *Clin. Vaccine Immunol.* 9 (3): 544–549.

29 Mukhopadhyay, S., Kim, B.S., Chipman, P.R. et al. (2003). Structure of West Nile virus. *Science* 302 (5643): 248–248.

30 Lindenbach, B.D., Thiel, H.J., and Rice, C.M. (2007). Flaviviridae: the viruses and their replication. In: *Fields Virology* (ed. D.M. Knipe and P.M. Howley), 1102–1153. Lippincott-Raven Publishers.

31 Mukhopadhyay, S., Kuhn, R.J., and Rossmann, M.G. A structural perspective of the flavivirus life cycle. *Nat. Rev. Microbiol.* 3: 13–22. https://www.nature.com/articles/nrmicro1067.

32 Brinton, M.A. (2002). The molecular biology of West Nile virus: a new invader of the western hemisphere. *Annu. Rev. Microbiol.* 56 (1): 371–402.

33 Deubel, V., Fiette, L., Gounon, P. et al. (2001). Variations in biological features of West Nile viruses. *Ann. N. Y. Acad. Sci.* 951 (1): 195–206.

34 Sampath, A. and Padmanabhan, R. (2009). Molecular targets for flavivirus drug discovery. *Antivir. Res.* 81 (1): 6–15.

35 Chung, K.M., Liszewski, M.K., Nybakken, G. et al. (2006). West Nile virus nonstructural protein NS1 inhibits complement activation by binding the regulatory protein factor H. *Proc. Natl. Acad. Sci.* 103 (50): 19111–19116.

36 Leung JY, Pijlman GP, Kondratieva N, Hyde J, Mackenzie JM, Khromykh AA, 2008. Role of nonstructural protein NS2A in flavivirus assembly | *J. Virol.*, 82:4731–41. https://doi.org/10.1128/jvi.00002-08

37 Mackenzie, J.M., Khromykh, A.A., Jones, M.K., and Westaway, E.G. (1998). Subcellular localization and some biochemical properties of the flavivirus kunjin nonstructural proteins NS2A and NS4A. *Virology* 245 (2): 203–215.

38 Egloff, M.-P., Benarroch, D., Selisko, B. et al. (2002). An RNA cap (nucleoside-2′-O-) -methyltransferase in the flavivirus RNA polymerase NS5: crystal structure and functional characterization. *EMBO J.* 21: 2757–2768. https://www.embopress.org/doi/full/10.1093/emboj/21.11.2757.

39 Khromykh, A.A., Kenney, M.T., and Westaway, E.G. (1998). trans-complementation of flavivirus RNA polymerase gene NS5 by using kunjin virus replicon-expressing BHK cells. *J. Virol.* 72: 7270–7279. https://journals.asm.org/doi/full/10.1128/jvi.72.9.7270-7279.1998.

40 Speight, G., Coia, G., Parker, M.D., and Westaway, E.G. (1988). Gene mapping and positive identification of the non-structural proteins NS2A, NS2B, NS3, NS4B and NS5 of the flavivirus kunjin and their cleavage sites. *J. Gen. Virol.* 69 (Pt 1): 23–34. https://www.microbiologyresearch.org/content/journal/jgv/10.1099/0022-1317-69-1-23.

41 Evans, J.D. and Seeger, C. (2007). Differential effects of mutations in NS4B on West Nile virus replication and inhibition of interferon signaling. *J. Virol.* 81 (21): 11809–11816.

42 Liu, W.J., Wang, X.J., Mokhonov, V.V. et al. (2005). Inhibition of interferon signaling by the New York 99 strain and kunjin subtype of West Nile virus involves blockage of STAT1 and STAT2 activation by nonstructural proteins. *J. Virol.* 79 (3): 1934–1942.

43 Muñoz-Jordán, J.L., Laurent-Rolle, M., Ashour, J. et al. (2005). Inhibition of alpha/beta interferon signaling by the NS4B protein of flaviviruses. *J. Virol.* 79 (13): 8004–8013.

44 Muñoz-Jordán, J.L., Sánchez-Burgos, G.G., Laurent-Rolle, M., and García-Sastre, A. (2003). Inhibition of interferon signaling by dengue virus. *Proc. Natl. Acad. Sci.* 100 (24): 14333–14338.

45 Khromykh, A.A., Sedlak, P.L., and Westaway, E.G. (2000). Cis- and trans-acting elements in flavivirus RNA replication. *J. Virol.* 74: 3253–3263. https://journals.asm.org/doi/full/10.1128/jvi.74.7.3253-3263.2000.

46 Berthet, F.X., Zeller, H.G., Drouet, M.T. et al. (1997). Extensive nucleotide changes and deletions within the envelope glycoprotein gene of Euro-African West Nile viruses. *J. Gen. Virol.* 78: 2293–2297. https://www.microbiologyresearch.org/content/journal/jgv/10.1099/0022-1317-78-9-2293.

47 Lanciotti, R.S., Ebel, G.D., Deubel, V. et al. (2002). Complete genome sequences and phylogenetic analysis of West Nile virus strains isolated from the United States, Europe, and the Middle East. *Virology* 298 (1): 96–105.

48 Lanciotti RS, Roehrig JT, Deubel V, et al. 1999. Origin of the West Nile virus responsible for an outbreak of encephalitis in the Northeastern United States | *Science*, 286:2333–7. https://doi.org/10.1126/science.286.5448.2333

49 Charrel, R.N., Brault, A.C., Gallian, P. et al. (2003). Evolutionary relationship between Old World West Nile virus strains: evidence for viral gene flow between Africa, the middle east, and Europe. *Virology* 315 (2): 381–388.

50 Romanca, C., Vladimirescu, A., Tsai, T.F. et al. (1999). Entomologic and avian investigations of an epidemic of West Nile fever in Romania in 1996, with serologic and molecular characterization of a virus isolate from mosquitoes. *Am. J. Trop. Med. Hyg.* 61 (4): 600–611.

51 Scherret, J.H., Poidinger, M., Mackenzie, J.S. et al. (2001). The relationships between West Nile and Kunjin viruses. *Emerg. Infect. Dis.* 7: 697–705. https://www.ncbi.nlm.nih.gov/pmc/articles/PMC2631745.

52 Murray, K.O., Mertens, E., and Desprès, P. (2010). West Nile virus and its emergence in the United States of America. *Vet. Res.* 41: 67. https://www.ncbi.nlm.nih.gov/pmc/articles/PMC2913730.

53 Hall, R.A., Scherret, J.H., and Mackenzie, J.S. (2001). Kunjin Virus. *Ann. N. Y. Acad. Sci.* 951 (1): 153–160.

54 Fall, G., Diallo, M., Loucoubar, C. et al. (2014). Vector competence of *Culex neavei* and *Culex quinquefasciatus* (Diptera: Culicidae) from Senegal for lineages 1, 2, Koutango and a putative new lineage of West Nile virus. *Am. J. Trop. Med. Hyg.* 90 (4): 747–754.

55 Papa, A., Bakonyi, T., Xanthopoulou, K. et al. (2011). Genetic characterization of West Nile virus lineage 2, Greece, 2010. *Emerg. Infect. Dis.* 17 (5): 920–922.

56 Bagnarelli, P., Marinelli, K., Trotta, D. et al. (2011). Human case of autochthonous West Nile virus lineage 2 infection in Italy, September 2011. *Euro Surveill.* 16 (43): 20002.

57 Botha, E.M., Markotter, W., Wolfaardt, M. et al. (2008). Genetic determinants of virulence in pathogenic lineage 2 West Nile virus strains. *Emerg. Infect. Dis.* 14: 222–230. https://www.ncbi.nlm.nih.gov/pmc/articles/PMC2600181.

58 Beasley, D.W.C., Li, L., Suderman, M.T., and Barrett, A.D.T. (2002). Mouse neuroinvasive phenotype of West Nile virus strains varies depending upon virus genotype. *Virology* 296: 17–23. https://www.sciencedirect.com/science/article/pii/S0042682202913723.

59 Hubálek, Z., Halouzka, J., Juricová, Z., and Sebesta, O. (1998). First isolation of mosquito-borne West Nile virus in the Czech Republic. *Acta Virol.* 42 (2): 119–120.

60 Bakonyi, T., Hubálek, Z., Rudolf, I., and Nowotny, N. (2005). Novel flavivirus or new lineage of West Nile virus, Central Europe. *Emerg. Infect. Dis.* 11 (2): 225–231.

61 Lvov, D.K., Butenko, A.M., Gromashevsky, V.L. et al. West Nile virus and other zoonotic viruses in Russia: examples of emerging-reemerging situations. In: *Emergence and Control of Zoonotic Viral Encephalitides* (ed. C.H. Calisher and D.E. Griffin), 85–96. Springer-Verlag Wien GmbH https://link.springer.com/chapter/10.1007/978-3-7091-0572-6_7.

62 Pachler, K., Lebl, K., Berer, D. et al. (2014). Putative new West Nile virus lineage in *Uranotaenia unguiculata* mosquitoes, Austria, 2013. *Emerg. Infect. Dis.* 20 (12): 2119–2122.

63 Schneider, B.S., Soong, L., Coffey, L.L. et al. (2010). Aedes aegypti saliva alters leukocyte recruitment and cytokine signaling by antigen-presenting cells during West Nile virus infection. *PLoS One* 5 (7): e11704.

64 Styer, L.M., Lim, P.Y., Louie, K.L. et al. (2011). Mosquito saliva causes enhancement of West Nile virus infection in mice. *J. Virol.* 85 (4): 1517–1527.

65 Lazear, H.M., Pinto, A.K., Ramos, H.J. et al. (2013). Pattern recognition receptor MDA5 modulates $CD8^+$ T cell-dependent clearance of West Nile virus from the central nervous system. *J. Virol.* 87 (21): 11401–11415.

66 Bai, F., Kong, K., Dai, J. et al. (2010). A paradoxical role for neutrophils in the pathogenesis of West Nile virus. *J. Infect. Dis.* 202 (12, 12): 1804.

67 Girard, Y.A., Klingler, K.A., and Higgs, S. (2004 Jun). West Nile virus dissemination and tissue tropisms in orally infected *Culex pipiens quinquefasciatus*. *Vector Borne Zoonotic Dis.* 4 (2): 109–122.

68 Moskalyk, L.A., Oo, M.M., and Jacobs-Lorena, M. (1996). Peritrophic matrix proteins of *Anopheles gambiae* and *Aedes aegypti*. *Insect Mol. Biol.* 5 (4): 261–268.

69 Cheng, G., Cox, J., Wang, P. et al. (2010). A C-type lectin collaborates with a CD45 phosphatase homolog to facilitate West Nile virus infection of mosquitoes. *Cell* 142 (5): 714–725.

70 Fredericksen, B.L., Keller, B.C., Fornek, J. et al. (2008). Establishment and maintenance of the innate antiviral response to West Nile virus involves both RIG-I and MDA5 signaling through IPS-1. *J. Virol.* 82 (2): 609–616.

71 Errett, J.S., Suthar, M.S., McMillan, A. et al. (2013). The essential, nonredundant roles of RIG-I and MDA5 in detecting and controlling West Nile virus infection. *J. Virol.* 87 (21): 11416–11425.

72 Daffis, S., Samuel, M.A., Suthar, M.S. et al. (2008). Toll-like receptor 3 has a protective role against West Nile virus infection. *J. Virol.* 82 (21): 10349–10358.

73 Szretter, K.J., Daffis, S., Patel, J. et al. (2010). The innate immune adaptor molecule MyD88 restricts West Nile virus replication and spread in neurons of the central nervous system. *J. Virol.* 84 (23): 12125–12138.

74 Yamamoto, M., Sato, S., Hemmi, H. et al. (2003). Role of adaptor TRIF in the MyD88-independent toll-like receptor signaling pathway. *Science* 301 (5633): 640–643.

75 Verma, S., Kumar, M., Gurjav, U. et al. (2010). Reversal of West Nile virus-induced blood–brain barrier disruption and tight junction proteins degradation by matrix metalloproteinases inhibitor. *Virology* 397 (1): 130–138.

76 Suthar, M.S., Ma, D.Y., Thomas, S. et al. (2010). IPS-1 is essential for the control of west nile virus infection and immunity. *PLoS Pathog.* 6 (2): e1000757.

77 Martina, B.E.E., Koraka, P., Van Den Doel, P. et al. (2008). DC-SIGN enhances infection of cells with glycosylated West Nile virus in vitro and virus replication in human dendritic cells induces production of IFN-α and TNF-α. *Virus Res.* 135 (1): 64–71.

78 Diamond, M.S. and Gale, M. (2012). Cell-intrinsic innate immune control of West Nile virus infection. *Trends Immunol.* 33 (10): 522–530.

79 Perwitasari, O., Cho, H., Diamond, M.S., and Gale, M. (2011). Inhibitor of κB kinase ϵ (IKKϵ), STAT1, and IFIT2 proteins define novel innate immune effector pathway against West Nile virus infection. *J. Biol. Chem.* 286 (52): 44412–44423.

80 Shrestha, B., Zhang, B., Purtha, W.E. et al. (2008). Tumor necrosis factor alpha protects against lethal West Nile virus infection by promoting trafficking of mononuclear leukocytes into the central nervous system. *J. Virol.* 82 (18): 8956–8964.

81 Shipley, J.G., Vandergaast, R., Deng, L. et al. (2012). Identification of multiple RIG-I-specific pathogen associated molecular patterns within the West Nile virus genome and antigenome. *Virology* 432 (1): 232–238.

82 Morgan, B. (2003). Complement therapeutics; history and current progress. *Mol. Immunol.* 40 (2–4): 159–170.

83 Fredericksen, B.L., Smith, M., Katze, M.G. et al. (2004). The host response to West Nile virus infection limits viral spread through the activation of the interferon regulatory factor 3 pathway. *J. Virol.* 78 (14): 7737–7747.

84 Keller, B.C., Fredericksen, B.L., Samuel, M.A. et al. (2006). Resistance to alpha/beta interferon is a determinant of West Nile virus replication fitness and virulence. *J. Virol.* 80 (19): 9424–9434.

85 Puig-Basagoiti, F., Deas, T.S., Ren, P. et al. (2005). High-throughput assays using a luciferase-expressing replicon, virus-like particles, and full-length virus for West Nile virus drug discovery. *Antimicrob. Agents Chemother.* 49 (12): 4980–4988.

86 Bouffard, J.P., Riudavets, M.A., Holman, R., and Rushing, E.J. (2004). Neuropathology of the brain and spinal cord in human West Nile virus infection. *Clin. Neuropathol.* 23 (2): 59–61.

87 Shieh, W.J. (2000). The role of pathology in an investigation of an outbreak of West Nile encephalitis in New York, 1999. *Emerg. Infect. Dis.* 6 (4): 370–372.

88 Guarner, J., Shieh, W.J., Hunter, S. et al. (2004). Clinicopathologic study and laboratory diagnosis of 23 cases with West Nile virus encephalomyelitis. *Hum. Pathol.* 35 (8): 983–990.

89 Omalu, B.I., Shakir, A.A., Wang, G. et al. (2006). Fatal fulminant pan-meningo-polioencephalitis due to West Nile virus. *Brain Pathol.* 13 (4): 465–472.

90 Khairallah, M., Benyahia, S., Ladjimi, A. et al. (2004). Chorioretinal involvement in patients with West Nile virus infection☆. *Ophthalmology* 111 (11): 2065–2070.

91 Garg, S. and Jampol, L.M. (2005). Systemic and intraocular manifestations of West Nile virus infection. *Surv. Ophthalmol.* 50 (1): 3–13.

92 Hayes, E.B. and O'Leary, D.R. (2004). West Nile virus infection: a pediatric perspective. *Pediatrics* 113 (5): 1375–1381.

93 Pealer, L.N., Marfin, A.A., Petersen, L.R. et al. (2003). Transmission of West Nile virus through blood transfusion in the United States in 2002. *N. Engl. J. Med.* 349 (13): 1236–1245.

94 Smithburn, K.C., Hughes, T.P., Burke, A.W., and Paul, J.H. (1940). A neurotropic virus isolated from the blood of a native of Uganda. *Am. J. Trop. Med. Hyg.* s1-20: 471–492. https://www.cabdirect.org/cabdirect/abstract/19412700112.

95 Habarugira, G., Suen, W.W., Hobson-Peters, J. et al. (2020). West Nile virus: an update on pathobiology, epidemiology, diagnostics, control and "one health" implications. *Pathogens* 9 (7): 589.

96 Song, B.-H., Yun, G.-N., Kim, J.-K. et al. (2012). Biological and genetic properties of SA14-14-2, a live-attenuated Japanese encephalitis vaccine that is currently available for humans. *J. Microbiol.* 50: 698–706. https://link.springer.com/article/10.1007/s12275-012-2336-6.

97 Komar, N., Langevin, S., Hinten, S. et al. (2003). Experimental infection of North American birds with the New York 1999 strain of West Nile virus. *Emerg. Infect. Dis.* 9 (3): 311–322.

98 Ulbert, S. (2011). West Nile virus: the complex biology of an emerging pathogen. *Intervirology* 54 (4): 171–184.

99 Roehrig, J.T. (2013). West Nile virus in the United States – a historical perspective. *Viruses* 5 (12): 3088–3108.

100 Sejvar, J.J. (2014). Clinical manifestations and outcomes of West Nile virus infection. *Viruses* 6 (2): 606–623.

101 Mostashari, F., Bunning, M.L., Kitsutani, P.T. et al. (2001). Epidemic West Nile encephalitis, New York, 1999: results of a household-based seroepidemiological survey. *Lancet* 358 (9278): 261–264.

102 Petersen, L.R. and Marfin, A.A. (2002). West Nile virus: a primer for the clinician. *Ann. Intern. Med.* 137 (3): 173–179.

103 Tyler, K.L. (2004). West Nile virus infection in the United States. *Arch. Neurol.* 61 (8): 1190–1195.

104 Bains, H.S., Jampol, L.M., Caughron, M.C., and Parnell, J.R. (2003). Vitritis and chorioretinitis in a patient with West Nile virus infection. *Arch. Ophthalmol.* 121 (2): 205–207.

105 Smith, R.D., Konoplev, S., DeCourten-Myers, G., and Brown, T. (2004). West Nile virus encephalitis with myositis and orchitis. *Hum. Pathol.* 35 (2): 254–258.

106 Fratkin, J.D., Leis, A.A., Stokic, D.S. et al. (2004). Spinal cord neuropathology in human West Nile virus infection. *Arch. Pathol. Lab. Med.* 128 (5): 533–537.

107 Anninger, W.V., Lomeo, M.D., Dingle, J. et al. (2003). West Nile virus-associated optic neuritis and chorioretinitis. *Am J. Ophthalmol.* 136 (6): 1183–1185.

108 Gorsche, R. and Tilley, P. (2005). The rash of West Nile virus infection. *CMAJ* 172 (11): 1440.

109 Anderson, R.C., Horn, K.B., Hoang, M.P. et al. (2004). Punctate exanthem of West Nile virus infection: report of 3 cases. *J. Am. Acad. Dermatol.* 51 (5): 820–823.

110 Ferguson, D.D., Gershman, K., LeBailly, A., and Petersen, L.R. (2005). Characteristics of the rash associated with West Nile virus fever. *Clin. Infect. Dis.* 41 (8): 1204–1207.

111 Emig, M. and Apple, D.J. (2004). Severe West Nile virus disease in healthy adults. *Clin. Infect. Dis.* 38 (2): 289–292.

112 O'Leary, D.R., Marfin, A.A., Montgomery, S.P. et al. (2004). The epidemic of West Nile virus in the United States, 2002. *Vector Borne Zoonotic Dis.* 4 (1): 61–70.

113 Sejvar, J.J., Lindsey, N.P., and Campbell, G.L. (2011). Primary causes of death in reported cases of fatal West Nile fever, United States, 2002–2006. *Vector Borne Zoonotic Dis.* 11 (2): 161–164.

114 Bode, A.V., Sejvar, J.J., Pape, W.J. et al. (2006). West Nile virus disease: a descriptive study of 228 patients hospitalized in a 4-county region of Colorado in 2003. *Clin. Infect. Dis.* 42 (9): 1234–1240.

115 Sejvar, J.J., Curns, A.T., Welburg, L. et al. (2008). Neurocognitive and functional outcomes in persons recovering from West Nile virus illness. *J. Neuropsychol.* 2 (2): 477–499.

116 Sejvar, J.J., Haddad, M.B., Tierney, B.C. et al. (2003). Neurologic manifestations and outcome of West Nile virus infection. *JAMA* 290 (4): 511–515.

117 Ceauşu, E., Erşcoiu, S., Calistru, P. et al. (1997). Clinical manifestations in the West Nile virus outbreak. *Rom. J. Virol.* 48 (1–4): 3–11.

118 Granwehr, B.P., Lillibridge, K.M., Higgs, S. et al. (2004). West Nile virus: where are we now? *Lancet Infect. Dis.* 4 (9): 547–556. https://pubmed.ncbi.nlm.nih.gov/15336221.

119 Murray, K.O., Resnick, M., and Miller, V. (2007). Depression after infection with West Nile Virus1. *Emerg. Infect. Dis.* 13 (3): 479–481.

120 Nolan, M.S., Hause, A.M., and Murray, K.O. (2012). Findings of long-term depression up to 8 years post infection from West Nile virus. *J. Clin. Psychol.* 68 (7): 801–808.

121 Sejvar, J.J., Bode, A.V., Marfin, A.A. et al. (2005). West Nile virus–associated flaccid paralysis. *Emerg. Infect. Dis.* 11 (7): 1021–1027.

122 Fan, E., Needham, D.M., Brunton, J. et al. (2004). West Nile virus infection in the intensive care unit: a case series and literature review. *Can. Respir. J.* 11 (5): 354–358.

123 Park, M., Hui, J.S., and Bartt, R.E. (2003). Acute anterior radiculitis associated with West Nile virus infection. *J. Neurol. Neurosurg. Psychiatry* 74 (6): 823–825.

124 Ahmed, S., Libman, R., Wesson, K. et al. (2000). Guillain–Barré syndrome: an unusual presentation of West Nile virus infection. *Neurology* 55 (1): 144–146.

125 Barzon, L., Pacenti, M., Ulbert, S., and Palù, G. (2015). Latest developments and challenges in the diagnosis of human West Nile virus infection. *Expert Rev. Anti-Infect. Ther.* 13 (3): 327–342.

126 Blitvich, B.J., Bowen, R.A., Marlenee, N.L. et al. (2003). Epitope-blocking enzyme-linked immunosorbent assays for detection of west nile virus antibodies in domestic mammals. *J. Clin. Microbiol.* 41 (6): 2676–2679.

127 Kitai, Y., Kondo, T., and Konishi, E. (2011). Non-structural protein 1 (NS1) antibody-based assays to differentiate West Nile (WN) virus from Japanese encephalitis virus infections in horses: effects of WN virus NS1 antibodies induced by inactivated WN vaccine. *J. Virol. Methods* 171 (1): 123–128.

128 Busch, M.P., Kleinman, S.H., Tobler, L.H. et al. (2008). Virus and antibody dynamics in acute West Nile virus infection. *J. Infect. Dis.* 198 (7): 984–993.

129 Higbie, C.T., Nevarez, J.G., Roy, A.F., and Piero, F.D. (2017). Presence of West Nile virus RNA in tissues of American alligators (*Alligator mississippiensis*) vaccinated with a killed West Nile virus vaccine. *J. Herpetol. Med. Surg.* 27: 18–21. https://scholar.google.com/scholar_lookup?title=Presence+of+West+Nile+Virus+RNA+in+Tissues+of+American+Alligators+%28Alligator+mississippiensis%29+Vaccinated+with+a+Killed+West+Nile+Virus+Vaccine&author=Higbie%2C+Christine+T.&publication_year=2017.

130 Boonham, N., Kreuze, J., Winter, S. et al. (2014). Methods in virus diagnostics: from ELISA to next generation sequencing. *Virus Res.* 186: 20–31.

131 Sun, W. (2010). Nucleic extraction and amplification. In: *Molecular Diagnostics* (ed. W.W. Grody and F.L. Kiechle), 35–47. Elsevier.

132 Barzon, L., Pacenti, M., Franchin, E. et al. (2014). Isolation of West Nile virus from urine samples of patients with acute infection. *J. Clin. Microbiol.* 52 (9): 3411–3413.

133 Lustig, Y., Sofer, D., Bucris, E.D., and Mendelson, E. (2018). Surveillance and diagnosis of West Nile virus in the face of flavivirus cross-reactivity. *Front. Microbiol.* 9: 2421.

134 Ziermann, R. and Sánchez-Guerrero, S.A. (2008). PROCLEIX West Nile virus assay based on transcription-mediated amplification. *Expert. Rev. Mol. Diagn.* 8 (3): 239–245.

135 Zhang, W., Wu, J., Li, Y. et al. (2009). Rapid and accurate in vitro assays for detection of West Nile virus in blood and tissues. *Transfus. Med. Rev.* 23 (2): 146–154.

136 Barzon, L., Lavezzo, E., Costanzi, G. et al. (2013). Next-generation sequencing technologies in diagnostic virology. *J. Clin. Virol.* 58 (2): 346–350.

137 Myhrvold, C., Freije, C.A., Gootenberg, J.S. et al. (2018). Field-deployable viral diagnostics using CRISPR-Cas13. *Science* 360 (6387): 444–448.

138 Wolk, D.M., Kaleta, E.J., and Wysocki, V.H. (2012). PCR-electrospray ionization mass spectrometry: the potential to change infectious disease diagnostics in clinical and public health laboratories. *J. Mol. Diagn.* 14 (4): 295–304.

139 Rossi, S.L., Ross, T.M., and Evans, J.D. (2010). West Nile virus. *Clin. Lab. Med.* 30 (1): 47–65.

140 Furuta, Y., Gowen, B.B., Takahashi, K. et al. (2013). Favipiravir (T-705), a novel viral RNA polymerase inhibitor. *Antivir. Res.* 100 (2): 446–454.

141 Chu, J.J.H. and Ng, M.L. (2004). Infectious entry of West Nile virus occurs through a clathrin-mediated endocytic pathway. *J. Virol.* 78 (19): 10543–10555.

142 Kuritzkes, D.R. (2009). HIV-1 entry inhibitors: an overview. *Curr. Opin. HIV AIDS* 4 (2): 82–87.

143 Garg, H., Viard, M., Jacobs, A., and Blumenthal, R. (2011). Targeting HIV-1 gp41-induced fusion and pathogenesis for anti-viral therapy. *Curr. Top. Med. Chem.* 11 (24): 2947–2958.

144 Modis, Y., Ogata, S., Clements, D., and Harrison, S.C. (2003). A ligand-binding pocket in the dengue virus envelope glycoprotein. *Proc. Natl. Acad. Sci.* 100 (12): 6986–6991.

145 Bressanelli, S., Stiasny, K., Allison, S.L. et al. (2004). Structure of a flavivirus envelope glycoprotein in its low-pH-induced membrane fusion conformation. *EMBO J.* 23 (4): 728–738.

146 Wang, Q.Y., Patel, S.J., Vangrevelinghe, E. et al. (2009). A small-molecule dengue virus entry inhibitor. *Antimicrob. Agents Chemother.* 53 (5): 1823–1831.

147 Vervaeke, P., Alen, M., Noppen, S. et al. (2013). Sulfated *Escherichia coli* K5 polysaccharide derivatives inhibit dengue virus infection of human microvascular endothelial cells by interacting with the viral envelope protein E domain III. *PLoS One* 8 (8): e74035.

148 Lin, L.T., Chen, T.Y., Lin, S.C. et al. (2013). Broad-spectrum antiviral activity of chebulagic acid and punicalagin against viruses that use glycosaminoglycans for entry. *BMC Microbiol.* 13: 187.

149 Lee, E., Pavy, M., Young, N. et al. (2006). Antiviral effect of the heparan sulfate mimetic, PI-88, against dengue and encephalitic flaviviruses. *Antivir. Res.* 69 (1): 31–38.

150 Acharya, D. and Bai, F. (2016). An overview of current approaches toward the treatment and prevention of West Nile virus infection. *West Nile Virus* 1435: 249–291.

10

Hantavirus Disease: A Global Update over the Last Decade

Anita Patel[1], Nisarg Patel[2], and Jayvadan K. Patel[3,4]

[1] R & D Department, Samrajya Aromatics Pvt. Ltd., Gandhinagar, Gujarat, India
[2] Department of Pharmacognosy, APMC College of Pharmaceutical Education and Research, Himatnagar, Gujarat, India
[3] R & D Department, Aavis Pharmaceuticals, Hoschton, GA, United States
[4] Faculty of Pharmacy, Sankalchand Patel University, Visnagar, Gujarat, India

10.1 Introduction

Small animals serve as hosts for Hantaviruses which are a member of the Bunyaviridae family. A virus can infect humans when they inhale contaminated aerosols or when they touch animal waste. Hantaviruses have the ability to cause two separate diseases in people: hantavirus pulmonary syndrome (HPS) and hemorrhagic fever with renal syndrome (HFRS.) Infections previously acknowledged as Korean hemorrhagic fever (KHF), epidemic hemorrhagic fever (EHF), and nephropathia epidemica (NE) are now collectively referred to as HFRS. These ailments are common across the Eurasian landmass and surrounding regions. Given that these syndromes share a common pathogenesis and clinical features, the word "Hantavirus disease/fever" may be used inclusively [1].

When 3200 cases of HFRS were reported among American troops serving in the Republic of Korea between 1951 and 1954, the disease caught the attention of the entire world [2]. The first Hantavirus type, known as Thottapalayam virus (TPMV), was found in the spleen of the insectivore *Suncus murinus* in Vellore, India, in 1964 [3]. After being isolated in 1978 from an infected rodent, it was given the moniker Hantaan virus (HTNV) in honor of the Han river basin where the rodent was captured, the Asian Hantavirus prototype [4]. In 1985, Schmaljohn, a pioneer in Hantavirus research, coined the name "Hantavirus." In the Four Corners region of the southwest United States, there was an outbreak of cardiac sickness in young, healthy people in 1993 (Mexico, Utah, Arizona, and Colorado). About 40% of the people died, and Sin Nombre virus (SNV) was the new pathogenic Hantavirus that caused it [5]. Hantaviruses, which were initially believed to trigger renal disease in Europe and Asia, altered their course as a result of this epidemic. In Peru, at the beginning of the twentieth century, human-to-human transmission of HCPS was allegedly caused by an isolate of the South American serotype, Andes virus (ANDV), from human serum. It was extracted from a patient's serum. A Hantavirus outbreak among tourists on their way to Yosemite National Park in California, USA, happened in 2012 [6]. Ten such instances were found after investigations. Eight patients were diagnosed with HCPS; five needed intensive care and ventilator support, and three passed away. Hantavirus infection was significantly correlated with staying the night in a "signature tent cabin" (nine case-patients; $P < 0.001$).

As of 2017, there were 41 known species of Hantaviruses; however, it is still debatable which viruses should be classified as separate species. Moreover, additional viruses have been discovered but have not yet been named.

10.2 Life Cycle

The family Bunyaviridae, which has over 300 viruses and affects people, plants, animals, and arthropods, contains the genus Hantavirus. As a result of the widespread distribution of their rodent reservoirs and the disease (HFRS or HPS) that results from the transmission to humans, Hantaviruses are commonly referred to as Old World and New World

Rising Contagious Diseases: Basics, Management, and Treatments, First Edition. Edited by Seth Kwabena Amponsah, Ranjita Shegokar, and Yashwant V. Pathak.

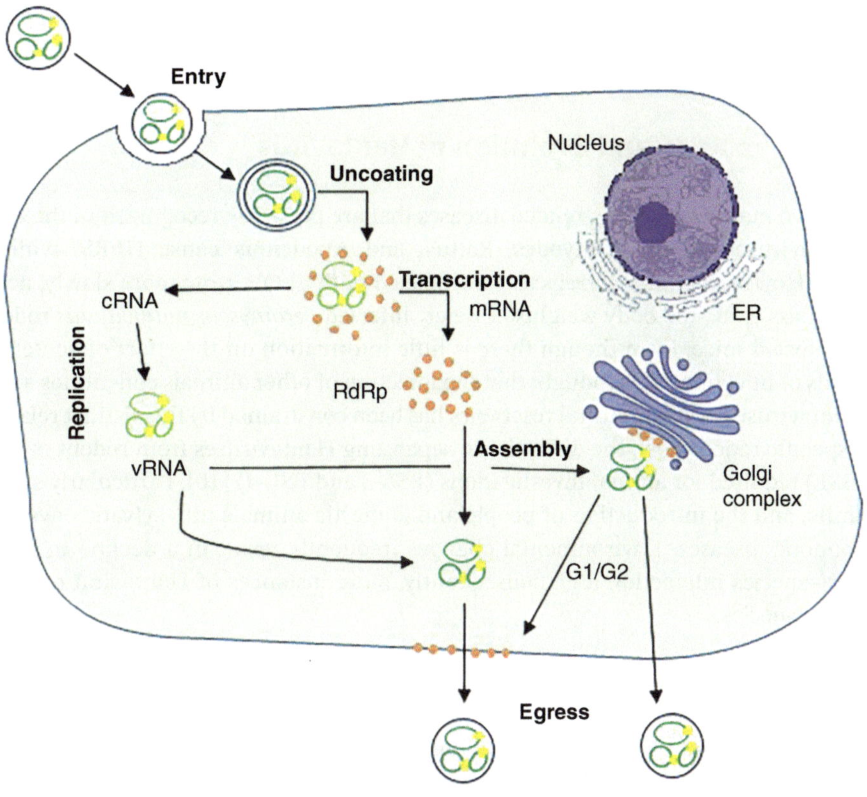

Figure 10.1 Life cycle of Hantavirus.

Hantaviruses [7]. The Old World and New World Hantaviruses show similar aspects of their life cycles (Figure 10.1) and exhibit high homology in the organization of their nucleic sequences, despite differences in geographical locations and diseases.

10.2.1 Genome Organization and Virion Structure

Hantavirus virions are spherical in form and range in size from 80 to 120 nm. Three single-stranded negative-sense RNA strands make up the Hantavirus genome. These three segments are designated as small (S), middle (M), and large (L) based on their sizes. The viral polymerase is encoded by the L segment, and the nucleocapsid (N) protein and two viral surface glycoproteins (G1 and G2) are encoded by the M and S segments, respectively. The quantity of genomic RNA in each virion is typically equimolar, and each segment of viral RNA is attached to a single molecule of the viral RNA-dependent RNA polymerase (RdRp). All viral RNA segments have N protein molecules on them, which come together to create ribonucleoproteins. G1 and G2 surface glycoproteins are anchored in the lipid bilayer that serves as their envelope [8].

Endocytosis occurs after the Hantavirus virion gets attached to the recipient cell via cellular receptors. Following pH-mediated membrane fusion, ribonucleic proteins (RNPs) are liberated from the late endosome and transported to the cytoplasm. Either at the location where RNPs are released or in the intermediary compartment between the endoplasmic reticulum and the Golgi, transcription and translation occur (ERGIC). In the latter scenario, the RNPs are delivered to ERGIC. RdRp, the viral polymerase, performs both virus transcription and reproduction, thanks to its endonuclease, replicase, and transcriptase activities. RdRp cleaves cellular mRNA to create bound primers, which begin the transcription process. Recently, it has also been proposed that cellular endonucleases contribute to the creation of capped primers by cleaving cellular mRNAs whose cap-structures are preserved from degradation by the specifically bound viral N protein. The viral mRNAs' transcription is started by these closed primers. The N protein is made using mRNA from the S section as a template, and it also results in the production of a nonstructural NSs protein for some specific Hantaviruses. Glycoprotein precursor (GPC) is produced on the ER membrane-bound ribosomes by mRNA taken from the M segment. The assembly process begins with the transfer of G1 and G2 glycoproteins from the ER to the Golgi complex or the plasma membrane. While New World Hantaviruses gather at the plasma membrane, Old World Hantaviruses congregate at the Golgi. Spike-like projections made by the tetramers of the viral surface glycoproteins are present in the newly formed envelope after assembly and appear to be crucial for both virus assembly and cell entrance. Through exocytosis, newly assembled virions are expelled [8].

10.3 Ecology and Evolution of Hantavirus

The two main Hantavirus-related diseases that are presently recognized in the Old World are HFRS and HPS (Table 10.1). Hantaviruses found in Myodes, Rattus, and Apodemus cause HFRS, while those found in sigmodontines cause HPS. However, *Rattus norvegicus* rats infected with SEOV grow more slowly, according to Childs et al. [9]. According to Douglass et al., the body weights of newly infected *Peromyscus maniculatus* rodents developed more slowly than those of uninfected mice. Even though there is little information on the effect of acute infection and the natural history of these kinds of infections, it is thought that the infection of other animals constitutes a spillover infection. Laboratory research on Hantaviruses in their natural reservoirs has been constrained by the distinct relationships between each viral genotype and a specific rodent host, the difficulty of separating Hantaviruses from rodent or human cases, as well as the biosafety level (BSL) required for animal investigations (BSL-3 and BSL-4) [10]. Particularly susceptible to ecological shifts, demographic shifts, and the introduction of people and domestic animals into sylvatic environments are the emergence and spread of zoonotic diseases. Environmental changes frequently result in a decline in rodent variety, which may promote greater host–species interaction and, consequently, more instances of Hantaviral transmission (aggressive encounters) within a single species.

10.3.1 Rodent Reservoirs of Old World Hantaviruses

These reservoirs include the following (i) HTNV and related *Apodemus* mouse-borne viruses in Asia (Amur virus [AMRV] and Soochong virus [SOOV]) and Europe (DOBV and Saaremaa virus [SAAV]), (ii) SEOV and related rat-borne viruses in Asia (Gou virus [GOUV] and Serang virus [SERV]) and bandicoot rat (genus *Bandicota*)-borne viruses (Thailand virus [THAIV]), and (iii) *Myodes* voleborne viruses, primarily Puumala virus (PUUV) in Europe. Tula virus (TULV),

Table 10.1 Old and new world Hantavirus strains' geographic distribution.

Group and subfamily	Rodent carrier	Virus strain	Abbreviation	Geographical distribution
Old world				
Murinae	*Apodemus agrarius*	Hantaan virus	HTNV	China, South Korea, Russia
	Apodemus flavicollis	Dobrava-Belgrade virus	DOBV	Balkans
	Rattus	Seoul virus	SEOV	Worldwide
	Apodemus peninsulae	Amur virus	AMRV	Far East Russia
Arvicolinae	*Clethrionomys glareolus*	Puumala virus	PUUV	Europe, Asia, and Americas
	Microtus fortis	Khabarovsk virus	KHAV	Far East Russia
	Myodes regulus	Muju virus	MUJV	South Korea
	Microtus arvalis	Tula virus	TULV	Russia/Europe
New world				
Sigmodontinae	*Peromyscus maniculatus*	Sin Nombre virus	SNV	North America
	Oligoryzomys longicaudatus	Andes virus	ANDV	Argentina, Chile
	Bolomys obscurus	Maciel virus	MCLV	Argentina
	Oligoryzomys longicaudatus	Oran virus	ORNV	Argentina
	Peromyscus maniculatus	Sin Nombre virus	SNV	North America
	Oligoryzomys nigripes	Juquitiba	JUQV	Brazil
	Peromyscus leucopus	New York virus	NYV	North America

Source: Adapted from Jonsson et al. [8].

Khabarovsk virus [KHAV], and Vladivostok virus – all capable of infecting humans – have not been definitively linked to human illnesses.

Apodemus mice, such as *Apodemus flavicollis*, are found throughout most of Europe, with the exception of northern Scandinavia, the British Isles, and western Europe close to the Atlantic Ocean. DOBV has been identified in the yellow-necked forest mouse (*A. flavicollis*) [11], which is linked to serious HFRS and high mortality rates in Europe. This virus, also known as the Belgrade virus and recovered from a human patient, shared the same genetic makeup as the DOBV isolate [12]. Later, a genetically similar virus known as Saaremaa virus (SAAV) was discovered in *Apodemus agrarius*, a subspecies of the same species but different from *A. Agrarius koreae*. The SAAV or DOBV-Aa lineage carrier in Europe, *A. agrarius*, is found in areas east of the former governmental "Iron Curtain." There appear to be several lineages in *A. apodemus* at the moment, and, additionally, a novel pathogenic lineage or virus has recently been detected in *Apodemus ponticus*, despite the viruses carried by *A. flavicollis* (DOBV, or DOBV-Af) and *A. apodemus* (SAAV, or DOBV-Aa) being able to be separated phylogenetically [13]. As first discovered in 1976, the *Apodemus Agrarius koreae* (Korean field mouse) is the vector of HTNV in the Far East and China. AMRV or SOOV, a similar genetic variant of HTNV, is present in *Apodemus peninsulae* [14]. In terms of the viruses transmitted by rats, *Rattus norvegicus* primarily harbors SEOV in the same geographic area [15]. *Rattus tanezumi*, commonly known as the Asian house rat, is the carrier of SERV and is found throughout Southeast Asia [16]. THAIV is also present in the lesser bandicoot rat (*Bandicota savilei*) and greater bandicoot rat (*Bandicota indica*), whose cross-neutralization tests have shown to be distinct species to SEOV [17]. The bank vole, *Myodes glareolus*, which is also the carrier of PUUV, is the most extensively dispersed rodent reservoir for Hantaviruses in Europe. Numerous nations in temperate Central Europe, the Balkans, and Scandinavia experience increases in rodent numbers, but these increases are attributed to various underlying causes. Rodent peaks are caused by mast years in northern Europe. A mast year is one in which many forest tree seeds gather on the ground, providing plenty of nutrients for forest rodents. High summer temperatures two-years prior, when flower buds form, and high autumn temperatures one-year prior, when the actual seed develops, correlate with and can be used to forecast high rodent density (and the associated epidemics) [18].

Increased surveillance as well as climatic variables including warmer summers triggering mast years are partly to blame for the rise in HFRS cases in both central and northern Europe. Additionally, the largest PUUV outbreak ever documented in northern Europe during the winter of 2006–2007 (incidence of 313 cases/100,000 people in 2007 in Vasterbotten, Sweden) may have been influenced by climatic variables [18]. Studies in China have demonstrated that a higher incidence of *A. agrarius*-borne HFRS cases was predicted by exceptional seed crops, little rain or no floods, and an abundance of field rodents, a result of the first two factors [19]. When closed, nonaerated, abandoned buildings (like barns and cabins) with access to rodents are cleaned up and reopened, contamination happens. Hantaviruses are most likely aerosol transmitted. At room temperature and below, PUUV is contagious for at least two weeks in bank vole excrement, and the absence of UV light indoors appears to help the virus survive [20]. Instead of using traps, poison can be used to control rodent populations inside the house; but this method increases exposure to possibly infectious excreta.

Hantavirus presence has been positively correlated with anthropogenic land cover disturbance in a number of ecosystems, according to a recent study. Infected rodents can spread the virus horizontally to uninfected rodents in the same cage or through contaminated bedding, as shown by the viral transmission between rodents of the same species [21]. Old World Hantavirus HTNV spreads through the *Apodemus* host's saliva, urine, and excrement, but it survives the longest in the urine.

10.3.2 Rodent Reservoirs of New World Hantaviruses

Over 30 new Hantaviral strains or genetic lineages from various rodent species throughout the American continent have been discovered since the discovery of HPS in the southwestern United States in 1993 [5] and the characterization of its pathogenic agent, SNV, from deer mice [22]. In the United States and Canada, Hantaviruses that cause HPS include Monongahela virus (MNGV), harbored by *Peromyscus maniculatus* [23]; New York virus (NYV), harbored by *Peromyscus leucopus* [24]; Bayou virus (BAYV), harbored by *Oryzomys palustris* [25]; and BCCV [26], harbored by *Sigmodon hispidus*. In Latin America, the primary HPS-causing Hantaviruses include Andes virus (ANDV), harbored by *Oligoryzomys longicaudatus* in Argentina [27] and Chile [28]; Araraquara virus (ARAV), harbored by *Necromys lasiurus* in Brazil [29]; Leguna Negra virus (LANV), harbored by *Calomys laucha* in Paraguay [30]; and Choclo virus, harbored by *Oligoryzomys fulvescens* in Panama [31]. Rio Mamore's virus has been identified in the Bolvian pygmy rice rodent *Oligoryzomys microtis* [32] but has not yet been connected to HPS. Rio Mamore's virus variant HTN-007 was discovered in Peru and recovered from the pygmy rice rat [33].

Oligoryzomys longicaudatus and *Oligoryzomys flavescens* are the main rodent hosts for the strains of ANDV, Ora'n virus (ORNV), Lechigunas virus (LECV), and Hu39694 virus, respectively, in Argentina. In Chile, *O. longicaudatus*, one of the most prevalent rat species, also serves as an ANDV reservoir [28]. According to surveillance studies, multiple rodent species in Argentina have a notable variety of genotypes. In Argentina, Maciel virus (MACV) and Bermejo virus (BMJV) are two new genotypes that have been identified as causing HPS. *Necromys benefactus* from central Argentina was used to identify MACV, whereas one *Oligoryzomys chacoensis* rodent that was caught close to Ora'n was used to identify the BMJV genotype [30]. The central Argentina-based *Akodon azarae*, which carries the Pergamino virus (PRGV), has not yet been connected to HPS [33]. The Araucaria virus and Juquitiba virus (JUQV) are carried by *Oligoryzomys nigripes* – a rodent that lives in the southern temperate Arauca'ria pine forests along with the Atlantic rain forest in Brazil. The rodent *Necromys lasiurus*, which carries ARAV, is found in the cerrado areas of the south-eastern and central plateaus, which resemble savannas. The natural habitats of *Oligoryzomys nigripes* are found on the edges of cultivated regions [34]. The Midwestern region's Calomys species inhabit an environment of cerrado and Amazon jungle, where they harbor a virus similar to LANV. The rodent *Oligoryzomys moojeni*, which resides in a large area of the northern Amazon that has been deforested, is most likely the source for the Castelo dos Sonhos virus. In Brazil's southeast cerrado, rodent reserves include the species *Necromys lasiurus*, *Akodon* species, and *Calomys tener*. In addition to Bolivia and Paraguay, the pampas plains of Argentina, Uruguay, and southern Brazil are home to akodon rodents. *Calomys laucha* was found to be the rodent reservoir for LANV in Paraguay, where it caused the country's first recorded HPS outbreak in the western area [35].

Moreover, Paraguay is home to at least four other Hantaviruses: Alto Paraguay virus (ALPAV), harbored by *Holochilus chacarius* (Chacoan marsh rat); Ape Aime virus (AAIV), harbored by *Akodon montensis* (montane akodont); an Itapu'a37/JUQV-like virus, harbored by *Oligoryzomys nigripes* (black-footed colilargo); and BMJV-N˜ eembucu (BMJV-N˜EBU), harbored by *Oligoryzomys chacoensis* (Chacoan colilargo). The yellow pygmy rice rat (*O. flavescens*), a nearby Hantavirus linked to HPS in Uruguay's southern area, may be the host for the Andes Central Plata virus [36]. Rodent species found in the Mexican states of Oaxaca and Zacatecas, *Peromyscus maniculatus, Peromyscus melanotis, Peromyscus megalotis, and Peromyscus hylocetes,* have been found to have Hantaviral antibodies or RNA [37]. *Oryzomys couesi* [38], a novel genotype of Hantavirus, was recently discovered in Colimas, Mexico. In Honduras, *Oryzomys couesi* also carries the Catacamas virus (CATV) [39]. The rodent genus *Oryzomys* is a large, diverse, and broadly distributed group that is primarily located in the neotropics [40], with the exception of one species (*O. palustris*) that relocates northward into the United States [41].

HPS outbreaks have been linked to weather and climatic events, notably precipitation, in North America, especially in the southwest [42]. The microhabitat of its source plays a major role in Hantavirus ecology [43]. Additionally, it has been demonstrated that the interactions within the rodent population are a significant contributor to viral ecology.

10.3.3 Evolution of Hantaviruses

Hantaviruses have a lengthy history of coevolution with the main rodent carriers, according to phylogenetic analyses of Hantaviruses and their rodent hosts (Figure 10.2). The recent research by Ramsden et al. suggests that since Hantaviruses are about 900 years old and have temporary evolutionary rates similar to those seen in rapidly evolving RNA viruses, Hantavirus dispersion and outbreak theories need to be reevaluated in light of new information about their major rodent hosts [44].

Both during the development of the Hantavirus genus and the development of the insectivore-borne Hantaviruses, it is evident that there have been species' jumps (viruses switching to a new host) [45]. In the Old World, Hantavirus recombination happens spontaneously between lineages, within lineages, and even between different viruses. Within SNVs among *Peromyscus maniculatus* rodents [46], in cell culture experiments [47], as well as within the DOBVs between *A. agrarius* and *A. flavicollis*, two different rodent host species [48], natural and laboratory genetic reassortment of genomic RNA segments have been described. A genetic reassortment with the L, S, and M segments of ANDV and the L, S, and M segments of SNV were isolated in an *in-vitro* investigation with mixed infection of ANDV and SNV [47]. Hantavirus phylogenetic studies in Paraguay indicated that *Akodon montensis* had previously undergone natural reassortments [49].

PUUV exhibits significant regional variation. However, the Muju virus from the South Korean royal vole, *Myodes regulus*, is close to the PUUV phylogenetic group and most likely causes HFRS [50]. The Hokkaido virus (or lineage) in *Myodes rufocamus* is then mentioned. This species appears to transmit a unique virus in Japan, China, and Russian Siberia. The majority of PUUVs are topographical variants, with genetic clusters developing among the PUUV genotypes in each nation or region. Overall, the nucleotide level of genetic diversity (at the level of individual rodents, populations, PUUV lineages, and up to 17% between lineages) ranged from 0.2% to 1.2% in rodents to 1% to 2% in populations [51].

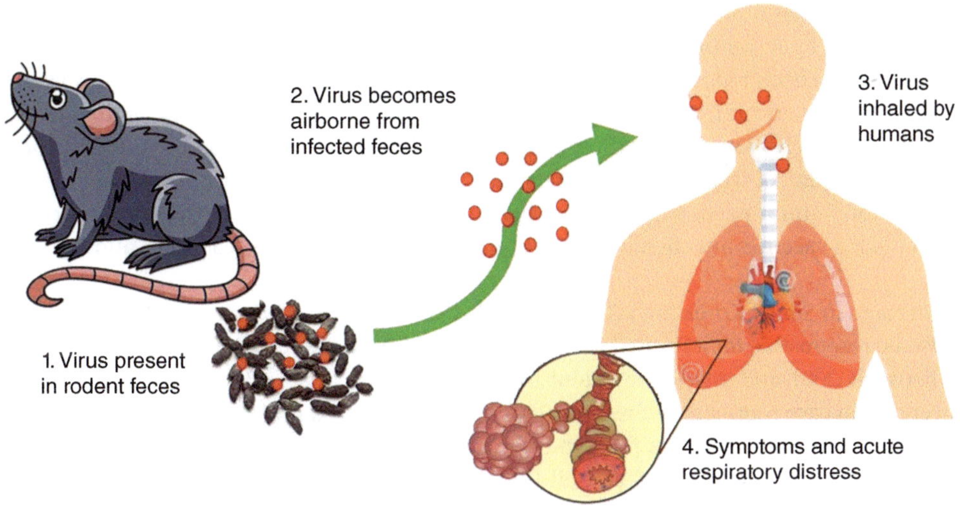

Figure 10.2 Hantavirus infection in humans.

10.4 Epidemiology of Hantavirus Infections

The prevalence of human peridomestic exposure to rodents in endemic regions is a major factor in the epidemiology of Hantavirus infections in human populations. Most of the time, people become infected after coming into contact with infected rodents or their excrement, usually by breathing in virus-contaminated aerosol. The bulk of Hantaviruses in the United States have not been associated with person-to-person transmission [52]; however, cases of ANDV person-to-person transmission have been documented, albeit infrequently, in Argentina and Chile. Compared to other Hantaviruses, ANDV is less susceptible in-vitro to the antiviral drugs found in saliva [53].

10.4.1 Epidemiology of Old World Hantaviruses

At first, only rural regions of Eurasia, particularly eastern Russia, northern Europe, South Korea, and China, were thought to be affected by HFRS [54]. Prudent surveillance showed that SEOV-caused HFRS could arise in many different locations and in urbanized areas.

With 35,424 cases reported by the end of 2006 [55], PUUV is the primary cause of HFRS in Europe; 95% of the cases were documented after 1990. Generally speaking, PUUV infections happen all over the continent within the *Myodes glareolus* habitat area. Although the host species is found throughout a much wider region, only the Balkan region has so far been associated with DOBV (derived from *A. flavicollis*) cases [56]. In Estonia, Russia, Denmark, Slovenia, and Slovakia [57, 58], SAAV (and/or DOBV-Aa) carried by *A. agrarius* has been found in field mice. It has also been linked to human disease outbreaks in Russia [59], as well as isolated instances in Germany [60] and Slovakia [58].

When rodent's population is high (and humans are exposed to more infected rodents in rural areas), epidemics in northern Europe usually reach their peak in November to December, with another peak occurring in August. Most of the cases in which PUUV infections take place in Baden-Wu¨rttemberg happen over the spring and summer months [61]. Research illustrates seroprevalences of 1.7% (SAAV moving eastward) [62] and 6.7% (PUUV seroprevalence in another prevalent area of Lower Bavaria) [63]. Northeastern France and Belgium's Ardennes area are two additional endemic regions.

The number of cases peaked in the summer during outbreaks that took place in the Balkans in 2002 and 2005, as well as in 1995 during the Bosnia–Herzegovina conflict, during which 354 cases were diagnosed. Between 1996 and 2006, 89,162 instances of HFRS were reported in Russia. From 1997 to 2007, the incidence averaged 5.8 cases per 100,000 people, with the peak incidence occurring in 1997 when there were 14.3 cases per 100,000 people. *A. ponticus* in the southern part of Sochi was recently found to contain another DOBV clade, which was connected to more commonly occurring severe HFRS cases (27% and 55%, correspondingly, in Lipetsk and Sochi) [13]. Fewer instances of HFRS are reported in Siberia, whereas rodents and cases of HFRS are reported more frequently in Russia's Far East than in South Korea and China, where severe HFRS outbreaks are brought on by *A. peninsulae*-carried viruses (AMRV/SOOV) [64].

Around 12,000 to 20,000 cases of HFRS have been reported to the China Center for Disease Control in the last few years, with a mortality incidence of about 1%. HTNV and HTNV-like viruses (Amur/Soochong virus) carried by apodemus, as well as rat-borne SEOV, are some of the viruses responsible for these occurrences. Living in a house on the outskirts of a town, camping, or living in huts in the fields, as well as owning cats, are risk factors, according to epidemiological studies conducted in China [65]. Traditional immunizations that have been formalin-inactivated are frequently used in China [66]. According to reports, inactivated vaccines are effective, elicit positive humoral immune reactions, and are well tolerated [67].

The most common causes of HFRS cases in South Korea are HTNV (AMRV/SOOV) and Muju virus (MUJUV), which is carried by *M. regulus*, which may also be responsible for a limited number of cases [50]. In the Republic of Korea, 323–450 HFRS cases were reported yearly between 2001 and 2008. The extent to which Hantavax use, which was prescribed to millions of individuals, caused a decrease in case numbers in the late 1990s has been debated [68].

Given that rodents and humans in Thailand [69], Indonesia [16], and India have been found to have THAIV antibodies, it is possible that other Asian countries will experience HFRS as well [70]. Since 1984, neither the SEOV, which is also prevalent in the region's rodents, nor the PUUV-like Hokkaido virus in *Myodes rufocanus* have been linked to disease in humans in Japan [71]. With a high fatality rate, the disease mostly affects males between the ages of 15 and 45.

10.4.2 Epidemiology of New World Hantaviruses

After getting laboratory diagnostic evidence for the HPS outbreak, Public Health Service researchers found the SNV hantavirus and its reservoir, *P. maniculatus*, about two weeks later. The New Mexico Office of the Medical Investigator and the Indian Health Service were the first to identify the HPS outbreak. In order to identify the broad sporadic geographic distribution of HPS cases across the United States, the Center for Disease Control and Prevention (CDC) established an emergency phone hotline [72]. With a mean patient age of 38 years old, the CDC received reports of 465 HPS cases in the United States since the outbreak began through March 2007. Of these, 35% (64% of the male cases and 37% of the female cases) ended in fatalities (range, 10–83 years). Although places outside the Four Corners region account for more than half of the confirmed cases, New Mexico encountered the most reports of cases [73].

Hantavirus risk has been linked to workplace risks where workers may come into contact with rodents or their droppings. Entering closed structures that are rodent-infested carry the highest danger [74]. SNV does not seem to be passed on to newborns through breast milk, does not appear to be transferred from mother to child, and does not appear to cause any unusual pathology in pregnant women [75]. There have been reports of HTNV's vertical spread [54]. Paraguay saw the first concentration of HPS cases outside the US from July 1995 to January 1996 [76]. The eastern department of Itapúa verified the first case in 2005. Although Hantaviruses that cause HPS have not yet been discovered, rodent surveillance data indicate that ANDV-like viruses are common in this area [48].

Between September and December 1996, a ski lodge in the southern Andes city of El Bolson experienced the country's first documented HPS outbreak. The Argentine Health Ministry recorded 324 cases of HPS between 1996 and 2001. AndV, Hu39694, and LECV viruses are endemic in the Santa Fe' Provinces and central Buenos Aires; pathogenic ANDV lineages are circulating in the southwestern provinces of Rio Negro, Chubut, and Neuquen; and ORNV that causes HPS is endemic in the northwest Salta Province.

A cluster of 25 HPS cases with two family clusters that happened between July 1997 and January 1998 in the Coyhaique Health District of Chile's Aysen region was identified soon after the HPS outbreak in Argentina [77]. The first human isolate of ANDV was found in the blood of a 10-year-old Chilean boy who had no symptoms and passed away 6 days later [78]. According to 485 cases and a 37% mortality rate recorded in various regions of Chile through 2006, the disease is likely to spread throughout the rodent reservoir's geographic distribution [79].

There were 12 cases and a 25% case fatality rate during the first HPS outbreak in Panama, which happened in Los Santos between 1999 and 2000 [31]. The three fatal instances, however, were not supported by serology. This virus was identified as the novel Hantavirus Choclo virus through sequence analysis of the virus genome.

The first three HPS cases in Uruguay were discovered in the Brazilian frontier region in 2004 (Ministry of Health of Uruguay, 2004). HPS is connected to LECV and Andes Central Plata virus – two closely linked Hantaviruses [36]. *Oligoryzomys flavescens* is the carrier of both of these viruses. The first HPS instances in Brazil were first noted in the southeast in 1993, and this small outbreak resulted in 66% of the patients dying [80].

10.5 The Clinical Course of Hantavirus Diseases and Pathology

10.5.1 European and Asian HFRS Disease Spectrum

In general, HFRS brought on by HTNV, the Amur virus, and DOBV are more serious, whereas SEOV and PUUV are less severe. Despite the apparent lack of data on patients infected with viruses derived from *A. agrarius*, the disease associated with *A. agrarius* in Europe appears to be less severe.

The five classic stages of HFRS are febrile, hypotensive, oliguric, polyuric, and convalescent. Around 11–40% of people who have a febrile sickness experience hypotension, and between 40% and 60% experience oliguria. The hypotensive stage, which is characterized by thirst, restlessness, nausea, and vomiting that lasts for hours or days at a time, ultimately augments the clinical symptoms of the febrile stage. Approximately one-third of all patients experience shock and confusion during the hypotensive episode of HFRS [81]. During this phase, tachycardia, stomach discomfort, and vascular leakage symptoms are seen. A one-third of all patients experience conjunctival, brain, and gastrointestinal (GI) hemorrhages [81]. About half of all Hantavirus-related fatalities occur during the oliguric stage. At this stage, patients are at the risk of developing high blood pressure, edema of the lungs, and kidney failure issues. About 40% of the HTNV patients and 20% of the SEOV patients need dialysis. Death is typically brought on by shock, hemorrhage, or problems from renal insufficiency.

Less than 0.1% of patients with PUUV infection die, whereas HFRS patients with HTNV infection have mortality rates as high as 15%. A less than 1% mortality incidence is linked to SEOV.

For PUUV patients, the illness typically develops suddenly and is accompanied by fever, headache, and abdominal discomfort, occasionally with vomiting or diarrhea [82]. There may be symptoms of central nervous system (CNS) involvement, such as drowsiness and somnolence. PUUV RNA has been discovered in cerebral fluid, despite the rarity of encephalitis [83]. In extreme circumstances, syncope or hypotension could occur [82]. On days 3–4, back pain, oliguria, anuria, and other renal involvement symptoms (such as an increased serum creatinine level and proteinuria) start to appear, and 5% of the hospitalized PUUV patients may need dialysis. A second week later, polyuria appears [82]. Patients in hospitals may develop pulmonary infiltrates or swelling that has a faint HPS-like appearance. According to a recent research, 57% of patients have acute electrocardiogram (ECG) changes [84]. The HLA B8-DR3 allele is connected to severe anuria and shock [85]. Hypophyseal bleeding, which leads to hypopituitarism, and mesangiocapillary glomerulonephritis are complications [86].

The clinical picture of DOBV resembles that of PUUV described above. Yet it is dangerous because abdominal and pleural effusions are seen, along with increased oliguric kidney injury (30–47%), bleeding issues, shock (21–28%) and other symptoms. DOBV mimics viruses like HTNV. About 20–30% of HTNV patients go through a severe course that includes shock, significant hemorrhages, electrolyte imbalance, renal failure, or pulmonary edema, while 30–40% of cases are deemed intermediate [22, 87].

Once more, the signs associated with SEOV infections are less severe. It is crucially important to avoid dangerous overhydration (for patients who are anuric and have leaky capillaries) when treating severe HFRS in order to maintain fluid balance and circulatory volume. Patients who have kidney insufficiency may require dialysis.

10.5.2 Spectrum of HPS Disease in the Americas

The serious, acute-illness HPS is linked to a sudden onset of cardiogenic shock and respiratory failure [22]. Similar to HFRS, HPS causes capillary leaks, but it affects the lungs instead of the kidneys [88, 89].

Because HPS causes a leaking syndrome in the lung capillaries 1–3 days after the onset of respiratory symptoms, which progresses to respiratory failure and cardiogenic shock for the patient, the result is typically severe. For instances of HPS brought on by ARAV, BAYV, SNV, and BCCV in the American continent [90, 91], renal failure symptoms and myositis have also been documented [91]. Generally, an incubation phase of 9–33 days precedes the onset of HPS symptoms (mean, 14–17 days).

After the incubation period, patients experience a febrile prodrome that includes temperature, myalgia, and thrombocytopenia that gets worse over time. Other symptoms include diarrhea, abdominal pain, back pain, and headache. Soon after the prodrome, IgM and IgG antibodies start showing up. According to data gathered from 70 Brazilian HPS cases, 79% of the cases begin with temperature, 60% with myalgia, and 21% with weakness. Additionally, 25% of the patients experience gastrointestinal complaints like diarrhea, vomiting, and nausea. Typically, these prodrome symptoms disappear after five days. After the second day of the sickness, patients start to experience breathlessness (87%) and cough (44%), then tachycardia (81%) and low arterial blood pressure levels (56%). Total 44% of patients experienced a drop in blood pressure, and 33% experienced shock 4–7 days after the start of the illness. In 51% of the cases, renal failure as indicated by elevated creatinine levels took place. Between days 3 and 5 after the start of the sickness, mild hemorrhagic disturbances like hematuria, hematemesis, intestinal bleeding, and metrorrhagia were noticed in 50% of the patients. Aspartate transaminase (AST) and alanine aminotransferase (ALT) blood levels were found to be elevated in 73% and 83% of the cases, respectively [92]. The majority of HPS patients were hospitalized between days 3 and 6 following the start of symptoms.

10.5.3 Pathogenesis

With the existence of the Hantavirus that causes infection and a strong immune activation that alters vascular permeability, HPS has a complicated pathogenesis.

Lung Hantaviral infection is brought on by target endothelium cells, macrophages, and platelets that have β3 integrin receptors on their cell membranes. These cells permit viral replication, which activates the immune system [93]. The development of respiratory failure and severe HPS may be influenced by immune activation, particularly, caused by macrophages and CD8 T cells [94]. Activated macrophages produce proinflammatory mediators such as IL-6, interleukin-1 (IL-1), and tumor necrosis factor alpha (TNF-α). The overproduction of Hantavirus-specific T cells and cytokines by macrophages in reaction to antigen recognition on infected pulmonary endothelial cells is likely to have an impact on the pathogenesis of HPS [94]. After recognizing an antigen, T cells can differentiate into at least two distinct types of helper cells called CD4 T helper 1 (Th1) and Th2 cells. TNF-β (also known as lymphotoxin-β) and Gamma Interferon (IFN-γ), which are responsible for cell-mediated immunity, are produced by Th1 cells, and IL-12 controls this differentiation. IL-4 and IL-5 are produced by Th2 cells, which support humoral and allergy reactions. On the other hand, inducible regulatory T cells play a significant role in controlling the immune response and infection-induced pathology by producing the immunosuppressive cytokines IL-10 and transforming growth factor β (TGF-β) [95].

In accordance with the strong humoral response seen in HPS patients with SNV [19], which contains Th2 antibody classes like IgA, IgG1, and IgG3, high levels of IL-6 and IL-5 in HPS patients point to some Th2 immune response activation in HPS. These findings imply that HPS may exhibit a mixed pattern of Th1 and Th2 reactions. Patients with HFRS showed a mix of Th1 and Th2 T-cell patterns but with preferential Th1 differentiation when their peripheral blood mononuclear cells (PBMC) were stained with IFN- and IL-4 intracellularly [96].

It is conceivable that Hantavirus infection inhibits the production of TGF-β. TGF-β-mediated immunosuppression is believed to be largely dependent on the downregulation of T cell function through the inhibition of their proliferation [97]. In contrast to other viral illnesses, HPS patients have higher levels of Hantavirus-specific CD8 T cells in their blood [94]. Although the underlying mechanism of this occurrence is unclear, it is tempting to make the assumption that the reduction of TGF-β synthesis caused by the Hantavirus affects the development of regulatory T cells in the early stages of HPS. A large number of IL-10-producing cells were observed in the hearts of fatal HPS patients during the acute stage of the illness [42].

Increased vascular permeability is a fundamental component of the pathogenesis of HFRS, and, as was already mentioned, the agents responsible for the disease attack endothelial cells without causing cytotoxic effects [98]. Animal histological investigations have shown that the kidney, spleen, and liver tissues contain viral antigen and RNA that can be detected by in situ hybridization, as well as nucleocapsid protein that can be detected by immune histochemical staining. Infiltrations of inflammatory cells and tubular injury colocalized with virus-infected cells in the kidneys, and these infiltrations primarily contained CD8-type T cells [99].

10.6 Laboratory Diagnosis of Hantavirus Infection

Laboratory tests and clinical and epidemiological data are used to diagnose Hantavirus infection in people. Clinical findings alone cannot provide a conclusive diagnosis, particularly when the illness is at the initial stage [100]. Samples from patients with respiratory distress, severe myalgia, undiagnosed fever, renal failure, or thrombocytopenia, as well as those who live in areas where the Hantavirus disease is prevalent or in those who recently engaged in outdoor activities that may have exposed them to rodents or their excrement, should be tested in a laboratory [101]. Because Hantaviruses vary in their geographic spread, mode of infection, and prognosis, it is crucial to use precise laboratory diagnostic tests [102].

10.6.1 Serologic Diagnosis

Serologic tests are the most feasible method for Hantavirus infection laboratory diagnosis [103, 104]. IgM antibodies, which are detectable at the start of symptoms, are induced to a high degree by the three structural proteins of Hantaviruses – Gn, Gc, and N [105].

10.6.2 Enzyme-Linked Immunosorbent Assay

The most popular serologic test for determining IgM class (acute phase) and IgG antibodies is the ELISA (convalescent stage). Hantavirus-specific IgG class antibodies are persistent throughout an individual's lifetime and can be used in retrospective research [106]. IgG or IgM ELISA can be used to identify seroconversion to a Hantavirus, but only IgM detection denotes recent exposure [107]. The timing of the appearance of IgM antibody varies based on the incubation period, dosage, immune status, and a number of other variables. IgM starts to decrease after a few weeks and continues for 3–6 months. IgG is visible a little later (and occasionally concurrently), but titers may continue to be elevated for many years [108]. The majority of Hantaviruses seem to have comparable dynamics. The majority of modern serologic tests use protein antigens obtained through recombinant DNA techniques as Hantavirus antigens. By using recombinant proteins, laboratory workers who would otherwise have to handle significant amounts of viruses to create conventional antigens are spared danger. Although Gn and Gc proteins have also been created, N proteins make up the majority of these antigens. A variety of recombinant expression systems, mainly bacterial ones, have been used to express and purify the N protein [109]. Hantavirus infections in rodents and human clinical samples were effectively examined using an ELISA technique based on a recombinant viral antigen. Recombinant antigens of ANDV (Instituto Carlos Malbra'n, Argentina) and SNV (CDC, USA) are utilized for the creation of ELISAs in the Americas through the cloning and expression of viral protein-encoding genes [108, 110]. A synthetic N protein from the Araraquara virus was created in Brazil by Figueiredo et al. [109] and is being used as an ELISA antigen. The nonpathogenic Maciel virus has been grown in Argentina for use in diagnosis based on blood sera; Maciel virus is cross-reactive with all the known New World Hantaviruses [111]. In 88 rural participants from northwest Colombia, by using ELISA, we discovered IgG antibodies reactive with SNV in 14% of serum samples in 2004 [112]. We found Maciel or Araraquara Hantavirus-specific IgG antibodies in 8.4% of 286 Colombian agricultural laborers in 2013. When compared to serosurveys conducted in North America, this prevalence is significantly greater. For instance, only 0.5% of North Americans had antibodies reactive with SNV 30 even when high-risk occupational groups (such as mammalogists) were selected. The Colombian participants had no unique occupational risks besides being rural laborers. This indicates that Hantavirus exposure among rural workers in northwestern Colombia may occur frequently [111].

10.6.3 Immunofluorescence Assay

Hantavirus fixed as an antigen on microscope plates was the first HFRS serodiagnosis test used in Asia and Europe. The need for biosafety level 3 labs prevents the widespread use of virus-infected cells for serologic tests (BSL-3) [112]. In Europe, the IFA is frequently used (for instance, Lubeck, Germany's Euroimmun Anti-Hantavirus-IIFT Mosaic II Test). On biochips that are being evaluated for Hantaviruses, reaction fields with uninfected cells or cells infected with one or more Hantaviruses are added with diluted serum samples.

10.6.4 Immunoblot Assay

Europe has also tried immunoblot tests, such as the recomLine Bunyavirus IgG/IgM test kit from Mikrogen in Neuried, Germany. Diluted sera are incubated on recomLine test strips containing six antigens that are either recombinant N-termini of the nucleocapsid antigens from PUUV, HTNV, DOBV, and SEOV or full nucleocapsid proteins from PUUV and HTNV. The antibody class is combined with a control band (IgG or IgM).

10.6.5 Focus Reduction Neutralization Test

The tests mentioned above can determine whether a patient has a Hantavirus infection, but none of them can identify which Hantavirus caused the infection because there is a substantial humoral cross-reaction between Hantavirus antibodies [113]. Focus reduction neutralization test (FRNT), the gold standard for Hantavirus diagnosis, can identify the particular infecting Hantavirus [112]. Although FRNT discriminates between Hantaviruses when testing serum from experimentally infected rats, it was less precise when testing serum from people in the acute phase [113]. FRNT can detect and quantify neutralizing antibodies by comparing serum titers with the relevant Hantaviruses. Since FRNT needs cell growth, it must be carried out in a BSL-3 environment. The test takes a lot of time and labor to complete.

10.6.6 Molecular Detection of Hantaviruses

Hantavirus illnesses are frequently diagnosed using molecular diagnostic techniques. The techniques are founded on practical, sensitive RT-PCR. Utilizing blood and tissue from fatal instances, postmortem diagnosis is frequently used [114]. The ability to sequence the PCR product in order to identify the virus and perform phylogenetic research is a clear benefit of PCR. Hantavirus RNA can be found using PCR 7–10 days after the start of symptoms. RT-PCR, which uses primers unique for the Hantavirus S and M segments, has also been used to diagnose Hantavirus infections. It is particularly helpful in instances where the disease is rapidly developing, and a febrile prodrome can turn into a severe respiratory illness in less than 24 hours in that situation, thanks to PCR's 24-hour turnaround time for results that allow for rapid detection of Hantavirus infection [115]. However, PCR using blood samples is only beneficial when patients are viremic and the infection is in its early phases [116]. Even before IgM antibodies show up, real-time RT-PCR is a reliable method for the early detection of Hantavirus RNA. Combining immunologic methods with Hantavirus real-time RTPCR can help with pathogenesis research and knowledge of antiviral therapies [116].

10.7 Treatment and Prevention

10.7.1 Treatment

As of right now, neither HFRS nor HCPS has a particular US Food and Drug Administration-approved therapy; instead, supportive care is the mainstay of treatment. Patients with severe HFRS and HCPS are advised to be transferred to an intensive care unit for careful observation and care. In order to prevent potentially harmful overhydration, it is crucial to maintain the fluid and electrolyte balance as well as circulatory volume. Based on the patient's fluid status, rate of diuresis, and kidney function, this needs to be carefully monitored (for patients that are anuric and with leaky capillaries). Dialysis may be necessary for HFRS patients with significant renal insufficiency, which is linked to pulmonary edema and extensive fluid retention. Platelet infusions may be used if there is significant bleeding and thrombocytopenia. Supplemental oxygen, when necessary, artificial ventilation, fluid control, and the proper application of pressors are essential in HCPS [117].

10.7.2 Ribavirin

Ribavirin has been shown to be effective against some Hantaviruses both *in vitro* and, to a lesser degree, in vivo. Various doses of ribavirin were administered to a suckling mouse model infected with the Hantaan virus in order to assess the effectiveness of ribavirin treatment against HFRS. Mice that received 50 mg of ribavirin/kg per day beginning on day 10 of the viral challenge survived longer than mice that received a placebo [118]. The effectiveness of ribavirin against HPS caused by the Andes virus has been demonstrated in vivo using a fatal hamster HPS model and in vitro using Vero E6 cells. Treatment with a dose as low as 5 mg/kg per day prevented lethal HPS disease in hamsters from day 1 through day 10 after infection when given intraperitoneally [119]. Up to 3 days after infection, intraperitoneal doses of ribavirin continued to offer 100% protection. Furthermore, ribavirin prevented HPS in hamsters without producing toxicity at doses of 50 mg/kg and 100 mg/kg, and it also provided a significant amount of protection against lethal HPS when administered 14 days after infection, according to another study [120]. These findings offer in vivo proof in favor of the possible use of ribavirin as a postexposure therapy to stop HPS brought on by ANDV. But according to two clinical studies, ribavirin was useless for treating HPS patients whose condition had advanced to the cardiopulmonary phase.

In the People's Republic of China, 242 patients with HRFS took part in a randomized, parallel, placebo-controlled clinical trial of intravenous ribavirin. A loading dose of 33 mg/kg was given to patients, followed by doses of 16 mg/kg every 6 hours for 4 consecutive days and 8 mg/kg at 8-hour intervals for the remaining 3 days of treatment. Patients receiving ribavirin experienced a sevenfold reduction in chances of mortality. On the other hand, a single, open-labeled clinical study carried out in Russia's European region demonstrated the insufficient effectiveness of IV ribavirin treatment for HFRS brought on by PUUV [121].

The majority of clinical trials that have been finished have demonstrated reversible anemia following the end of ribavirin therapy. Some individuals have also experienced rashes, sinus bradycardia, and hyperbilirubinemia.

10.7.3 Favipiravir

In vitro, SNV and ANDV are effectively inhibited by favipiravir (T-705) [122]. The 90% effective dose for both the viruses was calculated to be 5 μg/ml. Standard oral administration of T-705 at 50 or 100 mg/kg substantially increased survival in the lethal ANDV hamster model. When started before viremia appeared, oral T-705 therapy continued to guard against HPS. Daily oral T-705 administration significantly reduced the ability of tissue samples to detect SNV RNA and antigen when using a hamster-adapted SNV, suggesting that the medication may also be useful against HPS. T-705 is easily tolerated by humans, and no severe side effects associated with its use in clinical trials have been documented. The fact that T-705's 50% lethal dose (LD50) in hamsters is at least six times higher than that of ribavirin supports the notion that it has a low level of toxicity [123].

10.7.4 Lactoferrin

Lactoferrin, an iron-binding glycoprotein, has been shown to be effective against Hantavirus both in vitro and in vivo [124]. Both the studies' findings demonstrated lactoferrin's antiviral action against the Hantavirus by preventing the virus from adhering to cells. In vitro foci development with lactoferrin and ribavirin was completely inhibited, demonstrating the synergistic effects of both the medications. Prior to the Hantavirus viral challenge, suckling mice were given doses of 40 and 160 mg/kg of in vivo lactoferrin for evaluation. With a single administration, higher survival rates were found to be 15% and 70%, and they were shown to be 85% and 94% with a double dose, respectively.

10.7.5 Vandetanib

Higher vascular permeability plays a major role in the progression of the serious illness associated with viral hemorrhagic fevers. Vascular endothelial growth factor (VEGF) is produced by endothelial cells infected with ANDV, and VEGF stimulates endothelial cells' VEGFR2 receptors. Vascular endothelial cadherin is induced to internalize from adherent junctions, and paracellular permeability is increased by VEGFR2 stimulation. Additionally, patients with severe HPS have elevated circulating amounts of VEGF, suggesting that VEGF may be a crucial player in the pathogenesis of HPS [125]. A 3D human lung model has also shown increased amounts of VEGF-A and proinflammatory cytokines in response to the ANDV challenge. Vandetanib is a tyrosine kinase inhibitor that targets the stimulation of the VEGF-receptor 2. In vitro studies have shown that vandetanib can prevent the phosphorylation of the VEGFR-2 and the degradation of the VE-cadherin [126]. In addition, delayed lethality and improved total survival by 23% were seen in ANDV/hamster models with 10, 25, and 50 mg/kg/day of Vandetanib starting 5 days before the ANDV challenge. Although many early VEGFR-2 tyrosine kinase inhibitors, including Vandetanib, were developed for the treatment of cancer, large-scale human trials have revealed the possibility of severe adverse events, such as dermatologic reactions, high blood pressure, and other cardiovascular effects.

10.7.6 Immunotherapy

Numerous studies have demonstrated that neutralizing antibodies may prevent HPS in vivo, even though there is currently no confirmed cure for the Hantaviruses that cause HPS. Hamsters have been immunized with recombinant DNA vaccines and passively transfused with polyclonal blood from ducks, geese, and rabbits to prevent HPS. An open trial has been carried out in Chile to assess the effectiveness and security of using human immune sera as an HPS treatment alternative [127]. A 5000 U/kg dose of ANDV-neutralizing antibodies was administered intravenously to 32 individuals with suspected or confirmed HPS. When compared to the 32% case fatality rate in the remainder of the nation during the study period, the results have borderline statistical significance.

Hamsters have been defended against the deadly ANDV-HPS by two newly created recombinant monoclonal antibodies. For this, a source individual with high antibody titers was chosen after screening 27 ANDV convalescent HCPS sera. Using recombinant DNA technology, separated memory B cells have been used to produce recombinant monoclonal antibodies. JL16 and MIB22—the ensuing monoclonal antibody alternatives—were demonstrated to completely eliminate ANDV in vitro.

10.7.7 ETAR

The Hantaviruses HTNV and ANDV are resistant to the nucleoside analogue 1-β-D-Ribofuranosyl-3-ethynyl-[1,2,4]Triazole (ETAR) in vitro. ETAR was intraperitoneally given to suckling mice 10 days after HTNV exposure, with approximately 25% mortality at 12.5 and 25 mg/kg [128].

10.7.8 Corticosteroids

In Chile, a research study has been conducted to determine the effectiveness of intravenous methylprednisolone for HPS infection using a parallel-group, placebo-controlled clinical trial [129]. It comes to this conclusion, because there was no discernible difference in mortality between therapy groups, and because patients did not experience any appreciable clinical benefits. The use of methylprednisolone for HPS is also not supported by the findings.

10.7.9 Prevention

The risk category for Hantavirus infection includes people who come into contact with rodents or their excrement. Therefore, the most crucial stage in the disease's prevention is rodent control in homes and other places where human activities take place. In order to effectively manage rodent populations, rodent food sources should be eliminated from inside and around the house, entry points should be blocked, traps should be used, and potential nesting sites should be eliminated [130]. When cleaning up possibly rodent-infested spaces and rooms, it is important to ventilate the area before entering, and use personnel protection items and devices to prevent breathing in contaminated particles. Hantavirus vaccines are also required for the targeted control of human infections, particularly in risk groups. Strategies prevailing in the current era for the development of Hantavirus vaccine are tabulated in Table 10.2.

10.7.10 First-Generation Vaccines

10.7.10.1 Chinese and Korean Immunizations against the Inactivated HFRS

In China and Korea, vaccines made from cell culture and from rodent brains have both been created and tested on people. Between 1950 and 2007, China recorded more than 1.5 million human cases of HFRS and more than 45,000 (3%) fatalities.

Table 10.2 Multiple strategies for the development of a Hantavirus vaccine.

Vaccine-type	Vaccines/antigens	Animal models
Recombinant proteins	Nucleoprotein from ANDV, TOPV, DOBV, or PUUV	Bank voles
	Yeast-expressed DOBV nucleoprotein	Mice
	Truncated recombinant PUUV nucleoprotein linked to bacterial membrane protein	Mice
DNA vaccines	HTNV and ANDV M gene segments	Rhesus macaques
	HTNV M segment	Rhesus macaques
	SNV M gene segment	Syrian hamsters
	HTNV/PUUV/SNV/ANDV M gene segment mix	Rabbits
Virus vectored	Replication-competent VSV-vectored ANDV glycoproteins	Syrian hamsters
	Replication-competent VSV-vectored ANDV or SNV glycoproteins	Syrian hamsters
	Non-replicating Ad vector expressing N, Gn, Gc, or Gn/Gc	Syrian hamsters (protection studies) Mice (cytotoxicity assays)
Virus-like particles (VLPs)	HTNV-VLP with CD40L or GM-CSF incorporation	Mice
Inactivated virus	Hantavax (formalin-inactivated HNTV)	Humans

Source: Adapted from Saavedra et al. [131].

However, the prevalence of HFRS and the mortality rate have significantly dropped since the introduction of comprehensive preventive steps, such as vaccination. In areas where HFRS is extremely prevalent, inactivated versions of the HTNV and SEOV vaccines have been given since 1995. Since 2003, Vero cells have been used to create pure, formaldehyde-inactivated bivalent vaccines against HTNV and SEOV. In the blood of vaccine recipients, HTNV-NP-specific IgM and IgG antibodies as well as HTNV-neutralizing antibodies were significantly more prevalent and had higher positive rates. The findings show that high seropositive rates remained for 33 months after the vaccination, and that the peak levels of HTNV-NP-specific IgG and HTNV-neutralizing antibodies occurred 3 months after the vaccination [132].

South Korea has made extensive use of the formalin-inactivated HATN vaccine (Hanatvax) to prevent HFRS. With a three-dose vaccination schedule, seroconversion and high specific antibody titers in people were demonstrated (0 day, 01 month, and 12 months). Following the booster dose after 12 months, less than 50% of the sampled community produced neutralizing antibodies. Hantavax (three doses at 0, 1, and 13 months) immunogenicity and safety were examined in healthy adults as part of a phase III, multicenter clinical study [133]. With a three-dose regimen, Hantavax showed a two-year immunogenicity and booster effect. The neutralizing antibody response after two main doses was quite poor; consequently, a prompt booster shot at 2–6 months may be required to safeguard high-risk patients promptly. But even after the three doses of vaccination, a case–control study carried out in the Korean army failed to demonstrate statistically significant efficacy.

10.7.11 Second-Generation Vaccines

10.7.11.1 Virus-like Particle Vaccine

Although inactivated HFRS vaccines have been used to reduce the number of cases annually in China and Korea for years, frequent immunization is advised due to the loss of effective cell-mediated immunity. As a result, there is a growing interest in creating vaccine techniques that can produce longer-lasting, more widespread protection against the Hantavirus. Humans have developed high titer, reliable antibody levels against some viruses in response to a virus-like particle immunization [134]. In China, dihydrofolate reductase to create HTNV virus-like particles (VLPs) adorned with CD40L or GM-CSF—a vector bearing the S segment was cotransfected with Chinese hamster ovary cells (dhfr-CHO). *In vitro*, dendritic cells and macrophages exhibited increased activation in response to GM-CSF or CD40L-anchored HTNV VLP. These HTNV VLPs have provided rodents with stable, long-lasting protection six months after immunization, as well as a high titer of neutralizing antibodies and cellular immune responses specific to HTNV, such as higher expression of IFN-g and CTL responses. Mice responded better to CD40L or GM-CSF adorned VLPS than to unadorned VLPs or deactivated HTNV vaccines in terms of humoral and cellular immunity [135].

European researchers used a chimeric hepatitis B virus inner particle expressing a 45-amino acid fragment of PUU strain CG18-20 N spliced in the c/e1 region to demonstrate the bank vole model's high resistance to PUUV challenge [136]. It has been demonstrated that the nucleocapsid protein includes a second, small protective region through vaccinations using VLPs comprising the amino acids 75–119 of the PUU strain CG18-20 at the C-terminus of the HBV core.

In B ALB/c and C57BL/6 mice, HBV core particles expressing the N terminal 120 amino acids of the Dobrava, Hantaan, or Puumala nucleocapsid protein are powerfully immunogenic with or without adjuvant [137]. All IgG subtypes were represented in the induced nucleocapsid-specific antibodies, which were also highly cross-reactive. Furthermore, preexisting core-specific antibodies had no impact on the induction of an N-specific immune reaction.

10.7.11.2 Recombinant Proteins

There is enough proof to support the claims that the nucleocapsid protein and the membrane glycoproteins (Gn and Gc) have potent pathogenic properties and can activate defenses against Hantaviruses. Additionally, a number of methods have been put forth for the creation of Hantavirus synthetic protein vaccines. Higher antibody titer responses were elicited by baculovirus recombinants that expressed both Gn and Gc than by those that only expressed Gn or Gc [138]. Additionally, recombinant baculoviruses that only produce nucleocapsid protein or a Gn and Gc combination have completely protected hamsters from a challenge. The nucleocapsid protein of PUUV has an amino terminus that has been identified as the primary antigenic region and contains six of the seven B-cell epitopes that are recognized by infected bank vole monoclonal antibodies.

The use of adjuvant has increased the antigenicity and defensive potency of the Hantavirus recombinant proteins. Examples of genetically linked or complex proteins that have been used as adjuvants include the human *IL-2* gene, the outer membrane protein A of *Klebsiella pneumoniae* (rP40), and the heat-shock protein (not for use in humans). Different Hantavirus types show greater conservation of the nucleocapsid protein. As a result, the nucleocapsid protein causes strong antibody cross-reactions.

10.7.12 Third-Generation Vaccines

10.7.12.1 Recombinant Vector Vaccine

Hantavirus infection in animals has been demonstrated to be successfully prevented by recombinant vector vaccines. The M and S genome regions of HTNV's cDNA were inserted into the vaccinia virus to create a double-recombinant molecular vaccination. In contrast to PUUV, this vaccinia vector recombinant vaccination effectively protected hamsters from challenges with the Hantaan and Seoul viruses. Moreover, phase I and phase II clinical trials have been conducted to assess this vaccine. The phase I data show that the second inoculation raised neutralizing antibody titers to both the vaccinia virus and the HTNV. When two vaccination methods were compared, it was discovered that scarification effectively produced neutralizing antibodies in volunteers who had never been exposed to the vaccinia virus, but that subcutaneous inoculation was preferable to scarification in people who were already immune to the disease. According to the phase II trial's findings, only 26% of volunteers who had immunity to the vaccinia virus and 72% of participants who had never had the virus had neutralizing antibodies to the HTNV. Thus far, no efforts have been made to develop this vaccine.

Bearers of nonreplicating adenovirus encoding ANDV N, Gc, Gn, or Gn + Gc resulted in powerful immune responses, protecting hamsters from deadly ANDV infection and triggering powerful cytotoxic T cell responses in mice [139]. The requirement for the creation of vectors employing uncommon adenovirus serotypes is highlighted by the significant issue of preexisting immunity against adenovirus type 5.

The vesicular stomatitis virus (VSV) pseudotype that causes HTNV Gn and Gc was immunogenic and protective in rodents that had previously been exposed to HTNV three times. Employing a replication-competent recombinant vesicular stomatitis virus (VSV), which produces the ANDV Gn and Gc precursors, quick-acting pre- and post-exposure efficacy against lethal ANDV challenge was shown in the Syrian hamster model. To develop long-term immunity, nevertheless, a regimen of booster shots may be necessary [140].

10.7.12.2 Nucleic Acid–Based Molecular Vaccine

DNA vaccines containing either the M or S gene component of SEOV have been produced using either bare DNA expression vectors (pWRG7077), DNA-based Sindbis replicons (pSIN2.5), or packaged Sindbis replicon vectors (pSINrep5) [141]. The M segment vaccinations were linked to protection, and gene gun injection was preferable to needle inoculation. Hamsters were also shielded from HTNV and SEOV infections. As a result, a new DNA vaccine expressing the HTNV M gene segment was created. In hamsters, this vaccine provided sterilizing immunity against HTNV, SEOV, and DOBV, and it evoked extremely high levels of neutralizing antibodies in monkeys [142]. Three additional M gene segment DNA vaccines were created for ANDV, PUUV, and SNV. The PUUV DNA vaccine produced high-titer neutralizing antibodies in nonhuman primates and hamsters, but it did not protect them from other HFRS-associated Hantaviruses. Moreover, large titers of neutralizing antibodies were produced in rabbits and monkeys, respectively, by the ANDV and SNV DNA vaccines [143]. To assess the viability of creating a pan-Hantavirus vaccine, rabbits were given vaccinations with the HPS mix (ANDV and SNV plasmids), the HFRS mix (HTNV and PUUV plasmids), or the HPS/HFRS combination (all four plasmids). All the four viruses produced neutralizing antibodies in response to the HPS/HFRS mixture. The potential for protection from the multivalent Hantavirus immunizations was not investigated.

To evaluate the efficacy and security of HTNV and PUUV M segment DNA viruses in connection with the vaccine delivery method, two (phase I) clinical trials were conducted [144]. In both the trials, vaccines were deemed safe with no significant side effects for use in humans. When vaccines were administered via particle-mediated epidermal delivery (PMED), 56% of volunteers in the combination vaccine group and 30% and 44% of persons in the solo vaccine groups, respectively, generated neutralizing antibodies against one or both the viruses. The overall seroconversion rate with PMED administration, however (below 50%), was too low for continued research. About 56% and 78% of volunteers in the single vaccine groups employing the intramuscular electroporation delivery technique, as well as 78% of volunteers in the combined vaccine group, respectively, developed neutralizing antibodies to HTNV or PUUV. These results clearly demonstrate that the development of delivery technology can enhance immunogenicity.

Targeting the lysosome-associated membrane protein 1 or the HTNV Gn, two different DNA vaccines were developed (LAMP1). Strong humoral, cellular, and long-lasting immune responses can be produced by CD4+ T cells when LAMP 1 modifies the antigen-presenting pathway and activates them [145]. Long-term immune responses in vivo were shown for both the vaccinations.

10.8 Recent Outbreaks of Hantavirus

As the weather changes, residents begin cleaning their cabins or outside structures, which are typically infested with a variety of rodents and pose a greater risk of Hantavirus exposure. Fever, broad body and muscular aches, exhaustion, headache, light-headedness, chills, nausea, vomiting, and diarrhea are some of the early signs of this viral illness. Pneumonia, acute respiratory failure, and cardiogenic shock are all results of HPS's severe destruction of the lung parenchyma. With time, symptoms develop, and hospitalization rates for Hantavirus infection become fairly high. The genotypes of human populations as well as the strains of the viruses themselves have an impact on the pathophysiology of Hantavirus, which is rather diverse. Depending on a person's genetics, lifestyle, and other variables, the severity of the illness brought on by a Hantavirus infection can range from fairly mild to life-threatening [146]. During an outbreak of serious respiratory sickness in the Four Corners region, where Colorado, New Mexico, Arizona, and Utah meet, Hantavirus disease surveillance in the United States started in 1993. Since 1995, HPS has become an illness that must be reported through the Nationally Notifiable Disease Surveillance System (NNDSS) [147]. A total of 833 instances of Hantavirus infection had been documented by the year 2020, with a 35% fatality rate in the United States [148].

Hantavirus infections must be reported to the German health authorities as of 2001. With the exception of the outbreaks in 2007 and 2010, which saw 1688 and 2017 cases, respectively, fewer than 450 cases per year were reported during this time.

From January 2012 through July 2012 of this year, there have already been 2182 documented cases of Hantavirus sickness. From October 2011, the number of infections began to rise, and the number of cases has been quickly increasing since the spring of 2012. Almost 70% of instances involve men, which is an overrepresentation. Most affected individuals fall into the 40–49 age range. These traits are similar to those of earlier Hantaviral epidemics.

A 57-year-old man who had a verified Hantavirus infection passed away in April 2022. Following notification from the City of Buenos Aires's Ministry of Health, the control measures got underway. A new Hantavirus case was reported by the Esquel Health Program Directorate, with the Andean area of Chubut province being one likely source of infection. The fact that the patient had remained in the forest proved an epidemiological connection. No human connections, however, were seen to have contracted an infection [149]. The National Ministry of Health of Argentina, South America, reported on 14 July 2022, that an 8-year-old girl had passed away in Chubut province as a result of a Hantavirus infection. The young patient complained of fever, vomiting, and acute abdominal pain. After being admitted to the Zone Hospital in Esquel, the patient's health deteriorated, and she eventually passed away from a Hantavirus infection. A number of control techniques were used to investigate the close interactions. A total of 32 cases of Hantavirus infection have been verified in Argentina so far this year, including this index case in South America. Five persons have died as a result of the cases, which are from the provinces of Buenos Aires, the City of Buenos Aircs, Chubut, Entre Ros, Jujuy, Salta, and Tucum'an. The index case's Hantavirus infection's source could not be determined. The Andes virus, which was discovered in the long-tailed pygmy rice rat host, *Oligoryzomys longicaudatus*, in western Argentina, is the most likely to blame for the Hantavirus infection. There was no evidence that this was the method of transmission in the cases mentioned above, despite the fact that the Andes virus has been proven to be directly transmitted between people in regular communication. In the Panamanian province of Los Santos, there were a total of 17 cases of Hantavirus infestation reported in July 2022, of which 12 had HPS [149]. The North Dakota Department of Health (NDDoH) announced a pediatrics-confirmed case of HPS in a person in eastern North Dakota on 29 August 2022. The patient made a full recovery after being hospitalized. The two cases before this one were reported in 2018 and 2016, making this the first Hantavirus infection case in North Dakota in the year 2022. In Brazil, where wood mouse and rice mouse (*Akodon* spp. *and Oligoryzomys* spp.), found in farmlands, are recognized reservoirs of the virus, 22 cases of the Hantavirus were reported in September 2022. In the outbreak, 10 people died from the deadly sickness, marking a significant death rate. Between January and September 2022, 29 Hantavirus infections were reported in Los Santos, Panama. A total of 16 cases of the Hantavirus Cardiopulmonary Syndrome and 13 cases of the Hantavirus fever without fatalities were reported to the Department of Epidemiology, and they primarily originated from the Las Tabla, Los Santos, Macaracas, and Tonos district.

10.9 Summary and Conclusion

One of biology's greatest unanswered puzzles is how zoonotic infections first appeared. There have been several pathogen emergences and reemergence throughout the past century, with an estimated occurrence of one new pathogen every one or two years. Most of these novel infections are zoonotic, having spread from wildlife, and many of them are RNA viruses, like the Hantaviruses.

Hantaviral infections have become much better understood and recognized globally during the last few decades. The variety of Hantaviral illnesses and the number of known viruses both continue to grow. Hantavirus is an old disease, despite its recent discovery. The geographic range, quantity, and dynamics of the rodent carrier, and consequently the epidemiology of Hantavirus infections, may be impacted by environmental changes. Hantaviruses are widely believed to be present throughout the world, but further research is still needed to determine how each individual virus is dispersed. Human Hantaviral infections will likely be found in new locations with the advent of faster, more accurate testing and higher clinician knowledge, and new rodent species may be found to carry as-yet-undiscovered viruses. HFRS and HPS are two clinical conditions connected to Hantavirus infection. Depending on the specific Hantavirus implicated, the death rate can range from 0.1% to 40%. The N protein, G1 and G2 glycoproteins, and viral polymerase are all encoded by three negative-sense single-stranded RNA segments found in the Hantavirus genome. Both naturally occurring and in vitro Hantavirus genetic reassortment has been seen. The creation of novel Hantavirus strains with new virulence traits and/or new host ranges may be the outcome of such reassortment. The research reveals that the Hantavirus N protein is crucial for the establishment of an environment that is favorable for virus replication inside the host cell as well as for virus replication, transcription, and assembly. Compared to nonpathogenic strains, pathogenic Hantaviruses significantly alter the transcriptional activity of many cellular genes.

There are no approved treatment alternatives for Hantavirus infections. Unfortunately, there is only weak evidence that some antivirals are effective against these viruses. Animal models and in vitro research account for the majority of this evidence. Human studies are therefore necessary to get objective knowledge about the clinical effectiveness and safety of these medicines. Commercially accessible antivirals include ribavirin, which has demonstrated efficacy in a small number of human trials. In certain circumstances or when used out of compassion, these could be regarded as off-label therapy choices.

There are numerous vaccine options that have proof of providing sustained protection against Hantaviruses. Some of these had already undergone human testing in various countries. Greater regulatory agencies like the FDA and WHO have not yet prequalified a vaccine for the Hantavirus. For vaccines and medicines to be made available and used on a larger scale, additional clinical trial information is needed.

References

1 Clement, J., Maes, P., Lagrou, K. et al. (2012). A unifying hypothesis and a single name for a complex globally emerging infection: hantavirus disease. *Eur. J. Clin. Microbiol. Infect. Dis.* 31 (1): 1–5.

2 Hart, C.A. and Bennett, M. (1999). Hantavirus infections: epidemiology and pathogenesis. *Microbes Infect.* 1 (14): 1229–1237.

3 Carey, D.E., Reuben, R., Panicker, K.N. et al. (1971). Thottapalayam virus: a presumptive arbovirus isolated from a shrew in India. *Indian J. Med. Res.* 59 (11): 1758–1760.

4 Lee, H.W., Lee, P.W., and Johnson, K.M. (1978). Isolation of the etiologic agent of Korean Hemorrhagic fever. *J. Infect. Dis.* 137 (3): 298–308.

5 Nichol, S.T., Spiropoulou, C.F., Morzunov, S. et al. (1993). Genetic identification of a hantavirus associated with an outbreak of acute respiratory illness. *Science* 262 (5135): 914–917.

6 Núñez, J.J., Fritz, C.L., Knust, B. et al. (2014). Hantavirus infections among overnight visitors to Yosemite National Park, California, USA, 2012. *Emerg. Infect. Dis.* 20 (3): 386–393.

7 Schmaljohn, C.S. and Hooper, J.W. (2001). Bunyaviridae: the viruses and their replication. In: *Fields Virology*, 4ee (ed. D.M. Knipe, P.M. Howley, D.E. Griffin, et al.), 1581–1602. Philadelphia, PA: Lippincott Williams & Wilkins.

8 Jonsson, C.B., Figueiredo, L.T., and Vapalahti, O. (2010). A global perspective on hantavirus ecology, epidemiology, and disease. *Clin. Microbiol. Rev.* 23 (2): 412–441.

9 Childs, J.E., Glass, G.E., Korch, G.W., and LeDuc, J.W. (1989). Effects of hantaviral infection on survival, growth and fertility in wild rat (*Rattus norvegicus*) populations of Baltimore, Maryland. *J. Wildl. Dis.* 25 (4): 469–476.

10 Klein, S.L., Bird, B.H., and Glass, G.E. (2000). Sex differences in Seoul virus infection are not related to adult sex steroid concentrations in Norway rats. *J. Virol.* 74 (17): 8213–8217.

11 Xiao, S.Y., Diglisic, G., Avsic-Zupanc, T., and LeDuc, J.W. (1993). Dobrava virus as a new hantavirus: evidenced by comparative sequence analysis. *J. Med. Virol.* 39 (2): 152–155.

12 Taller, A.M., Xiao, S.Y., Godec, M.S. et al. (1993). Belgrade virus, a cause of hemorrhagic fever with renal syndrome in the Balkans, is closely related to Dobrava virus of field mice. *J. Infect. Dis.* 168 (3): 750–753.

13 Klempa, B., Tkachenko, E.A., Dzagurova, T.K. et al. (2008). Hemorrhagic fever with renal syndrome caused by 2 lineages of Dobrava hantavirus, Russia. *Emerg. Infect. Dis.* 14 (4): 617–625.

14 Jiang, J.F., Zhang, W.Y., Wu, X.M. et al. (2007). Soochong virus and Amur virus might be the same entities of hantavirus. *J. Med. Virol.* 79 (11): 1792–1795.

15 Zhang, Y.Z., Zhang, F.X., Wang, J.B. et al. (2009). Hantaviruses in rodents and humans, Inner Mongolia Autonomous Region, China. *Emerg. Infect. Dis.* 15 (6): 885–891.

16 Plyusnina, A., Ibrahim, I.N., and Plyusnin, A. (2009). A newly recognized hantavirus in the Asian house rat (*Rattus tanezumi*) in Indonesia. *J. Gen. Virol.* 90 (Pt 1): 205–209.

17 Hugot, J.P., Plyusnina, A., Herbreteau, V. et al. (2006). Genetic analysis of Thailand hantavirus in *Bandicota indica* trapped in Thailand. *Virol. J.* 5 (3): 72.

18 Pettersson, L., Boman, J., Juto, P. et al. (2008). Outbreak of Puumala virus infection, Sweden. *Emerg. Infect. Dis.* 14 (5): 808–810.

19 Hu, W., Mengersen, K., Bi, P., and Tong, S. (2007). Time-series analysis of the risk factors for haemorrhagic fever with renal syndrome: comparison of statistical models. *Epidemiol. Infect.* 135 (2): 245–252.

20 Kallio, E.R., Klingström, J., Gustafsson, E. et al. (2006). Prolonged survival of Puumala hantavirus outside the host: evidence for indirect transmission via the environment. *J. Gen. Virol.* 87 (Pt 8): 2127–2134.

21 Lee, H.W., Lee, P.W., Baek, L.J. et al. (1981). Intraspecific transmission of Hantaan virus, etiologic agent of Korean hemorrhagic fever, in the rodent *Apodemus agrarius*. *Am. J. Trop. Med. Hyg.* 30 (5): 1106–1112.

22 Peters, C.J. and Khan, A.S. (2002). Hantavirus pulmonary syndrome: the new American hemorrhagic fever. *Clin. Infect. Dis.* 34 (9): 1224–1231.

23 Song, J.W., Baek, L.J., Nagle, J.W. et al. (1996). Genetic and phylogenetic analyses of hantaviral sequences amplified from archival tissues of deer mice (*Peromyscus maniculatus nubiterrae*) captured in the eastern United States. *Arch. Virol.* 141 (5): 959–967.

24 Hjelle, B., Lee, S.W., Song, W. et al. (1995). Molecular linkage of hantavirus pulmonary syndrome to the white-footed mouse, *Peromyscus leucopus*: genetic characterization of the M genome of New York virus. *J. Virol.* 69 (12): 8137–8141.

25 Morzunov, S.P., Feldmann, H., Spiropoulou, C.F. et al. (1995). A newly recognized virus associated with a fatal case of hantavirus pulmonary syndrome in Louisiana. *J. Virol.* 69 (3): 1980–1983.

26 Ravkov, E.V., Rollin, P.E., Ksiazek, T.G. et al. (1995). Genetic and serologic analysis of Black Creek Canal virus and its association with human disease and *Sigmodon hispidus* infection. *Virology* 210 (2): 482–489.

27 Levis, S., Morzunov, S.P., Rowe, J.E. et al. (1998). Genetic diversity and epidemiology of hantaviruses in Argentina. *J. Infect. Dis.* 177 (3): 529–538.

28 Padula, P., Figueroa, R., Navarrete, M. et al. (2004). Transmission study of Andes hantavirus infection in wild sigmodontine rodents. *J. Virol.* 78 (21): 11972–11979.

29 Figueiredo, L.T., Moreli, M.L., Campos, G.M., and Sousa, R.L. (2003). Hantaviruses in São Paulo State, Brazil. *Emerg. Infect. Dis.* 9 (7): 891–892.

30 Padula, P., Martinez, V.P., Bellomo, C. et al. (2007). Pathogenic hantaviruses, northeastern Argentina and eastern Paraguay. *Emerg. Infect. Dis.* 13 (8): 1211–1214.

31 Vincent, M.J., Quiroz, E., Gracia, F. et al. (2000). Hantavirus pulmonary syndrome in Panama: identification of novel hantaviruses and their likely reservoirs. *Virology* 277 (1): 14–19.

32 Powers, A.M., Mercer, D.R., Watts, D.M. et al. (1999). Isolation and genetic characterization of a hantavirus (Bunyaviridae: Hantavirus) from a rodent, *Oligoryzomys microtis* (Muridae), collected in northeastern Peru. *Am. J. Trop. Med. Hyg.* 61 (1): 92–98.

33 Levis, S., Garcia, J., Pini, N. et al. (2004). Hantavirus pulmonary syndrome in northwestern Argentina: circulation of Laguna Negra virus associated with *Calomys callosus*. *Am. J. Trop. Med. Hyg.* 71 (5): 658–663.

34 Raboni, S.M., Probst, C.M., Bordignon, J. et al. (2005). Hantaviruses in Central South America: phylogenetic analysis of the S segment from HPS cases in Paraná, Brazil. *J. Med. Virol.* 76 (4): 553–562.

35 Johnson, A.M., Bowen, M.D., Ksiazek, T.G. et al. (1997). Laguna Negra virus associated with HPS in western Paraguay and Bolivia. *Virology* 238 (1): 115–127.

36 Delfraro, A., Clara, M., Tomé, L. et al. (2003). Yellow pigmy rice rat (*Oligoryzomys flavescens*) and hantavirus pulmonary syndrome in Uruguay. *Emerg. Infect. Dis.* 9 (7): 846–852.

37 Mantooth, S.J., Milazzo, M.L., Bradley, R.D. et al. (2001). Geographical distribution of rodent-associated hantaviruses in Texas. *J. Vector Ecol.* 26 (1): 7–14.

38 Chu, Y.K., Owen, R.D., Sánchez-Hernández, C. et al. (2008). Genetic characterization and phylogeny of a hantavirus from Western Mexico. *Virus Res.* 131 (2): 180–188.

39 Milazzo, M.L., Cajimat, M.N., Hanson, J.D. et al. (2006). Catacamas virus, a hantaviral species naturally associated with *Oryzomys couesi* (Coues' oryzomys) in Honduras. *Am. J. Trop. Med. Hyg.* 75 (5): 1003–1010.

40 Musser, G.G. and Carleton, M.D. (2005). Superfamily Muroidea. In: *Mammal Species of the World: A Taxonomic and Geographic Reference*, 3rde (ed. D.E. Wilson and D.M. Reeder), 894–1531. Baltimore, MD: The Johns Hopkins University Press.

41 Hall, E.R. (1981). *The Mammals of North America*, 2nde, vol. vol. 176, 1–5. New York: Wiley.

42 Guidotti, L.G. and Chisari, F.V. (2001). Noncytolytic control of viral infections by the innate and adaptive immune response. *Annu. Rev. Immunol.* 19: 65–91.

43 Mills, J.N., Ksiazek, T.G., Peters, C.J., and Childs, J.E. (1999). Long-term studies of hantavirus reservoir populations in the southwestern United States: a synthesis. *Emerg. Infect. Dis.* 5 (1): 135–142.

44 Ramsden, C., Holmes, E.C., and Charleston, M.A. (2009). Hantavirus evolution in relation to its rodent and insectivore hosts: no evidence for codivergence. *Mol. Biol. Evol.* 26 (1): 143–153.

45 Kang, H.J., Bennett, S.N., Sumibcay, L. et al. (2009). Evolutionary insights from a genetically divergent hantavirus harbored by the European common mole (*Talpa europaea*). *PLoS One* 4 (7): e6149.

46 Li, D., Schmaljohn, A.L., Anderson, K., and Schmaljohn, C.S. (1995). Complete nucleotide sequences of the M and S segments of two hantavirus isolates from California: evidence for reassortment in nature among viruses related to hantavirus pulmonary syndrome. *Virology* 206 (2): 973–983.

47 McElroy, A.K., Smith, J.M., Hooper, J.W., and Schmaljohn, C.S. (2004). Andes virus M genome segment is not sufficient to confer the virulence associated with Andes virus in Syrian hamsters. *Virology* 326 (1): 130–139.

48 Klempa, B., Schmidt, H.A., Ulrich, R. et al. (2003). Genetic interaction between distinct Dobrava hantavirus subtypes in *Apodemus agrarius* and *A. flavicollis* in nature. *J. Virol.* 77 (1): 804–809.

49 Chu, Y.K., Milligan, B., Owen, R.D. et al. (2006). Phylogenetic and geographical relationships of hantavirus strains in eastern and western Paraguay. *Am. J. Trop. Med. Hyg.* 75 (6): 1127–1134.

50 Song, K.J., Baek, L.J., Moon, S. et al. (2007). Muju virus, a novel hantavirus harboured by the arvicolid rodent *Myodes regulus* in Korea. *J. Gen. Virol.* 88 (Pt 11): 3121–3129.

51 Plyusnin, A., Vapalahti, O., Lehväslaiho, H. et al. (1995). Genetic variation of wild Puumala viruses within the serotype, local rodent populations and individual animal. *Virus Res.* 38 (1): 25–41.

52 Wells, R.M., Young, J., Williams, R.J. et al. (1997). Hantavirus transmission in the United States. *Emerg. Infect. Dis.* 3 (3): 361–365.

53 Hardestam, J., Lundkvist, A., and Klingström, J. (2009). Sensitivity of Andes hantavirus to antiviral effect of human saliva. *Emerg. Infect. Dis.* 15 (7): 1140–1142.

54 Lee, H.W. (1989). Hemorrhagic fever with renal syndrome in Korea. *Rev. Infect. Dis.* 11 (Suppl 4): S864–S876.

55 Heyman, P., Vaheri, A., Lundkvist, A., and Avsic-Zupanc, T. (2009). Hantavirus infections in Europe: from virus carriers to a major public-health problem. *Expert Rev. Anti-Infect. Ther.* 7 (2): 205–217.

56 Papa, A., Johnson, A.M., Stockton, P.C. et al. (1998). Retrospective serological and genetic study of the distribution of hantaviruses in Greece. *J. Med. Virol.* 55 (4): 321–327.

57 Plyusnin, A., Mustonen, J., Asikainen, K. et al. (1999). Analysis of Puumala hantavirus genome in patients with *Nephropathia epidemica* and rodent carriers from the sites of infection. *J. Med. Virol.* 59 (3): 397–405.

58 Sibold, C., Meisel, H., Lundkvist, A. et al. (1999). Short report: simultaneous occurrence of Dobrava, Puumala, and Tula Hantaviruses in Slovakia. *Am. J. Trop. Med. Hyg.* 61 (3): 409–411.

59 Plyusnin, A., Nemirov, K., Apekina, N. et al. (1999). Dobrava hantavirus in Russia. *Lancet* 353 (9148): 207.

60 Klempa, B., Meisel, H., Räth, S. et al. (2003). Occurrence of renal and pulmonary syndrome in a region of northeast Germany where Tula hantavirus circulates. *J. Clin. Microbiol.* 41 (10): 4894–4897.

61 Brummer-Korvenkontio, M., Vapalahti, O., Henttonen, H. et al. (1999). Epidemiological study of *Nephropathia epidemica* in Finland 1989-96. *Scand. J. Infect. Dis.* 31 (5): 427–435.

62 Zöller, L., Faulde, M., Meisel, H. et al. (1995). Seroprevalence of hantavirus antibodies in Germany as determined by a new recombinant enzyme immunoassay. *Eur. J. Clin. Microbiol. Infect. Dis.* 14 (4): 305–313.

63 Mertens, M., Wölfel, R., Ullrich, K. et al. (2009). Seroepidemiological study in a Puumala virus outbreak area in South-East Germany. *Med. Microbiol. Immunol.* 198 (2): 83–91.

64 Lokugamage, K., Kariwa, H., Hayasaka, D. et al. (2002). Genetic characterization of hantaviruses transmitted by the Korean field mouse (*Apodemus peninsulae*), Far East Russia. *Emerg. Infect. Dis.* 8 (8): 768–776.

65 Xu, Z.Y., Tang, Y.W., Kan, L.Y., and Tsai, T.F. (1987). Cats—source of protection or infection? A case-control study of hemorrhagic fever with renal syndrome. *Am. J. Epidemiol.* 126 (5): 942–948.

66 Hooper, J.W. and Li, D. (2001). Vaccines against hantaviruses. *Curr. Top. Microbiol. Immunol.* 256: 171–191.

67 Lu, Q., Zhu, Z., and Weng, J. (1996). Immune responses to inactivated vaccine in people naturally infected with hantaviruses. *J. Med. Virol.* 49 (4): 333–335.

68 Park, K., Kim, C.S., and Moon, K.T. (2004). Protective effectiveness of hantavirus vaccine. *Emerg. Infect. Dis.* 10 (12): 2218–2220.

69 Suputthamongkol, Y., Nitatpattana, N., Chayakulkeeree, M. et al. (2005). Hantavirus infection in Thailand: first clinical case report. *Southeast Asian J. Trop. Med. Public Health* 36 (1): 217–220.

70 Chandy, S., Boorugu, H., Chrispal, A. et al. (2009). Hantavirus infection: a case report from India. *Indian J. Med. Microbiol.* 27 (3): 267–270.

71 Arikawa, J., Yoshimatsu, K., and Kariwa, H. (2001). Epidemiology and epizootiology of hantavirus infection in Japan. *Jpn. J. Infect. Dis.* 54 (3): 95–102.

72 Tappero, J.W., Khan, A.S., Pinner, R.W. et al. (1996). Utility of emergency, telephone-based national surveillance for Hantavirus pulmonary syndrome. Hantavirus Task Force. *JAMA* 275 (5): 398–400.

73 Chu, Y.K., Owen, R.D., Gonzalez, L.M., and Jonsson, C.B. (2003). The complex ecology of hantavirus in Paraguay. *Am. J. Trop. Med. Hyg.* 69 (3): 263–268.

74 Armstrong, L.R., Zaki, S.R., Goldoft, M.J. et al. (1995). Hantavirus pulmonary syndrome associated with entering or cleaning rarely used, rodent-infested structures. *J. Infect. Dis.* 172 (4): 1166.

75 Howard, M.J., Doyle, T.J., Koster, F.T. et al. (1999). Hantavirus pulmonary syndrome in pregnancy. *Clin. Infect. Dis.* 29 (6): 1538–1544.

76 Williams, R.J., Bryan, R.T., Mills, J.N. et al. (1997). An outbreak of hantavirus pulmonary syndrome in western Paraguay. *Am. J. Trop. Med. Hyg.* 57 (3): 274–282.

77 Toro, J., Vega, J.D., Khan, A.S. et al. (1998). An outbreak of hantavirus pulmonary syndrome, Chile, 1997. *Emerg. Infect. Dis.* 4 (4): 687–694.

78 Galeno, H., Mora, J., Villagra, E. et al. (2002). First human isolate of Hantavirus (Andes virus) in the Americas. *Emerg. Infect. Dis.* 8 (7): 657–661.

79 Navarrete, M., Barrera, C., Zaror, L., and Otth, C. (2007). Rapid immunochromatographic test for hantavirus Andes contrasted with capture-IgM ELISA for detection of Andes-specific IgM antibodies. *J. Med. Virol.* 79 (1): 41–44.

80 Vasconcelos, M.I., Lima, V.P., Iversson, L.B. et al. (1997). Hantavirus pulmonary syndrome in the rural area of Juquitiba, São Paulo metropolitan area, Brazil. *Rev. Inst. Med. Trop. Sao Paulo* 39 (4): 237–238.

81 Lee, H.W. (1989). Clinical manifestations of HFRS. In: *Manual of Hemorrhagic Fever with Renal Syndrome* (ed. H.W. Lee and J.M. Dalrymple), 19–38. Seoul, South Korea: WHO Collaborating Center for Virus Reference and Research (Hemorrhagic fever with renal syndrome) Institute for Viral Diseases, Korea University.

82 Mustonen, J., Brummer-Korvenkontio, M., Hedman, K. et al. (1994). *Nephropathia epidemica* in Finland: a retrospective study of 126 cases. *Scand. J. Infect. Dis.* 26 (1): 7–13.

83 Mähönen, S.M., Sironen, T., Vapalahti, O. et al. (2007). Puumala virus RNA in cerebrospinal fluid in a patient with uncomplicated *Nephropathia epidemica*. *J. Clin. Virol.* 40 (3): 248–251.

84 Makela, S., Kokkonen, L., Ala-Houhala, I. et al. (2009). More than half of the patients with acute Puumala hantavirus infection have abnormal cardiac findings. *Scand. J. Infect. Dis.* 41 (1): 57–62.

85 Mustonen, J., Partanen, J., Kanerva, M. et al. (1996). Genetic susceptibility to severe course of *Nephropathia epidemica* caused by Puumala hantavirus. *Kidney Int.* 49 (1): 217–221.

86 Mustonen, J., Mäkelä, S., Helin, H. et al. (2001). Glomerulonephritis caused by Puumala hantavirus infection. *Nephron* 89 (4): 402–407.

87 LeDuc, J.W., Childs, J.E., and Glass, G.E. (1992). The Hantaviruses, etiologic agents of hemorrhagic fever with renal syndrome: a possible cause of hypertension and chronic renal disease in the United States. *Annu. Rev. Public Health* 13: 79–98.

88 Mertz, G.J., Hjelle, B., Crowley, M. et al. (2006). Diagnosis and treatment of new world hantavirus infections. *Curr. Opin. Infect. Dis.* 19 (5): 437–442.

89 Peters, C.J., Simpson, G.L., and Levy, H. (1999). Spectrum of hantavirus infection: hemorrhagic fever with renal syndrome and hantavirus pulmonary syndrome. *Annu. Rev. Med.* 50: 531–545.

90 Padula, P.J., Colavecchia, S.B., Martínez, V.P. et al. (2000). Genetic diversity, distribution, and serological features of hantavirus infection in five countries in South America. *J. Clin. Microbiol.* 38 (8): 3029–3035.

91 Khan, A.S., Khabbaz, R.F., Armstrong, L.R. et al. (1996). Hantavirus pulmonary syndrome: the first 100 US cases. *J. Infect. Dis.* 173 (6): 1297–1303.

92 Campos, G.M., Borges, A.A., Badra, S.J. et al. (2009). Síndrome pulmonar e cardiovascular por hantavírus: aspectos clínicos de uma doença emergente no sudeste brasileiro [Pulmonary and cardiovascular syndrome due to hantavirus: clinical aspects of an emerging disease in southeastern Brazil]. *Rev. Soc. Bras. Med. Trop.* 42 (3): 282–289.

93 Gavrilovskaya, I., LaMonica, R., Fay, M.E. et al. (1999). New York 1 and Sin Nombre viruses are serotypically distinct viruses associated with hantavirus pulmonary syndrome. *J. Clin. Microbiol.* 37 (1): 122–126.

94 Kilpatrick, E.D., Terajima, M., Koster, F.T. et al. (2004). Role of specific CD8+ T cells in the severity of a fulminant zoonotic viral hemorrhagic fever, hantavirus pulmonary syndrome. *J. Immunol.* 172 (5): 3297–3304.

95 Borges, A.A., Campos, G.M., Moreli, M.L. et al. (2006). Hantavirus cardiopulmonary syndrome: immune response and pathogenesis. *Microbes Infect.* 8 (8): 2324–2330.

96 Takala, A., Jousela, I., Jansson, S.E. et al. (1999). Markers of systemic inflammation predicting organ failure in community-acquired septic shock. *Clin. Sci. (Lond.)* 97 (5): 529–538.

97 Nakao, A., Miike, S., Hatano, M. et al. (2000). Blockade of transforming growth factor beta/Smad signaling in T cells by overexpression of Smad7 enhances antigen-induced airway inflammation and airway reactivity. *J. Exp. Med.* 192 (2): 151–158.

98 Vapalahti, O., Mustonen, J., Lundkvist, A. et al. (2003). Hantavirus infections in Europe. *Lancet Infect. Dis.* 3 (10): 653–661.

99 Sironen, T., Klingström, J., Vaheri, A. et al. (2008). Pathology of Puumala hantavirus infection in macaques. *PLoS One* 3 (8): e3035.

100 Figueiredo, L.T., Souza, W.M., Ferrés, M., and Enria, D.A. (2014). Hantaviruses and cardiopulmonary syndrome in South America. *Virus Res.* 187: 43–54.

101 Fulhorst, C.H., Koster, F., Enrı'a, D. et al. (2011). Hantavirus in tropical infectious diseases. *Infec. Chapter* 71: 470–480.

102 Engler, O., Klingstrom, J., Aliyev, E. et al. (2013). Seroprevalence of hantavirus infections in Switzerland in 2009: difficulties in determining prevalence in a country with low endemicity. *Euro Surveill.* 18 (50): 20660.

103 Bayard, V., Kitsutani, P.T., Barria, E.O. et al. (2004). Outbreak of hantavirus pulmonary syndrome, Los Santos, Panama, 1999-2000. *Emerg. Infect. Dis.* 10 (9): 1635–1642.

104 Kruger, D.H., Figueiredo, L.T., Song, J.W., and Klempa, B. (2015). Hantaviruses—globally emerging pathogens. *J. Clin. Virol.* 64: 128–136.

105 Figueiredo, L.T., Moreli, M.L., Borges, A.A. et al. (2008). Diagnosis of hantavirus relies on demonstration of immune response, which becomes evident within 2 to 8 days of acute onset. *Braz. J. Med. Biol. Res.* 41: 596–599.

106 Lundkvist, A., Hörling, J., and Niklasson, B. (1993). The humoral response to Puumala virus infection (*Nephropathia epidemica*) investigated by viral protein specific immunoassays. *Arch. Virol.* 130 (1–2): 121–130.

107 Mattar, S., Garzon, D., Tadeu, L. et al. (2014). Serological diagnosis of hantavirus pulmonary syndrome in a febrile patient in Colombia. *Int. J. Infect. Dis.* 25: 201–203.

108 Padula, P.J., Rossi, C.M., Valle, M.O.D. et al. (2000). Development and evaluation of a solid-phase enzyme immunoassay based on Andes hantavirus recombinant nucleoprotein. *J. Med. Microbiol.* 49 (2): 149–155.

109 Figueiredo, L.T., Moreli, M.L., Borges, A.A. et al. (2009). Evaluation of an enzyme-linked immunosorbent assay based on Araraquara virus recombinant nucleocapsid protein. *Am. J. Trop. Med. Hyg.* 81 (2): 273–276.

110 Ksiazek, T.G., Peters, C.J., Rollin, P.E. et al. (1995). Identification of a new North American hantavirus that causes acute pulmonary insufficiency. *Am. J. Trop. Med. Hyg.* 52 (2): 117–123.

111 Guzmán, C., Mattar, S., Levis, S. et al. (2013). Prevalence of antibody to hantaviruses in humans and rodents in the Caribbean region of Colombia determined using Araraquara and Maciel virus antigens. *Mem. Inst. Oswaldo Cruz* 108 (2): 167–171.

112 Máttar, S. and Parra, M. (2004). Serologic evidence of hantavirus infection in humans, Colombia. *Emerg. Infect. Dis.* 10 (12): 2263–2264.

113 Bharadwaj, M., Botten, J., Torrez-Martinez, N., and Hjelle, B. (1997). Rio Mamore virus: genetic characterization of a newly recognized hantavirus of the pygmy rice rat, *Oligoryzomys microtis*, from Bolivia. *Am. J. Trop. Med. Hyg.* 57 (3): 368–374.

114 Noh, J.Y., Cheong, H.J., Song, J.Y. et al. (2013). Clinical and molecular epidemiological features of hemorrhagic fever with renal syndrome in Korea over a 10-year period. *J. Clin. Virol.* 58 (1): 11–17.

115 Ramanathan, H.N. and Jonsson, C.B. (2008). New and Old World hantaviruses differentially utilize host cytoskeletal components during their life cycles. *Virology* 374 (1): 138–150.

116 Willemann, M.C.A. and de Oliveira, S.V. (2014). Risk factors associated with hantavirosis fatality: a regional analysis from a case-control study in Brazil. *Rev. Soc. Bras. Med. Trop.* 47 (1): 47–51.

117 Dheerasekara, K., Sumathipala, S., and Muthugala, R. (2020). Hantavirus infections-treatment and prevention. *Curr. Treat. Options Infect. Dis.* 12 (4): 410–421.

118 Huggins, J.W., Kim, G.R., Brand, O.M., and McKee, K.T. Jr. (1986). Ribavirin therapy for Hantaan virus infection in suckling mice. *J. Infect. Dis.* 153 (3): 489–497.

119 Safronetz, D., Haddock, E., Feldmann, F. et al. (2011). In vitro and in vivo activity of ribavirin against Andes virus infection. *PLoS One* 6 (8): e23560.

120 Ogg, M., Jonsson, C.B., Camp, J.V., and Hooper, J.W. (2013). Ribavirin protects Syrian hamsters against lethal hantavirus pulmonary syndrome—after intranasal exposure to Andes virus. *Viruses* 5 (11): 2704–2720.

121 Malinin, O.V. and Platonov, A.E. (2017). Insufficient efficacy and safety of intravenous ribavirin in treatment of haemorrhagic fever with renal syndrome caused by Puumala virus. *Infect. Dis. (Auckl)* 49 (7): 514–520.

122 Safronetz, D., Falzarano, D., Scott, D.P. et al. (2013). Antiviral efficacy of favipiravir against two prominent etiological agents of hantavirus pulmonary syndrome. *Antimicrob. Agents Chemother.* 57 (10): 4673–4680.

123 Vaheri, A., Henttonen, H., Voutilainen, L. et al. (2013). Hantavirus infections in Europe and their impact on public health. *Rev. Med. Virol.* 23: 35–49.

124 Gowen, B.B., Wong, M.H., Jung, K.H. et al. (2007). In vitro and in vivo activities of T-705 against arenavirus and bunyavirus infections. *Antimicrob. Agents Chemother.* 51 (9): 3168–3176.

125 Shrivastava-Ranjan, P., Rollin, P.E., and Spiropoulou, C.F. (2010). Andes virus disrupts the endothelial cell barrier by induction of vascular endothelial growth factor and downregulation of VE-cadherin. *J. Virol.* 84 (21): 11227–11234.

126 Bird, B.H., Shrivastava-Ranjan, P., Dodd, K.A. et al. (2016). Effect of Vandetanib on Andes virus survival in the hamster model of hantavirus pulmonary syndrome. *Antivir. Res.* 132: 66–69.

127 Vial, P.A., Valdivieso, F., Calvo, M. et al. (2015). A non-randomized multicentre trial of human immune plasma for treatment of hantavirus cardiopulmonary syndrome caused by Andes virus. *Antivir. Ther.* 20 (4): 377–386.

128 Chung, D.H., Kumarapperuma, S.C., Sun, Y. et al. (2008). Synthesis of 1-beta-D-ribofuranosyl-3-ethynyl-[1,2,4]triazole and its in vitro and in vivo efficacy against Hantavirus. *Antivir. Res.* 79 (1): 19–27.

129 Vial, P.A., Valdivieso, F., Ferres, M. et al. (2013). High-dose intravenous methylprednisolone for hantavirus cardiopulmonary syndrome in Chile: a double-blind, randomized controlled clinical trial. *Clin. Infect. Dis.* 57 (7): 943–951.

130 Krüger, D.H., Ulrich, R., and Lundkvist, Å. (2001). Hantavirus infections and their prevention. *Microbes Infect.* 3 (13): 1129–1144.

131 Saavedra, F., Díaz, F.E., Retamal-Díaz, A. et al. (2021). Immune response during hantavirus diseases: implications for immunotherapies and vaccine design. *Immunology* 163 (3): 262–277.

132 Li, Z., Zeng, H., Wang, Y. et al. (2017). The assessment of Hantaan virus-specific antibody responses after the immunization program for hemorrhagic fever with renal syndrome in northwest China. *Hum. Vaccin. Immunother.* 13 (4): 802–807.

133 Song, J.Y., Woo, H.J., Cheong, H.J. et al. (2016). Long-term immunogenicity and safety of inactivated Hantaan virus vaccine (Hantavax™) in healthy adults. *Vaccine* 34 (10): 1289–1295.

134 Mohsen, M.O., Zha, L., Cabral-miranda, G., and Bachmann, M.F. (2017). Seminars in immunology major findings and recent advances in virus—like particle (VLP) -based vaccines. *Semin. Immunol.* 34 (July): 123–132.

135 Dong, Y., Ma, T., Zhang, X. et al. (2019). Incorporation of CD40 ligand or granulocyte macrophage colony stimulating factor into Hantaan virus (HTNV) virus-like particles significantly enhances the long-term immunity potency against HTNV infection. *J. Med. Microbiol.* 68 (3): 480–492.

136 Ulrich, R., Koletzki, D., Lachmann, S. et al. (1999). New chimaeric hepatitis B virus core particles carrying hantavirus (serotype Puumala) epitopes: immunogenicity and protection against virus challenge. *J. Biotechnol.* 73 (2–3): 141–153.

137 Geldmacher, A., Skrastina, D., Borisova, G. et al. (2005). A hantavirus nucleocapsid protein segment exposed on hepatitis B virus core particles is highly immunogenic in mice when applied without adjuvants or in the presence of pre-existing anti-core antibodies. *Vaccine* 23 (30): 3973–3983.

138 Schmaljohn, C.S., Chu, Y.K., Schmaljohn, A.L., and Dalrymple, J.M. (1990). Antigenic subunits of Hantaan virus expressed by baculovirus and vaccinia virus recombinants. *J. Virol.* 64 (7): 3162–3170.

139 Safronetz, D., Hegde, N.R., Ebihara, H. et al. (2009). Adenovirus vectors expressing hantavirus proteins protect hamsters against lethal challenge with Andes virus. *J. Virol.* 83 (14): 7285–7295.

140 Prescott, J., DeBuysscher, B.L., Brown, K.S., and Feldmann, H. (2014). Long-term single-dose efficacy of a vesicular stomatitis virus-based Andes virus vaccine in Syrian hamsters. *Viruses* 6 (2): 516–523.

141 Kamrud, K.I., Hooper, J.W., Elgh, F., and Schmaljohn, C.S. (1999). Comparison of the protective efficacy of naked DNA, DNA-based Sindbis replicon, and packaged Sindbis replicon vectors expressing Hantavirus structural genes in hamsters. *Virology* 263 (1): 209–219.

142 Hooper, J.W., Custer, D.M., Thompson, E., and Schmaljohn, C.S. (2001). DNA vaccination with the Hantaan virus M gene protects Hamsters against three of four HFRS hantaviruses and elicits a high-titer neutralizing antibody response in Rhesus monkeys. *J. Virol.* 75 (18): 8469–8477.

143 Hooper, J.W., Josleyn, M., Ballantyne, J., and Brocato, R. (2013). A novel sin Nombre virus DNA vaccine and its inclusion in a candidate pan-hantavirus vaccine against hantavirus pulmonary syndrome (HPS) and hemorrhagic fever with renal syndrome (HFRS). *Vaccine* 31 (40): 4314–4321.

144 Hooper, J.W., Moon, J.E., Paolino, K.M. et al. (2014). A Phase 1 clinical trial of Hantaan virus and Puumala virus M-segment DNA vaccines for haemorrhagic fever with renal syndrome delivered by intramuscular electroporation. *Clin. Microbiol. Infect.* 20 (Suppl 5): 110–117.

145 Jiang, D.B., Zhang, J.P., Cheng, L.F. et al. (2018). Hantavirus Gc induces long-term immune protection via LAMP-targeting DNA vaccine strategy. *Antivir. Res.* 150: 174–182.

146 Transmission. https://www.cdc.gov/hantavirus/hps/transmission.html.

147 Reported Cases of Hantavirus Disease. https://www.cdc.gov/hantavirus/surveillance/index.html.

148 North Dakota Department of Health Reports Hantavirus Pulmonary Syndrome in a Pediatric Patient. https://www.health.nd.gov/news/north-dakota-department-health-reports-hantavirus-pulmonary-syndrome-pediatric-patient.

149 Hantavirus: A Total of 17 Cases Have Been Reported in Panama in 2022. http://outbreaknewstoday.com/hantavirus-a-total-of-17-cases-have-been-reported-in-panama-in-2022-2022

11

Current Advances in Marburg Virus Disease

Emmanuel K. Ofori and Eric N.Y. Nyarko

Department of Chemical Pathology, University of Ghana Medical School, Accra, Ghana

11.1 Introduction

11.1.1 General Properties of Marburg Virus (MARV)

11.1.1.1 Classification and Taxonomy

MARV belongs to the order *Mononegavirales*, the family *Filoviridae*, and the genus *Marburgvirus*. The genera *Marburgvirus, Ebolavirus, Cuevavirus, Striavirus,* and *Thamnovirus* are all members of the *Filoviridae* family [1]. Only three of the genera – Cuevavirus, Striavirus, and Thamnovirus – have not been linked to human or non-human primate (NHP) illnesses [2]. There are just two species in the genus *Marburgviruses* which are human pathogens: the MARV (formerly known as Lake Victoria Marburgvirus) and the *Ravn virus* [3]. They both have a genetic difference of about 20%. Both cause the Marburg Virus Disease (MVD) and are regarded as highly pathogenic [4]. After the original epidemic in Frankfurt and Yugoslavia, the virus was named the "Marburg virus" (after the German city Marburg), when it was discovered and described using electron microscopy on plasma from infected guinea pigs [5].

11.1.1.2 Genetics, Molecular, and Chemical Composition with Their Characteristics

MARV is an RNA virus with a single-strand that is filamentous, enveloped, non-segmented, negative-sense, and known to produce viral hemorrhagic fever with characteristics similar to those of the Ebola virus disease [6, 7]. Its virions have a variety of shapes; they might appear as U- or 6-shaped, toroidal, or branched. The virions contain ribonucleoprotein (RNP) complexes which are normally made up of genomic RNA (Figure 11.1). Underneath the viral envelope, it is typically composed of structural proteins such as nucleoprotein (NP) and polymerase co-factor (VP35). Other components include the transcriptional activator (VP30), RNP-associated protein (VP24), and RNA-dependent RNA polymerase (L). The matrix protein (VP40) is responsible for the formation of a regular layer. Glycoproteins (GP) are responsible for the creation of spikes on the surface of the virion. These spikes are approximately 7 nm in diameter and are spaced approximately 10 nm apart [1].

Filovirus genomes are generally 15–19 kb in size without a 5′-cap or 3′-poly (A). The replication and transcription promoters are located in the terminal leader and trailer sequences. The MARV genome is composed of five distinct, continuous open reading frames (ORFs), with an additional two (VP30 and VP24) linked together. These continuous ORFs are surrounded by non-coding 3′- and 5′-terminal sections that include transcription, initiation, and termination sites. The structural proteins of the virion are encoded by the ORFs [1, 8]. The seven structural proteins in the MARV genome – Nucleoprotein (NP), Viral protein 35 (VP35), Viral protein (VP40), Glycoprotein (GP), Viral protein 30 (VP30), Viral protein 24 (VP24), and Large Protein (L) – are all encoded by the seven genes present in the genome [3]. The MARV genome is encased by a nucleocapsid complex composed of four structural proteins: NP, VP35, VP30, and L. The four nucleocapsid complex proteins are all required for viral genome transcription and replication [9].

Rising Contagious Diseases: Basics, Management, and Treatments, First Edition. Edited by Seth Kwabena Amponsah, Ranjita Shegokar, and Yashwant V. Pathak.

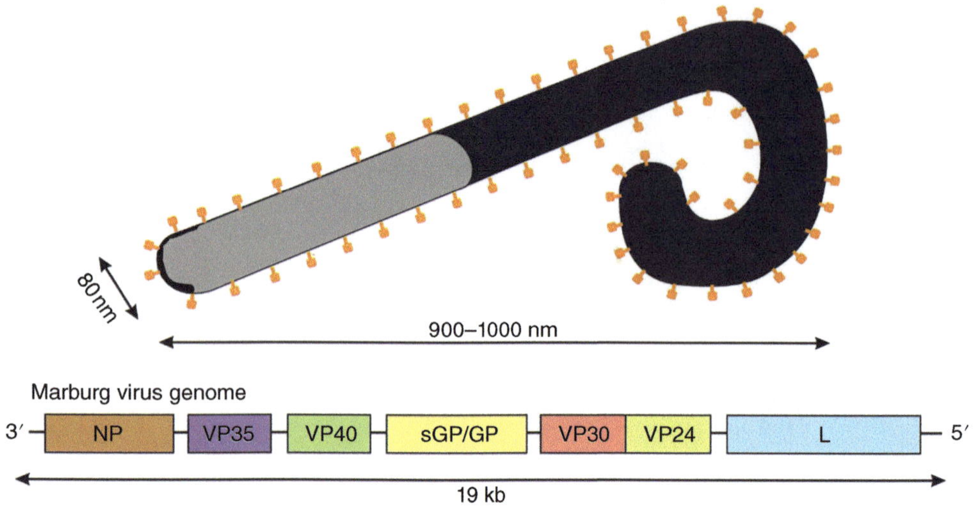

Figure 11.1 Protein (top) and genomic (bottom) characteristics of MARV. NP is nucleoprotein, VP is viral protein, GP is glycoprotein, and L is large protein.

Nucleoprotein (NP) The most abundant protein in the nucleocapsid is called NP, and it is responsible for the formation of a tubular helical shape. This structure interacts with VP35 and the complex that results interact with L. It can attach to the VP35, VP40, VP30, and VP24 proteins. It plays an important role in the development of the nucleocapsid, the cellular inclusion body, the encapsidation of the RNA genome as well as the antigenome, and the replication, transcription, and budding processes [10].

Viral Protein 35 (VP35) VP35 serves as an RNA-dependent RNA polymerase (RdRp) cofactor, a replicase transcriptase cofactor, and an antagonist for interferon (IFN). It can bind to both NP and L, as well as double-stranded RNA (dsRNA). It plays a role in the process of constructing the nucleocapsid [11].

Viral Protein 30 (VP30) It binds to single-stranded RNA (ssRNA), NP, and L. It plays a role in the creation of nucleocapsid, as well as the initiation, re-initiation, and antitermination of transcription, as well as amplification of the process [9].

Large Protein (L) The L protein performs the role of RNA-dependent RNA polymerase (RdRp). Binds to VP35 and VP30, as well as genomic and antigenomic RNA. Functions as an mRNA capping enzyme and the RdRp catalytic domain. Involved in genome replication and mRNA transcription [12].

Glycoprotein (GP) MARV has a membrane layer produced from the host that is spiked at regular intervals. These spikes are composed of highly glycosylated proteins (GP), which are required for attachment to vulnerable host cells. GP is in charge of determining cell and tissue tropism. It is a Class I fusion protein and a type I transmembrane protein. It works as a tetherin antagonist. It generates heterodimers by combining the GP1 and GP2 subunits. The mature protein is present in the form of a trimer of GP1,2 heterodimers. ADAM Metallopeptidase Domain (ADAM17) converts GP1,2 to soluble GP1,2 (GP1,2Δ) [9].

Viral Protein (VP40) VP40 is the protein that creates the virion's inner matrix, and it is this matrix that is responsible for budding. There is a connection between it and the matrix and the nucleocapsid. There is a motif of the late budding present here. Single-stranded RNA (ssRNA) and VP35 are both targets of its binding. It is the protein that is found in the greatest abundance in viral particles and infected cells. It does not like water and is related to membranes. It acts as a negative regulator of the replication and transcription processes. It is in charge of regulating the morphogenesis of the virion as well as the process of egress from the cell. It interferes with the Janus kinase-signal transducer and activator of transcription (JAK–STAT) [13].

Viral Protein 24 (VP24) The discharge of progeny virions from the cell is linked to VP24's ability to interact with the NP and other cellular membranes. Nucleocapsids require VP24 presence during their development and maturity. The activation of cytoprotective responses, the inhibition of transcription, the control of replication, the regulation of virion morphogenesis, and the management of viral egress are all functions that it plays [6].

11.1.1.3 Etiology

The main natural host of the MARV is *Rousettus aegyptiacus*, a member of the *Pteropodidae* family, popularly known as the Egyptian Rousette Bat (ERB) (Figure 11.2), an African fruit bat [14]. MARV RNA and antibodies were first identified in ERBs in Gabon and the Democratic Republic of Congo (DRC) [15]. These fruit bats live in colonies of multiple individuals and construct roosts in caves [16]. No illness symptoms are present in the infected fruit bats. NHPs (non-human primates), such as chimpanzees and monkeys, and humans can contract MARV [14].

During the course of conducting research on imported Ugandan African green monkeys (*Chlorocebus aethiops*) in facilities in Germany (Marburg and Frankfurt) and Serbia (Belgrade), to produce the polio vaccine, the first outbreak of MARV occurred in 1967 [5]. Although the exact cause of the monkeys' MARV infection has not been determined, it has been speculated that they may have been exposed while in Uganda [15]. The infected scientists were reported to be exposed to the tissues of the monkeys [5]. However, many additional outbreaks started in the mines and caves that were home to fruit bats. Epidemiological investigations conducted on these outbreaks revealed that the MVD outbreaks began when infected patients entered a cave or mine, and the exact sites of these caves and mines were determined [17]. Some of the MVD outbreaks, including the numerous outbreaks that occurred in Angola in 2004–2005 have not been able to pinpoint the origins of the human infections that caused them. However, infected ERBs have the potential to introduce MARV into remote locations because the maximum distance that can be relocated is proven to be 500 km and the long-distance flight of ERBs is known to range between 11 and 32 km [16]. Studies have revealed that other bat species, such as *Hipposideros caffer* and a few other *Chiroptera,* can act as natural sources of infection with MARV, in addition to the *Rousettus aegyptiacus* bats that are the predominant natural sources of infection [9].

11.1.1.4 Evolution

Studies have concluded that there are five separate lineages ofMARV based on the evolutionary analysis of genomic sequence information obtained from samples collected throughout a variety of outbreaks. This information was obtained from samples taken at various times during outbreaks. These lineages have been reclassified into two separate viruses: the *Ravn virus* (RAVV) and the MARV [9]. In yet another study, the full-length genome sequence was analyzed, and the results showed that there were two separate evolutionary clades. The first clade consisted of the isolates from Uganda, the Poppinga isolate from the 1967 German outbreak, and the bat isolates from the python cave in Queen Elizabeth National Park (the location in which the Dutch and American tourists likely acquired the MVD), the solitary specimen from Durba (the Democratic Republic of the Congo, in 1999), the Kenyan Musoke isolate from 1980, and all available specimens from the 2004 to 2005 Angola epidemic. The second clade is made up of a single isolate that was found in Zimbabwe in 1975 (the Ozolin isolate), the majority of the isolates that were found during the outbreak that occurred in the DRC from 1998 to 2000, recent isolates from ERBs that were found in South Africa and Zambia, specimens that were acquired in Uganda from 2007 to 2009, as well as the human cases that occurred in Uganda in 2014 and 2017 [7]. It is believed that the rate of molecular evolution for MARV is 5.67×10^{-4} nucleotide substitutions per site per year. This is a slower pace than that shown in Ebola and Reston viruses, but a faster rate than that observed in the Sudan strain of the Ebola virus [7].

Figure 11.2 *Rousettus aegyptiacus* bat, the vector of the Marburg virus.

11.1.2 Laboratory Characteristics of the MARV

11.1.2.1 Cultivation and Detection of MARV

Both the cultivation of MARV as well as its detection take place within the host bat. There are two methods for capturing bats, both of which adhere to the guidelines for the safe handling and sampling of mammals that carry the risk of infectious disease [18]. First, mist nets can be used to catch bats when they are flying through certain areas. The second step is to look for caves and then place nets at the cave entrances and in the area surrounding them to catch the bats that live there [16]. The bats that have been captured should be transported to a biosafety laboratory level 4 (BSL-4) in cotton bags that are breathable and should be moistened to prevent the bats from becoming dehydrated [16]. Blood from bats that have been euthanized can have their hearts punctured, and the blood can then be collected in sterile tubes containing 0.5 M EDTA. After that, samples of blood are put through a centrifuge to separate the plasma from the cells in the blood. Urine can also be collected postmortem with a sterile syringe from the bladder or antemortem in sterile Petri plates. Urine can also be collected antemortem in sterile Petri dishes [16]. During a necropsy, samples of skin, brain slices, lymph nodes, salivary glands, liver, spleen, kidney, lung, and intestinal tissue can be taken and stored at a temperature of −20 °C in RNA tissue-protect buffers (such as those manufactured by Qiagen in Hilden, Germany). Immediately following the completion of the necropsy, pools of bats' internal organs (liver, spleen) and lymph nodes can be homogenized with a Tissue Lyser LT (Qiagen, Hilden, Germany) using 0.5 ml of 0.15 M sodium chloride solution [16].

After this, a RNeasy Plus kit (Qiagen, Hilden, Germany) is used to extract total RNA from a volume of 50 l containing a 10% suspension. An Amplisens FiloA-Fl commercial kit (Central Research Institute of Epidemiology, Moscow, Russia) created for the detection of *Zaire ebolavirus*, *Sudan ebolavirus*, and MARV can be utilized for the detection of filovirus RNA [16]. The diagnosis of MVD can be accomplished with a high degree of accuracy by isolating the virus. The most common cell types used are Vero cells and Vero E6 cells; however, MARV can grow well in a variety of other cell lines as well. It is recommended that viral isolation and all other processes related to viruses be performed in a BSL-4 laboratory [19].

11.1.2.2 Reaction of MARV to Chemical and Physical Agents

Reaction to Chemicals (Chemical Inactivation) Sodium hypochlorite, -propiolactone, 3% acetic acid with a pH of 2.5, phenolic disinfectants, formaldehyde and paraformaldehyde, 1% glutaraldehyde, formalin, lipid solvents, and detergents are all able to kill filovirus [20]. For surfaces that can tolerate stronger bleach solutions such as cement and metal, the World Health Organization (WHO) recommends flooding the area with a 1 : 10 dilution of 5.25% household bleach for 10 minutes after a spill of blood or body fluids. This means that one part household bleach should be diluted in nine parts water, or 0.525% sodium hypochlorite. It is advised that surfaces that have the potential to corrode or discolor be cleaned with a 1 : 100 dilution of 5.25% household bleach for more than 10 minutes. This would be equivalent to using one component household bleach diluted in 99 parts water, or 0.0525% sodium hypochlorite [21].

Reaction to Physical Agents (Physical Inactivation) On contaminated surfaces, filoviruses can last up to four to five days, and they can also endure for several days in liquids or dried materials. It has been reported that 60 °C heating for 30–60 minutes, 5 minutes of boiling water, γ radiation, and UV radiation are all effective against them [19, 22].

11.1.2.3 Laboratory Safety of MARV

MARV has been categorized as a risk group 4 pathogen because of its highly infectious nature, propensity for human-to-human transmission, high risk for laboratory-acquired infections, and the absence of a vaccination that is both specific and safe. This classification mandates that MARV be contained at a biosafety level of 4, often known as BSL-4 [23]. Therefore, to safely confirm an infection, testing should only be performed in a BSL-4 facility, also known as a high-containment laboratory, that employs good laboratory practices [24].

11.1.3 Mode of Transmission, Replication, and Intracellular Life Cycle

11.1.3.1 Mode of Transmission

Since it was discovered that ERBs are the natural hosts of the MARV virus, the majority of the research efforts have been directed toward determining how the virus is passed on to humans. Amman et al. and Paweska et al. published their first experimental research using ERBs that had been bred in captivity [25, 26]. The research that was conducted by Amman and colleagues involved the use of a natural bat isolate of MARV. The researchers found the highest levels of viral shedding

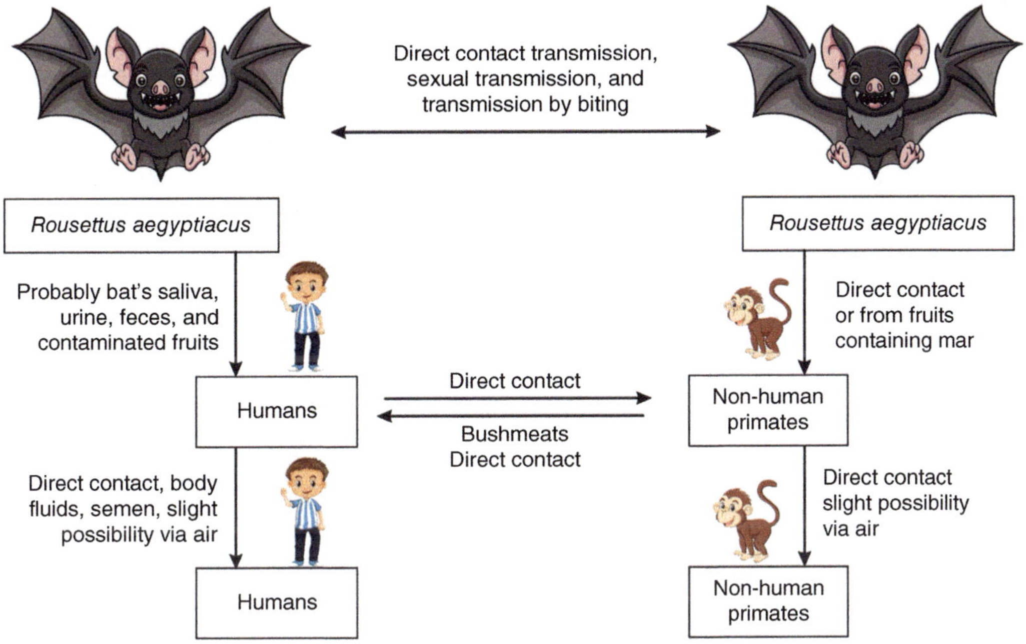

Figure 11.3 Cycle of transmission of the MARV between the bat (*Rousettus aegyptiacus*) host, a non-human primate (monkey), and humans.

in the oral secretions of the bats, which suggests that mucosal contact with the oral secretions of infected ERBs could be a potential mode of transmission. Paweska et al. also discovered MARV RNA in the urine of experimentally infected ERBs, which indicates that potential contact with infectious urine could be another route of transmission. This is based on the fact that it is common knowledge that bats urinate and defecate while flying. A further study that was carried out by Amman et al. to evaluate viral persistence or viral environmental stability and viability on fruits and the potential of MARV transmission to humans through environmental contamination, specifically contaminated fruits over 24 hours post-inoculation found that the MARV is stable for at least 6 hours on fruits, particularly mangoes and bananas, which are routinely consumed by both ERBs and primates [15].

The majority of MARV gearbox occurs close to where ERBs have been discovered [15]. These ERBs may be taken by hand, hunted opportunistically with sticks or machetes, or even shot with firearms. Eating contaminated fruits and coming into touch with sick primates that have become unwell as a result of eating contaminated fruits are the two ways in which the MARV can be passed on. ERBs typically do not consume the entirety of the fruits that they eat; rather, they chew the fruits to extract the juice and then throw away the pulp. In addition, ERBs are known to sample fruits by biting them, urinating on them, or dropping fruits they are in the process of eating [15]. This feeding behavior, in conjunction with their capacity to excrete the virus through secretions such as pee and saliva [16], has been posited as a potential mode of MARV transmission and is being investigated. There have been reports of cases involving male survivors who had documented ZEBOV RNA persistence in their sperm for longer than a year; as a result, MARV RNA might potentially be transmitted in the same manner [27]. An annotated cycle of how the MARV is transmitted between three species (including humans) is demonstrated in Figure 11.3.

Direct contact with infected body fluids or close, physical contact with an infected animal or human is the most typical method by which MARV infections are contracted (Figure 11.3), either by exposing mucosa or abraded skin or by being pierced by a needle or another sharp object [28], which has been exposed to infected fluids and secretions from any of these sources.

11.1.3.2 Cellular Life Cycle and Replication of MARV

The MARV, upon entering the body (especially the blood system), infects several cells including the macrophages (and monocytes), liver cells (or hepatocytes), dendritic cells (DCs), fibroblasts, endothelial cells and others, via its interaction with cell surface glycoproteins [29]. First, in order to initially connect to its host cells, MARV engages in interactions with cell surface molecules, the majority of which are glycoproteins (GP). Endosomal proteases will cut GP1 into smaller pieces after it has been endocytosed. This cleavage makes it easier for the molecule to attach to the receptor and transporter for endosomal cholesterol. The fusion process is then mediated by GP2 in a pH-dependent manner. When a virus replicates, its nucleocapsid is discharged into the cytosol of its host cell. The transcription of the viral genome begins after the viral nucleocapsid is degraded and released into the cytoplasm. The mRNA is then translated by the machinery of the host cell. The endoplasmic reticulum (ER) is responsible for the synthesis of GP, and as it moves along the conventional secretory route, GP goes through a number of posttranslational modifications. The newly arrived viral genomes are included in the production of the positive sense antigenomes. After that, these intermediate products are put to use in the process of replicating new genomes with a negative sense. After the Golgi cleavage, the GP is moved to the multivesicular bodies (MVB) and the cell membrane, both of which are locations where budding takes place. At places where viral budding is taking place, nucleocapsids and viral protein 24 (VP24) are also brought in. Matrix protein, also known as VP40, is the molecule that is in charge of promoting recruitment to viral budding sites. Matrix protein (VP40)-induced budding at the plasma membrane results in the assembly of the virions, which includes the acquisition of the GP1,2-containing lipid envelope. This is followed by the release of the virions extracellularly, which concludes the process [1, 6].

11.2 Pathogenesis and Pathophysiology of MARV

As was said previously, when MARV is exposed to the human blood system, the GP on the surface of the virus mediates binding and entry into the host cells. This occurs after the virus has been exposed to the human blood system. The GP surface unit, also known as GP1, connects to cellular receptors, whereas the internal fusion loop, also known as GP2, inserts itself into the cell membrane [30]. Targets of the virus include dendritic cells, monocytes, parenchyma cells in the liver, adrenocortical cells, and several lymphoid organs. The virus enters the body through the blood or lymphatic system [9]. The GP that is found on the surface of MARV acts as a mediator for the virus's ability to bind to and enter host cells. The GP surface unit, also known as GP1, connects to cellular receptors, whereas the internal fusion loop, also known as GP2, inserts itself into the cell membrane [30]. The virus enters the body through the bloodstream or the lymphatic system, and it attacks approximately five primary types of cells, including dendritic cells, monocytes, parenchyma cells of the liver, adrenocortical cells, and many lymphoid tissues [9]. These cells are where the first replication of the virus takes place, and they are also the ones responsible for its further dissemination to hepatocytes, endothelial cells, and epithelial cells [31]. Although the virus does not directly infect lymphocytes, dendritic cell infection results in poor stimulation of T lymphocytes, which results in natural killer and lymphocyte apoptosis [31]. Because of this, the body's immunity is suppressed as a result of the depletion of lymphoid cells in the spleen, liver, lymph nodes, and thymus. This uncontrolled viral replication influences the ability to mount an adaptive immune response which then leads to increased cytokines/chemokines numbers (cytokines-chemokines storm) and impaired humoral response, ultimately leading to shock as well as multi-organ injuries [9, 31].

A considerable number of monocytes and macrophages can be found in the majority of organs and tissues. It is possible to acquire a systemic infection as a consequence of the movement of infected monocytes and macrophages into neighboring tissues or as a consequence of the movement of free viruses through the lymphatic or circulatory systems. Infection of macrophages or monocytes causes uncontrolled activation of cytokines and chemokines, which in turn induces bystander death in the lymphocyte population, which in turn leads to immunosuppression and causes damage to the endothelium cells [9]. Endothelial cell infection, through different pathways, causes immune dysfunction which results in an increase of blood vessels permeability influenced by TNF-alpha (Tumor Necrosis Factor-alpha), nitric oxide, and other vasoactive compounds, and disseminated intravascular coagulopathy (DIC) influenced by abundant tissue factor expressed by macrophages. Both increased vascular permeability and DIC lead to hemorrhages. Systemic replication would continue because of the infection of endothelial cells [9, 31].

The presence of parenchymal cell infection in the liver has been shown to have an effect on the synthesis of clotting factors and contribute to the aberrant coagulation patterns seen in MVD. In the long run, bleeding may be a consequence of it as well [9]. Infection of the adrenocortical cells of the adrenal gland by the MARV virus can result in metabolic abnormalities as well as dysregulation of blood pressure (hypotension). The MARV infection causes hemorrhaging to emerge at a later point in the disease's progression [9]. Tissue necrosis and defective adaptive response result from MARV infection of lymphoid tissues in the lymphatic system, particularly the lymph nodes and spleen. Necrotic lesions were seen throughout the body, but they were most severe in the lymph nodes, liver, and spleen. Because these organs (mentioned above) have high concentrations of reticuloendothelial cells, the movement of infected cells makes it easier for the infection to spread to multiple organs, which can lead to the development of systemic infection shock and lymphoid organ damage can occur in the later stages of the systemic infection [32].

11.3 Marburg Virus Disease (MVD)

MARV is the principal agent in the development of MARV disease (MVD). MVD is a form of hemorrhagic fever that is brought on by a genetically distinct animal-borne RNA virus that is a member of the filovirus family. In humans and non-human primates (NHPs), it is fatal and frequently becomes untreatable as the disease progresses. A hemorrhagic fever, organ dysfunctions such as liver failure, the infection of the spleen, brain, and renal/adrenal tissues, and coagulation problems throughout the body are all symptoms that can be caused by the MARV virus, much as they can be caused by the Ebola virus [9]. There is evidence that the African fruit bat acts as a reservoir for this virus. It is generally believed that MVD is a zoonotic illness that is transmitted from animals to humans. The virus can be passed on to people from bats (specifically the *Rousettus aegyptiacus*) by extended exposure to mines and caves that are inhabited by these bat colonies as well as through contact with bat saliva, feces, and contaminated fruits [16].

Transmission from one individual to another can also take place by direct contact (with damaged skin or mucous membranes), with the secretions or bodily fluids of a person who is infected with the virus [9]. It is also possible for the MARV virus to be spread by fluids and secretions that have contaminated surfaces and materials (such as bed linens and clothing). During the process of preparing an infected body for burial, there is a possibility of viral transmission [33]. The virus has been found in the sperm of a patient who was previously infected, indicating that it can also be passed on through sexual contact. Male survivors of MVD are encouraged to engage in safer sexual activity for one year, during which time their sperm must be tested negative [27]. The history of MVD epidemics in the past demonstrates that they can be extremely lethal, with a death rate that can reach up to 90% [24]. The progression of the disease can be broken down into three separate stages: the generalization stage, the early organ stage, and the late organ stage, often known as the convalescence stage. The incubation period in humans can be anywhere from 2 to 21 days, with 5 to 10 days being the most common range [24]. At this time, there is neither a vaccine nor a medication specifically designed to combat MARV illness. The therapy consists solely of supportive care, such as rehydration, the replacement of lost blood and clotting factors, the management of symptoms, and the diagnosis and treatment of exacerbating co-infections [33]. However, there is currently research being conducted on several antivirals that are in the process of being developed to combat the virus. These antivirals, which include Galidesivir, Favipiravir, Remdesivir, Polyclonal concentrated IgG, and others, have the potential to offer a specific pharmacological treatment for MVD [31].

11.3.1 Historical Epidemiology of MVD

There have been around 13 outbreaks of MARV infection across the globe (so far). While scientists were conducting research on an imported Ugandan African green monkey called *Chlorocebus aethiops* in 1967, which ultimately led to the discovery of MARV, the first outbreak of MARV occurred simultaneously in laboratories in Marburg and Frankfurt, Germany as well as in Belgrade, Serbia. It was discovered that the infected individuals had been in contact with the monkeys' tissues [5]. The second epidemic was the very first to take place on the African continent, and it happened in South Africa in the year 1975. The information that was gathered revealed that the index patient, who was a tourist, went to the Sinoia caves in Zimbabwe and may have had direct contact with bats while he was there. In addition to this, he slept in rooms that had insectivorous bats and where there was the possibility of coming into direct touch with bat discharges [17]. Transmission from human to human resulted in the infection of his traveling companion and a nurse [9]. In the year 1980, Kenya was the location of the third MARV outbreak. The index patient was said to have visited the Kitum cave in Mount Elgon National Park, which is known to be a MARV reservoir. He was likely exposed to the virus there. Following the initial outbreak that took place in Kenya in 1980, a second outbreak took place there in 1987. According to reports, the index patient had been to the Kitum cave in question [17].

In 1988, 1990, 1991, and 1995, researchers in Russia reported incidences of laboratory contamination caused by MARV [9]. Ten years after the last outbreak in Kenya, in 1998, gold miners in the Democratic Republic of the Congo (DRC) contracted the disease after coming into touch with the fauna of the mine and their waste secretions, such as bat excreta. This outbreak persisted until late 2000. Some of the miners were said to have had contact with infected people, including non-mining relatives of infected miners. Exposure to the fauna and the waste secretions of the mine may have caused the epidemic, as it subsided after the gold mine was flooded in late 2000 [17]. During this epidemic, nine MARV lineages that were shown to be genetically distinct were discovered. After four years, a further outbreak was recorded in Angola during the years of 2004 and 2005. This new outbreak was the greatest MARV outbreak (252 cases) ever documented as of the present day. On the other hand, there was an insufficient epidemiological correlation between cases to determine where the illness originated [14]. In addition, it was hypothesized that the first outbreak could have been caused by the death of a hospital worker in the province of Uige in Angola and that the disease then went on to spread to other regions. Clinical investigations indicated genetic sequence similarities between previously identified MARV isolates and those connected with the MARV outbreak in Angola between 2004 and 2005 [9].

In 2007, Uganda was the site of an outbreak, however, there were only four confirmed cases. It was reported that all of the patients were workers at the Kitaka mine, which is located in the Ibanda district. It was said that bats roosted in the tunnels of the mine where the workers did their jobs without wearing masks, respirators, or goggles; they simply wore gloves. The index patient, who was infected, shared a tent with two other patients at the time of the outbreak. It was reported that the fourth patient became infected after entering the mine without wearing any type of personal protective equipment (PPE) at the time of the outbreak. After ecological sampling of bats from the mine was carried out to isolate the MARV virus, researchers were able to identify a definitive filovirus reservoir for the very first time. It was determined that the host reservoir for MARV is the *Rousettus aegypti* species [17].

In 2008, there were reports of two incidents in Uganda involving visitors from the United States and the Netherlands. It was said that the American tourist went camping, visited local towns, and observed animals, while the Dutch tourist was said to have explored two caves – one with bats and another without them – and came within five meters of wild gorillas. It was also reported that the American tourist camped there. The American tourist contracted the disease while she was out camping or seeing wildlife, activities in which she came into contact with wild animals. It is also possible that the Dutch tourist contracted the disease after being exposed to bats in the Python cave, which have been recognized as reservoirs of the MARV virus [4]. The American patient made it through the disease's aftermath, while the Dutch one did not [9]. As a result of this event, Uganda was chosen as the location for the subsequent three epidemics in the years 2012, 2014, and 2017. The first of these three epidemics took place in 2012, however, the patient who served as the index case was never located. However, after conducting contact tracing, the earliest confirmed case was found to have originated from the Ibanda district, which was the location of the MARV outbreak in 2007, indicating that the patient may have been exposed to ERBs [18]. In the year 2014, Uganda witnessed a case of MARV exposure that resulted in the death of a healthcare worker. Studies on secondary infection and probable contact with the *Rousettus aegyptiacus* bat were conducted epidemiologically and ecologically, respectively, and the results were negative [17]. In another case, the MVD claimed the lives of three out of four members of one family in the Kween district of Uganda in the year 2017. Because of the timely detection of new cases and the availability of a well-trained response team, the outbreak was brought under control in less than a month. The original patient was a herdsman who hunted game in a sub-county that has a big population of ERBs hiding in caves [18]. A case of MARV exposure occurred in Guinea in 2021 for the first time in West Africa, in a patient who was diagnosed after death. After monitoring more than 170 high-risk contacts for 21 days, epidemiological investigations found no evidence of secondary infection [24].

After confirming the presence of MARV in a 26-year-old index patient using RT-PCR at the Noguchi Memorial Institute for Medical Research (NMIMR), Ghana (which is also located in West Africa) declared its first epidemic of MVD on 7 July 2022, in the Ashanti area. On 22 June, he started experiencing symptoms, and soon after that, he started bleeding from his lips and nose. The patient was taken to the hospital on the 26th of June 2022, but he or she did not survive the night. The second patient, a kid who had just turned 14 months old, started showing signs and passed away on the third day after admission. On 26 July 2022, a female patient who was 24 years old was checked into a facility that had been classified by the government as an isolation ward. Each of the three cases originated in the same family home. The outbreak was brought under control as a result of the prompt identification of new cases and the presence of skilled response teams. After conducting epidemiological research on a total of 198 identified contacts who were separated and monitored for 42 days (twice the maximum incubation period), Ghana's Ministry of Health announced on 16 September that the outbreak had come to an end. This came after the contacts had been quarantined for the full maximum incubation period. Because the epidemiological investigation has not yet located the origin of this outbreak, community-based surveillance must be increased. The results of genetic sequencing, on the other hand, suggested that the MARV genomes from Ghana are related to the sequence from the outbreak in 2021 and, more generally, to groups with sequences recovered from bats in Sierra Leone and an outbreak that occurred in Angola in 2004–2005 [34]. Table 11.1 shows the history of the worldwide outbreak of the MVD pandemic.

Table 11.1 History of worldwide outbreaks of the Marburg virus disease.

Year of outbreak	Suffering country	Suspected country of origin	Number of cases	Number of deaths	Case fatality rate (CFR) (%)
1967	Serbia and Germany	Uganda	31	7	23
1975	South Africa	Zimbabwe	3	1	33
1980	Kenya	Kitum Cave, Kenya	2	1	50
1987	Kenya	Kenya	1	1	100
1988–1995	Russia	Russia	4	2	50
1998–2000	Democratic Republic of Congo (DRC)	Durba, DRC	154	128	83
2004–2005	Angola	Uige, Angola	252	227	90
2007	Uganda	Ibanda District (Kitaka mine), Uganda	4	1	25
2008	United States	Cave in Maramagambo forest, Uganda	1	0	0
2008	Netherlands	Cave in Maramagambo forest, Uganda	1	1	100
2012	Uganda	Ibanda, Kabale and Kamwenge, Uganda	15	4	27
2014	Uganda	Kampala, Uganda	1	1	100
2017	Uganda	Kween, Uganda	4	3	75
2021	Guinea	Gueckedou, Guinea	1	1	100
2022	Ghana	Gueckedou, Guinea	3	2	67

11.3.2 Immunity, Clinical Features, and Findings in MVD

According to experimental data, mononuclear phagocytic cells are the virus's principal target, with epithelial cells in several organs coming in a close second. Variables such as the virulence of the infecting strain can affect how MARV manifests in a patient's body. Other factors include the patient's physical status, their susceptibility to the virus, and their need for ongoing medical care [9].

11.3.2.1 Immune Evasion Mechanism of MARV

Within 24–48 hours of infection, the host's gene expression profile is altered by the MARV infection, which evades the immune system in large part by changing genes involved in coagulation, apoptosis, and immunoregulation. Interferon-stimulated gene (ISG) synthesis in hepatocytes is associated with altered gene expression and can function as a potent antiviral suppression mediator [9]. According to the findings, MARV is capable of effectively inhibiting interferon (IFN) reactions, including type 1 and type 2 IFN signaling, and eliciting an inflammatory cytokine response.

Table 11.1 shows a brief overview of the history of the Marburg virus disease around the world.

It has been demonstrated that the action of EBOV VP35 suppresses the development of IFN by blocking the activation of IFN-regulatory factor 3 (IRF3), but this response has not yet been explicitly established for MARV. On the other hand, there is currently no evidence that can be used to explain the acquired immune responses following recovery from Marburg hemorrhagic fever (MHF). During the outbreak in the DRC, after the beginning of symptoms, MARV survivors showed signs of having developed MARV-specific antibodies, as evidenced by the presence of both IgM and IgG antibodies [9]. Hematological changes, such as early leukopenia, neutrophilia, and monocytosis, along with moderate eosinophilia, have also been noted. Patients suffering from MVD experience immunosuppression as a direct result of these alterations, which makes them more susceptible to secondary bacterial infections. It has also been observed that increased amounts of soluble nitric oxide and proinflammatory cytokine in the blood can promote intravascular apoptosis [9]. The observed incubation period in humans spans from 2 to 21 days, with an average of 5–10 days, and is probably influenced by factors including infectious dosage and perhaps infection route [24].

11.3.2.2 Clinical Features and Findings in MVD

According to the most recent clinical evidence, the progression of MVD may be broken down into three stages, each of which is characterized by a unique set of symptoms: the first stage, known as the initial generalization phase, is followed by the early organ phase, and the third stage, known as the late organ phase or the convalescent phase, comes last [9].

1) **Phase 1 (Generalization phase):** General flu-like symptoms, a high fever (39–40 °C), weariness, loss of appetite, abdominal discomfort, severe weight loss, severe nausea, vomiting, watery diarrhea, and anorexia; severe headaches, myalgia, chills, and malaise, and so on are all symptoms that may appear during this stage. It is common for symptoms such as conjunctivitis, dysphasia, enanthem, and pharyngitis to appear toward the conclusion of this first stage [9]. A distinguishing sign of filovirus infection appears around the fifth day when a peculiar maculopapular rash may appear conspicuously over the trunk (neck, back, chest, and stomach) [24]. In addition to this, patients may experience lymphadenopathy, leukopenia, and thrombocytopenia [9].
2) **Phase 2 (Early organ phase):** This stage is characterized by a persistent high temperature and nonspecific symptoms. Patients may experience conjunctivitis infection, prostration, shortness of breath (dyspnea), viral exanthema, aberrant vascular permeability, and edema in addition to neurological symptoms like disorientation, encephalitis, irritability, psychosis, and aggressiveness [9]. Hemorrhagic symptoms are present in around 75% of patients. These symptoms include mucosal bleeding, melena, petechiae, bloody diarrhea, visceral hemorrhagic effusions, uncontrolled leakage from venipuncture sites, hematemesis, and ecchymoses. Other symptoms include ecchymoses and hematemesis. There have also been reports of bleeding from the mouth, the nose, and the vagina. As a result of the presence of hemorrhagic symptoms in certain individuals, the MARV infection is commonly referred to as MHF, which stands for Marburg Hemorrhagic Fever, in certain instances; however, this does not guarantee a positive outcome in all cases. During this stage of the infection, several organs, including the kidney, the liver, and the pancreas, are compromised [9].
3) **Phase 3 (Late organ phase or convalescent phase):** Late-stage MARV infection can have one of two unique outcomes: the infection can either be fatal or non-fatal, and patients enter a protracted phase of restoration. After the onset of symptoms, death typically occurs anywhere between 8 and 16 days later, on average. Generally speaking, shock and the failure of many organs are the leading causes of mortality [9]. The case fatality rate for MHF is between 23% and 90% [24]. During the disease, the number of cases that do not result in death begins on day 13 and continues until day 20 and beyond. At this stage, severe metabolic abnormalities, such as convulsions and severe dehydration, are the cause of multi-organ dysfunction and anuria. Anuria is the absence of urine production. Neurological symptoms are still prevalent during this phase of the condition. Orchitis is a condition that can affect men, and unplanned pregnancies can result in spontaneous abortions in women. During this drawn-out period, symptoms such as myalgia, weariness, partial amnesia, sweating, peeling skin in rash-affected areas, and secondary infections are present and noticeable. During the stage of the infection known as convalescence, symptoms such as arthralgia, hepatitis, asthenia, ophthalmic disease, and psychosis are common side effects [9]. There have also been reports of other particular symptoms that are associated with the temporary persistence of the virus in immune-privileged areas such as the testicles and the eyes. For example, the testicles are more susceptible to the effects of the virus than the eyes. During the MARV illness outbreak that took place in the Democratic Republic of the Congo between the years 1998 and 2000, researchers found that patients who exhibited conjunctival injection and hiccups had larger rates of fatality [35].

11.3.3 Laboratory Diagnosis of MVD

The diagnosis of MVD can be made using a variety of methods, such as electron microscopy, culture, RT-PCR, antigen testing, serology, and immunohistochemistry. These methods are used in conjunction with one another and are dependent on the progression of the infection. The direct detection of viral particles, proteins, or particular RNA in a suspected case can be used to arrive at a diagnosis of MARV. Diagnostic materials such as blood, other body fluids, and tissue recovered at autopsy can be used to make this determination [31].

The NP, L, and GP genes are the primary focus of diagnostic tests that make use of molecular techniques. These genes have been demonstrated to have high sensitivity and specificity when it comes to the diagnosis of MVDs. In addition, because it was shown that the GP gene is unique to each strain of the virus, tests that focus on the GP gene can be used to identify the specific strain of the virus that was responsible for the infection. On the other hand, antigen testing is successful in the early stages of MVD, when the viral load in blood and tissues is high. This is the case because antigens can detect the virus before it spreads throughout the body [8].

There are a few different serological methods, the most common of which are antigen-capture ELISA and indirect immunofluorescence assay (IFA). IgM-capture ELISA may detect IgM antibodies indicating current infection as early as 2 to 4 days after the onset of symptoms indicating the beginning of symptoms, in contrast to IgG-capture ELISA, which can identify IgG antibodies indicating prior infection. ELISA can detect IgM antibodies indicating current infection as early as 2 to 4 days after the onset of symptoms indicating the beginning of symptoms. 8 to 10 days following the initial manifestation of symptoms [8]. It is possible to make a retrospective diagnosis of MVD in individuals who have passed away by using immunohistochemistry, virus isolation, or RT-PCR on blood or tissue specimens [24].

11.3.4 Treatment, Control, and Prevention of MVD

11.3.4.1 Treatment

At this time, neither vaccines nor direct antiviral medicines have been authorized for the treatment of MVD. On the other hand, it has been observed that providing supportive care in the form of rehydration through oral or intravenous fluids and treatment of specific symptoms can improve a patients' chances of survival [14].

Pre-Exposure Prophylaxis (Potential Vaccines) The three main candidates of MARV vaccines (cAd3, MVA BN Filo, and MARV DNA) are in phase 1 clinical studies [36]. Multiple Marburg candidate platforms (Adenovirus, DNA, VLP, rVSV) have demonstrated protection in non-human primates (NHPs) [37]. Below are some vaccines currently being developed for use against MARV:

- **Adenovirus-based vaccine:** Several adenovirus-based vaccines have been studied for the Ebolavirus (EBOV), but studies are limited to MARV [31]. The most commonly used vector for GP vaccines is recombinant adenovirus serotype 5 (rAd5). Five filovirus antigens are used in the complex adenovirus (CAdVax) platform. These include three GPs: EBOV, Sudan virus (SUDV), and MARV (Ravn, Musoke and Ci67 strains), and two NPs: EBOV and MARV-Musoke [4]. A phase 1 clinical trial for the chimpanzee adenovirus serotype 3 vector vaccine, or (cAd3) MARV vaccine, is presently underway. The combination of a cAd3 prime and a Modified Vaccinia Ankara boost exhibited 100% protection against the Ebola virus and may do the same for MARV. The use of MARV in comparable research in NHPs has not yet been done [31].

Phase 2 clinical studies are scheduled for the modified vaccinia Ankara vector vaccine (MVA-BN-Filo), which contains GPs for EBOV, SUDV, MARV, and Tai Forest NP [36]. Ad26.ZEBOV or MVA-BN-Filo was administered to 87 participants in phase 1 clinical trial of the vaccines. Twenty-eight days following primary immunization, 97% of Ad26. ZEBOV recipients and 23% of MVA-BN-Filo recipients had measurable IgG responses. An alternate vaccination was given to the participants at 14, 28, and 56 days following their initial immunization. There were no severe adverse reactions to the vaccines. All recipients had detectable IgG levels at 21 days and 8 months following the alternative vaccine boost [31]. Following this, in May 2020, the European Medicines Agency (EMA) granted marketing authorization to Zabdeno (Ad26.ZEBOV) and Mvabea (MVA-BN-Filo) to produce vaccines against EVD [14]. The vaccine could potentially protect against MVD too, but its efficacy has not been proven in clinical trials yet [14].

- **DNA-based vaccines:** Greater success has been achieved with DNA-based vaccinations, such as the MARV DNA plasmid vaccine (VRC-MARDNA025-00-VP), which has finished phase 1b clinical testing and expresses MARV Angola DNA. 90% of the 10 recipients of the vaccine showed antibody responses at, 4, and 8 weeks. Seven recipients received a fourth dose at 12 weeks, which increased fading antibody titers [31].
- **Virus-like particles (VLP) based vaccines:** Using MARV VP40, GP, and NP, MARV virus-like particles (mVLP) vaccines have been developed. A mVLP vaccination against the MARV-Musoke, Ravn, and Ci67 isolates was evaluated in non-human primates, and antibody responses to all three strains were observed. Four weeks later, when exposed to MARV, all showed cross-protection [4].
- **Recombinant vesicular stomatitis virus (rVSV):** MARV Recombinant Vesicular Stomatitis Virus (rVSV) vaccines contain MARV GP in place of its innate surface membrane glycoprotein and are replication-competent. The use of three VSV vectors containing MARV, EBOV, and SUDV GP in a vaccine that was administered to NHPs resulted in antibody responses to all three components and 100% cross-protection against MARV, EBOV, and SUDV 28 days later [31]. A ring vaccination trial using the rVSV-EBOV vaccine was successful during the West African EBOV pandemic that lasted from 2014 to 2016 [38]. It has since been employed during recent EBOV outbreaks in the Democratic Republic of the Congo

(DRC). The vaccine is now approved in both the United States and Europe [31]. Although there are currently no phase I clinical trials underway, the rVSV platform seems promising for MARV.

Post-Exposure Prophylaxis There are monoclonal antibodies (mAbs) under development for post-exposure prophylaxis [14]. After challenging Rhesus macaques (*Macaca mulatta*) with Marburg Musoke, there were reported reduced death when a VSV-vectored vaccine was used as a post-exposure prophylaxis (PEP) administered within 20–30 minutes which demonstrated 5/5 protection, administered at 24 hours demonstrated 5/6 protection and administered at 48 hours demonstrated 2/6 protection [31]. Depending on the dose (high 1000 pfu, low 50 pfu), Rhesus macaques were granted post-exposure protection against the more virulent Marburg Angola variant when they were vaccinated against it within 20–30 minutes of being challenged with the virus [39].

Antivirals Antivirals such as Remdesivir and Favipiravir, which have been used in clinical studies for Ebola Virus Disease (EVD), could also be tested for MVD, or they could be administered under compassionate use or expanded access provisions [14]. Multiple anti-MARV pharmaceutical active agents are currently in the research and development stage [31]. However, the following antivirals have been used so far.

- **Galidesivir:** It is a synthetic nucleoside analogue that works as a non-obligate RNA chain terminator to inhibit viral RNA polymerase from being active. Six cynomolgus macaque groups were MARV-challenged, and 1, 24, and 48 hours later, Galidesivir (15 mg/kg) twice-daily IM injections were administered. Within the first hour, 5/6 animals recovered. Both the 24- and 48-hour groups had 6/6 survivors. There were no overt toxicity indicators found. Animals receiving treatment showed improved liver enzyme levels, decreased clotting times, and lower viremia levels. Galidesivir has also been shown in rhesus macaques to provide post-exposure defense against the Ebola virus. A phase 1 safety study for testing in humans was completed in 2016, however, the findings have not yet been made public [31].
- **Favipiravir:** It is a synthetic guanidine nucleoside analogue that possesses significant antiviral activity against a wide variety of RNA virus families. On five out of six Cynomolgus macaques that were given, the Marburg Angola virus survived after receiving favipiravir intravenously on a twice-daily basis for 14 days, beginning on the day of the challenge. Oral administration of the drug had no therapeutic effects [40]. Despite the absence of a significant survival effect, the second study carried out in Guinea revealed a tendency toward improved survival in the population that had been treated [41].
- **Remdesivir:** It has an in vitro Marburg activity and is the prodrug of an adenosine analogue. It has been used to successfully treat EVD in nonhuman primates [42] and has shown promise in the treatment of EVD in humans after being administered to a preterm infant born to an infected mother. The baby was given leukocytes and ZMapp and was able to recover enough to go home [43]. Treatment of Marburg-infected cynomolgus macaques four to five days after exposure with once-daily doses of either 5 or 10 mg for 12 days has also recently shown promise, with 50% and 83% survival rates, respectively [42]. Following the terms of the emergency use permission, remdesivir has been put to extensive use as a prophylactic measure during the 2019–2020 COVID-19 outbreak, and several clinical investigations are still being carried out with it [44].

Other Treatment Options

- **Interferon-beta:** Studies found that administering interferon beta after a Marburg infection in macaques enhanced mean survival time, but did not affect mortality, suggesting that it may be used as adjuvant therapy [31].
- **Concentrated (polyclonal) IgG:** Concentrated (polyclonal) IgG was derived from vaccinated NHPs that had survived the challenge with MARV. Following a MARV challenge, rhesus macaques were given three doses (100 mg/kg) that completely protected the animals from viremia and clinical sickness, but they did develop an IGM response. MARV challenge issued again demonstrated full protection 77 days later [31].
- **Monoclonal antibodies(mAbs):** A panel of monoclonal antibodies was given to guinea pigs and nonhuman primates, and all of them were able to survive after demonstrating protection against the MARV challenge in mice [45]. During the Ebola virus disease (EVD) outbreak that occurred in West Africa, a three-antibody cocktail known as ZMapp was tested on humans in a randomized controlled trial (RCT). The patients who were given ZMapp fared significantly better than the patients who were only given supportive therapy. During the 2018–2019 Ebola outbreak that occurred in the Democratic Republic of the Congo (DRC), ZMapp was also evaluated as part of a four-drug RCT known as the PALM study [46]. One particular formulation of mAbs (mAb114 and REGN-EB3) performed exceptionally well in reducing

mortality. There is some indication that treating VHF illnesses like MVD with a single monoclonal preparation may be sufficient. This is even though ZMapp and REGN-EB3 are triclonal cocktail products, whereas the mAb1114 product is a single monoclonal antibody [31].

- **Phosphorodiamidate morpholino oligomers (PMOs) with positive charges:** PMOs prevent the translation of mRNA by steric hindrance [39]. The translational machinery is prevented from reaching the morpholino group, which is analogous to a ribose base in RNA, by a linking molecule known as methylene phosphorodiamidate. This molecule physically attaches to mRNA. It is believed that the inclusion of a piperazine residue leads to the insertion of a positive charge (denoted by PMO plus), which improves binding to negatively charged mRNA and prevents the generation of resistance mutations. Antisense PMOs are extremely stable and soluble once they have bound to the mRNA target, which enables them to achieve substantial levels of suppression while maintaining consistently low levels of toxicity [31].

11.3.4.2 Control and Prevention

Infection Prevention and Control (IPC) at Health Care Facilities Healthcare providers should follow infection prevention and control (IPC) procedures when tending to patients who have MVD, whether they are suspected or confirmed cases, to prevent any exposure to blood, body fluids, and/or potentially contaminated environments. Aerosol-generating procedures should be avoided, or where necessarily unavoidable, well-contained. The usage of needles and other sharp objects ought to be minimized [4]. It is recommended that patients suspected of having MVD be isolated in a single room with a separate bathroom immediately, as per the Viral Hemorrhagic Fever (VHF) protocol, after they have been identified through screening or triage [4]. Appropriate isolation infrastructure and human resources should be made available. Hand hygiene resources (i.e. soap and water or alcohol-based hand rub) and appropriate personal protective equipment (PPE) should be made accessible to healthcare workers. Healthcare staff should have access to and be given the proper resources for decontaminating, sterilizing, and disposing of infectious waste. Within affected communities, integrated disease surveillance and response efforts should be made [34].

The production of aerosols is the determining factor in the availability of various treatment options. These alternatives may include non-invasive positive pressure ventilation (NPPV), masking up, or careful intubation. However, careful handling is required in every scenario. It is recommended to use NPPV rather than mask therapies, but intubation should be done instead of NPPV if it is at all possible. Furthermore, manual ventilation procedures, particularly during resuscitation, should be kept to a minimum as much as possible. Endotracheal intubation should only be done by the most qualified medical professionals, who should wear the appropriate personal protective equipment (PPE), and ventilated patients diagnosed with HID are required to have a highly precautionary tracheostomy and bronchoscopy, either with or without bronchoalveolar lavage. The pediatric population that is infected with MARD needs to be handled with extreme caution as well because there is a risk of infection spreading due to inadequate precautionary methods [47, 48].

Infection Prevention and Control (IPC) in the Community The nature of the disease, the risk factors connected with the sickness, and the precautionary measures that persons may take to reduce human exposure to the virus should be brought to people's attention through education so that efforts can be made to reduce the number of human infections and fatalities. The goal of education should be to reduce the risk of human-to-human transmission in the community, which typically occurs as a result of close contact with infected individuals, particularly when their bodily fluids are involved. Additionally, education should focus on reducing the risk of wildlife-to-human transmission, such as coming into contact with monkeys, apes, and fruit bats. This educational awareness should include the following: avoiding physical contact with individuals who have MVD; any suspected case should not be managed at home, but rather immediately notified and transferred to a health center for isolation and treatment (where necessary). It is important to promote safe burial practices, such as decontaminating the body with a bleach solution of 1 : 10 and placing it in a body bag before interment. After that, the corpse bag is sealed, and the exterior of the bag undergoes the same decontamination process [4]. The practice of stigmatizing members of the community need to be avoided. It is important to encourage people to wash their hands frequently, especially after seeing a sick family member in the hospital. When working with wild animals, it is important to wear gloves and any other protective gear that is applicable. Before consuming any products derived from animals, including blood and meat, ensure that they have been thoroughly cooked, and steer clear of eating raw meat. Masks, gloves, and any other protective gear deemed necessary should be worn by tourists and researchers alike while venturing into mines or caverns that are inhabited by ERBs.

Since sexual transmission is still possible even after clinical recovery, male survivors should abstain or engage in safe sex practices (until semen repeat tests negative on two separate occasions) [4]. There have been reports of sexual transmission episodes that originated from male survivors who had verified ZEBOV RNA persistence in their semen for more than a year [27]. Therefore, it is an absolute requirement to have a negative test result for the virus in the sperm after achieving clinical recovery.

11.4 Conclusion

Since the very first documented case of MARD in 1967, which occurred as a direct result of human-to-animal contact leading to the transmission of the disease, there have been many documented instances of the disease spreading to other populations. Although a wide variety of medications have been investigated up to this point, there has been no progress made. Therefore, the most effective method of treatment continues to be supportive care, which entails close monitoring of the patient as well as isolating them whenever possible. Although disease-modifying medications and inhibitors of viral protein have shown some positive outcomes in a large number of patients and can be given to people who are afflicted, these treatments are not the definitive solution to the fatal virus. There is an immediate need for the development of an advanced strategy, such as a vaccine that is risk-free and dependable, to protect those people who are at an especially high risk of contracting the disease. We need to focus not only on the advancements that are related to MARD but also on the effects that it has had on the healthcare system. This will be accomplished by presenting an in-depth analysis of the MARD. The authorities in charge of public health need to come up with a strategy while bearing in mind the current climate, in which an unchecked endemic or pandemic can rapidly develop into an epidemic, wreaking havoc on the already inadequate number of healthcare facilities and posing an overwhelming number of challenges. In light of the recent Ebola outbreak in the Democratic Republic of the Congo, the government agencies in charge of public health need to formulate a strategy while keeping the current situation in mind.

References

1 Kuhn, J.H., Amarasinghe, G.K., Basler, C.F. et al. (2019). ICTV virus taxonomy profile: filoviridae. *J. Gen. Virol.* 100 (6): 911–912.

2 Kuhn, J.H., Adachi, T., Adhikari, N.K.J. et al. (2019). New filovirus disease classification and nomenclature. *Nat. Rev. Microbiol.* 17 (5): 261–263.

3 Shifflett, K. and Marzi, A. (2019). Marburg virus pathogenesis – differences and similarities in humans and animal models. *Virol. J.* 16 (1): 1–12.

4 European Centre for Disease Prevention and Control. Factsheet About Ebola Virus Disease. 2022.

5 Ristanović, E.S., Kokoškov, N.S., Crozier, I. et al. (2020). A forgotten episode of marburg virus disease: Belgrade, Yugoslavia, 1967. *Microbiol. Mol. Biol. Rev.* 84 (2): e00095-19.

6 Brauburger, K., Hume, A.J., Mühlberger, E., and Olejnik, J. (2012). Forty-five years of Marburg virus research. *Viruses* 4 (10): 1878–1927.

7 Woolsey, C., Geisbert, T.W., and Cross, R.W. (2021). Marburg and Ravn viruses (Filoviridae). In: *Encyclopedia of Virology*, 608–618. Elsevier.

8 Emperador, D.M., Mazzola, L.T., Trainor, B.W. et al. (2019). Diagnostics for filovirus detection: impact of recent outbreaks on the diagnostic landscape. *BMJ Glob. Health* 4 (Suppl 2): e001112.

9 Abir, M.H., Rahman, T., Das, A. et al. (2022). Pathogenicity and virulence of Marburg virus. *Virulence* 13 (1): 609–633.

10 Dolnik, O. and Becker, S. (2022). Assembly and transport of filovirus nucleocapsids. *PLoS Pathog.* 18 (7): 1–15. https://doi.org/10.1371/journal.ppat.1010616.

11 Edwards, M.R., Liu, G., Mire, C.E. et al. (2016). Differential regulation of interferon responses by Ebola and Marburg virus VP35 proteins. *Cell Rep.* 14 (7): 1632–1640. https://doi.org/10.1016/j.celrep.2016.01.049.

12 Basler, C.F., Krogan, N.J., Leung, D.W., and Amarasinghe, G.K. (2019). Virus and host interactions critical for filoviral RNA synthesis as therapeutic targets. *Antivir. Res.* 162 (404): 90–100.

13 Kolesnikova, L., Bugany, H., Klenk, H.-D., and Becker, S. (2002). VP40, the matrix protein of Marburg virus, is associated with membranes of the late endosomal compartment. *J. Virol.* 76 (4): 1825–1838.

14 World Health Organisation. Marburg Virus Disease. 2021.

15 Amman, B.R., Schuh, A.J., Albariño, C.G., and Towner, J.S. (2021). Marburg virus persistence on fruit as a plausible route of bat to primate filovirus transmission. *Viruses* 13 (12): 2394.

16 Makenov, M., Boumbaly, S., Tolno, F.R. et al. (2022). Investigating the zoonotic origin of the marburg virus outbreak in guinea in 2021. *PLoS Negl. Trop. Dis. 17* (4): 1–16, e0011279.

17 Languon, S. and Quaye, O. (2019). Filovirus disease outbreaks: a chronological overview. *Virology* 10: 1178122X19849927.

18 Nyakarahuka, L., Shoemaker, T.R., Balinandi, S. et al. (2019). Marburg virus disease outbreak in Kween District Uganda, 2017: epidemiological and laboratory findings. *PLoS Negl. Trop. Dis.* 13 (3): e0007257.

19 Miraglia, C.M. (2019). Marburgviruses: an update. *Lab. Med.* 50 (1): 16–28.

20 Canada Laboratory Biosafety and Biosecurity (2018). *Pathogen Safety Data Sheets: Infectious Substances – Marburg Virus.* Government of Canada.

21 WHO. Interim Infection Prevention and Control Guidance for Care of Patients with Suspected or Confirmed Filovirus Haemorrhagic Fever in Health-Care Settings, with Focus on Ebola. 2019.

22 Feldmann, F., Shupert, W.L., Haddock, E. et al. (2019). Gamma irradiation as an effective method for inactivation of emerging viral pathogens. *Am. J. Trop. Med. Hyg.* 100 (5): 1275–1277.

23 Zhao, F., He, Y., and Lu, H. (2022). Marburg virus disease: a deadly rare virus is coming. *Biosci. Trends* 16 (4): 312–316.

24 CDC. 2021. Factsheet-Marburg Hemorrhagic Fever. https://www.cdc.gov/vhf/marburg/pdf/factsheet.pdf

25 Amman, B.R., Jones, M.E.B., Sealy, T.K. et al. (2015). Oral shedding of Marburg virus in experimentally infected Egyptian fruit bats (*Rousettus aegyptiacus*). *J. Wildl. Dis.* 51 (1): 113–124.

26 Paweska, J.T., Jansen van Vuren, P., Masumu, J. et al. (2012). Virological and serological findings in *Rousettus aegyptiacus* experimentally inoculated with vero cells-adapted Hogan strain of Marburg virus. *PLoS One* 7 (9): 1–11.

27 Den Boon, S., Marston, B.J., Nyenswah, T.G. et al. (2019). Ebola virus infection associated with transmission from survivors. *Emerg. Infect. Dis.* 25 (2): 240.

28 Araf, Y., Maliha, S.T., Zhai, J., and Zheng, C. (2022). Marburg virus outbreak in 2022: a public health concern. *Lancet Microbe* 4: e9.

29 Olejnik, J., Ryabchikova, E., Corley, R.B., and Mühlberger, E. (2011). Intracellular events and cell fate in filovirus infection. *Viruses* 3 (8): 1501–1531.

30 Hoffmann, M., Crone, L., Dietzel, E. et al. (2017). A polymorphism within the internal fusion loop of the Ebola virus glycoprotein modulates host cell entry. *J. Virol.* 91 (9): e00177-17.

31 Kortepeter, M.G., Dierberg, K., Shenoy, E.S., and Cieslak, T.J. (2020). Marburg virus disease: a summary for clinicians. *Int. J. Infect. Dis.* 99: 233–242.

32 Odendaal, L. and Davis, A.S.V.E. (2021). Insights into the pathogenesis of viral haemorrhagic fever valley fever. *Viruses* 13 (4): 709.

33 Sah, R., Mohanty, A., Reda, A. et al. (2022). Marburg virus re-emerged in 2022: recently detected in Ghana, another zoonotic pathogen coming up amid rising cases of Monkeypox and ongoing COVID-19 pandemic – global health concerns and counteracting measures. *Vet. Q.* 42 (1): 167–171.

34 WHO. Marburg Virus Disease – Ghana. 2022.

35 Nyakarahuka, L., Schafer, I.J., Balinandi, S. et al. (2020). A retrospective cohort investigation of seroprevalence of Marburg virus and ebolaviruses in two different ecological zones in Uganda. *BMC Infect. Dis.* 20 (1): 1–20.

36 Finch, C.L., Martinez, C., Leffel, E. et al. (2022). Vaccine licensure in the absence of human efficacy data. *Vaccine* 10 (3): 368.

37 Reynolds, P. and Marzi, A. (2017). Ebola and Marburg virus vaccines. *Virus Genes* 53 (4): 501–515.

38 Henao-Restrepo, A.M., Camacho, A., Longini, I.M. et al. (2017). Efficacy and effectiveness of an rVSV-vectored vaccine in preventing Ebola virus disease: final results from the Guinea ring vaccination, open-label, cluster-randomised trial (Ebola Ça Suffit!). *Lancet* 389 (10068): 505–518.

39 Woolsey, C., Geisbert, J.B., Matassov, D. et al. (2018). Postexposure efficacy of recombinant vesicular stomatitis virus vectors against high and low doses of Marburg virus variant angola in nonhuman primates. *J. Infect. Dis.* 218 (Suppl. 5): S582.

40 Bixler, S.L., Bocan, T.M., Wells, J. et al. (2018). Efficacy of favipiravir (T-705) in nonhuman primates infected with Ebola virus or Marburg virus. *Antivir. Res.* 151: 97–104.

41 Kerber, R., Lorenz, E., Duraffour, S. et al. (2019 Jun). Laboratory findings, compassionate use of favipiravir, and outcome in patients with Ebola virus disease, Guinea, 2015 – a retrospective observational study. *J. Infect. Dis.* 220 (2): 195–202.

42 Porter, D.P., Weidner, J.M., Gomba, L. et al. (2020). Remdesivir (GS-5734) is efficacious in cynomolgus macaques infected with Marburg virus. *J. Infect. Dis.* 222 (11): 1894–1901.

43 Dörnemann, J., Burzio, C., Ronsse, A. et al. (2017). First newborn baby to receive experimental therapies survives Ebola virus disease. *J. Infect. Dis.* 215 (2): jiw493.

44 Beigel, J.H., Tomashek, K.M., Dodd, L.E. et al. (2020). Remdesivir for the treatment of Covid-19 - final report. *N. Engl. J. Med.* 383 (19): 1813–1826.

45 Cross, R.W., Mire, C.E., Feldmann, H., and Geisbert, T.W. (2018). Post-exposure treatments for Ebola and Marburg virus infections. *Nat. Rev. Drug Discov.* 17 (6): 413–434.

46 Mulangu, S., Dodd, L.E., Davey, R.T. et al. (2019). A randomized, controlled trial of Ebola virus disease therapeutics. *N. Engl. J. Med.* 381 (24): 2293–2303.

47 Gayer, M., Legros, D., Formenty, P., and Connolly, M.A. (2007). Conflict and emerging. *Infect. Dis.* 13 (11): 1625–1631.

48 Schmidt, K.M. and Mühlberger, E. (2016). Marburg virus reverse genetics systems. *Viruses* 8: 6, 178.

12

Recent Advances in Combating Nipah Virus Disease

Mehul Chorawala[1], Aanshi Pandya[1], Ishika Shah[1], Bhupendra G. Prajapati[2], Nirjari Kothari[1], and Aayushi Shah[1]

[1] *Department of Pharmacology and Pharmacy Practice, L. M. College of Pharmacy, Gujarat University, Ahmedabad, Gujarat, India*
[2] *Department of Pharmaceutics and Pharmaceutical Technology, Shree S. K. Patel College of Pharmaceutical Education & Research, Ganpat University, Mehsana, Gujarat, India*

12.1 Introduction

The Nipah virus (NiV) and Hendra virus (HeV) are newly emerged and immensely dangerous RNA virus entities that are carried by bats and hold origin from the Henipavirus genus which is a part of the Paramyxoviridae family. It is responsible for causing severe respiratory illness along with encephalitis in animals as well as humans. It is considered a Biosafety Level-4 (BSL-4) pathogen and transmitted by specific types of fruit bats, mainly *Pteropus* spp. which are natural reservoir hosts [1]. It is a concern for public health and safety due to a possible high mortality among people and a lack of vaccines and therapies to counteract it [2]. NiV was first identified in 1998–1999 during a brain inflammation outbreak among pig farmers in Malaysia, which led to the extermination of over one million pigs and resulted in a considerable economic effect. Since then, NiV has caused outbreaks almost every year in Bangladesh and India [3].

There are two identified distinct strains of the Nipah virus namely NiV Malaysia (NiV-M) and NiV Bangladesh (NiV-B) of which the latter strain is currently suspected to be in circulation [4]. Humans infected with this virus are susceptible to severe breathing problems and acute brain inflammation (encephalitis), with a death rate ranging from 40% to 100% [5]. The main method of treating patients is through supportive care, which involves ensuring proper airways, breathing, circulation, and balance of fluids and electrolytes. Accurate early diagnosis of NiV infection is crucial for preventing the spread of outbreaks and providing appropriate medical care since the symptoms of this infection resemble those observed in other febrile diseases. Various laboratory tests are available for detecting the Nipah virus, including nucleic acid amplification tests (NAAT) such as PCR and sequencing, IgG/IgM/antigen ELISA, virus isolation, immunofluorescence, and histopathology and neutralization [6, 7]. Currently, most international laboratories use their own NiV assays, but the validation levels defined for these tests vary greatly. Only a handful commercial PCR kits are available, and only a single commercial reagent source is used in ELISA testing (not a fully assembled and validated kit) [8]. Polymerase chain reaction (PCR) is considered to be a highly sensitive method for diagnosing NiV infection present in active form, but NiV-specific IgM ELISA can be used as an alternative when PCR is not available. In cases of fatal NiV infections, histopathology (immunohistochemistry) is performed after death to confirm the diagnosis. PCR based tests have the ability to detect the virus nucleic acid in blood or tissue samples, however, they are not widely accessible and can be costly. Therefore, there remains a requirement for improved diagnostic tools for NiV infection that can clearly distinguish between the various phases of infection or progression and accurately diagnose the disease across all populations [9]. Virus isolation and neutralization techniques are potentially utilized for confirmatory purposes, but they are limited to Biosafety Level 4 facilities with stricter safety measures.

As of now, there are no approved antivirals or vaccines available for the management of NiV pathogenesis in humans. The available treatment alternatives are restricted to provision of supportive care in infected patients. Despite significant advances met during the development of experimental vaccines, progress in the creation of drug treatments has been

Rising Contagious Diseases: Basics, Management, and Treatments, First Edition. Edited by Seth Kwabena Amponsah, Ranjita Shegokar, and Yashwant V. Pathak.

slower [10]. Although there have been some indications that the drug ribavirin may reduce mortality, its effectiveness against NiV has not been established. In 1998, a study involving the use of ribavirin including a population of 140 patients during the initial outbreak of Nipah in Malaysia was carried out, however it was an open-label trial and not a controlled clinical study [11], thus it is difficult to make definitive conclusions based on the results. To combat NiV, several vaccine alternatives are currently in development phase which utilize NiV glycoprotein (G) and fusion (F) proteins in order to trigger a protective immune arsenal [12]. Some of these approaches aim to create specific neutralizing antibody activities, while other approaches are under evaluation for both immune response and efficacy. In Australia, the vaccine targeting Hendra G protein has been created to provide protection in horses from Hendra virus which might also offer protection against NiV [13, 14].

This article summarizes recent reports on pathogenesis of NiV. It also shows how the responses are modulated which are induced by NiV along with the development of new diagnostic tools with preventative and therapeutic perspectives required to confine this highly life-threatening infection.

12.1.1 Virus Structure and Biology

NiV is an enveloped virus which has a single-stranded negative-sense RNA genome and pleomorphic shape (40–1900 nm). It possesses six genes encoding for glycoprotein (G), nucleocapsid (N), phosphoprotein (P), matrix protein (M), fusion protein (F), and polymerase protein (L). From these above genes, P encodes some significant accessory proteins such as C, V, and W [15]. C protein regulates synthesis of viral RNA and virulence factor, while V and W proteins are critical for virulence and preventing interferon-inducible-promoter activation [16]. Human NiV isolates which were collected from the outbreaks in India, Malaysia, and Bangladesh have shown strain variations. Also, bat isolates collected from different geographical areas have shown genomic variations. NiV showed 68–92% homology in the protein-coded region and 40–60% in the non-translated region, with HeV [17]. Antibodies to NiV show reactivity with HeV contrasting to other members of the family [18]. The contrasting feature of both viruses to other paramyxoviruses includes much longer genomes (18.2 kb for NiV and 15.5 kb for HeV) and a broad host range. The common features of both viruses with other members of paramyxoviruses include an absence of haemagglutinin and neuraminidase activities [19]. The detailed structural aspects of NiV are described in Figure 12.1.

12.1.2 Transmission

From the outbreak in Malaysia and Singapore, it was identified that humans were infected by the virus through direct contact with infected pigs or their contaminated tissues. Nasal secretion of pigs and respiratory droplets and open contact with the throat was considered to be a source of transmission [20]. Liability to infection is high when close proximity, touching, feeding, or contact with NiV-infected person is frequent [21]. Also, the virus was transmitted from fruit bats which are considered natural hosts to amplification hosts, and then to humans in Malaysia. While in India and Bangladesh, urine or saliva of bats contaminate fruits and are responsible for the spread of infection [22]. Others were infected via occupational work with trees [23]. The best example of it is date palm sap in Bangladesh [24]. Data reveals that date palm tree is often surrounded by bats and during sap collection, bats slurp them [25]. NiV from *Pteropus* sp. can transmit infection from single to multiple individuals and thus might scourge infection from person-to-person [26]. Risk factors in this situation will include patient operation and association with the secretions of susceptible individual. There is a record showing virus transmission through infected cows in Bangladesh in 2001 [27]. The transmission cycle has been depicted in Figure 12.2.

12.1.3 Epidemiology

During 1988–1999, there was a disruption of respiratory illness and encephalitis in Malaysia and Singapore among pig farmers and individuals who were in accord with pigs. The first incident of the city of Ipoh located in the state of Perak, West Malaysia, took place in late September 1998. In December 1998 and January 1999, the second outburst transpired in a small town named Sikamat situated in the state of Negri Sembilan. The third and the largest outburst, in the same year 1998, took place in the city of Bukit Pelandok [28]. The name was given after the name of a village, Sungai Nipah, located in Malaysia. There was another outbreak in Bangladesh in the year 2010 [29].

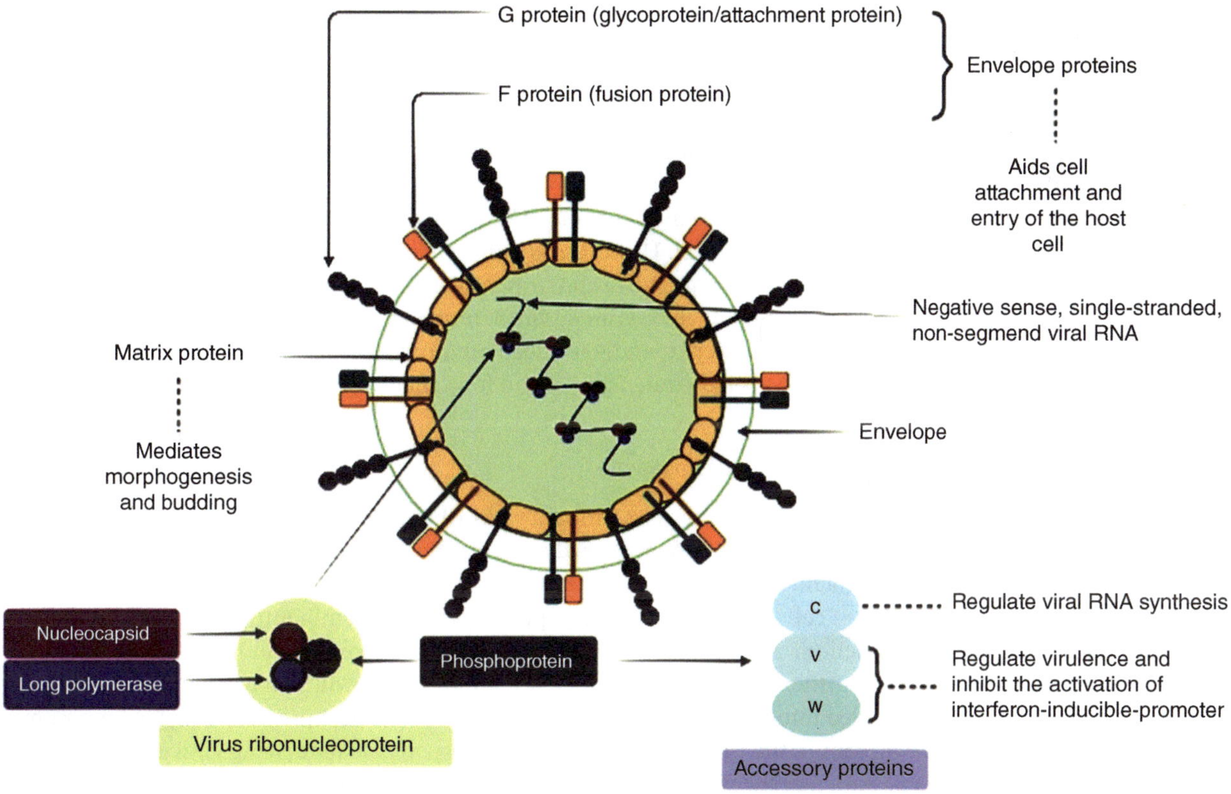

Figure 12.1 Structural biology of NiV. This illustration demonstrates the intricate structure of the NiV. With the core structure surrounded by a lipid bi-layered envelope, the viral assembly constitutes five major proteins with two of them being riveted on envelope (F and G). Three other proteins (L, P, and N) participate in regulating virulence and replication-associated processes. Other than these five, viral body also comprises of matrix proteins that play role in morphogenesis and budding.

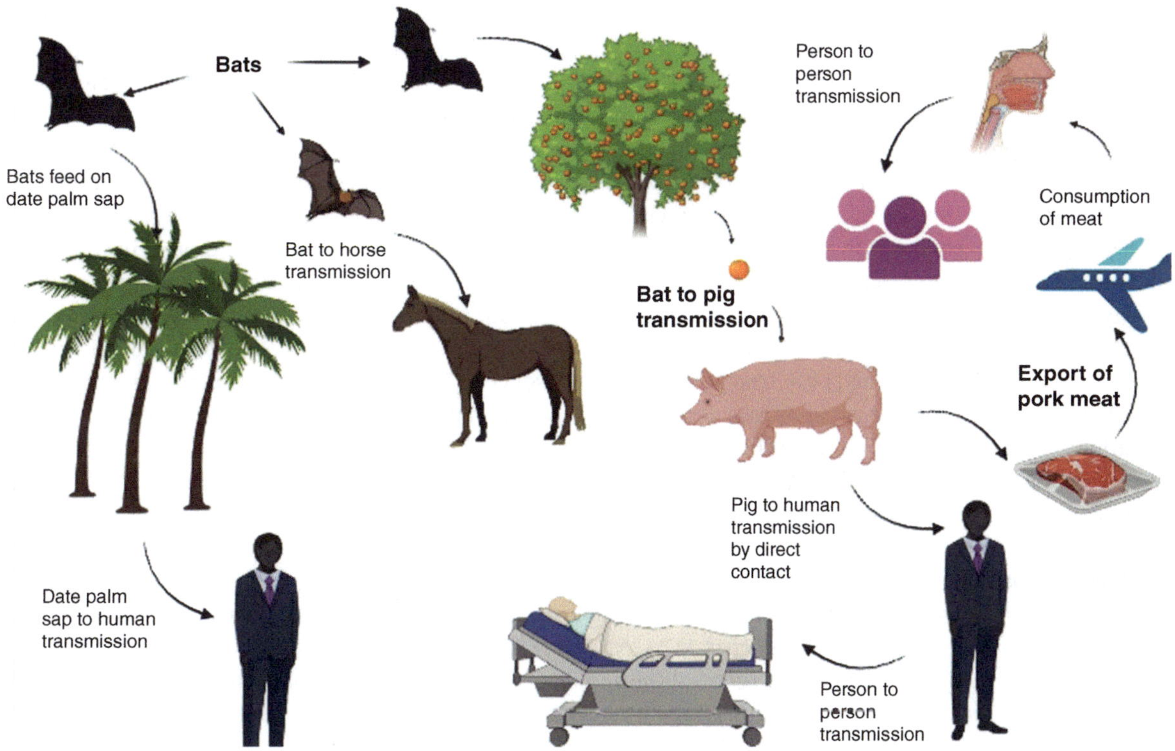

Figure 12.2 Transmission and life-cycle of Nipah virus. This figure depicts spread of NiV from bats to human population. Bats feed on date palm sap and other trees from which virus is transmitted through the consumption of products. Virus gets transmitted to pigs and through the consumption of meat products it gets transmitted to human. Also, when the same meat is exported, the virus gets transmitted to other population.

In India, the first case was found in the Siliguri district of West Bengal in 2001, due to transmission from bats to human population contrasting to the Malaysian outbreak as it involved the pigs. There was another outbreak in 2007 in the Nadia district of West Bengal. From 2001 to 2015, there had been cases from Bangladesh. Recently, in Kerala, India, an outburst took place resulting in the death of 17 individuals, in 2018 [30].

Antibodies against NiV were found in Lyle's flying foxes in Cambodia. Also, antibodies were identified in Indonesia from the sera of *P. vampyrus* which were detected by ELISA [31]. However, surveillance and sero-surveillance preformed in bat population showed the presence of NiV RNA and IgG in bats' saliva and urine, and serum, respectively which indicated long-term NiV persistence in them in Thailand [32]. In 2014, there was an outbreak in the Philippines in which 17 incidents were confirmed and the mortality rate was 28%. Ten horses were found to be dead out of which nine showed neurological symptoms. Five patients were infected through person-to-person transmission and from that two were health care personnel [33].

12.1.4 Immunopathological Mystery of Nipah Virus Disease

Three proteins namely, N, P, and L get affinated to viral RNA and form a virus ribonucleoprotein while the other two proteins namely, F and G are accountable for cellular adjunction to capsid and virus host-cell penetration [34]. The tetrameric G protein will bind with homotrimeric F protein before receptor adherence [35]. G protein binds with the globular head of the ephrin B2/3 receptor present on the host cell [36]. Ephrin B2 and Ephrin B3 are predominantly manifested in the vasculature of mammals, along with endothelium and nervous system [37]. Following receptor binding, a conformational change is induced in the G protein which further leads to F protein refolding [38]. There will be a production of precursor F protein (F0) which will further get divided into two subunits namely, F1 and F2 by a host protease enzyme, endosomal cathepsin L. There is a fusion peptide present in the F1 subunit which will be responsible for driving viral and human cell membranes fusion, supporting the penetration of the virus [39]. This will occur after initial exposure of the virus at the cell surface followed by an internalization process. The disulfide-linked polypeptides travel back to the cell surface where it gets assimilated into budding virions, facilitating the cell–cell spread of the virus, and forming multinucleated syncytia [40]. M protein will be responsible for budding and morphogenesis [41]. It is believed that for greater pathogenicity of the virus, replication and the fusion of ephrin receptors are accountable [42].

Recently, it was revealed that viral modulation of host cell machinery targets the nucleolar DNA damage response (DDR) pathway by inhibiting nucleolar *Treacle* protein which is responsible for the production of the virus [43]. There is a rapid spread of the virus among humans, affecting their airway epithelia having major entry receptors, ephrin B2 [44]. NiV virus upraises IFN-γ in respiratory epithelial cells of human. Prior treatment with IFN-γ will hinder NiV replication and demonstrate antiviral activity and thus there will be variation in receptor expression which ultimately will contribute to NiV replication kinetics [45]. NiV V protein, an accessory protein, which is encoded by viral P protein, has an important role in its pathogenesis. NiV V protein increases the proportion of host protein UBXN1 (UBX domain-containing protein 1, which is a negative controller of RIG-I-like receptor signaling). It does so, by inhibiting host protein proteolysis and thus repressing the enrollment of innate interferons [46]. It's important to pin down that molecular details of the virus are sustained among the M protein [47].

Due to the up regulation of the nucleocapsid protein of NiV, the production of viral RNA and also the regulation of viral polymerase activity would be increased. Due to such protein overexpression, there is inhibition of transcription but the conglomeration of the genome of the virus is increased. As there is a bias of polymerase activity toward the concoction of the genome, lineage of the virus is restrained [48]. Irrespective of the presence of M protein, F and G proteins are randomly distributed on the NiV plasma membrane which is revealed by super-resolution microscopy. Virus-like particles (VLPs) are produced as a result of fabrication of M protein at the plasma membrane [49]. G protein gets recruited into VLPs and this is supplemented by viral particles guided by F, M as well as M/F [50]. The microRNA processing machinery and PRP19 complexes are the host recipient of a virus. W protein of a virus alters p53 control along with the genesis [51].

12.1.4.1 Innate Immunity and Interferon Type I Signaling

There is a major role of innate immunity in host defense and modulation against virus which have been depicted by many reports [52]. Strong impression of anti-viral genes MX1, RSAD2, ISG15, and OAS1 was seen during former stages of infection in ferrets, however, it was not enough to control the spread of the infection [53]. Disruption of the first line of defense is seen due to suppression of IFN-1 production which leads to death. The structural and non-structural proteins of NiV are associated with IFN-1 signaling blockade [54, 55].

There are several in vitro models present, showing processes of modulation of IFN-1 associated pathway by viral proteins. Caspase activation and recruitment domain of RIG-1 and SPRY domain of TRIM25 gets linked with the C-terminal domain leading to the prevention of ubiquitination of RIG-1 and its downstream signaling [56]. There are also some reports that shows involvement of M protein in disrupting IFN-1 signaling. It does so by interacting with E3ubiquition ligase TRIM6, setting of its degradation and further preventing IKKε kinase-mediated IFN-1 response which was evident only when M was expressed and showcased decrease in TRIM6 level [57].

Averting the nuclear transport of signal transducer and activator of transcription 1 (STAT1) and STAT2, IFN-stimulated genes are hampered due to the presence of NiV protein N [58]. All these directions of the virus protein indicate a viral evasion mechanism that affects IFN-1 signaling.

12.1.4.2 Adaptive Immunity

NiV infection affects both cell-mediated and humoral immunity [59]. There is distinct elevation of CD8+ T-cells, co-expressing granzyme B and PD-1 which suggests elimination of infected cells due to increased lymphocyte population. There is also increased level of NiV-specific IgG and IgM antibodies with increased B-lymphocytes counts. Survivors of NiV infection showed more stable immune reaction compared to mortalities. Three cases support resistance following infection through respiratory route in peripheral blood and tissues of ferrets and monkeys [53, 60, 61]. Among ferrets, following infection with NiV Bangladesh strain, there was time-dependent elevation in macrophage markers and consistent level of lymphocyte markers in lungs depicted by gene analysis which was characterized by host gene expression during HeV infection [53]. A data shows the production of Th1-associated cytokines, increase in natural killer (NK)-cells during both acute and convalescent phases, and further increase in CD8+ T-cell activation markers in lung, brain, and blood in NiV infected AGM, depicting a cell-mediated immune response, although neutralizing anti-bodies were not initiated within 10 days of infection [60]. Figure 12.3 shows immune response to NiV infection.

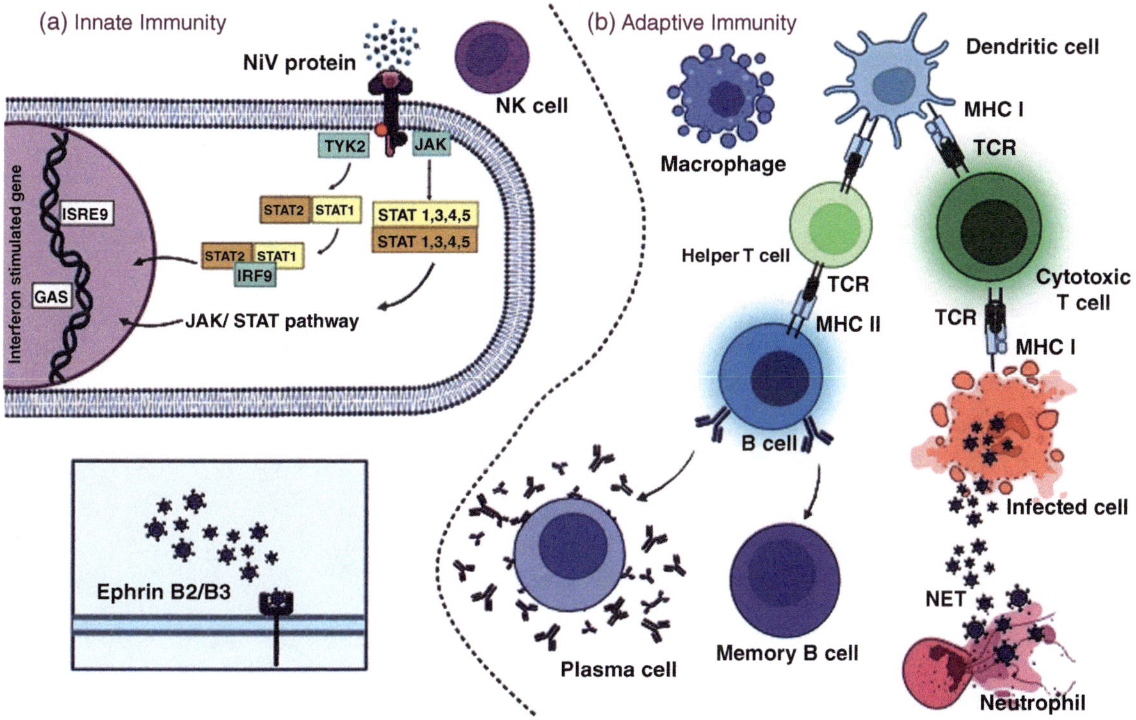

Figure 12.3 Immune response to NiV infection. (a) Response of innate immunity against NiV infection: The response initiates with recruitment of neutrophils at the site of infection. The NiV proteins inhibit the IFN activity through interaction with JAK/STAT pathway. (b) Acquired immune response against NiV infection: After the entry of viral particles through Ephrin- B2 or -B3 receptors present on the host cell membrane, engulfment, and lysis occurs, turning them into viral peptides by antigen-presenting cells such as macrophages and dendritic cells. These viral peptides presented on MHC molecules are recognized which leads to an activation of the T cells via TCRs. Activated helper T cells subsequently stimulate B cells and allow B cells to proliferate and form mature antibodies against NiV. Later on, plasma cells and memory B cells are formed which produce NiV-specific antibodies to combat the infection. Secondly, the activation of cytotoxic T cells directly aims and kill the NiV-infected cells.

12.1.4.3 In Humans

In the initial stages of illness, the epithelium of bronchi and type-II pneumocytes gets infected by viral antigens being their first target. It leads to the induction of inflammatory cytokines which further recruit cells of the immune system and eventually leading to acute respiratory distress syndrome (ARDS)-like condition [27]. Also, when respiratory epithelium gets infected, various inflammatory mediators such as interleukins viz. granulocyte-colony stimulating factor (G-CSF), IL-1α, IL-6, IL-8, C-X-C motif chemokine 10 (CXCL10) gets induced. From the epithelium of the lungs, NiV gets propagated to endothelial cells of the lungs in the subsequent stages of the disease. Finally, NiV enters the bloodstream either freely or in association with the host leukocyte. Not only the lung, but the virus also infects other organs such as the brain, spleen, and kidneys leading to multiple organ failures [62, 63].

The entry of the virus in the CNS is mediated in two ways: one is a hematogenous route which involves the choroid plexus or the blood vessels of the cerebrum and another is olfactory nerves [64]. Eventually, there is a disordering of the blood–brain barrier (BBB) and articulation of IL-1β and tumor necrosis factor (TNF-α) showing neurological signs [27]. Inclusion bodies may also be present in the infected person. Both the gray and white matter plaques might see escorted by necrosis [62]. Figure 12.4 depicts the overall pathogenesis of NiV in the body.

12.1.4.4 In Animals

In pigs, the respiratory system and meninges were involved. Pneumonia, peri-bronchial inflammation, and tracheitis were found. Multi-nucleated cells, alveolar macrophages, and neutrophils were found in the lungs. Even though vasculitis was rare, it showed syncytial cells in the small blood vessels. Meningitis detected to be infiltrated with lymphocytes, plasma cells, and macrophages while characterized by edema and vasculitis. NiV-infected dogs, cats, and horses were not much investigated as there was lesser number of pathological specimens for evaluation. However, in dogs and horses, non-dischargeable meningitis and bran parenchymal rarefaction were revealed [27].

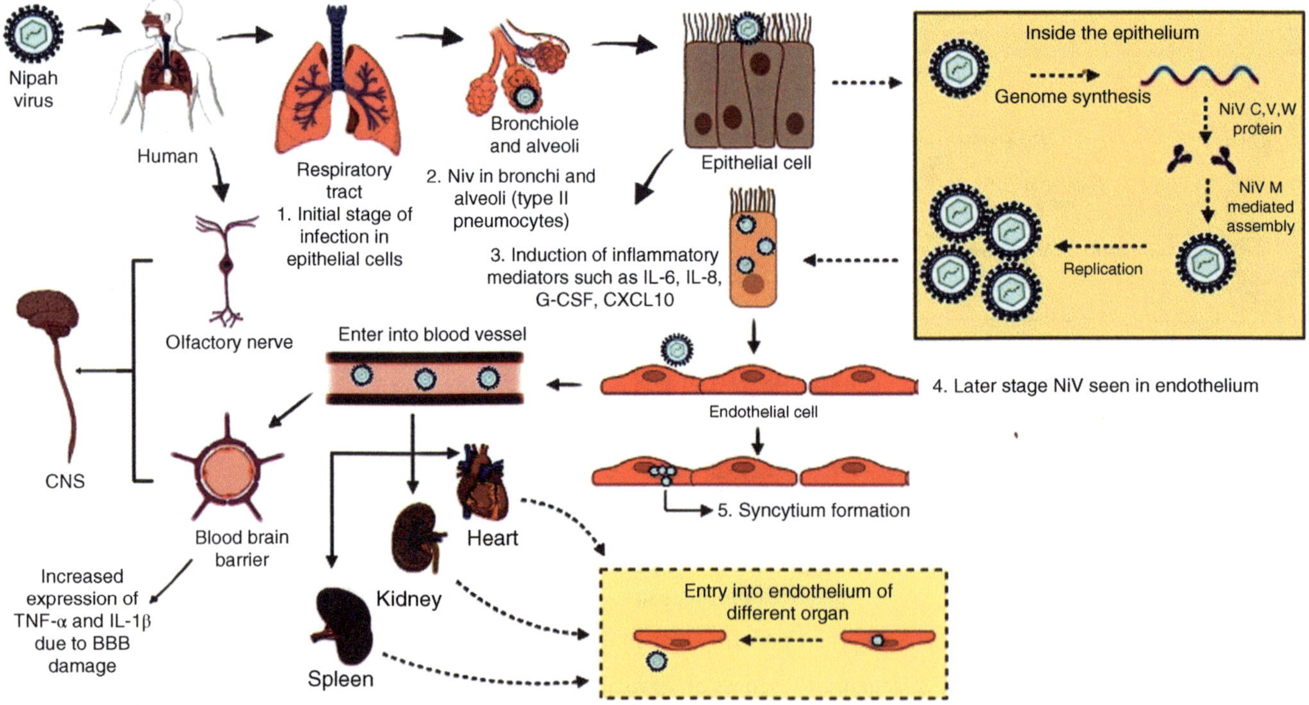

Figure 12.4 Pathological role of NiV virus. The figure depicts the transmission of the virus and its pathogenesis in the body. Firstly, the virus infects the respiratory tract of the person where alveoli and bronchiole get affected. It enters the epithelium of the respiratory tract where genome synthesis starts subsequently leading to the replication of the virus. From the epithelium, the virus infects the endothelium and syncytium formation takes place indicating the end stage of infection. From endothelial cells, the virus enters into blood stream affecting other organs. The virus damages BBB and through direct entry by the olfactory route it enters CNS and leads to destruction.

12.1.5 Clinical Presentation

HeV virus incubation period gets extended from four days to two weeks and may stay up to four months. Milder symptoms shown included fever, headache, sore throat, and influenza-like drowsiness during the early stages of the illness. Neurological manifestations included seizures, confusion, and motor deficit. The chest X-ray showed hypoxemia, influenza-like illness, and dispersed alveolar shadowing in pulmonary syndrome [65]. The clinical development starts with hyperthermia followed by convulsions and ultimately encephalitis and respiratory syndrome leading to death [66]. A study showed clinical features such as headache (74.9%), altered sensorium (71.6%), segmental myoclonus (49.3%), and segmental sweating (26.2%). Due to the involvement of the brain stem, noticeable vasomotor changes such as tachycardia (43.3%), and hypertension (42.3%) were seen indicating the presence of the medullary vasomotor part of the brain stem. Also, some incidents of seizures (23.2%) were seen in the Malaysian outbreak. Hyporeflexia (26.3%) and limb weakness (10.5%) were also present [67]. Dizziness and vomiting were observed in one-third of the patients, while non-productive cough and myalgia were seen in 14% and 12% of patients respectively. The patient experiencing relapse or late-onset encephalitis had a lower mortality rate than patients suffering from acute encephalitis [68].

The cerebrospinal fluid examination (CSF) indicated elevated protein levels and an increased white blood cell count; however, glucose levels remained unaffected. Electrolyte levels were within the normal range, except for hyponatremia. The electroencephalography revealed continuous diffuse and symmetrical slowing, either with or without focal discharges, in cases of acute encephalitis caused by NiV (Nipah virus) [69]. Seroconversion showed that Anti-NiV IgM appeared early and stayed for more than three months, while IgG persisted for more than years which is seen by 100% in three weeks [70]. Potentially threatening NiV encephalitic syndrome reflects fever, headache, reduced consciousness, and dizziness. Tachycardia, areflexia, meningism, abnormal pupillary, and doll's eye reflex, hypotonia, hypertension, and seizures arise as some of the clinical signs [71]. A study showed that in 160 cases of patients who survived NiV encephalitis, 7.5% had a relapse and 3.4% had late-onset encephalitis [72]. Diffuse bilateral opacities consistent with ARDS were reported in the Bangladesh NiV outbreak with the help of a chest radiograph indicating edema, inflammation in the lung parenchyma, and interstitial thickening.

12.2 Recent Advancements in Nipah Virus Diagnostic Tools

The virus is first isolated from mammalian tissues during the first outbreak highlighting the idea of a new infectious etiology. The diagnosis includes a wide array of methods [73]. As NiV is a BSL-4 agent, handling and sample collection should be done with great care [74]. Throat swabs and tissue specimens can be stored in a viral transport medium, 10 ml of urine in a universal sterile container, 5 ml of blood in a plain vial, and 1–2 ml of CSF in a sterile container. From the above samples, virus seclusion and Real Time-Polymerase Chain Reaction (RT-PCR) should be performed. From CSF or serum samples, convalescent-phase antibody detection by enzyme-linked immunoassay test (ELISA) can also be done which is having a high specificity for diagnosis [75]. Although the presence of antibodies detection is the best method, it cannot be performed as it's a BSL-4 agent. A conventional polymerase chain reaction had been used in the past for diagnosis, however these are considered substandard due to advancements in rapid detection methods [74]. A few of the serological, molecular, and tissue culture methods have been described below:

12.2.1 Serological Methods

ELISA is a technique that detects antigens and also evaluates antibody response. Virus detection and differentiation of the virus from HeV has been apprehended by a monoclonal antibody-based N protein [76]. Polyclonal antibody immunized with NiV G protein derived from rabbits develops an antigen capture sandwich ELISA [77]. If we use native antigens then biosafety level 4 facilities must be there, so normally we utilize recombinant proteins in the guise of alternative antigens instead [76]. Antibodies are detected in pigs sera against truncated soluble HeV G or NiV G protein in *L. tarentolae* and sub-cloned N protein expressed in *E. coli* using indirect ELISA [78]. For rapid detection of viruses and as an alternative to PCR, such methods provide cheap and high output methods for sample winnowing [77]. However, false positive results can lead to high economic loss and false negatives can aggravate public health, thus remaining a major concern.

Virus neutralization tests (VNTs) had been soon developed following a commencing outbreak in Malaysia and are taken into account as the citation for serology testing. As an alternative to a conventional method, plaque neutralization assays have been found. For mitigating the requirement of handling live viruses, a variety of pseudotyped viruses having NiV F and G glycoproteins have been instituted in high containment [73]. This approach correlated well with the conventional VNTs featuring the capability of low containment tools together with high sensitivity and safety [78]. Serum neutralization tests have been considered a gold standard but require PC4 facilities in the laboratory owing to the BSL4 risk of the virus [79]. Cell fusion assay is another method for diagnosis in which the effector cells demonstrating NiV F and G proteins, accompanied by a T7 RNA polymerase, encrypt a T7 promoter-driven *E. coli* lacZ cassette cultured with target cell. By measuring beta-galactosidase activity, we can compute NiV-mediated cell fusion when the cytoplasm of fused cells mixes. Before target cell addition, serial dilution of sera incubated with target cells demonstrated inhibition of cell–cell fusion [80]. This technique could be put together nicely with classical VNTs to scrutinize both virus-cell and cell–cell fusion inhibition [80].

12.2.2 Molecular Method

TaqMan RT-PCR detects specimens and laboratory materials rapidly which is depended on the N gene sequence. A real-time TaqMan array card can simultaneously detect 26 acute febrile illness-associated organisms, inclusive of NiV. These when set side by side to standard real-time PCR, showed vulnerability and reliable rapid sensing of NiV for diagnosis [81]. These help in the generation of phylogenetic analysis [82]. While it was the first line of diagnostic testing, as a second line of testing, the detection of anti-NiV IgG and IgM was carried out by CDC [75].

12.2.3 Tissue-Culture Method

Virus isolation is attempted in newer cases where a sample has been collected from the kidney, spleen, and/or lungs. NiV flourish in Vero cells and shows a cytopathic effect inside the limit of three days of culture in the form of syncytia and characteristic plaques in the cell monolayer. Electron-microscopy and immune-electron microscopy are some useful tools for structure identification and detection of virus-antibody interactions, respectively. According to CDC guidelines, virus isolation techniques and RT-PCR from the nasal swabs and throat, blood, CSF, and urine should be used during acute phase while antibody spotting by ELISA can be used in the later stages [83]. A brain magnetic resonance imaging (MRI) scan has shown focal lesions in the subcortical and deep white matter which were less than 5 mm without surrounding edema. This aids in the diagnosis of acute encephalitis. These lesions were also visible in the later stages of uncomplicated acute NiV encephalitis. These do not get enlarged, however, there may be a size reduction. Cerebral cortex lesions may become more prominent [84].

Immunohistochemistry previously used the convalescent human serum. Now, it uses formalin-fixed tissue for the detection of viral antigens. An immense scope of tissues is used including the brain, spleen, lungs, lymph nodes, and kidney as the site for viral replication is vascular endothelium [79]. For pregnant animals, the uterus, placenta, and products of conception are used [75]. For pinpointing NiV-associated pathogenesis including lesions, necrosis, and vasculitis, analysis of the tissue section is important [85].

12.3 Recent Development in Therapeutics and Treatment Modalities Against NiV Disease

World Health Organization (WHO) has listed NiV as one of the eight dangerous pathogens in the priority list of pathogens, in efficiently monitoring and preventing a pandemic [3]. The infection caused by NiV may either occur as a severe and acute respiratory manifestation or acute, late-onset, or relapsed encephalitis. Adding to this, the fatality rates of the infection are spread over the spectrum of 40–100% [5]. The treatment strategy against NiV mainly focuses on supportive care, coupled with the thorough conservation of the "ABC" concept of resuscitation that includes airway, breathing, and respiration. This also involves the maintenance of patency of the fluid and electrolyte concentrations in the bodily compartments. Due to factors such as extreme pathogenicity, intricate and complex transmission cycle, unavailability of vaccine, unpredictable viral course, and a difficult management strategy, NiV presents a substantial threat to the world, thus necessitating the treatment protocols to be advanced efficiently [86]. However, before moving forward with the advancements in treatment strategies, the current treatment plans must be grasped well. As of now, no licensed therapeutic interventions have been licensed, although an agriculture-based subunit vaccine has been approved for NiV-related

Hendra virus (HeV) [87, 88]. The preclinical studies are still investigating the therapeutic drugs and vaccines for fabricating an optimum strategy for NiV and those have been introduced in Tables 12.1 and 12.2. Along with these preclinical assessments, two vaccines have successfully entered the clinical phase and the same have been described in Table 12.3. Figure 12.5 depicts novel therapeutic approaches against NiV. Antiviral treatment remains the apparent choice, which includes agents like Ribavirin, Favipiravir, Remdesivir and Acyclovir.

12.3.1 Antivirals

12.3.1.1 Ribavirin

Being the sole antiviral to treat NiV infection in humans, ribavirin acts as a guanosine analog and it possesses a wide spectrum of activity against the viruses, including DNA as well as RNA viruses. Also, it has showcased good in-vitro effects of inhibiting the replication of the HeV viruses [10]. Observational studies point out Ribavirin to be efficacious in reducing the mortality rates by 36% in patients suffering from encephalitis caused by the Nipah virus, except for its use as a prophylactic [11]. A second trial was conducted in the patients during the Malaysian outbreak which involved the administration of ribavirin in 78% of the patients, nevertheless with no significant reduction in mortality rates. Furthermore, a study using hamster models to evaluate efficacy of ribavirin, either alone or in combination with chloroquine, did not yield positive results [88]. However, their ultimate efficacy as well as the dosage protocol for NiV has not been definitively established yet. The 2008 study by the Infectious Diseases Society of America has suggested the short-term treatment for NiV encephalitis, as the long-term treatment in animals, has led to teratogenic effects and numerous other adverse effects [108]. With the paucity of any other effective treatment protocol, the National Centre for Disease Control, India has recommended ribavirin for the mitigation of NiV infection, although not as post-exposure prophylactic [10].

A 28-year-old male Keralian was presented to the hospital with fever, vomiting, and altered mental sensorium on 17 May 2018. He had the signs and symptoms of diaphoresis, restlessness, and confusion; Hypertension (180/100 mmHg), tachypnoea (36/min), and tachycardia (180/min). The examination of the nervous system detected hypotonia of all four limbs with the diminishment of tendon reflexes and limb myoclonus. Chest examination revealed bilateral diffuse fine crackles. The CBC, LFT, RFT, and coagulation particulars were normal, except for mild thrombocytopenia (140,000 cells/mm^3). The arterial blood gas profile and ECG were normal. However, the CT scan revealed cerebral edema. The supportive treatment included elective intubation and intermittent positive-pressure ventilation. IV mannitol was administered, and a lumbar puncture was performed. The CSF study showcased high sugar and protein levels with cell count being normal. The empirical treatment was initiated with IV acyclovir and ceftriaxone for a suspected central nervous system infection. However, the patient died on 18 May 2018, and detailed microbiological and virological analysis results proved the infection.

Insufficient ventilation, overcrowding, and inefficacious infection control practices in the ERs, ward, and corridors of a few public hospitals led to the rapid transmission of infection to the bystanders, caregivers, and ultimately other hospital-associated patients. For a general view, patients suffering from respiratory distress received the aid of non-invasive ventilation as well as intermittent positive pressure ventilation. The majority of the patients were administered IV antibiotics with the aim to prevent secondary bacterial infections. Seizure-experiencing patients were given antiepileptic agents. Brain edema and elevated intracranial tension were therapeutically managed with the help of dexamethasone and mannitol. Antihypertensive agents were provided to the patients who experienced hypertensive crises. Crystalloids and vasopressors were administered to patients with septic shock. To conclude, of the 19 patients, 8 (42.1%) were administered Ribavirin. Initially, intravenous Acyclovir was given to three of the patients and later, was changed to Ribavirin with the dose of 1200 mg/day in six divided doses. Two of the patients survived on the receipt of 14 days of Ribavirin, which was initiated in the very early stage of the disease [67].

12.3.1.2 Favipiravir

Favipiravir, a small molecule, acts as a purine derivative and performs inhibition of RNA-dependent RNA polymerase. It has a characteristic wide-spectrum activity in fighting RNA viruses. The drug has been approved for its use in Japan for the treatment of various influenza virus strains [93]. Also, a few recent studies have recommended Favipiravir, for it possesses the highest antiviral activity against mitigating NiV [108]. Preclinical studies focused on evaluating Favipiravir were performed in the Syrian hamsters against NiV-M, and the study concluded that the animal models were efficiently treated with 14 days of daily administration of the drug, yet further clinical studies need to be established to optimize the therapeutically relevant doses, route of administration, and duration of the treatment following infection [93].

Table 12.1 Preclinical studies evaluating therapeutic drugs and biologics against NiV.

Sr. no	Drug class	Intervention	Mechanism of action	Animals used	Results attained	Reference
1	Monoclonal antibodies	A. Human cross-reactive mAb m102.4	Antibody has enough potential to neutralize both NiV and HeV by blocking the interaction of glycoprotein G with Ephrin B2 and B3 entry receptors	Ferrets	Full protection observed in ferrets after single IV infusion of m102.4 administered 10 h after development of NiV intranasal infection	[89]
				African Green Monkeys (AGM)	Foremost pilot study conducted on ferrets subjected the animals to m102.4 against NiV immediately after contracting the infection. IV dose of m102.4 at 10 h was repeated three days after the infection. Treated animals showed a healthy improvement while the control group succumbed to infectious pathology on the eighth day of infection	[90]
		B. Humanized, cross-reactive mAb h5B3.1	mAb with neutralizing capacity which targets the fusion glycoprotein F of NiV and HeV. It further blocks the conformational change necessary for membrane fusion and progression of infection	Ferrets	A cohort of ferrets were exposed to antibody through i.p. route on day 1 following the lethal i.n. exposure of NiV with a subsequent dosage of antibody on day 3 after the challenge (D1/D3 cohort) with another cohort of D3/D5. Animals from both cohorts survived until the end of study, gained weight during the study and showed minor clinical manifestations of NiV-driven disease without any observable morphological changes upon necropsy at the study end point	[91]
2	Antivirals	Remdesivir	Remdesivir's active form behaves as a nucleoside analogue and thus inhibits the viral replication and ultimately the viral load	African Green Monkeys (AGM)	The AGMS were given dose intranasally and intratracheally in addition to a lethal dose of NiV-B. With the two groups including the remdesivir and vehicle alone group, two out of four members of remdesivir were positive to mildly involve respiratory system which eradicated 12–14 dpi. Remdesivir group clinical scored resolved to baseline point along with the survival of all models until the end of the study (92 dpi). No changes in respiratory rate were encountered in the remdesivir group throughout the study, compared to the vehicle-alone group	[92]
		Favipiravir	This small molecule purine analog inhibits the viral replication and transcription	Syrian Golden Hamsters	Favipiravir proffered full protection against the lethally NiV-M-challenged animals, when administered immediately after the challenge for 14 consecutive days. No clinical manifestations were observed in the models through the course of study and no viral RNA or pathological alterations were noted in the tissues	[93]
		Griffithsin (GRFT) Monomeric GRFT (mGRFT) mGRFT tandemers (2mG-Dimer; 3mG-Trimer) Oxidation-resistant GRFT (Q-GRFT)	GRFT interacts with viral and cellular glycoproteins and blocks the entry and spread of NiV. The proteins present on the envelope of numerous viruses showcase rich glycosylation of mannose, that can bind with the monosaccharide binding sites of the GRFT, and this renders interference in host cell invasion	Syrian Golden Hamsters	Due to the enhanced ability of trimeric tandemer (3mG) to block NiV glycoprotein-induced syncytia formation, it has comparatively greater potency compared to GRFT Also, the prophylactic evaluation of Q-GRFT preferred significant protection against a lethal challenge of NiV	[94]
		Ribavirin	It modifies the cellular nucleotide pools and interferes with the viral mRNA synthesis, thus inhibiting replication	Hamsters	Out of the five animals who were administered ribavirin individually at a dose of 30 mg/kg every 12 h, two hamsters survived till the end of the experiment (day 21). Ribavirin did delay Niv-induced death by approximately five days, however with no prevention of mortality	[95]

Table 12.2 Preclinical studies evaluating vaccines against NiV.

Vaccines	Animal models	References
Recombinant subunit	Ferret, Cat, AGM	[90, 96–98]
Recombinant vaccinia virus	Hamster	[80]
Recombinant VSV	Ferret, Hamster, AGM	[96, 99–101]
Recombinant measles virus	Hamster, AGM	[27]
Recombinant Newcastle disease virus	Pig, Balb/c rodents	[102]
Recombinant canarypox virus	Pig	[103]
Recombinant adeno-associated virus	Hamster, Balb/c rodents	[103]
Virus-like particles (VLPs)	Hamster	[104]
Venezuelan equine encephalitis virus replicon particles	C3H/He mice	[105]

Table 12.3 Ongoing clinical trials evaluating therapeutic strategies against NiV.

Sr. no	NCT no.	Intervention	Arms	Phase	Status	Outcomes to be measured	Reference
1	NCT05178901	rVSV-Nipah Virus Vaccine Candidate PHV02 versus Placebo in healthy adult volunteers	1) PHV02 2×10^5 pfu 2) PHV02 2×10^6 pfu 3) PHV02 2×10^7 pfu 4) Placebo	1	Recruiting	To assess the number of participants with treatment-related adverse events (TRAEs)	[106]
2	NCT05398796	mRNA-1215 NiV vaccine in healthy adults	1) mRNA −1215 – 25 mcg IM 2) mRNA −1215 – 50 mcg IM 3) mRNA −1215 – 100 mcg IM 4) mRNA −1215 – X mcg IM (based on evaluation of arms 1–3)	1	Recruiting	To evaluate the safety and tolerability of mRNA 1215 in a 2-dose vaccine regimen	[107]

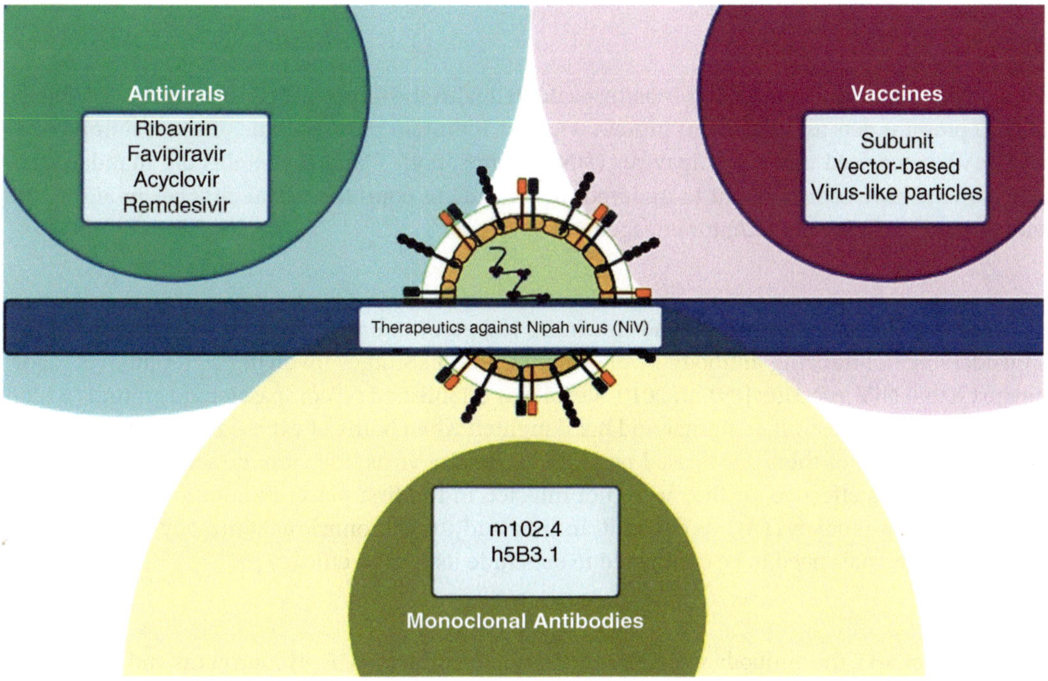

Figure 12.5 Novel therapeutic approaches against NiV. The sphere of therapeutic approaches against NiV embodies three major divisions, them being the antiviral agents, vaccines and monoclonal antibodies against specified targets (biologics).

12.3.1.3 Remdesivir

Remdesivir is also a nucleoside derivative, the antiviral activity of which is broad-spectrum. It has been reported to inhibit the replicative cycle of paramyxovirus, filovirus, and coronavirus. The National Institutes of Health (NIH) has presented reports about remdesivir being the drug that has led to 100% protection of African green monkeys (AGMs) from the lethal infection of the Nipah virus, when it was started 24 hours after being infected with NiV-B and continued daily for 12 days, through an intravenous route. Though the results of this study were promising, future campaigns need to be executed to establish the assured efficacy of remdesivir. This drug has entered clinical trials for evaluating its efficacy against the Ebola virus and has been reported to be safe, though not for the Nipah virus study [109]. Also, recently, remdesivir has gained a good deal of attention as a repurposed antiviral against Covid infection.

12.3.1.4 Acyclovir

Another nucleoside analogue, acyclovir incorporates itself into the viral DNA and prevents further synthesis. Acyclovir was used along with ceftriaxone in nine patients, in the March 1999 outbreak of NiV. Out of these nine patients, eight of them survived. However, whether the reason for survival was acyclovir is yet to be discovered [10]. The studies incorporating the evaluation of acyclovir against NiV in vitro or in additional *in vivo* have not been published yet.

12.3.1.5 4′azidocytidine (R1479)

This cytidine analogue has a wide spectrum activity against pneumo- and flavi- viruses, along with a promising antiviral effect against in vitro NiV in low micromolar ranges [110]. The derivatives of this agent such as 2′-monofluoro- and 2′-difluoro have shown more encouraging results. These derivatives showcased upto 20 times enhanced antiviral activity against NiV than 4′azidocytidine in vitro and reported comparatively less cytotoxicity [111]. These characteristics lead us toward a promising therapeutic evaluation. Additionally, Balavir, a prodrug of R1479, had also been developed. Nevertheless, clinical trials assessing its efficacy against flavivirus-induced infections did not present promising results coupled with negative side effects and this led to the discontinuation of its trials [112].

12.3.1.6 4′-Chloromethyl-2′-Deoxy-2′-Fluorocytidine (ALS-8112)

This drug is another cytidine analogue with being a parent drug of lumicitabine and has undergone clinical trials to get evaluated for its efficacy in treating respiratory syncytial virus-induced infections [113]. The profile of ALS-8112 has been strong in the in vitro NiV field at low micromolar ranges coupled with minimal toxic effects, excluding neutropenia and lymphopenia [114].

12.3.2 Monoclonal Antibodies

Passive immunization through two monoclonal antibodies is being evaluated with the target antigens being the NiV membrane glycoproteins F and G. G proteins acts as attachment protein whereas F protein plays the role of the fusion protein. For NiV and Hendra virus (HeV), comprised under Henipavirus (HNV) genus, both F and G proteins are mandatory in causing infection and these proteins have been thought to undergo the cascade of conformational changes in aiding the promotion of host-virus membrane fusion for the receptor engagement [115].

12.3.2.1 m102.4

This experimental monoclonal antibody sets the ephrin-B2 and ephrin-B3 receptor binding domain of the G protein of NiV as a target and acts as a considerable neutralizing antibody in the in vitro profile. Studies involving ferret models report m102.4 to be efficacious against lethal NiV infection [89]. In 2010, a trial was established on compassionate grounds which involved administration of this mAb to an Australian mother and her daughter, when both got exposed to the Hendra virus through their already infected horse. Both of them showcased no signs of Hendra virus infection, nonetheless it remains unclear as to whether the treatment was effective, or they were not infected in the first place. In addition to the ferrets, m102.4 was tested in the African Green monkeys (AGMs) as well. In this study, this monoclonal antibody demonstrated 100% efficacy, however, further human trials need to be conducted to conclude its overall efficacy [90].

12.3.2.2 h5B3.1

This mAb is the humanized version of 5B3, the antibody which is targeted against the HNV F glycoproteins and thus inhibits the infections of NiV and HeV. A preclinical trial was performed in ferrets to assess the therapeutic profile of h5B3.1 and the results proved it promisingly efficacious [91, 115]. Though these encouraging results demonstrated by monoclonal antibodies paved the way either in therapeutic or prophylactic capacity, studies still face the paucity of providing a comprehensive understanding of G and F protein epitopes when fought against, by these antibodies.

12.3.3 Vaccines

Numerous vaccines are in the developmental stage with an increasing understanding of the molecular biology of NiV. Initially, recombinant vaccines were focused upon with the evidence that enveloped glycoprotein-mediated vaccination could trigger an immune response and thus safety against the virus. With the establishment of this knowledge, several vaccination strategies had been proposed. Some of these approaches involved vaccinating the horses and swine herds in order to avoid the spread to human populations. Nevertheless, these animal vaccinations would fail in countries like Bangladesh, wherein the food sources are contaminated by fruit bats too. Ergo, this necessitated the development of human vaccinations as well.

12.3.3.1 Subunit Vaccines

These vaccines, which are fabricated with setting an oligomeric and soluble form of G glycoprotein (sG) of NiV as a base, has been the most studied strategy. sG is capable of providing a complete protection from high dose NiV challenge coupled with zero signs of pathological disease or viral replication. Vaccinated rodents, AGMs, and cats did successfully produce good levels of NiV-specific neutralizing antibodies (nAb) and IgGs [96, 116]. Equivac is one such vaccine that has been developed for horses against HeV, in Australia [116]. Nonetheless, another study reported that HeV sG vaccinated pigs produced low nAbs to NiV post-challenge, which explains the requirement of the cell-mediated immune response as well, for protection [99].

12.3.3.2 Vector-Based Vaccines

Another approach involved virus vector-based recombinant vaccines, which expressed the F or G proteins on their surface. Three main vectors have been used for the vaccination namely, Poxvirus vectors, Vesicular stomatitis virus (VSV), and Paramyxovirus vectors.

Poxvirus Vectors A canarypox vaccine vector (ALVAC) has been fabricated to aid the expression of NiV-G or NiV-F or both of them, and thus immunologically sensitize the pigs in an experimental model which resulted in the presence of no viral shedding in the vaccinated models. Co-expression of both NiV-G and NiV-F gave the best response with the production of highest nAb [80].

Vesicular Stomatitis Virus (VSV) Vector-Based Vaccines This technology involves pseudotyping the replication-defective VSV, which lacks the envelope G protein, with that of NiV-G or NiV-F. This could also involve live attenuated recombinant VSV, also known as rVSV. That expresses NiV-G or NiV-F [102, 103, 117]. These vaccines have been developed to neutralize both the NiV-B and M strains. Additionally, only a single-dose inoculation of the vaccine was capable of providing full protection to Syrian hamster models, AGMs, and ferrets that were previously inoculated with lethal doses of NiV [102, 103, 118]. The results presented low viral antigens, high nAb titers, and absence of disease pathology in these animals.

Paramyxovirus Vectors A complete protection to hamster models has been showcased by a recombinant-based measles virus vaccine that expressed the NiV envelope glycoproteins. Nevertheless, when two doses of the same vaccine were administered to the AGMs, they did display the clinical signs of neurological disease, which provides the evidence of partial protection [119]. Another such vaccine that made use of the Newcastle disease virus strain, expressing the NiV-F or NiV-G proteins, is being developed as well [120].

Other than these three vectors, few other vectors such as Venezuelan equine virus vector and adeno-associated virus vector are also being used in the developmental stages.

12.3.3.3 Virus-Like Particles (VLPs)

The VLPs have been created with the help of simultaneous expression of NiV F or G proteins in mammalian cells [121]. These vaccines have been investigated in the Balb/c rodents with the results showcasing the induction of nAb response. Also, these vaccines have been tested in hamsters to provide protection against NiV challenge, with a single-shot as well as 3-dose regime [80].

12.4 Conclusion and Future Directions

Having witnessed a recent death-dealing Covid-19 pandemic, worldwide scientific researchers and health care workers are now attentive for such viral global breakouts. As NiV exists as an RNA virus, it is plausible to accept that the strains could undergo constant mutations and create a deadlier version. This characteristic, coupled with a wide host range and high transmission, could lead to NiV being one such threat in the near future. NiV is a highly infectious virus that has the potential to cause pandemics with a high fatality rate. Its natural host is fruit bats, and its ability to spread via the respiratory route raises the risk of future pandemics. NiV proteins effectively interact with the immune response, disabling the establishment of a protective anti-viral immunity. Understanding the host-pathogen relationship and developing innovative anti-viral strategies is necessary for the development of efficient anti-viral approaches. A multidisciplinary approach is required, putting together virologists, immunologists, epidemiologists, veterinarians, and physicians within a "one health approach" to understand and control NiV infections.

Like pathogenesis and management, diagnostic tests for NiV, a highly prevalent cause of febrile encephalitic disease, have a vital role in identifying and controlling outbreaks. Furthermore, these tests provide valuable insights into the timing and duration of infection, transmission risks, and risk factors associated with the severity of the disease. Through diagnostic testing, a better understanding of the window of positivity can be obtained, which helps to inform disease management strategies. To achieve the objectives of the R&D blueprint, diagnostic tests play a crucial role, and therefore, the World Health Organization (WHO) is working to coordinate research and funding efforts. To ensure the development, evaluation, and distribution of high-quality, affordable NiV diagnostic tests, WHO is collaborating with product development partnerships, such as Foundation for Innovative New Diagnostics (FIND).

The Coalition for Epidemic Preparedness Innovations (CEPI) has included NiV as a priority for vaccine development. Multi sectoral and interdisciplinary approaches are essential in preventing the emergence of NiV. Additionally, intricate understanding regarding the excretions of virus by bats could be established with a strong coordination among the international virologists and ecologists. Platforms such as Global Outbreak Alert and Response Network (GOARN) have shown keen interest in fabricating a communication between the veterinary and medical segments. Proper hygienic environment should be maintained while slaughtering the livestock.

In a nutshell, human and animal healthcare should be considered one, especially when NiV is considered. Shifting focus toward the advancements, the diagnostic sphere is surely evolving with a liberal speed; however, their availability still remains a concern. Where differential diagnosis stands up to be a hindrance, low containment biosafe assays and tool could be the savior. The low-income regions facing paucity of the diagnostic tools, particulars like surveillance, properly designed quarantine protocols, and adequate training campaigns could be the only alternative. A majority of the therapeutic approaches, including the biologics and vaccines, are rapidly progressing in the preclinical stages and with the continuance of this momentum, this world would surely be able to defend itself against the forthcoming pandemics.

Acknowledgment

The authors are grateful to **Prof. Gaurang B. Shah**, Department of Pharmacology, L. M. College of Pharmacy, Ahmedabad, Gujarat, India for kind support and guidance in manuscript preparation. The authors also extend their appreciation to L. M. College of Pharmacy, Ahmedabad, India for providing continuous library resources support throughout literature survey and data collection.

References

1 Halpin, K., Young, P.L., Field, H.E., and Mackenzie, J.S. (2000). Isolation of Hendra virus from pteropid bats: a natural reservoir of Hendra virus. *J. Gen. Virol.* 81 (Pt. 8): 1927–1932.

2 Epstein, J.H., Field, H.E., Luby, S. et al. (2006). Nipah virus: impact, origins, and causes of emergence. *Curr. Infect. Dis. Rep.* 8 (1): 59–65.

3 Johnson, K., Vu, M., and Freiberg, A.N. (2021). Recent advances in combating Nipah virus. *Fac. Rev.* 10: 74.

4 Soman Pillai, V., Krishna, G., and Valiya, V.M. (2020). Nipah virus: past outbreaks and future containment. *Viruses* 12 (4): 465.

5 Kenmoe, S., Demanou, M., Bigna, J.J. et al. (2019). Case fatality rate and risk factors for Nipah virus encephalitis: a systematic review and meta-analysis. *J. Clin. Virol.* 117: 19–26.

6 Hooper, P., Zaki, S., Daniels, P., and Middleton, D. (2001). Comparative pathology of the diseases caused by Hendra and Nipah viruses. *Microbes Infect.* 3 (4): 315–322.

7 Escaffre, O., Borisevich, V., Carmical, J.R. et al. (2013). Henipavirus pathogenesis in human respiratory epithelial cells. *J. Virol.* 87 (6): 3284–3294.

8 Mazzola, L.T. and Kelly-Cirino, C. (2019). Diagnostics for Nipah virus: a zoonotic pathogen endemic to Southeast Asia. *BMJ Glob. Health* 4 (Suppl 2): e001118.

9 Nipah virus. https://www.who.int/news-room/fact-sheets/detail/nipah-virus.

10 Hauser, N., Gushiken, A.C., Narayanan, S. et al. (2021). Evolution of Nipah virus infection: past, present, and future considerations. *Trop. Med. Infect. Dis.* 6 (1): 24.

11 Chong, H.T., Kamarulzaman, A., Tan, C.T. et al. (2001). Treatment of acute Nipah encephalitis with ribavirin. *Ann. Neurol.* 49 (6): 810–813.

12 Guillaume, V., Aslan, H., Ainouze, M. et al. (2006). Evidence of a potential receptor-binding site on the Nipah virus G protein (NiV-G): identification of globular head residues with a role in fusion promotion and their localization on an NiV-G structural model. *J. Virol.* 80 (15): 7546–7554.

13 Geisbert, T.W., Bobb, K., Borisevich, V. et al. (2021). A single dose investigational subunit vaccine for human use against Nipah virus and Hendra virus. *NPJ Vaccines* 6 (1): 23.

14 Walpita, P., Cong, Y., Jahrling, P.B. et al. (2017). A VLP-based vaccine provides complete protection against Nipah virus challenge following multiple-dose or single-dose vaccination schedules in a hamster model. *NPJ Vaccines* 2: 21.

15 Harcourt, B.H., Lowe, L., Tamin, A. et al. (2005). Genetic characterization of Nipah virus, Bangladesh, 2004. *Emerg. Infect. Dis.* 11 (10): 1594–1597.

16 Harcourt, B.H., Tamin, A., Ksiazek, T.G. et al. (2000). Molecular characterization of Nipah virus, a newly emergent paramyxovirus. *Virology* 271 (2): 334–349.

17 Goldsmith, C.S., Whistler, T., Rollin, P.E. et al. (2003). Elucidation of Nipah virus morphogenesis and replication using ultrastructural and molecular approaches. *Virus Res.* 92 (1): 89–98.

18 Zhu, Z., Dimitrov, A.S., Bossart, K.N. et al. (2006). Potent neutralization of Hendra and Nipah viruses by human monoclonal antibodies. *J. Virol.* 80 (2): 891–899.

19 Wang, L.F., Yu, M., Hansson, E. et al. (2000). The exceptionally large genome of Hendra virus: support for creation of a new genus within the family Paramyxoviridae. *J. Virol.* 74 (21): 9972–9979.

20 Luby, S.P., Rahman, M., Hossain, M.J. et al. (2006). Foodborne transmission of Nipah virus, Bangladesh. *Emerg. Infect. Dis.* 12 (12): 1888–1894.

21 Escaffre, O., Hill, T., Ikegami, T. et al. (2018). Experimental infection of Syrian hamsters with aerosolized Nipah virus. *J. Infect. Dis.* 218 (10): 1602–1610.

22 Luby, S.P., Gurley, E.S., and Hossain, M.J. (2009). Transmission of human infection with Nipah virus. *Clin. Infect. Dis.* 49 (11): 1743–1748.

23 Montgomery, J.M., Hossain, M.J., Gurley, E. et al. (2008). Risk factors for Nipah virus encephalitis in Bangladesh. *Emerg. Infect. Dis.* 14 (10): 1526–1532.

24 Ali, M.Y., Fattah, S., Islam, M.M. et al. (2011). Outbreak of Nipah encephalitis in Greater Faridpur district. *Faridpur Med. Coll. J.* 5: 63–65.

25 Fogarty, R., Halpin, K., Hyatt, A.D. et al. (2008). Henipavirus susceptibility to environmental variables. *Virus Res.* 132 (1–2): 140–144.

26 Gurley, E.S., Montgomery, J.M., Hossain, M.J. et al. (2007). Person-to-person transmission of Nipah virus in a Bangladeshi community. *Emerg. Infect. Dis.* 13 (7): 1031–1037.

27 Singh, R.K., Dhama, K., Chakraborty, S. et al. (2019). Nipah virus: epidemiology, pathology, immunobiology and advances in diagnosis, vaccine designing and control strategies – a comprehensive review. *Vet. Q.* 39 (1): 26–55.

28 Ang, B.S.P., Lim, T.C.C., and Wang, L. (2018). Nipah virus infection. *J. Clin. Microbiol.* 56 (6): e01875-17.

29 Dabra, A., Arya, P., and Guarve, K. (2018). Nipah virus (NiV): diagnosis, pathophysiology and treatment. *Asian J. Pharm. Pharmacol.* 4: 739–743.

30 Sharma, V., Kaushik, S., Kumar, R. et al. (2019). Emerging trends of Nipah virus: a review. *Rev. Med. Virol.* 29 (1): e2010.

31 Reynes, J.M., Counor, D., Ong, S. et al. (2005). Nipah virus in Lyle's flying foxes, Cambodia. *Emerg. Infect. Dis.* 11 (7): 1042–1047.

32 Wacharapluesadee, S., Lumlertdacha, B., Boongird, K. et al. (2005). Bat Nipah virus, Thailand. *Emerg. Infect. Dis.* 11 (12): 1949–1951.

33 Ching, P.K., de los Reyes, V.C., Sucaldito, M.N. et al. (2015). Outbreak of henipavirus infection, Philippines, 2014. *Emerg. Infect. Dis.* 21 (2): 328–331.

34 Skowron, K., Bauza-Kaszewska, J., Grudlewska-Buda, K. et al. (2021). Nipah virus-another threat from the world of zoonotic viruses. *Front. Microbiol.* 12: 811157.

35 Stone, J.A., Vemulapati, B.M., Bradel-Tretheway, B., and Aguilar, H.C. (2016). Multiple strategies reveal a bidentate interaction between the Nipah virus attachment and fusion glycoproteins. *J. Virol.* 90 (23): 10762–10773.

36 Bowden, T.A., Aricescu, A.R., Gilbert, R.J. et al. (2008). Structural basis of Nipah and Hendra virus attachment to their cell-surface receptor ephrin-B2. *Nat. Struct. Mol. Biol.* 15 (6): 567–572.

37 Steffen, D.L., Xu, K., Nikolov, D.B., and Broder, C.C. (2012). Henipavirus mediated membrane fusion, virus entry and targeted therapeutics. *Viruses* 4 (2): 280–308.

38 Bose, S., Jardetzky, T.S., and Lamb, R.A. (2015). Timing is everything: fine-tuned molecular machines orchestrate paramyxovirus entry. *Virology* 479–480: 518–531.

39 Liu, Q., Bradel-Tretheway, B., Monreal, A.I. et al. (2015). Nipah virus attachment glycoprotein stalk C-terminal region links receptor binding to fusion triggering. *J. Virol.* 89 (3): 1838–1850.

40 Eaton, B.T., Broder, C.C., Middleton, D., and Wang, L.F. (2006). Hendra and Nipah viruses: different and dangerous. *Nat. Rev. Microbiol.* 4 (1): 23–35.

41 Aguilar, H.C., Aspericueta, V., Robinson, L.R. et al. (2010). A quantitative and kinetic fusion protein-triggering assay can discern distinct steps in the Nipah virus membrane fusion cascade. *J. Virol.* 84 (16): 8033–8041.

42 Marsh, G.A. and Wang, L.F. (2012). Hendra and Nipah viruses: why are they so deadly? *Curr. Opin. Virol.* 2 (3): 242–247.

43 Rawlinson, S.M., Zhao, T., Rozario, A.M. et al. (2018). Viral regulation of host cell biology by hijacking of the nucleolar DNA-damage response. *Nat. Commun.* 9 (1): 3057. https://doi.org/10.1038/s41467-018-05354-7.

44 Sauerhering, L., Zickler, M., Elvert, M. et al. (2016). Species-specific and individual differences in Nipah virus replication in porcine and human airway epithelial cells. *J. Gen. Virol.* 97 (7): 1511–1519.

45 Sauerhering, L., Müller, H., Behner, L. et al. (2017). Variability of interferon-λ induction and antiviral activity in Nipah virus infected differentiated human bronchial epithelial cells of two human donors. *J. Gen. Virol.* 98 (10): 2447–2453.

46 Uchida, S., Horie, R., Sato, H. et al. (2018). Possible role of the Nipah virus V protein in the regulation of the interferon beta induction by interacting with UBX domain-containing protein1. *Sci. Rep.* 8 (1): 7682.

47 Watkinson, R.E. and Lee, B. (2016). Nipah virus matrix protein: expert hacker of cellular machines. *FEBS Lett.* 590 (15): 2494–2511.

48 Ranadheera, C., Proulx, R., Chaiyakul, M. et al. (2018). The interaction between the Nipah virus nucleocapsid protein and phosphoprotein regulates virus replication. *Sci. Rep.* 8 (1): 15994.

49 Liu, Q., Chen, L., Aguilar, H.C., and Chou, K.C. (2018). A stochastic assembly model for Nipah virus revealed by super-resolution microscopy. *Nat. Commun.* 9 (1): 3050.

50 Johnston, G.P., Contreras, E.M., Dabundo, J. et al. (2017). Cytoplasmic motifs in the Nipah virus fusion protein modulate virus particle assembly and egress. *J. Virol.* 91 (10): e02150-16.

51 Gao, Z., Li, T., Han, J. et al. (2022). Assessment of the immunogenicity and protection of a Nipah virus soluble G vaccine candidate in mice and pigs. *Front. Microbiol.* 13: 1031523.

52 Mathieu, C., Guillaume, V., Volchkova, V.A. et al. (2012). Nonstructural Nipah virus C protein regulates both the early host proinflammatory response and viral virulence. *J. Virol.* 86 (19): 10766–10775.

53 Leon, A.J., Borisevich, V., Boroumand, N. et al. (2018). Host gene expression profiles in ferrets infected with genetically distinct henipavirus strains. *PLoS Negl. Trop. Dis.* 12 (3): e0006343.

54 Andrejeva, J., Childs, K.S., Young, D.F. et al. (2004). The V proteins of paramyxoviruses bind the IFN-inducible RNA helicase, mda-5, and inhibit its activation of the IFN-beta promoter. *Proc. Natl. Acad. Sci. U. S. A.* 101 (49): 17264–17269.

55 Shaw, M.L., Cardenas, W.B., Zamarin, D. et al. (2005). Nuclear localization of the Nipah virus W protein allows for inhibition of both virus- and toll-like receptor 3-triggered signaling pathways. *J. Virol.* 79 (10): 6078–6088.

56 Sánchez-Aparicio, M.T., Feinman, L.J., García-Sastre, A., and Shaw, M.L. (2018). Paramyxovirus V proteins interact with the RIG-I/TRIM25 regulatory complex and inhibit RIG-I signaling. *J. Virol.* 92 (6).

57 Bharaj, P., Wang, Y.E., Dawes, B.E. et al. (2016). The matrix protein of Nipah virus targets the E3-ubiquitin ligase TRIM6 to inhibit the IKKε kinase-mediated type-I IFN antiviral response. *PLoS Pathog.* 12 (9): e1005880.

58 Sugai, A., Sato, H., Takayama, I. et al. (2017). Nipah and Hendra virus nucleoproteins inhibit nuclear accumulation of signal transducer and activator of transcription 1 (STAT1) and STAT2 by interfering with their complex formation. *J. Virol.* 91 (21).

59 Satter, S.M., Nazneen, A., Aquib, W.R. et al. (2023). Vertical transfer of humoral immunity against Nipah virus: a novel evidence from Bangladesh. *Trop. Med. Infect. Dis.* 8 (1): 16.

60 Cong, Y., Lentz, M.R., Lara, A. et al. (2017). Loss in lung volume and changes in the immune response demonstrate disease progression in African green monkeys infected by small-particle aerosol and intratracheal exposure to Nipah virus. *PLoS Negl. Trop. Dis.* 11 (4): e0005532.

61 Lara, A., Cong, Y., Jahrling, P.B. et al. (2019). Peripheral immune response in the African green monkey model following Nipah-Malaysia virus exposure by intermediate-size particle aerosol. *PLoS Negl. Trop. Dis.* 13 (6): e0007454.

62 Escaffre, O., Borisevich, V., and Rockx, B. (2013). Pathogenesis of Hendra and Nipah virus infection in humans. *J. Infect. Dev. Ctries.* 7 (4): 308–311.

63 Arunkumar, G., Devadiga, S., McElroy, A.K. et al. (2019). Adaptive immune responses in humans during Nipah virus acute and convalescent phases of infection. *Clin. Infect. Dis.* 69 (10): 1752–1756.

64 Weingartl, H., Czub, S., Copps, J. et al. (2005). Invasion of the central nervous system in a porcine host by nipah virus. *J. Virol.* 79 (12): 7528–7534.

65 Selvey, L.A., Wells, R.M., McCormack, J.G. et al. (1995). Infection of humans and horses by a newly described morbillivirus. *Med. J. Aust.* 162 (12): 642–645.

66 Chua, K.B. (2003). Nipah virus outbreak in Malaysia. *J. Clin. Virol.* 26 (3): 265–275.

67 Pallivalappil, B., Ali, A., Thulaseedharan, N.K. et al. (2020). Dissecting an outbreak: a clinico-epidemiological study of Nipah virus infection in Kerala, India, 2018. *J. Global Infect. Dis.* 12 (1): 21–27.

68 Goh, K.J., Tan, C.T., Chew, N.K. et al. (2000). Clinical features of Nipah virus encephalitis among pig farmers in Malaysia. *N. Engl. J. Med.* 342 (17): 1229–1235.

69 Lee, K.E., Umapathi, T., Tan, C.B. et al. (1999). The neurological manifestations of Nipah virus encephalitis, a novel paramyxovirus. *Ann. Neurol.* 46 (3): 428–432.

70 Shete, A.M., Radhakrishnan, C., Pardeshi, P.G. et al. (2021). Antibody response in symptomatic & asymptomatic Nipah virus cases from Kerala, India. *Indian J. Med. Res.* 154 (3): 533–535.

71 Chong, H.T., Kunjapan, S.R., Thayaparan, T. et al. (2002). Nipah encephalitis outbreak in Malaysia, clinical features in patients from Seremban. *Can. J. Neurol. Sci.* 29 (1): 83–87.

72 Tan, C.T., Goh, K.J., Wong, K.T. et al. (2002). Relapsed and late-onset Nipah encephalitis. *Ann. Neurol.* 51 (6): 703–708.

73 Wang, L.F. and Daniels, P. (2012). Diagnosis of henipavirus infection: current capabilities and future directions. *Curr. Top. Microbiol. Immunol.* 359: 179–196.

74 Jose, B. (2018). Nipah: the deadly menace. *J. Acad. Clin. Microbiol.* 20 (2): 66–73.

75 Daniels, P., Ksiazek, T., and Eaton, B.T. (2001). Laboratory diagnosis of Nipah and Hendra virus infections. *Microbes Infect.* 3 (4): 289–295.

76 Chiang, C.F., Lo, M.K., Rota, P.A. et al. (2010). Use of monoclonal antibodies against Hendra and Nipah viruses in an antigen capture ELISA. *Virol. J.* 7: 115.

77 Kulkarni, D.D., Tosh, C., Venkatesh, G., and Senthil, K.D. (2013). Nipah virus infection: current scenario. *Indian J. Virol.* 24 (3): 398–408.

78 Fischer, K., Diederich, S., Smith, G. et al. (2018). Indirect ELISA based on Hendra and Nipah virus proteins for the detection of henipavirus specific antibodies in pigs. *PLoS One* 13 (4): e0194385.

79 Ong, K.C. and Wong, K.T. (2015). Henipavirus encephalitis: recent developments and advances. *Brain Pathol.* 25 (5): 605–613.

80 Thakur, N. and Bailey, D. (2019). Advances in diagnostics, vaccines and therapeutics for Nipah virus. *Microbes Infect.* 21 (7): 278–286.

81 Guillaume, V., Lefeuvre, A., Faure, C. et al. (2004). Specific detection of Nipah virus using real-time RT-PCR (TaqMan). *J. Virol. Methods* 120 (2): 229–237.

82 Liu, J., Ochieng, C., Wiersma, S. et al. (2016). Development of a TaqMan array card for acute-febrile-illness outbreak investigation and surveillance of emerging pathogens, including Ebola virus. *J. Clin. Microbiol.* 54 (1): 49–58.

83 Diagnosis. https://www.cdc.gov/vhf/nipah/diagnosis/index.html.

84 Lim, C.C., Lee, K.E., Lee, W.L. et al. (2002). Nipah virus encephalitis: serial MR study of an emerging disease. *Radiology* 222 (1): 219–226.

85 Ranawaka, U.K. (2022). Emerging tropical neurological infections. *Clin. Med. (Lond.)* 22 (1): 18–20.

86 Clayton, B.A., Middleton, D., Bergfeld, J. et al. (2012). Transmission routes for Nipah virus from Malaysia and Bangladesh. *Emerg. Infect. Dis.* 18 (12): 1983–1993.

87 Middleton, D., Pallister, J., Klein, R. et al. (2014). Hendra virus vaccine, a one health approach to protecting horse, human, and environmental health. *Emerg. Infect. Dis.* 20 (3): 372–379.

88 Ochani, R.K., Batra, S., Shaikh, A., and Asad, A. (2019). Nipah virus – the rising epidemic: a review. *Infez. Med.* 27 (2): 117–127.

89 Bossart, K.N., Zhu, Z., Middleton, D. et al. (2009). A neutralizing human monoclonal antibody protects against lethal disease in a new ferret model of acute Nipah virus infection. *PLoS Pathog.* 5 (10): e1000642.

90 Geisbert, T.W., Mire, C.E., Geisbert, J.B. et al. (2014). Therapeutic treatment of Nipah virus infection in nonhuman primates with a neutralizing human monoclonal antibody. *Sci. Transl. Med.* 6 (242): 242ra82.

91 Mire, C.E., Chan, Y.P., Borisevich, V. et al. (2020). A cross-reactive humanized monoclonal antibody targeting fusion glycoprotein function protects ferrets against lethal Nipah virus and Hendra virus infection. *J. Infect. Dis.* 221 (Suppl. 4): S471–S479.

92 Lo, M.K., Feldmann, F., Gary, J.M. et al. (2019). Remdesivir (GS-5734) protects African green monkeys from Nipah virus challenge. *Sci. Transl. Med.* 11 (494).

93 Dawes, B.E., Kalveram, B., Ikegami, T. et al. (2018). Favipiravir (T-705) protects against Nipah virus infection in the hamster model. *Sci. Rep.* 8 (1): 7604.

94 Lo, M.K., Spengler, J.R., Krumpe, L.R.H. et al. (2020). Griffithsin inhibits Nipah virus entry and fusion and can protect Syrian golden hamsters from lethal Nipah virus challenge. *J. Infect. Dis.* 221 (Supplement_4): S480–S492.

95 Freiberg, A.N., Worthy, M.N., Lee, B., and Holbrook, M.R. (2010). Combined chloroquine and ribavirin treatment does not prevent death in a hamster model of Nipah and Hendra virus infection. *J. Gen. Virol.* 91 (Pt 3): 765–772.

96 Pallister, J.A., Klein, R., Arkinstall, R. et al. (2013). Vaccination of ferrets with a recombinant G glycoprotein subunit vaccine provides protection against Nipah virus disease for over 12 months. *Virol. J.* 10: 237.

97 Tanimura, N., Imada, T., Kashiwazaki, Y., and Sharifah, S.H. (2006). Distribution of viral antigens and development of lesions in chicken embryos inoculated with Nipah virus. *J. Comp. Pathol.* 135 (2): 74–82.

98 Zhu, Z., Bossart, K.N., Bishop, K.A. et al. (2008). Exceptionally potent cross-reactive neutralization of Nipah and Hendra viruses by a human monoclonal antibody. *J. Infect. Dis.* 197 (6): 846–853.

99 Pickering, B.S., Hardham, J.M., Smith, G. et al. (2016). Protection against henipaviruses in swine requires both, cell-mediated and humoral immune response. *Vaccine* 34 (40): 4777–4786.

100 Guillaume, V., Contamin, H., Loth, P. et al. (2004). Nipah virus: vaccination and passive protection studies in a Hamster model. *J. Virol.* 78 (2): 834–840.

101 Broder, C.C., Xu, K., Nikolov, D.B. et al. (2013). A treatment for and vaccine against the deadly Hendra and Nipah viruses. *Antivir. Res.* 100 (1): 8–13.

102 Mire, C.E., Versteeg, K.M., Cross, R.W. et al. (2013). Single injection recombinant vesicular stomatitis virus vaccines protect ferrets against lethal Nipah virus disease. *Virol. J.* 10: 353.

103 Lo, M.K., Bird, B.H., Chattopadhyay, A. et al. (2014). Single-dose replication-defective VSV-based Nipah virus vaccines provide protection from lethal challenge in Syrian hamsters. *Antivir. Res.* 101: 26–29.

104 Valbuena, G., Halliday, H., Borisevich, V. et al. (2014). A human lung xenograft mouse model of Nipah virus infection. *PLoS Pathog.* 10 (4): e1004063.

105 Broder, C.C. and Wong, K.T. (2016). Henipaviruses. In: *Neurotropic Viral Infections: Volume 1: Neurotropic RNA Viruses* (ed. C.S. Reiss), 45–83. Springer International Publishing.

106 A Phase 1 Study to Evaluate Safety & Immunogenicity of rVSV-Nipah Virus Vaccine Candidate PHV02 in Healthy Adult Subjects. https://clinicaltrials.gov/ct2/show/record/NCT05178901?cond=Nipah+virus&draw=2&rank=1.

107 Loomis, R.J., Stewart-Jones, G.B.E., Tsybovsky, Y. et al. (2020). Structure-based design of Nipah virus vaccines: a generalizable approach to paramyxovirus immunogen development. *Front. Immunol.* 11: 842.

108 Banerjee, S., Gupta, N., Kodan, P. et al. (2019). Nipah virus disease: a rare and intractable disease. *Intractable Rare Dis. Res.* 8 (1): 1–8.

109 Lipin, R., Dhanabalan, A.K., Gunasekaran, K., and Solomon, R.V. (2021). Piperazine-substituted derivatives of favipiravir for Nipah virus inhibition: what do in silico studies unravel? *SN Appl. Sci.* 3 (1): 110.

110 Sen, N., Kanitkar, T.R., Roy, A.A. et al. (2019). Predicting and designing therapeutics against the Nipah virus. *PLoS Negl. Trop. Dis.* 13 (12): e0007419.

111 Lo, M.K., Jordan, P.C., Stevens, S. et al. (2018). Susceptibility of paramyxoviruses and filoviruses to inhibition by 2′-monofluoro- and 2′-difluoro-4′-azidocytidine analogs. *Antivir. Res.* 153: 101–113.

112 Nguyen, N.M., Tran, C.N., Phung, L.K. et al. (2013). A randomized, double-blind placebo controlled trial of balapiravir, a polymerase inhibitor, in adult dengue patients. *J. Infect. Dis.* 207 (9): 1442–1450.

113 Patel, K., Kirkpatrick, C.M., Nieforth, K.A. et al. (2019). Respiratory syncytial virus – a dynamics and the effects of lumicitabine, a nucleoside viral replication inhibitor, in experimentally infected humans. *J. Antimicrob. Chemother.* 74 (2): 442–452.

114 Lo, M.K., Amblard, F., Flint, M. et al. (2020). Potent in vitro activity of β-d-4′-chloromethyl-2′-deoxy-2′-fluorocytidine against Nipah virus. *Antivir. Res.* 175: 104712.

115 Dang, H.V., Cross, R.W., Borisevich, V. et al. (2021). Broadly neutralizing antibody cocktails targeting Nipah virus and Hendra virus fusion glycoproteins. *Nat. Struct. Mol. Biol.* 28 (5): 426–434.

116 Bhalla, S., Melnekoff, D.T., Aleman, A. et al. (2021). Patient similarity network of newly diagnosed multiple myeloma identifies patient subgroups with distinct genetic features and clinical implications. *Sci. Adv.* 7 (47): eabg9551.

117 DeBuysscher, B.L., Scott, D., Marzi, A. et al. (2014). Single-dose live-attenuated Nipah virus vaccines confer complete protection by eliciting antibodies directed against surface glycoproteins. *Vaccine* 32 (22): 2637–2644.

118 Prescott, J., DeBuysscher, B.L., Feldmann, F. et al. (2015). Single-dose live-attenuated vesicular stomatitis virus-based vaccine protects African green monkeys from Nipah virus disease. *Vaccine* 33 (24): 2823–2829.

119 Yoneda, M., Georges-Courbot, M.C., Ikeda, F. et al. (2013). Recombinant measles virus vaccine expressing the Nipah virus glycoprotein protects against lethal Nipah virus challenge. *PLoS One* 8 (3): e58414.

120 Kong, D., Wen, Z., Su, H. et al. (2012). Newcastle disease virus-vectored Nipah encephalitis vaccines induce B and T cell responses in mice and long-lasting neutralizing antibodies in pigs. *Virology* 432 (2): 327–335.

121 Vera-Velasco, N.M., García-Murria, M.J., Sánchez Del Pino, M.M. et al. (2018). Proteomic composition of Nipah virus-like particles. *J. Proteome* 172: 190–200.

13

Middle East Respiratory Syndrome (MERS)

Orhan E. Arslan

Department of Cellular Biology and Pharmacology, Florida International University Herbert Wertheim College of Medicine, Miami, FL, United States

13.1 Introduction

Middle East respiratory syndrome (MERS) is a highly fatal respiratory zoonotic disease first reported at Jeddah, Saudi Arabia, in 2012 by a patient diagnosed with pneumonia and renal failure [1]. Later, MERS spread to Qatar, a bordering country then to Iran, and eventually to the rest of the world [2]. The largest outbreak recorded outside the Middle East was in South Korea, which rapidly spread to other parts of the region. This disease is caused by a novel single-stranded, positive-sense ribonucleic acid (RNA) beta coronavirus (MERS-CoV), which was first isolated from a patient deceased subsequent to hypoxemic respiratory distress and multi-organ failure [1].

Reports have indicated that bats and dromedary camels were the natural hosts of MERS-CoV, and the spread of this viral infection to humans occurred due to either direct or indirect means, but the mechanism and route of transmission to humans remained uncertain. MERS cases were related to several bat CoVs, including HKU4 and HKU5 [3]. Although Dromedary camels were considered the source of animal-to-human transmission, outbreaks of human-to-human transmission in health care settings have also been reported.

Coronaviruses (CoVs) are a group of enveloped viruses with a positive single-stranded pathogenic RNA genome [4]. MERS-CoVs, the largest member of the family Coronaviridae with genomes of ~30 kb, are encapsulated, positive-single-stranded RNA viruses. They are categorized on the basis of phylogenetic clustering into alpha, beta, gamma, and delta genera and further subdivided into clearly marked lineages. Although MERS-CoV and SARS-CoV belong to lineages C and B, respectively, both MERS and SARS are caused by β-CoVs. CoVs not only cause the previously listed diseases, but they are also responsible for myriad of conditions with varying severity in humans and domesticated animals, including enteric, respiratory, hepatic, and neurologic abnormalities.

Interestingly, CoVs have a highly preserved genome structural organization consisting of a single large 5′ open reading frame (ORF) encoding a replicase (an enzyme that catalyzes the production of a complementary RNA molecule) polyprotein trailed by various supplementary ORFs that encode structural and accessory genes scattered through the 3′ end of the genome [5]. Accessory genes modulate the innate immune response to CoVs.

Facilitation of virus replication occurs through the nonstructural proteins that are formed by the cleavage of the polyprotein. Several virion structural proteins, including the nucleocapsid (N), matrix (M), and envelope (E) proteins, as well as the spike (S) protein, are encoded at the 3′ end of the genome encode. Virulence is primarily determined by the S protein, which enables binding of the virus to the host cell receptor. As indicated earlier, the MERS-CoV genome encodes exclusive accessory proteins, of which at least two, 4A and 4B, modulate immune reactions by producing interferon (IFN), particularly when in isolation [6].

Dipeptidyl peptidase 4 (DPP4), an ectopeptidase found on the surfaces of several other cells, is regarded as the receptor that mediates MERS-CoV cell entry [7]. It is a key factor distinguishing MERS-CoV from SARS-CoV, as the latter utilizes angiotensin-converting enzyme 2 (ACE2), another large ectopeptidase receptor [8]. There is no evidence that the MERS-CoV has underwent mutations, to enhance binding to DPP4 in spite of documented substantial recombination of the MERS-CoV in camel, and genetic changes in the virus [9]. Reduced affinity of the virus for the receptor has been a rare

Rising Contagious Diseases: Basics, Management, and Treatments, First Edition. Edited by Seth Kwabena Amponsah, Ranjita Shegokar, and Yashwant V. Pathak.

occurrence, with the exception of only one recorded instance [10]. Nevertheless, the spread of the reduced affinity human CD26 mutant MERS-CoV has been reported during the South Korean outbreak, which suggests that the MERS-CoV spike protein is unable to adapt to human DPP4. Yet, SARS-CoV retained the capability to adapt to the human ACE2 protein through extensive mutation [11]. Despite the compiled data, there is an urgent need to holistically explore the virus, fully determine its pathogenesis and virulence, and eventually develop effective therapeutic strategies.

Following entry into host cells, MERS-CoV binds to the carcinoembryonic antigen-related cell-adhesion molecule 5 (CEACAM5) cofactor ligand and the dipeptidyl peptidase 4 (DPP-4) receptor and begins to replicate in the human respiratory epithelium. After nearly two weeks of incubation period, infected patients may show signs and symptoms of hypoxemic respiratory failure accompanied by dysfunction in other organs that requires immediate intervention.

MERS' presentations are commonly subclinical and in mild form, although infection with this virus can also produce severe manifestations with grave consequences, such as multi-organ failure and signs of acute respiratory distress, particularly in patients with comorbidities including diabetes and renal disease. High mortality rates have been reported in outbreaks in either family or hospital settings, particularly at sites where appropriate care and management are lacking. In the meantime, reports of nosocomial epidemics, sporadic clinical cases in poorly infection-controlled environments, and community-based outbreaks of MERS-CoV continue to provide information about this disease. Human-to-human transmission on a large scale has been documented in healthcare settings, the most noticeable of which occurred in the capital of Saudi Arabia.

Since its emergence, the endemic nature of the virus in dromedary camels in Saudi Arabia, Jordan, Qatar, Oman, Kuwait, and the United Arab Emirates, and Yemen, as well as in Africa is responsible for the recurrence of transmission of infection from animal to human and human to human. The ability of the infecting virus to mutate can augment transmission among human beings. MERS Coronavirus (MERS-CoV) potential to infect different species of animals and humans in wide regions of the world necessitates clear mapping of the geographic distribution of this disease.

Lack of Federal Food and Drug Administration (FDA) approved vaccines necessitated the urgent development of effective therapies. In view of the variance in immune responses recorded in animal models and the lack of concrete data on the epidemiology of MERS in human population had delayed to a large extent the progress toward cure and development of vaccines. Prevention of epidemics may require robust infection control in healthcare delivery systems and long-term public health scrutiny measures.

In order to limit the spread of MERS virus, a number of diagnostic methodologies, therapeutic strategies, and preventive measures have been put into effect, but with the exception of palliative management of disease manifestations, no clear success in vaccine development or efficient therapy was possible. Diagnostic measures for MERS primarily included Real-time Reverse-Transcription–Polymerase-Chain-Reaction (RT-PCR) testing of respiratory secretions and examination of samples obtained from the lower respiratory tract, particularly from highly advanced cases.

Because of the severity of patients' manifestations, high mortality rate, hazards posed to the public, epidemic potential, and absence of real-time treatment or preventive measures, MERS-CoV is considered a priority and high-threat pathogen by the World Health Organization (WHO). Thus, to keep the rate of spread of MERS under control, healthcare organizations must institute rigid prevention measures to curtail the high threat of the pathogen to public health.

Presence of wide knowledge gap about transmission of the MERS virus from animal to human and from human to human, and about disease pathogenesis, therapy, and a vaccine development had hindered the establishment of a robust antiviral therapy regiments and efficacious prevention protocols. However, the door remains wide open for active research work toward an in-depth understanding of the transmission of the virus and disease course both in humans and in animals.

13.2 Discussion

13.2.1 Epidemiology

As indicated earlier, the first reported case ever of MERS-CoV occurred in 2012 in Jeddah, Saudi Arabia, upon the death of an individual from severe pneumonia [12]. Majority of MERS-CoV cases were tracked to individuals residing in the Kingdom of Saudi Arabia or had recently traveled from this region. The virus continued to spread and was later identified in other countries on the Arabian Peninsula. Two years after isolating the virus, a substantial outbreak occurred in Saudi Arabia caused by poor infection control standards in hospitals and outpatient clinics, which was eventually halted by a public awareness campaign, the establishment of a reporting system, and strict infection-control measures.

South Korea MERS outbreak in 2015 was the largest outside the Middle East and was also caused by a returning traveler from a visit to Bahrain, the United Arab Emirates, Qatar, and Saudi Arabia [13].

Reports confirmed that this particular patient visited several hospitals before receiving a diagnosis and treatment [14].

Several factors were thought to have contributed to the South Korean outbreak, which included delayed diagnosis, overcrowded health facilities, the desire by patients to seek medical care in multiple facilities, "doctor shopping," the tendency of visitors to stay with the patients for prolonged periods, limitations in the national quarantine system, and the lack of infectious disease specialists [15, 16].

This outbreak resulted in a number of deaths before ceasing subsequent to the establishment of the Rapid Response Team (RRT) and placement of the population under stringent quarantine. These drastic measures reduced the case fatality rate to 20% far less than the fatality rate of 40% documented earlier in the country of Saudi Arabia [17].

Strict testing was conducted to identify all affected patients, including those within the subclinical stage, and established a 20/80 (20/800 standard, where it is predicted that 20% of the affected population is responsible for 80% of the transmission [18]. These interventions played an important role in realizing a lower fatality rate. Nevertheless, the spread of the infection to a large population outside the Middle East was a threatening development. Patient–patient transmission was subsequently identified in Iran [19].

Another case of MERS-CoV was reported in England at a local hospital after a Qatari national received treatment for the condition. When this patient was identified, immediate, stringent public health measures, including enhanced monitoring systems, were put into effect, particularly among travelers from the Middle East [20]. A case of MERS-CoV was identified in Italy when a traveler returned from a trip to Jordan and was successfully treated and recovered [21]. Similarly, imported cases of MERS were also identified in the Netherlands and Greece [22]. In China, a MERS-CoV infected male traveler in his 40s was identified when he returned to China through Hong Kong after visiting a MERS-CoV infected father in South Korea [23]. Also, a case of MERS-CoV was identified in the state of Indiana in the United States upon returning from a recent trip to Saudi Arabia. A second, but unrelated, case was also reported in the state of Florida.

Although the source of infection in patients with primary disease contracted from other than MERS patients is unclear, camels remain the apparent source for transmission of the virus to humans. Cases also develop subsequent to the spread of infection in households and poorly controlled healthcare settings, although to a limited extent [24]. Interhuman transmission may occur as a result of contact with contaminated fomite, spray, or large droplets, particularly in a close-knit environment such as overcrowded hospital facilities, considering the fact that the virus can survive for up to a day in such an environment [25]. Further, intrinsic disorder analysis indicated that, due to its lengthy existence in the environment, MERS-CoV maintains a high risk of oral-fecal, oral-urine, and salivary transmissibility [26].

It is evident from random MERS cases in past outbreaks that inter-human transmission of the infection is relatively slow when compared to SARS infection, a fact that provides an understanding of the dynamic of virus transmission [27]. Also, the challenge encountered when tracing individual cases that lasted only for a short period of time imposed limitations on the identification of transmission mode [28]. A consistent finding with human-to-human virus transmission is the similarity of the genetic sequences in MERS-CoV obtained from a single epidemic.

Unlike the genetic sequencing of SARS-CoV, the MER-CoV obtained from patients infected at various times and locations showed no controlled mutation, and thus the virus lacks the ability to adapt to human populations [29]. The transmission rate of infection among humans due to household interactions remained minimal, inefficient, and estimated to be around 4% [30]. Patients affected by this mode of infection exhibited mild symptoms or were completely asymptomatic, and the contagiousness and transmissibility of the virus (R_0 factor) through house contacts is around 0.5, significantly lower than the mark of epidemic potential of R_0 1 [28].

Serologic measurement studies have estimated that 45,000 patients in Saudi Arabia were infected with MERS-COV, and that infected individuals were able to spread the virus and show clinical manifestations [31].

Demographic studies show that although males are most commonly affected and are prone to infection at a higher rate than females, in two-thirds of MERS-CoV cases no difference was detected in the fatality rate between males and females, a fact that may be linked to the degree of exposure to the virus [32–34]. The gender disparity with regard to infection rate is linked to male-dominant Arabian culture and customs that enable males to have more contact with camels than females [35]. However, in South Korea, the opposite gender ratio has been observed among hospital employees, as females are dominant workers in healthcare delivery systems [36].

In addition to gender, age has been an additional risk factor, as older patients commonly represent the primary cases in addition to clinical cases acquired as a result of poor health standards in hospitals and medical clinics. In contrast, younger patients tend to be infected by secondary household members and healthcare practitioners. It has been reported that the average age of infection for males and females was 55 and 57 years, respectively. Irrespective of gender, the elderly suffer a higher mortality rate [37–39]. Chan JF et al. reported that patients aged 65 and older experience 10 times higher mortality as compared to others [40]. When considering mortality rate, it is much greater, approximately 90% in the elderly patients, particularly those exceeding the eighth decade of their lives, when compared with 10% in teenage patients. Despite the emphasis placed on MERS-CoV infection in the adult population, infected juvenile cases were also recorded [41, 42]. In the absence of co-morbidities, infected children endure a lower mortality rate when compared to adult patients [43].

Above all the underling comorbidities considered scientifically the most reliable risk factors occurring in 75% of patients diagnosed with MERS-CoV. These comorbidities include approximately 70% of patients with diabetes, 50% with chronic renal disease, 30% with hypertension and obesity, and 30% with chronic cardiac pathologies, pulmonary conditions, including asthma, and chronic obstructive pulmonary disease (COPD) [44]. The fatality rate increased by seven to eight times when the MERS-CoV was associated with co-morbidities [45].

13.2.2 Disease Development

The attachment of MERS-CoV to the host cell receptor identified as the dipeptidyl peptidase DPP4 (CD 26) is critical for infection initiation [46–48]. DPP4, an element responsible for species MERS tropism, is selectively expressed in certain species. Yet, it is widely found in T and B lymphocytes, hepatic parenchyma, renal and intestinal tissues [49]. MERS-CoV infection in transgenic mice with hCD26/DPP4 receptors exhibited severe respiratory compromise and multi-organ damage [50–52], indicating that DPP4 is critical for the onset of MERS-CoV infection. The presence of MERS antigens is also linked to DPP4 receptors located in type I and type II pneumocytes, ciliated bronchial epithelium, non-ciliated terminal bronchioles, and alveolar macrophages [53–56].

The spike glycoprotein (S) is a cell entry protein that bounds to the cell surface receptor DPP4 on host cells. It is a crown-like peplomer lodged on the surface of the enveloped coronavirus particle, mediating cell entry of the virus and initiating the release of neutralizing antibodies from the host [57, 58]. The S protein is considered a type I membrane glycoprotein with two terminals: N-terminal (S1 residues) and the C-terminal (S2 residues). The S1 subunit comprises the receptor-binding domain (RBD), whereas the S2 subunit includes fusion peptide and heptad repeats 1 and 2 (HR1 and HR2) [59, 60].

Cell entry of the virus is enabled by a lock-and-key interaction of the S protein with the host receptors subsequent to the S1 and S2 junctions under protease cleavage action. Cleavage at the S2 position by Furin, a universally expressed protease, has also been claimed [10]. RBD in the S1 subunit binds to the cell receptor, initiating the conformational changes in the S2 subunit. Facilitation of virus entry into the cell occurs upon formation of axis-helix bundle fusion core between heptad repeats 1 and 2 (HR1 and HR2) [61]. When RBD, HR1, and HR2 bind together, the stalk of the S protein is formed, enabling the attachment of the virus to host cell receptors. Accordingly, termination of an infection or limiting the fusion of viruses into cells can be achieved with minor mutations on the S protein or DPP4 receptor residues. An interesting observation in the South Korean outbreak is the reported mutation in the RBD of the viral spike (S) protein of MERS-CoV.

Endosomal membrane fusion and endocytosis are the two pathways that govern MERS-CoV cell entry. As indicated earlier, endosomal membrane fusion is mediated by the cleavage of MERS-CoV spike protein, whereas virus entry via endocytosis is a cathepsin L-dependent process that excludes S protein cleavage. As a result, viral genomic RNA is released into the host cell cytoplasm.

Spike (S) protein interaction with cellular receptor DPP4 signals the beginning of MERS pathogenesis and entry of the virus into the upper respiratory tract. DPP4 expression occurs on several respiratory tract cells, such as pneumocytes type I and II, endothelial cells, hematopoietic cells, and non-ciliated bronchial epithelial cells [55]. DPP4 is highly expressed on the epithelial cells of distal parts of the respiratory tract and alveoli, but remains scanty on the surface of the epithelia of the upper respiratory tract, including the nasal cavity, nasopharynx, and trachea, to conduct airways [62]. The limitation of interhuman transmission may be attributed to the relatively narrow expression of DPP4 in the epithelia of the upper part of the respiratory tract. Epithelial cells of several other organs show high expression of DPP4, such as the intestine, kidney, thymus, and bone marrow.

Increased DPP4 expression on pneumocyte type I and II cells and alveolar macrophages compared to control groups has been documented in immunohistochemical research studies conducted on patients with COPD, which supports the notion that comorbidities, such as preexisting pulmonary disease, cause heightened DPP4 expression and subsequently render individuals with these comorbidities prone to more severe forms of MERS.

More detailed data on MERS pathogenesis have been limited because of the difficulty of accessing autopsy or surgical specimens from patients infected from the Middle East or the South Korean outbreak. This limitation diverted research effort toward in vitro studies on experimental animals, infected human lung explants, and autopsy findings [63]. The in vitro tissue culture research demonstrated that MERS-CoV replication is vigorous in differentiated and undifferentiated human primary epithelial cells as well as in other tissue cell lines [64–66].

In vitro examination of infected human pulmonary explants revealed Mers-CoV antigen in pneumocytes type I and type II, in non-ciliated and also in the ciliated bronchial and bronchiolar epithelial cells, endothelial cells, and alveolar macrophages [53, 67]. The presence of MERS-CoV antigen in pneumocytes (types I and II), epithelial and endothelial cells, as well as macrophages, as shown in the immunohistochemical studies on the MERS autopsy case, is also confirmed by ex vivo infected lung tissue [68–70].

13.2.3 Pathogen Transmission

Data on human-to-human transmission has been inadequate, and the results of studies conducted in 2013 showed that the R_0 (basic reproductive number) was less than 1, a signal that the virus transmission resolved spontaneously [71]. However, self-limitation of transmission appeared inapplicable when analyzing the MERS-CoV infection of human clusters after the enforcement of infection control [72].

A broad view of human-to-human transmission demonstrates that it is linked to two major factors: family and healthcare. A report by Memish ZA et al. suggested that the risk of transmission among family members is low [73]. Another study indicated that silently/and or mildly-infected human patients exist in which an individuals can carry low viral loads of MERS-CoV RNA without obvious symptoms. The existence of MERS-CoV infection in these individuals had been confirmed by RT-PCR assay. The presence of MERS-CoV was reported in the upper respiratory tracts of the study group, which included seven household contacts [30]. However, the rate of transmission within families remained obscure [74]. Therefore, the introduction of serologic testing may be useful in determining the transmission rate among close-knit family members.

MERS-CoV infection subsequent to healthcare related factors was first reported retrospectively from Jordan, Saudi Arabia, in 2012 and later in 2013 and cases were confirmed by the Reverse Transcription-Polymerase Chain Reaction (RT-PCR) technique [44, 75]. Infected health care practitioners during the 2012 outbreak in Jordan represented only 10% of the total workforce exposed [76].

Human-to-human transmission of MERS-CoV through the nosocomial route has been reported among a number of healthcare practitioners in France, Italy, Tunisia, and England, representing additional evidence of hospital-associated infection [77, 78]. Considerable morbidity among healthcare workers due to human-to-human transmission has been reported. It has also been demonstrated by phylogenetic and epidemiologic analyses that human-to-human transmission is the route responsible for the MERS-CoV outbreak [78].

Several factors have been attributed to outbreaks among healthcare workers, including ineffective and/or prevention and control measures, and improper monitoring [79]. It is a significant fact that healthcare employees may play a role in the spread of the MERS-CoV to patients, particularly when infected workers are underdiagnosed, asymptomatic, and/or at a subclinical stage of the disease [80]. Therefore, transmission of infection to individuals seeking medical help within hospital settings from healthcare practitioners has been identified as the most striking cause [81], emphasizing the importance of enforcing targeted infection control standards, self-protection, and the development of specific diagnostic analyses to identify MERS-CoV infection [82].

Nosocomial transmission has been recorded as the responsible route for most confirmed cases of the MERS-CoV outbreak in South Korea. The transmission occurred among patients in the same hospital facility and spread from one health facility to another. It was estimated that this route of transmission was responsible for virus transmission to 20 hospitals [83]. Also, the higher the number of contacts with the virus, the higher the recorded infection rate with MERS-CoV [84].

13.2.4 Immune Response to Pathogen

MERS-CoV consists of structural and accessory proteins, including protein M, ORF 4b, and ORF 5, which function as potent IFN antagonists. ORF4 in particular is considered the most powerful suppressor of IFN antiviral activity and production. A study conducted by Yang et al. reaffirmed the role of MERS-CoV ORF4b as IFN antagonist [85]. Ex vivo study of human lung cell lines demonstrated that induction of antiviral pro-inflammatory cytokines, which form the first-line antiviral defense in humans, is inhibited during MERS-CoV infection [86]. A delay in the induction or activation of pro-inflammatory cytokines such as IL-1β, IL-8, and IL-6 has been detected in a study conducted on the respiratory epithelial cell line Calu-3 cells [87].

The presence of papain-like proteases (PLpro), an IFN antagonist, in MERS-CoV may be responsible for blocking the expression of IFN-α and IFN-β, which occurs upon cell entry of the virus. The PLpro had the potential to block the upregulation of cytokines and alter the host innate immune reaction in response to viral infection in the host. The mechanism that mediates PLpro suppression of IFN-β includes inhibition of the phosphorylation and nuclear translocation of IFN regulatory factor 3 (IRF3) [88, 89].

Since human macrophages and monocyte-derived dendritic cells (mDCs) allow replication of MERS-CoV [90, 91], expression of viral antigen and viral RNA heighten when MERS-CoV infection occurs, causing an increase in the IFN-λ1, CXCL10, and MxA mRNAs in both macrophages and dendritic cells. An in vitro study revealed increased expression of type I and III IFNs primarily in human plasmacytoid dendritic cells (pDCs) and in the titer of the infected cell culture media.

T cells (both $CD4^+$ and $CD8^+$) may be targeted by MERS-CoV as they express DPP4 receptors [92].

Apoptosis induced in the host T cells by MERS-CoV infection is mediated through intrinsic and extrinsic pathways that produce robust viral replication. To elucidate and analyze host responses to MERS-CoV infection, mice models or transduced mice models with Ad/hDPP4 were used. These animal-based research studies showed that depletion of $CD8^+$ T cells and macrophages by MERS-CoV infection compromised lung function [93]. Another mice model showed that infection triggered critical changes to counter MERS-CoV, including increased expression of $D8^+$ T cells and neutralizing antibodies combined with a type I IFN response [94].

Initial viral infection usually triggers a rapid and well-coordinated innate immune response as a first-line defense mechanism. This innate immune response, usually robust but unrestrained, is associated with immunopathological reactions. In particular, MERS-CoV infection of human epithelial cells induces significant delayed type I, II, and III IFN responses [95]. MERS-CoV infection produces delayed induction of pro-inflammatory cytokines and chemokines, including interleukin (IL)-1β, IL-6, and IL-8 [86].

MERS-CoV not only infects epithelial cells within the lower respiratory tract but also hematopoietic cells such as monocyte–macrophages, dendritic cells, and activated T cells, as shown in in vitro studies [91, 96]. Whereas SARS-CoV infect monocyte–macrophages and dendritic cells without producing any progeny virus (abortive infection) [97]. Elevated levels of cytokines and chemokines such as CCL-2, CCL-3, RANTES (**R**egulated upon **A**ctivation, **N**ormal **T** Cell **E**xpressed and Presumably **S**ecreted, is a chemokine secreted by platelets that mediates monocyte/macrophage infiltration), IL-2, and IL-8 in the human macrophages and dendritic cells have also been induced by MERS-CoV infection, similar to that of the epithelial cells immune response.

Plasmacytoid dendritic cells produce type I and III IFN in high concentration upon infection, unlike the induction of type 1 IFN by monocyte–macrophages and dendritic cells, which remains meager. Both intrinsic and extrinsic pathways engage in the apoptosis induced by MERS-CoV infection of the activated T cells, which may be responsible for lymphopenia associated with poor consequences [98].

In severely-ill MERS-CoV patients, data have been insufficient on elevated CXCL-10, IL-17, and IL-23 expression, with scanty IFN-α and IFN-γ expression. Insufficient data and limited available autopsy specimens made the determination of factors actively involved in protective versus pathologic immune reactions a major challenge [70].

13.2.5 Clinical Signs and Symptoms

The median incubation period of one week, with a range of two days to two weeks, has been reported by studies on inter-human MERS-CoV transmission [99]. It has also been estimated that the median interval from the onset of the disease to hospital admission was four days, whereas the median period from the start of symptoms to intensive care admission was around five days [100].

MERS-CoV infection produces clinical symptoms and signs that range from mild to severe. Some patients remain asymptomatic yet are able to infect others. Most illnesses involve pathologies of the lower respiratory tract, but other internal organ injuries have also been documented. Respiratory illnesses include acute pneumonia that can rapidly worsen to acute lung injury, then to acute respiratory distress syndrome, progressing to organ failure and eventual death [101–104]. Mild respiratory clinical signs and symptoms of MERS-CoV include cough, dyspnea, pneumonitis, and pneumonia, leading rapidly to acute respiratory distress syndrome and pulmonary hypoxemia [105]. Multi-organ injury, which also develops as a result of immunosuppression, comprises renal failure and hepatitis, particularly in critically ill MERS-CoV patients, which may be associated with elevated levels of alanine aminotransferase and aspartate aminotransferase [106].

According to a research study conducted during the outbreak in South Korea in 2015, transmission of MERS-CoV is intricately linked to the delay in patients' isolation, the presence of high fever, and extensive chest infiltrates of the patients, dictating a rigorous quarantine during the early-stage MERS outbreak [107]. Hospitalized severely-ill MERS-CoV patients have had mild initial flu-like symptoms, such as dry cough, chest pain, fever, sore throat, headache, muscle pain, and dyspnea that rapidly progress to pneumonia, but also involve other systems, including the gastrointestinal tract, causing crampy abdominal pain, diarrhea, and vomiting.

13.2.6 Laboratory and Radiological Findings

Laboratory tests conducted on samples from the upper respiratory tract of MERS-CoV patients revealed that the titers and RNA copy numbers were less as compared to those of samples obtained from the lower respiratory tract. Examination of blood, stool, and urine samples from some patients revealed low levels of MERS-CoV genomic RNA [108, 109]. The blood sample analysis of MERS patients shows characteristic lymphocytopenia and thrombocytopenia, but also laboratory values that point to renal and liver dysfunctions, such as increased levels of lactate dehydrogenase, creatinine, alanine, and aspartate aminotransferase, have been affirmed.

Findings of plain chest radiographs and CT scans of MERS-CoV patients demonstrated unilateral and bilateral pathologic changes that ranged from minor to extensive tissue damage of the airspace produced nodular or reticular opacities with reticulonodular shadowing and pleural effusions. Additional radiographic findings encompassed airspace consolidations, patchy infiltrates, and enhanced bronchovascular markings [110].

13.2.7 Pathological Findings

Although MERS-CoV infection was associated with outbreaks in wide areas of Saudi Arabia, South Korea, and certain parts of the Arabian Peninsula, macroscopic and histopathologic changes in humans caused by MERS-CoV, which included edema, generalized congestion, and consolidation of all pulmonary lobes, were based primarily on a single autopsy.

The list of microscopic lesions includes exudative diffuse alveolar damage with necrosis of alveolar epithelial cells, hyperplasia of type II pneumocytes, sloughing of necrotic bronchiolar epithelium, alveolar edema, hyaline membrane formation, marked thickening of alveolar septa, and infiltration of monocyte–macrophages, neutrophils, and lymphocytes, as well as alveolar fibrin deposition. The latter is initiated by heightened activity of the extrinsic coagulation cascade pathway related to factor VII. To substantiate if the changes observed in the heart, kidney, bone marrow, and primary and secondary lymphoid organs are the results of direct virus infection or if they are nonspecific findings at the terminal stage, further investigation is required.

13.2.8 Zoonotic Infection

Dromedary, a common breed of camel throughout Africa and the Arabian Peninsula, was seropositive to MERS-CoV at a high nearing 80% in some populations [111].

Data generated from serum samples of individuals from Saudi Arabia that date back to three or four decades appeared to be seropositive for MERS-CoV antibodies, indicating that MERS-CoV has in fact infected camels in the past years, but for unknown reasons, the infection did not produce noticeable manifestations prior to the outbreaks. In contrast, camels from Australia, Canada, the United States, Germany, the Netherlands, and Japan were seronegative for MERS-CoV during the corresponding periods. These findings account for the belief that transmission of MERS-CoV from bats to camels in Africa occurred years ago and then spread to the Middle East.

Although cases of MERS-CoV transmission from camels to humans have been reported, the bulk of primary human cases were not confined to camel exposure. However, a sequence analysis conducted on the samples obtained from humans and camels at a barn in Qatar revealed that the same strain of virus infected both camels and humans. Research studies conducted in Qatar and Saudi Arabia indicate that individuals in close contact with camels are more likely to test positive in serological tests compared to the general population [112].

The possible role of camels in human transmission correlates with the findings of studies showing that MERS-CoVs infected camels in Saudi Arabia were linked to different genetic lineages, and that β-CoVs 1-HKU23 and camelid CoVs were associated with human CoV 229E. Additionally, the evidence that provides credence to the link between camel and human infection stems from the recombinant nature of MERS-CoV strains, which are analogous to those in human populations.

Due to the high expression of DPP4 in the upper respiratory tract and the increased concentration of MESR-CoV in nasal discharge, transmission of the infection between camels is more favorable when compared with transmission to humans. However, the same explanation does not account for the mode of transmission from camels to humans, as this zoonotic transmission appears to be multifactorial. Despite the primary role of MERS-CoV in zoonotic infection, direct contact with infected animals is thought to be an important avenue of transmission [113]. There is no concrete indication that transmission to humans is linked to occupational exposure, as is the case with slaughterhouse employees, notwithstanding the heightened rate of seropositivity among this population that supports the viewpoint that transmission from camels to humans is tenuous [114].

Lack of sanitation and the substandard pasteurization of the milk obtained from Camels are responsible for the detection of MERS-CoV RNA in raw, unpasteurized milk collected in a marketplace in many parts of the Middle East [115]. This observation has highlighted that the transmission of the virus through contaminated camel milk and meat, in addition to respiratory mode, is highly likely. Since the virus may naturally be present in milk and meat or introduced as a result of poor hygiene, prevention of camel-to-human transmission may require extensive precautionary actions such as a concerted public awareness campaign and the use of personal protective equipment. Historically, camels were and remain an integral part of Arab culture and traditions, and thus changing local customs, although an overarching challenge may help overcome the threat of MERS-CoV.

Transmission of MERS-CoV from animals to humans is not common, and there is no strong evidence to the contrary. This observation does not negate the fact that zoonotic transmission is likely responsible for most of the epidemic associated with the virus [116]. The absence of serologic evidence of human infections among workers in close contact with camels in southern Saudi Arabia indicated that zoonotic transmission was uncommon [117]. This finding does not mean that camels are unable to transmit the virus, and the close resemblance of the genetic sequence between MERs-CoV derived from camels from Qatar and Oman to that of human MERS-CoV imparts support to the zoonotic transmission. The likelihood of transmission through the respiratory path of camels is reinforced by the observation that nasal and conjunctival swabs contained viral shedding. Also, several additional researches on MERS-CoV RNA in infected camels' milk may have augmented the evidence in support of the notion that transmission in humans indeed occurs through consumption of milk and uncooked meat [118].

After a series of molecular research in 2013, Ziad et al. revealed that bats in Saudi Arabia were infected by several alpha-coronaviruses and beta-coronaviruses and that the infected bats were from the *Vespertilionidae* family, also known as European Pipistrellus bats [119, 120]. Bats have been viewed as the ultimate source for MERS-CoV since the latter is closely related to the Tylonycteris and Pipistrellus bat CoVs HKU4 and HKU5, respectively [121]. Further, β-CoV, a closely related CoV to MERS-CoV, was obtained from an adult female *Neormocia* cf. *zuluensis* bat from South Africa, a finding that corroborates the concept that this virus harbors the root of the phylogenetic tree of MERS-CoV, adding strength to the view that MERS-CoV distantly originated in African bats [122].

A genetic characterization study of beta-coronavirus lineage C viruses in bats indicates that the RNA polymerase of Pi-BatCoV HKU5 was more closely linked to the MERS-CoV, whereas the spike gene of MERS-CoV was more closely related to Ty-BatCoV HKU4. Also, Pi-BatCoV HKU5 displayed a remarkable divergence in the spike protein, enabling adaptation to a host habitat [123]. Phylogenetic analysis suggests that recombination of S1 subunits of the spike gene produces the variation in spike phenotype necessary for the pathogenesis of MERS-CoV in humans [124]. Furthermore, the detection of cell surface CD26/DPP4 expression in bats added credence to the finding that bats are the origin point of MERS-CoV [125].

An important reason for considering bats as the main reservoir for the MERS-CoV is the phylogenetic analogy of CoVs identified in bats from South Africa to that of human MERS-CoV. Some research scientists reported that bats pose a greater threat to human health and have the capability to unleash devastation through infection transmission when contrasted with other infected animals [126]. Widagdo et al. indicated that the predilection site for the receptor DPP4 was in the intestine of insectivorous bats, further evidence that MERS-CoV transmission occurred via the fecal–oral path [127]. Lack of direct contact between bats and humans renders a direct transmission of the virus to humans highly unlikely. In view of this fact, serologic tests were applied to domestic animals, such as camels, horses, sheep, goats, and chickens, to ascertain MERS-CoV positivity. Research data have shown that isolates from horses and camels contain potent MERS-CoV receptors for the entry of viruses compared to those from bats and goats, and that the genetic sequence of MERS-CoV obtained from camels matches closely with that of human MERS-CoV [128].

13.2.9 Diagnostic Methodologies

Because of the similarity of the presentations of patients with MERS-CoV to those of other common respiratory viral infections, the diagnosis of MERS on a clinical basis may pose a great deal of difficulty, requiring extensive diagnostic tests. As a result, rigid criteria have been established to confirm the MERS-CoV diagnosis. Detection of either an infectious virus or viral RNA in respiratory samples during the acute stage of the disease helps confirm the diagnosis of the disease.

To determine a higher viral load level, diagnostic sampling should be obtained from the lower respiratory tract. Two positive qRT-PCR samples on two separate specific genomic segments or a single positive qRT-PCR outcome with a sequence of another positive genomic segment may be needed to establish a definite diagnosis of MERS-CoV infection using viral RNA detection.

It is a commonly accepted practice that targeting upstream segment of E gene (UpE) and ORF1ab may enable qRT-PCR detection of MERS-CoV genomic mRNA in the upper and lower airway samples [129]. Serum samples containing virus-specific antibodies obtained within two to three weeks after disease onset are utilized in the diagnosis, with the condition that a positive immunofluorescence and/or microneutralization test (a highly specific confirmatory test used to measure neutralizing antibodies) be preceded by a positive enzyme-linked immunosorbent assay (ELISA). When performing these assays, consideration must be given to the fact that MERS-CoV antibody titers decline with time, especially in patients with a mild form of the condition [130].

13.2.10 Vaccination

A series of experimental vaccines were developed using inactivated virus, live attenuated virus, viral vectored, inactivated subunit, and DNA vaccines but failed to reach the stage of clinical trials, and thus to date no specific vaccine against MERS-CoV has been approved for public immunization [131, 132]. The use of live but attenuated viruses in the development of vaccines offers the advantages of strong and prolonged antibody and T cell responses but also holds the propensity to recombine with circulating CoVs to produce novel strains of pathogenic CoVs, which renders the clinical use of the vaccine not plausible [133].

Most of the vaccines that are under development utilize the S protein, or the domain of the S protein, which is essential for binding to host DPP4 as an immunogen since the neutralizing antibodies are primarily driven toward the RBD.

13.2.11 Therapeutic Regimen

At present, no specific approved medication exists to treat MERS, so reliance rests primarily on palliative care. However, the combined timely administration of type I IFNs, an inhibitor of viral replication, with ribavirin, an antiviral drug used to treat hepatitis C, is believed to lessen lung injury through a reduction of pulmonary viral load [134]. Experimental administration of these combined regiments to infected rhesus macaques shortly after infection produced a perceptible reduction of pulmonary immunopathologic lesions and MRS-CoV titers [135].

The same drug combination also produced longer survival in critically ill patients when administered at the end of two weeks, but lacked the same effect when administered at the end of the fourth week post-infection [136]. These findings were supported by another study, which revealed that late administration of the combined viral treatment regiments to patients in the intensive care unit had no noticeable impact on mortality rates, and the variable results of this study were attributed to the challenges encountered in the early administration of medications, a critical factor in the success of therapeutic measures [137].

The introduction of plasma with high titers of neutralizing antibodies or MERS-CoV-specific antibodies from recovering patients has been a promising development that has invigorated the effort to treat MERS patients, yet it has been challenged by the inability to locate MERS survivors with high-titer antibodies [138]. The efficacy of neutralizing antibodies in inhibiting MERS-CoV replication was affirmed experimentally in a mouse model in which researchers were able to isolate the antibody from memory B cells of a MERS patient and effectively inoculate the animal model to impede replication [139]. Ohnuma et al. reported that anti-CD26 monoclonal antibodies inhibit MERS-CoV infection, a finding that supports the utilization of monoclonal antibodies as a first-line treatment for MERS infection [140].

A research study identified m336, m337, and m338 monoclonal antibodies that appear highly effective in neutralizing MERS-CoV by targeting the CD26/DPP4-binding domain (RBD) of the MERS-CoV spike glycoprotein [141, 142]. Reports also identified neutralizing monoclonal antibodies that specifically target the N-terminal domain of the residual region of the MERS-CoV spike protein. Efforts to construct a recombinant Modified Vaccine Virus Ankara (MVA) that expressed the MERS-CoV spike (S) protein (MVA-MERS-S) were successful [143, 144].

MERS-CoV RBD is considered an effective inhibitor of MERS-CoV entry into the cell, and also a potential candidate for MERS-CoV subunit vaccines because of its ability to interact effectively with human DPP4 [145]. Reductions in pulmonary pathology and virus replication have been achieved experimentally in inoculated rhesus macaques via neutralizing antibodies developed using naïve human antibody-phage display libraries, which are usually harvested from unimmunized individuals [140, 146].

Another promising immunotherapeutic approach that drastically inhibited MERS-CoV replication both in vitro and in vivo in a mouse model was the administration of polyclonal antibodies obtained from transchromosomic bovines (developed by chromosomal transfer) [147].

The fact that antibodies act on the RBD of the MERS-CoV S protein is a characteristic that may be utilized most efficiently for prophylaxis in high-risk hospital and healthcare settings. Yet, efficacy in the utilization of neutralizing antibodies for therapeutic purposes appears to be linked to the rapid and immediate administration at the beginning of disease manifestations.

Recent clinical investigations also included other antiviral drugs than IFN and neutralizing antibodies [148]. For instance, the in vitro action of MERs-CoV-specific peptide fusion inhibitors in curtailing viral entry and replication has been experimentally shown to reduce pulmonary viral loads introduced prophylactically or as therapeutic agents in mouse and macaque models [149, 150].

Because of the specific function of viral enzymes and their ability to inhibit replication, drugs like papain-like protease and 3C-like protease can be advantageous alternatives for bringing down viral loads in patients. Drugs with anti-HIV protease activity, such as lopinavir, have also been introduced as therapies but without proven efficacy. Although there is no solid evidence to support effectiveness, other antiviral drugs, including chloroquine, chlorpromazine, mycophenolic acid, and nitazoxanide, have also been utilized as therapeutic agents in MERS patients.

Analysis of a retrospective study revealed that the use of macrolide therapy did not produce a reduction in the mortality rate of MERS-CoV patients, nor did it achieve a quicker clearance of MERS-CoV RNA as compared to the control group [151]. Reports indicate that critically ill patients due to MERS-CoV infection showed that viral clearance was delayed by systemic corticosteroids, but no direct improvement in patients' clinical outcomes was accomplished with this type of therapy [152].

13.2.12 MERS-COV Animal Models

In order to conduct studies on the pathogenesis of MERS-CoV and evaluate antiviral therapeutics and vaccines, the use of animal models may be essential. It has also been reported that the inability of the MERS-CoV S protein to bind to DPP4 orthologs in small laboratory animals may prevent it from enhancing MERS-CoV replication. An innovative method to transduce or sensitize mice to MERS-CoV infection by intranasal transduction of human DPP4 has been applied to MERS-CoV mice models that exhibited mild to moderate signs and symptoms of pneumonia after MERS-CoV experiment. After these successful experimental transductions, several human DPP4 transgenic mouse models were created, but with fatal consequences due to encephalitis [153], necessitating a robust assessment of the developed vaccines and anti-MERS-CoV therapeutic agents. When rabbits were utilized as animal model to mimic human MERS-CoV infection, their manifestations remained incomplete at the subclinical level [154], whereas rhesus macaques exhibited manifestations that can be described as respiratory diseases with a mild to moderate nature. However, severe symptoms were seen in marmosets, which appeared highly susceptible to infection [54].

It is unclear how these conflicting outcomes of therapeutic approaches arise, and the explanations may lie in the manner of drug administration, variance in viral load, sources and/or ages of animal models, findings that may be a useful basis for future studies on inter-camel and camel-to-human zoonotic transmission as well as the basis for mild nasal inflammation in infected camels, but with no signs of pneumonia [155].

13.2.13 Infection Control

The spread of MERS-CoV cases among healthcare practitioners is common and attributed to poor prevention standards, and a downward trend in infection rate has been observed subsequent to the application of robust quarantine standards. Due to the close resemblance of early manifestations of MERS-CoV to other respiratory disorders, healthcare practitioners and staff, particularly in the Middle East, are required to exercise extreme care, adhering to the highest rules of infection control.

Precautionary measures when in contact with infected patients or handling droplets must consistently be put into practice when dealing with verified MERS-CoV patients. Since patients with comorbidities such as diabetes, renal failure, chronic lung disease, or incompetent immune systems are more prone to develop severe forms of MRS-COV, they should refrain or be prevented from making visits to farms or markets where close contact may be established and the potential transmission of the virus exists. General hygiene practice through regular hand washing is likely to reduce infection. Additionally, consuming raw camel milk, under-cooked meat, or medicinally used camel urine should be avoided. When these measures were combined with other World Health Organization recommendations, an overall positive impact was achieved, which is evident in the consistent reduction in the number of secondary cases of MERS.

13.3 Conclusions

The past decades have seen significant progress in medical science research, yet humanity remains vulnerable to newly emerging diseases and infections. We have attempted in this review to demonstrate the impact of understanding pathogenesis, mode of transmission, and significance of infection-control in curtailing future MERS-CoV outbreaks. Lack of effective therapy and reliance on only supportive treatment crystallize the slow advancement in this field and the enormous task ahead toward disease eradication. A pressing need to identify and develop advanced targeted therapies to efficiently offset future threats of pandemic outbreaks became of outmost urgency. Fortunately, work is under way on the development of a number of vaccines for MERS-CoV disease.

After decades of observations, trials and errors, laboratory tests, and robust clinical and immunologic research on MERS-CoV and its potential epidemic threats, it became clear that the dromedary camel is the zoonotic source of the virus. Increased screening and close monitoring of MERS-CoVs for mutations are also crucial, particularly when COVs have a tendency for recombination that boosts virulence and infection transmission in humans. In order to prevent virus outbreaks, health practitioners must establish strict infection control standards and pay careful attention to patients with fever and respiratory manifestations, particularly those who have had a travel history to the Middle East.

Although massive efforts and resources continue to be diverted toward developing vaccines and drugs to control and treat infection, no perceptible success has been recorded. In this regard, specific emphasis may be given to animal models that mimic manifestations of severe human MERS-CoV cases, enabling proper assessment of vaccine efficacy and studying the effects of novel and repurposed drugs as well as antibodies. As part of this approach, development of new drugs and vaccines may require the determination of additional targets, a new host, and viral pathogenesis.

Transportation may play an important role in the spread of MERS-CoV infection, a finding that could serve as a guide in directing detection and intervention measures to control endemics and block a potential pandemic. Since healthcare facilities are considered high-risk sites for the transmission of MERS-CoV infection to patients and to other healthcare personnel due to the imbalance in infection control and delayed isolation of suspected, but undiagnosed cases, robust precautions and intervention tools have to be instituted at these sites. Health practitioners should anticipate and prioritize the management of pulmonary function abnormalities, protracted psychological impairment, including depression, anxiety, and reduced musculoskeletal capacity.

An increased awareness and understanding of the clinical features of MERS-CoV can be enhanced by early screening and management of this infection, an approach that paves the way toward effective therapeutic standards.

References

1 Moh Zaki, A., Van Boheemen, S., Bestebroer, T.M. et al. (2012). Isolation of a novel coronavirus from a man with pneumonia in Saudi Arabia. *N. Engl. J. Med.* 367: 1814–1834.

2 Alyami, M.H., Alyami, H.S., and Warraich, A. (2020). Middle East respiratory syndrome (MERS) and novel coronavirus disease-2019 (COVID-19): from causes to preventions in Saudi Arabia. *Saudi Pharm. J.* 28: 1481–1491.

3 de Groot, R.J., Baker, S.C., Baric, R.S. et al. (2013). Middle East respiratory syndrome coronavirus (MERS-CoV): announcement of the Coronavirus Study Group. *J. Virol.* 87: 7790–7792.

4 Xu, Z., Shi, L., Wang, Y. et al. (2020). Pathological findings of COVID-19 associated with acute respiratory distress syndrome. *Lancet Respir. Med.* 8: 420–422.

5 Masters, P.S. and Perlman, S. (2013). Coronaviridae. In: *Fields Virology*, 6ee, vol. 6 (ed. D.M. Knipe and P.M. Howley), 825–858. Philadelphia, PA: Lippincott Williams & Wilkins.

6 Yang, Y., Zhang, L., Geng, H. et al. (2013). The structural and accessory proteins M, ORF 4a, ORF 4b, and ORF 5 of Middle East respiratory syndrome coronavirus (MERS-CoV) are potent interferon antagonists. *Protein Cell* 4: 951–961.

7 Raj, V.S., Mou, H., Smits, S.L. et al. (2013). Dipeptidyl peptidase 4 is a functional receptor for the emerging human coronavirus-EMC. *Nature* 495: 251–254.

8 Li, W., Moore, M.J., Vasilieva, N. et al. (2003). Angiotensin-converting enzyme 2 is a functional receptor for the SARS coronavirus. *Nature* 426: 450–454.

9 Sabir, J.S., Lam, T.T., Ahmed, M.M. et al. (2016). Co-circulation of three camel coronavirus species and recombination of MERS-CoVs in Saudi Arabia. *Science* 351: 81–84.

10 Kim, Y., Cheon, S., Min, C.K. et al. (2016). Spread of mutant Middle East Respiratory Syndrome coronavirus with reduced affinity to human CD26 during the South Korean Outbreak. *MBio* 7: e00019–e00016.

11 Chinese SARS Molecular Epidemiology Consortium (2004). Molecular evolution of the SARS coronavirus during the course of the SARS epidemic in China. *Science* 303 (5664): 1666–1669.

12 Balkhair, A., Al Maamari, K., and Alawi, F.B. (2013). The struggle against MERS-CoV (the novel coronavirus). *Oman Med. J.* 28 (4): 226–227.

13 Cowling, B.J., Park, M., Fang, V.J. et al. (2015). Preliminary epidemiological assessment of MERS-CoV outbreak in South Korea, May to June 2015. *Euro Surveill.* 20 (25): 7–13.

14 Park, J.W., Lee, K.J., Lee, K.H. et al. (2017). Hospital outbreaks of Middle East respiratory syndrome, Daejeon, South Korea, 2015. *Emerg. Infect. Dis.* 23 (6): 898–905.

15 Choi, J.H., Yoo, B., Lee, S.Y. et al. (2015). Epidemiological investigation of the 119th confirmed Middle East respiratory Syndrome coronavirus case with an indefinite mode of transmission during the Pyeongtaek outbreak in Korea. *Epidemiol. Health* 37: e2015054.

16 Ki, M. (2015). MERS outbreak in Korea: hospital-to-hospital transmission. *Epidemiol. Health* 37: e2015033.

17 Korean Society of Infectious Diseases and Korean Society for Healthcare-associated Infection Control and Prevention An unexpected outbreak of Middle East respiratory syndrome coronavirus infection in the Republic of Korea, 2015. *Infect. Chemother.* 2015, 47: 120–122.

18 Yong, C.J. (2015). An outbreak of Middle East respiratory syndrome coronavirus infection in South Korea. *Yonsei Med. J.* 56: 1174–1176.

19 Gautret, P., Benkouiten, S., Al-Tawfiq, J.A., and Memish, Z.A. (2016). The spectrum of respiratory pathogens among returning Hajj pilgrims: myths and reality. *Int. J. Infect. Dis.* 47 (Supplement C): 83–85.

20 Puzelli, S., Azzi, A., Santini, M.G. et al. (2013). Investigation of an imported case of Middle East respiratory syndrome coronavirus (MERS-CoV) infection in Florence, Italy. *Euro Surveill.* 18 (34): 20564.

21 Kraaij-Dirkzwager, M., Timen, A., Dirksen, K. et al. (2014). Middle East respiratory syndrome coronavirus (MERS-CoV) infections in two returning travellers in the Netherlands, May 2014. *Euro Surveill.* 19 (21): 20817.

22 Tsiodras, S., Baka, A., Mentis, A. et al. (2014). A case of imported Middle East respiratory syndrome coronavirus infection and public health response, Greece, April 2014. *Euro Surveill.* 19 (16): 20782.

23 Su, S., Wong, G., Liu, Y. et al. (2015). MERS in South Korea and China: a potential outbreak threat? *Lancet* 385 (9985): 2349–2350.

24 Alsahafi, A.J. and Cheng, A.C. (2016). The epidemiology of Middle East respiratory syndrome coronavirus in the Kingdom of Saudi Arabia, 2012–2015. *Int. J. Infect. Dis.* 45: 1–4.

25 Hunter, J.C., Nguyen, D., Aden, B. et al. (2016). Transmission of Middle East respiratory syndrome coronavirus infections in healthcare settings, Abu Dhabi. *Emerg. Infect. Dis.* 22 (4): 647–656.

26 Goh, G.K., Dunker, A.K., and Uversky, V. (2013). Prediction of intrinsic disorder in MERS-CoV/HCoV-EMC supports a high oral-fecal transmission. *PLoS Curr.* 5.

27 Hui, D.S. and Zumla, A. (2014). Advancing priority research on the Middle East respiratory syndrome coronavirus. *J. Infect. Dis.* 209 (2): 173–176.

28 Poletto, C., Pelat, C., Levy-Bruhl, D. et al. (2014). Assessment of the Middle East respiratory syndrome coronavirus (MERS-CoV) epidemic in the Middle East and risk of international spread using a novel maximum likelihood analysis approach. *Euro Surveill.* 19 (23): 20824.

29 Cotten, M., Watson, S.J., Zumla, A.I. et al. (2014). Spread, circulation, and evolution of the Middle East respiratory syndrome coronavirus. *MBio* 5 (1): e01062–e01013.

30 Drosten, C., Meyer, B., Muller, M.A. et al. (2014). Transmission of MERS-coronavirus in household contacts. *N. Engl. J. Med.* 371: 828–835.

31 Muller, M.A., Meyer, B., Corman, V.M. et al. (2015). Presence of Middle East respiratory syndrome coronavirus antibodies in Saudi Arabia: a nationwide, cross-sectional, serological study. *Lancet Infect. Dis.* 15 (5): 559–564.

32 Kim, K.H., Tandi, T.E., Choi, J.W. et al. (2017). Middle East respiratory syndrome coronavirus (MERS-CoV) outbreak in South Korea, 2015: epidemiology, characteristics and public health implications. *J. Hosp. Infect.* 95 (2): 207–213.

33 Jansen, A., Chiew, M., Konings, F. et al. (2015). Sex matters—a preliminary analysis of Middle East respiratory syndrome in the Republic of Korea, 2015. *West. Pac. Surveill. Response J.* 6 (3): 68–71.

34 Shapiro, M., London, B., Nigri, D. et al. (2016). Middle East respiratory syndrome coronavirus: review of the current situation in the world. *Disaster Mil. Med.* 2: 1–5.

35 Sha, J., Li, Y., Chen, X. et al. (2017). Fatality risks for nosocomial outbreaks of Middle East respiratory syndrome coronavirus in the Middle East and South Korea. *Arch. Virol.* 162 (1): 33–44.

36 Feikin, D.R., Alraddadi, B., Qutub, M. et al. (2015). Association of higher MERS-CoV virus load with severe disease and death, Saudi Arabia, 2014. *Emerg. Infect. Dis.* 21 (11): 2029–2035.

37 Thabet, F., Chehab, M., Bafaqih, H., and AlMohaimeed, S. (2015). Middle East respiratory syndrome coronavirus in children. *Saudi Med. J.* 36 (4): 484–486.

38 Majumder, M.S., Kluberg, S.A., Mekaru, S.R., and Brownstein, J.S. (2015). Mortality risk factors for Middle East respiratory syndrome outbreak, South Korea, 2015. *Emerg. Infect. Dis.* 21 (11): 2088–2090.

39 Mizumoto, K., Endo, A., Chowell, G. et al. (2015). Real-time characterization of risks of death associated with the Middle East respiratory syndrome (MERS) in the Republic of Korea. *BMC Med.* 13: 2282015.

40 Chan, J.F., Lau, S.K., To, K.K. et al. (2015). Middle East respiratory syndrome coronavirus: another zoonotic betacoronavirus causing SARS-like disease. *Clin. Microbiol. Rev.* 28 (2): 465–522.

41 Momattin, H., Mohammed, K., Zumla, A. et al. (2013). Therapeutic options for Middle East respiratory syndrome coronavirus (MERS-CoV)—lessons from a systematic review of SARS-CoV therapy. *Int. J. Infect. Dis.* 17 (10): e792–e798.

42 Alraddadi, B.M., Watson, J.T., Almarashi, A. et al. (2016). Risk factors for primary Middle East respiratory syndrome coronavirus illness in humans, Saudi Arabia, 2014. *Emerg. Infect. Dis.* 22 (1): 49–55.

43 Kim, S.W., Park, J.W., Jung, H.D. et al. (2017). Risk factors for transmission of Middle East respiratory syndrome coronavirus infection during the 2015 outbreak in South Korea. *Clin. Infect. Dis.* 64 (5): 551–557.

44 Assiri, A., Al-Tawfiq, J.A., Al-Rabeeah, A.A. et al. (2013). Epidemiological, demographic, and clinical characteristics of 47 cases of Middle East respiratory syndrome coronavirus disease from Saudi Arabia: a descriptive study. *Lancet Infect. Dis.* 13: 752–761.

45 Alraddadi, B.M., Al-Salmi, H.S., Jacobs-Slifka, K. et al. (2016). Risk factors for Middle East respiratory syndrome coronavirus infection among healthcare personnel. *Emerg. Infect. Dis.* 22 (11): 1915–1920.

46 Kim, S.H., Ko, J.H., Park, G.E. et al. (2017). Atypical presentations of MERS-CoV infection in immunocompromised hosts. *J. Infect. Chemother.* 23 (11): 769–773.

47 Lu, G., Hu, Y., Wang, Q. et al. (2013). Molecular basis of binding between novel human coronavirus MERS-CoV and its receptor CD26. *Nature* 500 (7461): 227–231.

48 Kandeel, M., Elaiziz, M.A., Kandeel, A. et al. (2014). Association of host tropism of Middle East syndrome coronavirus with the amino acid structure of host cell receptor dipeptidyl peptidase 4. *Acta Virol.* 58 (4): 359–363.

49 Boonacker, E. and Van Noorden, C.J. (2003). The multifunctional or moonlighting protein CD26/DPPIV. *Eur. J. Cell Biol.* 82 (2): 53–73.

50 Zhao, G., Jiang, Y., Qiu, H. et al. (2015). Multi-organ damage in human dipeptidyl peptidase 4 transgenic mice infected with Middle East respiratory syndrome-coronavirus. *PLoS One* 10 (12): e0145561.

51 Xia, Z.-Q., Zhang, J., Xue, Y.-K. et al. (2015). Modeling the transmission of middle east respirator syndrome corona virus in the Republic of Korea. *PLoS One* 10 (12): e0144778.

52 Agrawal, A.S., Garron, T., Tao, X. et al. (2015). Generation of a transgenic mouse model of Middle East respiratory syndrome coronavirus infection and disease. *J. Virol.* 89 (7): 3659–3670.

53 Hocke, A.C., Becher, A., Knepper, J. et al. (2013). Emerging human middle East respiratory syndrome coronavirus causes widespread infection and alveolar damage in human lungs. *Am. J. Respir. Crit. Care Med.* 188 (7): 882–886.

54 Baseler, L.J., Falzarano, D., Scott, D.P. et al. (2016). An acute immune response to Middle East respiratory syndrome coronavirus replication contributes to viral pathogenicity. *Am. J. Pathol.* 186 (3): 630–638.

55 Meyerholz, D.K., Lambertz, A.M., and McCray, P.B. (2016). Dipeptidyl peptidase 4 distribution in the human respiratory tract implications for the Middle East respiratory syndrome. *Am. J. Pathol.* 186 (1): 78–86.

56 Yu, P., Xu, Y., Deng, W. et al. (2017). Comparative pathology of rhesus macaque and common marmoset animal models with Middle East respiratory syndrome coronavirus. *PLoS One* 12 (2): e0172093.

57 Bosch, B.J., Raj, V.S., and Haagmans, B.L. (2013). Spiking the MERS-coronavirus receptor. *Cell Res.* 23 (9): 1069–1070.

58 Xia, S., Liu, Q., Wang, Q. et al. (2014). Middle East respiratory syndrome coronavirus (MERS-CoV) entry inhibitors targeting spike protein. *Virus Res.* 194 (Supplement C): 200–210.

59 Lu, L., Liu, Q., Zhu, Y. et al. (2014). Structure-based discovery of Middle East respiratory syndrome coronavirus fusion inhibitor. *Nat. Commun.* 5: 3067.

60 Gao, J., Lu, G., Qi, J. et al. (2013). Structure of the fusion core and inhibition of fusion by a heptad repeat peptide derived from the S protein of Middle East respiratory syndrome coronavirus. *J. Virol.* 87 (24): 13134–13140.

61 Millet, J.K. and Whittaker, G.R. (2014). Host cell entry of Middle East respiratory syndrome coronavirus after two-step, furin-mediated activation of the spike protein. *Proc. Natl. Acad. Sci. U. S. A.* 111 (42): 15214–15219.

62 Widagdo, W., Raj, V.S., Schipper, D. et al. (2016). Differential expression of the Middle East respiratory syndrome coronavirus receptor in the upper respiratory tracts of humans and dromedary camels. *J. Virol.* 90: 4838–4842.

63 Ng, D.L., Al Hosani, F., Keating, M.K. et al. (2016). Clinicopathologic, immunohistochemical, and ultrastructural findings of a fatal case of Middle East respiratory syndrome coronavirus infection in the United Arab Emirates. *Am. J. Pathol.* 186: 652–658.

64 Zielecki, F., Weber, M., Eickmann, M. et al. (2013). Human cell tropism and innate immune system interactions of human respiratory coronavirus EMC compared to those of severe acute respiratory syndrome coronavirus. *J. Virol.* 87: 5300–5304.

65 Chan, J.F., Chan, K.H., Choi, G.K. et al. (2013). Differential cell line susceptibility to the emerging novel human betacoronavirus 2c EMC/2012: implications for disease pathogenesis and clinical manifestation. *J. Infect. Dis.* 207: 1743–1752.

66 Dijkman, R., Jebbink, M.F., Koekkoek, S.M. et al. (2013). Isolation and characterization of current human coronavirus strains in primary human epithelial cell cultures reveal differences in target cell tropism. *J. Virol.* 87: 6081–6090.

67 Chan, R.W., Chan, M.C., Agnihothram, S. et al. (2013). Tropism of and innate immune responses to the novel human betacoronavirus lineage C virus in human ex vivo respiratory organ cultures. *J. Virol.* 87: 6604–6614.

68 Law, H.K., Cheung, C.Y., Ng, H.Y. et al. (2005). Chemokine up-regulation in SARS-coronavirus-infected, monocyte-derived human dendritic cells. *Blood* 106: 2366–2374.

69 Scheuplein, V.A., Seifried, J., Malczyk, A.H. et al. (2015). High secretion of interferons by human plasmacytoid dendritic cells upon recognition of Middle East respiratory syndrome coronavirus. *J. Virol.* 89: 3859–3869.

70 Faure, E., Poissy, J., Goffard, A. et al. (2014). Distinct immune response in two MERS-CoV-infected patients: can we go from bench to bedside? *PLoS One* 9: e88716.

71 Breban, R., Riou, J., and Fontanet, A. (2013). Interhuman transmissibility of Middle East respiratory syndrome coronavirus: estimation of pandemic risk. *Lancet* 382 (9893): 694–699.

72 Cauchemez, S., Fraser, C., Van Kerkhove, M.D. et al. (2014). Middle East respiratory syndrome coronavirus: quantification of the extent of the epidemic, surveillance biases, and transmissibility. *Lancet Infect. Dis.* 14 (1): 50–56.

73 Memish, Z.A., Zumla, A.I., Al-Hakeem, R.F. et al. (2013). Family cluster of Middle East respiratory syndrome coronavirus infections. *N. Engl. J. Med.* 368 (26): 2487–2494.

74 Memish, Z.A., Al-Tawfiq, J.A., Makhdoom, H.Q. et al. (2014). Screening for Middle East respiratory syndrome coronavirus infection in hospital patients and their healthcare worker and family contacts: a prospective descriptive study. *Clin. Microbiol. Infect.* 20 (5): 469–474.

75 Hijawi, B., Abdallat, M., Sayaydeh, A. et al. (2013). Novel coronavirus infections in Jordan: epidemiological findings from a retrospective investigation. *East Mediterr. Health J.* 19 (Suppl 1): S12–S18.

76 Al-Abdallat, M.M., Payne, D.C., Alqasrawi, S. et al. (2014). Hospital-associated outbreak of Middle East respiratory syndrome coronavirus: a serologic, epidemiologic, and clinical description. *Clin. Infect. Dis.* 59 (9): 1225–1233.

77 Guery, B., Poissy, J., and el Mansouf, L. (2013). Clinical features and viral diagnosis of two cases of infection with Middle East Respiratory Syndrome coronavirus: a report of nosocomial transmission. *Lancet* 381 (9885): 2254–2265.

78 Harriman, K., Brosseau, L., and Trivedi, K. (2013). Hospital-associated Middle East respiratory syndrome coronavirus infections. *N. Engl. J. Med.* 369 (18): 1761.

79 Sampathkumar, P. (2014). Middle East respiratory syndrome: what clinicians need to know. *Mayo Clin. Proc.* 89 (8): 1153–1158.

80 Al-Gethamy, M., Corman, V.M., Hussain, R. et al. (2015). A case of long-term excretion and subclinical infection with Middle East respiratory syndrome coronavirus in a healthcare worker. *Clin. Infect. Dis.* 60 (6): 973–974.

81 Lee, S.S. and Wong, N.S. (2015). Probable transmission chains of Middle East respiratory syndrome coronavirus and the multiple generations of secondary infection in South Korea. *Int. J. Infect. Dis.* 38: 65–67.

82 Memish, Z.A., Zumla, A.I., and Assiri, A. (2013). Middle East respiratory syndrome coronavirus infections in health care workers. *N. Engl. J. Med.* 369 (9): 884–886.

83 Banik, G.R., Khandaker, G., and Rashid, H. (2015). Middle East respiratory syndrome coronavirus "MERS-CoV": current knowledge gaps. *Paediatr. Respir. Rev.* 16 (3): 197–202.

84 Omrani, A.S. and Shalhoub, S. (2015). Middle East respiratory syndrome coronavirus (MERS-CoV): what lessons can we learn? *J. Hosp. Infect.* 91 (3): 188–196.

85 Yang, Y., Ye, F., Zhu, N. et al. (2015). Middle East respiratory syndrome coronavirus ORF4b protein inhibits type I interferon production through both cytoplasmic and nuclear targets. *Sci. Rep.* 5: 17554.

86 Lau, S.K., Lau, C.C., Chan, K.H. et al. (2013). Delayed induction of proinflammatory cytokines and suppression of innate antiviral response by the novel Middle East respiratory syndrome coronavirus: implications for pathogenesis and treatment. *J. Gen. Virol.* 94 (Pt 12): 2679–2690.

87 Subbaram, K., Kannan, H., and Khalil, G.M. (2017). Emerging developments on pathogenicity, molecular virulence, epidemiology and clinical symptoms of current Middle East respiratory syndrome coronavirus (MERS-CoV). *Hayati J. Biosci.* 24 (2): 53–56.

88 Mielech, A.M., Kilianski, A., Baez-Santos, Y.M. et al. (2014). MERS-CoV papain-like protease has deISGylating and deubiquitinating activities. *Virology* 450–451: 64–70.

89 Yang, X., Chen, X., Bian, G. et al. (2014). Proteolytic processing, deubiquitinase and interferon antagonist activities of Middle East respiratory syndrome coronavirus papain-like protease. *J. Gen. Virol.* 95 (Pt 3): 614–626.

90 Chu, H., Zhou, J., Wong, B.H. et al. (2014). Productive replication of Middle East respiratory syndrome coronavirus in monocyte-derived dendritic cells modulates innate immune response. *Virology* 454–455: 197–205.

91 Tynell, J., Westenius, V., Ronkko, E. et al. (2016). Middle East respiratory syndrome coronavirus shows poor replication but significant induction of antiviral responses in human monocyte-derived macrophages and dendritic cells. *J. Gen. Virol.* 97 (2): 344–355.

92 Zhong, J., Rao, X., and Rajagopalan, S. (2013). An emerging role of dipeptidyl peptidase 4 (DPP4) beyond glucose control: potential implications in cardiovascular disease. *Atherosclerosis* 226 (2): 305–314.

93 Coleman, C.M., Sisk, J.M., Halasz, G. et al. (2017). CD8+ T cells and macrophages regulate pathogenesis in a mouse model of Middle East respiratory syndrome. *J. Virol.* 91 (1): e01825–e01816.

94 Zhao, J., Li, K., Wohlford-Lenane, C. et al. (2014). Rapid generation of a mouse model for Middle East respiratory syndrome. *Proc. Natl. Acad. Sci. U. S. A.* 111 (13): 4970–4975.

95 Menachery, V.D., Eisfeld, A.J., Schafer, A. et al. (2014). Pathogenic influenza viruses and coronaviruses utilize similar and contrasting approaches to control interferon-stimulated gene responses. *MBio* 5: e01174–e01114.

96 Zhou, J., Chu, H., Li, C. et al. (2014). Active replication of Middle East respiratory syndrome coronavirus and aberrant induction of inflammatory cytokines and chemokines in human macrophages: implications for pathogenesis. *J. Infect. Dis.* 209: 1331–1342.

97 Cheung, C.Y., Poon, L.L., Ng, I.H. et al. (2005). Cytokine responses in severe acute respiratory syndrome coronavirus-infected macrophages in vitro: possible relevance to pathogenesis. *J. Virol.* 79: 7819–7826.

98 Chu, H., Zhou, J., Wong, B.H. et al. (2015). Middle East respiratory syndrome coronavirus efficiently infects human primary T lymphocytes and activates the extrinsic and intrinsic apoptosis pathways. *J. Infect. Dis.* 213 (6): 904–914.

99 Virlogeux, V., Park, M., Wu, J.T., and Cowling, B.J. (2016, 2014). Association between severity of MERS-CoV infection and incubation period. *Emerg. Infect. Dis.* 22 (3): 526–528.

100 Al-Tawfiq, J.A., Hinedi, K., Ghandour, J. et al. (2014). Middle East respiratory syndrome coronavirus: a case-control study of hospitalized patients. *Clin. Infect. Dis.* 59: 160–165.

101 Zumla, A., Hui, D.S., and Perlman, S. (2015). Middle East respiratory syndrome. *Lancet* 386: 995–1007.

102 Arabi, Y.M., Arifi, A.A., Balkhy, H.H. et al. (2014). Clinical course and outcomes of critically ill patients with Middle East respiratory syndrome coronavirus infection. *Ann. Intern. Med.* 160: 389–397.

103 Assiri, A., McGeer, A., Perl, T.M. et al. (2013). Hospital outbreak of Middle East respiratory syndrome coronavirus. *N. Engl. J. Med.* 369: 407–416.

104 Saad, M., Omrani, A.S., Baig, K. et al. (2014). Clinical aspects and outcomes of 70 patients with Middle East respiratory syndrome coronavirus infection: a single-center experience in Saudi Arabia. *Int. J. Infect. Dis.* 29: 301–306.

105 Xie, Q., Cao, Y., Su, J. et al. (2017). Two deletion variants of Middle East respiratory syndrome coronavirus found in a patient with characteristic symptoms. *Arch. Virol.* 162 (8): 2445–2449.

106 Omrani, A.S., Matin, M.A., Haddad, Q. et al. (2013). A family cluster of Middle East respiratory syndrome coronavirus infections related to a likely unrecognized asymptomatic or mild case. *Int. J. Infect. Dis.* 17 (9): e668–e672.

107 Kang, C.K., Song, K.-H., Gyun Choe, P. et al. (2017). Clinical and epidemiologic characteristics of spreaders of Middle East respiratory syndrome coronavirus during the 2015 outbreak in Korea. *J. Korean Med. Sci.* 32 (5): 744–749.

108 Poissy, J., Goffard, A., Parmentier-Decrucq, E. et al. (2014). Kinetics and pattern of viral excretion in biological specimens of two MERS-CoV cases. *J. Clin. Virol.* 61: 275–278.

109 Memish, Z.A., Al-Tawfiq, J.A., Makhdoom, H.Q. et al. (2014). Respiratory tract samples, viral load, and genome fraction yield in patients with Middle East respiratory syndrome. *J. Infect. Dis.* 210: 1590–1594.

110 Ajlan, A.M., Ahyad, R.A., Jamjoom, L.G. et al. (2014). Middle East respiratory syndrome coronavirus (MERS-CoV) infection: chest CT findings. *Am. J. Roentgenol.* 203: 782–787.

111 Omrani, A.S., Al-Tawfiq, J.A., and Memish, Z.A. (2015). Middle East respiratory syndrome coronavirus (MERS-CoV): animal to human interaction. *Pathog. Glob. Health* 109: 354–362.

112 Reusken, C.B., Raj, V.S., Koopmans, M.P., and Haagmans, B.L. (2016). Cross host transmission in the emergence of MERS coronavirus. *Curr. Opin. Virol.* 16: 55–62. It has also been demonstrated.

113 Adney, D.R., van Doremalen, N., Brown, V.R. et al. (2014). Replication and shedding of MERS-CoV in upper respiratory tract of inoculated dromedary camels. *Emerg. Infect. Dis.* 20: 1999–2005.

114 Hemida, M.G., Al-Naeem, A., Perera, R.A. et al. (2015). Lack of Middle East respiratory syndrome coronavirus transmission from infected camels. *Emerg. Infect. Dis.* 21: 699–701.

115 Reusken, C.B., Farag, E.A., Jonges, M. et al. (2014). Middle East respiratory syndrome coronavirus (MERS-CoV) RNA and neutralizing antibodies in milk collected according to local customs from dromedary camels, Qatar. *Euro Surveill.* 19 (23): 20829.

116 Memish, Z.A., Cotten, M., Meyer, B. et al. (2014). Human infection with MERS coronavirus after exposure to infected camels, Saudi Arabia, 2013. *Emerg. Infect. Dis.* 20 (6): 1012–1015.

117 Aburizaiza, A.S., Mattes, F.M., Azhar, E.I. et al. (2014). Investigation of anti-middle East respiratory syndrome antibodies in blood donors and slaughterhouse workers in Jeddah and Makkah, Saudi Arabia, fall 2012. *J. Infect. Dis.* 209 (2): 243–246.

118 Yang, Y., Liu, C., Du, L. et al. (2015). Two mutations were critical for bat-to-human transmission of Middle East respiratory syndrome coronavirus. *J. Virol.* 89 (17): 9119–9123.

119 Memish, Z.A., Mishra, N., Olival, K.J. et al. (2013). Middle East respiratory syndrome coronavirus in Bats, Saudi Arabia. *Emerg. Infect. Dis.* 19 (11): 1819–1823.

120 Ithete, N.L., Stoffberg, S., Corman, V.M. et al. (2013). Close relative of human Middle East respiratory syndrome coronavirus in bat, South Africa. *Emerg. Infect. Dis.* 19 (10): 1697–1699.

121 Mizutani, T. (2013). A novel coronavirus, MERS-CoV. *Uirusu* 63 (1): 1–6.

122 Corman, V.M., Ithete, N.L., Richards, L.R. et al. (2014). Rooting the phylogenetic tree of Middle East respiratory syndrome coronavirus by characterization of a conspecific virus from an African bat. *J. Virol.* 88: 11297–11303.

123 Lau, S.K.P., Li, K.S., Tsang, A.K. et al. (2013). Genetic characterization of betacoronavirus lineage C viruses in bats reveals marked sequence divergence in the spike protein of pipistrellus bat coronavirus HKU5 in Japanese pipistrelle: implications for the origin of the novel Middle East respiratory syndrome coronavirus. *J. Virol.* 87 (15): 8638–8650.

124 Anthony, S.J., Gilardi, K., Menachery, V.D. et al. (2017). Further evidence for bats as the evolutionary source of Middle East respiratory syndrome coronavirus. *MBio* 8 (2): e00373–e00317.

125 Cai, Y., Yu, S.Q., Postnikova, E.N. et al. (2014). CD26/DPP4 cell-surface expression in bat cells correlates with bat cell susceptibility to Middle East respiratory syndrome coronavirus (MERS-CoV) infection and evolution of persistent infection. *PLoS One* 9 (11): e112060.

126 Kupferschmidt, K. (2013). Link to MERS virus underscores bats' puzzling threat. *Science* 341: 948–949.

127 Widagdo, W., Begeman, L., Schipper, D. et al. (2017). Tissue distribution of the MERS-coronavirus receptor in bats. *Sci. Rep.* 7 (1): 1193.

128 Nowotny, N. and Kolodziejek, J. (2014). Middle East respiratory syndrome coronavirus (MERS-CoV) in dromedary camels, Oman, 2013. *Euro Surveill.* 19 (16): 20781.

129 Corman, V.M., Muller, M.A., Costabel, U. et al. (2012). Assays for laboratory confirmation of novel human coronavirus (hCoV-EMC) infections. *Euro Surveill.* 17 (49): 20334.

130 Alshukairi, A.N.K.I., Ahmed, W.A., Dada, A.M. et al. (2016). Antibody response and disease severity in healthcare worker MERS survivors. *Emerg. Infect. Dis.* 22 (6): 1113–1115.

131 Graham, R.L., Donaldson, E.F., and Baric, R.S. (2013). A decade after SARS: strategies for controlling emerging coronaviruses. *Nat. Rev. Microbiol.* 11: 836–848.

132 Perlman, S. and Vijay, R. (2016). Middle East respiratory syndrome vaccines. *Int. J. Infect. Dis.* 16: 31021–31029.

133 Haagmans, B.L., van den Brand, J.M., Raj, V.S. et al. (2016). An orthopoxvirus-based vaccine reduces virus excretion after MERS-CoV infection in dromedary camels. *Science* 351: 77–81.

134 Kindler, E., Jonsdottir, H.R., Muth, D. et al. (2013). Efficient replication of the novel human betacoronavirus EMC on primary human epithelium highlights its zoonotic potential. *MBio* 4: e00611–e00612.

135 Falzarano, D., de Wit, E., Rasmussen, A.L. et al. (2013). Treatment with interferon-alpha2b and ribavirin improves outcome in MERS-CoV-infected rhesus macaques. *Nat. Med.* 19: 1313–1317.

136 Omrani, A.S., Saad, M.M., Baig, K. et al. (2014). Ribavirin and interferon alfa-2a for severe Middle East respiratory syndrome coronavirus infection: a retrospective cohort study. *Lancet Infect. Dis.* 14: 1090–1095.

137 Al-Tawfiq, J.A., Momattin, H., Dib, J., and Memish, Z.A. (2014). Ribavirin and interferon therapy in patients infected with the Middle East respiratory syndrome coronavirus: an observational study. *Int. J. Infect. Dis.* 20: 42–46.

138 Mair-Jenkins, J., Saavedra-Campos, M., Baillie, J.K. et al. (2015). The effectiveness of convalescent plasma and hyperimmune immunoglobulin for the treatment of severe acute respiratory infections of viral etiology: a systematic review and exploratory meta-analysis. *J. Infect. Dis.* 211: 80–90.

139 Corti, D., Zhao, J., Pedotti, M. et al. (2015). Prophylactic and postexposure efficacy of a potent human monoclonal antibody against MERS coronavirus. *PNAS* 112: 10473–10478.

140 Ying, T., Du, L., Ju, T.W. et al. (2014). Exceptionally potent neutralization of Middle East respiratory syndrome coronavirus by human monoclonal antibodies. *J. Virol.* 88 (14): 7796–7805.

141 Ohnuma, K., Haagmans, B.L., Hatano, R. et al. (2013). Inhibition of Middle East respiratory syndrome coronavirus infection by anti-CD26 monoclonal antibody. *J. Virol.* 87 (24): 13892–13899.

142 Ying, T., Li, H., Lu, L. et al. (2015). Development of human neutralizing monoclonal antibodies for prevention and therapy of MERS-CoV infections. *Microbes Infect.* 17 (2): 142–148.

143 Du, L., Zhao, G., Yang, Y. et al. (2014). A conformation-dependent neutralizing monoclonal antibody specifically targeting receptor-binding domain in Middle East respiratory syndrome coronavirus spike protein. *J. Virol.* 88 (12): 7045–7053.

144 Song, F., Fux, R., Provacia, L.B. et al. (2013). Middle East respiratory syndrome coronavirus spike protein delivered by modified vaccinia virus ankara efficiently induces virus-neutralizing antibodies. *J. Virol.* 87 (21): 11950–11954.

145 Chen, Y., Rajashankar, K.R., Yang, Y. et al. (2013). Crystal structure of the receptor-binding domain from newly emerged Middle East respiratory syndrome coronavirus. *J. Virol.* 87 (19): 10777–10783.

146 Tang, X.C., Agnihothram, S.S., Jiao, Y. et al. (2014). Identification of human neutralizing antibodies against MERS-CoV and their role in virus adaptive evolution. *Proc. Natl. Acad. Sci.* 111: E2018–E2026.

147 Luke, T., Wu, H., Zhao, J. et al. (2016). Human polyclonal immunoglobulin G from transchromosomic bovines inhibits MERS-CoV in vivo. *Sci. Transl. Med.* 8: 326ra21.

148 Zumla, A., Chan, J.F., Azhar, E.I. et al. (2016). Coronaviruses—drug discovery and therapeutic options. *Nat. Rev. Drug Discov.* 15 (5): 327–347.

149 Lan, J., Yao, Y., Deng, Y. et al. (2015). Recombinant receptor binding domain protein induces partial protective immunity in rhesus macaques against Middle East. *EBioMedicine* 2 (10): 1438–1446.

150 Channappanavar, R., Lu, L., Xia, S. et al. (2015). Protective effect of intranasal regimens containing peptidic Middle East respiratory syndrome coronavirus fusion inhibitor against MERS-CoV infection. *J. Infect. Dis.* 2015 (212): 1894–1903.

151 Arabi, Y.M., Deeb, A.M., Al-Hameed, F. et al. (2019). Macrolides in critically ill patients with Middle East respiratory syndrome. *Int. J. Infect. Dis.* 81: 184–190.

152 Arabi, Y.M., Mandourah, Y., Al-Hameed, F. et al. (2018). Corticosteroid therapy for critically ill patients with middle east respiratory syndrome. *Am. J. Respir. Crit. Care Med.* 197 (6): 757–767.

153 Li, K., Wohlford-Lenane, C., Perlman, S. et al. (2016). Middle East respiratory syndrome coronavirus causes multiple organ damage and lethal disease in mice transgenic for human dipeptidyl peptidase 4. *J. Infect. Dis.* 213: 712–722.

154 Houser, K.V., Gretebeck, L., Ying, T. et al. (2016). Prophylaxis with a Middle East respiratory syndrome coronavirus (MERS-CoV)-specific human monoclonal antibody protects rabbits from MERS-CoV infection. *J. Infect. Dis.* 213: 1557–1561.

155 Johnson, R.F., Bagci, U., Keith, L. et al. (2016). 3B11-N, a monoclonal antibody against MERS-CoV, reduces lung pathology in rhesus monkeys following intratracheal inoculation of MERS-CoV Jordan-n3/2012. *Virology* 490: 49–58.

14

Chikungunya Fever: Epidemiology, Clinical Manifestation, and Management

Chandrakanta[1], Bhupendra G. Prajapati[2], and Archana N. Sah[1]

[1] *Department of Pharmaceutical Sciences, Faculty of Technology, Sir J. C. Bose Technical Campus Bhimtal, Kumaun University Nainital, Nainital, Uttarakhand, India*
[2] *Department of Pharmaceutics and Pharmaceutical Technology, Shree S K Patel College of Pharmaceutical Education and Research, Ganpat University, Mehsana, Gujarat, India*

14.1 Introduction

Chikungunya fever (CHIKF) is an arthropod-borne disease, caused by the CHIKV, endemic to tropical regions. It is spread by female mosquitos *Aedes aegypti* and *Aedes albopictus*. This disease has become a public health issue in countries where it has spread. CHIKV is an RNA virus that is a member of the genus alphavirus in the Togaviridae family [1, 2]. The word "chikungunya" originates from the Makonde language, which is spoken by the Makonde ethnic group of Southern Tanzania and Northern Mozambique. Its meaning is "that which bends" which describes the hunched posture of people who have arthralgias' distinctive painful symptoms [3]. CHIKV was first isolated from patients with symptomatic picture of fever and arthralgia in Newala, Tanzania in 1952 [4]. CHIKV has three genotypes known as West African (WA), Asian, and East Central South Africa (ECSA) [5].

14.1.1 Structure

The CHIKV is an enveloped alphavirus of approximately 60–70 nm in diameter. A positive-sense single-stranded RNA CHIKV consists of 11.8 kilo base pairs along with a capsid and a phospholipid envelope [6], and its genome consists of two open reading frames– the 5ORF, that is translated from genomic RNA and encodes non-structural proteins such as nsP1, nsP2, nsP3, and nsP4, and 3ORF, that is translated from sub-genomic RNA and encodes structural proteins such as capsid (C), envelope (E1 and E2), and two peptides (E3 and 6K) [7]. The association of these proteins results in the formation of a viral replication complex, which is necessary for viral genome replication downstream [7, 8]. However, the icosahedral fenestrated nucleocapsid of a mature virus, which will house the viral RNA genome, is formed with the help of the structural capsid protein [9]. The trimeric spikes formed by the E1 and E2 glycoproteins, which first form a heterodimer before being incorporated onto the surface of the mature virion, are crucial for the virion's attachment to and entry into susceptible target cells during a subsequent infection. A mature virion has a total of 80 trimeric spikes on its surface [10]. The alpha virus 6K protein is involved in virus assembly and budding during infections, and it is the component of a cation selective ion channel with the potential to influence membrane permeability [11], through appropriate interactions with E2 glycoprotein, E3 proteins play a crucial role in directing structural proteins to the endoplasmic reticulum for proper assembly and spike formation [12]. Additionally, it has been reported that alphavirus E3 protein may contribute to pH defense during virus formation and budding. Non-structural proteins are not packaged into the final virions, and hence structural proteins, mainly surface glycoproteins E2 and E1, are the key targets of the host humoral immune response and of most anti-CHIKV vaccines [13].

Rising Contagious Diseases: Basics, Management, and Treatments, First Edition. Edited by Seth Kwabena Amponsah, Ranjita Shegokar, and Yashwant V. Pathak.

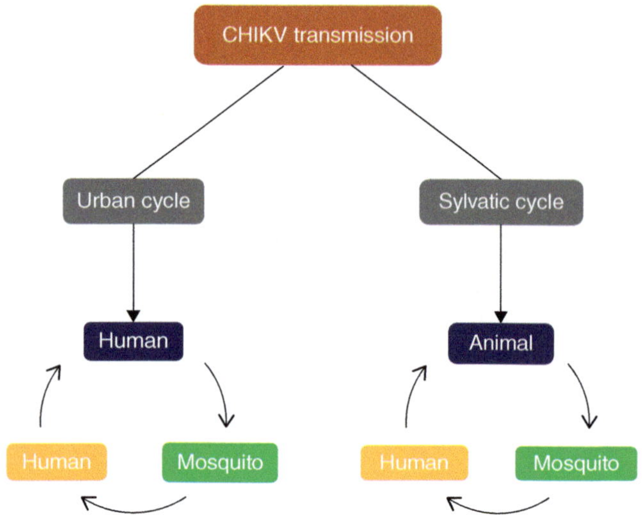

Figure 14.1 CHIKV transmission in sylvatic and urban cycle.

14.1.2 Target Cell

CHIKV has a well-established route of infection in humans, with certain cell types being particularly vulnerable to the infection. Epithelial, endothelial, primary fibroblast, and macrophages derived from monocytes are human cell types that are vulnerable to infection and viral production. After the initial replication, the host immune system reacts, but the virus still spreads to the lymph nodes before moving on to other tissues via the circulatory system. The disease's viremic phase, during which mosquitoes can spread through blood, is caused by replication in other tissues [14].

14.1.3 Transmission

CHIKV spreads in urban and sylvatic cycles. Urban cycle refers to the transmission from human to mosquito to human, whereas sylvatic transmission is from an animal to a mosquito, and then to a human (see Figure 14.1). In Africa, the sylvatic cycle is the principal form of maintenance [15]. CHIKV is predominantly maintained in more densely populated regions by an urban cycle in which people serve as primary hosts and mosquitoes of the genus Aedes serve as vectors [16]. *A. aegypti* and *A. albopictus* are the primary vectors of the virus. In the past, *A. aegypti* served as the main vector for the spread of CHIKV, but in several recent outbreaks in Europe, Gabon, and La Reunion, *A. albopictus* served as the main vector. Nevertheless, *A. aegypti* continues to be a significant viral vector, as evidenced by the Caribbean outbreak in 2013. A mutation in the E1 envelope protein enhanced CHIKV fitness in *A. albopictus* and increased transmission to vertebrate species, which is thought to be the cause of the adaptation to *A. albopictus* due to a lack of *A. aegypti* vectors [17].

14.1.4 Replication

Upon viral entry into target cells, the acidic endosomal environment triggers conformational changes in the viral envelope. These confirmational changes will expose the E1 glycoprotein, which will then mediate the fusion of the CHIKV membrane with the host endosomal membrane and the release of nucleocapsid into the cytoplasm. Following translation into a polyprotein, the free viral RNA is then cleaved into the non-structural proteins 1–4. The functional viral replication complex formed by these non-structural proteins will go on to produce the full-length negative sense RNA intermediate. The 26s sub genomic mRNA and the 49s genomic RNA will both be synthesized later using this intermediate RNA as a template. To release the capsid into the cytoplasm, serine proteases further cleave the Capsid-pE2-6K-E1 polyprotein obtained from the translation of the 26s sub-genomic mRNA. The remaining proteins are then sent to the endoplasmic reticulum, where they will experience additional post translational alterations. The Golgi body will then cleave pE2 to produce E2 and E3 glycoproteins. The E1 and E2 glycoproteins will heterodimerize and go to the host cell plasma membrane, where they will be integrated as trimeric spikes on the virion surface. In the cytoplasm, capsid protein combines to create the icosahedral nucleocapsid, which houses the 49s genomic RNA. After that, the mature virions assemble at the plasma membrane, and rupture the infected host cell and bud out (see Figure 14.2), during which the mature virions will acquire a membrane bilayer from the host cell plasma membrane [6, 9, 18, 19].

14.1.5 Sign and Symptoms of CHIKF

People of all ages and genders are equally affected by CHIKF. The signs and symptoms of the virus infection can be divided into two phases– the acute phase, which lasts for about a week, and the chronic phase, also referred to as the persistent phase, which can last for months to years [20, 21].

14.1.6 Acute Phase Symptoms

In this stage, fever can be constant or intermittent and typically lasts for a few days to a week. However, a drop in temperature is not accompanied by an aggravation of the symptoms. Rarely, a relative bradycardia occurs along with the fever.

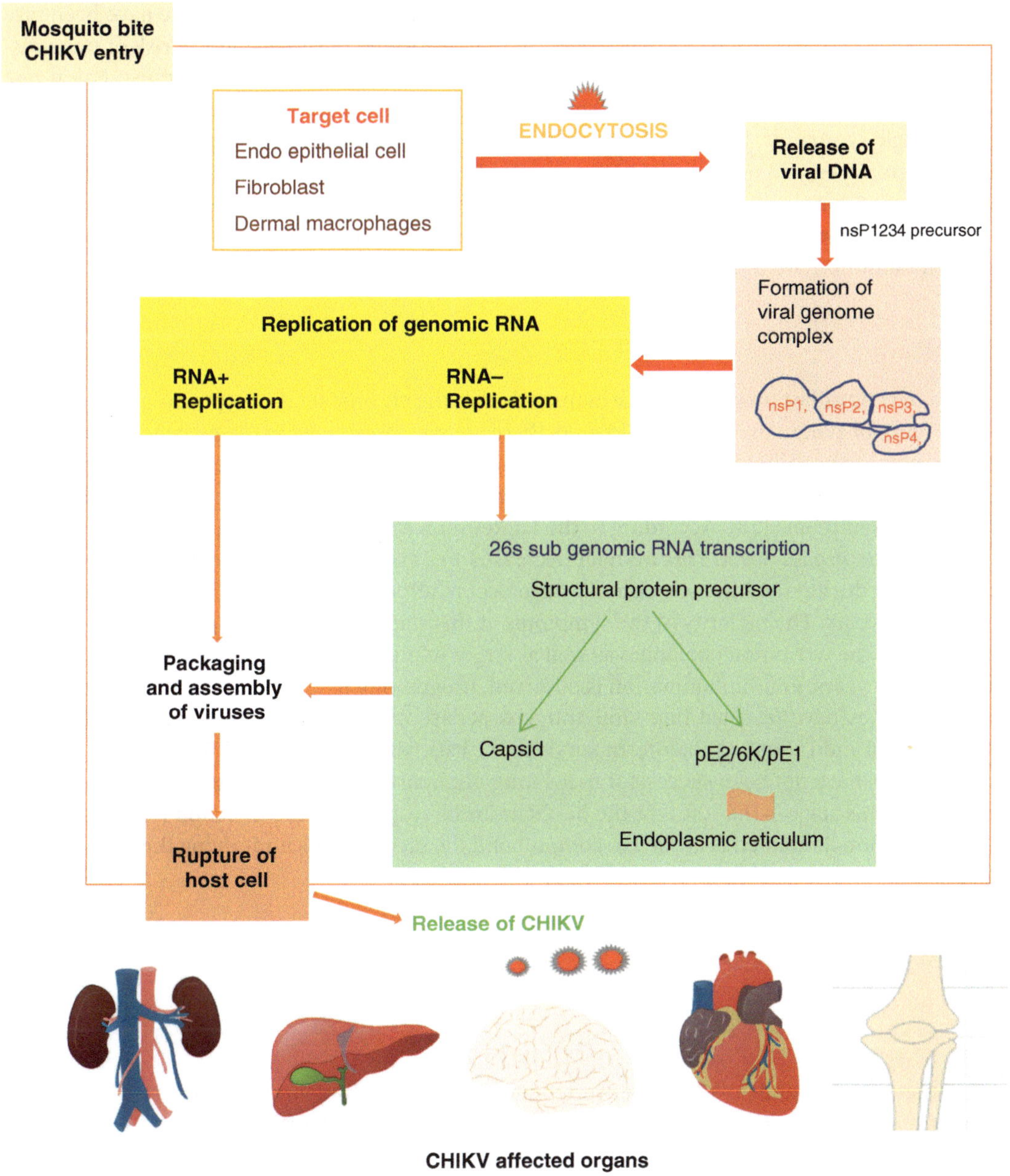

Figure 14.2 Mechanism of chikungunya virus in human cells and affected organs by CHIKV.

CHIKF has an attack rate of 40–85% in susceptible groups [22]. Joint symptoms are usually symmetric and mostly affect the hands and feet, but they can also impact more proximal joints. Patients are frequently rendered immobile because of pain, edema, and stiffness of joints. Many people may be unable to conduct normal duties, and many will be restricted to bed as result of these symptoms [23]. Rashes occur in nearly half of all patients 2–5 days following the onset of fever. It is usually maculopapular and affects the trunk and extremities, although it can also affect the palms, soles, and face. The rash might also appear as a widespread erythema that blanches when touched. Fluid filled lesions are the most prevalent cutaneous presentation in neonates. According to serosurveys, between 3% and 28% of those with CHIKV antibodies had asymptomatic illnesses. If the vectors that are capable of transmitting the virus are actively present in the same area where people are acutely infected with CHIKV, even if clinically evident or asymptomatic, it can help in the spread of the disease [24].

Most CHIKV infections that occur during pregnancy do not spread the virus to the fetus. However, there have been rare and isolated reports of spontaneous abortions caused by CHIKV infection in the mother [25]. The highest risk of transmission appears to be when women become infected during pregnancy [26]. During this time, the vertical transmission rate reaches 49%. Infants are typically asymptomatic after birth until they develop fever, peripheral edema, and rash. Intrapartum infections can result in hemorrhagic symptoms, heart disease, and neurologic disease, for example meningoencephalitis, brain swelling, white matter abnormalities, and intracranial hemorrhage. Elderly people are more prone to die from severe and unusual diseases. Persons over 65 have a 50-fold higher mortality risk than people under 45. Although the cause of older people's increased risk for more severe disease is unknown, it may be because of the prevalence of underlying conditions that coexist or a weakened immune system [27].

14.1.7 Chronic Phase Symptoms

The symptoms of chikungunya infections last more than three months during the chronic phase. Depending on the study, and the time span between the onset of symptoms and the follow-up, there are wide variations in the percentage of participants reporting persistent symptoms. According to research conducted in South Africa, 12–18% people will experience chronic symptoms at 18 months and for up to 3 years after that. Most recent studies conducted in India found that after 10 months, 49% of patients still had symptoms [28]. According to the La Reunion study, up to 93% patients complained of chronic symptoms three months after disease onset. This number decreased to 57% in 15 months and 47% in 2 years. The development of inflammatory joints during the disease's chronic stage occurs after the acute stage has subsided, and in some patients, it can last for several years. The majority of the symptoms at this stage are polyarthralgia and polyarthritis, which mostly affect minor joints like the wrists and phalanges as well as larger joints like the shoulders, knees, and ankles, limiting the mobility of patients. For years, joint inflammation is observed, progressing to chronic inflammatory rheumatism [29–31]. Other alphavirus studies have revealed that viral antigens persist, causing joint inflammation [32]. Some investigations on CHIKV have found indications of long-term survival of virus, viral proteins, and RNA in synovial tissues [28]. In contrast, several studies have not been successful in locating viral particles in CHIKV-infected joints during the chronic phase, contending that this stage is the result of the host's immune response to the infection [33]. These contradicting findings suggest that additional research is needed to comprehend the mechanisms underlying the development of CHIKF into a chronic condition.

14.2 Epidemiology

14.2.1 Worldwide

Africa: Several nonhuman primates (NHPs) and possibly other vertebrates are used by the zoonotic virus CHIKV as amplification hosts, and these animals may also act as virus reservoirs during interepidemic intervals [34]. Aedes species mosquitoes that live in forests, and nonhuman primates, have been discovered to have a sylvatic cycle with the virus, which is thought to have its origins in Central/East Africa. Human instances are occasional in these places, although significant human epidemics are rare [35]. However, the virus can spread between mosquitoes and unaware human hosts in urban areas of Africa and Asia in a cycle resembling that of dengue viruses. The primary vectors of CHIKV transmission in cities are the mosquitoes *A. aegypti* and *A. albopictus*. In Tanzania, in 1952, the first human case of CHIKV infection was documented [36]. Over the following 30 years, there were sporadic minor outbreaks, but no significant outbreaks were noted until 2004 when a large epidemic began on the coast of Kenya. Kenya witnessed two epidemics of CHIKF beginning in 2004. The first outbreak in Lamu resulted in an estimated 13,500 cases, the outbreak was previously considered to be caused by malaria, but testing eventually revealed that CHIKV was responsible. A few months after the Lamu outbreak, a second outbreak occurred in the city of Mombasa [37]. An outbreak of dengue-like illness was discovered in the union of the Comoros in 2005. About 40,000 cases were reported per week during the peak of the outbreak in La Reunion in the same year. It was estimated that 266,000 instances occurred in total. This was the first time that CHIKV was linked to a variety of neurological symptoms, prenatal infections, and mortality. Chikungunya epidemics in Africa have historically been related to spill over from natural reservoir vectors, *Aedes Stegomyia* spp. mosquitoes, into humans near sylvatic transmission cycles. Nonetheless, there has been an increase in large-scale outbreaks associated with African urban areas in recent years, La Reunion in 2005–2006 [38], or in Gabon in 2006–2007 [39], and Ethiopia in 2019 [40].

In Ethiopia, 311 cases, including three confirmed cases of CHIKV infection and no deaths, have been reported as of 8 May 2022. In Kenya as of 11 December 2022, 291 cases have been reported, including five confirmed cases and one death, and in Sudan as of October 2022, eight cases have been reported with no deaths [41].

Asia: The first substantial urban CHIKF epidemics were recorded in Bangkok in the early 1960s [42]. Following several decades of no documented outbreaks, the arrival of the ECSA genotype to India through the Indian Ocean Islands in 2005 resulted in large-scale transmission and multiple outbreaks in India and Southeast Asia over the next decade. Chikungunya epidemics have been observed in India from 1963 to 1973 and again from 2005 to 2022 [43, 44]. In 1963, the first chikungunya outbreak was recorded in Kolkata, West Bengal. While outbreaks were still being reported between 2005 and 2022. A reported outbreak that was ultimately identified as a CHIKV infection occurred in 2000–2007 in the general hospital in Peradeniya, Sri Lanka. Moreover, CHIKV outbreaks occurred in Bangladesh in 2008, 2011, and 2017 [45, 46]. CHIKV was first documented in Bhutan in 2012 [47]. Also, Nepal reported an increase in CHIKV cases from 2013 to 2015 [48]. Although the first case was discovered in Pakistan in December 2016, anti-CHIKV antibodies in human serum and rodents were discovered as early as the 1980s. This virus was introduced into Zhejiang and Tianjin provinces of China in 2017 by migrants from Bangladesh to Myanmar [49–51].

In 2022, 108,957 cases of CHIKV disease were recorded in India, with 5320 confirmed cases and no deaths. There have been 814 confirmed instances of CHIKV sickness in Malaysia, with no deaths reported. The Republic of Philippines, in 2022, confirmed 600 cases of CHIKV disease were, and no deaths were reported, while 1370 instances of CHIKV disease occurred in Thailand and no deaths were reported [41].

Americas: Imported cases of CHIKF were first reported in the Americas, primarily in the Caribbean Islands, and later developed into endemic patterns in several countries of the region due to suitable climatic circumstances and the emergence of vector mosquito. The reemergence of CHIKV in the Americas was announced in December 2013, with the first cases diagnosed on the island of Saint Martin, from which the virus spread to several Caribbean countries, including Martinique, Guadeloupe, the Dominican Republic, and the British Virgin Islands, with notification of approximately 1.1 million cases of CHIKF within a year, including severe cases with deaths [52–54].

Suriname had the greatest prevalence with 760 suspected cases per 10,000 population recorded in 2014–2015 [55]. The WHO Pan American Health Organization (PAHO) recorded 271,006 cases of CHIKV infection and 75 deaths across the Americas, upto 31 December 2022. Brazil (265,289), Paraguay (2443), Guatemala (1933), Peru (595), and Bolivia (232), reported the most cases [41].

Europe: Chikungunya epidemiology in Europe is characterized by small outbreaks involving local transmission because of the arrival of an imported index case. Between June and September 2007, the first European outbreak was observed in Northeastern Italy [36]. In Europe, five outbreaks involving local transmission of chikungunya by *A. albopictus* were observed in 2010–2017. Two of these occurred in Italy and three in France, each with a small number of patients. Except for France in 2010, where no distinct CHIKV lineage was found, the ECSA genotype was identified in all European outbreaks [56]. In Europe, no autochthonous cases of the chikungunya virus disease have been documented as of 2022 [41].

14.2.2 In India

The first CHIKF outbreak in India was reported in Kolkata (Calcutta), West Bengal, in 1963. CHIKV resurfaced in India 32 years after first appearing on the Indian Ocean Islands in 2005 [57]. In South India, the virus infected Hyderabad and the Anantapur district of Andhra Pradesh, affecting 1.4 million people across 13 states [58]. The 2005–2006 outbreak in India was caused by the ECSA genotype [59]. In 2009, the infection spread to 15 Indian states. By 2015, the number of states reporting CHIKF had risen to 23, 28 in 2016, 25 in 2017 [60], 29 in 2018, and 21 states and 3 union territories in 2019. In 2020, a total of 43,424 cases had been clinically suspected of CHIKV, with 6324 laboratory-confirmed cases reported in 20 Indian states and three Union territories. Karnataka had the most CHIKV cases (1326), followed by Gujarat (1061), Rajasthan (1015), and the least number of cases were reported from Jammu and Kashmir [1]. In 2021, a total of 119,070 cases were clinically suspected of CHIKV, and 11,890 cases were laboratory-confirmed as CHIKF. In 2022, a total of 108,957 cases were clinically suspected of CHIKV and 5320 confirmed chikungunya cases were reported, with the highest number of confirmed cases reported from Karnataka (1789) and the least from Assam [2], till October 2022 (see Figure 14.3, Table 14.1). Before December 2022, Jammu and Kashmir, Mizoram, and Manipur had reported ≤10 laboratory-confirmed CHIKF cases, but in 2022, 51 confirmed cases reported in Mizoram. CHIKF instances have not been reported in the union territories of Dadra and Nagar Haveli and Lakshadweep until now [61].

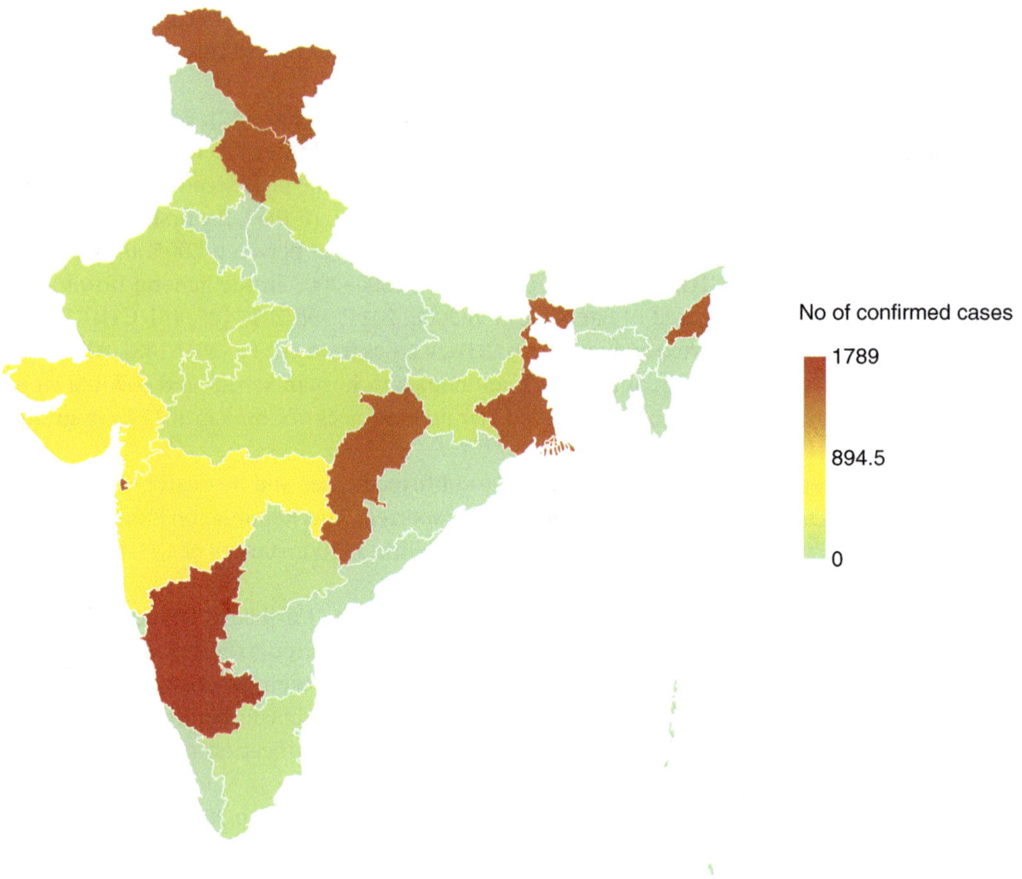

Figure 14.3 State wise chikungunya fever cases in India (till October 2022). *Source:* Geonames / CC BY 4.0.

14.3 Clinical Manifestation

14.3.1 Rheumatic Manifestation

About 75% of patients infected with CHIKV develop musculoskeletal signs of the disease. Arthralgia typically affects multiple joints in a bilateral and symmetric pattern, most notably the knees, ankles, hands, and wrists [62]. Over 50% of patients in certain cohorts developed arthralgia and clinically evident joint swelling three years after their initial infection. Javelle et al. conducted a six-year retrospective analysis in La Reunion on patients referred to a rheumatologist for rheumatic symptoms persisting more than four months after CHIKV infection. They discovered that 59% of 159 cases fit the criteria for de novo chronic inflammatory rheumatism such as rheumatoid arthritis, spondyloarthropathy, and undifferentiated polyarthritis, while 31% had pre-existing rheumatic musculoskeletal diseases [63].

14.3.2 Ophthalmic Manifestation

Another well-known consequence of CHIKV infection is ocular manifestation. Ocular involvement in chikungunya infection manifests in a variety of ways, ranging from anterior segment compromise to the development of lesions in the posterior pole. Some of the target cells involved in the onset of ocular disease [64]. Its high rate of replication results in a large production of infectious viral particles. CHIKV preferentially infects epithelium and endothelium cells in the cornea. The fibroblast is the ideal cell for the virus in the scleral connective tissues, the stroma of smooth muscle in ciliary bodies, and the stroma of the iris. Ocular manifestations can appear during or after the systemic illness. The first ophthalmological manifestations of the disease include photophobia retroocular pain and enlargement of conjunctival vessels. Other

Table 14.1 Chikungunya cases in India 2017–2022.

	2017		2018		2019		2020		2021		2022 (31 October)	
Affected states and union territories	**No. of suspected case**	**No. of confirmed cases**	**No. susp. cases**	**No. of conf. cases**	**No. susp. cases**	**No. of conf. cases**	**No. susp. cases**	**No. of conf. cases**	**No. susp. cases**	**No. of conf. cases**	**No. susp. cases**	**No. of conf. cases**
Andhra Pradesh	1162	108	622	79	832	88	318	28	753	48	1527	17
Arunachal Pradesh	133	0	507	1	332	55	0	0	0	0	0	0
Assam	41	41	3	3	0	0	0	0	2	2	2	2
Bihar	1251	1251	156	156	594	594	38	38	40	40	56	56
Goa	509	48	455	77	867	366	64	15	114	12	891	106
Gujarat	7953	1363	10,601	1290	8084	669	8120	1061	32,372	4044	16,350	852
Haryana	220	6	62	3	0	0	930	14	765	21	701	29
Jammu and Kashmir	0	0	1	1	0	0	1	1	7	7	0	0
Jharkhand	269	17	3405	851	1691	169	627	157	1064	215	861	139
Karnataka	32,831	3511	20,411	2546	43,698	3664	16,111	1326	40,134	2188	53,201	1789
Kerala	78	74	77	77	109	109	2302	752	3030	334	1131	63
Madhya Pradesh	2477	858	3211	1609	2749	756	1871	133	3822	397	3285	301
Meghalaya	236	45	44	2	114	48	3	0	0	0	0	0
Mizoram	0	0	93	10	0	0	34	0	0	0	160	51
Maharashtra	8110	1438	9884	1009	5158	1646	4258	782	21,680	2526	12,047	865
Manipur	0	0	2	0	40	3	0	0	0	0	0	0
Odisha	0	0	0	0	242	21	15	2	0	0	0	0
Punjab	3251	201	736	25	372	11	25	0	644	144	443	192
Rajasthan	1612	1612	254	254	365	365	1015	1015	1157	1157	193	193
Sikkim	130	8	384	28	1126	95	253	13	457	32	1627	4
Tamil Nadu	131	131	284	284	623	623	1461	224	3654	153	149	149
Telangana	1277	58	1954	489	5352	1358	364	183	220	76	6169	104
Tripura	574	64	683	75	1105	125	454	27	948	20	826	18
Uttar Pradesh	103	103	58	58	72	72	53	53	70	70	13	13
Uttarakhand	0	0	29	7	1	1	0	0	3	1	572	154
West Bengal	2103	577	52	23	NR	NR	391	82	154	20	—	—
Andaman & Nicobar Island	93	17	205	27	702	53	610	16	187	0	5	0
Chandigarh	1810	54	357	4	82	0	0	0	397	7	32	32
Dadra & Nagar Haveli	0	0	0	0	0	0	0	0	0	0		
Delhi	940	940	407	407	520	520	126	126	112	112	48	48
Lakshadweep	0	0	NR	NR	NR	NR	—	—	—	—	105	0
Puducherry	475	23	2876	361	7084	794	3980	276	7284	264	8480	141
Total	67,769	12,548	57,813	9756	81,914	12,205	43,424	6324	119,070	11,890	108,957	5320

Source: Adapted from Ref. [61].

symptoms such as blurred vision, eye discharge, eye irritation, and diplopia are common after a month to a year of the latent period during the chronic phase. The most frequent ocular problems of CHIKF includes retinitis, conjunctivitis, optic neuritis, iridocyclitis, episcleritis, and uveitis [65].

14.3.3 Renal Manifestation

In a retrospective analysis, it was revealed that there were 120 cases of prerenal failure out of 610 patients with atypical manifestations during acute infection of CHIKV [27]. One-third of the patients with prerenal failure had pre-existing renal disease, the majority of which was chronic kidney disease [65]. During an epidemic in Mayotte, Comoros, two more cases of acute renal failure were reported [66].

14.3.4 Dermatological Manifestation

Dermatological symptoms of the disease are possible in 40–50% of all CHIKF patients [67]. The most common dermatological symptom of a typical CHIKV infection is a maculopapular rash that begins on the upper limbs, followed by face, neck, trunk and ear lobes [68]. It typically appears 3–5 days after the onset of the fever and goes away without any long-term consequences in 3–4 days. The rashes are asymptomatic in approximately 80% of patients, with the rest complaining of minor itchiness. Skin hypermelanosis may develop in several patients soon after the rashes are completely resolved. According to a study conducted in India, intertriginous aphthous-like ulcers (21.37%), maculopapular eruption (33%), and pigmentary changes (42% of all cases) were the most frequent skin manifestations. Only infants were found to have widespread vesiculobullous eruption (2.75%), which is unusual [69–71].

14.3.5 Neurological Manifestation

The most frequent cerebral complications of CHIKV are encephalitis, encephalopathy, and peripheral neuropathy. On average, 10 days after the first symptoms appear, neurological symptoms appear. However, CHIKV appears to be more common in the elderly, newborns, and critically ill people. Neurological symptoms were found in 9% of the confirmed CHIKV cases and 3.14% of the suspected cases, according to a screening of 405 suspected CHIKF patients in India in 2006. Encephalitis, which has been observed in cases from India, Italy, Thailand, La Reunion Island, and the Caribbean, appears to be one of the most prevalent neurologic manifestations of CHIKV. Encephalitis was reported in 9% of the patients with neurologic disease in a study conducted in India on 300 patients who had neurological symptoms following a possible CHIKV infection [72]. In a study of individuals with severe CHIKV symptoms in Ahmedabad and Pune, encephalitis accounted for 15.4% of the cases, and it was also discovered that anti-CHIKV IgM was present in 91–95% of their serum and 79.7% of their cerebral spinal fluid. CHIKV RNA was shown to be less sensitive. The mortality rate in these instances was high. The prevalence of encephalitis in the Caribbean outbreak seems to be lower than that of India and La Reunion Island. Only 3/65 of the severely ill CHIKV patients in the Caribbean met the criteria for encephalitis, according to an analysis [73, 74]. Another reported CNS manifestation of CHIKV is myelopathy. According to an analysis of 300 cases of suspected CHIKV infection and neurologic disease in India, 5% of the patients had myelopathy or myeloneuropathy, which accounted for 27% of the cases with neurologic involvement. Electromyography (EMG) was abnormal in every patient, while only 3/14 patients' MRI signals showed signal changes that indicated spinal cord involvement. Half of the patients had these myelopathic symptoms that only gradually appeared after infection. Myelopathy and myeloneuropathy from CHIKV may occur more frequently in critically ill patients. It was also reported that a patients with myeloneuropathy in Thailand also experienced a delayed onset of symptoms [75].

Guillain–Barre syndrome is another important clinical manifestation of CHIKV infection. In 2006, there was an increase in Guillain–Barre syndrome cases on La Réunion Island which increased to 22% on the island in 2006, before returning to baseline in 2007 [36]. The CHIKV outbreak in French Polynesia in 2013–2014 was accompanied by an increase in the number of cases of Guillain–Barre syndrome, implying a connection with the virus [76, 77].

14.3.6 Cardiovascular Manifestation

Geographically, the Caribbean, parts of South America, and India are the regions most affected by CHIKV-related cardiac problems. Since these locations have been linked to Asian CHIKV circulation, particularly in South America and the Caribbean, it may initially appear that these issues are caused by Asian ancestry [78–80]. Although there have only been a

Table 14.2 Clinical manifestation of chikungunya fever.

CHIKV affected systems	Manifestations	References
Ocular	Optic neuritis, conjunctivitis Anterior and posterior uveitis Retinitis, pan uveitis, choroiditis	[64, 65]
Neurological	Cerebral syndrome, encephalopathy, generalized encephalitis, meningoencephalitis, acute demyelinating encephalomyelitis, myelopathy, polyneuropathy, Guillain–Barre syndrome, brainstem encephalitis, cognitive and psychiatric	[93–96]
Respiratory	Respiratory failure, pneumonia	[27, 87]
Rheumatic	Arthralgia, polyarthralgia, rheumatic arthritis, fibromyalgia, spondyloarthropathies, synovitis, joint effusions	[62–63, 68]
Cardiovascular	Myocarditis, heart failure, arrhythmias, unstable blood pressure, ischemic heart disease, disseminate intravascular coagulation	[81–87]
Dermatologic	Maculopapular, eruption, hyperpigmentation penoscrotal ulcers, bullous dermatosis	[67–71]
Renal	Acute renal failure Nephritis Hepatitis	[65, 66, 97]

few cases reported in the literature, myocarditis is mentioned as a potential side effect of CHIKV infection. In the 1970s, a Reunion Island outbreak and a number of cases were both reported. According to a retrospective data analysis of the Reunion Island epidemics, 13% individuals suffered heart failure after acute infection, despite 59% of them already having a heart problem. Similar to arrhythmias, which occurred in 7% of the cases but in 36% of them due to a cardiac or systemic cause, they were more likely to occur. There were 6% incidences of myocarditis or pericarditis that were linked to CHIKV infection [81]. Myocarditis is a CHIKV infection symptom that affects people of all ages. Myocarditis is an inflammation of the heart tissue caused by a viral infection, which leads to viral myocarditis. People who arrived with tachycardia, or a heartbeat that was greater than 60–100 beats per minute, frequently experienced arrhythmias [82, 83]. Despite the fact that there was no conclusive evidence of heart disease, some individuals had electrocardiograms that revealed repolarization abnormalities, right bundle branch blockage, and poor R wave progression. Older age and cardiac comorbidities like hypertension and congestive heart failure are risk factors for older patients, and these comorbidities are uncommon in children. Any of these disorders can result in death [84–86].

For CHIKV-positive patients admitted to hospitals with serious conditions, the mortality rate ranges from 26% to 48%. Another study estimated that 10% of their 32 patients had cardiac complications, despite the fact that only a few studies have examined the frequency of CHIKV patients who arrive with cardiac complications, including the use of echocardiograms to confirm arrhythmias [87–90]. The signs of these illnesses may worsen when CHIKV coexists with other arboviruses like dengue and zika [91, 92].

These clinical manifestations may be brought on by the virus itself, an immune response to the virus, or drug toxicity (see Table 14.2).

14.4 Diagnosis of Chikungunya Fever

CHIKV infection is diagnosed using three types of laboratory tests– virus isolation, serology, and reverse transcriptase polymerase chain reaction (RT-PCR) [98, 99].

14.4.1 Virus Isolation

The gold standard for CHIKV diagnosis for many years was virus isolation; however, this method is no longer used in everyday diagnosis. Approximately two to six days after the onset of symptoms, the virus can be found in high titers in serum or plasma samples from the patient. But after five days of symptom onset, the viral load starts to decrease. As a result, the test's diagnostic use is restricted to the acute phase of the infection. This test has the advantage of allowing virus amplification and the characterization of viral strains which can be defined for further epidemiology and research studies [100].

14.4.2 Serology

For serological diagnosis of CHIKV infection, serum collected from whole blood is used in enzyme-linked immunosorbent assay (ELISA) and plaque reduction neutralization testing (PRNT). The serum should not be frozen and should be delivered at a temperature of 2–8 °C. A serologic diagnosis can be made using acute and convalescent specimens with IgM antibodies specific for CHIKV or a fourfold increase in PRNT titer. To demonstrate IgM antibodies specific for CHIKV, the IgM antibody capture ELISA is used first, followed by the PRNT [101, 102].

14.4.3 Reverse Transcriptase Polymerase Chain Reaction

Acute phase molecular diagnosis CHIKV detection with RT-PCR is a widely used method nowadays. It is a more specific and less time-demanding procedure. Serum and plasma are the preferred samples; however, saliva and urine can also be utilized, and cerebrospinal fluid can be used in case of meningitis and encephalitis. Practically all viral genes are used for diagnostic purposes [103, 104].

14.5 Management of Chikungunya Fever

Chikungunya infection does not have a specific antiviral treatment or vaccination till now. After the confirmation of the diagnosis, the patient can begin symptomatic treatment with nonsteroidal anti-inflammatory drugs and joint pain physiotherapy.

14.5.1 For Acute Phase

During the acute phase of CHIKV infection, dehydration prevention, rest, and analgesics, are the only treatments available and recommended. Depending on how severe the pain is, these medications are used. Due to the risk of hemorrhage and renal insufficiency, non-steroidal anti-inflammatory drugs (NSAIDs) should be completely avoided during the acute phase [105].

14.5.1.1 First-Line Treatment

Paracetamol is used as the first line of treatment for the acute phase of CHIKV infection. When patient's pain is mild to moderate, Paracetamol should be recommended at 500–750 mg every six hours. However, the total daily dose should not be more than 4 g due to the risk of hepatotoxicity [106].

14.5.1.2 Second Line Treatment

If first line treatment is ineffective, then opioid analgesics like tramadol and codeine are recommended because these drugs are effective and safe in acute pain. Tramadol should be taken every six hours at a dose of 50–100 mg. Codeine is an opioid that should be taken every six hours in a dose of 30 mg and can be combined with other analgesics [107].

14.5.2 For Subacute and Chronic Phase

14.5.2.1 Corticosteroids

Corticosteroids were suggested for the treatment of 27.7% of arthralgia cases by the Reunion Island hospital staff, particularly for invalidating forms during the chikungunya infection [108]. Prednisolone and aceclofenac therapy reduced pain and enhanced the quality of life in patients with chikungunya arthritis when compared to aceclofenac therapy given alone in the treatment of early chikungunya fever, according to a study carried out in South India during the 2007 chikungunya epidemic with the aim of developing a standard treatment protocol. Their research suggests that the most effective treatment option for sub-acute chikungunya cases, that also have arthralgia, is the co-administration of low-dose systemic corticosteroids with NSAIDS [109]. Another patient who received steroid treatment for post-chikungunya reversible demyelinating encephalitis with vertigo, dysarthria, and ataxia showed full clinical and radiological recovery [110]. Nevertheless, corticosteroid usage during acute viral arthritis is regarded as inappropriate due to the possibility of immunosuppression, which can lead to increased infection and disease worsening [111]. Systemic glucocorticoids may be used to treat chronic joint pain with the dose determined by the degree of severity of symptoms. Because post chikungunya chronic arthritis sometimes resembles well-diagnosed rheumatic presentations, it has frequently been misdiagnosed and treated as such with acceptable results [106].

14.6 Disease Modifying Antirheumatic Drugs

14.6.1 Chloroquine

In the chronic stage of CHIKV infection, chloroquine is used. The mechanism behind chloroquine's ability to effectively treat chronic CHIKV-induced arthralgia is most likely due to its anti-inflammatory properties, which are also used to treat a number of autoimmune diseases [112]. In the chronic stage, hydroxychloroquine at a dose of 6 mg/kg/day given orally for six weeks can be used to commence therapy. The patients should be reviewed after six weeks. If the patient is no longer in rheumatic pain, the treatment should be stopped and if the patient does not recover, the course of treatment should be continued for an additional six weeks, with the addition of analgesics as needed [113].

14.6.2 Sulfasalazine

If the pain persists after six weeks of hydroxychloroquine treatments, sulfasalazine at a dose of 1000 mg in 12 hours intervals should be administered, with the usage of hydroxychloroquine maintained for another six weeks. This combination has been shown to be useful in managing joint problems in patients who have previously failed treatment. After six weeks, the patient should be evaluated again, if the patient is no longer in pain after this period, the drug should be discontinued and if the condition does not improve, the combination should be continued for another six weeks [114].

14.6.3 Methotrexate

If the combination of hydroxychloroquine and sulfasalazine fails to improve the pain, sulfasalazine should be replaced with methotrexate at an initial dose of 10 mg/week, orally; this quantity may be increased to 25 mg/week. Folic acid should always be administered in combination with methotrexate, at a dose of 5 mg/week, the day after methotrexate [115, 116]. During the 2005–2006 La Reunion outbreak, several individuals who developed chronic rheumatic symptoms following CHIKV infection were successfully treated with methotrexate [28].

14.7 Non-Pharmacological Treatment

Physiotherapy should be considered in all three stages of CHIKV infection, although it should be prioritized in the subacute and chronic stages. As an adjuvant measure, the manufacturing and use of orthoses may be required; this may result in more quick pain relief and more importantly, may help to prevent muscular atrophy caused by the disuse of the affected joints. In the subacute and chronic phases, other therapies, such as acupuncture, may be used [117].

14.8 Medicinal Plants for CHIKV and Its Associated Symptoms

14.8.1 *Persicaria odorata*

Chan et al. investigated the anti-CHIKV potential of six different solvent extracts of the aerial part of *P. odorata*. The cytopathic effect of CHIKV was significantly reduced by ethanolic extract, which had a cell viability of more than 70%, while metabolic extract had a moderate suppression with cell viability of 50%. The neutral red uptake test was used to determine cell viability. In quantitative RT-PCR, ethanol extract reduced the chikungunya viral load by 90% at 160 g/ml when compared to the virus control. As a result, ethanol-soluble extractives may be a valuable source of anti-CHIKV drugs [118].

14.8.2 *Picrorhiza kurroa*

It is commonly known as Katuki. The primary active components of katuki are kutkin and apocynin. The rhizomes of this plant's aqueous extract at a dose of 10 g/ml, suppressed CHIKV by more than 80% in a plaque reduction assay without causing cytotoxicity, proving that the extract is a viral attachment inhibitor. The plant extract suppressed the activity of the helicase enzyme in a helicase assay, suggesting that it is a viral replication inhibitor. Virus replication inhibition assay results showed that it inhibited viral replication by 38.3% and 67% at dosages of 100 g/ml and 200 g/ml, respectively [119].

14.8.3 *Andrographis paniculata*

It is also known as the Kalmegh. By measuring the cell viability of virus-infected Vero cells after treatment with extract nanoparticles, it was determined in a study that aqueous leaf extract silver nanoparticles had anti-CHIKV properties. In this study, silver nanoparticles made from the plant's aqueous leaf extract increased the viability of infected Vero cells from 25% to around 81% when used at the highest nontoxic dosage of 31.25 g/ml. It has strong anti-CHIKV potential, according to the findings [120].

14.8.4 *Ipomoea aquatica*

A study investigated the anti-CHIKV activity of the aerial part of *I. aquatica* extracted with six different polarity solvents. Only ethyl acetate extract, with 82% cell viability at a dose of 320 g/ml, significantly inhibited the cytopathic effect of CHIKV in Vero cell lines, as determined by the neutral red uptake assay. When compared to virus control, a quantitative RT-PCR analysis of the extract showed a 65% decrease in viral load at a dose of 320 g/ml [118].

14.8.5 *Azadirachta indica*

At a dosage of 62.5 and 32.25 g/ml, aqueous and aqueous ethanolic extracts of leaves suppressed the development of Asian CHIKV in Vero cell lines. The extract's selectivity scores were 1 and 2. In the same experimental conditions, these extracts had no effect on the proliferation of the ECSA lineage of CHIKV [121].

14.8.6 *Terminalia chebulla*

The high medicinal value and pharmacological activity of this plant make it effective in treating a wide range of diseases. This includes anti-inflammatory, hepatoprotective, anti-mutagenic, anti-proliferative, radioprotective, cardioprotective, anti-arthritic, gastrointestinal motility, wound healing, analgesic, anti-allergic, anti-carcinogenic, cytroprotective, renoprotective, anti-viral, adaptogenic, and anti-anaphylactic properties. The symptoms of chikungunya are lessened by *T. chebula*'s anti-inflammatory, anti-arthritic, analgesic, anti-allergic, anti-viral, and immunomodulatory properties [122]. In another study. it is demonstrated that the *T. chebulla* fruit extract has 100% antiviral activity. According to the results, this extract inhibited CHIKV adhesion, entry, and multiplication by more than 80% in helicase, protease, and plaque reduction assays, demonstrating comprehensive antiviral activity against CHIKV. At concentrations of 25, 50, 100, and 200 g/ml, the extract inhibited CHIKV by 23%, 51.6%, 83.3%, and 100%, respectively, in the replication inhibition assay [119].

14.8.7 *Alpinia officinarum*

It is reported that *A. officinarum* rhizomes' aqueous, ethanolic, and aqueous ethanolic extracts all demonstrated antiviral activity against Asian CHIKV in Vero cell lines at doses of 72.5, 125, and 72.5 g/ml. The ECSA CHIKV lineage's growth was not inhibited by the extracts, despite their selectivity indices of 3.44, 2, and 6 [123].

14.8.8 *Zingiber officinalis*

It is the most significant plant with nutritional and therapeutic characteristics and belongs to the Zingiberaceae family. There have been reports of essential oils, phenolic compounds, flavonoids, carbohydrates, proteins, alkaloids, glycosides, saponins, steroids, terpenoids, and tannin as the main active ingredients present as phytochemical groups. Its pharmacological significance includes its use in the treatment of diabetes, obesity, diarrhea, allergies, pain, fever, rheumatoid arthritis, inflammation, and different types of cancer. It has a variety of antioxidant, anti-microbial, anti-diabetic, anti-cancer, anti-inflammatory, analgesic, antipyretic, immunomodulatory, anti-platelet aggregation, anti-angiogenic, and hepatoprotective potential [124]. Some of the specific qualities that ease chikungunya symptoms include anti-pyretic, anti-arthritic, analgesic, immunomodulatory, and anti-allergic properties. The aqueous extracts of fresh dried rhizomes reportedly exhibited anti-chikungunya potential by reducing the cytopathic effect of CHIKV in Vero cell lines and increasing cell survival by 51.05% and 35.10% when pre-treated with maximum non-toxic dose (MNTD) or half of MNTD, respectively. Similar to this, Vero cell survival increases by 52.90% and 49.02% when co-administered with MNTD or Half of MNTD, respectively [125].

14.8.9 *Pisidium guajava*

It is reported that the aqueous leaf extracts of *P. guajava* and their greenly synthesized silver nanoparticles decreased the cytopathic effect of CHIKV in Vero cell lines and improved cell survival by 60% and 40%, respectively, in comparison to the viral control. This demonstrates the plant's ability to combat CHIKV viruses. When tested using the MTT assay in Vero cells, the extract and the nanoparticles both demonstrated MNTD 15 g/ml. Longifolene and quercetin interact with the CHIKV cysteine protease (nsP2), which is essential for replication, according to docking studies. This interaction may be the antiviral mechanism [126].

14.9 Future Prospectives

CHIKF disease is a major public health concern in Africa, Asia, and the Americas. Despite producing a self-limited, acute disease with a low mortality rate, CHIKF poses a challenge to the human population because of the debilitating nature of the disease's progression and its long-lasting side effects that negatively affect the quality of life. As a result, it is critical to implement prevention and control strategies. The year-wise growth in CHIKV outbreaks throughout the world is worsening, a sign that has encouraged us to focus on virus biology, virus–host interaction, transmission, strain mutation, and circulation. The development of new therapies, such as vaccines and antivirals, that can treat the illness and potentially lower the high viremia and significant morbidity linked to CHIKV infection, as well as research into the pathophysiology of chronic arthralgia, should continue.

14.10 Conclusion

In conclusion, CHIKF poses a serious threat to public health, especially in tropical regions where the main CHIKV transmitters, *A. aegypti* and *A. albopictus*, are heavily endemic. After being bitten by a female mosquito carrying the virus, the virus enters the host's body through the skin and travels to various organs, including the brain. In addition to vector transmission, vertical transmission and transmission through blood transfusions are both possibilities. Neurological, cardiac, renal, cutaneous, and ophthalmic clinical manifestations of CHIKF can have serious effects on the patient. The majority of these problems seem to be made worse by pre-existing co-morbidities. Adults over 65 tend to exhibit more atypical symptoms, and people over 40 are more likely to experience severe disease. It is unclear whether atypical symptoms are brought on by mutant CHIKV strains or an unexpected host response to the virus. Women who experience intrapartum viremia have a higher risk of vertical transmission, which may cause infection in newborns. With the beginning of the twenty-first century, CHIKV has reemerged as a serious viral danger. The virus has been responsible for outbreaks all over the world and has proven to be capable of spreading locally in regions where it is not endemic. The possibility of a more widespread epidemic will only increase as global travel and migration rates increase. Finding efficient CHIKV infection treatment and preventive alternatives becomes crucial in this situation because efforts to combat vectors have been ineffective.

References

1 Kraemer, M., Sinka, M.E., Duda, K.A. et al. (2015). The global distribution of the arbovirus vectors *Aedes aegypti* and *Ae. albopictus*. *elife* 4: 1–18.

2 Mayer, S.V., Tesh, R.B., and Vasilakis, N. (2017). The emergence of arthropod-borne viral diseases: a global prospective on dengue, chikungunya and zika fevers. *Acta Trop.* 166: 155–163.

3 Benjamin, M. (2012). Chikungunya is not a Swahili word, it is from the Makonde language. *J. Ophthalmic Inflamm. Infect.* 3: 35.

4 Lumsden, W.H.R. (1955). An epidemic of virus disease in Southern Province, Tanganyika Territory, in 1952–53: II. General description and epidemiology. *Trans. R. Soc. Trop. Med. Hyg.* 49 (1): 33–57.

5 Volk, S.M., Chen, R., Tsetsarkin, K.A. et al. (2010). Genome-scale phylogenetic analyses of chikungunya virus reveal independent emergences of recent epidemics and various evolutionary rates. *J. Virol.* 84: 6497–6504.

6 Straus, J.H. and Straus, E.G. (1994). The alphaviruses: gene expression, replication, and evolution. *Microbiol. Rev.* 58 (3): 491–562.

7 Solignat, M., Gay, B., Higgs, S. et al. (2009). Replication cycle of 993 chikungunya: a re-emerging arbovirus. *Virology* 393 (2): 183–197.

8 Schwartz, O. and Albert, M.L. (2010). Biology and pathogenesis of chikungunya virus. *Nat. Rev. Microbiol.* 8 (7): 491–500.

9 Rashad, A.A., Mahalingam, S., and Keller, P.A. (2014). Chikungunya virus: emerging 938 targets and new opportunities for medicinal chemistry. *J. Med. Chem.* 57 (4): 1147–1166.

10 Voss, J.E., Vaney, M.C., Duquerroy, S. et al. (2010). Glycoprotein organization of chikungunya virus particles revealed by X-ray crystallography. *Nature* 468 (7324): 709–712.

11 Melton, J.V., Ewart, G.D., Weir, R.C. et al. (2022). Alphavirus 874 6K proteins form ion channels. *J. Biol. Chem.* 277 (49): 46923–46931.

12 Snyder, A.J. and Mukhopadhyay, S. (2012). The alphavirus E3 glycoprotein functions in a clade-specific manner. *J. Virol.* 86 (24): 13609–13620.

13 Uchime, O., Fields, W., and Kielian, M. (2013). The role of E3 in pH protection during alphavirus assembly and exit. *J. Virol.* 87 (18): 10255–10262.

14 Thiberville, S.D., Moyen, N., Dupuis-Maguiraga, L. et al. (2013). Chikungunya fever: epidemiology, clinical syndrome, pathogenesis and therapy. *Antivir. Res.* 99 (3): 345–370.

15 Mavale, M., Parashar, D., Sudeep, A. et al. (2010). Venereal transmission of chikungunya virus by *Aedes aegypti* mosquitoes (Diptera: Culicidae). *Am. J. Trop. Med. Hyg.* 83 (6): 1242–1244.

16 Vazeille, M., Moutailler, S., Coudrier, D. et al. (2007). Two chikungunya isolates from the outbreak of La Reunion (Indian Ocean) exhibit different patterns of infection in the mosquito, *Aedes albopictus*. *PLoS One* 2 (11): e1168.

17 Chhabra, M., Mittal, V., Bhattacharya, D. et al. (2008). Chikungunya fever: a re-emerging viral infection. *Indian J. Med. Microbiol.* 26 (1): 5–12.

18 Leung, J.Y.S., Ng, M.M.L., and Chu, J.J.H. (2011). Replication of alphaviruses: a review on the entry process of alphaviruses into cells. *Adv. Virol.* 2011: 1–9.

19 Ong, R.Y., Lum, F.M., and Ng, L.F. (2014). The fine line between protection and pathology in neurotropic flavivirus and alphavirus infections. *Future Virol.* 9 (3): 313–330.

20 Rahman, S., Suchana, S.A., Rashid, S.S., and Pavel, O.F. (2017). A review article on chikungunya virus. *World J. Pharma. Res.* 6 (13): 100–107.

21 Sharma, S.K. and Jain, S. (2018). Chikungunya: a rheumatologist's perspective. *Int. J. Rheum. Dis.* 21 (3): 584–601.

22 Sam, I.C. and AbuBakar, S. (2006). Chikungunya virus infection. *Med. J. Malaysia* 61: 264–269.

23 Sissoko, D., Malvy, D., Ezzedine, K. et al. (2009). Post-epidemic chikungunya disease on Reunion Island: course of rheumatic manifestations and associated factors over a 15-month period. *PLoS Negl. Trop. Dis.* 3: e389.

24 Robinson, M.C. (1955). An epidemic of virus disease in Southern Province, Tanganyika Territory, in 1952–53; I. Clinical features. *Trans. R. Soc. Trop. Med. Hyg.* 49: 28–32.

25 Touret, Y., Randrianaivo, H., Michault, A. et al. (2006). Early maternal-fetal transmission of the chikungunya virus. *Presse Med.* 35 (11): 1656–1658.

26 Gerardin, P., Barau, G., Michault, A. et al. (2008). Multidisciplinary prospective study of mother-to-child chikungunya virus infections on the island of La Réunion. *PLoS Med.* 5 (3): e60.

27 Economopoulou, A., Dominguez, M., Helynck, B. et al. (2009). Atypical chikungunya virus infections: clinical manifestations, mortality and risk factors for severe disease during the 2005–2006 outbreak on Réunion. *Epidemiol. Infect.* 137 (4): 534–541.

28 Hoarau, J.J., Jaffar Bandjee, M.C., Trotot, P.K. et al. (2010). Persistent chronic inflammation and infection by chikungunya arthritogenic alphavirus in spite of a robust host immune response. *J. Immunol.* 184 (10): 5914–5927.

29 Rajapakse, S., Rodrigo, C., and Rajapakse, A. (2010). Atypical manifestation of chikungunya infection. *Trans. R. Soc. Trop. Med. Hyg.* 104 (2): 89–96.

30 Simon, F., Javelle, E., Oliver, M., and Marimoutou, C.L. (2011). Chikungunya virus infection. *Curr. Infect. Dis. Rep.* 13 (3): 218–228.

31 Suhrbier, A., Bandjee, C.M., and Gasque, P. (2012). Arthritogenic alphaviruses – an overview. *Nat. Rev. Rheumatol.* 8 (7): 420–429.

32 Chang, Y.A., Martins, A.K., Encinales, L. et al. (2018). Chikungunya arthritis mechanism in the Americas: a cross sectional analysis of chikungunya arthritis patients twenty two months after infection demonstrating no detectable viral persistence in synovial fluid. *Arthritis Rheum.* 70 (4): 585–593.

33 Pardigon, N. (2009). The biology of chikungunya: a brief review of what we still do not know. *Pathol. Biol.* 57 (2): 127–132.

34 Tsetsarkin, K.A., Chen, R., and Weaver, S.C. (2016). Interspecies transmission and chikungunya virus emergence. *Curr. Opin. Virol.* 16: 143–150.

35 Ross, R.W. (1956). A laboratory technique for studying the insect transmission of animal viruses, employing a bat-wing membrane, demonstrated with two African viruses. *J. Hyg. (Lond).* 54 (2): 192–200.

36 Powers, A.M. and Logue, C.H. (2007). Changing patterns of chikungunya virus: re-emergence of a zoonotic arbovirus. *J. Gen. Virol.* 88: 2363–2377.

37 Sergon, K., Njuguna, C., Kalani, R. et al. (2008). Seroprevalence of chikungunya virus (CHIKV) infection on Lamu Island, Kenya, October 2004. *Am. J. Trop. Med. Hyg.* 78: 333–337.

38 Renault, P., Solet, J.L., Sissoko, D. et al. (2007). A major epidemic of chikungunya virus infection on Reunion Island, France, 2005–2006. *Am. J. Trop. Med. Hyg.* 77 (4): 727–731.

39 Peyrefitte, C.N., Bessaud, M., Pastorino, B.A. et al. (2008). Circulation of chikungunya virus in Gabon, 2006–2007. *J. Med. Virol.* 80 (3): 430–433.

40 World Health Organization. 2019. Chikungunya Outbreak in Ethiopia. WHO Regional Office for Africa. https://www.afro.who.int/health-topics/chikungunya.

41 European Center for Disease Prevention and Control. Chikungunya Virus Disease Cases January–December 2022. https://www.ecdc.europa.eu/en/chikungunya-monthly.

42 Hammon, W.M., Rudnick, A., and Sather, G.E. (1960). Viruses associated with epidemic hemorrhagic fevers of the Philippines and Thailand. *Science* 131 (3407): 1102–1103.

43 Yergolkar, P.N., Tandale, B.V., Arankalle, V.A. et al. (2006). Chikungunya outbreaks caused by African genotype, India. *Emerg. Infect. Dis.* 12 (10): 1580–1583.

44 Padbidri, V.S. and Gnaneswar, T.T. (1979). Epidemiological investigations of chikungunya epidemic at Barsi, Maharashtra state, India. *J. Hyg. Epidemiol. Microbiol. Immunol.* 23 (4): 445–451.

45 Khatun, S., Chakraborty, A., Rahman, M. et al. (2015). An outbreak of chikungunya in rural Bangladesh, 2011. *PLoS Negl. Trop. Dis.* 9: e0003907.

46 Rahman, M., Yamagishi, J., Rahim, R. et al. (2019). East/central/south African genotype in a chikungunya outbreak. Dhaka, Bangladesh, 2017. *Emerg. Infect. Dis.* 25: 370–372.

47 Wangchuk, S., Chinnawirotpisan, P., Dorji, T. et al. (2013). Chikungunya fever outbreak, Bhutan, 2012. *Emerg. Infect. Dis.* 19: 1681–1684.

48 Pandey, K., Pandey, B.D., Chaurasiya, R.R. et al. (2017). Evidence of chikungunya virus circulation in the Terai region of Nepal in 2014 and 2015. *Trans. R. Soc. Trop. Med. Hyg.* 111: 294–299.

49 Darwish, M.A., Hoogstraal, H., Roberts, T.J. et al. (1983). A sero-epidemiological survey for certain arboviruses (Togaviridae) in Pakistan. *Trans. R. Soc. Trop. Med. Hyg.* 77: 442–445.

50 Pan, J., Fang, C., Yan, J. et al. (2019). Chikungunya fever outbreak, Zhejiang province, China, 2017. *Emerg. Infect. Dis.* 25: 1589–1591.

51 Xie, T., Lyu, L.K., Tan, Z.L. et al. (2020). Genotyping on one case with chikungunya infection introduced into Tianjin in China from Myanmar. *Zhonghua Liu Xing Bing Xue Za Zhi* 41: 2131–2134.

52 Powers, A.M., Brault Robert, B.A.C., Tesh, R.B. et al. (2000). Reemergence of chikungunya and o'nyong-nyong viruses: evidence for distinct geographical lineages and distant evolutionary relationships. *J. Gen. Virol.* 81: 471–479.

53 Cassadou, S., Boucau, S., Huc, M.P. et al. (2014). Emergence of chikungunya fever on the French side of Saint Martin Island, October to December 2013. *Euro Surveill.* 19 (13): 20752.

54 Halstead, S.B. (2015). Reappearance of chikungunya, formerly called dengue, in the Americas. *Emerg. Infect. Dis.* 21 (4): 557–561.

55 Van Genderen, F.T., Krishnadath, I., Sno, R. et al. (2016). First chikungunya outbreak in suriname; clinical and epidemiological features. *PLoS Negl. Trop. Dis.* 10 (4): e0004625.

56 Vairo, F., Mammone, A., Lanini, S. et al. (2018). Local transmission of chikungunya in Rome and the Lazio region, Italy. *PLoS One* 13 (12): e0208896.

57 Lahariya, C. and Pradhan, S.K. (2006). Emergence of chikungunya virus in Indian subcontinent after 32 years: a review. *J. Vector Borne Dis.* 43 (4): 151–160.

58 Schuffenecker, I., Iteman, I., Michault, A. et al. (2006). Genome microevolution of chikungunya viruses causing the Indian Ocean outbreak. *PLoS Med.* 3 (7): e263.

59 Kalantri, S.P., Joshi, R., and Riley, L.W. (2006). Chikungunya epidemic: an Indian perspective. *Natl Med. J. India* 19 (6): 315–322.

60 The Translational Research Consortia (TRC) for Chikungunya Virus in India (2021). Current status of chikungunya in India. *Front. Microbiol.* 12: 695173.

61 National Vector Borne Disease Control Programme (NVBDCP), Ministry of Health & Family Welfare, Government of India. Chikungunya Situation in India. https://ncvbdc.mohfw.gov.in/index4.php?lang=1&level=0&linkid=486&lid=3765. Accessed February 2023.

62 Sam, I.C., Kümmerer, B.M., Chan, Y.F. et al. (2015). Updates on chikungunya epidemiology, clinical disease, and diagnostics. *Vector Borne Zoonotic Dis.* 15: 223–230.

63 Javelle, E., Tiong, T.H., Leparc-Goffart, I. et al. (2014). Inflammation of the external ear in acute chikungunya infection: experience from the outbreak in Johor Bahru, Malaysia, 2008. *J. Clin. Virol.* 59: 270–273.

64 Mittal, A., Mittal, S., Bharati, M.J. et al. (2007). Optic neuritis associated with chikungunya virus infection in South India. *Arch. Ophthalmol.* 125: 1381–1386.

65 Mahendradas, P., Ranganna, S.K., Shetty, R. et al. (2008). Ocular manifestations associated with chikungunya. *Ophthalmology* 115: 287–291.

66 Sissoko, D., Malvy, D., Giry, C. et al. (2008). Outbreak of chikungunya fever in Mayotte, Comoros archipelago, 2005–2006. *Trans. R. Soc. Trop. Med. Hyg.* 102: 780–786.

67 Staikowsky, F., Talarmin, F., Grivard, P. et al. (2009). Prospective study of chikungunya virus acute infection in the Island of La Reunion during the 2005–2006 outbreak. *PLoS One* 28: 4.

68 Simon, F., Parola, P., Grandadam, M. et al. (2007). Chikungunya infection: an emerging rheumatism among travelers returned from Indian Ocean islands. Report of 47 cases. *Medicine (Baltimore)* 86: 123–137.

69 Bandyopadhyay, D. and Ghosh, S.K. (2008). Mucocutaneous features of chikungunya fever: a study from an outbreak in West Bengal, India. *Int. J. Dermatol.* 47: 1148–1152.

70 Inamadar, A.C., Palit, A., Sampagavi, V.V. et al. (2008). Cutaneous manifestations of chikungunya fever: observations made during a recent outbreak in south India. *Int. J. Dermatol.* 47: 154–159.

71 Mishra, K. and Rajawat, V. (2008). Chikungunya-induced genital ulcers. *Indian J. Dermatol. Venereol. Leprol.* 74: 383–384.

72 Chandak, N.H., Kashyap, R.S., Kabra, D. et al. (2009). Neurological complications of chikungunya virus infection. *Neurol. India* 57 (2): 177–180.

73 Crosby, L., Perreau, C., Madeux, B. et al. (2016). Severe manifestations of chikungunya virus in critically ill patients during the 2013–2014 Caribbean outbreak. *Int. J. Infect. Dis.* 48: 78–80.

74 Tandale, B.V., Sathe, P.S., Arankalle, V.A. et al. (2009). Systemic involvements and fatalities during chikungunya epidemic in India, 2006. *J. Clin. Virol.* 46 (2): 145–149.

75 Chusri, S., Siripaitoon, P., Hirunpat, S., and Silpapojakul, K. (2011). Case reports of neuro-chikungunya in southern Thailand. *Am. J. Trop. Med. Hyg.* 85 (2): 386–389.

76 Brasil, P., Sequeira, P.C., Freitas, A.D. et al. (2016). Guillain-Barré syndrome associated with Zika virus infection. *Lancet* 387 (10026): 1482.

77 Lebrun, G., Chadda, K., Reboux, A.H. et al. (2009). Guillain-Barré syndrome after chikungunya infection. *Emerg. Infect. Dis.* 15 (3): 495–496.

78 Zeller, H., Van Bortel, W., and Sudre, B. Chikungunya: its history in Africa and Asia and its spread to new regions in 2013–2014. *J. Infect. Dis.* 214: S436–S440.

79 Cunha, M.S., Costa, P.A.G., Correa, I.A. et al. (2020). Chikungunya virus: an emergent arbovirus to the South American continent and a continuous threat to the world. *Front. Microbiol.* 11: 1297.

80 Villero-Wolf, Y., Mattar, S., Puerta-González, A. et al. (2019). Genomic epidemiology of chikungunya virus in Colombia reveals genetic variability of strains and multiple geographic introductions in outbreak, 2014. *Sci. Rep.* 9: 9970.

81 Farias, B.G., Beserra, C.N., Fernandes, L. et al. (2019). Myocarditis following recent chikungunya and dengue virus coinfection: a case report. *Arq. Bras. Cardiol.* 113: 783–786.

82 Al Amin, M., Ibrahim, S., Balkhair, A. et al. (2019). Chikungunya masquerading as acute rheumatism in an Omani traveler. *Oman Med. J.* 34: 63–65.

83 Rahim, M.A. and Uddin, K.N. (2017). Chikungunya: an emerging viral infection with varied clinical presentations in Bangladesh: reports of seven cases. *BMC. Res. Notes* 10: 410.

84 Colavita, F., Vita, S., Lalle, E. et al. (2018). Overproduction of IL-6 and type-I IFN in a lethal case of chikungunya virus infection in an elderly man during the 2017 Italian outbreak. *Open Forum Infect. Dis.* 5: ofy276.

85 Villamil-Gómez, W.E., Ramirez-Vallejo, E., Cardona-Ospina, J.A. et al. (2016). Electrocardiographic alterations in patients with chikungunya fever from Sucre, Colombia: a 42-case series. *Travel Med. Infect. Dis.* 14: 510–512.

86 Nasci, R.S. (2014). Movement of chikungunya virus into the western hemisphere. *Emerg. Infect. Dis.* 20: 1394–1395.

87 Lemant, J., Boisson, V., Winer, A. et al. (2008). Serious acute chikungunya virus infection requiring intensive care during the Reunion Island outbreak in 2005–2006. *Crit. Care Med.* 36: 2536–2541.

88 Thiery, G., Crosby, L., Perreau, C. et al. (2015). Critically ill patients with chikungunya virus infection during the Caribbean outbreak 2013–2014. *Intensive Care Med. Exp.* 3: A348.

89 Nazneen, S., Saha, M., Hossain, R. et al. (2017). Chikungunya viral infection requiring intensive care–experience in a tertiary care hospital in Bangladesh. *BIRDEM Med. J.* 8: 16.

90 Torres, J.R., Leopoldo, C.G., Castro, J.S. et al. (2015). Chikungunya fever: atypical and lethal cases in the western hemisphere: a Venezuelan experience. *ID Cases* 2: 6–10.

91 Gonzalez, K.A., Mendoza, I.J., Finizola, V. et al. (2016). Abstract 13292: bradycardia as a manifestation of chikungunya myocarditis. A new threat to America. *Circulation* 134: A13292.

92 Aletti, M., Lecoules, S., Kanczuga, V. et al. (2017). Transient myocarditis associated with acute Zika virus infection. *Clin. Infect. Dis.* 64: 678–679.

93 Rampal, Sharda, M., and Meena, H. (2007). Neurological complications in chikungunya fever. *J. Assoc. Physicians India* 55: 765–769.

94 Robin, S., Ramful, D., Le Seach, F. et al. (2008). Neurologic manifestations of pediatric chikungunya infection. *J. Child Neurol.* 23: 1028–1035.

95 Musthafa, A.K., Abdurahiman, P., and Jose, J. (2008). Case of ADEM following chikungunya fever. *J. Assoc. Physicians India* 56: 473.

96 Lewthwaite, P., Vasanthapuram, R., Osborne, J.C. et al. (2009). Chikungunya virus and central nervous system infections in children. *India Emerg. Infect. Dis.* 15: 329–331.

97 Solanki, B.S., Arya, S.C., and Maheshwari, P. (2007). Chikungunya disease with nephritic presentation. *Int. J. Clin. Pract.* 61: 1941.

98 Jayakeerthi, R.S., Potula, R.V., Srinivasan, S., and Badrinath, S. (2006). Shell vial culture assay for the rapid diagnosis of Japanese encephalitis, West Nile and Dengue-2 viral encephalitis. *Virol. J.* 3: 2.

99 Lanciotti, R.S., Kosoy, O.L., Laven, J.J. et al. (2007). Chikungunya virus in US travelers returning from India, 2006. *Emerg. Infect. Dis.* 13 (5): 764–767.

100 Natrajan, M.S., Rojas, A., and Waggoner, J.J. (2019). Beyond fever and pain: diagnostic methods for chikungunya virus. *J. Clin. Microbiol.* 57: e00350-19.

101 Álvarez-Argüelles, M.E., Alba, S.R., Pérez, M.R. et al. (2019). Diagnosis and molecular characterization of chikungunya virus infections. In: *Current Topics in Neglected Tropical Diseases* (ed. A.J.R. Morales), 28–51. Germany: BoD – Books on Demand.

102 Weaver, S.C., Forrester, N.L., Liu, J., and Vasilakis, N. (2021). Population bottlenecks and founder effects: implications for mosquito-borne arboviral emergence. *Nat. Rev. Microbiol.* 19: 184–195.

103 Acevedo, N., Waggoner, J., Rodriguez, M. et al. (2017). Zika virus, chikungunya virus, and dengue virus in cerebrospinal fluid from adults with neurological manifestations, Guayaquil, Ecuador. *Front. Microbiol.* 8: 42.

104 Musso, D., Teissier, A., Rouault, E. et al. (2016). Detection of chikungunya virus in saliva and urine. *Virol. J.* 13: 102.

105 Mardekian, S.K. and Roberts, A.L. (2015). Diagnostic options and challenges for dengue and chikungunya viruses. *Biomed. Res. Int.* 2015: 834371.

106 Krutikov, M. and Manson, J. (2016). Chikungunya virus infection: an update on joint manifestations and management. *Rambam Maimonides Med. J.* 7: e0033.

107 Da Cunha, R.V. and Trinta, K.S. (2017). Chikungunya virus: clinical aspects and treatment – a review. *Mem. Inst. Oswaldo Cruz* 112: 523–531.

108 Staikowsky, F., Le Roux, K., Schuffenecker, I. et al. (2008). Retrospective survey of chikungunya disease in Reunion Island hospital staff. *Epidemiol. Infect.* 136 (2): 196–206.

109 Padmakumar, B., Jayan, J.B., Menon, R.M.R. et al. (2009). Comparative evaluation of four therapeutic regimes in chikungunya arthritis: a prospective randomized parallel-group study. *Indian J. Rheumatol.* 4 (3): 94–101.

110 Gauri, L.A., Ranwa, B.L., Nagar, K. et al. (2012). Post chikungunya brain stem encephalitis. *J. Assoc. Physicians India* 60 (4): 68–69.

111 Mylonas, A.D., Harley, D., Purdie, D.M. et al. (2004). Corticosteroid therapy in an alphaviral arthritis. *J. Clin. Rheumatol.* 10 (6): 326–330.

112 Brighton, S.W. (1984). Chloroquine phosphate treatment of chronic chikungunya arthritis. An open pilot study. *S. Afr. Med. J.* 66 (6): 217–218.

113 Ben-Zvi, I., Kivity, S., Langevitz, P., and Shoenfeld, Y. (2012). Hydroxychloroquine: from malaria to autoimmunity. *Clin. Rev. Allergy Immunol.* 42: 145–153.

114 Arroyo-Ávila, M. and Vilá, L.M. (2015). Rheumatic manifestations in patients with chikungunya infection. *P. R. Health Sci. J.* 34: 71–77.

115 Simon, F., Javelle, E., Cabie, A. et al. (2015). French guidelines for the management of chikungunya (acute and persistent presentations). *Med. Mal. Infect.* 45: 243–263.

116 Javelle, E., Ribera, A., Degasne, I. et al. (2015). Specific management of post-chikungunya rheumatic disorders: a retrospective study of 159 cases in Reunion Island from 2006–2012. *PLoS Negl. Trop. Dis.* 9: e0003603.

117 Khoury, V.J. and Camilo, P.R. (2016). Chikungunya virus (CHIKV): what can be expected after the acute phase? *Reumatol. Clin.* 12: 1–3.

118 Chan, Y.S., Khoo, K.S., and Sit, N.W. (2016). Investigation of twenty selected medicinal plants from Malaysia for anti-chikungunya virus activity. *Int. Microbiol.* 19: 175–182.

119 Raghavendhar, S., Tripati, P.K., Ray, P., and Patel, A.K. (2019). Evaluation of medicinal herbs for anti-CHIKV activity. *Virology* 533: 45–49.

120 Sharma, V., Kaushik, S., Pandit, P. et al. (2019). Green synthesis of silver nanoparticles from medicinal plants and evaluation of their antiviral potential against chikungunya virus. *Appl. Microbiol. Biotechnol.* 103: 881–891.

121 Peng, Q., Shang, X., Zhu, C. et al. (2019). Qualitative and quantitative evaluation of *Oroxylum indicum* (L.) Kurz by HPLC and LC-qTOF-MS/MS. *Biomed. Chromatogr.* 33: 4657.

122 Bag, A., Bhattacharyya, S.K., and Chattopadhyay, R.R. (2013). The development of *Terminalia chebula* Retz. (Combretaceae) in clinical research. *Asian Pac. J. Trop. Biomed.* 3 (3): 244–252.

123 Sangeetha, K. and Rajarajan, S. (2015). In-vitro antiviral activity of Indian medicinal plants to Asian and east central south African lineage of chikungunya virus. *Int. J. Pharm. Sci. Res.* 6: 692–697.

124 Dhanik, J., Arya, N., and Nand, V.A. (2017). Review on *Zingiber officinale*. *J. Pharmacogn. Phytochem.* 6 (3): 174–184.

125 Kaushik, S., Jangra, G., Kundu, V. et al. (2020). Anti-viral activity of *Zingiber officinale* (ginger) ingredients against the chikungunya virus. *Virusdisease* 31: 1–7.

126 Sharma, Y., Kawatra, A., Sharma, V. et al. (2021). In-vitro and in-silico evaluation of the anti-chikungunya potential of *Psidium guajava* leaf extract and their synthesized silver nanoparticles. *Virusdisease* 32: 1–6.

15

Lassa Fever: Recent Clinical Reports and Management Update

Abigail Aning[1], Kwasi A. Bugyei[2], Bismarck A. Hottor[3], and Seth K. Amponsah[2]

[1] *Department of Clinical Pathology, Noguchi Memorial Institute for Medical Research, University of Ghana, Accra, Ghana*
[2] *Department of Medical Pharmacology, University of Ghana Medical School, Accra, Ghana*
[3] *Department of Anatomy, University of Ghana Medical School, Accra, Ghana*

15.1 Background

Lassa fever is a severe viral hemorrhagic illness caused by the Lassa virus (LASV), a member of the Arenaviridae family [1]. Genetic dating suggests that zoonotic transfers of LASV to humans have been occurring for centuries. However, it was not until 1969 that the first patient with Lassa fever was described. The patient was a missionary nurse infected while working in a rural clinic near Lassa, Nigeria [2]. According to the Centers for Disease Control and Prevention (CDC), there are approximately 100,000–300,000 cases of Lassa fever annually, resulting in 5000 deaths [3]. However, due to underreporting and limited surveillance systems in the region, the actual incidence and prevalence of Lassa fever may be higher than documented. Over the years, there have been numerous clinical reports and research aimed at enhancing the understanding of the disease and improving its management.

15.2 Geographical Distribution

Lassa fever is primarily endemic to several countries in West Africa, including Nigeria, Sierra Leone, Liberia, Guinea, and Benin. Nigeria bears the highest burden of Lassa fever, with reported cases from various states across the country [4]. The distribution of Lassa fever is influenced by ecological factors, such as the presence of the multimammate rat (*Mastomys natalensis*), which serves as the primary reservoir for the virus, and agricultural practices that facilitate rodent-human contact [5].

15.3 Mode of Transmission

The primary mode of transmission of Lassa fever to humans is through direct or indirect contact with the urine or feces of infected rodents (Figure 15.1), particularly the multimammate rat (*Mastomys natalensis*) [6]. Inhalation of aerosolized particles contaminated with rodent excreta is also a potential route of transmission. Additionally, human-to-human transmission can occur through direct contact with blood, tissue, or other body fluids of infected individuals, particularly in healthcare settings where infection control measures may be lacking [3].

15.4 Reservoir Hosts

The multimammate rat (*Mastomys natalensis*) is considered the primary reservoir host for the LASV. These rats are highly adaptable and can be found in various ecotypes, including rural and urban areas. They are known to shed the virus asymptomatically, which contributes to the persistence of the virus in the environment and potential transmission to humans [5].

Rising Contagious Diseases: Basics, Management, and Treatments, First Edition. Edited by Seth Kwabena Amponsah, Ranjita Shegokar, and Yashwant V. Pathak.

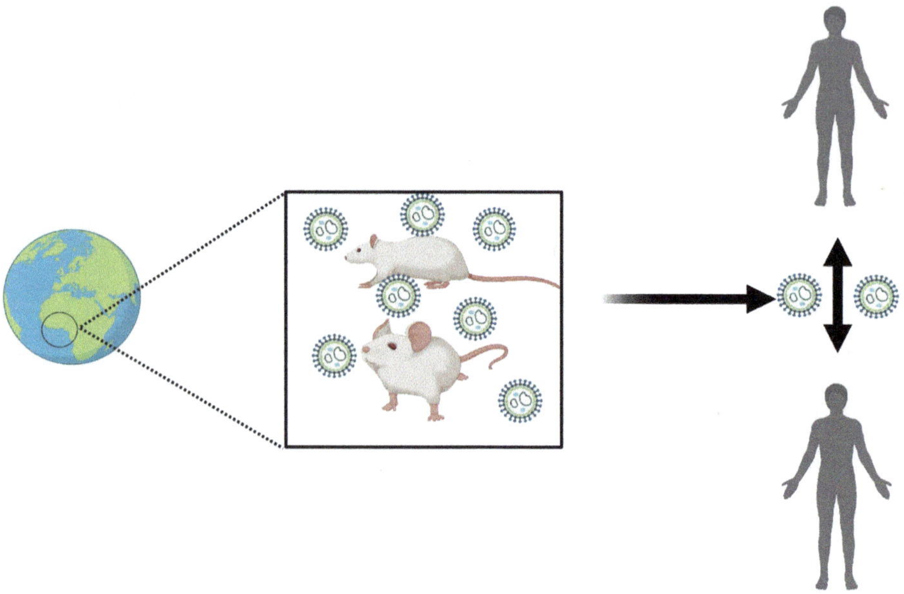

Figure 15.1 Primary mode of transmission of Lassa virus.

Infected rodents shed the virus in their urine and feces, thereby contaminating the environment and facilitating human exposure [6]. Other rodent species, such as the African soft-furred rat (Praomys spp.) and the Guinea multimammate rat (*Mastomys erythroleucus*), may also play a role as secondary or occasional reservoir hosts [7].

15.5 Incidence and Prevalence

Determining the precise incidence and prevalence of Lassa fever is challenging due to underreporting and limited surveillance systems in endemic regions. The CDC estimates that there are approximately 100,000–300,000 cases of Lassa fever annually, resulting in 5000 deaths [3]. However, these figures likely underestimate the true burden of the disease. The incidence of Lassa fever exhibits seasonal variation, with higher transmission observed during the dry season when rodents migrate and come into closer proximity to humans [8]. The disease can affect individuals of all ages and both sexes, with healthcare workers being at high risk due to their occupational exposure to infected patients [6].

15.6 Clinical Presentation

Lassa fever presents with a diverse range of clinical manifestations, from mild to severe. The incubation period of the virus typically ranges from 3–21 days [9]. The initial symptoms of Lassa fever are often nonspecific and can resemble other febrile illnesses, making diagnosis challenging [10]. However, as the disease progresses, more severe symptoms may develop. Patients with Lassa fever commonly experience fever, malaise, headache, and myalgia during the initial phase of the illness [11]. Gastrointestinal symptoms, such as nausea, vomiting, and diarrhea, are also frequently reported [11]. These early symptoms are often indistinguishable from those of other common infectious diseases, making it important to consider Lassa fever in the differential diagnosis, especially in regions where the disease is endemic [12].

As the disease advances, more serious symptoms may emerge. Respiratory distress, accompanied by cough and chest pain, can occur as the virus spreads to the lungs [13]. Hemorrhagic manifestations, such as petechiae (small red or purple spots), ecchymosis (bruises), bleeding from mucosal surfaces (nose, gums, gastrointestinal tract), and in some cases, hematemesis (vomiting blood) or melena (blood in stool), may be observed in severe cases [13].

Lassa fever can lead to complications affecting various organ systems. Neurological involvement may occur, resulting in encephalopathy, seizures, and coma [14]. Cardiovascular manifestations, including hypotension and shock, can develop due to fluid loss, vascular leakage, and sepsis [15].

Specific populations may experience unique clinical presentations. Pregnant women infected with the LASV may have an increased risk of spontaneous abortion, stillbirth, or maternal death [16]. Furthermore, hearing loss is a notable complication associated with Lassa fever, affecting up to one-third of the survivors [17].

15.7 Diagnostic Methods

Accurate and timely diagnosis of Lassa fever is crucial for patient management, implementation of infection control measures, and public health interventions. Several diagnostic methods are available to detect the presence of the LASV. Detection of LASV antigens or genetic material in patient samples is a definitive diagnostic approach. Various laboratory techniques are employed, including enzyme-linked immunosorbent assay (ELISA), immunofluorescence assay (IFA), and reverse transcription-polymerase chain reaction (RT-PCR) [18].

15.7.1 Serological Assays

Serological assays are commonly used to detect antibodies produced by the host in response to LASV infection. Immunoglobulin M (IgM) and immunoglobulin G (IgG) antibodies can be detected in a patient's serum using ELISA or IFA methods. IgM antibodies are typically present during the acute phase of the infection, while IgG antibodies indicate past or recent exposure to the virus [18].

15.7.2 Polymerase Chain Reaction (PCR)

PCR is a highly sensitive and specific molecular diagnostic technique used to detect the genetic material, ribonucleic acid (RNA), of LASV in patient samples. RT-PCR, a variant of PCR, is specifically used to amplify and detect viral RNA. This method enables rapid and accurate diagnosis, especially during the early stages of LASV infection when viral load is high [19].

15.7.3 Imaging Techniques

Imaging techniques are primarily used to assess complications and organ involvement in severe cases of Lassa fever. Chest X-rays may reveal pulmonary infiltrates and respiratory distress, while abdominal ultrasound or computed tomography (CT) scans can help identify organ-specific abnormalities such as hepatomegaly or splenomegaly [20, 21].

It is important to note that diagnostic testing for Lassa fever should be performed in well-equipped laboratories with appropriate biosafety measures to prevent laboratory-acquired infections. Due to potential cross-reactivity with other related viruses, confirmatory testing and consultation with reference laboratories may be necessary for accurate diagnosis [18].

15.8 Recent Clinical Reports

Recent clinical reports on Lassa fever provide valuable insights into the evolving understanding of the disease, its clinical presentation, and management strategies. These reports contribute to our understanding of the current trends and challenges associated with Lassa fever management.

Since 2016, there has been a rise in the number of West African cases of Lassa fever, particularly in Nigeria, Liberia, Benin, and Togo [22–24]. In 2018 and 2019, Nigeria saw many outbreaks of Lassa fever [25]. During this period, Benin recorded 54 cases and 28 deaths, Liberia recorded 7 cases and 3 deaths, whiles Togo and Sierra Leone each reported 2 cases [26]. Additionally, outside West Africa, there have been cases reported in Sweden and Germany; mainly travelers from Africa [26].

A study conducted by Merson et al. [27] showed that at baseline, fever was the most reported symptom of Lassa fever. This was identified in 88% of individuals who had the disease. Other common symptoms included headache, vomiting, abdominal pain, and cough [27]. A smaller number of patients presented with clinically severe or life-threatening symptoms such as shock, breathing difficulty, and seizure [27].

Okokhere et al. [28] conducted a retrospective, observational cohort study of consecutive patients in Nigeria with Lassa fever. The study detailed clinical and laboratory data of Nigerian patients with Lassa fever and provided strong evidence of intrinsic renal dysfunction among some of the patients [28]. Early recognition and treatment of acute kidney injury would significantly reduce mortality.

A case study highlighted the occurrence of bilateral sensorineural hearing loss as a complication of Lassa fever in a patient [29]. The report emphasized the need for early identification and prompt management of hearing loss in Lassa fever survivors to mitigate its long-term impact on the quality of life.

A retrospective cohort study described a high rate of maternal and fetal death among 30 pregnant women with Lassa fever, highlighting the increased risk of adverse outcomes in this population [30]. The report emphasized the importance of timely diagnosis, aggressive supportive care, and close monitoring of pregnant women with Lassa fever to optimize maternal and fetal outcomes.

These clinical reports provide valuable insights into the current understanding of Lassa fever, and its clinical presentation. These, among others, contribute to ongoing efforts to improve diagnosis, prevention, and treatment of this potentially life-threatening disease. These case reports contribute to the expanding knowledge of Lassa fever and its varied clinical presentations.

15.9 Management and Treatment

15.9.1 Antiviral Therapy

Antiviral therapy is a cornerstone in Lassa fever treatment. Ribavirin, a broad-spectrum antiviral drug, has shown efficacy in reducing morbidity and mortality when administered early in the course of the disease [31]. Ribavirin inhibits viral replication and is typically administered intravenously or orally depending on the severity of the illness and the patient's condition. Ribavirin has only been tested as a Lassa fever treatment in one nonrandomized clinical study and in a retrospective field study [32]. Nonetheless, there is a need for better treatment options because of the high death rates seen in both ribavirin-treated and untreated individuals during LASV infection [33].

The recommended duration and dosage of ribavirin vary based on disease severity and patient characteristics. Intravenous ribavirin is usually administered for a period of 10 days, while oral ribavirin may be continued for up to 14 days or longer [8].

15.9.2 Supportive Care

Supportive care is an important aspect of managing Lassa fever patients. It involves addressing symptoms, maintaining fluid and electrolyte balance, and managing complications [34]. Patients with Lassa fever often require close monitoring of vital signs, such as blood pressure, heart rate, and oxygen saturation. Patients who are admitted to specialized Lassa fever treatment units that have resources for supportive care, such as dialysis and respiratory support, have a high chance of survival [26].

15.9.3 Experimental Therapies

There have been a number of experimental therapies that are being explored to expand treatment options for Lassa fever. Favipiravir, an antiviral agent with activity against RNA viruses, has shown promise in preclinical studies and is being evaluated in clinical trials [35]. Favipiravir has been shown to decrease LASV viremia in animal models [25]. Other investigational therapies include monoclonal antibodies, small interfering RNA (siRNA), and antiviral peptides. These are being studied for their efficacy against LASV infections [36]. Monoclonal antibodies specific for LASV that have been cloned from West African Lassa fever survivors seem to bind to individual or combined Lassa GP protein subunits, which can potently neutralize all LASV lineages – an early start to immunotherapeutic development [37]. Other reports also suggest that human monoclonal antibody therapy protects non-human primates against advanced Lassa fever [38].

15.9.4 Immune-Based Therapies

Immunotherapy approaches, such as passive immunization with convalescent plasma or monoclonal antibodies, are being explored as potential treatment options for Lassa fever. Convalescent plasma obtained from recovered Lassa fever patients contains antibodies that can neutralize the virus and may improve patient outcomes [39].

It is important to note that although progress has been made in Lassa fever management, challenges still exist, especially in resource-limited settings where access to timely diagnosis, appropriate supportive care, and effective antiviral therapy may be limited. Efforts to strengthen healthcare systems, enhance diagnostic capabilities, and develop more accessible and affordable treatment options remain critical in improving the management and outcomes of patients with Lassa fever.

15.10 Future Prospects and Research Directions

Lassa fever remains a significant public health concern in endemic regions, and ongoing research has focused on advancing preventive, diagnostic, therapeutic, and surveillance strategies. Several areas of future prospects and research directions are being explored to combat Lassa fever effectively, some of which include vaccine development,

15.10.1 Vaccine Development

Vaccine development for Lassa fever is a critical area of research. Several vaccine candidates have shown promise in preclinical studies, including DNA-based vaccines, viral vectors expressing LASV antigens, and virus-like particle (VLP) vaccines [40, 41]. Future research aims to advance these candidates into clinical trials to assess their safety and efficacy in generating protective immune responses against LASV infection. Scientists are also making progress in developing vaccines that can make humans have protective immune responses against LASV, providing long-term immunity and reducing the risk of infection and transmission.

15.10.2 Therapeutic Interventions

Further research is needed to explore and develop novel therapies for Lassa fever. This includes the evaluation of new antiviral drugs with enhanced efficacy, improved delivery methods, and reduced toxicity profiles. Additionally, the identification of host factors that contribute to disease severity and the development of host-targeted therapies hold promise for future therapy against Lassa fever infections [42].

15.10.3 Improved Diagnostic Tools

Rapid and accurate diagnosis of Lassa fever is crucial for early treatment and effective control. Currently, research efforts are focused on the development of improved diagnostic tools, such as point-of-care tests that are affordable, sensitive, specific, and easy to use in resource-limited settings. This includes the exploration of novel diagnostic platforms, such as portable molecular assays and serological tests with enhanced performance characteristics [43]. Integration of new technologies, such as loop-mediated isothermal amplification (LAMP) and portable sequencing platforms, can enhance the speed and accuracy of Lassa fever diagnosis.

15.10.4 Strengthening Surveillance

Strengthening surveillance systems is essential for timely detection, monitoring, and response to Lassa fever outbreaks. Research is aimed at improving disease surveillance through the integration of advanced technologies, including real-time data collection, geographic information systems (GIS), and modeling approaches. This will facilitate early detection of cases, better understanding of disease transmission dynamics, and implementation of targeted control measures.

15.10.5 Understanding Viral Pathogenesis

A deeper understanding of LASV pathogenesis is necessary to identify critical host-virus interactions and develop targeted interventions. Research focused on unraveling the mechanisms of viral entry, immune evasion, and host immune response to LASV infection is relevant. This knowledge can contribute to the development of therapies that modulate host immune response and limit viral replication.

15.10.6 Public Health Interventions

Effective control and prevention of Lassa fever require comprehensive approaches that include community engagement, health education, improved infection prevention, and control measures in healthcare settings. There have been a number of studies conducted to evaluate the impact of public health interventions (behavioral change communication campaigns, vector control strategies, and strengthening of healthcare infrastructure) on the spread of Lassa fever.

15.10.7 One Health Approach

Lassa fever research is increasingly adopting a "One Health Approach," recognizing the interconnectivity of human, animal, and environmental health. Future research aims to investigate the role of animal reservoirs, transmission dynamics, and ecological factors that influence LASV spillover into human populations. Collaboration between human health, veterinary, and environmental sectors can contribute to a better understanding of Lassa fever epidemiology and inform comprehensive prevention and control strategies.

15.11 Conclusions

Lassa fever remains a significant public health concern in endemic regions, and recent clinical reports have shed light on its clinical presentation, management, and challenges. The disease poses a threat to human health, with potentially severe outcomes and long-term complications. However, advancements in diagnostic methods, treatment strategies, and ongoing research provide hope for improved outcomes and control of this disease.

Key points discussed in this chapter include the geographical distribution of Lassa fever, its mode of transmission, reservoir hosts, incidence, and prevalence. The clinical presentation of Lassa fever varies from mild to severe manifestations, highlighting the need for early diagnosis and appropriate management.

Diagnostic methods, such as laboratory testing, serological assays, PCR, and imaging techniques, play a crucial role in confirming Lassa fever and guiding treatment decisions. Supportive care, antiviral therapy with ribavirin, experimental therapies, and immune-based therapies are integral components of Lassa fever management.

Going forward, the research focus should be on vaccine development, therapeutic interventions, improved diagnostic tools, enhanced surveillance systems, understanding viral pathogenesis, and public health interventions. Addressing the challenges and seizing the opportunities in Lassa fever management requires collaboration, funding, and resource allocation.

A call to action is needed to strengthen healthcare systems, improve access to timely diagnosis and treatment, enhance surveillance systems, foster research collaborations, and prioritize community engagement. By taking collective action, we can make significant progress in mitigating the impact of Lassa fever, improving patient outcomes, and preventing future outbreaks. Lassa fever serves as a reminder of the ongoing threats posed by emerging infectious diseases and the importance of a comprehensive, multidisciplinary approach to their management.

References

1 Yun, N.E. and Walker, D.H. (2012). Pathogenesis of Lassa fever. *Viruses* 4 (10): 2031–2048.
2 Garry, R.F. (2023). Lassa fever – the road ahead. *Nat. Rev. Microbiol.* 21 (2): 87–96.
3 CDC. 2022. Lassa Fever. http://cdc.gov/vhf/lassa/index.html (accessed 07 April 2023).
4 Denue, B., Stephen, M., and Dauvoux, I. (2017). The unending threat of Lassa fever in Nigeria, what can be done; what should be done. *Port Harcourt Med. J.* 11 (3): 113–121.
5 Fichet-Calvet, E. and Rogers, D.J. (2009). Risk maps of Lassa fever in West Africa. *PLoS Negl. Trop. Dis.* 3 (3): e388.
6 McCormick, J.B. et al. (1987). A prospective study of the epidemiology and ecology of Lassa fever. *J. Infect. Dis.* 155 (3): 437–444.
7 Monath, T.P. et al. (1974). Lassa virus isolation from *Mastomys natalensis* rodents during an epidemic in Sierra Leone. *Science* 185 (4147): 263–265.

8 Bausch, D.G. et al. (2010). Review of the literature and proposed guidelines for the use of oral ribavirin as postexposure prophylaxis for Lassa fever. *Clin. Infect. Dis.* 51 (12): 1435–1441.
9 Atkin, S. et al. (2009). The first case of Lassa fever imported from Mali to the United Kingdom, February 2009. *Euro Surveill.* 14 (10): 19145.
10 Raabe, V. and Koehler, J. (2017). Laboratory diagnosis of Lassa fever. *J. Clin. Microbiol.* 55 (6): 1629–1637.
11 Mustapha, A. (2017). Lassa fever: unveiling the misery of the Nigerian health worker. *Ann. Niger. Med.* 11: 1.
12 Murphy, H.L. and Ly, H. (2021). Pathogenicity and virulence mechanisms of Lassa virus and its animal modeling, diagnostic, prophylactic, and therapeutic developments. *Virulence* 12 (1): 2989–3014.
13 Olayemi, A. et al. (2016). New hosts of the Lassa virus. *Sci. Rep.* 6 (1): 1–6.
14 Abdulraheem, I. (2002). Public health importance of Lassa fever epidemiology, clinical features and current management review of literature. *Afr. J. Clin. Exp. Microbiol.* 3 (1): 33–37.
15 World Health Organization. Lassa Fever. Key Facts 2017 (accessed 30 June 2023).
16 Adewole, A.A. et al. (2022). Lassa fever in pregnancy: report of two maternal deaths in a tertiary center in the middle-belt region of Nigeria. *Afr. J. Reprod. Health* 26 (2): 137–143.
17 Okokhere, P.O., Ibekwe, T.S., and Akpede, G.O. (2009). Sensorineural hearing loss in Lassa fever: two case reports. *J. Med. Case Rep.* 3: 36.
18 Branco, L.M. et al. (2011). Emerging trends in Lassa fever: redefining the role of immunoglobulin M and inflammation in diagnosing acute infection. *Virol. J.* 8: 1–15.
19 Günther, S. et al. (2000). Imported Lassa fever in Germany: molecular characterization of a new Lassa virus strain. *Emerg. Infect. Dis.* 6 (5): 466.
20 Erameh, C.O. et al. (2023). Focussed assessment with sonography in acute Lassa fever (FASLa): development of a point-of-care protocol and description of common ultrasound findings. *J. Infect.* 87 (1): 27–33.
21 Eze, K.C., Salami, T.A., and Kpolugbo, J.U. (2014). Acute abdominal pain in patients with Lassa fever: radiological assessment and diagnostic challenges. *Niger. Med. J.* 55 (3): 195.
22 Whitmer, S.L. et al. (2018). New lineage of Lassa virus, Togo, 2016. *Emerg. Infect. Dis.* 24 (3): 599.
23 Kafetzopoulou, L. et al. (2019). Metagenomic sequencing at the epicenter of the Nigeria 2018 Lassa fever outbreak. *Science* 363 (6422): 74–77.
24 Akpede, G.O. et al. (2019). Caseload and case fatality of Lassa fever in Nigeria, 2001–2018: a specialist center's experience and its implications. *Front. Public Health* 7: 170.
25 NCDC. 2019. Lassa Fever. (accessed 4 July 2023).
26 Asogun, D.A. et al. (2019). Lassa fever: epidemiology, clinical features, diagnosis, management and prevention. *Infect. Dis. Clin.* 33 (4): 933–951.
27 Merson, L. et al. (2021). Clinical characterization of Lassa fever: a systematic review of clinical reports and research to inform clinical trial design. *PLoS Negl. Trop. Dis.* 15 (9): e0009788.
28 Okokhere, P. et al. (2018). Clinical and laboratory predictors of Lassa fever outcome in a dedicated treatment facility in Nigeria: a retrospective, observational cohort study. *Lancet Infect. Dis.* 18 (6): 684–695.
29 Ibekwe, T. et al. (2011). Early-onset sensorineural hearing loss in Lassa fever. *Eur. Arch. Otorhinolaryngol.* 268: 197–201.
30 Okogbenin, S. et al. (2019). Retrospective cohort study of Lassa fever in pregnancy, Southern Nigeria. *Emerg. Infect. Dis.* 25 (8): 1494.
31 McCormick, J. and Fisher-Hoch, S. (2002). Lassa fever. In: *Arenaviruses I: The Epidemiology, Molecular and Cell Biology of Arenaviruses* (ed. M.B.A. Oldstone), 75–109. Springer.
32 McCormick, J.B. et al. (1986). Lassa fever. *N. Engl. J. Med.* 314 (1): 20–26.
33 Shaffer, J.G. et al. (2014). Lassa fever in post-conflict Sierra Leone. *PLoS Negl. Trop. Dis.* 8 (3): e2748.
34 Akpede, G.O. et al. (2018). Lassa fever outbreaks in Nigeria. *Expert Rev. Anti-Infect. Ther.* 16 (9): 663–666.
35 Hansen, F. et al. (2021). Lassa virus treatment options. *Microorganisms* 9 (4): 772.
36 Hodgson, J. (2020). The pandemic pipeline. *Nat. Biotechnol.* 38 (5): 523–532.
37 Robinson, J.E. et al. (2016). Most neutralizing human monoclonal antibodies target novel epitopes requiring both Lassa virus glycoprotein subunits. *Nat. Commun.* 7 (1): 11544.
38 Mire, C.E. et al. (2017). Human-monoclonal-antibody therapy protects nonhuman primates against advanced Lassa fever. *Nat. Med.* 23 (10): 1146–1149.

39 Akpede, G.O. and Lawal, W. (2016). Part nine: chapter 61. Viral haemorrhagic fevers (VHF) with particular focus on Lassa and Ebola. In: *Paediatric and Child Health in a Tropical Region* (ed. J.C. Azubuike and K. Nkanginieme), 625–656. Educational Printing and Publishing.

40 Grant-Klein, R.J., Altamura, L.A., and Schmaljohn, C.S. (2011). Progress in recombinant DNA-derived vaccines for Lassa virus and filoviruses. *Virus Res.* 162 (1–2): 148–161.

41 Medugu, N. et al. (2023). A review of the recent advances on Lassa fever with special reference to molecular epidemiology and progress in vaccine development. *Afr. J. Clin. Exp. Microbiol.* 24 (2): 130–146.

42 Meganck, R.M. and Baric, R.S. (2021). Developing therapeutic approaches for twenty-first-century emerging infectious viral diseases. *Nat. Med.* 27 (3): 401–410.

43 Dhillon, R.S., Srikrishna, D., and Garry, R.F. (2018). Early detection of Lassa fever: the need for point-of-care diagnostics. *Lancet Infect. Dis.* 18 (6): 601–602.

16

Lyme Disease Management: Antibiotics and Beyond

Aparoop Das[1], Kalyani Pathak[1], Manash P. Pathak[2], Urvashee Gogoi[1], and Riya Saikia[1]

[1] *Department of Pharmaceutical Sciences, Dibrugarh University, Dibrugarh, Assam, India*
[2] *Faculty of Pharmaceutical Sciences, Assam Down Town University, Guwahati, Assam, India*

16.1 Introduction to Lyme Disease

Lyme disease is a tick-borne bacterial infection caused by the spirochete *Borrelia burgdorferi*. The disease was first identified in Lyme, Connecticut, in 1975, and has since become the most common vector-borne illness in the United States, with an estimated 300,000 cases reported annually [1]. The bacteria are primarily transmitted to humans through the bite of infected black-legged ticks, also known as deer ticks, which are commonly found in grassy and wooded areas in the northeastern and midwestern regions of the United States, as well as in parts of Europe and Asia. Lyme disease symptoms can include fever, fatigue, headache, and a characteristic skin rash called erythema migrans (EM), which appears at the tick bite site and expands over time [2]. If left untreated, the infection can spread to the joints, heart, and nervous system, causing more severe symptoms such as arthritis, heart palpitations, and meningitis [3]. The diagnosis of Lyme disease can be challenging, as the symptoms are often non-specific and may mimic other conditions. The most reliable method of diagnosis is laboratory testing of blood or cerebrospinal fluid for the presence of antibodies to the bacteria [4]. The treatment of Lyme disease typically involves a course of antibiotics, such as doxycycline, amoxicillin, or cefuroxime, depending on the severity and stage of the infection [5]. Preventing Lyme disease involves measures to avoid tick bites, such as wearing protective clothing, using insect repellent, and performing thorough tick checks after spending time outdoors. A vaccine is also available for Lyme disease, but it is not currently recommended for widespread use [6].

Lyme disease is caused by the bacterium *B. burgdorferi*, which is primarily transmitted to humans through the bite of infected black-legged ticks. The causes of Lyme disease are described as follows [7–10]–

I. *Borrelia Burgdorferi* Bacteria
Borrelia burgdorferi is a spirochete bacterium that causes Lyme disease. It was first identified in the mid-1980s by Dr. Willy Burgdorfer, who discovered the bacteria in the midgut of ticks collected in the woods of Connecticut. The bacteria are transmitted to humans through the bite of infected black-legged ticks (also known as deer ticks) that have fed on infected small mammals, such as mice and chipmunks.

II. Tick Vectors
Ticks are the primary vectors for Lyme disease. Black-legged ticks are the most common carriers of *B. burgdorferi*, but other species of ticks can also transmit the bacteria. In the United States, most Lyme disease cases are reported in the northeast and upper midwest regions, where black-legged ticks are most prevalent. However, cases of Lyme disease have also been reported in other parts of the country where other species of ticks are present.

III. Host Reservoirs
Small mammals, such as mice and chipmunks, serve as the primary host reservoirs for *B. burgdorferi*. These animals are important in transmitting the bacteria to ticks because they are often infested with ticks and can carry high levels of the

Rising Contagious Diseases: Basics, Management, and Treatments, First Edition. Edited by Seth Kwabena Amponsah, Ranjita Shegokar, and Yashwant V. Pathak.

bacteria in their blood. Ticks feeding on infected small mammals can acquire the bacteria and transmit it to other hosts, including humans.

IV. Risk Factors
Individuals who spend time in wooded or grassy areas are at the most significant risk of contracting Lyme disease. Other risk factors include living in or visiting areas where Lyme disease is common, such as the northeast and upper midwest regions of the United States. Outdoor activities such as camping, hiking, and gardening, can also increase the risk of exposure to infected ticks. Individuals who have pets that spend time outdoors may also be at risk, as ticks can attach to pets and be brought inside the home.

V. Genetic Factors
Genetic factors may also play a role in an individual's susceptibility to Lyme disease. Studies have shown that certain genes involved in the immune response may influence an individual's ability to clear the bacteria and prevent the disease. For example, a study published in the journal *Infection and Immunity* found that individuals with certain genetic variations in the IL-10 gene were more likely to develop Lyme disease after exposure to infected ticks.

VI. Coinfections
In addition to *B. burgdorferi*, ticks can also transmit other pathogens that can cause coinfections. These coinfections can complicate the diagnosis and treatment of Lyme disease and can lead to more severe symptoms. Common coinfections include *Anaplasma phagocytophilum*, which causes anaplasmosis, and *Babesia microti*, which causes babesiosis.

VII. Tick Life Cycle
The life cycle of a black-legged tick has four stages– egg, larva, nymph, and adult. Ticks can become infected with *B. burgdorferi* at any stage of their life cycle, but the highest risk of infection occurs during the nymph stage. Nymphs are the most common stage at which ticks are encountered by humans, and they are small enough to be easily missed or mistaken for a freckle.

VIII. Climate Change
Climate change may also contribute to the spread of Lyme disease. As the temperature is rising, tick populations are expanding into new areas, and the tick season is becoming longer. A study published in the journal *Environmental Health Perspectives* found that the incidence of Lyme disease in the United States associated with particular environmental circumstances like temperature, measurements of land cover, and the diversity of vertebrate species.

16.2 Modes of Transmission

Lyme disease is transmitted to humans through the bite of infected black-legged ticks. The black-legged ticks, also known as deer ticks, feed on the blood of animals infected with *B. burgdorferi*, the bacterium that causes Lyme disease, and then transmit the bacterium to humans when they bite them. The transmission of the bacterium usually occurs during the nymphal stage of the tick's development, which is during the spring and summer months.

Ticks are parasites that live by feeding on the blood of mammals, birds, and reptiles. In order for the tick to transmit the bacterium that causes Lyme disease to a human, the tick must remain attached to the human for at least 24 hours. This is because the bacterium needs to travel from the tick's gut to its salivary glands before it can be transmitted to the human.

Ticks are commonly found in wooded and grassy areas, where they wait on the tips of grass or on the edges of leaves for an animal or human to pass by. When the tick senses a host, it crawls onto the host's body and begins to search for a suitable place to bite. Once it finds a suitable spot, it attaches itself to the host's skin and begins to feed on the host's blood.

The black-legged tick is the most common tick species that transmits Lyme disease in the United States, although other species of ticks can also transmit the disease in other parts of the world. The black-legged tick is found in the northeastern and midwestern United States, as well as in parts of the Pacific Coast. In recent years, the range of the black-legged tick has been expanding, and it is now found in more areas than it was in the past.

The black-legged tick can also transmit other diseases besides Lyme disease, such as anaplasmosis and babesiosis. The bacterium *A. phagocytophilum* causes anaplasmosis, and babesiosis is caused by the parasite *B. microti*. Both of these diseases can be transmitted to humans when they are bitten by an infected tick.

In addition to tick bites, there has been some evidence of transmission of Lyme disease through blood transfusions, although this is considered a rare transmission mode. Research study reported that the polymerase chain reaction (PCR) test can detect *B. burgdorferi* in blood and urine samples, indicating that the bacterium can be present in the bloodstream of infected individuals [11]. However, Hinckley et al. [9] report that large commercial laboratories in the United States do not routinely test blood for Lyme disease, indicating that transmission through blood transfusions is not a major concern.

There have also been some reports of Lyme disease transmission from an infected mother to her unborn child. Steere et al. [7] report that in some cases, the bacterium that causes Lyme disease has been found in the placenta, amniotic fluid, and cord blood of infected pregnant women. However, they also note that the risk of transmission from an infected mother to her unborn child is low.

In conclusion, Lyme disease is primarily transmitted to humans through the bite of an infected black-legged tick. Other modes of transmission, such as transmission through blood transfusions or from an infected mother to her unborn child, are considered to be rare. It is important for individuals to take precautions to avoid tick bites when spending time in wooded or grassy areas, such as wearing long sleeves and pants, using insect repellent, and checking for ticks after being outdoors. Early detection and treatment of Lyme disease is important to prevent the development of more severe symptoms.

16.3 Recent Trends in Lyme Disease Management

16.3.1 Significance of Early Detection

Early detection and diagnosis of Lyme disease is crucial for several reasons. Lyme disease is highly treatable in its early stages. Antibiotics are the primary treatment for Lyme disease and, when started early, they can effectively eliminate the infection. However, if left untreated, the disease can progress to more advanced stages, becoming harder to treat and leading to long-term health problems [12–15]. Lyme disease can cause long-term complications if left untreated or if treatment is delayed. These can include joint pain, neurological symptoms, and heart problems. Early diagnosis and treatment can prevent these complications from developing countries. Reduced healthcare costs– Early detection and treatment of Lyme disease can reduce healthcare costs associated with prolonged illness and long-term complications [16, 17]. Lyme disease can significantly impact a person's quality of life, especially if it progresses to more advanced stages. Early diagnosis and treatment can reduce the risk of long-term complications and improve overall health and well-being [18]. Lyme disease is a public health concern, especially in areas where it is endemic. Early detection and diagnosis can help identify outbreaks and prevent the spread of the disease to others [19]. Overall, early detection and diagnosis of Lyme disease is essential for effective treatment, long-term complication prevention, healthcare costs, quality of life, and public health concerns [19].

16.3.2 The Role of Education and Awareness

Education and awareness of Lyme disease is critical in preventing and managing the disease. Educating people about Lyme disease, its symptoms, and how to prevent it can help reduce the number of cases. People who are aware of the risk factors are more likely to take preventive measures, such as wearing protective clothing, using insect repellent, and checking for ticks after spending time outdoors [20, 21]. Awareness of Lyme disease symptoms can help people seek medical attention early, leading to timely diagnosis and treatment. Educating people about the early signs of Lyme disease, such as bullseye rash, flu-like symptoms, and joint pain, can help them recognize the disease and seek medical attention promptly [22].

Educating people about the importance of seeking treatment early can lead to more effective treatment and a better prognosis. When people are aware of the risks and symptoms of Lyme disease, they are more likely to seek medical attention early and follow the prescribed treatment regimen [23]. Educating the public about Lyme disease can help raise awareness of the disease, its prevalence, and its associated risks. This can help public health officials develop strategies for the prevention, surveillance, and control of the disease [24]. There are many misconceptions about Lyme disease, such as the belief that it is only prevalent in specific regions or that it can only be contracted from deer ticks. Educating the public about the true nature of the disease can help dispel these myths and ensure that people take appropriate preventive measures [21, 22]. Education and awareness of Lyme disease are essential for prevention, early detection, effective treatment, public health, and addressing misconceptions.

16.3.3 Non-Pharmacologic Approaches

16.3.3.1 Lifestyle Changes and Self-Care

Lyme disease is a tick-borne illness that can cause various symptoms including fever, fatigue, joint pain, and a characteristic "bull's-eye" rash. While antibiotic treatment can help alleviate the symptoms, lifestyle changes and self-care can also play an important role in managing the condition [25, 26]. Here are some lifestyle changes and self-care strategies that may help manage Lyme disease–

- Get plenty of rest: Lyme disease can cause fatigue, so it is important to get plenty of rest. Make sure you get enough sleep at night and consider taking naps during the day, if needed [25].
- Eat a healthy diet: Eating a healthy diet can help support your immune system and give you the energy you need to fight off the infection. Focus on eating plenty of fruits, vegetables, lean protein, and whole grains [26].
- Exercise regularly: While it may be tempting to avoid exercise when you are feeling fatigued, regular exercise can actually boost your energy levels and support your immune system. Start with gentle exercises like walking or yoga, and gradually increase the intensity as you feel better [27].
- Manage stress: Stress can worsen Lyme disease symptoms, so it is crucial to find ways to manage stress [28].

16.3.3.2 Complementary and Alternative Therapies

Complementary and alternative therapies (CAM) may be helpful for some people in managing the symptoms of Lyme disease. However, it is essential to keep in mind that these therapies should not be used as a substitute for conventional medical treatment. Always talk to your healthcare provider before trying any new CAM therapies [29, 30]. Here are some examples of CAM therapies that may help manage Lyme disease–

- Acupuncture: Acupuncture involves the insertion of fine needles into specific points on the body. Some people find that acupuncture can help alleviate pain and other symptoms associated with Lyme disease [29].
- Herbal remedies: Certain herbs, such as garlic, oregano, and cat's claw, may have antimicrobial properties that can help fight the infection. However, caution is essential when using herbal remedies, as they can interact with medications and may have side effects [29].
- Massage: Massage therapy can help alleviate muscle tension and reduce pain and stiffness associated with Lyme disease. However, it is important to choose a massage therapist who has experience working with people with Lyme disease, as some types of massages may exacerbate symptoms [30].
- Homeopathy: Homeopathy involves using highly diluted substances to stimulate the body's natural healing processes. Some people find that homeopathy can help alleviate Lyme disease symptoms, but there is limited scientific evidence to support its effectiveness [31].
- Mind–body therapies: Mind–body therapies like meditation, yoga, and tai chi can help reduce stress and promote relaxation. Some people find that these therapies can help manage Lyme disease symptoms. It is important to work closely with your healthcare provider to develop a comprehensive treatment plan for Lyme disease. While CAM therapies may be helpful for some people, they should not be used as a substitute for conventional medical treatment [32].

16.3.3.3 Pharmacologic Management

Antibiotic Therapy Early diagnosis and antibiotic treatment of Lyme disease are critical after the discovery of an EM lesion. The recommended treatment for patients in the early or early disseminated phase without any neurological involvement is 14–21 days of doxycycline or amoxicillin administration [5]. Antibiotic treatment for Lyme disease depends upon the symptoms and types, i.e. EM rash, Neurologic Lyme (NL) disease, Lyme carditis (LC) disease, and Lyme arthritis (LA) disease.

Antibiotics for EM rash: Early-stage EM may get reduced on its own without any antibiotic intervention, however, antibiotic therapy may shorten the duration of the disease and reduces the chances of occurrence of the late-stage Lyme disease variants [33]. Adults showing symptoms of EM are prescribed doxycycline – 100 mg twice per day orally for 10–14 days, amoxicillin – 500 mg thrice per day orally for 14 days, or cefuroxime – 500 mg twice daily for 14 days. For children, the doses of antibiotics for EM are doxycycline – 44 mg/kg per day (maximum 100 mg per dose) divided into two doses orally for 10–14 days or amoxicillin – 50 mg/kg per day (maximum 500 mg per dose) divided into three doses orally for 14 days, or cefuroxime – 30 mg/kg per day (maximum 500 mg per dose) divided into two doses orally for 14 days [34]. In the case of pregnant women suspected of having ER due to Lyme disease, the treatment strategy is similar to that of a non-pregnant adult. Antibiotics other than doxycycline may be used in the pregnant woman because the administration of the latter may result in the discoloration of teeth permanently and may impact the formation of bones in the fetus [35]. Doxycycline comes under the category of tetracyclines, which are reportedly bacteriostatic in nature and inhibit the protein synthesis of bacteria [36]. Apart from their protein inhibition properties, tetracyclines are also reported to demonstrate anti-inflammatory efficacy against Lyme disease-causing bacteria *B. burgdorferi*. Both doxycycline and minocycline are reported to reduce the production of inflammatory markers like TNF-alpha, IL-6, and IL-8 as well as reduce the expression of *NFκB, IKKα*, phosphorylated IκBα in *B. burgdorferi* stimulated human monocyte THP-1, rhesus monkey brain astrocytes, and microglia [36]. It is noted that most of the antibiotics prescribed for early-stage EM are administered orally.

NL disease: NL disease or neuroborreliosis is a type of **CLD** characterized by peripheral facial nerve palsy (droop), radiculoneuropathy, Lyme meningitis followed by numbness, tingling, shooting pain, and weakness in the limbs due to the involvement of cranial nerve, peripheral nerve, and the central nervous system [37, 38]. Based on the occurrences, NL has been designated as early or acute and late manifestations. As per the report of a study conducted in Eastern Germany, Lyme neuroborreliosis (3.3%) is found to be the second most frequent clinical manifestation after EM (95.4%) [39]. Evidence suggested NL results from many tick-borne coinfections like Babesia spp, *A. phagocytophilum*, and *Bartonella henselae* [39]. A study conducted in Southern Sweden reported that children are more prone to Lyme neuroborreliosis than adults because infected ticks bite them on their heads [38]. Adults showing symptoms of facial palsy are prescribed doxycycline – 100 mg twice per day orally for 14–21 days, while children are prescribed doxycycline – 4.4 mg/kg per day (maximum 100 mg per dose) divided into two doses orally for 14–21 days [40]. Adults showing symptoms of Lyme meningitis or radiculoneuritis are prescribed doxycycline 200 mg per day divided into one or two doses orally for 14–21 days, or ceftriaxone 2 mg per day through intravenous route for 14–21 days. For children, the doxycycline dose is 4.4 mg/kg per day (maximum 100 mg per dose) divided into 1 or 2 doses orally for 14–21 days, or ceftriaxone 50–75 mg/kg per day (maximum 100 mg per dose) through intravenous route for 14–21 days [41]. Doxycycline and ceftriaxone are the first-line drugs for Lyme neuroborreliosis where the former is administered orally, and the latter is via the *i.v* route. Based on the route of administration, there was speculation on the effectiveness of doxycycline compared to ceftriaxone. However, a recent study found that oral doxycycline is as effective as intravenously administered ceftriaxone in adults [42], however studies on children are underway [43].

LC disease: LC occurs in the second (early disseminated stage) of the three stages of Lyme disease manifested by abnormalities in the conduction system and myocardium [44]. LC is characterized by the infiltration of inflammatory markers in different heart components, including the base of the heart, atrial and ventricular epicardium, endocardium, myocardium, and so on. The predominance of the chemokines-activated macrophage is found to be responsible for inflammation [45]. As per guidelines, patients showing symptoms like syncope, chest pain, or having first-degree blockage of the heart with a PR interval greater than or equal to 300 ms or patients with third-degree heart block should be immediately hospitalized and put under continuous monitoring [5]. A prophylactic antibiotic regimen of a single dose of oral doxycycline (200 mg for adults and 4.4 mg/kg for children) is recommended [46]. Adults suspected with mild LC and having first-degree AV block and PR above 300 ms are suggested to be prescribed doxycycline 100 mg twice per day for 10–14 days, or amoxicillin 500 mg thrice per day orally for 14–21 days, or cefuroxime 500 mg twice per day orally for 14–21 days. For children, the dose of doxycycline is 4.4 mg/kg per day (maximum 100 mg per dose) divided into two doses orally for 14–21 days, or amoxicillin 50 mg/kg per day (maximum 500 mg per dose) divided into three doses orally for 14–21 days, or cefuroxime 30 mg/kg per day (maximum 500 mg per dose) divided into two doses orally for 14–21 days [47]. However, for severe LC having first-degree, second-degree, or third-degree AV block with PR interval greater than 300 ms, IV administration of ceftriaxone for both adult and children are recommended [47].

LA disease: Lyme disease, when left untreated, may develop into LAand may persist for many years. Over the course of many months and years, Lyme disease bacteria enter the larger joints, especially the knees, that causes inflammation, followed by intermittent and persistent pain and swelling that may persist for many years [48, 49]. Apart from joint swelling, LA is also characterized by the infiltration of immune cell followed by the activation of the innate and adaptive immune system, the production of antibodies specific to the disease-causing antigens, synovial hypertrophy, and vascular proliferation [50, 51]. Many factors like the patient's age, previous and current medical history, pregnancy, or allergies decide the preferred choice of antibiotics and the route of administration. Initially, LA patients are administered oral antibiotics, however, when patients did not respond to oral antibiotics or when they become refractory to oral antibiotics, they were administered with the second regimen of parenteral antibiotics. Adults suspected of LA are initially suggested to prescribe doxycycline 100 mg twice per day orally for 28 days, amoxicillin 500 mg thrice per day orally for 28 days, or cefuroxime 500 mg twice daily for 28 days. For children with age eightyears or above, doxycycline at the dose of 4.4 mg/kg per day (maximum 100 mg per dose) divided into two doses orally for 28 days or amoxicillin 50 mg/kg per day (maximum 500 mg per dose) divided into three doses orally for 28 days, or cefuroxime 30 mg/kg per day (maximum 500 mg per dose) divided into two doses orally for 28 days are suggested. For children with age eightyears or below, amoxicillin 50 mg/kg per day (maximum 500 mg per dose) divided into three doses orally for 28 days or cefuroxime 30 mg/kg per day (maximum 500 mg per dose) divided into two doses orally for 28 days are suggested. When LA patients do not respond to the initial oral antibiotics, they are administered parenteral antibiotics as the second regimen. Adults are suggested to administer 2 g of ceftriaxone through *IV* once a day for 14–28 days, and children are suggested to be administered 50–75 mg/kg of ceftriaxone *IV* once a day for 14–28 with a maximum dose of 2 g per day [52].

Management of Post-Treatment Lyme Disease Syndrome Most of the patients recover from Lyme disease within 3–4 weeks of antibiotic treatment, however, nearly 2–40% of the patients continue to suffer from somatic as well as neurocognitive symptoms for a prolonged period. The symptoms include musculoskeletal and neurocognitive symptoms like fatigue, myalgia, arthralgia, sleep apnea, headache, and cognitive impairment which may persist for months to years. These symptoms are characterized by post-Lyme disease syndrome (PLDS) or post-treatment Lyme disease syndrome (PTLDS) [53, 54]. Currently, there is no FDA-approved treatment therapy for PTLDS, and the efficacy of the off-label or complementary therapies is not scientifically validated, and neither is their acceptance in the scientific community [53]. Treatment is based on a case-by-case basis where analgesics, anti-psychotics, anti-depressant, anti-convulsants, and alternative therapies are practiced [55].

Management of Co-Infections in Patients with Lyme Disease As per reports from a few New York state areas like Dutchess County and Hudson Valley, *Ixodes scapularis* (deer tick or black-legged tick), in addition to *B. burgdorferi*, can transmit four other pathogens: *A. phagocytophilum*, *B. microti*, *Borrelia miyamotoi*, and the deer tick virus subtype of Powassan virus [56]. Some previous research also reported a few Bartonella species such as *Yersinia enterocolitica*, *Chlamydophila pneumoniae*, *Chlamydia trachomatis*, and *Mycoplasma pneumoniae* to be the causative agents of co-infections in patients with Lyme disease [57]. Co-infecting agents can be transferred by tick bite alongside *B. burgdorferi*, resulting in multiple infections, or a subset of co-infections can arise independently from the tick bite. Co-infecting agents are reported to exacerbate the pathological conditions posed by Lyme disease by exacerbating the symptoms and may be responsible for symptoms similar to that of Lyme disease. These coinfections may pose a hindrance to accurate diagnosis and treatment as they may present nonspecific or overlapping symptoms. If not appropriately diagnosed by an experienced physician, the patients may receive insufficient treatment that may result in driving the disease to a chronic stage [58]. Third-generation cephalosporins are the preferred antibiotics for Lyme disease with co-infections, followed by some specific antibiotics targeting some specific agents responsible for transmitting co-infections. Carbapenems are used in infections caused by *Y. enterocolitica*, in combination with rifampicin for *B. henselae*, *C. trachomatis*, and *C. pneumoniae*, erythromycin for *Campylobacter jejuni*, and quinolones like gemifloxacin for alternative treatment [57]. Lyme disease, when associated with co-infections, makes the CLD/PTLDS more complicated and does not respond to the conventional antibiotic regimen. A recent study reported the effectiveness of dapsone combination therapy in CLD/PTLDS patients. A short-term (seven to eight weeks) high-dose pulsed dapsone combination therapy resulted in the remission of symptoms in almost 50% of patients within one year of treatment [59].

16.3.4 Novel Approaches to Lyme Disease Treatment

16.3.4.1 Vaccination

Lyme borreliosis cases are on the rise, especially in northern latitudes and high-altitude regions of Europe and North America [60, 61]. Early in the 1990s, two pharmaceutical companies developed Lyme disease vaccines, LYMErix (SmithKlineBeecham) and ImuLyme (PasteurMérieux-Connaught). Vaccines based on outer surface protein A (OspA) of B burgdoferi went on to phase III trials [62]. On 21 December 1998, the US Food and Drug Administration (FDA) approved LYMErix for use in people aged 15–70. The primary vaccination was administered in three doses, with the second booster dose being given 1 month and 12 months later. According to the phase III trial, 76% of those who finished the three-dose series prevented laboratory-confirmed Lyme borreliosis, and 100% prevented asymptomatic infection [63]. GlaxoSmithKline (formerly SmithKlineBeecham) voluntarily discontinued LYMErix distribution in February 2002 due to low sales, media attention, and concerns about possible side effects. Later, they revoked their license [64].

Multiple research investigations have indicated that the persistent autoimmune effects caused by Lyme disease are controlled by the interaction between the outer surface protein A (OspA) of *B. burgdorferi* and the human leukocyte function antigen (LFA) type 1 via cross-reactivity [65, 66]. Following the 1998 meeting of FDA's Advisory Committee on "Vaccines and Related Biological Products", scientists began devising plans to re-engineer recombinant OspA (rOspA) as a potential vaccine frontrunner. The goal was to alter the rOspA epitope that was discovered to be a prospective analog of human leukocyte function-associated antigen-1 (hLFA1) while maintaining the protein's immunogenicity and structural integrity [67].

The mv rOspA LB vaccine for Lyme borreliosis developed in Vienna, Austria, Baxter Healthcare. This vaccine comprises three recombinant OspA antigens and is formulated to protect against most of the *B. burgdorferi* sensu lato strains associated with human diseases worldwide [68]. Using information about protective OspA epitopes, researchers created a novel recombinant OspA molecule that merged the protective traits of both OspA antigens into a single entity. The Phase I/II study showed the candidate vaccine to induce significant antibody titers toward all targeted species of *Borrelia*, the causative agent of Lyme disease [69]. Following that, Valneva Austria GmbH used an identical strategy to create a composition that included only the modified C-terminus domains of six rOspA serotypes to create a subunit multivalent broadly protective vaccine (VLA15) [70, 71]. VLA15 is presently being tested in two Phase III clinical trials on 18,000 healthy volunteers aged 5 and up to determine its effectiveness, safety, tolerability, immunogenicity, and lot uniformity (http://ClinicalTrials.gov). Pfizer purchased VLA15 technology in April 2020, and the two businesses are working together to codevelop and launch a Lyme disease vaccine [67].

Although several vaccine candidates are being developed, no licensed Lyme disease vaccines are currently available in the United States. Lyme pre-exposure prophylaxis (Lyme PrEP) is a different strategy developed by the University of Massachusetts Medical School. Unlike a vaccine, Lyme PrEP delivers a single defensive antibody. The vaccine targets OspA, a protein found on the surface of the Lyme bacterium. The vaccine has been shown in preclinical studies to protect against multiple strains of the bacterium. A Phase I clinical trial of a preventive Lyme disease vaccine designed by MassBiologics at UMass Chan Medical School is approaching completion, and the next trial phase could start immediately.

Researchers are also looking at developing a vaccine that targets the tick-produced protein Subolesin, which is critical to the growth of Lyme. Subolesin (SUB), a tick antigen, has demonstrated potential as a vaccine in the management of numerous tick species in cattle. Furthermore, SUB-protective epitopes and peptide sequences linked with protein–protein interactions in this tick antigen have now been identified using new quantum vaccinomics approaches. Therefore, a novel approach for developing vaccines may require the identification of B-cell-reactive epitopes through epitope mapping using a subpeptide array. To assess the efficacy of these methods for identifying protective epitopes connected to vaccine protection and efficacy, Subolesin can be used as a model. The vaccine has shown promising results in preclinical studies [72].

It is critical to note that the process of creating vaccines is complicated, and any of these vaccines may not be approved for use for several years. Until then, precautions such as wearing safety garments, using repellents, and checking for ticks after being outside can help reduce the chance of developing Lyme disease.

16.3.4.2 Other Emerging Therapies

Stem Cell Therapy CLD is frequently attributed to the immune system's inability to eliminate the bacteria, resulting in chronic inflammation and tissue damage. In recent years, stem cell therapy has become an exciting new therapeutic approach for Lyme disease. Specifically, mesenchymal stem cells possess characteristics that can reduce inflammation and modulate the immune system, improving Lyme disease symptoms and accelerating the recovery process [59]. Shroff [73] used hESC therapy to regenerate damaged cells in two patients with multiple sclerosis (MS) and Lyme disease, injecting the hESCs via intramuscular, intravenous, and auxiliary routes (e.g. deep spinal, caudal, and intercostal via eye drops). Both patients' cognitive ability, physical strength, functional abilities, and general stamina improved significantly [74]. Horowitz and Freeman [72] reported the first successful treatment of Lyme disease-related common variable immunodeficiency with human embryonic stem cell therapy [75]. Stem cell-based therapies are particularly promising for patients who are resistant to antibiotics. MSCs can develop into supportive cells capable of repairing damaged nerve, muscle, immune, and other cells and tissue types [76]. The effectiveness of stem cell therapy in treating Lyme disease differs significantly according to the patient and the stage of the illness. DVC stem reports that 85% of their patients experienced greater quality of life improvements within three months of receiving stem cell therapy [77].

Despite being in the early stages of study, stem cell treatment for Lyme disease has demonstrated encouraging results in animal tests and clinical studies. Continued study in this area may open a new path for treating Lyme disease patients and, ultimately, help in finding a cure for this persistent condition.

Immunotherapy Low-dose immunotherapy (LDI) is a radically new treatment option that has been shown to provide significant symptom relief and better health outcomes for patients suffering from a variety of chronic illnesses, including Lyme disease [78]. LDI works by reducing inflammation brought on by foreign invaders or allergens, assisting the immune system in regaining proper balance. CLD typically correlates with a persistent and active infection, harm to various bodily systems including the endocrine and central nervous systems, continuous inflammation, and frequently, autoimmune disorders. In autoimmune conditions, the immune system mistakenly targets and attacks the patient's tissues, such as the joints, thyroid, or brain. LDI is a treatment method that combines particular antigens with an enzyme named beta-glucuronidase. This enzyme attracts T cells, a specialized white blood cell type that plays a crucial role in the immune response and signals them to pay attention to the included antigens. The enzyme subsequently triggers the downregulation or desensitization of TH2, the immune system component that was previously overreacting to the antigens. This process reduces TH2 activity, enhancing the weakened TH1 immunity, which can effectively combat intracellular infections (Low Dose Immunotherapy Bloomington MN | One Agora Health).

Repurposing Available Drugs

- **Disulfiram:** Disulfiram, also known as tetraethylthiuram disulfide, is a medication that the FDA has authorized for treating alcohol abuse. Disulfiram monotherapy has been found useful in the treatment of Lyme disease [63]. Three patients with chronic relapsing neurological Lyme disease and relapsing babesiosis, who were receiving open-ended antibiotic treatment to control symptoms, were finally able to halt treatment temporarily and stayed symptom-free for 6–23 months after finishing a limited course of treatment using only disulfiram, as reported in a case study [79]. Disulfiram is being repurposed as a Lyme antibiotic, and some patients have found it to be a game-changer in treating CLD. Although the exact mechanism of Disulfiram's action in treating Lyme disease is unknown, several hypotheses have been made regarding its mode of action. Some believe that Disulfiram may affect the flexibility and function of microbial surface membranes thereby eliciting its anti-microbial effect. *B. burgdorferi* requires zinc and manganese as cofactors for key biological functions [80–82]. Due to its high affinity for metal ions, Disulfiram may inhibit microbial metabolism [83]. Disulfiram monotherapy has been useful in treating persistent symptoms after initial Lyme disease treatment, but regular laboratory monitoring and close clinical follow-up are necessary. According to clinical trials, Disulfiram is deemed safe and well-tolerated and has been found to enhance fatigue levels and quality of life in patients experiencing persistent symptoms [84, 85].
- **Dapsone:** Dapsone, a sulfur-based antibiotic that has been conventionally used to treat leprosy, is emerging as a potential treatment option for Lyme disease. While antibiotics are typically prescribed as a primary treatment for Lyme disease, relapses are common, and ongoing symptoms may be attributed to the presence of an antibiotic-resistant form known as biofilms. Persisters, which are multi-drug tolerant cells found in substantial numbers in biofilms, have been used to emphasize the significance of *Borrelia* biofilms [73, 86] in postmortem tissues from a well-known Lyme disease patient [87]. Fend et al. have found in their study that among different sulfa drugs, dapsone was the most active against *B. burgdorferi* persisters" [88]. Dr. Horowitz, a pioneering researcher in the field of Lyme disease, has employed dapsone as a treatment option for the disease due to its anti-inflammatory properties, ability to target persister cells, and capacity to penetrate the central nervous system while enhancing patients' immune responses. Two separate retrospective case series involving a total of 300 patients have demonstrated that the use of Dapsone combination therapy (DDS CT) can significantly alleviate eight major Lyme disease symptoms and improve treatment outcomes in patients with CLD/PTLDS and concurrent co-infections that have failed to respond to conventional antibiotic therapy [89, 90]. In patients with CLD/PTLDS, a treatment known as high-dose pulsed dapsone combination therapy (DDDCT) can result in symptom remission for a minimum of a year in 50% of the cases. This therapy is also effective against resistant biofilm forms of *B. burgdorferi*. Patients experienced improvement in eight key Lyme symptoms, such as pain, fatigue, headaches, sweating, cognition, insomnia, and neuropathy, and the dosage of dapsone had a positive impact on the treatment's effectiveness. Although DDDCT shows potential as a new treatment for resistant CLD/PTLDS, clinical trials are necessary to confirm this [91].

Overall, these trends reflect a growing understanding of the complexities of Lyme disease and the need for a multifaceted approach to management, including prevention, early diagnosis, and effective treatment of both Lyme disease and co-infections.

16.4 Future Directions and Challenges in Lyme Disease Management

The management of Lyme disease is a rapidly evolving field, and there are several areas of ongoing research and future directions that are worth discussing. Development of a Lyme disease vaccine – There have been previous vaccines for Lyme disease, but they were discontinued due to low demand and concerns about side effects. However, there is currently research underway to develop a new vaccine that could be more effective and safer. While current diagnostic tests have improved, there is still room for improvement in terms of accuracy and reliability, particularly in detecting early-stage Lyme disease. New tests and diagnostic approaches are being developed to address this. There is still much to be learned about the behavior and life cycle of *B. burgdorferi*, including how it interacts with the immune system and how it persists in the body despite antibiotic treatment. A better understanding of these factors could lead to more effective treatments. Development of personalized treatment plans – Lyme disease can present in many different ways and can affect different people in different ways. Developing personalized treatment plans that take into account factors such as age, co-infections, and immune status could improve outcomes for patients. While awareness of Lyme disease has increased in recent years, there is still a need for more education and awareness campaigns to help people understand the risks and take appropriate preventative measures. Additionally, efforts to reduce the spread of tick populations and to improve tick control measures could help to reduce the incidence of Lyme disease. In summary, the future direction of Lyme disease management is focused on improving prevention, diagnosis, and treatment options, as well as better understanding the biology of the bacteria and the disease. With continued research and public education efforts, it is hoped that the burden of Lyme disease can be reduced in the coming years.

16.5 Conclusion

Lyme disease is a tick-borne illness caused by the bacterium *B. burgdorferi*, which can cause a range of symptoms including fever, rash, and joint pain. In recent years, there have been several notable trends in Lyme disease management. With the rise in Lyme disease cases in many parts of the world, there has been a concerted effort to increase public awareness about tick-borne illnesses and prevention measures, such as wearing long clothing, using tick repellent, and performing tick checks after spending time outdoors. There have been significant advances in the development of diagnostic tests for Lyme disease, including more sensitive and specific blood tests and point-of-care tests that can provide rapid results. In addition to the standard course of antibiotics, new treatment options have emerged for Lyme disease, including longer courses of antibiotics, immunotherapy, and herbal remedies. However, the effectiveness of these treatments is still being studied. Lyme disease is often accompanied by other tick-borne illnesses, known as co-infections. There is increasing recognition of the importance of testing for and treating these co-infections, which can affect the severity and duration of Lyme disease symptoms. There is ongoing research into the potential long-term effects of Lyme disease, including chronic symptoms such as fatigue, joint pain, and neurological issues. This has led to a greater understanding of the need for early diagnosis and treatment to prevent these long-term effects. Overall, the chapter emphasizes the importance of early diagnosis and treatment of Lyme disease and the need for a multidisciplinary approach to managing the disease and its associated symptoms.

References

1 Mead, P.S. (2015). Epidemiology of Lyme disease. *Infect. Dis. Clin. N. Am.* 29 (2): 187–210. https://doi.org/10.1016/j.idc.2015.02.010.

2 Centers for Disease Control and Prevention. (2021). Tickborne Diseases of the United States: A Reference Manual for Health Care Providers. 5th ed. https://www.cdc.gov/ticks/tickbornediseases/tickID.html

3 Steere, A.C. (2001). Lyme disease. *N. Engl. J. Med.* 345 (2): 115–125. https://doi.org/10.1056/nejm200107123450207.

4 Centers for Disease Control and Prevention. (2021). Signs and Symptoms of Lyme Disease. https://www.cdc.gov/lyme/signs_symptoms/index.html

5 Wormser, G.P., Dattwyler, R.J., Shapiro, E.D. et al. (2006). The clinical assessment, treatment, and prevention of Lyme disease, human granulocytic anaplasmosis, and babesiosis: clinical practice guidelines by the Infectious Diseases Society of America. *Clin. Infect. Dis.* 43 (9): 1089–1134. https://doi.org/10.1086/508667.

6 Sanchez, E., Vannier, E., Wormser, G.P., and Hu, L.T. (2016). Diagnosis, treatment, and prevention of lyme disease, human granulocytic anaplasmosis, and babesiosis: a review. *JAMA* 315 (16): 1767–1777. https://doi.org/10.1001/jama.2016.2884.

7 Steere, A.C., Strle, F., Wormser, G.P. et al. (2016). Lyme borreliosis. *Nat. Rev. Dis. Primers* 2: 16090. https://doi.org/10.1038/nrdp.2016.90.

8 McGuire, B.S., Chandler, F.W., Felz, M.W. et al. (1992). Detection of *Borrelia burgdorferi* in human blood and urine using the polymerase chain reaction. *Pathobiology* 60 (3): 163–167. https://doi.org/10.1159/000163717.

9 Hinckley, A.F., Connally, N.P., Meek, J.I. et al. (2014). Lyme disease testing by large commercial laboratories in the United States. *Clin. Infect. Dis.* 59 (5): 676–681. https://doi.org/10.1093/cid/ciu393.

10 Johnson, L., Aylward, A., and Stricker, R.B. (2013). Healthcare access and burden of care for patients with Lyme disease: a large United States survey. *Health Policy* 112 (1–2): 131–138. https://doi.org/10.1016/j.healthpol.2013.06.015.

11 Liebling, M.R., Nishio, M.J., Rodriguez, A. et al. (1993). The polymerase chain reaction for the detection of Borrelia burgdorferi in human body fluids. *Arthritis and Rheumatism* 36 (5): 665–675. https://doi.org/10.1002/art.1780360514.

12 Kuehn, B.M. (2013). CDC estimates 300,000 US cases of Lyme disease annually. *JAMA* 310 (11): 1110. https://doi.org/10.1001/jama.2013.278331 [PubMed] [CrossRef] [Google Scholar].

13 Henry, B., Roth, D., Reilly, R. et al. (2011). How big is the Lyme problem? Using novel methods to estimate the true number of Lyme disease cases in British Columbia residents from 1997 to 2008. *Vector Borne Zoonotic Dis.* 11 (7): 863–868. https://doi.org/10.1089/vbz.2010.0142 [PubMed] [CrossRef] [Google Scholar].

14 Owens, B. (2013). Risk of tick-borne infections on the rise. *CMAJ* 185 (15): E711. https://doi.org/10.1503/cmaj.109-4593 [PMC free article] [PubMed] [CrossRef] [Google Scholar].

15 Ogden, N.H., Koffi, J.K., Pelcat, Y., and Lindsay, L.R. (2014). Environmental risk from Lyme disease in central and eastern Canada: a summary of recent surveillance information. *Can. Commun. Dis. Rep.* 40 (5): 74–83. [PMC free article] [PubMed] [Google Scholar].

16 Ogden, N.H., Lindsay, L.R., Morshed, M. et al. (2009). The emergence of Lyme disease in Canada. *CMAJ* 180 (12): 1221–1224. https://doi.org/10.1503/cmaj.080148 [PMC free article] [PubMed] [CrossRef] [Google Scholar].

17 Lindsay, L.R., Bernat, K., and Dibernardo, A. (2014). Laboratory diagnostics for Lyme disease. *Can. Commun. Dis. Rep.* 40: 209–217. [PMC free article] [PubMed] [Google Scholar].

18 Huegli, D., Moret, J., Rais, O. et al. (2011). Prospective study on the incidence of infection by *Borrelia burgdorferi* sensu lato after a tick bite in a highly endemic area of Switzerland. *Ticks Tick Borne Dis.* 2 (3): 129–136. https://doi.org/10.1016/j.ttbdis.2011.05.002 [PubMed] [CrossRef] [Google Scholar].

19 Costello, C.M., Steere, A.C., Pinkerton, R.E., and Feder, H.M. Jr. (1989). A prospective study of tick bites in an endemic area for Lyme disease. *J. Infect. Dis.* 159 (1): 136–139. https://doi.org/10.1093/infdis/159.1.136 [PubMed] [CrossRef] [Google Scholar].

20 Canada Centre for Occupational Health and Safety (2016). Lyme disease: a warm weather hazard. *Health Saf. Rep.* 12: https://www.ccohs.ca/newsletters/hsreport/issues/2014/07/ezine.html#hsreport-tipstools (accessed 4 January 2020) [Google Scholar].

21 Cisak, E., Zajac, V., Wojcik-Fatla, A., and Dutkiewicz, J. (2012). Risk of tick-borne diseases in various categories of employment among forestry workers in eastern Poland. *Ann. Agric. Environ. Med.* 19: 469–474. [PubMed] [Google Scholar].

22 Kurnatowski, P., Warpechowska, M., and Kurnatowska, A.J. (2011). Knowledge on Lyme disease among foresters. *Int. J. Occup. Med. Environ. Health* 24: 78–93. https://doi.org/10.2478/s13382-011-0001-0. [PubMed] [CrossRef] [Google Scholar].

23 Thorin, C., Rigaud, E., Capek, I. et al. (2008). Séroprévalence de la borréliose de Lyme et de l'encéphalite à tiques chez des professionnels exposés dans le Grand Est de la France. *Med. Mal. Infect.* 38: 533–542. https://doi.org/10.1016/j.medmal.2008.06.008. [PubMed] [CrossRef] [Google Scholar].

24 Valente, S.L., Wemple, D., Ramos, S. et al. (2015). Preventive behaviors and knowledge of tick-borne illnesses: results of a survey from an endemic area. *J. Public Health Manag. Pract.* 21: E16–E23. https://doi.org/10.1097/PHH.0000000000000098. [PubMed] [CrossRef] [Google Scholar].

25 Abbott, A. (2006). Lyme disease: uphill struggle. *Nature* 439: 524–525. [PubMed] [Google Scholar].

26 Adelson, M.E., Rao, R.V.S., Tilton, R.C. et al. (2004). Prevalence of *Borrelia burgdorferi*, *Bartonella* spp., *Babesia microti*, and *Anaplasma phagocytophila* in *Ixodes scapularis* ticks collected in Northern New Jersey. *J. Clin. Microbiol.* 42 (6): 2799–2801. [PMC free article] [PubMed] [Google Scholar].

27 Ajzen, I. (1991). The theory of planned behaviour. *Organ. Behav. Hum. Decis. Process.* 50: 179–211. [Google Scholar].

28 Ajzen, I. (2002). Perceived behavioural control, self-efficacy, locus of control, and the theory of planned behaviour. *J. Appl. Soc. Psychol.* 32: 665–683. [Google Scholar].

29 Allan, B., Keesing, F., and Ostfield, R. (2003). Effect of forest fragmentation on Lyme disease risk. *Conserv. Biol.* 17 (1): 267–272. [Google Scholar].

30 Barbour, A., Maupin, G., Teltow, G. et al. (1996). Identification of an uncultivable Borrelia species in the hard tick *Amblyomma americanum*: possible agent of a Lyme disease-like illness. *J. Infect. Dis.* 173 (2): 403–409. [PubMed] [Google Scholar].

31 Bartosik, K., Sitarz, M., Szymańska, J., and Buczek, A. (2011). Tick-bites on humans in the agricultural and recreational areas in south-eastern Poland. *Ann. Agric. Environ. Med.* 18 (1): 151–157. [PubMed] [Google Scholar].

32 Bayles, B., Evans, G., and Allan, B. (2013). Knowledge and prevention of tick-borne diseases vary across an urban-to-rural human land-use gradient. *Ticks Tick Borne Dis.* 4 (4): 352–358. [PubMed] [Google Scholar].

33 Schoen, R.T. (2020). Lyme disease: diagnosis and treatment. *Curr. Opin. Rheumatol.* 32 (3): 247–254.

34 Centre for Disease Control and Prevention. 2022. Erythema Migrans Rash. https://www.cdc.gov/lyme/treatment/erythema-migrans-rash.html (accessed 20 March 2023).

35 Smith, G.N., Gemmill, I., and Moore, K.M. (2012). Management of tick bites and Lyme disease during pregnancy. *J. Obstet. Gynaecol. Can.* 34 (11): 1087–1091.

36 Franklin, T.J. (1963). The inhibition of incorporation of leucine into protein of cell-free systems from rat liver and *Escherichia coli* by chlortetracycline. *Biochem. J.* 87 (3): 449.

37 Bernardino, A.L., Kaushal, D., and Philipp, M.T. (2009). The antibiotics doxycycline and minocycline inhibit the inflammatory responses to the Lyme disease spirochete *Borrelia burgdorferi*. *J. Infect. Dis.* 199 (9): 1379–1388.

38 Centre for Disease Control and Prevention. 2022. Neurologic Lyme Disease. https://www.cdc.gov/lyme/treatment/NeurologicLyme.html#:~:text=What%20is%20neurologic%20Lyme%20disease,both%20sides%20of%20the%20face. (accessed 20 March 2023).

39 Lantos, P.M. (2015). Chronic Lyme disease. *Infect. Dis. Clin.* 29 (2): 325–340.

40 Wilking, H. and Stark, K. (2014). Trends in surveillance data of human Lyme borreliosis from six federal states in eastern Germany, 2009–2012. *Ticks Tick Borne Dis.* 5 (3): 219–224.

41 Berglund, J., Eitrem, R., Ornstein, K. et al. (1995). An epidemiologic study of Lyme disease in southern Sweden. *N. Engl. J. Med.* 333 (20): 1319–1324.

42 Kortela, E., Kanerva, M.J., Puustinen, J. et al. (2021). Oral doxycycline compared to intravenous ceftriaxone in the treatment of Lyme neuroborreliosis: a multicenter, equivalence, randomized, open-label trial. *Clin. Infect. Dis.* 72 (8): 1323–1331.

43 Nigrovic, L.E., Chun, T.H., Vargas, S.E. et al. (2023). Comparative effectiveness and complications of intravenous ceftriaxone compared with oral doxycycline in Lyme meningitis in children: a multicentre prospective cohort study. *BMJ Open* 13 (2): e071141.

44 Fish, A.E., Pride, Y.B., and Pinto, D.S. (2008). Lyme carditis. *Infect. Dis. Clin. N. Am.* 22 (2): 275–288.

45 Robinson, M.L., Kobayashi, T., Higgins, Y. et al. (2015). Lyme carditis. *Infect. Dis. Clin.* 29 (2): 255–268.

46 Centre for Disease Control and Prevention. 2022. Lyme Carditis. https://www.cdc.gov/lyme/treatment/lymecarditis.html (accessed 20 March 2023).

47 Radesich, C., Del Mestre, E., Medo, K. et al. (2022). Lyme carditis: from pathophysiology to clinical management. *Pathogens* 11 (5): 582.

48 Steere, A.C., Schoen, R.T., and Taylor, E. (1987). The clinical evolution of Lyme arthritis. *Ann. Intern. Med.* 107 (5): 725–731.

49 Miller, J.B. and Aucott, J.N. (2021). Stages of Lyme arthritis. *J. Clin. Rheumatol.* 27 (8): e540–e546.

50 Codolo, G., Bossi, F., Durigutto, P. et al. (2013). Orchestration of inflammation and adaptive immunity in *Borrelia burgdorferi*–induced arthritis by neutrophil-activating protein A. *Arthritis Rheum.* 65 (5): 1232–1242.

51 Arvikar, S.L. and Steere, A.C. (2015). Diagnosis and treatment of Lyme arthritis. *Infect. Dis. Clin.* 29 (2): 269–280.

52 Centre for Disease Control and Prevention. 2022. Lyme Arthritis. https://www.cdc.gov/lyme/treatment/LymeArthritis.html (accessed 25 March 2023).

53 Ścieszka, J., Dąbek, J., and Cieślik, P. (2015). Post-Lyme disease syndrome. *Reumatologia* 53 (1): 46–48.

54 Nemeth, J., Bernasconi, E., Heininger, U. et al. (2016). Update of the Swiss guidelines on post-treatment Lyme disease syndrome. *Swiss Med. Wkly.* 146 (4950): w14353.

55 Rebman, A.W. and Aucott, J.N. (2020). Post-treatment Lyme disease as a model for persistent symptoms in Lyme disease. *Front. Med.* 7: 57.

56 Wormser, G.P., McKenna, D., Scavarda, C. et al. (2019). Co-infections in persons with early Lyme disease, New York, USA. *Emerg. Infect. Dis.* 25 (4): 748.

57 Berghoff, W. (2012). Suppl 1: chronic Lyme disease and co-infections: differential diagnosis. *Open Neurol. J.* 6: 158.

58 Caulfield, A.J. and Pritt, B.S. (2015). Lyme disease coinfections in the United States. *Clin. Lab. Med.* 35 (4): 827–846.

59 Horowitz, R.I. and Freeman, P.R. (2022). Efficacy of short-term high dose pulsed dapsone combination therapy in the treatment of chronic Lyme disease/post-treatment Lyme disease syndrome (PTLDS) and associated co-infections: a report of three cases and literature review. *Antibiotics* 11 (7): 912.

60 ClinicalTrials.gov An Efficacy, Safety, Tolerability, Immunogenicity, and Lot-Consistency Clinical Trial of a 6-Valent OspA-Based Lyme Disease Vaccine (VLA15) - Ongoing clinical trial. https://clinicaltrials.gov/ct2/show/NCT05477524.

61 Barth, K. and Malcolm, R. (2012). Disulfiram: an old therapeutic with new applications. *CNS Neurol. Disord. Drug Targets* 9: 5–12. https://doi.org/10.2174/187152710790966678.

62 Comstedt, P., Schüler, W., Meinke, A., and Lundberg, U. (2017). The novel Lyme borreliosis vaccine VLA15 shows broad protection against Borrelia species expressing six different OspA serotypes. *PLoS One* 12: e0184357. https://doi.org/10.1371/journal.pone.0184357.

63 Llen, A., Teere, C.S., Ikand, I.K.S. et al. (1998). Vaccination against Lyme disease with recombinant *Borrelia burgdorferi* outer-surface lipoprotein A with adjuvant. *N. Engl. J. Med.* 339: 209–215. https://doi.org/10.1056/NEJM199807233390401.

64 Contreras, M., Kasaija, P.D., Kabi, F. et al. (2022). The correlation between subolesin-reactive epitopes and vaccine efficacy. *Vaccines* 10: 1327. https://doi.org/10.3390/VACCINES10081327/S1.

65 Custodio, M.M., Sparks, J., and Long, T.E. (2022). Disulfiram: a repurposed drug in preclinical and clinical development for the treatment of infectious diseases. *Antiinfect. Agents* 20: e040122199856. https://doi.org/10.2174/2211352520666220104104747.

66 Dattwyler, R.J. and Gomes-Solecki, M. (2022, 2022). The year that shaped the outcome of the OspA vaccine for human Lyme disease. *npj Vaccines* 71 (7): 1–5. https://doi.org/10.1038/s41541-022-00429-5.

67 Feng, J., Auwaerter, P.G., and Zhang, Y. (2015). Drug combinations against *Borrelia burgdorferi* persisters in vitro: eradication achieved by using daptomycin, cefoperazone and doxycycline. *PLoS One* 10: e0117207. https://doi.org/10.1371/journal.pone.0117207.

68 Feng, J., Zhang, S., Shi, W., and Zhang, Y. (2017). Activity of sulfa drugs and their combinations against stationary phase b. Burgdorferi in vitro. *Antibiotics* 6: 10. https://doi.org/10.3390/antibiotics6010010.

69 Gao, J., Gong, Z., Montesano, D. et al. (2020). “Repurposing” disulfiram in the treatment of Lyme disease and Babesiosis: retrospective review of first 3 Years’ experience in one medical practice. *Antibiotics* 9: 868. https://doi.org/10.3390/antibiotics9120868.

70 Goren, A., Viljugrein, H., Rivrud, I.M. et al. (2023). The emergence and shift in seasonality of Lyme borreliosis in northern Europe. *Proc. R. Soc. B Biol. Sci.* 290: 20222420. https://doi.org/10.1098/RSPB.2022.2420/.

71 Hitt, E. (2002). Poor sales trigger vaccine withdrawal. *Nat. Med.* 8: 311–312. https://doi.org/10.1038/nm0402-311b.

72 Horowitz, R. and Freeman, P.R. (2018). Improvement of common variable immunodeficiency using embryonic stem cell therapy in a patient with Lyme disease: a clinical case report. *Clin. Case Rep.* 6: 1166. https://doi.org/10.1002/CCR3.1556.

73 Shroff, G. (2016). Transplantation of human embryonic stem cells in patients with multiple sclerosis and Lyme disease. *Am. J. Case Rep.* 17: 944–949. https://doi.org/10.12659/AJCR.899745.

74 Horowitz, R.I., Murali, K., Gaur, G. et al. (2020). Effect of dapsone alone and in combination with intracellular antibiotics against the biofilm form of *B. burgdorferi*. *BMC Res. Notes* 13: 1–7. https://doi.org/10.1186/S13104-020-05298-6/TABLES/2.

75 Hsieh, Y.F., Liu, H.W., Hsu, T.C. et al. (2007). Serum reactivity against *Borrelia burgdorferi* OspA in patients with rheumatoid arthritis. *Clin. Vaccine Immunol.* 14: 1437–1441. https://doi.org/10.1128/CVI.00151-07.

76 Horowitz, R.I. and Freeman, P.R. (2016). The use of dapsone as a novel “persister” drug in the treatment of chronic Lyme disease/post treatment Lyme disease syndrome. *J. Clin. Exp. Dermatol. Res.* 7: 1–7. https://doi.org/10.4172/2155-9554.1000345.

77 Koide, S., Yang, X., Huang, X. et al. (2005). Structure-based design of a second-generation Lyme disease vaccine based on a C-terminal fragment of *Borrelia burgdorferi* OspA. *J. Mol. Biol.* 350: 290–299. https://doi.org/10.1016/J.JMB.2005.04.066.

78 Liegner, K.B. (2019). Disulfiram (tetraethylthiuram disulfide) in the treatment of Lyme disease and babesiosis: report of experience in three cases. *Antibiotics* 8: 72. https://doi.org/10.3390/antibiotics8020072.

79 One Agora Health *Low Dose Immunotherapy*. Bloomington, MN: One Agora Health https://www.oneagorahealth.com/immunotherapy-ldi-lda.html.

80 MassBiologics Research into Preventive Shot for Lyme Disease Continues to Move Forward. https://www.umassmed.edu/news/news-archives/2022/05/massbiologics-research-into-preventive-shot-for-lyme-disease-continues-to-move-forward

81 Melia, M.T. and Auwaerter, P.G. (2016). Time for a different approach to Lyme disease and long-term symptoms. *N. Engl. J. Med.* 374: 1277–1278. https://doi.org/10.1056/NEJME1502350.

82 Rizzoli, A., Hauffe, H.C., Carpi, G. et al. (2011). Lyme borreliosis in Europe. *Eurosurveillance* 16: 19906. https://doi.org/10.2807/ESE.16.27.19906-EN/CITE/PLAINTEXT.

83 Sapi, E., Kasliwala, R.S., Ismail, H. et al. (2019). The long-term persistence of *Borrelia burgdorferi* antigens and DNA in the tissues of a patient with Lyme disease. *Antibiotics* 8: 183. https://doi.org/10.3390/antibiotics8040183.

84 Shahani, L., Singh, S., and Khardori, N.M. (2012). Immunotherapy in clinical medicine: historical perspective and current status. *Med. Clin. North Am.* 96: 421–431. https://doi.org/10.1016/j.mcna.2012.04.001.

85 Sharma, A., Chakraborty, A., and Jaganathan, B.G. (2021). Review of the potential of mesenchymal stem cells for the treatment of infectious diseases. *World J. Stem Cells* 13: 568–593. https://doi.org/10.4252/wjsc.v13.i6.568.

86 Shen, A.K., Mead, P.S., and Beard, C.B. (2011). The Lyme disease vaccine – a public health perspective. *Clin. Infect. Dis.* 52: s247–s252. https://doi.org/10.1093/CID/CIQ115.

87 Steere, A.C. and Livey, I. (2013). Lyme disease vaccines. In: *Vaccines*, 6the (ed. S.A. Plotkin, W.A. Orenstein, and P.A. Offit), 1122–1132. Elsevier Inc 10.1016/B978-1-4557-0090-5.00055-0.

88 Van Den Berk, L.C.J., Jansen, B.J.H., Snowden, S. et al. (2014). Cord blood mesenchymal stem cells suppress DC-T cell proliferation via prostaglandin B2. *Stem Cells Dev.* 23: 1582–1593. https://doi.org/10.1089/scd.2013.0433.

89 Wagh, D., Pothineni, V.R., Inayathullah, M. et al. (2015). Borreliacidal activity of Borrelia metal transporter A (BmtA) binding small molecules by manganese transport inhibition. *Drug Des. Devel. Ther.* 9: 805–816. https://doi.org/10.2147/DDDT.S77063.

90 Willett, T.A., Meyer, A.L., Brown, E.L., and Huber, B.T. (2004). An effective second-generation outer surface protein A-derived Lyme vaccine that eliminates a potentially autoreactive T cell epitope. *Proc. Natl. Acad. Sci. U. S. A.* 101 (5): 1303–1308.

91 Wressnigg, N., Noel Barrett, P., Pöllabauer, E.M. et al. (2014). A novel multivalent OspA vaccine against Lyme Borreliosis is safe and immunogenic in an adult population previously infected with *Borrelia burgdorferi* Sensu Lato. *Clin. Vaccine Immunol.* 21: 1490. https://doi.org/10.1128/CVI.00406-14.

17

Chagas Disease: Historical and Current Trends

Vivek Patel[1], Dhara Patel[2], Grishma Patel[2], and Jayvadan K. Patel[3]

[1] Sun Pharmaceutical Industries Ltd., Vadodara, Gujarat, India
[2] Pioneer Pharmacy College, Vadodara, Gujarat, India
[3] Aavis Pharmaceuticals, Hoschton, GA, United States

17.1 Introduction

Chagas disease, also known as American trypanosomiasis caused by the protozoan flagellate *Trypanosoma cruzi* (*T. cruzi*), is an endemic illness that can be fatal and initially was identified in 1909 by Brazilian physician Carlos Ribeiro Justiniano Chagas (1879–1934) [1]. *T. cruzi* is a parasite that affects an estimated 6–7 million individuals globally and is responsible for about 10,000 disease-related fatalities, 30,000–40,000 new cases each year, and 8600 newborns become infected during pregnancies. Around 75 million additional people worldwide reside in places where they are susceptible to infection [2]. According to WHO, it was one among the world's top 13 most neglected tropical diseases (NTDs) in 2005 [3]. Despite being identified more than 110years ago, Chagas disease is still one of the public health concerns for 21 countries in continental Latin America [4]. Due to population movement, urbanization, and emigration, the epidemiological spectrum of the disease has shifted in the previous 10 years from a rural to an urban illness. Due to the ongoing migration of individuals from endemic locations, Chagas disease has become an alarm for non-endemic areas such as Japan, Canada, Europe, the United States, and Australia [5]. It is usually referred to as a "silent and silenced disease" because the majority of persons who are affected show no symptoms or just very minor signs [6]. Despite advancements, there are still a number of issues with dealing with this illness: in addition to the absence of a vaccine, it is thought that less than 10% of those affected by Chagas disease worldwide receive a diagnosis, and only about 1% have access to specialized care [7]. The currently available treatments, Benznidazole and Nifurtimox (the latter being used on a smaller scale), have questionable efficacy and demand constant observation. Furthermore, the need for long-term administration of these compounds results in a number of side effects that may necessitate early treatment discontinuation. These compounds also have significant manufacturing costs and are exceedingly hazardous [8, 9]. When medical attention is given right away after an infection, the disease can be cured. Chagas disease, which has a propensity for digestive clinical symptoms, thrombotic vascular accidents, and neurological sequelae, can cause arrhythmias and dilated cardiomyopathy, which can result in sudden death or heart failure if undetected and untreated in its chronic phase. The most significant mortality and morbidity among parasite infections, chronic Chagas disease, is a disabling disorder that costs USD $627.5 million in medical expenses worldwide each year [8]. The cost per patient is estimated to be $200 in the early stages of the infection, but it may rise to $400–$6000 throughout the chronic symptomatic phase [10]. Taking into account the fact that Chagas disease is now impacting non-endemic nations as well, in order to increase public awareness of the disease, the seventy-second World Health Assembly decided in May 2019 to establish World Chagas Disease Day, which will be observed on 14 April (the day in 1909 when Carlos Chagas diagnosed the first human case of the illness, a two-year-old girl named Berenice). In order to meet the targets for eliminating Chagas disease by 2030, the road map for neglected tropical diseases 2021–2030 lists primary goals; such as confirming the cessation of vectorial domestic transmission; interruption of blood transfusions transmission and organ transplants; eradicating the disease congenital form; and ensuring that 75% of the eligible population receives antiparasitic treatment [11].

Rising Contagious Diseases: Basics, Management, and Treatments, First Edition. Edited by Seth Kwabena Amponsah, Ranjita Shegokar, and Yashwant V. Pathak.

17.2 Discovery of Chagas Disease

The name "Chagas disease" was given in honor of the disease's discoverer, Carlos Ribeiro Justiniano Chagas, who was born on a coffee estate in Oliveira, Minas Gerais, Brazil, on 9 July 1878. In 1903, Chagas received his medical degree [12]. Oswaldo Cruz then invited him to serve as a doctor at the ministry of Public Health and Hygiene in Brazil, where he initially implemented the intra household vector control against malaria [13]. As a result of the success of his work, Chagas joined the National Academy of Medicine of Brazil and was given a number of honors and titles by organizations in Paris, United States, Belgium, and Lima, including the title of Doctor Honoris Causa from Harvard University. In order to safeguard the lives of those who worked on the construction of the central do Brasil railway, Carlos Chagas was proposed in 1907 to battle paludism in the Rio das Velhas river basin. Chagas discovered a hematophagous bug (*Panstrongylus megistus*) carrying a brand-new trypanosomatide, which he named *T. cruzi* during his field research investigation. The same parasite was discovered in the blood of two-year-old Berenice on 14 April 1909, who also had a high fever, hepatosplenomegaly, and face edema. Chagas reported his findings to the scholarly journal Brasil Medico. Throughout her entire life, Berenice experienced no symptoms, and she passed away at the age of 73 due to other medical conditions. Berenice was included in a number of clinical research of Chagas disease from the age of 55–71 years old [14].

The study of Chagas disease in Latin America then intensified, with the first cases being reported in El Salvador (1913) [15], Peru [16] and Venezuela [17] (1919), Costa Rica (1922) [18], Guatemala (1924) [19], Mexico (1938) [20], Chile (1937) [21], Bolivia (1942) [22], Colombia (1947) [23], Argentina and Nicaragua (1949) [24], and Honduras (1960) [19]. Recent discoveries of *T. cruzi* DNA in mummies from Brazil and Peru/Chile, which date to 7050 and 2500–5000 years BCE, respectively, show that the illness has been present in Latin America for more than 9000 years [25, 26]. The history of Chagas disease, according to Lannes-Vieira et al. can be split into three significant phases. The outstanding work of Chagas and the debate over how to define and accept the disease as a reality of science and a social issue serve to describe the first stage, which lasted from 1909 until 1934. The second stage, which spans the years 1935–1960 and occurs after Chagas's death, was marked by the discovery of the acute form of the illness in Argentina by Mazza and Romana and the endemic and chronic nature of the disease by Evandro Chagas and Emmanuel Dias. Last, but not the least, the third stage, which spans from 1961 to the present, poses a challenge for Chagas disease management and comprehension in a variety of ways, making it necessary to continuously put national and international health policies into practice [27]. Nifurtimox (NFX), the initial medication to fight Chagas disease, was not developed until 1960. In 1972, the second and final medication to be developed was Benznidazole (BZN). Even though Chagas disease is listed by the WHO as the third most widely disseminated infectious illness in Latin America, it is also the most common NTD that causes and promotes poverty. Only these two nitro heterocyclic medications, created more than 50 years ago, are currently approved treatments. Serious downsides include protracted treatment times, harmful side effects, and frequently unsuccessful treatments [28, 29]. There have been numerous attempts in recent decades to develop novel medicines; efficient treatment or vaccines have proven challenging due to the lack of understanding the biology of the infection, the intricate pathology, and the persistent nature of Chagas disease. Table 17.1 displays the significant turning point in Chagas disease history.

Table 17.1 Important landmark in the history of Chagas disease.

Anos	Landmark
1909	Discovery of Chagas Disease (Carlos Chagas)
1922	Cardiac form description by Chagas and Villela
1942	Cardiac form modifications by Laranja and Associates
1943	Study Centre in Bambui, MG
1968–1985	Research into the autonomic nerve system
1993–1996	Chronic parasite (Higuchi, Bocchi, and Belloti)
2012	Chronic Chagas heart disease and cellular therapy
2013–2015	London Declaration on the genome and TcKAP7 protein demonstration
2015	Benefits of using Benznidazole for chronic Chagas heart illness
2018	Creating scenarios for the peninsula of Yucatán
2020 to till date	Aspects of new medications and education

17.3 Etiology of Chagas Disease

Chagas disease is a vector-borne illness that affects humans, more than 150 different domestic animals species, such as cats, dogs, and guinea pigs, as well as wild mammals, such as rats, armadillos, and marsupials. It is mostly transmitted by huge, blood-sucking, or kissing reduviid bugs of the subfamily triatominae that occur during three overlapping cycles: domestic, peridomestic, and sylvatic [30]. There are more than 130 different species of triatominae bugs, but a few of them are capable of transmitting *T. cruzi*. The three vector species that play the most significant roles in transmitting *T. cruzi* to humans are *Triatoma infestans, Triatoma dimidiate*, and *Rhodnius prolixus. T. infestans* has long been the most important vector, serving as the main vector in sub-Amazonian endemic regions of southern South America. *R. prolixus* is commonly found in northern South and Central America, and *T. dimidiate* also occurs there but extends further north into Mexico. The blood meal is taken at night by reduviid bugs of both sexes, although female bugs are only able to lay eggs after blood meals. Reduviid bugs discharge parasites in their feces and urine near mucous surfaces, most commonly the mouth or eyes, as they consume the food or even afterward. Following that, the parasite enters via a bite wound or mucosal membrane. The most prevalent source of the contagious trypomastigotes form is the feces of infected reduviid bugs. Chagas disease can also spread by transfusions of blood, bone marrow or organ transplants, vertical transmission from mother to child that results in congenital Chagas disease, consuming tainted food or drink, and accidental exposure. The main risk factors for contracting the disease are thought to be ignorance of the condition, poor hygiene practices, living with animals, and poor housing conditions. Less than 10–20% of recipients of one unit of blood from infected donors may get Chagas disease; this risk depends on a variety of factors, including the blood component transfused, amount of parasites in the donor's blood, and possibly the parasite strain. The likelihood of transmitting Chagas disease after receiving one unit of blood from an infected donor is less than 10–20%, and it relies on a number of variables, including the blood component transfused, the parasite concentration in the donor's blood, and maybe the parasite strain [31]. Figure 17.1 depicts the several routes of transmission for the Chagas disease.

17.4 Lifecycle of *T. cruzi*

Triatomine insects (the invertebrate vector) and humans (the vertebrate host) are the two intermediate hosts in *T. cruzi's* classical life cycle, which also includes three developmental stages called trypomastigotes, amastigotes, and epimastigotes. Figure 17.2 depicts the overall perspective of *T. cruzi's* life cycle. *Triatoma infestans*, the vector, consumes blood and excretes metacyclic trypomastigotes. Trypomastigotes enter bite wounds and circulate through the blood. Later, they can infect different cell types and change into amastigotes, which reproduce via binary fission. Finally, the amastigote nest-containing cells detonate, and the parasites transform back into trypomastigotes, continuing the cycle in the vertebrate host. A triatomine bug eventually consumes fresh blood, and the parasites change into epimastigotes in the midgut where they proliferate through binary fission. In the gastrointestinal tract, they change into infectious metacyclics [32].

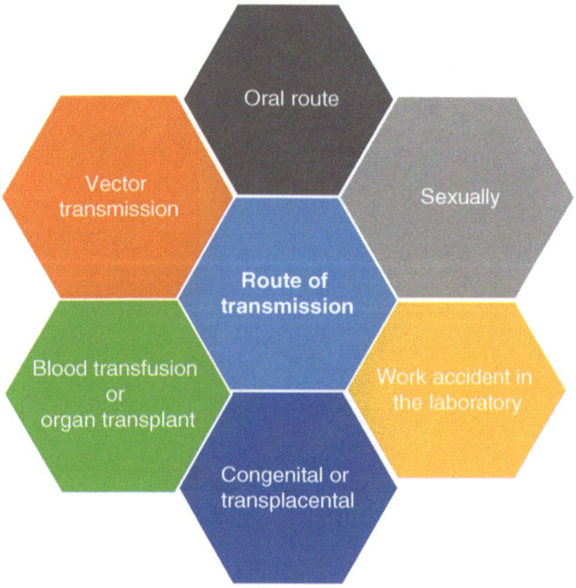

Figure 17.1 Route of transmission in Chagas disease.

Recent developments in *T. cruzi* stage forms and its life cycle include: (1) uncovered a novel morphological form of *T. cruzi*, which has a substantial impact on the new understanding of the life cycle and its stages, and (2) found crucial stages that permit genetic exchange across *T. cruzi* strains, thereby enhancing genetic diversity of these parasites. Therefore, these discoveries advance the search for novel treatments and contribute to our understanding of the illness. The classical approach is very superficial, and advanced research has shown that the process is undeniably more complicated in mammalian hosts. Two fundamental morphologies of blood cells, slender and broad BTs, were found in a pleomorphic population in 1963. In 2003, it was discovered that intracellular epimastigote-like forms

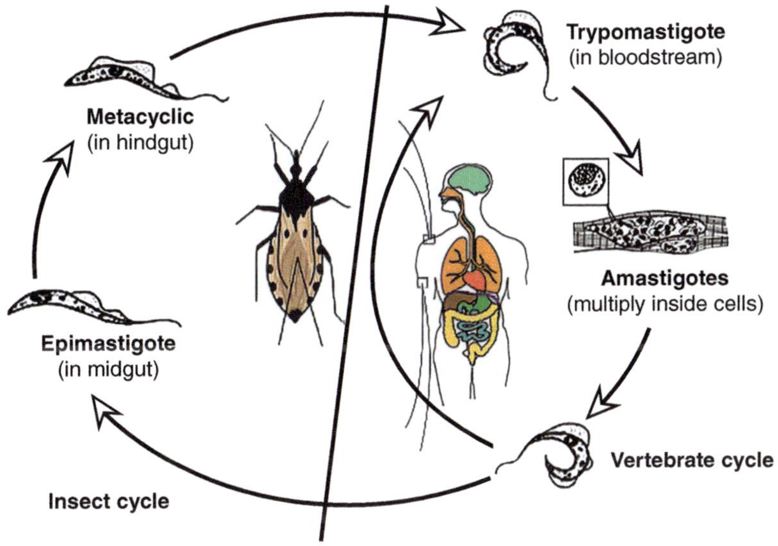

Figure 17.2 The *T. cruzi* life cycle demonstrates the many protozoan forms found in the hosts, which are invertebrate (triatomines) and vertebrate (mammals).

existed; however, it is not yet apparent if these forms constitute a crucial intracellular stage of the life cycle or merely a step in the transformation of amastigotes into trypomastigotes. Year 2014 saw the discovery of a brand-new intracellular morphology known as the zoid. One amastigote and one zoid (a cell with a kinetoplast but no nucleus, which immediately dies and is eliminated by the host cell) occur from the initial differentiation of the metacyclic trypomastigotes into this zoid through asymmetric cell division. In 2017, new clinically noteworthy findings were announced. An interesting feature regarding treatment resistance in Chagas disease is that some amastigotes have been documented to become metabolically quiescent (also known as latent amastigotes). Even after therapy, this form has the ability to impulsively restart cell cycle and regain infection in chronically infected tissues of mammalian hosts. *T. cruzi* modifies its gene expression quickly and significantly during its life cycle in response to environmental changes. The majority of these modifications are made at the posttranscriptional level by altering the stability of messenger RNA (mRNA) and translational control mechanisms. *T. cruzi* undergoes severe structural and metabolic changes to adapt to the many environmental conditions; they encounter in one or more host species. It is also able to differentiate between proliferative and non-proliferative cells within the same host. The differential expression of stage-specific genes serves as evidence for all these changes [28].

17.5 Clinical Forms of Chagas Disease

Human Chagas Disease has a complicated etiology, and it has two sequential clinical manifestations. After *T. cruzi* infection, the acute phase is characterized by high-grade parasitemia, which is frequently asymptomatic or oligosymptomatic with fever, tachycardia, and anorexia. Due to a deficiency in clinical symptoms or their absence, the acute phase is frequently missed. The main symptoms of symptomatic infection includes persistent fever, lymph node enlargement, hepatomegaly, splenomegaly, subcutaneous edema, indications of the parasite's portal of entry (inoculation chagoma or Roma's sign), and in severe cases, meningoencephalitis and/or myocarditis. Even without the use of trypanocidal medications, symptomatic signs of the acute phase gradually resolve in 90% of cases within 60 days, and it is likely that 60–70% of infected people never have any symptom or sign associated with Chagas disease. The infection then transitions into the chronic phase, which typically begins with a protracted, asymptomatic, or latent clinical form (known as the indeterminate form), which can be identified by normal findings on X-rays of the heart, esophagus, and colon as well as no alternation on the electrocardiogram. In 60–70% of instances, the infection is clinically quiet throughout the duration of the chronic phase. However, over the course of 10–30 years, 30–40% of asymptomatic patients will eventually develop chronic diseases of various organs, primarily the heart and digestive system (cardiac, digestive, and cardio digestive diseases). The three main cardiac syndromes of cardiac involvement – cardiac arrhythmias, heart failure, and thromboembolism – are the most severe and prevalent manifestations of chronic Chagas disease. The most common terms for the digestive form of Chagas disease are megaesophagus and/or megacolon; however, involvement of the entire digestive system can occur [33, 34] (See Figure 17.3).

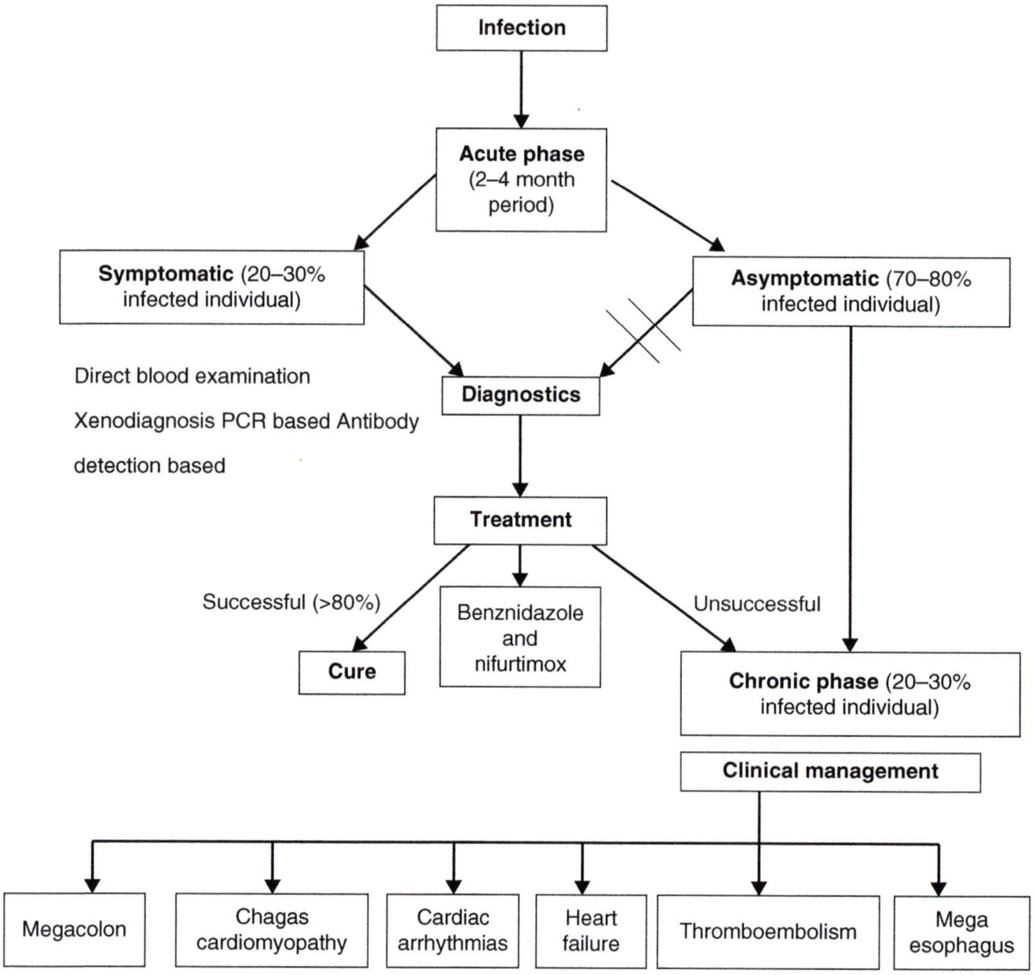

Figure 17.3 Overview of Chagas disease clinical manifestations, diagnosis, management, and treatment.

17.6 Diagnosis and Management of Chagas Disease

In endemic areas with advantageous sociocultural, political, and economic factors, Chagas disease cases typically occur in certain environments. Chagas disease has similar characteristics, such as the absence of symptoms in most cases, the inability to recognize and/or celebrate the condition, and the ignorance of accessible resources and immigration laws [35]. In non-endemic nations, it is in fact too high to be useful for gaining disease preventive and control strategies. In the blood, parasites may be seen circulating while in the acute stage of Chagas disease. By observing the parasite in a blood smear under a microscope, the diagnosis of Chagas disease can be made. For the purpose of visualizing parasites, a thin and thick blood smear are generated and stained. Three major categories can be used to categorize the modern Chagas-complaint scenario: parasitological, serological, and molecular system [36]. (See Figure 17.4).

The presence of parasites is visualized using parasitological methods, and the perceptivity of the parasites changes with the infection stage. Immunological (serological) techniques are based primarily on the search for G antibodies (IgG) anti-*T. cruzi* in patient blood and their colorimetric response visible in the case that the patient's blood contains the antibodies to identify the complaint in the stage where the parasitemia is actually less. The three methods with the highest usage rates are indirect immunofluorescence (IFI, perceptivity = 98%, specificity = 98%), indirect hemagglutination (HAI), and the ELISA test (perceptivity = 94–100%, specificity = 96–100%). Serological testing for malignancy can have some cross-reactivity despite being mainly sensitive and specific. Another method based on the search for anti-*T. cruzi* antibodies is the Western blot, which has been used in the opinion of Chagas complaint and commonly uses recombinant proteins and considerable excretion-secretion antigens [37]. However, owing to a deficit of

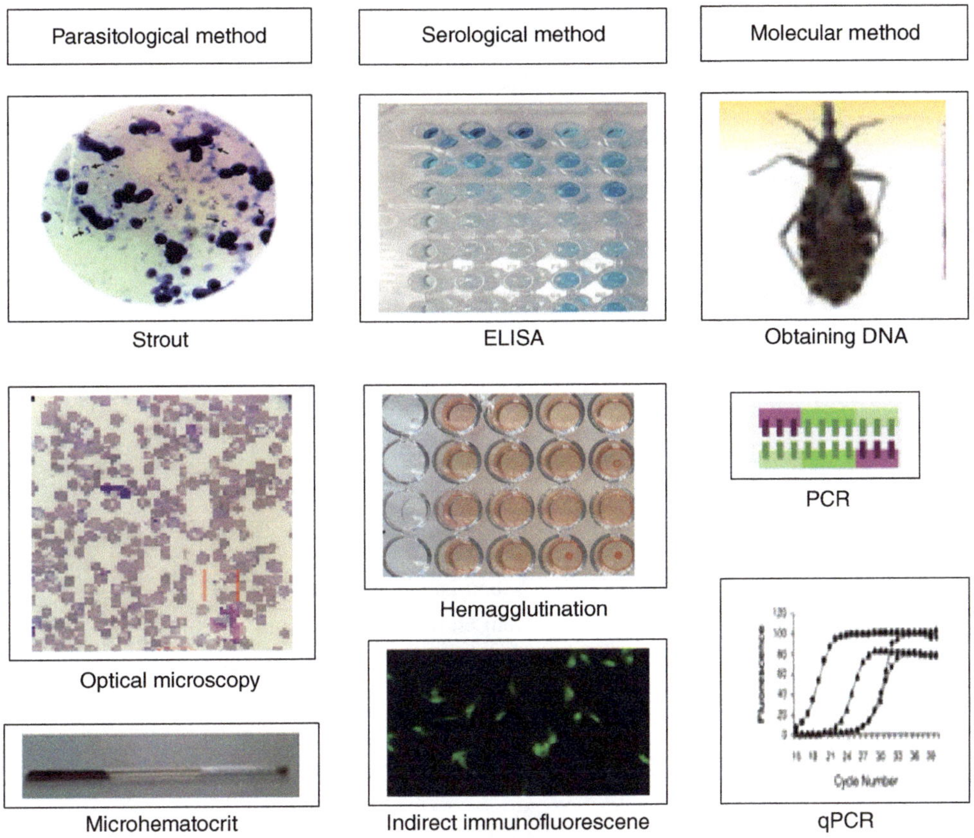

Figure 17.4 Parasitological, serological, and molecular diagnostic methods for Chagas disease.

standardization, complexity, clinical evidence, and application costs, its use in aboriginal regions is still limited. Molecular individual aids may serve as an alternative to or a supplement to conventional individual styles in the finding of parasite DNA. As a result of molecular natural processes, recombinant antigens and synthetic peptides, among others, can now be produced in significant numbers for use in immunologic procedures. Polymerase chain reaction (PCR), a method with a high degree of specificity and perceptivity, has been employed in hybridization approaches to identify distinct DNA fractions of the parasite genome, particularly during the acute phase. The poor perceptivity in the habitual phase of the PCR technique, which results from the parasites' genuinely low position in the bloodstream since they are restricted to tissues, is one of the biggest drawbacks of the method in relation to Chagas disease. However, it is crucial to remember that the DNA birth mechanism and the amount of blood used for DNA birth also provide information about how well PCR works. In chronic infections where parasites hide within organs, parasite detection via organ necropsies is more successful; nevertheless, this method is problematic because necropsies are not routinely obtained. Some researchers believe that circulating parasite antigens could be used as highly specific markers of *T. cruzi* infection, as seen in mouse models. Additionally, suggested as a marker for Chagasic patients are serum proteins. However, the detection of antibodies against the parasite continues to dominate diagnosis. The identification of the antigens used in anti-*T. cruzi* assays is essential for the specificity of the assay, especially in the case of persons infected with related protozoan parasites similar to Leishmania. Chagas disease is currently diagnosed using a variety of serologic assays based on whole parasite extracts and/or recombinant antigens (Table 17.2) [38]. The first screening test for blood donations received FDA approval in December 2006, and a number of blood-collecting organizations immediately began checking donated blood for serologic evidence of *T. cruzi* infection. In 2010, a different screening test produced by a different manufacturer received approval, and in 2017, the FDA released updated recommendations on whether donors who tested falsely positive for a screening test could still donate blood [39, 40]. These serological assays for identifying *T. cruzi* antibodies are typically categorized as confirmatory assays (Table 17.2).

Table 17.2 List of available commercial diagnostic tests for the serological identification of *T. cruzi* (Chagas disease).

Test	Antigen	Manufacturer
Immunoenzymatic assay		
AccuDiag™ Chagas ELISA kit	NI	Diagnostic Automation/Cortez diagnostics, Inc. USA
ImmunoComb II Chagas Ab Kit	RA	Alere Inc. Germany
Anti-ChagasIgG ELISA kit	NI	Abcam, UK
Abbott ES Chagas	Purified antigens	Abbott Laboratories, USA
EIAgen *Trypanosoma cruzi* Ab	Total extract	Adaltis, Italy
Chagas ELISA IgG + IgM	RA	Vircell, Spain
Pathozyme Chagas	RA	Omega Diagnostics Limited, Scotland
NovaLisa Chagas	NI	NovaTec, Germany
Premier Chagas IgG ELISA test	Purified antigens	Meridian Diagnostics, USA
Bio-Manguinhos EIA	Total extract/RP	Bio-Manguinhos, Brazil
Cellabs *T. cruzi* IgG CELISA	NI	Cellabs Pty Ltd., Australia
Abbott Chagas Anticorpos EIA	RA	Abbott Laboratories, USA
Cruzi TEST ELISA	NI	GenCell Biosystems, Ireland
Chagas test IICS, ELISA	Total extract (Y strain)	IICS Univ de Asuncion, Paraguay
Immunochromatographic assay		
OnSite Chagas Ab Rapid test	RA	CTK Biotech, USA
Chagas AB rapid	RA	Standard Diagnostics, Korea
WL Check Chagas	RA	Wiener Lab, Argentina
Chagas Instantest	Antigen attached to colloidal gold	Silanes, Mexico
Hemagglutination assays		
Hemacruzi	Erythrocytes sensitized with parasite lysate	Biolab-Mérieux, Brazil
Immunofluorescence assays		
Chagas IFA	NI	Vircell, Spain
Immunofluor Chagas Kit	Epimastigotes	Biocientifica S.A, Argentina
Chemiluminescent immunoassay		
Architect Chagas assay (prototype) immunoparticles	RA	Abbott Laboratories, Spain
PCR Assays		
RealStar Chagas PCR Kit RUO	NA	Altona Diagnostics GmbH, Germany
VIASURE *Trypanosoma cruzi* real-time PCR Detection Kit	NA	Certets Biotec, Spain
Confirmatory assays		
Radioimmunoprecipitation analysis [RIPA]	Radiolabeled *T. cruzi* surface antigens	University of Iowa, USA

Abbreviations: RA, recombinant antigen; NI, not indicated; NA, not applicable.

17.6.1 Treatment of Chagas Disease

The medications that were authorized to treat *T. cruzi* infection more than 50 years ago are still in use today. These medications include Nifurtimox, a nitrofuran chemical that Bayer released in 1965, and Benznidazole, a nitroimidazole derivative that Roche (Roche 7-501, Rochagan) introduced in 1971. Most experts regard Benznidazole to be the most preferred treatment for Chagas disease because of its superior safety profile and tolerability. The FDA approved the use of Benznidazole in children between the ages of 2 and 12 on 29 August 2017, while children under 2 may still receive it off-label. The website

Table 17.3 Overview of vital strength and weakness of Benznidazole and Nifurtimox from the view of pharmaceutical company.

Drugs	Benznidazole	Nifurtimox
Mechanism of action	By the action of the trypanosome-specific nitro reductase (TcNTR-1), a prodrug is converted into toxic, highly reactive intermediates that bind to DNA, proteins, and small molecules to form adducts.	
Drug resistance	Multifactorial pathways, inducible in-vitro, and TcNTR mutation. Extremely varied intrinsic drug sensitivity across parasite subtypes	
Formulation and regimen	2.5 and 100 mg oral tablets Weight-based dosing, tablet slurry for children (typically 2–3 × day × 60 d)	30 and 120 mg oral tablets Weight-based dosing, tablet slurry for children (typically 3 × day × 60 d)
Pharmacokinetic characteristics	Wide tissue distribution and trans placental transit C_{max} is 2–3 h post dose elimination half-life 13 h Unknown metabolic pathways	C_{max} 4 h after the dose-wide tissue dispersion; elimination half-life of 3 h; and transplacental passage. Nitroreductases are used in metabolism
Efficacy	In the acute period, both medicines have a >70% parasitological cure rate. Limited efficacy in treating chronic, unknown diseases in children and adults (10–35% cure according to seroreversion);	
Safety	Hypersensitivity reactions (30% of patients develop dermatitis, for example); depression of the bone marrow Neuropathy in the periphery risk of embryofetal harm; mutagenic and clastogenic	Anorexia/weight loss/nausea/vomiting Psychic alterations Excitability or sleepiness mutagenic and clastogenic Embryofetal toxicity risk
Drug interaction	Alcohol, Disulfiram	Alcohol
Tolerability	Because of its superior safety profile, BZN is frequently chosen over NFX. Children tolerate acute illness and tolerability better. Early therapy withdrawal (within 19 d) due to side effects is typical. 12–24% of patients stop receiving treatment.	

https://www.benznidazoletablets.com/en allows customers to order Benznidazole from a single distributor. The FDA approved Lampit (Bayer) on 6 August 2020, for the treatment of Chagas disease in patients 18 years of age and less who weighed less than 5.5 lbs. (2.5 kg). The trypomastigotes and amastigote forms of the parasite are both susceptible to the effects of these nitro heterocyclic prodrugs, which prevent DNA replication. While Benznidazole should be administered to adults at a dosage of 5–7 mg/kg/day in two divided doses for 60 days, Nifurtimox should be administered at a dosage of 8–10 mg/kg/day in three divided doses for 90 days. For congenital infections (10 mg/kg/day for Benznidazole and 15–20 mg/kg/day for Nifurtimox) and acute meningoencephalitis (up to 25 mg/kg/day for Benznidazole), higher doses are preferred. Table 17.3 provides an overview of the major advantages and disadvantages of Benznidazole and Nifurtimox from the standpoint of a pharmaceutical company [41].

Several reviews have been written about the experimental anti-trypanosomal drug candidates in the preclinical and development pipeline [42–45]. Phase II studies have only been conducted on a small number of novel medications, most notably azoles like Posaconazole and Fosravuconazole (E1224). Unfortunately, none has advanced past BZN in terms of producing long-lasting parasite suppression [46–49]. Two phase II trials of Chagas disease (NCT 02498782, NCT 03587766) used Fexinidazole, a 5-nitroimidazole authorized for the treatment of Human African Trypanosomiasis (HAT). A second research with modified dose regimens was carried out as a result of the previous study's poor tolerability [50], but the findings have not yet been published. Oxaboroles, Nitroimadazoles, and proteosome inhibitors are three further innovative medication classes being investigated for different trypanosomal disorders such as leishmaniasis and HAT may also have capability to treat Chagas disease [51]. These drug replacement and new drug studies offer the potential to find additional therapeutic approaches to the current Chagas disease treatment (Table 17.4).

17.6.2 Diagnosis, Management, and Treatment of Chronic Chagas Cardiomyopathy (CCC) and Gastrointestinal Complication

About 30% of patients with *T. cruzi* infection experience CCC, which typically results in malignant ventricular arrhythmias (MVA), apical aneurysms, left ventricular dilation, congestive cardiac failure, and sudden death. Patients with CCC have a greater death rate than those with non-Chagas disease CCC, and MVA are more frequent in Chagas disease than in other

Table 17.4 Summarizes the key discovery in chagas disease treatment nowadays.

Drugs	Formulation	Study phase or model	Comments	References
Nitro heterocyclic				
Benznidazole	Monotherapy	Human in chronic phase	According to the patients' geographic region, Benznidazole decreased the identification of circulating parasites but did not decrease cardiac clinical progression (The BENEFIT study).	[52]
	Pediatric monotherapy	Implementation in humans	The dosage in this formulation is suitable for young children.	[53]
Fexinidazole21,2	Monotherapy	Murine model of acute and chronic infection	Using strains of mice that were sensitive to varying doses of Benznidazole, fexinidazole proved very successful in treating both acute and chronic mouse infections.	[54, 55]
	Monotherapy	Phase II proof-of-concept study	Results are still pending.	[56]
Triazole				
Itraconazole	Monotherapy	Humans in chronic phase and mice in acute and chronic phase	Both therapeutic activity in chronic Chagas disease in humans and action in animals infected with various *T. cruzi* genotypes.	[57–59]
	Combined with Benznidazole	Mice in acute phase	More effectively than either medicine alone removes parasites from the blood and decreases damage in heart muscle tissue.	[60]
Ketoconazole	Monotherapy	In-vitro and in-vivo in humans with chronic phase	Active in-vitro but not in patients in the chronic phase.	[61]
	Combination with Benznidazole	Mice in acute phase	Effect of complementary medications.	[62]
Voriconazole	Monotherapy	Mice in acute phase	Decreased parasitemia and mortality, however Benznidazole performed well.	[63]
	Combined with Benznidazole	In-vitro and in-vivo with mice in acute phase	Although it was effective in-vitro, Benznidazole was still the better option in-vivo.	[64]
Posaconazole	Monotherapy	In-vitro and in-vivo with mice in acute and chronic phase	Strong and specific anti-*T. cruzi* action.	[65]
	Combination with Benznidazole	Mice in acute phase	Together, these medications were more potent than they were alone.	[66]
	Monotherapy	Phase II clinical trials in chronic phase	In a human study called the "CHAGASAZOL Trial," this medication did not perform as intended.	[67]
	Monotherapy and combined with Benznidazole	Phase II clinical trials in chronic phase	In the STOP-CHAGAS Trial, there were no benefits to combination therapy over Benznidazole monotherapy.	[68]
Ravuconazole	Monotherapy	In-vitro and in-vivo with murine and canine model in acute phase	Activity is modest in-vivo but is present in-vitro.	[69]
	Monotherapy (E1224)	Phase II clinical trials in chronic phase	E1224 caused the blood to be cleared of parasites, but later, parasitemia returned.	[70]

Table 17.4 (Continued)

Drugs	Formulation	Study phase or model	Comments	References
Other drugs				
Allopurinol	Combination with Benznidazole and Nifurtimox	In-vitro and in-vivo with mice in acute phase	Combining the medications proved more efficient than administering them separately.	[71]
	Monotherapy	Humans in acute and chronic phase	Effective in lowering the incidence of positive xenodiagnosis and ECG abnormalities, but ineffectual in causing a parasitological cure.	[72]
Oxaborole AN4169	Monotherapy	Mice in acute phase	When given in continuous doses rather than intermittently, AN4169 completely cured sick mice. Results from another in-vivo trial are not yet available.	[73]
	Monotherapy	In-vitro	In-vitro testing revealed that oxaborole was effective against several *T. cruzi* strains and clones	[74]
Azabenzoxazole GNF6702	Monotherapy	In-vitro and in-vivo with mice in chronic phase	Because it particularly inhibited the parasite proteasome, it had substantial effectiveness against *T. cruzi*.	[75]
Tris (2-aminomethyl) amine	Monotherapy	*Vero cells* and *Balb/c mice*	Acute and chronic Chagas disease trypanocidal efficacy with less harmful side effects than Benznidazole.	[76]
Estafietin (Sesquiterpene lactones derivatives)	Monotherapy	*Vero cells*	Epoxyestafietin, which has trypanocidal activity, was the substance that was most effective against trypomastigotes and amastigotes of *T. cruzi*.	[77]
Natural compounds isolated from cashew nut (*Anacardium occidentale*, L. Anacardiaceae)	Monotherapy	*hFIB cells*	Sirtuin inhibition has anti-*T. cruzi* effects similar to those of Benznidazole. Strong in-vitro trypanocidal action. Limited curative effect in-vivo, nevertheless.	[78]
AS-48 (Bacteriocin)	Monotherapy	Vero cells	Effective against all *T. cruzi* morphological forms, showing lesser cytotoxicity than benznidazole	[79]
Cordycepin Pentostatin	Monotherapy	*HeLa cells* and *swiss mice*	Strong in-vitro trypanocidal action, limited curative effect in-vivo, nevertheless.	[80]
AN15368	Monotherapy	*Mice and non-human primates*	The parasite that causes Chagas disease, *Trypanosoma cruzi*, has been found in mice and non-human primates, and researchers have discovered a brand-new drug that is 100% effective in treating it. The substance, known as AN15368, seems to be secure, causes no noticeable side effects, and is more efficient than current medications.	[81]

types of cardiac illness. MVA are more frequent in Chagas disease than with other types of cardiac illness, and CCC patients die more frequently than people with non-Chagas disease CCC. 19% of Latin American-born migrants with nonischemic cardiomyopathy tested positive for Chagas disease, according to a study, and those in this category had significantly higher mortality or transplant rates than those without Chagas disease. All patients with a proven *T. cruzi* infection should undergo baseline cardiac examinations, as shown in the Table 17.5. If CCC is present, the patient should be sent to a cardiologist because antitrypanosomal therapy may be less successful. The American College of Cardiology's (ACC) and American Heart Association's (AHA) guidelines for cardiac failure are applicable to CCC, whereas the Brazilian guidelines now in effect state clinical care of CCC. Additionally, as is mentioned below, emphasis should be paid to the distinct clinical

Table 17.5 Recommended surveillance in patients with confirmed *T. cruzi* infection.

Test	Observations	Frequency
ECG (electrocardiogram)	With a normal ECG, patients with Chagas disease have a good prognosis. Conduction irregularities may be a sign of future cardiac damage. Right bundle branch block, a characteristic of CCC, commonly coexists with left anterior fascicular block. It has been shown that the quantity of ECG anomalies correlates with the severity of cardiac damage.	Baseline and annually
Echocardiogram	Fibrosis-induced left ventricular systolic impairment is a typical feature of CCC. CCC frequently includes apical aneurysms and aberrant wall motion.	Baseline and annually
24-h Holter	To determine the possibility of arrhythmias and autonomic dysfunction, which raise the risk of unexpected death.	Baseline
Chest X-ray	In order to assess cardiomegaly	Baseline

manifestation of CCC in comparison to non-Chagas disease CCC. One of the most frequent clinical outcomes of chronic determined Chagas disease are arrhythmias, which can develop in bradyarrhythmias, atrial and ventricular fibrillation, sick sinus syndrome, and ventricular tachycardia. In a study conducted in Los Angeles, 7.5% of Latin American-born patients with pacemaker were found to have Chagas disease, indicating that Chagas disease may be a serious cause of cardiomyopathy in the country [82, 83].

The prevention of sudden death is one of the main goals of treating CCC. Latin American guidelines prescribe amiodarone for individuals at risk of sudden death from nonsustained ventricular tachycardia with symptoms of myocardial dysfunction because treatment may improve outcomes. It has been demonstrated that angiotensin converting enzyme (ACEs) inhibitors lower death rate in CCC patients. Taking beta-blockers could make Bradycardia in patients with Chagas disease worse. However, compared to untreated individuals, beta-blocker medication enhanced survival in CCC patients in a Brazilian randomized trial. Another randomized experiment concluded that Carvedilol was safe and increased LVEF in CCC patients. In the later study, patients' clinical complaints were first treated with ACE inhibitors, and only then were beta-blockers prescribed [83, 84]. Table 17.6 displays the list of research on some immunomodulatory agents used in Chronic Chagas Cardiomyopathy.

Table 17.6 Immunomodulatory agents used in Chronic Chagas Cardiomyopathy.

Drug	Route/Dose	*T. cruzi* strain animal mode	Main findings	References
Granulocyte colony-stimulating factor	I.P./200 μg/kg	Colombian C57 BL/6	Reduction of myocarditis with a rise in inflammatory cells that are undergoing apoptosis and an improvement in heart function	[85]
Acetyl salicylic acid	Oral/2 or 40 mg/kg	Dm28c BALB/c	Decreasing the amount of inflammatory infiltrates in the heart and increasing endothelial function	[86]
Pentoxifyline	I.P./20 mg/kg	Colombian C57 BL/6	Fibrosis, electrical alteration, and myocrditis all decreased	[87]
Fenofibrate + benzindazole	Oral/50– 300 mg/kg	K-98 and RA BALB/c	Reversal of cardiac dysfunction and a decrease in pro-inflammatory chemicals are both associated with a reduction in myocarditis	[88]
Simvastatin + benznidazole	Ora /5–40 mg/kg	Dm28c BALB/c and Sv/129	Reduced endothelial activation, cardiac fibrosis, and inflammation associated with 15-epi-lipoxin A4.	[89]
BA5 (semisynthetic derivative from betulinic acid)	Oral/1 or 10 mg/kg	Colombian C57 BL/6	Decrease in IL-10 production and M2 polarization-related cardiac fibrosis and inflammation.	[90]

The most frequent gastrointestinal symptoms are megacolon and megaesophagus, but these conditions are less frequently seen in patients from Mexico and Central America and more frequently in those from Brazil, Argentina, Bolivia, Chile, and Paraguay (about 15% of patients). Of these, 30% of the patients had both gastric and heart problems. Regurgitation and difficulty swallowing may be signs of esophageal involvement, whereas colon injury may cause constipation, volvulus, or irregular bowel motions. In some circumstances, a barium enema and radiographic examination are advised when there is a clinical suspicion of gastrointestinal issues resulting from chagas disease. Early-stage gastrointestinal indication is not usually a contraindication for etiological treatment, even though more advance cases of megaesophagus or megacolon may require surgical repair before any use of antitrypanosomal medicine [91].

17.6.3 Public Health Initiatives, Local and International Associations, Awareness Campaigns, Regulatory Health Perspectives, and Chagas Disease Control Programmes

Both endemic and non-endemic nations must prioritize public health, no-shame, awareness, and education initiatives involving patients and the general public in order to: (1) normalize the diagnosis and eliminate embarrassment, (2) promote asymptomatic screening among those at risk, and (3) inform the public and healthcare professionals about the condition, available therapies, and available preventive measures. Public health, no-shame, awareness, and education initiatives involving patients and the general public are crucial in both endemic and non-endemic countries in order to: (1) normalize the diagnosis and eradicate embarrassment, (2) promote the risk population's asymptomatic screening, and (3) educate the general public and healthcare professionals about the illness, its treatments, and its choices for prevention. Chagas disease control and eradication plans were released by the PAHO [92] and WHO [93] in 2009 and 2010, respectively. The 2009 WHO study provided the first definition of the rules and procedures to prevent disease transmission in Europe [94]. Recently, a five-year, US $2.6 million partnership between the PAHO and Unitaid (a worldwide health organization that seeks out creative ways to prevent, diagnose, and treat illness in low- and middle-income nations more effectively, quickly, and affordably) was established to intensify local and national efforts to end mother-to-child transmission of the Chagas disease [94]. There is numerous nonprofit, patient-focused organizations that provide assistance and information in both endemic and non-endemic nations. These organizations include some that are a part of Findechagas (A nonprofit organization aims to develop, promote, and disseminate the guidelines outlined in the Declaration of Uberaba, which was adopted at the first gathering of associations of Chagas disease sufferers in the Americas, Europe, and the Western Pacific [Minas Gerais, Brazil, October 2009]). More than 30 associations from countries including Argentina, Australia, Bolivia, Brazil, Colombia, Italy, Mexico, Spain, Switzerland, the United States of America, and Venezuela are members of the International Federation of Associations of People Affected by Chagas Disease (FINDECHAGAS), which was established in October 2010 [95]. The Global Chagas Coalition is a multidisciplinary collaborative group of funders, researchers, patients, funders, and health professionals that focuses on advocacy, increasing policy makers' understanding of the condition, and mobilizing financial resources [96]. The WHO, PAHO, Mundo Sano (family foundation whose goal is to improve the lives of those who suffer from neglected disease), the Oswaldo Cruz Foundation (FIOCRUZ, regarded as one of the premier research centers for public health worldwide), Global Chagas' Coalition, the Barcelona Institute for Global Health (ISGlobal) (Its ultimate objective is to contribute to the reduction of health disparities both within and between various geographical areas), DNDi (The Drugs for Neglected Diseases initiative), and numerous international organizations committed to eradicating this illness. Control policies of Chagas disease must combine two general strategies for action: (1) prevention of transmission to stop the development of new cases, but these actions are cost-effective; and (2) prompt diagnosis and treatment of infected people to stop the clinical progression of the illness and to enable them to regain their health. To maximize the impact, every activity should be started as thoroughly and uniformly as feasible (Figure 17.5).

17.7 Conclusion

Chagas disease spread across the entire world 113 years after it was first identified and described in Latin America. Infecting millions of individuals worldwide, the Chagas disease is one of the neglected tropical diseases. In Latin America, it is regarded as the fourth most common disability and is considered the sickness that is passed from animal to human. A wide range of topics, from the disease's epidemiology to its diagnosis, have been thoroughly studied, which has increased our understanding of the condition and our understanding of how to treat it. The disease and its severity are still largely unknown to the general public. It is ignored by those who are aware of it, and those who are affected by it but are not

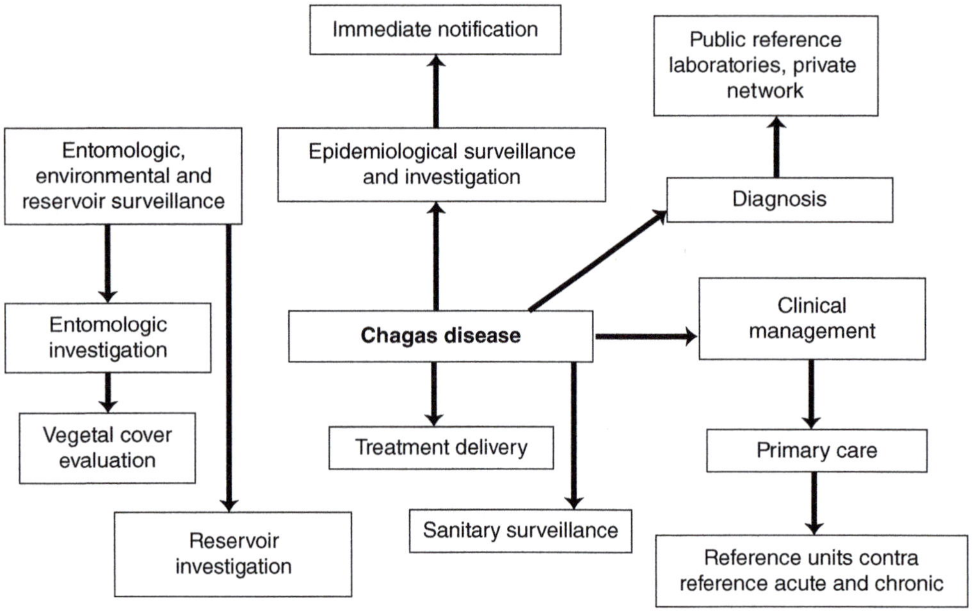

Figure 17.5 Flowchart for an organized approach to Chagas disease prevention.

promptly diagnosed or treated. A vaccination for populations at risk of contracting the disease has not yet been developed, and the present treatments for Chagas disease are only partially effective. In both endemic and non-endemic nations, Chagas disease is still a major concern, and there are still some goals to be accomplished in terms of curing or completely eradicating this illness. Given the existing situation, a multidisciplinary strategy is required to address this difficult condition in order to improve control techniques, create new diagnostic methods and drugs, and investigate and treat comorbidities linked to chronic Chagas disease.

References

1 Lynn, M.K., Bossak, B.H., Sandifer, P.A. et al. (2020). Contemporary autochthonous human Chagas disease in the USA. *Acta Trop.* 205: 105361.
2 World Health Organization *Chagas Disease (also known as American Trypanosomiasis).* World Health Organization Press https://www.who.int/news-room/fact-sheets/detail/chagas-disease-(american-trypanosomiasis).
3 Engels, D. and Zhou, X.N. (2020). Neglected tropical diseases: an effective global response to local poverty-related disease priorities. *Infect. Dis. Poverty* 9 (1): 10.
4 Lidani, K.C.F., Andrade, F.A., Bavia, L. et al. (2019). Chagas disease: from discovery to a worldwide health problem. *Front. Public Health* 7: 166.
5 Pan American Health Organization *Neglected Infectious Diseases in the Americas: Success Stories and Innovation to Reach the Neediest.* Pan American Health Organization Press https://iris.paho.org/handle/10665.2/31250.
6 World Health Organization *World Chagas Disease Day*. World Health Organization Press https://www.who.int/campaigns/world-chagas-disease-day/2022.
7 Sangenito, L.S., Branquinha, M.H., and Santos, A.L.S. (2020). Funding for chagas disease: a 10-year (2009–2018) survey. *Trop. Med. Infect. Dis.* 5 (2): 88.
8 Jackson, Y., Wyssa, B., and Chappuis, F. (2020). Tolerance to nifurtimox and benznidazole in adult patients with chronic Chagas' disease. *J. Antimicrob. Chemother.* 75 (3): 690–696.
9 Francisco, A.F., Jayawardhana, S., Olmo, F. et al. (2020). Challenges in chagas disease drug development. *Molecules* 25 (12): 2799.
10 Suárez, C., Nolder, D., García-Mingo, A. et al. (2022). Diagnosis and clinical management of chagas disease: an increasing challenge in non-endemic areas. *Res. Rep. Trop. Med.* 13: 25–40.

11 World Health Organization *World Chagas Disease Day: Finding and Reporting Every Case.* World Health Organization Press https://www.who.int/news/item/14-04-2022-world-chagas-disease-day-bringing-a-forgotten-disease-to-the-fore-of-global-attention.

12 Chagas, C. (1909). Nova tripanozomiaze humana: estudos sobre a morfolojia e o ciclo evolutivo do Schizotrypanum cruzi n. gen, n sp, ajente etiolojico de nova entidade morbida do homem. *Mem. Inst. Oswaldo Cruz* 1: 159–218.

13 Kropf, S.P. and Sá, M.R. (2009). The discovery of Trypanosoma cruzi and Chagas disease (1908–1909): tropical medicine in Brazil. *Hist. Cienc. Saude Manguinhos* 1: 13–34.

14 Sosa-Estani, S. and Segura, E.L. (2015). Integrated control of Chagas disease for its elimination as public health problem—a review. *Mem. Inst. Oswaldo Cruz* 110 (3): 289–298.

15 Segovia, J.C. (1913). Un caso de trypanosomiasis. *Arch. Hosp. Rosales En San Salvador* 8: 249–254.

16 Ayulo, V.M. and Herrer, A. (1944). Estudios sobre trypanosomiasis americana en el Perú: I. Observaciones en el departamento de Arequipa. *Rev. Peru. Med. Exp. Salud Publica* 3 (2): 96–117.

17 Tejera, E. (1919). La trypanosomose americaine ou maladie de Chagas au Venezuela. *Bull. Soc Pathol. Exot.* 12: 509–513.

18 De Araujo-Jorge, T.C., Telleria, J., and Rios-Dalenz, J. (2010). History of the discovery of American trypanosomiasis (Chagas disease). In: *American Trypanosomiasis*, 1ee (ed. J. Telleira and M. Tibayrenc), 3–23. Amsterdam: Elsevier Inc.

19 León Gómez, A., Flores Fiállos, A., Reyes, Q.L. et al. (1960). La Enfermad de Chagas en Honduras. *Rev. Med. Hondur.* 28: 78–83.

20 Gasic, G. and Bertin, V. (1940). Epidemiologia de la enfermedad de chagas en Chile. *Rev. Chil. Pediatr.* 11: 561–584.

21 Mazzotti, L. (1940). Dos casos de enfermedad de Chagas en el estado de Oaxaca. *Gac. Med. Mex.* 70: 417–420.

22 Mazza, S. (1942). Consideraciones sobre la enfermedad de Chagas en Bolivia. *Prensa Med. Argent.* 29: 51.

23 Florez, F.S. (2000). Historia de la Tripanosomiasis Americana en Colombia. *Medicina* 22 (2): 75–77.

24 Mazza, S. (1949). La enfermedad de chagas em la republica Argentina. *Mem. Inst. Oswaldo Cruz* 47: 273–288.

25 Aufderheide, A.C., Salo, W., Madden, M. et al. (2004). A 9,000-year record of Chagas' disease. *Proc. Natl. Acad. Sci. U. S. A.* 101 (7): 2034–2039.

26 Araújo, A., Jansen, A.M., Reinhard, K., and Ferreira, L.F. (2009). Paleoparasitology of Chagas disease—a review. *Mem. Inst. Oswaldo Cruz* 104 (1): 9–16.

27 Lannes-Vieira, J., de Araújo-Jorge, T.C., de NC, S.M. et al. (2010). The centennial of the discovery of Chagas disease: facing the current challenges. *PLoS Negl.Trop. Dis.* 4 (6): e645.

28 Martín-Escolano, J., Marín, C., Rosales, M.J. et al. (2022). An updated view of the Trypanosoma cruzi life cycle: intervention points for an effective treatment. *ACS Infect. Dis.* 8 (6): 1107–1115.

29 Rassi, F.M., Minohara, L., Rassi, A. Jr. et al. (2019). Systematic review and meta-analysis of clinical outcome after implantable cardioverter-defibrillator therapy in patients with Chagas heart disease. *JACC Clin. Electrophysiol.* 5 (10): 1213–1223.

30 Bern, C., Kjos, S., Yabsley, M.J., and Montgomery, S.P. (2011). Trypanosoma cruzi and Chagas' disease in the United States. *Clin. Microbiol. Rev.* 24 (4): 655–681.

31 Flores-Ferrer, A., Marcou, O., Waleckx, E. et al. (2018). Evolutionary ecology of Chagas disease; what do we know and what do we need? *Evol. Appl.* 11 (4): 470–487.

32 Nogueira, N.F.S., Gonzalez, M.S., Gomes, J.E. et al. (2007). Trypanosoma cruzi: involvement of glycoinositolphospholipids in the attachment to the luminal midgut surface of Rhodnius prolixus. *Exp. Parasitol.* 116 (2): 120–128.

33 Francisco, A.F., Jayawardhana, S., Lewis, M.D. et al. (2017). Biological factors that impinge on Chagas disease drug development. *Parasitology* 144 (14): 1871–1880.

34 Sánchez-Valdéz, F.J., Padilla, A., Wang, W. et al. (2018). Spontaneous dormancy protects Trypanosoma cruzi during extended drug exposure. *Elife* 7: e34039.

35 Suárez, C., Nolder, D., García-Mingo, A. et al. (2022). Diagnosis and clinical management of Chagas disease: an increasing challenge in non-endemic areas. *Res. Rep. Trop. Med.* 13: 25–40.

36 Ribeiro, A.L., Nunes, M.P., Teixeira, M.M., and Rocha, M.O.C. (2012). Diagnosis and management of Chagas disease and cardiomyopathy. *Nat. Rev. Cardiol.* 9 (10): 576–589.

37 Morais, M.C.C., Silva, D., Milagre, M.M. et al. (2022). Automatic detection of the parasite Trypanosoma cruzi in blood smears using a machine learning approach applied to mobile phone images. *PeerJ* 10: e13470.

38 Pereiro, A.C. (2019). Guidelines for the diagnosis and treatment of Chagas disease. *Lancet* 393: 1486–1487.

39 Forsyth, C.J., Manne-Goehler, J., Bern, C. et al. (2022). Recommendations for screening and diagnosis of Chagas disease in the United States. *J. Infect. Dis.* 225 (9): 1601–1610.

40 López-Monteon, A., Dumonteil, E., and Ramos-Ligonio, A. (2019). More than a hundred years in the search for an accurate diagnosis for Chagas disease: current panorama and expectations. In: *Current Topics in Neglected Tropical Diseases*. Alfonso J. Rodriguez-Morales, IntechOpen.

41 Man, A. and Segal, F. (2022). New therapeutics for Chagas disease: charting a course to drug approval. In: *Chagas Disease—From Cellular and Molecular Aspects of Trypanosoma cruzi-Host Interactions to the Clinical Intervention* (ed. R. Menna-Barreto). Intechopen.

42 Sales, P.A., Molina, I., and Murta, S. (2017). Experimental and clinical treatment of Chagas disease: a review. *Am. J. Trop. Med. Hyg.* 97 (5): 1289–1303.

43 Rao, S.P.S., Barrett, M.P., Dranoff, G. et al. (2019). Drug discovery for kinetoplastid diseases: future directions. *ACS Infect. Dis.* 5 (2): 152–157.

44 Ribeiro, V., Dias, N., and Taís, P. (2020). Current trends in the pharmacological management of Chagas' disease. *Int. J. Parasitol. Drugs Drug Resist.* 2: 7–17.

45 Martínez-Peinado, N., Cortes-Serra, N., Losada-Galvan, I. et al. (2020). Emerging agents for the treatment of Chagas disease: what is in the preclinical and clinical development pipeline? *Expert Opin. Investig. Drugs* 29 (9): 947–959.

46 Torrico, F., Gascón, J., and Barreira, F. (2021). New regimens of benznidazole monotherapy and in combination with fosravuconazole for treatment of Chagas' disease [BENDITA]: a phase 2, double-blind, randomised trial. *Lancet Infect. Dis.* 46: 1129–1140.

47 Molina, I., Gómez i Prat, J., Salvador, F. et al. (2014). Randomized trial of posaconazole and benznidazole for chronic Chagas' disease. *N. Engl. J. Med.* 370 (20): 1899–1908.

48 Morillo, C.A., Waskin, H., and Sosa-Estani, S. (2017). Benznidazole and Posaconazole in eliminating parasites in Asymptomatic T. cruzi carriers. The STOP-CHAGAS' trial. *J. Am. Coll. Cardiol.* 69 (8): 939–947.

49 Torrico, F., Gascon, J., and Ortiz, L. (2018). Treatment of adult chronic indeterminate Chagas disease with benznidazole and three E1224 dosing regimens: a proof-of-concept, randomized, placebo-controlled trial. *Lancet Infect. Dis.* 18: 419–430.

50 DNDi fexinidazole for Chagas Disease. 2020. https://dndi.org/research-development/portfolio/fexinidazole-chagas.

51 Drugs for Neglected Diseases Initiative, R&D Portfolio. 2021. https://dndi.org/wp-content/uploads/2021/12/DNDi-RD-Portfolio-December-2021.pdf.

52 Morillo, C.A., Marin-Neto, J.A., Avezum, A. et al. (2015). Randomized trial of benznidazole for chronic Chagas' cardiomyopathy. *N. Engl. J. Med.* 373 (14): 1295–1306.

53 Drugs for Neglected Diseases initiative (DNDi) (2020). *Portfolio*. DNDi [cited 2023 Apr 28] https://dndi.org/research-development/portfolio.

54 Bahia, M.T., de Andrade, I.M., Martins, T.A.F. et al. (2012). Fexinidazole: a potential new drug candidate for Chagas disease. *PLoS Negl.Trop. Dis.* 6 (11): e1870.

55 Bahia, M.T., Nascimento, A.F.S., Mazzeti, A.L. et al. (2014). Antitrypanosomal activity of fexinidazole metabolites, potential new drug candidates for Chagas disease. *Antimicrob. Agents Chemother.* 58 (8): 4362–4370.

56 DNDi, Drugs for Neglected Diseases Initiative Research and Development Portfolio (2020). *Fexindazole*. Drugs for Neglected Diseases Initiative https://dndi.org/research-development/portfolio.

57 Apt, W., Aguilera, X., Arribada, A. et al. (1998). Treatment of chronic Chagas' disease with itraconazole and allopurinol. *Am. J. Trop. Med. Hyg.* 59 (1): 133–138.

58 Apt, W., Arribada, A., Zulantay, I. et al. (2013). Treatment of Chagas' disease with itraconazole: electrocardiographic and parasitological conditions after 20 years of follow-up. *J. Antimicrob. Chemother.* 68 (9): 2164–2169.

59 Toledo, M.J.d.O., Bahia, M.T., Carneiro, C.M. et al. (2003). Chemotherapy with benznidazole and itraconazole for mice infected with different Trypanosoma cruzi clonal genotypes. *Antimicrob. Agents Chemother.* 47 (1): 223–230.

60 Martins, A.F., De Figueiredo, D.T., and Mazzeti, L. (2015). Benznidazole/itraconazole combination treatment enhances anti-Trypanosoma cruzi activity in experimental Chagas disease. *PLoS One* 10 (6): e0128707.

61 Brener, Z., Cançado, J.R., Galvão, L.M. et al. (1993). An experimental and clinical assay with ketoconazole in the treatment of Chagas disease. *Mem. Inst. Oswaldo Cruz* 88 (1): 149–153.

62 Araújo, M.S., Martins-Filho, O.A., Pereira, M.E., and Brener, Z. (2000). A combination of benznidazole and ketoconazole enhances efficacy of chemotherapy of experimental Chagas' disease. *J. Antimicrob. Chemother.* 45 (6): 819–824.

63 Gulin, J.E.N., Eagleson, M.A., Postan, M. et al. (2013). Efficacy of voriconazole in a murine model of acute Trypanosoma cruzi infection. *J. Antimicrob. Chemother.* 68 (4): 888–894.

64 Gulin, J.E.N., Eagleson, M.A., López-Muñoz, R.A. et al. (2020). In vitro and in vivo activity of voriconazole and benznidazole combination on trypanosoma cruzi infection models. *Acta Trop.* 211 (105606): 105606.

65 Urbina, J.A., Payares, G., Contreras, L.M. et al. (1998). Antiproliferative effects and mechanism of action of SCH 56592 against Trypanosoma (Schizotrypanum) cruzi: in vitro and in vivo studies. *Antimicrob. Agents Chemother.* 42 (7): 1771–1777.

66 Diniz Lde, F., Urbina, J.A., and De Andrade, I.M. (2013). Benznidazole and posaconazole in experimental Chagas disease: positive interaction in concomitant and sequential treatments. *PLoS Negl.Trop. Dis.* 7 (8): 2367.

67 Sosa Estani, S., Segura, E.L., Ruiz, A.M. et al. (1998). Efficacy of chemotherapy with benznidazole in children in the indeterminate phase of Chagas' disease. *Am. J. Trop. Med. Hyg.* 59 (4): 526–529.

68 Morillo, C.A., Waskin, H., and Sosa-Estani, S. (2017). STOP-CHAGAS investigators. Benznidazole and Posaconazole in eliminating parasites in asymptomatic T. Cruzi carriers: the STOP-CHAGAS trial. *J. Am. Coll. Cardiol.* 69 (8): 939–947.

69 Diniz Lde, F., Caldas, I.S., and Guedes, P.M. (2010). Effects of ravuconazole treatment on parasite load and immune response in dogs experimentally infected with Trypanosoma cruzi. *Antimicrob. Agents Chemother.* 54 (7): 2979–2986.

70 Torrico, F., Gascon, J., Ortiz, L. et al. (2018). Treatment of adult chronic indeterminate Chagas disease with benznidazole and three E1224 dosing regimens: a proof-of-concept, randomised, placebo-controlled trial. *Lancet Infect. Dis.* 18 (4): 419–430.

71 Mazzeti, A.L., Diniz, L.d.F., Gonçalves, K.R. et al. (2019). Synergic effect of allopurinol in combination with nitroheterocyclic compounds against Trypanosoma cruzi. *Antimicrob. Agents Chemother.* 63 (6): E02264-18.

72 Apt, W., Arribada, A., Zulantay, I. et al. (2003). Itraconazole or allopurinol in the treatment of chronic American trypanosomiasis: the regression and prevention of electrocardiographic abnormalities during 9 years of follow-up. *Ann. Trop. Med. Parasitol.* 97 (1): 23–29.

73 Moraes, C.B., Giardini, M.A., Kim, H. et al. (2014). Nitroheterocyclic compounds are more efficacious than CYP51 inhibitors against Trypanosoma cruzi: implications for Chagas disease drug discovery and development. *Sci. Rep.* 4 (1): 4703.

74 Khare, S., Nagle, A.S., Biggart, A. et al. (2016). Proteasome inhibition for treatment of leishmaniasis, Chagas disease and sleeping sickness. *Nature* 537 (7619): 229–233.

75 Martín-Escolano, R., Cebrián, R., Martín-Escolano, J. et al. (2019). Insights into Chagas treatment based on the potential of bacteriocin AS-48. *Int. J. Parasitol. Drugs Drug Resist.* 10: 1–8.

76 Sülsen, V.P., Lizarraga, E.F., Elso, O.G. et al. (2019). Activity of estafietin and analogues on Trypanosoma cruzi and Leishmania braziliensis. *Molecules* 24 (7): 1209.

77 Matutino Bastos, T., Mannochio Russo, H., Silvio Moretti, N. et al. (2019). Chemical constituents of Anacardium occidentale as inhibitors of Trypanosoma cruzi sirtuins. *Molecules* 24 (7): 1299.

78 Adade, C.M., Oliveira, I.R.S., Pais, J.A.R., and Souto-Padrón, T. (2013). Melittin peptide kills Trypanosoma cruzi parasites by inducing different cell death pathways. *Toxicon* 69: 227–239.

79 Carmo, G.M., De Sá, M.F., and Grando, T.H. (2019). Cordycepin (3′-deoxyadenosine) and pentostatin (deoxycoformycin) against Trypanosoma cruzi. *Exp. Parasitol.* 199: 47–51.

80 Dalla Rosa, L., da Silva, A.S., Gressler, L.T. et al. (2013). Cordycepin (3′-deoxyadenosine) pentostatin (deoxycoformycin) combination treatment of mice experimentally infected with Trypanosoma evansi. *Parasitology* 140 (5): 663–671.

81 Yancy, C., Jessup, M., and Butler, J. (2013). ACCF/AHA. Guideline for the management of heart failure. *J. Am. Coll. Cardiol.* 81: e147–e239.

82 Bestetti, R.B., Otaviano, A.P., Cardinalli-Neto, A. et al. (2011). Effects of B-Blockers on outcome of patients with Chagas' cardiomyopathy with chronic heart failure. *Int. J. Cardiol.* 151 (2): 205–208.

83 Botoni, F.A., Poole-Wilson, P.A., Ribeiro, A.L.P. et al. (2007). A randomized trial of carvedilol after renin-angiotensin system inhibition in chronic Chagas cardiomyopathy. *Am. Heart J.* 153 (4): 544.e1–544.e8.

84 Macambira, S.G., Vasconcelos, J.F., Costa, C.R. et al. (2009). Granulocyte colony-stimulating factor treatment in chronic Chagas disease: preservation and improvement of cardiac structure and function. *FASEB J.* 23 (11): 3843–3850.

85 Molina-Berríos, A., Campos-Estrada, C., Henriquez, N. et al. (2013). Protective role of acetylsalicylic acid in experimental Trypanosoma cruzi infection: evidence of a 15-epi-lipoxin A_4-mediated effect. *PLoS Negl.Trop. Dis.* 7 (4): e2173.

86 Pereira, I.R., Vilar-Pereira, G., Moreira, O.C. et al. (2015). Pentoxifylline reverses chronic experimental Chagasic cardiomyopathy in association with repositioning of abnormal CD8+ T-cell response. *PLoS Negl.Trop. Dis.* 9 (3): e0003659.

87 Cevey, Á.C., Mirkin, G.A., Donato, M. et al. (2017). Treatment with Fenofibrate plus a low dose of Benznidazole attenuates cardiac dysfunction in experimental Chagas disease. *Int. J. Parasitol. Drugs Drug Resist.* 7 (3): 378–387.

88 González-Herrera, F., Cramer, A., Pimentel, P. et al. (2017). Simvastatin attenuates endothelial activation through 15-epi-Lipoxin A4 production in murine chronic Chagas cardiomyopathy. *Antimicrob. Agents Chemother.* 61 (3): e02137–e02116.

89 Meira, C.S., do Espírito Santo, R.F., dos Santos, T.B. et al. (2017). Betulinic acid derivative BA5, a dual NF-kB/calcineurin inhibitor, alleviates experimental shock and delayed hypersensitivity. *Eur. J. Pharmacol.* 815: 156–165.

90 Hugo García Orozco, V., Enrique Villalvazo Navarro, J., Solar Aguirre, C. et al. (2022). Digestive disorders in Chagas disease: megaesophagus and chagasic megacolon. In: *Chagas Disease—From Cellular and Molecular Aspects of Trypanosoma cruzi-Host Interactions to the Clinical Intervention* (ed. R. Menna-Barreto). IntechOpen https://doi.org/10.5772/intechopen.102871.

91 Pan American Health Organization (2016). Chagas Disease. http://Paho.org, https://www.paho.org/en/topics/chagas-disease (accessed 29 April 2023).

92 World Health Organization (2010). Chagas disease: control and elimination. In: *63rd World Health Assembly, Provisional agenda item 11.14 reported by the Secretariat*. World Health Organization Press 4: 56.

93 World Health Organization (2009). *Report of a WHO Informal Consultation (Jointly Organized by WHO Headquarters and the WHO Regional Office for Europe)*. World Health Organization Press. Geneva, Switzerland.

94 PAHO and Unitaid Launch Collaboration. Advance the Elimination of Mother-to-Child Transmission of Chagas Disease. http://Paho.org. https://www.paho.org/en/news/30-6-2022-paho-and-unitaid-launch-collaboration-advance-elimination-mother-child-transmission (accessed 29 April 2023).

95 Home EN—FINDECHAGAS (2022). http://Findechagas.org, https://findechagas.org/home-en (accessed 29 April 2023).

96 Chagas Coalition (2023). http://Coalicionchagas.org, http://www.coalicionchagas.org (accessed 29 April 2023).

18

Legionnaires' Disease: Current Trends in Microbiology and Pharmacology

Julia Wang[1], Neha Chintapally[1], Meera Nagpal[1], Anjali Mahapatra[1], and Charles Preuss[2]

[1] *University of South Florida Morsani College of Medicine, Tampa, FL, United States*
[2] *Department of Molecular Pharmacology & Physiology, University of South Florida Morsani College of Medicine, Tampa, FL, United States*

18.1 History of *Legionella* spp.

In the summer of 1979, in Philadelphia, Pennsylvania, United States, attendees at an American Legion conference experienced a severe pneumonia outbreak. This unusual respiratory disease affected 221 attendees with 34 fatal cases [1]. The Center for Disease Control and Prevention (CDC) identified the source of the infection as a gram-negative aerobic rod-shaped bacterium: a new bacterium genus named *Legionella*, and the specific species as *L. pneumophila,* after the conference at which it was discovered [1]. After its discovery, it was determined that outbreaks of previously unexplained flu-like illnesses, such as the one that occurred in Pontiac, Michigan 1968, clinically known as Pontiac fever, were brought on by *Legionella* species [2].

18.1.1 Characteristics

Legionella is a Gram-negative rod-shaped proteobacteria that is found in freshwater environments and moist soil [3]. They exist in the environment as either free-living biofilm-associated bacteria or infect aquatic amoeba, replicating within these protozoan hosts. Studies confirm that this capacity to infect protozoa allows the bacteria to also replicate within macrophages in the human respiratory system [3].

The transmission of *Legionella* to humans is primarily through aerosols generated by contaminated man-made water sources, such as air-conditioning systems, plumbing, showers, and more. Human inhalation of aerosols as well as aspiration of contaminated water directly into the pulmonary system [3]. Though, because human-to-human occurrence rarely occurs, human infection is a dead-end for bacteria and is only incidental.

L. pneumophila is a common cause of community-acquired pneumonia, accounting for 90% of *Legionella* cases [4]. It consists of 15 serogroups, with the first serogroup LP1 responsible for majority of the reported cases.

Legionella longbeachae is a major cause of disease only in Australia and New Zealand, though cases have been rising across Europe, the United States, Canada, Thailand, and Taiwan. It is unclear whether this increased reporting of *L. longbeachae* is due to higher incidence or increased clinical awareness and improved detection methods. Only two serogroups are recognized for *L. longbeachae*. While *Legionella* species are found in aquatic environments, *L. longbeachae* is found primarily in moist soil and potting mixes, when contaminated particles are aerosolized upon soil bags opening, handling potting mix, or when plants are watered [4]. Therefore associated with an increased frequency of infection in the spring and summer when the unique risk factors of gardening and using potting soil are at their highest.

Rising Contagious Diseases: Basics, Management, and Treatments, First Edition. Edited by Seth Kwabena Amponsah, Ranjita Shegokar, and Yashwant V. Pathak.

18.1.2 Epidemiology

The incidence of *Legionella* is thought to be underestimated due to misdiagnosis, lack of reporting and testing, variability in reporting systems, and the asymptomatic nature of some infections. As a result, the global incidence of *Legionella* is difficult to quantify. The CDC estimates that between 8000–18000 hospitalizations occur in the United States due to Legionnaires' disease each year [5]. In Europe, the average estimated incidence of *Legionella* is 9.2 per million people in 2011, with variability among different countries, which is slightly less than the estimated incidence of 10.8 per million people in the United States [6]. Around 75–80% of all reported cases are among individuals over 50 years old, with males accounting for 60–70% of the reported cases [7]. The incidence of *Legionella* has been increasing in many countries over the past two decades. Specifically, rates of *Legionella* are increasing in the United States and Europe annually, possibly due to seasonal changes that may impact incidence, an increasing number of susceptible individuals (elderly and immunocompromised individuals), and better surveillance and reporting systems [6].

Legionnaires' disease can be fatal. Most infected individuals have a typical mortality rate of 8–12%, but this rate may be higher in those who smoke, acquire the disease in healthcare settings, are elderly, have comorbidities, or receive a delay in treatment [6]. According to the World Health Organization (WHO), the reported mortality rate of *Legionella* ranges from 5–30% depending on the population affected and the specific outbreak [7]. Untreated immunocompromised patients may experience a death rate as high as 40–80%, but this rate can be reduced to 5–30% with appropriate management [7]. On average, the case-fatality rate for Legionnaires' disease is 10% in Europe and 8% in the United States, but these rates increase for nosocomial cases [5].

18.2 Bacterial Life Cycle

In the environmental stage of the virus, *Legionella* survives and replicates within free-living amoeba, a reservoir for the species, allowing it to inhabit freshwater environments and biofilms that surround plumbing. After being aerosolized, it can be inhaled into human lungs and infect human alveolar macrophages. *L. pneumophila* has a biphasic life cycle, switching between a non-virulent, replicative form and a virulent, transmissive form. This biphasic life cycle relies upon the metabolic state of the cell. The non-virulent, replicative form occurs in a phase of exponential growth in nutrient-rich conditions. In an environment with limited nutrients and increased bacterial density, *L. pneumophila* stops replicating in a stationary growth phase and begins to express transmission traits of motility, osmotic- and acid-resistance, and cytotoxicity [8]. After programmed cell death of the alveolar macrophage, these traits allow the bacteria to move extracellularly. The released bacteria can then infect new macrophages to begin the cycle anew (Figure 18.1).

18.3 Pathogenicity

Effector proteins are proteins that are usually essential for a pathogen's ability for virulence, survival, replication, or immunosuppression. The expression of Dot/Icm effector proteins in particular is crucial for the infective capacity of *L. pneumophila*. Dot stands for the defect in organelle trafficking, and Icm stands for intracellular multiplication. The Dot/Icm proteins encode a Type 4B secretion system (T4BSS). The T4BSS is a bacterial transport system, closely related to conjugation systems, of the uptake of macromolecules such as proteins and DNA [8]. *L. pneumophila* uses the Dot/Icm system to transfer circular pieces of bacterial DNA called *IncL* plasmids into the host cell, alongside at least 40 effector proteins into the host cell [8]. *Legionella* uses a unique method of motility called "gliding motility." Gliding motility is a mechanism of movement that allows bacteria to move smoothly over surfaces without the use of flagella. In the case of *L. pneumophila*, it uses the movement of a specialized organelle called the Type IV secretion system (T4SS) to achieve gliding motility. The T4SS is a complex structure that spans the bacterial membrane and facilitates the transfer of bacterial DNA, proteins, and other molecules between bacteria and host cells. In *L. pneumophila*, the T4SS plays a crucial role in its virulence and ability to cause disease by allowing the bacterium to move around and invade host cells [9].

Legionella pneumophila enters the host cell through their mechanisms of phagocytosis, which is the process of engulfing the microbe. An effector protein known as SidF is a crucial effector protein to manipulate the host cell cytoskeleton and promote bacterial entry into host cells [9]. Then, they carve out an intracellular niche in a membrane-bound replicative compartment known as *Legionella*-containing vacuole (LCV), using effector proteins such as SidJ [9]. After bacterial uptake

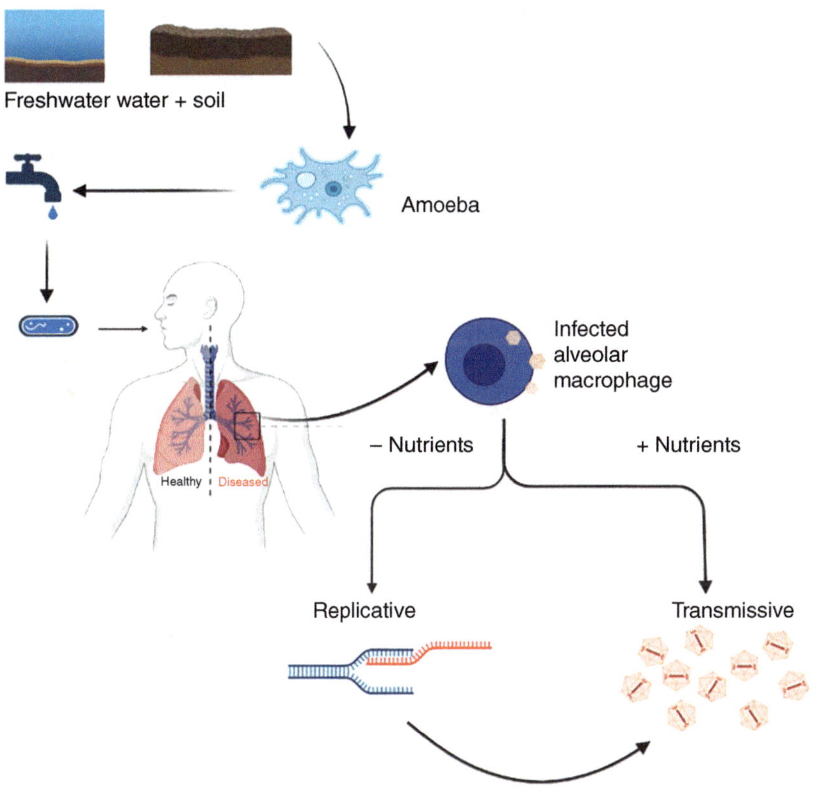

Figure 18.1 *Legionella pneumophila* life cycle.

into the cell, *Legionella* shelters within the LCV and thus evades phagolysosomal degradation and intercepts nutrients to support their transmissive phase [8]. The bacteria then secretes many effector proteins to alter the host cell signaling pathways and ensure a safe environment to replicate [8]. Alongside the T4BSS, *Legionella* also employs a Type 2 secretion system (T2SS) that serves to shuttle proteins from the inside of the cell to the outside in the extracellular matrix and plays an essential role in infection [8]. *L. pneumophila* in particular also employs a Type 1 secretion system (T1SS) for the secretion of important enzymes such as proteases, lipases, and other pore-forming enzymes into the extracellular space. A functional T1SS is a requirement for entry into the host cell [8]. The T2SS system allows for the delivery of over 25 effector proteins, first across the inner membrane of a cell into the periplasm, then through a T2SS-formed pore in the outer membrane, into the cytoplasm [8]. The effector proteins assist with intracellular replication within the amoeba and other functions such as a chitinase that prolongs bacterial survival within the lungs [8].

18.4 Alterations to Host Cell Pathways

The endoplasmic reticulum (ER) in a cell is where transmembrane proteins and lipids are produced for most of the cell's organelles, including secretory vesicles, the plasma membrane, the ER itself, and much more. *L. pneumophilia* interacts with the ER by intercepting vesicle traffic that is constantly entering and exiting the ER, and instead re-purposing them for their pathogenicity by establishing their replication vacuole [8]. The LCV is located by the host cell's ER sites of vesicle exit. Arf1, Sar1, and Rab1 are host enzymes that cleave GTP (GTPases) to produce energy, that regulate the transport of host cell vesicles to the membrane. *L. pneumophilia* hijacks their original purpose and instead uses them to recruit vesicles from the ER to the LCV membrane [8]. Rab1 is recruited to the LCV through a T4SS-dependent process and facilitates the recruitment of vesicles to the LCV membrane, while Sar1 and Arf1 work to recruit and tether ER vesicles to the membrane. Effector proteins SidE and SidF targets Rab in order to control *Legionella* replication, and LegB targets Arf GTPases for controlling vesicle trafficking and cytoskeletal organization. The LCV uses these SNARE proteins, host proteins that fuse membranes, to pair the LCV with the ER [8]. *L. pneumophilia* genes encode for an effector protein LseA that acts as a protein to mediate the fusion of the vesicle to the cell membrane and exocytose the contents. The LCV can move along the multitude of microtubules that surround the ER through T4SS effector protein and GTPase LegG1, thus promoting LCV motility through microtubule stabilization [8].

Legionella pneumophila not only hijacks the vesicular trafficking from the ER for the LCV to ensure a safe environment to replicate, but also alters and regulates host cell signaling pathways. The host cell phagosome, the organelle that allows macrophages to engulf and induce apoptosis in intracellular pathogens, is converted into a vacuole derived from the ER to resemble an immature autophagosome; the LCV has markers of this autophagosome that allows for it to survive undetected by host cell degradation pathways [8]. *L. pneumophila* also hijacks the NF-κB pathway, a pathway crucial for host cell innate immune response, apoptosis, and inflammation, to its advantage, using effector proteins such as AnkB [4]. Depending on the stage of infection, *L. pneumophila* secretes different effectors to either counteract the effects of NF-κB activation at the beginning of the infection cycle or strongly activate the NF-κB transcription factors, at the end of the replication cycle, such as effector protein LegK1 [4, 8]. Similarly, the MAPK pathway, a signaling cascade that relays signals to elicit cellular proliferation, differentiation, and apoptosis in host cells, is activated in a T4SS-dependent manner by *L. pneumophila* to activate MAPK and shape what the host cells can transcribe. Protein effector RomA methylates histones to directly affect what genes are silenced and what genes can be transcribed [8]. Through alterations of the host cell pathways, *L. pneumophila* is able to not only infect the cell but also use all of its innate machinery to its advantage for survival, replication, and infectivity.

18.5 Ubiquitin Pathways

Ubiquitination is when a molecule of ubiquitin protein is attached to a target protein to label it for degradation by a protease, thus regulating the functional activity, differentiation, maturation, and stability of proteins. It is essential for the host's immune response to extracellular pathogens. In *L. pneumophila*, several of the T4SS effector proteins are markedly similar in structure and function to typical eukaryotic E3 ubiquitin ligases, the primary enzyme that catalyzes the process of ubiquitination [8]. In particular, one *L. pneumophila* effector protein SidC that enhances the ER recruitment to the LCV is structurally similar to the E3 ubiquitin ligase family; SidF tags for degradation of the pro-apoptotic BNIP3 and Bcl-rambo proteins; SdhA tags an unknown substrate protein but prevents cell death. In addition, many other effector proteins can induce pro-apoptotic caspase-3 activity to promote cell death [8]. In a tightly controlled balance, *L. pneumophila* is able to secrete effector proteins that are both pro-apoptotic and anti-apoptotic to suit its needs to support both bacterial replication and survival initially, to bacterial release from the host cell at the end of the infection cycle (Table 18.1).

Table 18.1 Effectors and cellular targets.

SidC	Enhance ER recruitment to the LCV
SidE	Promote replication by interacting with host cell GTPase Rab and ubiquitin ligases
SidF	Promote bacterial entry into host cell through manipulation of the host cell microtubule cytoskeleton, recruit ER proteins to LCV, interact with host GTPase Rab1, and opposes the function of pro-apoptotic Bcl-rambo protein
SidJ	Modifies other effector proteins, and is necessary for the establishment of *Legionella*-containing vacuoles (LCVs) within host cells
SdhA	Inhibit host cell protein synthesis, which helps to prevent the host cell from mounting an effective immune response while preventing cell death
AnkB	Ubiquinate proteins to promote bacterial survival within host cells and supply nutrients to LCV
LepA	Inhibit host cell autophagy, which is a process that normally functions to clear intracellular pathogens
LepB	Target GTPases Arf1 and Arf6 to interrupt vesicular trafficking and cytoskeletal organization
LegG1	Activate GTPase for mitochondrial fragmentation in inducing cell apoptosis
LegK1	Phosphorylate proteins to interrupt NF-κB pathway and autophagy
LseA	Mediate membrane fusion for LCV creation
RomA	Changes histone marks for epigenetic regulation

18.6 *Legionella*-Amoeba Interactions

Legionella replicates and proliferates within amoeba, free-living protozoans within aquatic environments and soil. Amoebas are multicellular organisms that consume microorganisms. *Legionella* has evolved over millennia to become one of many bacteria that have learned to escape intracellular degradation and survive within the protozoa. *Legionella* infects amoeba the same way it infects human macrophages– lysosomal degradation of the phagocytized material, and replication within the organism while being protected from hostile environments. With this ability, the free-living amoeba can transport *Legionella* to new environments before expelling them through vesicles or releasing from the host cell through apoptosis [8]. Other bacteria known to be associated with amoeba include *Listeria*, *Mycobacterium*, and *Chlamydia*, as well as giant viruses Mimiviridae and Marseilleviridae [8]. The microorganisms can all reside within the amoeba simultaneously, leading to the movement of genetic material in multiple directions. While many bacteria tend to compress their genome when hiding inside of a host cell, *Legionella* and other amoeba-resistant bacteria do the opposite and instead expand their genome considerably by continuously acquiring more genes to create a modern *Legionella* that is considerably larger than their ancestral genome [8]. *Legionella* and other intracellular amoeba-resistant pathogens have been also seen to contain eukaryotic genes, of which the method of acquisition and integration is still unknown. In all, the amoeba represents a specialized host that both houses and protects many human pathogens, including *Legionella*. More research will be crucial to develop new technologies for disease prevention, control, and treatment.

18.7 Risk Factors

The bacterium *L. pneumophila* can be found in natural water sources and man-made water systems. Although anyone may contract the disease through the inhalation of contaminated aerosols or aspiration of contaminated water, certain factors increase the likelihood of developing Legionnaires' disease, including age above 50, cigarette smoking, excessive alcohol consumption, chronic lung diseases, occupational exposure (cooling towers or other water systems), and travel history to high-risk areas such as certain cruise ships or hotels [8]. Importantly, individuals with weakened immune systems, such as those having HIV/AIDS, cancer, or people taking immunosuppressive medications, are more likely to contract the disease [10, 11]. There is also a seasonal prevalence of transmission with most people getting infected during the summer and early fall seasons [12]. Prevention measures such as proper maintenance and disinfection of water systems, especially in high-risk settings, and avoiding high-risk activities can reduce the risk of infection.

18.8 Detection/Diagnostic Testing

For patients with Legionnaires' disease, the time taken to diagnose the disease can play a big role in determining clinical outcomes. The diagnosis of Legionnaires' disease requires a detailed patient history, clinical presentation, and diagnostic testing. The classical clinical presentation of Legionnaires' disease includes pneumonia with a productive cough. Patients may also experience extrapulmonary manifestations including neurological and gastrointestinal symptoms (Figure 18.2) [13]. These symptoms can help healthcare providers identify Legionnaires' disease in patients with community-acquired pneumonia.

Since *L. pneumophila* is a faintly staining, facultative intracellular, gram-negative bacterium, gram stain typically shows few organisms relative to neutrophils [8]. The gold standard for diagnosing Legionnaires' disease is the isolation and culture of the bacteria from clinical samples. *L. pneumophila* bacteria grow on a selective medium containing buffered charcoal yeast extract [14]. Serological tests detect the antibodies produced by the host in response to the infection. The two types of serological tests used for *L. pneumophila* are indirect fluorescent antibody (IFA) and enzyme-linked immunosorbent assay (ELISA) [15]. In addition, NAATs detects the DNA or RNA of the bacteria in clinical or environmental samples. Polymerase chain reaction (PCR) is the most commonly used NAAT for the detection of *L. pneumophila* [16]. Another method of detection is next-generation sequencing (NGS), a high-throughput sequencing technology that can sequence millions of DNA fragments simultaneously. Metagenomic sequencing using NGS has been shown to be a useful tool for the detection of *L. pneumophila* in hospitalized patients [17].

In clinical practice, diagnosis typically involves the combined use of imaging, urinary antigen testing, and blood culturing. Chest X-ray may show evidence of pneumonia, including infiltrates and consolidation [6]. The urinary antigen test detects a specific antigen produced by *L. pneumophila* in urine samples. This test is rapid, non-invasive, and has high sensitivity

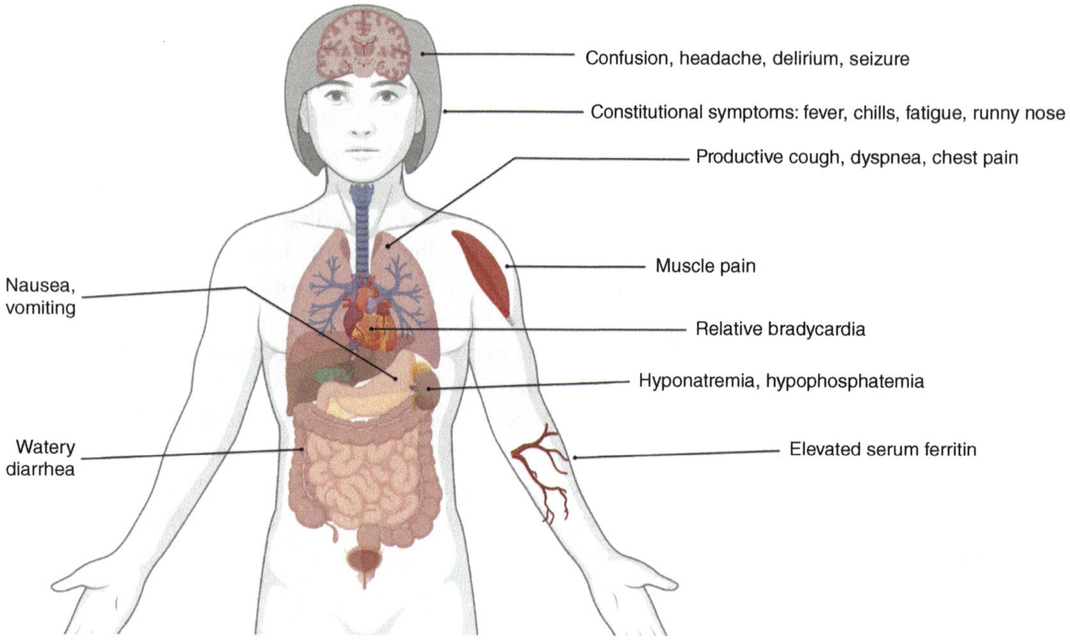

Figure 18.2 Clinical presentation of Legionnaires' disease.

and specificity [6]. Lastly, blood culture can be used to diagnose Legionnaires' disease in patients with severe illness. However, the bacteria grow slowly, and it can take more than three days for visible growth to appear [18].

Early detection and diagnosis of *L. pneumophila* are essential for preventing outbreaks and ensuring timely treatment. A combination of culture-based methods, serological tests, NAATs, and NGS can improve the accuracy of detection, while chest X-ray, urinary antigen tests, and blood culture are commonly used for the diagnosis of Legionnaires' disease.

18.9 Symptoms

The symptoms of Legionnaires' disease typically begin to appear 2–10 days after exposure to the bacteria [19]. The initial symptoms of the disease can be similar to those of the flu, including fever, chills, cough, and muscle aches [20]. However, unlike the flu, Legionnaires' disease can also cause gastrointestinal symptoms such as nausea, vomiting, and watery diarrhea [13].

A classical finding in patients with Legionnaires' disease is hyponatremia caused by salt wasting [21, 22]. Hyponatremia occurs because the bacteria can release toxins that damage the cells in the lungs and cause inflammation; the resulting increase in microvascular permeability causes fluid to leak into the lungs which consequently causes dilutional hyponatremia, as the body tries to maintain the proper balance of fluid in and around the cells [13]. Hyponatremia in Legionnaires' disease may also be caused by inappropriate secretion of antidiuretic hormone (ADH), also known as vasopressin which can lead to dilutional hyponatremia [23]. There is, however, conflicting information on the role of ADH in Legionnaires' disease hyponatremia. Another possible mechanism behind hyponatremia in Legionnaires' disease is a renal tubulointerstitial disease, which can impair sodium reabsorption by the kidneys and lead to excessive loss of sodium in the urine [24]. Hyponatremia may manifest as symptoms of nausea, vomiting, confusion, headache, fatigue, muscle weakness, seizures, and delirium [25, 26]. These neurological symptoms can be particularly dangerous in older adults and individuals with weakened immune systems.

Patients with Legionnaires' disease typically have a high fever, above 39 °C, and a heart rate that is slower than expected [13]. This condition, where the heart rate is slower than expected relative to the severity of the fever, is known as relative bradycardia and can be a helpful diagnostic clue for physicians. Other laboratory findings associated with *L. pneumophila* disease include hypophosphatemia. This electrolyte imbalance can be caused by a combination of factors, including the release of phosphorus from damaged cells and the renal wasting of phosphorus [27, 28]. In addition, patients with Legionnaires' disease may have increased serum ferritin levels. Ferritin is a protein that binds to iron and stores it in the body [13]. The mechanism behind this finding is not fully understood, but it may be related to the inflammation and tissue damage caused by the infection.

As the disease progresses, patients may experience severe community-acquired pneumonia with shortness of breath, chest pain, and a persistent productive cough [13]. In some cases, the disease can cause exudative pleural effusion or, rarely, acute respiratory distress syndrome (ARDS), a life-threatening condition that can lead to respiratory failure [13, 29]. In summary, Legionnaires' disease can cause a wide range of symptoms, including fever, chills, cough, muscle aches, neurological symptoms, gastrointestinal symptoms, relative bradycardia, high fever, hyponatremia, hypophosphatemia, and increased serum ferritin levels. Early recognition and treatment of Legionnaires' disease are essential to prevent severe complications such as ARDS.

18.10 Long-Term Effects and Co-morbidities

Several studies have suggested that Legionnaires' disease can have negative long-term effects on respiratory health related to post-pneumonic pulmonary fibrosis [30, 31]. Patients with Legionnaires' disease were also found to have a higher risk of developing rhabdomyolysis and renal failure [32].

In addition to respiratory health, Legionnaires' disease has also been associated with neurological and cognitive deficits [13]. Some studies have also drawn relationships between Legionnaires' disease and neurological dysfunction, often manifesting as ataxic motor symptoms [33]. Other studies have suggested that Legionnaires' disease can cause cognitive impairment, including memory loss and difficulty with attention and concentration [13]. Fatigue is a common symptom reported by patients recovering from Legionnaires' disease [13]. A study published in the *European Journal of Clinical Microbiology & Infectious Diseases* found that 44% of patients with Legionnaires' disease reported fatigue at six months post-discharge [34]. Another study published in the *Journal of Infection* in 2016 found that 64% of patients with Legionnaires' disease reported fatigue at six months post-discharge [35]. Some studies have also suggested that Post Traumatic Stress Disorder (PTSD) may occur in patients who have experienced severe illnesses, including those requiring hospitalization and intensive care unit (ICU) treatment, such as Legionnaires' disease. One study published in the *European Journal of Psychotraumatology* in 2015 found that 19% of patients who had been hospitalized with severe community-acquired pneumonia, including cases of Legionnaires' disease, developed PTSD symptoms during the first year after hospital discharge [36]. Another study published in the journal *Intensive Care Medicine,* in 2012, found that ICU survivors had an increased risk of developing PTSD, with symptoms appearing up to two years after hospital discharge [37]. While this study did not specifically focus on patients with Legionnaires' disease, it did include patients with severe respiratory infections.

In addition to the long-term effects of Legionnaires' disease, patients with this disease may also be at increased risk for comorbidities [13]. Patients with Legionnaires' disease were more likely to have cardiovascular disease and diabetes than the general population [6]. Other studies have suggested that Legionnaires' disease may increase the risk of developing autoimmune disorders and certain cancers [13]. The exact mechanisms underlying these comorbidities are not well understood, but it is thought that the inflammatory response to *L. pneumophila* infection may play a role [13]. *L. pneumophila* bacteria can trigger a strong immune response, which can lead to tissue damage and inflammation which may contribute to the development of chronic disease [38].

18.11 Prevention

There are several strategies for the prevention of Legionnaires' disease caused by *L. pneumophila*. Routine monitoring and maintenance of water systems is crucial for preventing the growth and spread of *Legionella*. This includes regular cleaning, disinfection, and temperature/pH monitoring of water systems, such as cooling towers, hot tubs, water tanks, and hot tubs.

Advanced water treatment technologies can help control the growth and spread of *Legionella* in water systems. A 2016 systematic review found evidence suggesting that both UV radiation and copper-silver ionization are techniques effective at reducing *Legionella* proliferation in environmental samples and hospital supplies of water [38]. UV disinfection is a simple strategy proven to disinfect water, but is only effective at the point of application, leaving downstream areas in water systems susceptible to *Legionella* growth [39]. Meanwhile, copper-silver ionization may be a preferable alternative to hyper chlorinating water to disinfect water systems of *L. pneumophila*.

Public education and awareness campaigns play a crucial role in raising awareness about the risk factors associated with Legionnaires' disease and the importance of maintaining water systems. These campaigns aim to educate the public, healthcare professionals, and at-risk populations about the disease, its transmission, and preventive measures. By disseminating information about the disease and preventive strategies, these campaigns contribute to reducing the incidence and severity of Legionnaires' disease.

Overall, prevention of Legionnaires' disease requires a multifaceted approach, including regular monitoring and maintenance of water systems, implementation of *Legionella*-specific regulations and guidelines, use of advanced water treatment technologies, and public education and awareness campaigns.

18.12 Pharmacology

Due to the nonspecific nature of *Legionella* infection's clinical and radiological symptoms, an empiric, broad-spectrum antibiotic treatment is recommended to treat the wide range of potential pneumonia pathogens if *L. pneumophila* is suspected [40, 41]. According to current American and European guidelines, macrolides and fluoroquinolones are recommended as first-line treatment for moderate and severe *Legionella* infection.

For healthy outpatient adults without comorbidities or risk factors for antibiotic-resistance pathogens, the American Thoracic Society (ATS) and Infectious Disease Society (IDS) recommend amoxicillin 1 g three times daily, doxycycline 100 mg twice daily, or a macrolide (i.e. azithromycin 500 mg on the first day then 250 mg daily) for empiric treatment of community-acquired pneumonia, which includes *Legionella* infection [42]. For outpatient adults with comorbidities like chronic heart, lung, renal, and liver diseases, the ATS and IDS recommend a combination therapy consisting of amoxicillin/clavulanate (i.e. 500 mg/125 mg three times daily), or a cephalosporin (i.e. cefpodoxime 200 mg twice daily), and a macrolide (i.e. azithromycin 500 mg on the first day then 250 mg daily) [42]. The duration for treatment should not exceed eight days in a responding patient [43]. ATS and IDS recommend a combination of B-lactam (i.e. ampicillin + sulbactam 1.5–3 g every six hours) and a macrolide (i.e. azithromycin 500 mg daily) or monotherapy of respiratory fluoroquinolone (i.e. levofloxacin 750 mg daily) for non-severe inpatient pneumonia. For severe inpatient pneumonia, ATS/IDS recommends either a B-lactam and macrolide or B-lactam and respiratory fluoroquinolone (Table 18.2).

Generally, the more severe the patient presents, the stronger the indication for combination therapy. Respiratory quinolones and macrolides have potent antimicrobial effects on both extracellular and intracellular *Legionella* [44]. Quinolones, especially levofloxacin, offer advantages over macrolides, such as shorter hospital stays [45], and they are efficacious at treating Legionnaires disease [46]. If a macrolide is used, azithromycin is the superior choice for *Legionella* specifically [43].

Table 18.2 Medication guidelines for Legionnaires' disease.

Outpatient adults with no comorbidities or risk factors for antibiotic-resistant pathogen	– Amoxicillin 1 g 3×/d or – Doxycycline 100 mg 2×/d or – Macrolide (e.g. azithromycin 500 mg on the first day then 250 mg daily)
Outpatient adults with comorbidities (e.g. chronic heart, lung, liver diseases)	– Amoxicillin/clavulanate (e.g. 500 mg/125 mg 3×/d) or – Cephalosporin (e.g. cefpodoxime 200 mg 2×/d) AND a macrolide (e.g. azithromycin 500 mg on the first day then 250 mg daily)
Inpatient adults with non-severe pneumonia	– B-lactam (e.g. ampicillin + sulbactam 1.5–3 g every 6 h) AND a macrolide (e.g. azithromycin 500 mg daily) or – Fluoroquinolone (e.g. levofloxacin 750 mg daily)
Inpatient adults with severe pneumonia	– B-lactam (e.g. ampicillin + sulbactam 1.5–3 g every 6 h) AND macrolide (azithromycin 500 mg daily) or – B-lactam (e.g. ampicillin + sulbactam 1.5–3 g every 6 h) and respiratory fluoroquinolone (e.g. levofloxacin 750 mg daily)

18.13 Treatments in Development

There are several new treatments in development for Legionnaires' disease. This includes antibiotics and specific therapies targeting different stages of the bacterial life cycle. Antibiotics are currently the primary treatment for Legionnaires' disease. These antibiotics help to reduce the severity of an infection and prevent complications. However, some strains of *Legionella* have become resistant to antibiotics, and new drugs are needed to treat these strains.

Researchers are currently developing new antibiotics, such as quinolones, ketolides, and glycylcyclines, that may be effective against antibiotic-resistant strains of *Legionella*. While levofloxacin is the preferred treatment among fluoroquinolones for Legionnaire's disease, there have been new developments in quinolones which include gatifloxacin, gemifloxacin, and moxifloxacin. These fourth-generation agents have an enhanced affinity for binding both DNA topoisomerase IV and gyrase, with their dual action amplifying the mechanism of action of fluoroquinolones. These treatments have shown to be as effective in treating Legionnaires' disease when compared to ciprofloxacin and levofloxacin, which are earlier fluoroquinolones [47]. Of these, gemifloxacin is indicated for atypical pneumonia such as Legionnaire's disease, due to its activity against atypical pathogens and aerobic gram-negative bacteria with minimal side effects and once daily oral dosing [48]. Third-generation macrolides, termed "ketolides," have been created in an effort to treat macrolide-resistant strains of bacteria. Ketolides are essentially a derivative of the second-generation macrolide erythromycin, created by replacing L-cladinose with a 3-keto group. This modification allows ketolides to bind to ribosomes with a higher affinity than macrolides, consequently inhibiting protein synthesis more effectively. Ketolides, including telithromycin, cethromycin, and solithromycin, have shown strong effectiveness in treating Legionnaires' disease [49]. Lastly, glycylcyclines are drugs derived from tetracyclines created to overcome pathogen resistance to tetracycline antibiotics, such as doxycycline. A particular glycylcycline, tigecycline, has shown promise in treating Legionnaire's disease, having shown 100% efficacy in treating *L. pneumophila* in guinea pigs [50]. However, further confirmation with clinical trials is needed to determine the efficacy of tigecycline in treating humans with Legionnaire's disease. It is evident that there are a myriad of first-line and newer antibiotics to choose from when treating *L. pneumophila*. Ultimately, the choice of antibiotic should be based off several factors including the specific strain of *Legionella* causing the infection, the severity of infection, antibiotic susceptibility testing, and individual patient factors.

Legionnaire's disease triggers an inflammatory response in the body. Some newer treatments focus on harnessing or modulating the body's immune system to fight infection, termed immunotherapies. Researchers are currently investigating the use of drugs that modulate the host immune response to reduce inflammation and improve the body's ability to fight infection. Specifically, the use of corticosteroids in Legionnaire's disease is a subject of debate and ongoing research. Corticosteroids are a class of immunomodulatory drugs that have anti-inflammatory properties; some studies suggest that combining such steroids with antibiotics may reduce the inflammatory response, therefore preventing complications and improving outcomes in community-acquired pneumonia [51]. However, this cannot be generalized to Legionnaire's disease without further clinical studies. Researchers are also currently investigating the use of monoclonal antibodies that target specific proteins on the surface of *Legionella*, but mainly for detection and diagnostic purposes.

Researchers are also investigating vaccines that can prevent infection or reduce the severity of symptoms. In particular, a recent study demonstrated that vaccination of mice with DNA vaccines expressing *Legionella* peptidoglycan-associated lipoprotein (PAL) successfully inducted PAL-specific $CD8^+$ T cells, producing a strong immunogenic response [52]. PAL is a protein present in the outer membrane of gram-negative bacteria such as the *Legionella* species that is also responsible for bacterial pathogenesis. Due to this induction of PAL-specific cytotoxic T cells, mice were protected to a greater extent when exposed to *L. pneumophila* when compared to control mice [52]. This may be one promising avenue for the development of future vaccines against *L. pneumophila*.

Phage therapy involves using bacteriophages, which are viruses that infect and kill bacteria, to treat bacterial infections. While phage therapy shows promise, it is still in the experimental stage for serving as a potential treatment for Legionnaire's disease. Researchers must identify bacteriophages that are capable of killing *L. pneumophila*, isolate and purify them, and then apply them at the site of infection in various forms, such as topically, orally, or through inhalation. In the context of Legionnaire's disease, inhalation would allow the bacteriophages to reach bacteria present in the respiratory system. From there, bacteriophages may attach to receptors on the surface of *L. pneumophila* bacteria, inject their genetic material, and ultimately cause lysis of the bacteria. This is the mechanism through which phage therapy may have the potential to prevent proliferation of bacteria. Studies have shown the promising identification of such bacteriophages that are active against *L. pneumophila*, but the later stages of isolation and purification have not been done successfully, so more research is needed on novel phage therapy [53]. Additionally, phage therapy may have the potential to not only treat humans but

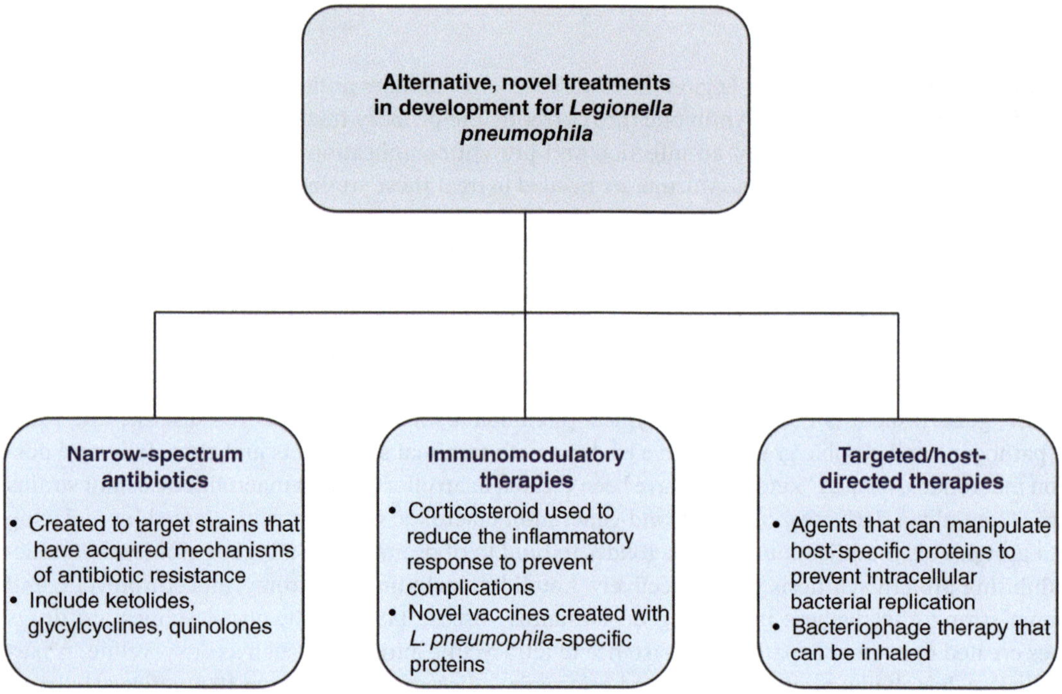

Figure 18.3 Treatments in development for *Legionella pneumophila*.

also serve as a means of decontamination in water systems. Another avenue for treating Legionnaire's disease may be through autophagy modulation, a type of host-directed therapy, which is a therapy that targets the host immune response rather than the bacteria itself. During pathogenesis, *L. pneumophila* inhibits host cell autophagy, which is a process that normally functions to clear intracellular pathogens. Therefore, drugs that modulate this autophagy process may have potential as a host-directed therapy to manipulate host response against *L. pneumophila* [53].

During pathogenesis, *L. pneumophila* prevents its own phagosome from fusing with a lysosome, effectively escaping degradation and therefore replicating intracellularly inside a host. One study found that a specific protein, termed an NLR protein Ipaf, is critical for restricting *Legionella* replication in mice by regulating this phagosomal maturation [54]. Essentially, in mice who lacked the NRL protein Ipaf, a *Legionella*-containing phagosome did not fuse with a lysosome, allowing the bacteria to escape degradation [7]. Findings like this may yield inspiration for new host-directed therapies, such as an agent or mechanism to increase this type of protein in host organisms to prevent *Legionella* bacterial growth.

Overall, these new treatments in development offer promising avenues for the prevention and treatment of Legionnaires' disease caused by *L. pneumophila* (Figure 18.3). However, more research is needed to determine their safety and efficacy in clinical settings.

18.14 New Technologies

There are several new technologies and approaches that are being developed to combat Legionnaires' disease caused by *L. pneumophila*. *Legionella*, identified 40 years ago, has been the focus of extensive research, revealing molecular tools that manipulate host pathways and uncovering mechanisms in eukaryotic cells. For example, *L. pneumophila*'s RomA was found to methylate Lys14 of histone H3, a modification unknown in mammalian cells [55]. Recent research shows that this epigenetic modification naturally occurs in eukaryotic cells, highlighting how bacterial research can advance our understanding of basic cellular processes. In the future, it is anticipated that epigenetic studies will involve comprehensive analysis of all identified histone modifications throughout an infection using genome-wide techniques [8]. These methods will be augmented with advanced tools such as Internal Standard Calibrated Chromatin Immunoprecipitation, which enables precise epigenomic profiling [56]. Additionally, new tools will facilitate the investigation of microRNA regulation in host cells during infection, as well as the study of nucleosome positioning in cells infected with *Legionella* [57]. Genome-wide techniques also include whole-genome sequencing, a technique that allows scientists to read the entire genetic code of an organism, including bacteria like *Legionella*. By sequencing the genomes of different strains of *Legionella*, researchers can identify unique genetic markers that can be used to track outbreaks, monitor the spread of the bacteria, and develop more effective treatments [58].

NGS of the lung microbiome during *Legionella* infection is another promising approach. It was originally believed that the lungs did not have any bacteria, but recent advances in sequencing technologies have revealed that they actually have their own microbiome, much like other parts of the body [59]. As a result, by using NGS on clinical lung samples like bronchoalveolar lavage fluids and sputum, it will be possible to analyze the microbiome of the lungs during a *Legionella* infection [60]. As the disease progresses, *Legionella* may displace the lung bacteria, akin to gut microbiome disruptions. Importantly, tracking microbiome changes will have implications for the development of new approaches for *Legionella* prevention, diagnosis, and control, as well as the creation of novel therapies [8].

Although we have gained greater insight into the biology and pathogenicity of *L. pneumophila*, there is still a lack of understanding regarding *L. longbeachae*'s infection mechanisms. Predictions of specific effector proteins in this species indicate that *L. longbeachae* can manipulate host cell pathways using different cellular mechanisms than those employed by *L. pneumophila*. The high prevalence of *L. longbeachae* in Australia and New Zealand is thought to be linked to the bacterium's presence in potting soils, which are primarily composed of composted pine bark or sawdust in these regions, unlike Europe. This implies that *L. longbeachae* may be associated with plants and trees, and that bacterial growth occurs during composting [61]. The *L. longbeachae* genome analysis showed that this species encodes a set of enzymes dedicated to using plant cell wall components for energy [62]. This supports the notion that *L. longbeachae* may be connected to, or even infecting, plants. Thus, it may be useful to further investigate whether other organisms, besides protozoa, may also serve as hosts for *Legionella* species [8]. *L. pneumophila* is known to hijack well-established immune pathways in macrophages that are not present in amoebae, such as caspase-mediated apoptosis or the NF-κB pathway, leading to speculation that interactions between *L. pneumophila* and other hosts closer to higher eukaryotes may have also played a role in molding this bacterium's effector range. Several findings support this hypothesis, including the observation that *L. pneumophila* can colonize and persist in the digestive tract of the nematode *Caenorhabditis elegans*, cause natural pneumonia in cattle, and be found in the microbiome of the gastrointestinal tract in *Panaque nigrolineatus*, a tropical freshwater fish [63–65]. The discovery of *Legionella* hosts other than protozoa will expand our knowledge and open up new avenues of research into *Legionella*-host interactions.

Advanced water treatment systems are more effective than traditional disinfection methods and can help prevent the transmission of *Legionella* in healthcare facilities, hotels, and other buildings [66]. Owing to the limitations associated with chemical and physical technologies, constant efforts are being made to develop alternatives. One such technology is UV irradiation. UV light induces the formation of pyrimidine dimers within the DNA strands. These dimers hinder proper DNA replication, leading to cell death [66]. A 2019 review article noted that "UV disinfection is a highly effective method for inactivating *Legionella*, and has been shown to reduce the risk of Legionnaires' disease in healthcare facilities" [67]. One study illustrated the effectiveness of UV irradiation in inactivating four different strains of *L. pneumophila* and *L. longbeachae*. At a fluence of $50\,J/m^2$ or $5.0\,mJ/cm^2$, a minimum reduction of three logs was achieved for all tested strains and species, which exhibited similar sensitivity to UV irradiation [66]. Limited research has focused on the effectiveness of UV irradiation for treating plumbing systems in hospitals [66]. In a study by Hall et al., the efficacy of UV irradiation was examined in a recently constructed hospital. Over a period of 13 years, regular sampling and *Legionella* testing yielded no positive environmental samples [68]. In contrast, data from other hospitals showed a 51% positivity rate, which was significantly higher than the hospital equipped with a UV lamp at the point-of-entry that had a 0% positivity rate. Additionally, the facility located across the street had positive environmental samples for *Legionella*, indicating the presence of this pathogen in the city's water supply. This study indicates that UV irradiation may effectively prevent the entry of pathogens into premise plumbing systems. However, it should be noted that UV lamps do not leave behind any residual disinfectant, making them ineffective for already colonized systems. Apart from the absence of residual disinfection, the current technology of mercury lamps used for UV irradiation has several drawbacks [68].

A major concern is the hazardous nature of improperly disposed mercury lamps, posing risks to both the environment and public health. Additionally, mercury lamps exhibit low energy efficiency with a wall-plug efficiency ranging from 15–35%. Their lifespan is also relatively short, lasting approximately 10,000 hours [69]. These limitations contribute to the infrequent utilization of UV lamps as a disinfection technology [68]. LEDs have recently emerged as a water treatment technology that can replace mercury lamps. LEDs can emit at particular wavelengths, allowing them to better target specific parts of a cell [70]. In addition, they are made of nontoxic and nonhazardous materials, like gallium/aluminum nitride or aluminum nitride, and are more energy efficient than mercury lamps [68]. One study found that specific wavelengths (i.e. 265 nm) are better at inactivating *L. pneumophila*, but further studies need to be conducted to better compare LED wavelengths to the current water treatment technologies [71]. Other proactive measures to improve water quality is the utilization of point-of-use (POU) filters, such as shower filters, especially in high-risk areas. These filters act as a physical barrier, protecting susceptible individuals from waterborne pathogens and playing a critical role in preventing Legionnaires' disease. Moreover, several review studies have shown that POU filters are effective in preventing exposure to *Legionella* spp. [72, 73]. A recent innovation of an electrically heatable carbon nanotube POU filter illustrated a 100% efficacy in inactivating *Legionella* on the membrane surface within 60 seconds. These filters can serve as an integral barrier to remove pathogens and eliminate microorganisms in both public and private water supplies [68].

Rapid diagnostic tests that can quickly and accurately detect *Legionella* infections are being developed. These tests can help doctors diagnose and treat Legionnaires' disease more quickly, which can improve patient outcomes and reduce the risk of transmission. For example, Matrix-Assisted Laser Desorption/Ionization Time-of-Flight Mass Spectrometry is a technique that can rapidly identify microorganisms based on their protein profiles. It has shown promise in identifying *Legionella* species directly from clinical samples, enabling quicker and more accurate diagnosis [74]. Another emerging technique is isothermal nucleic acid amplification, which refers to the amplification of DNA or RNA at a constant temperature, without the need for cycling as in PCR. This approach offers the advantage of rapid target detection (typically within 15–60 minutes) without requiring expensive and complex thermal cyclers that consume a lot of energy. Two commonly used isothermal techniques for detecting *Legionella* DNA in clinical and environmental samples are nucleic acid sequence-based amplification and loop-mediated isothermal amplification [75].

Microbiome-based approaches involve modifying the microbial communities in water systems to prevent the growth of *Legionella*. By introducing beneficial bacteria or modifying the environment to favor the growth of beneficial bacteria, researchers hope to create a hostile environment for *Legionella* and reduce the risk of Legionnaires' disease. For example, altering the water filter microbial community can seed the water treatment further downstream, allowing healthier drinking water [76].

These new technologies and approaches offer promising ways to prevent and treat Legionnaires' disease caused by *L. pneumophila*. However, more research is needed to determine their effectiveness and safety in clinical and real-world settings.

18.15 Conclusion

In conclusion, *L. pneumophila* is a gram-negative bacterium that can cause a severe form of pneumonia known as Legionnaires' disease. It exists as free-living biofilm-associated bacteria or infects aquatic amoeba, allowing it to replicate within these hosts. This capacity to infect protozoa also enables *Legionella* to replicate within macrophages in the human respiratory system. The transmission of *Legionella* to humans primarily occurs through contaminated man-made water sources, such as air-conditioning systems, showers, and plumbing, which generate aerosols.

L. pneumophila has a biphasic life cycle, switching between a replicative form and a transmissive form. The pathogenesis of Legionnaires' disease is complex and involves several virulence factors, including the Dot/Icm and T4BSSs, formation of a LCV, interactions with amoeba, and a multitude of effector proteins that manipulate host cell functions. *Legionella* also alters host cell vesicle trafficking, interacts with the endoplasmic reticulum, and hijacks host cell signaling pathways to evade degradation and ensure a favorable environment for replication. The bacteria employ ubiquitin pathways to regulate protein degradation and induce cell death. *Legionella*'s ability to infect amoeba plays a crucial role in its survival and dissemination. It can replicate within amoeba, which act as reservoirs and transport *Legionella* to new environments. Other bacteria and even viruses can coexist with *Legionella* inside amoeba, facilitating the exchange of genetic material.

Risk factors for Legionnaires' disease include age above 50, cigarette smoking, alcohol consumption, chronic lung diseases, occupational exposure, and travel to high-risk areas. Weakened immune systems increase susceptibility. Timely diagnosis relies on patient history, clinical presentation, and various tests such as gram stain, culture, and serological tests. Symptoms include flu-like manifestations, gastrointestinal symptoms, hyponatremia, relative bradycardia, and respiratory complications. Long-term effects may involve respiratory dysfunction, pulmonary fibrosis, increased risk of COPD and asthma, neurological deficits, cognitive impairment, fatigue, and potential comorbidities such as cardiovascular disease, diabetes, autoimmune disorders, and certain cancers. Inflammatory response to infection may contribute to chronic diseases.

Early recognition and treatment are crucial to prevent complications. Treatment for Legionnaires' disease typically involves fluoroquinolone and macrolide antibiotics, but the emergence of antibiotic-resistant strains highlights the need for alternative treatment options. Prevention of Legionnaires' disease requires a multifaceted approach, including regular monitoring and maintenance of water systems, implementation of *Legionella*-specific regulations and guidelines, and public education and awareness campaigns. Several new treatments are currently in development, including use of advanced water treatment technologies, genome-wide sequencing to target specific virulence factors of *Legionella*, rapid diagnostic tests, and microbiome-based approaches. Overall, a better understanding of the microbiology and pathogenesis of *Legionella*, coupled with improved prevention and treatment strategies, can help reduce the incidence and severity of Legionnaires' disease.

References

1 Fraser, D.W., Tsai, T.R., Orenstein, W. et al. (1977). Legionnaires' disease: description of an epidemic of pneumonia. *N. Engl. J. Med.* 297 (22): 1189–1197.

2 Glick, T.H., Gregg, M.B., Berman, B. et al. (1978). Pontiac fever: an epidemic of unknown etiology in a health department: I. Clinical and epidemiologic aspects. *Am. J. Epidemiol.* 107 (2): 149–160.

3 Cunha, B.A., Burillo, A., and Bouza, E. (2016). Legionnaires' disease. *Lancet* 387 (10016): 376–385. https://doi.org/10.1016/S0140-6736(15)60078-2.

4 Gonçalves, I.G., Simões, L.C., and Simões, M. (2021). *Legionella pneumophila*. *Trends Microbiol.* 29 (9): 860–861. https://doi.org/10.1016/j.tim.2021.04.005.

5 Hicks, L.A. et al. (2011). *Legionellosis – United States, 2000-2009*. Centers for Disease Control and Prevention https://www.cdc.gov/mmwr/preview/mmwrhtml/mm6032a3.htm.

6 Phin, N., Parry-Ford, F., Harrison, T. et al. (2014). Epidemiology and clinical management of Legionnaires' disease. *Lancet Infect. Dis.* 14 (10): 1011–1021. https://doi.org/10.1016/S1473-3099(14)70713-3. Epub 2014 Jun 23. PMID: 24970283.

7 World Health Organization (2022). Legionellosis. https://www.who.int/news-room/fact-sheets/detail/legionellosis.

8 Mondino, S., Schmidt, S., Rolando, M. et al. (2020). Legionnaires' disease: state of the art knowledge of pathogenesis mechanisms of *Legionella*. *Annu. Rev. Pathol.* 15: 439–466.

9 Ensminger, A.W. (2016). *Legionella pneumophila*, armed to the hilt: justifying the largest arsenal of effectors in the bacterial world. *Curr. Opin. Microbiol.* 29: 74–80. https://doi.org/10.1016/j.mib.2015.11.002. Epub 2015 Dec 19. PMID: 26709975.

10 Lanternier, F., Tubach, F., Ravaud, P. et al. (2013). Incidence and risk factors of *Legionella pneumophila* pneumonia during anti-tumor necrosis factor therapy: a prospective French study. *Chest* 144 (3): 990–998. https://doi.org/10.1378/chest.12-2820. PMID: 23744173.

11 Kao, A.S., Myer, S., Wickrama, M. et al. (2021). Multidisciplinary management of *Legionella* disease in immunocompromised patients. *Cureus* 13 (11): e19214. https://doi.org/10.7759/cureus.19214. PMID: 34873543; PMCID: PMC8638927.

12 Cunha, C.B. and Cunha, B.A. (2017). Legionnaire's disease since Philadelphia: lessons learned and continued progress. *Infect. Dis. Clin. North Am.* 31 (1): 1–5.

13 Cunha, B.A. (2010). Legionnaires' disease: clinical differentiation from typical and other atypical pneumonias. *Infect. Dis. Clin.* 24 (1): 73–105. https://doi.org/10.1016/j.idc.2009.10.006.

14 Pierre, D.M., Baron, J., Yu, V.L., and Stout, J.E. (2017). Diagnostic testing for Legionnaires' disease. *Ann. Clin. Microbiol. Antimicrob.* 16 (1): 59. https://doi.org/10.1186/s12941-017-0229-6. PMID: 28851372; PMCID: PMC5576257.

15 Rojas, A., Navarro, M.D., Fornés, F.E. et al. (2005). Value of serological testing for diagnosis of legionellosis in outbreak patients. *J. Clin. Microbiol.* 43 (8): 4022–4025. https://doi.org/10.1128/JCM.43.8.4022-4025.2005. PMID: 16081945; PMCID: PMC1233976.

16 Chen, D.J., Procop, G.W., Vogel, S. et al. (2015). Utility of PCR, culture, and antigen detection methods for diagnosis of legionellosis. *J. Clin. Microbiol.* 53 (11): 3474–3477. https://doi.org/10.1128/JCM.01808-15. Epub 2015 Aug 19. PMID: 26292304; PMCID: PMC4609676.

17 Huang, P.H., Huang, Y.T., Lee, P.H. et al. (2023). Diagnosis of Legionnaires' disease assisted by next-generation sequencing in a patient with COVID-19. *Infect. Drug Resist.* 16: 355–362. https://doi.org/10.2147/IDR.S396254.

18 Barth Reller, L., Weinstein, M.P., and Murdoch, D.R. (2003). Diagnosis of *Legionella* infection. *Clin. Infect. Dis.* 36 (1): 64–69. https://doi.org/10.1086/345529.

19 Edens, W. (2020). Legionnaires' disease & pontiac fever. In: *CDC Yellow Book 2020: Health Information for International Travelers*, 390–392. New York: Oxford University Press.

20 Centers for Disease Control and Prevention (CDC). (2021). *Signs and Symptoms of Legionnaires' Disease*. Retrieved from: https://www.cdc.gov/legionella/about/signs-symptoms.html.

21 Puri, S., Kelly, M.B., Walker, J.D. et al. (2017). Clinical and laboratory findings between *Legionella* and non-*Legionella pneumonia* in a veteran population. *Open Forum Infect. Dis.* 4 (Suppl 1): S585. https://doi.org/10.1093/ofid/ofx163.1530.

22 Saraya, T., Nunokawa, H., Ohkuma, K. et al. (2018). A novel diagnostic scoring system to differentiate between *Legionella pneumophila* pneumonia and *Streptococcus pneumoniae* pneumonia. *Intern. Med.* 57 (17): 2479–2487. https://doi.org/10.2169/internalmedicine.0491-17. Epub 2018 Mar 30. PMID: 29607950; PMCID: PMC6172550.

23 Schuetz, P., Haubitz, S., Christ-Crain, M. et al. (2013). Hyponatremia and anti-diuretic hormone in Legionnaires' disease. *BMC Infect. Dis.* 13: 585. https://doi.org/10.1186/1471-2334-13-585. PMID: 24330484; PMCID: PMC3880094.

24 Daumas, A., El-Mekaoui, F., Bataille, S. et al. (2012). Acute tubulointerstitial nephritis complicating Legionnaires' disease: a case report. *J. Med. Case Rep.* 6: 100. https://doi.org/10.1186/1752-1947-6-100. PMID: 22475340; PMCID: PMC3359167. https://pubmed.ncbi.nlm.nih.gov/10732834/.

25 Douglas, I. (2006). Hyponatremia: why it matters, how it presents, how we can manage it. *Cleve. Clin. J. Med.* 73 (Suppl 3): S4–S12. https://doi.org/10.3949/ccjm.73.suppl_3.s4. PMID: 16970147.

26 Weismann, D., Schneider, A., and Höybye, C. (2016). Clinical aspects of symptomatic hyponatremia. *Endocr. Connect.* 5 (5): R35–R43. https://doi.org/10.1530/EC-16-0046. Epub 2016 Sep 8. PMID: 27609587; PMCID: PMC5314806.

27 Watanabe, S., Kono, K., Fujii, H. et al. (2016). Two cases of hypophosphatemia with increased renal phosphate excretion in *Legionella pneumonia*. *Case Rep. Nephrol. Dial.* 6 (1): 40–45. https://doi.org/10.1159/000444875. PMID: 27066493; PMCID: PMC4821156.

28 Mouton, W.J., Boshuizen, H.C., Kuipers, S. et al. (2021). A prospective cohort study of long-term health effects of *Legionella pneumophila* infection. *BMC Infect. Dis.* 21 (1): 128. https://doi.org/10.1186/s12879-021-05838-7. PMID: 33531034; PMCID: PMC8797088.

29 Kashif, M., Patel, R., Bajantri, B., and Diaz-Fuentes, G. (2017). *Legionella pneumonia* associated with severe acute respiratory distress syndrome and diffuse alveolar hemorrhage – a rare association. *Respir. Med. Case Rep.* 21: 7–11. https://doi.org/10.1016/j.rmcr.2017.03.008.

30 Cunha, B.A. (2014). Legionnaires' disease: clinical differentiation from typical and other atypical pneumonias. *Infect. Dis. Clin. North Am.* 28 (1): 105–118. https://doi.org/10.1016/j.idc.2013.10.002.

31 Blackmon, J.A., Harley, R.A., Hicklin, M.D., and Chandler, F.W. (1979). Pulmonary sequelae of acute Legionnaires' disease pneumonia. *Ann. Intern. Med.* 90 (4): 552–554. https://doi.org/10.7326/0003-4819-90-4-552. PMID: 434633.

32 Soni, A.J. and Peter, A. (2019). Established association of *Legionella* with rhabdomyolysis and renal failure: a review of the literature. *Respir. Med. Case Rep.* 28: 100962. https://doi.org/10.1016/j.rmcr.2019.100962. PMID: 31720209; PMCID: PMC6838801.

33 de Lau, L.M.L., Siepman, D.A.M., Remmers, M.J.M. et al. (2010). Acute disseminating encephalomyelitis following Legionnaires disease. *Arch. Neurol.* 67 (5): 623–626. https://doi.org/10.1001/archneurol.2010.75.

34 Cao, B., Ren, L.L., Zhao, F. et al. (2013). Discharge criteria for COVID-19 patients. *Lancet Infect. Dis.* 20 (5): 540–548.

35 Zhou, L., Zhou, J., Sun, D. et al. (2016). Clinical features of pneumonia caused by *Mycoplasma pneumoniae* in children. *Chin. J. Pediatr.* 54 (8): 584–589.

36 Vythilingam, M., Vermetten, E., Anderson, G.M. et al. (2015). Hippocampal volume, memory, and cortisol status in major depressive disorder: effects of treatment. *Biol. Psychiatry* 77 (3): 285–293. https://doi.org/10.1016/j.biopsych.2014.08.009.

37 Parker, A.M., Sricharoenchai, T., Raparla, S. et al. (2012). Posttraumatic stress disorder in critical illness survivors: a metaanalysis. *Crit. Care Med.* 40 (9): 616–623. https://doi.org/10.1097/CCM.0b013e31823e99e5.

38 Almeida, D., Cristovam, E., Caldeira, D. et al. (2016). Are there effective interventions to prevent hospital-acquired Legionnaires' disease or to reduce environmental reservoirs of *Legionella* in hospitals? A systematic review. *Am. J. Infect. Control* 44 (11): e183–e188. https://doi.org/10.1016/j.ajic.2016.06.018. PMID: 27524259.

39 Bartram, J., Chartier, Y., Lee, J.V. et al. (2007). *Legionella and the Prevention of Legionellosis*. World Health Organization https://apps.who.int/iris/handle/10665/43233.

40 Sharma, L., Losier, A., Tolbert, T. et al. (2017). Atypical pneumonia: updates on *Legionella*, *Chlamydophila*, and *Mycoplasma pneumonia*. *Clin. Chest Med.* 38: 45–58.

41 Carratalà, J. and Garcia-Vidal, C. (2010). An update on *Legionella*. *Curr. Opin. Infect. Dis.* 23: 152–157.

42 Metlay, J.P., Waterer, G.W., Long, A.C. et al. (2019). Diagnosis and treatment of adults with community-acquired pneumonia. An official clinical practice guideline of the American Thoracic Society and Infectious Diseases Society of America. *Am. J. Respir. Crit. Care Med.* 200 (7): e45–e67. https://doi.org/10.1164/rccm.201908-1581ST.

43 Plouffe, J.F., Breiman, R.F., Fields, B.S. et al. (2003). Azithromycin in the treatment of *Legionella pneumonia* requiring hospitalization. *Clin. Infect. Dis.* 37 (11): 1475–1480. https://doi.org/10.1086/379329.

44 Miyashita, N., Kobayashi, I., Higa, F. et al. (2018). In vitro activity of various antibiotics against clinical strains of *Legionella* species isolated in Japan. *J. Infect. Chemother.* 24 (5): 325–329. https://doi.org/10.1016/j.jiac.2018.01.018.

45 Sabrià, M., Pedro-Botet, M.L., Gómez, J. et al. (2005). Fluoroquinolones vs macrolides in the treatment of Legionnaires disease. *Chest* 128 (3): 1401–1405. https://doi.org/10.1378/chest.128.3.1401.

46 Yu, V.L., Greenberg, R.N., Zadeikis, N. et al. (2004). Levofloxacin efficacy in the treatment of community-acquired legionellosis. *Chest* 125 (6): 2135–2139. https://doi.org/10.1378/chest.125.6.2135.

47 Saravolatz, L.D. and Leggett, J. (2003). Gatifloxacin, gemifloxacin, and moxifloxacin: the role of 3 newer fluoroquinolones. *Clin. Infect. Dis.* 37 (9): 1210–1215. https://doi.org/10.1086/378809.

48 Amitabh, V., Singhal, A., Kumar, S. et al. (2012). Efficacy and safety of oral gemifloxacin for the empirical treatment of pneumonia. *Lung India* 29 (3): 248–253. https://doi.org/10.4103/0970-2113.99109. PMID: 22919164; PMCID: PMC3424864.

49 Dinos, G.P. (2017). The macrolide antibiotic renaissance. *Br. J. Pharmacol.* 174 (18): 2967–2983. https://doi.org/10.1111/bph.13936. Epub 2017 Aug 10. PMID: 28664582; PMCID: PMC5573421.

50 Edelstein, P.H., Weiss, W.J., and Edelstein, M.A. (2003). Activities of tigecycline (GAR-936) against *Legionella pneumophila* in vitro and in guinea pigs with *L. pneumophila* pneumonia. *Antimicrob. Agents Chemother.* 47 (2): 533–540. https://doi.org/10.1128/AAC.47.2.533-540.2003. PMID: 12543655; PMCID: PMC151731.

51 Confalonieri, M., Urbino, R., Potena, A. et al. (2005). Hydrocortisone infusion for severe community-acquired pneumonia. *Am. J. Respir. Crit. Care Med.* 171 (3): 242–248. https://doi.org/10.1164/rccm.200406-808oc.

52 Kim, S.J., Sin, J.-I., and Kim, M.J. (2020). $CD8^+$ T cells directed against a peptide epitope derived from peptidoglycan-associated lipoprotein of *Legionella pneumophila* confer disease protection. *Front. Immunol.* 11: 604413. https://doi.org/10.3389/fimmu.2020.604413.

53 Lammertyn, E., Voorde, J.V., Meyen, E. et al. (2008). Evidence for the presence of *Legionella* bacteriophages in environmental water samples. *Microb. Ecol.* 56 (1): 191–197. http://www.jstor.org/stable/40343358.

54 Amer, A., Franchi, L., Kanneganti, T.D. et al. (2006). Regulation of *Legionella* phagosome maturation and infection through flagellin and host Ipaf. *J. Biol. Chem.* 281 (46): 35217–35223. https://doi.org/10.1074/jbc.M604933200. Epub 2006 Sep 19. PMID: 16984919.

55 Rolando, M., Sanulli, S., Rusniok, C. et al. (2013). *Legionella pneumophila* effector RomA uniquely modifies host chromatin to repress gene expression and promote intracellular bacterial replication. *Cell Host Microbe* 13 (4): 395–405.

56 Grzybowski, A.T., Chen, Z., and Ruthenburg, A.J. (2015). Calibrating ChIP-Seq with nucleosomal internal standards to measure histone modification density genome wide. *Mol. Cell* 58 (5): 886–899.

57 Schones, D.E., Cui, K., Cuddapah, S. et al. (2008). Dynamic regulation of nucleosome positioning in the human genome. *Cell* 132 (5): 887–898.

58 Reuter, S., Harrison, T.G., Köser, C.U. et al. (2013). A pilot study of rapid whole-genome sequencing for the investigation of a *Legionella* outbreak. *BMJ Open* 3: e002175. https://doi.org/10.1136/bmjopen-2012-002175.

59 Dickson, R.P., Erb-Downward, J.R., and Huffnagle, G.B. (2015). Homeostasis and its disruption in the lung microbiome. *Am. J. Physiol. Lung Cell. Mol. Physiol.* 309 (10): L1047–L1055.

60 Pérez-Cobas, A.E. and Buchrieser, C. (2019). Analysis of the pulmonary microbiome composition of *Legionella pneumophila*-infected patients. *Methods Mol. Biol.* 1921 (4): 429–443.

61 Whiley, H. and Bentham, R. (2011). *Legionella longbeachae* and legionellosis. *Emerg. Infect. Dis.* 17 (4): 579–583.

62 Cazalet, C., Gomez-Valero, L., Rusniok, C. et al. (2010). Analysis of the *Legionella longbeachae* genome and transcriptome uncovers unique strategies to cause Legionnaires' disease. *PLos Genet.* 6 (2): e1000851.

63 Brassinga, A.K.C., Kinchen, J.M., Cupp, M.E. et al. (2010). *Caenorhabditis* is a metazoan host for *Legionella*. *Cell. Microbiol.* 12 (3): 343–361.

64 Fabbi, M., Pastoris, M.C., Scanziani, E. et al. (1998). Epidemiological and environmental investigations of *Legionella pneumophila* infection in cattle and case report of fatal pneumonia in a calf. *J. Clin. Microbiol.* 36 (7): 1942–1947.

65 McDonald, R., Schreier, H.J., and Watts, J.E.M. (2012). Phylogenetic analysis of microbial communities in different regions of the gastrointestinal tract in *Panaque nigrolineatus*, a wood-eating fish. *PLoS One* 7 (10): e48018.

66 Knudson, G.B. (1985). Photoreactivation of UV-irradiated *Legionella pneumophila* and other *Legionella* species. *Appl. Environ. Microbiol.* 49: 975–980.

67 Hall, K.K., Giannetta, E.T., Getchell-White, S.E. et al. (2003). Ultraviolet light disinfection of hospital water for preventing nosocomial *Legionella* infection: a 13-year follow-up. *Infect. Control Hosp. Epidemiol.* 24: 580–583.

68 Carlson, K.M., Boczek, L.A., Chae, S., and Ryu, H. (2020). Legionellosis and recent advances in technologies for *Legionella* control in premise plumbing systems: a review. *Water (Basel)* 12 (3): 1–676. https://doi.org/10.3390/w12030676. PMID: 32704396; PMCID: PMC7377215.

69 Song, K., Mohseni, M., and Taghipour, F. (2016). Application of ultraviolet light-emitting diodes (UV-LEDs) for water disinfection: a review. *Water Res.* 94: 341–349.

70 Vilhunen, S., Sarkka, J., and Silanpaa, M. (2009). Ultraviolet light-emitting diodes in water disinfection. *Environ. Sci. Pollut. Res.* 16: 439–442.

71 Rattanakul, S. and Oguma, K. (2018). Inactivation kinetics and efficiencies of UV-LEDs against *Pseudomonas aeruginosa*, *Legionella pneumophila*, and surrogate microorganisms. *Water Res.* 130: 31–37.

72 Sheffer, P.J., Stout, J.E., Wagener, M.M., and Muder, R.R. (2005). Efficacy of new point-of-use water filter for preventing exposure to *Legionella* and waterborne bacteria. *Am. J. Infect. Control* 33: S20–S25.

73 Baron, J.L., Peters, T., Shafer, R. et al. (2014). Field evaluation of a new point-of-use faucet filter for preventing exposure to *Legionella* and other waterborne pathogens in health care facilities. *Am. J. Infect. Control* 42: 1193–1196.

74 Pascale, M.R., Mazzotta, M., Salaris, S. et al. (2020). Evaluation of MALDI-TOF mass spectrometry in diagnostic and environmental surveillance of *Legionella* species: a comparison with culture and *Mip*-gene sequencing technique. *Front. Microbiol.* 11: 589369. https://doi.org/10.3389/fmicb.2020.589369. PMID: 33384668; PMCID: PMC7771186.

75 Mercante, J.W. and Winchell, J.M. (2015). Current and emerging *Legionella* diagnostics for laboratory and outbreak investigations. *Clin. Microbiol. Rev.* 28 (1): 95–133. https://doi.org/10.1128/CMR.00029-14. PMID: 25567224; PMCID: PMC4284297.

76 Wang, H. et al. (2013). Probiotic approach to pathogen control in premise plumbing systems? A review. *Environ. Sci. Technol.* 47 (18): 10117–10128.

19

Babesiosis: An Emerging Global Threat

Komal Parmar[1] and Jayvadan K. Patel[2]

[1] *Department of Pharmacy, ROFEL, Shri G.M. Bilakhia College of Pharmacy, Vapi, Gujarat, India*
[2] *Formulation and Development, Aavis Pharmaceuticals, Hoschton, GA, United States*

19.1 Introduction

Babesiosis is a serious, developing infectious illness that destroys the red blood cells, with widespread implications that affect both people and domestic animals. The multisystemic disease babesiosis, also known as piroplasmosis because of its pear-shape, is driven on by protozoan parasites of the genus *Babesia* [1–3]. Although additional methods of transmission exist, such as vertical transmission, transmission through blood transfusions, or organ donation, the primary method of transfer of babesiae to mammalian hosts is by a bite of ticks, which have diverse geographic ranges dependent on the presence of their capable natural animal hosts [4]. Many parasite species belonging to the genus *Babesia* are spread when ticks feed on the blood of their vertebrate hosts. Although it has long been known that these natural hosts can become infected with parasites, which can lead to diseases like bovine babesiosis and cause major economic costs, the seriousness of human infection is quickly becoming obvious whether the disease was originally transmitted by a tick bite or secondarily transferred through a blood transfusion with infected blood. *Babesia microti* (common in the United States) [5], *Babesia duncani* (formerly WA1 type) (across Canada) [6], *Babesia divergens* (restricted to Europe only) *B. divergens*-like (MO1 and EU1) [7], and *Babesia venatorum* (China) [8] are among the *Babesia* species that can infect people. Other genetically related pathogen sub-strains, such as *B. divergens*-like and *B. microti*-like pathogens, have been identified to infect people [9, 10]. Most cases of human babesiosis occur in the temperate zone. The most common species is *B. microti*, which is confined to southwestern China, the north-eastern and northern midwestern United States. Table 19.1 exhibits the first reports of babesia infections in humans worldwide (adapted from [5]). Table 19.2 tabulates the worldwide case distribution of babesiosis human infection. Figure 19.1 shows geographic distribution of human babesiosis transmission.

19.2 World-Wide Babesiosis Human Infection

19.2.1 Americas

Babesiosis was first discovered in the United States in California in 1966 [11]. Then, in 1970, *B. microti* was reported in Nantucket, Massachusetts [31], and with subsequent reports, the disease became infamous as Nantucket fever [32]. The frequency of cases from New Jersey, where babesiosis case reporting started in 1985, highlights the rise in human babesiosis in the northeast corridor [33]. Additionally, a recent *B. microti* infection in Canada [34] and instances recorded further east into Pennsylvania [35] demonstrate that the range of transmission is unmistakably growing. *B. duncani* was reported in western America, primarily in Washington and California [36]. Isolated cases of *B. divergens* are reported in various parts of America, such as Missouri, Kentucky, and Washington State [37–39]. Human babesios infections are also reported in some parts of South America, such as Columbia [40] and Mexico [41].

Rising Contagious Diseases: Basics, Management, and Treatments, First Edition. Edited by Seth Kwabena Amponsah, Ranjita Shegokar, and Yashwant V. Pathak.

Table 19.1 *Babesia* human infection first reported worldwide.

***Babesia* species infected**	**Year of report**	**Region of infection**	**Reference**
Babesia microti	1966	North-east USA	[11]
Babesia divergens	1957	Western Europe	[12]
Babesia duncani	1991	West USA	[13]
Babesia venatorum	2003	Europe	[14]
Babesia motasi	2007	South Korea	[15]
Babesia crassa-like agent	2018	China	[16]

Table 19.2 Worldwide case distribution of babesiosis human infection.

Continent/country	**Species of Babesiosis infection**	**References**
United States	*B. microti, B. divergens*–like	[9, 17, 18]
Canada	*B. duncani*	[6]
Spain	*B. microti, B. divergens*	[19–21]
Korea	*B. microti, B. motasi*	[22]
Russia	*B. divergens*	[23]
Europe	New Human Babesia sp. FR1	[24]
China	*B. venatorum, B. microti, B. divergens, B. crassa*	[8, 16, 25, 26]
India	*Babesia* species	[27]
Norway	*B. divergens*	[28]
Colombia	*Babesia* species	[29]
Ecuador	*Babesia* species	[30]

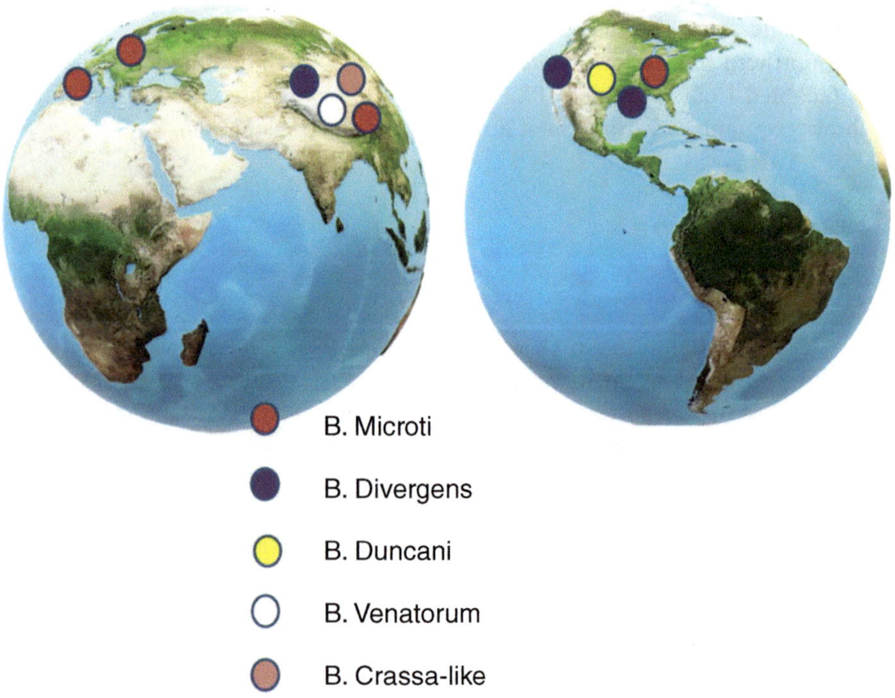

Figure 19.1 Schematic diagram showing global transmission of human babesiosis infection.

19.2.2 Europe

The first reported case of human babesiosis infection was in the former Yugoslavia, now Croatia, in 1957. Considering this, more than many instances of babesiosis have been documented on the European continent [42]. *B. divergens* is the main pathogen in Europe, but *B. microti* and *B. venatorum* have been found in a handful of cases [14, 43, 44]. Epidemiologic studies have revealed that *B. divergens* and its associated tick vector, *Ixodes ricinus*, are widely distributed throughout Europe [45].

19.2.3 Asia, Africa and Australia

Four Babesia spp. that have been linked to human infections in China are *B. microti*, *B. divergens*, *B. venatorum*, and *Babesia crassa*-like agent [16, 26, 46, 47]. In Taiwan, Japan, and South Korea, similar organisms to *B. microti* have been discovered [15, 48, 49]. Infections similar to *B. divergens* have been documented on the Canary Islands [50], and other, as of yet unidentified babesia species, have been reported in Egypt, Mozambique, and South Africa [51]. Whereas *B. microti* was documented in Australia [52].

19.3 Life Cycle and Transmission

Babesia species have intricate life cycles that go through various stages in both the tick and the mammalian host (Figure 19.2). When *Babesia* sporozoites are introduced into a mammalian host during a blood meal, they immediately invade and multiply in the erythrocytes. In the host or after being ingested by a tick, some of the parasite population changes into gametocytes and produces gametes in the tick's intestinal lumen. Diploid zygotes are created from gametes after fertilization. Kinetes are formed when zygotes enter the midgut epithelium and go through meiosis. The invasion and growth of kinetes in several organs, such as the salivary glands, leads to transstadial transfer. Except for *B. microti*, most *Babesia* species have kinetes that enter the ovaries and eggs, which causes the parasite to be passed on to the progeny [53]. During each of these stages – replication, egress, and invasion – the parasite makes use of extensive signaling networks [54].

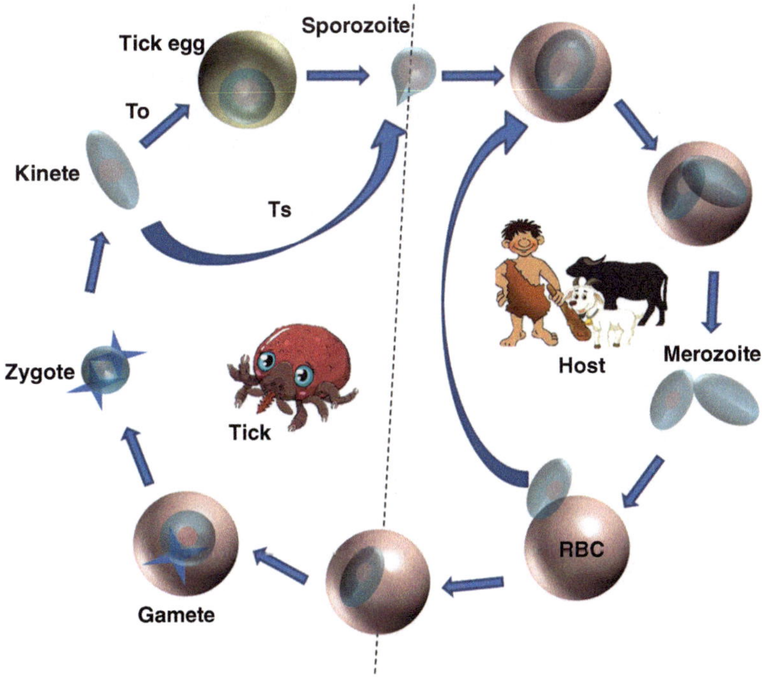

Figure 19.2 Life cycle and transmission of Babesia species.

19.4 Clinical Features, Pathogenesis, and Diagnosis of Babesiosis

Babesiosis symptoms can linger for six to eight weeks, and asymptomatic patients can go unnoticed for years. The clinical manifestations of *Babesia* species, intraerythrocytic protozoans that cause similar inflammatory reactions, include headaches, fever, chills, nausea, vomiting, myalgia, altered mental status, disseminated intravascular coagulation, anemia with hypotension, respiratory distress, hepatomegaly, and renal insufficiency [53, 55, 56]. The host's immune capacity frequently determines how serious an infection will be. It has been observed that the incubation phase normally lasts between one and six weeks. Only high-risk people, particularly those with a history of asplenia, are frequently affected by severe disease. These patients may suffer from respiratory distress, congestive heart failure, renal failure, splenic rupture, disseminated intravascular coagulation (DIC), hepatitis, or coma, among other multi-organ dysfunctions.

Giemsa or Wright staining is commonly used to detect the organism on a thin smear of peripheral blood, and the severity of parasitemia can be determined. It is advised to examine numerous thin smears in the early stages of infection because the parasite burden may be minimal. The most frequent shapes are rings, and each cell may include several rings (Figure 19.3). Maltese crosses, sometimes known as tetrad formations, are occasionally observed. In addition, reference facilities offer PCR testing, which is more accurate than peripheral smears. The indirect immunofluorescent antibody test used in serology is helpful for verifying the diagnosis. A fourfold increase in acute and convalescent titers confirms a recent infection, yet a single positive serology cannot discriminate between an acute and prior infection [57, 58]. Anemia, raised LDH, thrombocytopenia, transaminitis, proteinuria, elevated BUN, and elevated creatinine are a few of the laboratory abnormalities that might be present in babesiosis patients [59–61].

19.5 Management of Babesiosis

If a blood test or PCR has been positive for more than three months and the patient is symptomatic or asymptomatic, treatment is recommended. Atovaquone, azithromycin, clindamycin, and quinine make up the majority of the current therapy options for human babesiosis. Atovaquone specifically targets the mitochondrial electron transport chain's cytochrome bc_1 complex [62, 63]. Toxoplasmosis [64], malaria (when combined with proguanil) [65], *Pneumocystis jirovecii* pneumonia [66], and other human diseases are all treated with atovaquone. Azithromycin is a broad-spectrum antibiotic used to treat a variety of bacterial infections, including those spurred on by Staphylococcus species [67] and Legionella species [68]. An effective inhibitor of protein synthesis, azithromycin targets the translational apparatus in the apicoplast of apicomplexan parasites [69]. The "delayed death" effect of azithromycin, which prevents parasite division from producing viable daughter cells that cannot divide in the future cycle, has been observed [70]. Another frequent antibiotic for treating different bacterial infections is clindamycin, which is also used to treat parasitic infections. Clindamycin is used to treat babesiosis and malaria in conjunction with quinine [71]. Numerous findings claim that clindamycin targets protein synthesis in the apicoplast and functions similarly to azithromycin in this regard. A similar target was also shown by the identification of *T. gondii* parasites that were resistant to clindamycin and azithromycin [72]. Quinine is a popular antimalarial drug that is often given along with an antibiotic like clindamycin or doxycycline [73]. There have been various suggested mechanisms of action for quinine in malaria parasites. According to the most often described mechanism of action, hemozoin

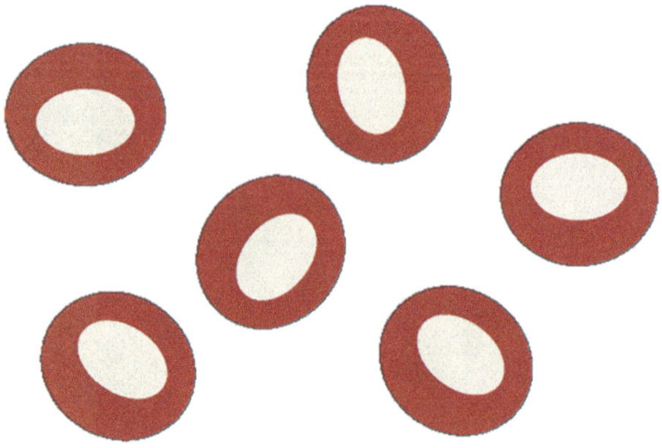

Figure 19.3 Schematic diagram showing Giemsa-stained thin blood smears ring form stage of Babesia species.

synthesis is disrupted, which leads to a buildup of free ferriprotoporphyrin IX, a byproduct of hemoglobin breakdown that is harmful to parasite growth [74, 75]. *Babesia* species differ from Plasmodium parasites in that they lack a digestive vacuole, do not break down hemoglobin, and do not create hemozoin. Quinine's method of action against *Babesia* parasites is therefore expected to be distinct from that of Plasmodium. Intriguingly, it was observed to bind to phospholipids and to accumulate in membrane-bound structures, such as the parasite plasma membrane, the endoplasmic reticulum, and the mitochondrion, indicating that quinine may inactivate particular biological processes in these organelles [76].

19.6 Conclusion

A significant disease impact is imposed by human babesiosis, a global health issue that is on the rise, particularly in the aging population and individuals with impaired immune systems. Many studies show that the actual number of *Babesia* cases is significantly underreported. Clindamycin plus quinidine, quinine, or atovaquone plus azithromycin are the only treatments now available for *Babesia* infections, but there are indications that drug resistance to atovaquone with azithromycin may already have been noticed in immunocompromised patients. Enhancing surveillance, creating novel medicines, providing supportive therapy, and developing a vaccine will all be crucial in reducing the effects of this illness.

References

1 Gagnon, J., Timalsina, S., Choi, J.Y. et al. (2022). Specific and sensitive diagnosis of *Babesia microti* active infection using monoclonal antibodies to the immunodominant antigen BmGPI12. *J. Clin. Microbiol.* 60 (9): e0092522.

2 Patel, K.M., Johnson, J.E., Reece, R. et al. (2019). Babesiosis-associated splenic rupture: case series from a hyperendemic region. *Clin. Infect. Dis.* 69 (7): 1212–1217.

3 Tonnetti, L., Townsend, R.L., Deisting, B.M. et al. (2019). The impact of *Babesia microti* blood donation screening. *Transfusion* 59 (2): 593–600.

4 Lantos, P.M. and Krause, P.J. (2002). Babesiosis: similar to malaria but different. *Pediatr. Ann.* 31 (3): 192–197.

5 Puri, A., Bajpai, S., Meredith, S. et al. (2021). *Babesia microti*: pathogen genomics, genetic variability, immunodominant antigens, and pathogenesis. *Front. Microbiol.* 12: 697669.

6 Scott, J.D. and Scott, C.M. (2018). Human Babesiosis caused by *Babesia duncani* has widespread distribution across Canada. *Healthcare (Basel)* 6 (2): 49.

7 Lobo, C.A., Cursino-Santos, J.R., Singh, M. et al. (2019). *Babesia divergens*: a drive to survive. *Pathogens* 8 (3): 95.

8 Zhao, L., Jiang, R., Jia, N. et al. (2020). Human case infected with *Babesia venatorum*: a 5-year follow-up study. *Open Forum Infect. Dis.* 7 (3): ofaa062.

9 Herc, E., Pritt, B., Huizenga, T. et al. (2018). Probable locally acquired *Babesia divergens*-like infection in woman, Michigan, USA. *Emerg. Infect. Dis.* 24 (8): 1558–1560.

10 Checa, R., López-Beceiro, A.M., Montoya, A. et al. (2018). *Babesia microti*-like piroplasm (syn. *Babesia vulpes*) infection in red foxes (*Vulpes vulpes*) in NW Spain (Galicia) and its relationship with *Ixodes hexagonus*. *Vet. Parasitol.* 252: 22–28.

11 Scholtens, R.G., Braff, E.H., Healey, G.A. et al. (1968). A case of babesiosis in man in the United States. *Am. J. Trop. Med. Hyg.* 17 (6): 810–813.

12 Skrabalo, Z. and Deanovic, Z. (1957). Piroplasmosis in man; report of a case. *Doc. Med. Geogr. Trop.* 9 (1): 11–16.

13 Quick, R.E., Herwaldt, B.L., Thomford, J.W. et al. (1993). Babesiosis in Washington state: a new species of Babesia? *Ann. Intern. Med.* 119 (4): 284–290.

14 Herwaldt, B.L., Cacciò, S., Gherlinzoni, F. et al. (2003). Molecular characterization of a non-*Babesia divergens* organism causing zoonotic babesiosis in Europe. *Emerg. Infect. Dis.* 9 (8): 942–948.

15 Kim, J.Y., Cho, S.H., Joo, H.N. et al. (2007). First case of human babesiosis in Korea: detection and characterization of a novel type of Babesia sp. (KO1) similar to ovine babesia. *J. Clin. Microbiol.* 45 (6): 2084–2087.

16 Jia, N., Zheng, Y.C., Jiang, J.F. et al. (2018). Human Babesiosis caused by a *Babesia crassa*-like pathogen: a case series. *Clin. Infect. Dis.* 67 (7): 1110–1119.

17 Swanson, M., Pickrel, A., Williamson, J. et al. (2023). Trends in reported babesiosis cases – United States, 2011–2019. *MMWR Morb. Mortal. Wkly Rep.* 72: 273–277.

18 Burgess, M.J., Rosenbaum, E.R., Pritt, B.S. et al. (2017). Possible transfusion-transmitted *Babesia divergens*-like/MO-1 infection in an Arkansas patient. *Clin. Infect. Dis.* 64 (11): 1622–1625. Erratum in: Clin Infect Dis. 2017; 65(8): 1431–1433.

19 Arsuaga, M., Gonzalez, L.M., Lobo, C.A. et al. (2016). First report of *Babesia microti*-caused babesiosis in Spain. *Vector Borne Zoonotic Dis.* 16 (10): 677–679.

20 Asensi, V., González, L.M., Fernández-Suárez, J. et al. (2018). A fatal case of *Babesia divergens* infection in Northwestern Spain. *Ticks Tick Borne Dis.* 9 (3): 730–734.

21 de Ramón, C., Cid, J., Rodríguez-Tajes, S. et al. (2016). Severe *Babesia microti* infection in an American immunocompetent patient diagnosed in Spain. *Transfus. Apher. Sci.* 55 (2): 243–244.

22 Hong, S.H., Kim, S.Y., Song, B.G. et al. (2019). Detection and characterization of an emerging type of Babesia sp. similar to *Babesia motasi* for the first case of human babesiosis and ticks in Korea. *Emerg. Microbes Infect.* 8 (1): 869–878.

23 Kukina, I.V., Guzeeva, T.M., Zelya, O.P. et al. (2018). Fatal human babesiosis caused by *Babesia divergens* in an asplenic host. *IDCases* 13: e00414.

24 Bonsergent, C., de Carné, M.C., de la Cotte, N. et al. (2021). The new human Babesia sp. FR1 is a European member of the Babesia sp. MO1 clade. *Pathogens* 10 (11): 1433.

25 Huang, S., Zhang, L., Yao, L. et al. (2018). Human babesiosis in Southeast China: a case report. *Int. J. Infect. Dis.* 68: 36–38.

26 Wang, J., Zhang, S., Yang, J. et al. (2019). *Babesia divergens* in human in Gansu province, China. *Emerg. Microbes Infect.* 8 (1): 959–961.

27 Godbole, R., Gaur, A., Nayar, P. et al. (2022). Case report: a fatal case of babesiosis in a splenectomized male patient from western India. *Am. J. Trop. Med. Hyg.* 106 (5): 1421–1425.

28 Mørch, K., Holmaas, G., Frolander, P.S. et al. (2015). Severe human *Babesia divergens* infection in Norway. *Int. J. Infect. Dis.* 33: 37–38.

29 Gonzalez, J., Echaide, I., Pabón, A. et al. (2018). Babesiosis prevalence in malaria-endemic regions of Colombia. *J. Vector Borne Dis.* 55 (3): 222–229.

30 Al Zoubi, M., Kwak, T., Patel, J. et al. (2016). Atypical challenging and first case report of babesiosis in Ecuador. *IDCases* 4: 15–17.

31 Western, K.A., Benson, G.D., Gleason, N.N. et al. (1970). Babesiosis in a Massachusetts resident. *N. Engl. J. Med.* 283 (16): 854–856.

32 Ruebush, T.K. 2nd, Juranek, D.D., Spielman, A. et al. (1981). Epidemiology of human babesiosis on Nantucket Island. *Am. J. Trop. Med. Hyg.* 30 (5): 937–941.

33 Apostolou, A., Sorhage, F., and Tan, C. (2014). Babesiosis surveillance, New Jersey, USA, 2006–2011. *Emerg. Infect. Dis.* 20 (8): 1407–1409.

34 Bullard, J.M., Ahsanuddin, A.N., Perry, A.M. et al. (2014). The first case of locally acquired tick-borne *Babesia microti* infection in Canada. *Can. J. Infect. Dis. Med. Microbiol.* 25 (6): e87–e89.

35 Acosta, M.E., Ender, P.T., Smith, E.M. et al. (2013). *Babesia microti* infection, eastern Pennsylvania, USA. *Emerg. Infect. Dis.* 19 (7): 1105–1107.

36 Swei, A., O'Connor, K.E., Couper, L.I. et al. (2019). Evidence for transmission of the zoonotic apicomplexan parasite *Babesia duncani* by the tick *Dermacentor albipictus*. *Int. J. Parasitol.* 49 (2): 95–103.

37 Herwaldt, B., Persing, D.H., Précigout, E.A. et al. (1996). A fatal case of babesiosis in Missouri: identification of another piroplasm that infects humans. *Ann. Intern. Med.* 124 (7): 643–650.

38 Herwaldt, B.L., de Bruyn, G., Pieniazek, N.J. et al. (2004). *Babesia divergens*-like infection, Washington State. *Emerg. Infect. Dis.* 10 (4): 622–629.

39 Beattie, J.F., Michelson, M.L., and Holman, P.J. (2002). Acute babesiosis caused by *Babesia divergens* in a resident of Kentucky. *N. Engl. J. Med.* 347 (9): 697–698.

40 Ríos, L., Alvarez, G., and Blair, S. (2003). Serological and parasitological study and report of the first case of human babesiosis in Colombia. *Rev. Soc. Bras. Med. Trop.* 36 (4): 493–498.

41 Kjemtrup, A.M. and Conrad, P.A. (2000). Human babesiosis: an emerging tick-borne disease. *Int. J. Parasitol.* 30 (12–13): 1323–1337.

42 Hildebrandt, A., Zintl, A., Montero, E. et al. (2021). Human babesiosis in Europe. *Pathogens* 10 (9): 1165.

43 Häselbarth, K., Tenter, A.M., Brade, V. et al. (2007). First case of human babesiosis in Germany—clinical presentation and molecular characterisation of the pathogen. *Int. J. Med. Microbiol.* 297 (3): 197–204.

44 Hildebrandt, A., Hunfeld, K.P., Baier, M. et al. (2007). First confirmed autochthonous case of human *Babesia microti* infection in Europe. *Eur. J. Clin. Microbiol. Infect. Dis.* 26 (8): 595–601.

45 Zintl, A., Mulcahy, G., Skerrett, H.E. et al. (2003). *Babesia divergens*, a bovine blood parasite of veterinary and zoonotic importance. *Clin. Microbiol. Rev.* 16 (4): 622–636.

46 Zhou, X., Li, S.G., Chen, S.B. et al. (2013). Co-infections with *Babesia microti* and Plasmodium parasites along the China-Myanmar border. *Infect. Dis. Poverty* 2 (1): 24.

47 Sun, Y., Li, S.G., Jiang, J.F. et al. (2014). *Babesia venatorum* infection in child, China. *Emerg. Infect. Dis.* 20 (5): 896–897.

48 Shaio, M.F. and Lin, P.R. (1998). A case study of cytokine profiles in acute human babesiosis. *Am. J. Trop. Med. Hyg.* 58 (3): 335–337.

49 Wei, Q., Tsuji, M., Zamoto, A. et al. (2001). Human babesiosis in Japan: isolation of *Babesia microti*-like parasites from an asymptomatic transfusion donor and from a rodent from an area where babesiosis is endemic. *J. Clin. Microbiol.* 39 (6): 2178–2183.

50 Vannier, E. and Krause, P.J. (2012). Human babesiosis. *N. Engl. J. Med.* 366 (25): 2397–2407.

51 El-Bahnasawy, M.M., Khalil, H.H., and Morsy, T.A. (2011). Babesiosis in an Egyptian boy aquired from pet dog, and a general review. *J. Egypt. Soc. Parasitol.* 41 (1): 99–108.

52 Mayne, P.J. (2015). Clinical determinants of Lyme borreliosis, babesiosis, bartonellosis, anaplasmosis, and ehrlichiosis in an Australian cohort. *Int. J. Gen. Med.* 8: 15–26.

53 Vannier, E.G., Diuk-Wasser, M.A., Ben Mamoun, C. et al. (2015). Babesiosis. *Infect. Dis. Clin. N. Am.* 29 (2): 357–370.

54 Elsworth, B. and Duraisingh, M.T. (2021). A framework for signaling throughout the life cycle of Babesia species. *Mol. Microbiol.* 115 (5): 882–890.

55 White, D.J., Talarico, J., Chang, H.G. et al. (1998). Human babesiosis in New York state: review of 139 hospitalized cases and analysis of prognostic factors. *Arch. Intern. Med.* 158 (19): 2149–2154.

56 Joseph, J.T., Roy, S.S., Shams, N. et al. (2011). Babesiosis in Lower Hudson Valley, New York, USA. *Emerg. Infect. Dis.* 17 (5): 843–847.

57 Krause, P.J. (2003). Babesiosis diagnosis and treatment. *Vector Borne Zoonotic Dis.* 3 (1): 45–51.

58 Parija, S.C., Dinoop, K.P., and Venugopal, H. (2015). Diagnosis and management of human babesiosis. *Trop. Parasitol.* 5 (2): 88–93.

59 Akel, T. and Mobarakai, N. (2017). Hematologic manifestations of babesiosis. *Ann. Clin. Microbiol. Antimicrob.* 16 (1): 6.

60 Narurkar, R., Mamorska-Dyga, A., Nelson, J.C. et al. (2017). Autoimmune hemolytic anemia associated with babesiosis. *Biomark. Res.* 5: 14.

61 Khangura, R.K., Williams, N., Cooper, S. et al. (2019). Babesiosis in pregnancy: an imitator of HELLP syndrome. *AJP Rep.* 9 (2): e147–e152.

62 Jacobsen, L., Husen, P., and Solov'yov, I.A. (2021). Inhibition mechanism of antimalarial drugs targeting the cytochrome bc1 complex. *J. Chem. Inf. Model.* 61 (3): 1334–1345.

63 Montazeri, M., Mehrzadi, S., Sharif, M. et al. (2018). Activities of anti-toxoplasma drugs and compounds against tissue cysts in the last three decades (1987 to 2017), a systematic review. *Parasitol. Res.* 117 (10): 3045–3057.

64 Dunay, I.R., Gajurel, K., Dhakal, R. et al. (2018). Treatment of toxoplasmosis: historical perspective, animal models, and current clinical practice. *Clin. Microbiol. Rev.* 31 (4): e00057–e00017.

65 Nixon, G.L., Moss, D.M., Shone, A.E. et al. (2013). Antimalarial pharmacology and therapeutics of atovaquone. *J. Antimicrob. Chemother.* 68 (5): 977–985.

66 Mantadakis, E. (2020). *Pneumocystis jirovecii* pneumonia in children with hematological malignancies: diagnosis and approaches to management. *J. Fungi (Basel)* 6 (4): 331.

67 Renteria, A.E., Maniakas, A., Mfuna, L.E. et al. (2021). Low-dose and long-term azithromycin significantly decreases *Staphylococcus aureus* in the microbiome of refractory CRS patients. *Int. Forum Allergy Rhinol.* 11 (2): 93–105.

68 Viasus, D., Gaia, V., Manzur-Barbur, C. et al. (2022). Legionnaires' disease: update on diagnosis and treatment. *Infect. Dis. Ther.* 11 (3): 973–986.

69 Chakraborty, A. (2016). Understanding the biology of the *Plasmodium falciparum* apicoplast; an excellent target for antimalarial drug development. *Life Sci.* 158: 104–110.

70 Burns, A.L., Sleebs, B.E., Siddiqui, G. et al. (2020). Retargeting azithromycin analogues to have dual-modality antimalarial activity. *BMC Biol.* 18 (1): 133.

71 Smith, R.P., Hunfeld, K.P., and Krause, P.J. (2020). Management strategies for human babesiosis. *Expert Rev. Anti-Infect. Ther.* 18 (7): 625–636.

72 Montazeri, M., Mehrzadi, S., Sharif, M. et al. (2018). Drug resistance in *Toxoplasma gondii*. *Front. Microbiol.* 9: 2587.

73 Talapko, J., Škrlec, I., Alebić, T. et al. (2019). Malaria: the past and the present. *Microorganisms* 7 (6): 179.

74 Tang, Y.Q., Ye, Q., Huang, H. et al. (2020). An overview of available antimalarials: discovery, mode of action and drug resistance. *Curr. Mol. Med.* 20 (8): 583–592.

75 Woodland, J.G., Hunter, R., Smith, P.J. et al. (2017). Shining new light on ancient drugs: preparation and subcellular localisation of novel fluorescent analogues of Cinchona alkaloids in intraerythrocytic Plasmodium falciparum. *Org. Biomol. Chem.* 15 (3): 589–597.

76 Woodland, J.G., Hunter, R., Smith, P.J. et al. (2018). Chemical proteomics and super-resolution imaging reveal that chloroquine interacts with *Plasmodium falciparum* multidrug resistance-associated protein and lipids. *ACS Chem. Biol.* 13 (10): 2939–2948.

20

Epidemiology and Current Trends in Malaria

Priya Patel[1], Arti Bagada[1], and Nasir Vadia[2]

[1] *Department of Pharmaceutical Sciences, Saurashtra University, Rajkot, Gujarat, India*
[2] *Department of Pharmaceutical Sciences, Faculty of Health Sciences, Marwadi University, Rajkot, Gujarat, India*

20.1 History of Malaria

Hippocrates and other ancient Greek physicians described the malaria that struck the country every year as an autumnal fever. According to certain scholarly theories, *P. vivax* and *P. malariae* were probably the causes of the malaria that existed in ancient Greece. The Pontine Marshes of the Roman Campagna were where malaria originally originated among the nations on the north shore of the Mediterranean Sea, and by the later classical age of the Roman Empire, it had become a significantly more serious condition. Bogs and marshes were associated with malarial fevers even in classical Greece, although mosquitoes were not thought to be a source of illness transmission. The ailment was later known as malaria, or "poor air," since the Romans believed it was caused by inhaling "miasmas," or vapors, that radiated from stagnant water bodies. Many early Greeks believed the illness was caused by swallowing marsh water. Draining swamps and stagnant marshes has been used to prevent malaria since Ancient Greece, but until Peruvian cinchona tree bark was brought into Spain in the 1630s, a particular malaria treatment was not made available in Europe [1].

According to the common miasma notion of humoral medicine, infectious diseases were frequently linked to the presence of impure or corrupt air that was poisoned by noxious vapor created by putrefying materials. Although the spread of the disease has frequently been influenced by environmental changes brought on by human activity, such as population changes and agricultural practices like irrigation and deforestation, malaria has changed the course of human history since ancient times. In the 1990s, archeologists found an infant burial in Lugnano that was buried about 450 CE, showing that malaria had advanced up the Tiber Valley. The discovery of such an infant cemetery is extremely exceptional, considering that infants were hardly ever given a respectable burial during the Roman era. In fact, baby remains were frequently found in sewers and waste bins throughout the Roman world [2]. The discovery of 47 newborns interred in the cemetery over the course of one summer by the excavation suggests a severe pandemic that afflicted nearly all pregnant women and the majority of the society as a whole [3, 4].

In order to protect the public health, this led to the removal of foul-smelling waste that may be a source of numerous diseases, as well as the draining of ditches and ponds with stagnant water. Since stagnant water is where mosquitoes lay their eggs and raise their larvae, it is a significant factor in the spread of malaria. There are about 30 of the 400 species of anopheline mosquitoes that transmit malaria with varied degrees of efficiency [5].

Due to the negative physical and mental impacts of recurrent malaria infections on affected populations, both in the tropics and more temperate locations, malaria has long been a barrier to social and economic growth. From Hippocrates onward, medical professionals have consistently described how the residents of dangerous locations are not only physically ill but also mentally distressed and depressed. Many malaria survivors still struggle with neurological or mental health issues, even in this day and age of better healthcare. Malaria also contributed to high infant mortality, dwindling populations, and even depopulation in places that were severely affected. By the Middle Ages, trade networks had allowed malaria to spread over northern, central, and eastern Europe, as well as into Scandinavia, the Balkans, Russia, and the Ukraine. In Britain, endemic malaria had a serious impact on marshy areas near river estuaries. However, the indigenous populations

Rising Contagious Diseases: Basics, Management, and Treatments, First Edition. Edited by Seth Kwabena Amponsah, Ranjita Shegokar, and Yashwant V. Pathak.

of Northern Europe lack the genetic variants selected by malaria, such as thalassemia and sickle-cell trait, suggesting that the disease did not become significant there until about 1000 years ago [3].

There are references to malaria in Chinese writings from around 2700 BCE, Egyptian papyri from 1570 BCE, and Hindu scriptures from as far back as the sixth century BCE. Although we should proceed with caution when reading these historical writings, as we go into more modern times, our foundation is beginning to feel more secure. The early Greeks, such as Homer in around 850 BCE, Empedocles of Agrigentum in around 550 BCE, and Hippocrates in around 400 BCE, were fully aware of the typical bad health, malarial fevers, and enlarged spleens seen in people living in marshy places [6]. Scientific studies were only made possible after the parasites were discovered by Charles Louis Alphonse Laveran in 1880 and the mosquitoes were recognized as the disease's carriers, first for avian malaria by Ronald Ross in 1897 and then for human malaria by Italian scientists between 1898 and 1900 [7–19].

A. Laveran, a French army commander stationed in Algeria, found the malaria parasite in the fresh blood of many feverish patients on November 8th, 1880, marking a significant advancement [20]. Despite the fact that Laveran's finding earned him the Nobel Prize in 1907, his observation was not well received at first. Klebs and Tommasi-Crudeli maintained the idea that malaria is brought on by a bacillus. Italian malariologists led by Marchiafava and Celli accurately identified the parasite as being inside the red blood cells without acknowledging Laveran's discovery. Italian researchers additionally, noted that *P. malariae* and *P. vivax* have different biological and physical traits from one another [21, 22].

20.1.1 Malaria and First World War

The First World War saw the unexpected arrival of malaria as an opponent. It affected all belligerent forces with disastrous effects for immense numbers of troops and massive civilian populations as a result of the environmental, civic, and demographic ramifications of troop dispersals and operations [23]. A malaria distribution map would have revealed that a significant number of the combatant nations were thought to be malaria-endemic when hostilities started in August 1914. They may be found in two bands, one on each side of the equator.

With the exception of areas with desert or high-altitude (2500 m+) topography, the inner tropical belt was around 2000 miles wide on either side of the equator. The Anopheles vector mosquitoes could not survive in these conditions due to a lack of water (for breeding grounds), an uncomfortably low temperature, and/or insufficient humidity. In this tropical belt, where the temperature ranged from 20° C to 30° C and the humidity level was at least 60%, *P. faciparum*, or tropical malaria, was therefore widespread [24]. Unawareness of the *P. vivax* malaria parasite's hepatic stage of growth, which was first identified in 1948 [25], made the problem worse. As a result, ineffective treatments were frequently prescribed, and a lot of sick soldiers had to be removed from battle zones [23]. All belligerent countries' militaries also had to deal with the problem of many conflict zones becoming more prone to malaria outbreaks as a result of the widespread introduction of soldiers carrying the malaria parasite from locations where the disease may spread. The same area contained a large number of non-immune soldiers who were highly susceptible to the disease. Following the First World War, 500 cases of malaria were recorded in Southern England's civilian population by British servicemen who had previously served in malarial areas [24].

20.2 Types of Malaria

An Anopheles mosquito bearing the Plasmodium parasite and infected with malaria will bite a person and transmit the parasitic sickness to them. Through a mosquito bite, this parasite enters your bloodstream and infects your red blood cells (RBCs).

Tropical and subtropical areas are often where you can find the Anopheles mosquito. Additionally, there are four primary kinds of malarial parasites that can infect people:

Plasmodium vivax (P.v.): The most prevalent. The least common form of malaria is *Plasmodium malariae* (P.m.), often known as *Plasmodium ovale* (P.o.). Most dangerous *Plasmodium falciparum* (P.f.)

20.2.1 *Plasmodium vivax* (P.v)

This type of malaria is the most prevalent. The most widely distributed variation is this one. More than half of the cases of malaria in India involve this type of infection. It is usually not fatal, although it can be very disabling. Fatigue, diarrhea, feverish spells, chills, and flu-like symptoms are among the symptoms of this particular kind of malaria.

20.2.2 *Plasmodium ovale* (P.o)

This is one of the rarest types of malaria. The main countries in western Africa where it is found are Ghana, Nigeria, and Liberia. Along with joint discomfort, anemia, vomiting, a sore back, disorientation, diarrhea, and body aches, high fevers and shivering chills are some symptoms. Malaria has a propensity to recur because the parasite can remain dormant in the patient's liver for up to four years. Any period within this interval can see a relapse. A new infection develops as a result of attacks on the red blood cells. Antimalarial drugs must be regularly administered to stop recurrence. Malaria caused by this specific strain can be lethal.

20.2.3 *Plasmodium falciparum* (P.f)

The most deadly strain of malaria now in use is this one. This parasite subspecies is present in Africa, South America, and South-East Asia. In addition to the nausea, tiredness, body aches, enlarged spleen, discomfort in the gut, muscles, and joints, fever, headaches, and anemia common with malaria, the patient also exhibits typical neurological symptoms such as confusion and seizures. Due to the seriousness and reputation of this malarial illness, it needs to be checked for, diagnosed, and treated right away in order to avoid fatalities. It particularly affects the brain and the nervous system. Additionally, malaria can cause seizures and paralysis. Non-treatment or postponed therapy can even have an effect on the brain and central nervous system due to a condition known as "cerebral malaria," leading to cognitive impairment.

20.2.4 *Plasmodium malariae* (P.m)

This type of malaria is the least common. Frequent signs of malaria include a fever and chills. This sort is less typical in the Indian subcontinent – less than 1%. South America, South-East Asia, and Africa all contain the malaria parasite that produces this particular variety of malaria. As part of the treatment, anti-malaria medication is taken in addition to other prescription drugs [26, 27].

20.2.5 Malaria and Plasmodium Biology

The two-part life cycle of Plasmodium is the same for all species. Malaria is spread from one infected vertebrate host to another by insect-mediated transmission and parasite infection in vertebrates.

The Plasmodium life cycle begins when sporozoites formed in the factor enter the blood of a vertebrate host following a mosquito bite (Figure 20.1). Deposited sporozoites move directly from the dermis to the liver, where they infect hepatocytes. Here, they engage in a process known as schizogeny when they multiply quickly and increase in number by thousands (asexual multiplication). To begin the malaria infection cycle, a female anopheline mosquito injects sporozoites into the skin [28]. Sporozoites are transferred from an infected anopheles mosquito into the human bloodstream when the human is bitten by the bug. The majority of the time, sporozoites penetrate human liver cells after entering the bloodstream. Upon entering the liver cell, sporozoites multiply asexually and give rise to schizonts. These rupture the liver cells, releasing merozoites [29]. The sporozoites enter the bloodstream, move to the liver, and then make their way out through the Kupffer cells and into a hepatocyte. The sporozoites move about the liver until they locate an appropriate hepatocyte before beginning their active invasion. They form a parasitophorous vacuole membrane (PVM) when they come into contact with one another, then go through schizogony until daughter merozoites are released into the vasculature in compartments of merosomes [28].

A continuous cycle of asexual reproduction starts in the bloodstream as soon as the merozoites interact with the erythrocytes. Due to genetic reprogramming, some of these asexually reproducing merozoites will go through gametocytogenesis. Parasites in the oocyte stage cause the disease's clinical symptoms [28].

The liver cells release merozoites into the vesicles. These merozoites depart from the heart and travel to the lungs, where they land in the capillaries. Vesicles start to break down as merozoites enter the blood phase, a stage in their development. Erythrocytes, generally known as red blood cells, are regularly attacked by merozoites as they travel through the bloodstream. Inside the red blood cell, they keep expanding until the cell erupts. The gametocytes that are consumed by a mosquito when it bites an infected individual eventually develop into gametes [29].

Gametocytes divide and continue to grow in the bone marrow within 15 days. When they are fully developed, they enter the peripheral circulation. When a mosquito eats them, they are subsequently carried to the midgut, where they develop

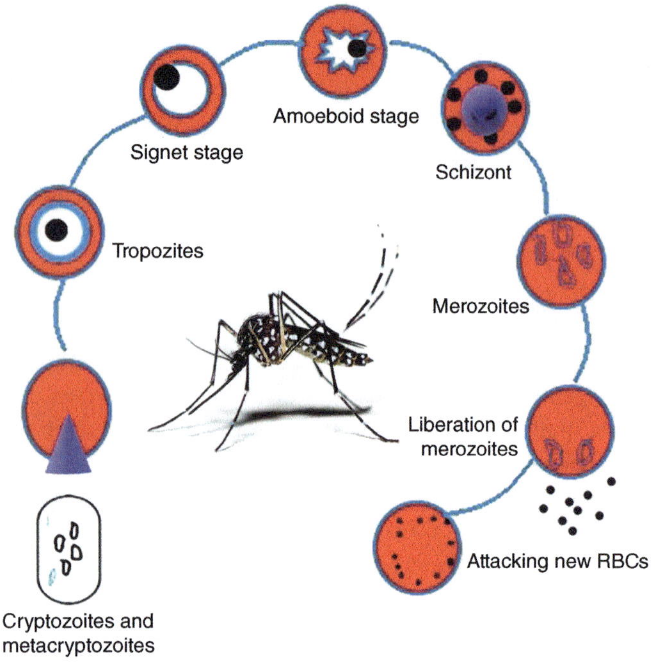

Figure 20.1 Plasmodium life cycle. *Source:* mailsonpignata/ Adobe Stock.

into extracellular male and female gametes. When the male microgamete and female macrogamete unite to produce a zygote, reproduction takes place. Over the course of 24 hours, this changes into an ookinete, which then passes through the mosquito's midgut epithelium to produce an oocyst. The sporozoites are released as the oocysts develop, burst, and travel to the salivary gland of the mosquito. The malaria life cycle is initiated when sporozoites are placed inside a new human host [28]. Male gametocytes are referred to as microgametocytes, whereas female gametophytes are referred to as microgametocytes. These microgametocytes are fertilized inside the anopheles mosquito. Inside the anopheles mosquito, gametocytes finally mate to produce a sporozoite. In between 15 and 18 days, a sporozoite develops. Sporozoite is a parasite [29].

20.3 Worldwide Trend of Malaria (Geographical and Country)/The Spatial Epidemiology of Malaria Globally

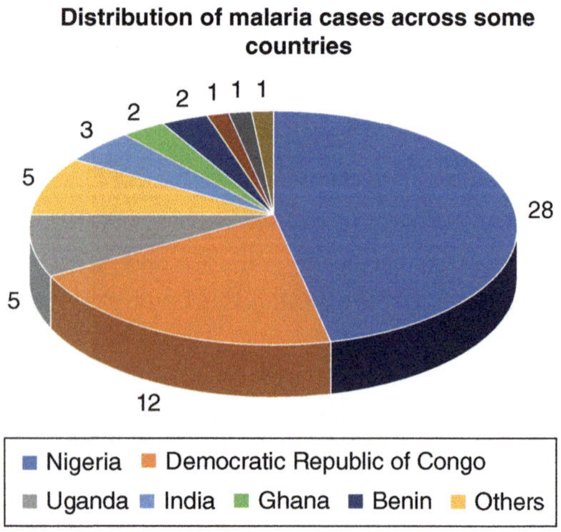

Figure 20.2 Distribution of malaria cases across some countries. *Source:* Data from WHO, 2021.

Despite tremendous advancements in medical technology, parasitic infectious illnesses continue to be a major global source of morbidity and mortality. More than 2 billion parasitic illness cases are thought to have affected the human population in 2013 alone [30]. Vector-borne parasite infections account for over 15% of all infectious diseases and more than 600,000 annual mortality [31]. One of the most common parasitic diseases spread by vectors, malaria claims the lives of more than 400,000 people annually [31, 32]. The majority of these deaths occur among children under the age of five.

Malaria has a unique place in the history books. Over the millennia, it has claimed the lives of people from the Neolithic period, ancient Chinese and Greeks, princes, and peasants. In the twentieth century alone, malaria claimed the lives of 150 million to 300 million people globally, accounting for 2–5% of all fatalities [33]. Despite the fact that the poor of sub-Saharan Africa, Asia, the Amazon basin, and other tropical regions continue to be its major victims today, 40% of the world's population still lives in malaria-endemic areas [34]. Figure 20.2 displays the distribution of malaria cases across several countries.

According to the most current World Malaria Report, there were 227 million cases of malaria in 2019 and 241 million cases in 2020. Malaria is expected to claim the lives of 627,000 individuals in 2020, up from 69,000 in 2019. About 2/3 of these deaths (47,000) were a result of the COVID-19 pandemic, while the remaining 1/3 (22,000) were due to a recent change in the WHO's methodology for calculating malaria mortality.

The updated cause-of-death method was applied to 32 countries in sub-Saharan Africa, which is home to 93% of all malaria deaths worldwide. The methodology's application revealed that, since 2000, malaria has killed significantly more African children each year than was previously thought.

In the WHO African Region, malaria still has a disproportionately high global burden. In the Region in 2020, 96% of all malaria deaths and 95% of all malaria infections occurred. Children under the age of five were responsible for slightly more than 80% of all malaria-related deaths in the area. Four African countries – Nigeria (31.9%), the Democratic Republic of the Congo (13.2%), the United Republic of Tanzania (4.1%), and Mozambique (3.8%) – accounted for just over half of all malaria-related deaths globally.

20.3.1 WHO Response

The WHO Global Technical Plan for Malaria 2016–2030, which was revised in 2021, provides a technical framework for all nations with a current malaria epidemic. It is intended to guide and support national and regional activities aimed at eradicating and controlling malaria. The strategy sets forth ambitious but achievable global objectives, such as preventing a resurgence of malaria in all malaria-free nations, reducing the incidence of malaria cases by at least 90% by 2030, reducing malaria mortality rates by at least 90% by 2030, and eradicating malaria in at least 35 nations by that year.

This plan directs the Global Malaria Programme, which also maintains an independent score of global progress, developing strategies for capacity building, system strengthening, and surveillance, and identifying new opportunities for action as well as risks to malaria control and eradication [35].

Even though comprehensive control and elimination strategies have been implemented through worldwide and national malaria control programmes, malaria remains the most serious parasitic disease in the world. Beginning in 1969, the Global Malaria Eradication Programme (GMEP) was a failure. It caused hundreds of millions of people to become infected, tens of millions of people to pass away (mostly in sub-Saharan Africa), hundreds of thousands of pregnant women to pass away during childbirth as a result of complications related to malaria, and millions of underweight children who later passed away or were disabled [36]. However, the first two decades of this century in the new millennium represent a golden era in the history of malaria control [4]. In 87 countries with a high incidence of malaria, there were an estimated 229 million cases in 2019, down 9 million cases from 2000, according to the World Health Organization's (WHO) most recent annual global malaria report. They exceeded the 218 million malaria cases projected in the Global Technical Strategy (GTS) for Malaria 2016–2030 for the base year of 2015 [32, 36].

Because of the high rates of morbidity and mortality associated with malaria, it is considered a severe public health concern. Despite being preventable and treatable, malaria continues to have a severe impact on people's health and way of life all across the world. About 100 nations and territories, which are home to about half of the world's population, are at risk from malaria [37]. Approximately 228 million cases of malaria were reported worldwide in 2018, according to the WHO, with 93% of those cases happening in the WHO Africa Region. Two WHO estimates that 405,000 people died from malaria-related causes worldwide in 2018, with children under 5 accounting for 67% of all malaria-related deaths worldwide [38, 39].

The international malaria community and WHO both share the objective of eradicating malaria from the planet. Part of this vision is the Global Technical Strategy for Malaria 2016–2030 (GTS), which establishes ambitious yet doable global goals for 2030 along with indicators of success for 2020 and 2025. These goals include a reduction of at least 40%, 75%, and 90% from 2015 in the worldwide fatality rates and case incidence of malaria by 2020, 2025, and 2030, respectively. However, according to WHO, the global incidence of malaria has either ceased or is declining. Even though 31 nations significantly lowered case incidence between 2015 and 2018 and were on target to cut incidence by 40% or more by 2020, the world is not on track to fulfill the 2020 milestones that will allow us to lower case incidence and mortality by 90% by 2030 [40].

20.3.2 High Burden to High Impact Approach

In the 11 countries with a high burden to high impact (HBHI), there were approximately 155 million cases of malaria in 2018, down from 177 million in 2010. 84 million (54%) of all cases were located in Nigeria and the Democratic Republic of the Congo. Of the 10 countries with the highest malaria burden in Africa, Ghana and Nigeria saw the largest absolute increases in malaria cases in 2018 compared to 2017. In all other countries, the burden of malaria was identical to that of 2017, with the exception of Uganda and India, where there were reported decreases in incidence of 1.5 and 2.6 million cases, respectively, in 2018 compared to 2017. Nigeria saw the largest drop, from almost 153,000 deaths in 2010 to about 95,000 in 2018, as the number of malaria deaths decreased from over 400,000 in 2010 to about 260,000 in 2018.

By 2018, at least 40% of the population at risk slept under long-lasting insecticidal nets (LLINs), with Uganda having the greatest percentage (80%) and Nigeria having the lowest (40%). In 2018, it was predicted that only Burkina Faso and the United Republic of Tanzania had more than 50% of pregnant women receiving three intermittent preventative therapy in pregnancy (IPTp3) doses. The estimated coverage in Cameroon, Nigeria, and Uganda was 30% or less.

In the Sahel sub-region of Africa, six nations conducted seasonal malaria chemoprevention (SMC) in 2018. Out of the 26 million children who were targeted, averages of 17 million children were treated per SMC cycle. From 58% in Mali to 82% in Uganda, children under the age of five who experienced fever and sought medical attention. In Mali and the Democratic Republic of the Congo, more than 40% of youngsters were never brought into care. Only 30% or less of the children who were brought for care in Cameroon, Nigeria, and the Democratic Republic of the Congo underwent testing, which is frighteningly low. Direct domestic investment in the HBHI countries, with the exception of India, is still incredibly low in contrast to international investment [41].

Table 20.1 lists the objectives, benchmarks, and targets for the Global Technical Strategy for Malaria from 2016 to 2030. The Worldwide Technical Strategy for Malaria 2016–2030 (GTS) sets four worldwide targets by 2030 along with checkpoints to track development. The 2030 goals include preventing the spread of malaria in all already malaria-free nations, reducing the frequency of malaria cases by at least 90%, reducing the mortality rates from the disease by at least 90%, and eradicating it in at least 35 nations.

One of the short-term GTS aims for 2020 is the eradication of malaria in at least 10 countries and a global reduction of at least 40% in malaria case incidence and fatality rates [42].

Every year on April 25, World Malaria Day is observed to encourage initiatives to eradicate the disease. The WHO is still urging more funding and coverage for tried-and-true malaria prevention, detection, and treatment methods.

On World Malaria Day, it is important to emphasize the need of ongoing financial support and political commitment to the prevention and control of malaria. The WHO Member States implemented it during the 2007 World Health Assembly.

The theme for World Malaria Day this year is "Utilize innovation to lessen the impact of malaria sickness and save lives." The subject emphasizes how important innovation is to achieving the aim of eradicating malaria worldwide.

While many various strategies are being employed internationally to combat malaria, no one tactic is adequate to eradicate the disease. To meet the goal of eradicating malaria worldwide, it is crucial to develop new tools in a variety of sectors, such as insecticides and vector control treatments, better diagnostics, and more potent antimalarial medications.

Malaria is an acute fever illness caused by Plasmodium parasites, which people catch through the bites of infected female anopheles mosquitoes. It is possible to both prevent and treat malaria. In 2022, 241 million malaria cases and 627,000 deaths worldwide were reported. The WHO African Region continues to represent a sizable percentage of the global malaria burden. In 2020, this region was home to 95% of all malaria cases and 96% of malaria-related deaths.

2% (5 million cases in 2020) of all malaria cases globally were found in the WHO South-East Asia Region, in contrast, where malaria cases had dramatically declined. In India, there were 2.09 million cases of malaria in 2001; by 2020, that number had dropped to 0.19 million [43].

Table 20.1 Goals, milestones, and targets for the Global Technical Strategy for malaria 2016–2030.

Goal	Milestone (2020)	Target (2030)
Reduce malaria mortality rates globally compared with 2015	At least 40%	At least 90%
Reduce malaria case incidence globally compared with 2015	At least 40%	At least 90%
Eliminate malaria from countries in which malaria was transmitted in 2015	At least 10 countries	At least 35 countries
Prevent re-establishment of malaria in all countries that are malaria-free	Re-establishment prevented	Re-establishment prevented

20.3.3 Current Malaria Burden in India

The nation has achieved significant progress in the fight against the eradication of malaria. 2020 saw 5.2 million fewer cases than 2017 and a 57% drop in the number of cases per 1000 people at risk, from 7.47 to 3.22. Deaths decreased by 56% during the same time frame, from 0.013 to 0.006 per 1000 at-risk individuals [44].

India's diverse geography and ecosystems add to complexity of the malaria epidemiology. The two most prevalent malaria parasites, *P. falciparum* and *P. vivax*, are the principal causes of malaria in India (although *P. ovale* and *P. malariae* instances have also been documented in specific regions of the nation). Six of the nine Anopheline species, or major vectors, spread the illness [45].

20.4 History of Malaria Control by Various Authorities/Agencies

Malaria is endemic in many nations and is especially prominent in tropical surroundings. The (WHO) estimates that in 2013, 3.3 billion individuals worldwide were at risk of contracting an infection, which would have caused over 200 million illnesses and more than 1 million fatalities. Incredibly, 90% of malaria infections are found in Africa, and more than two-thirds of those instances are in children under the age of five. The only infectious disease that claims more lives than any other is TB [46].

Malaria, a medieval Italian term for poor air quality, is the genesis of the term malaria. At the time, the miasma theory of humoral medicine, which was generally accepted, said that a number of infectious ailments were caused by noxious or corrupt air that was tainted by foul gases released by congealing substances. It further led to the disposal of foul-smelling refuse, which might be the wellspring of several pathogens, as well as the depletion of culverts and lagoons, which actually contain stagnant water. Stagnated waters are crucial in the propagation of malaria due to the fact that mosquitoes rest their own embryos and grow their own larval stage. Roughly 30 of the 400 anopheles mosquito vectors convey malaria transmission diseases with some success.

The illness is caused by Plasmodium protozoan parasites that are conveyed into the human blood system through the sting of female anopheles mosquitoes. *P. vivax*, *P. malariae*, *P. ovale*, *P. falciparum*, and *P. knowlesi* are the protozoa species responsible for human malaria infections [46]. The two parasitic genera *P. vivax* and *P. falciparum*, which together account for nearly the same percentage of predicted malaria cases globally, offer the biggest threat among the five pathogen species that cause malaria in humans [47]. Between 2010 and 2018, the prevalence of malaria in the vulnerable population was drastically reduced, dropping from 71 to 57 instances per 1000 individuals [48].

20.4.1 History of Malaria Control

The development of dichloro-diphenyl-trichloroethane (DDT) as a prototype residual pesticide led to a significant change in malaria control strategies in the early 1940s. Beginning in the 1930s, insecticides were sprayed on residential walls to kill inside resting adult mosquitoes. However, because of the lax requirements, this practice had little impact. Following multiple field tests, an increasing number of national control initiatives supported DDT spraying in the late 1940s and early 1950s. The aforementioned plans showed that if spraying was stopped, the spread could be stopped, and malaria might not necessarily return [49, 50]. History of malaria control around the globe is shown in Table 20.2.

The Swiss Tropical and Public Health Institute (Swiss TPH) has spent a lot of time researching malaria. In the early years of Swiss TPH, there's existed just a few instances involving malaria-related investigations and assistance. These cases involved *P. vivax* infections in Italian and Yugoslavian troops, many of whom had become plights of war. They received treatment in the Swiss Tropical Institute's "tropical clinic" toward the close of World War II [59, 60].

The GMEP of the WHO ran from 1955 to 1969. The 8th World Health Assembly (WHA) in Mexico in 1955 adopted the GMEP. Representatives from member nations decided to employ Dichlorodiphenyltrichloroethane (DDT) to eradicate the deadly disease during the WHO's 8th World Health Assembly in Mexico City in 1955 [61]. When Geigy AG produced it in Basel in 1939, legitimate hopes were raised. The GMEP had a mixed record of success outside of Africa. By 1970, 18 nations had completely wiped out the disease thanks to GMEP. A few years after the GMEP ended, malaria was eradicated in eight additional nations [51].

Undoubtedly, the WHA was inspired by the 1968–1969 outbreak of malaria in Sri Lanka (formerly Ceylon), a country that had previously been regarded as a model for malariologist training. This country's monitoring system had not yet

Table 20.2 History of malaria control globally.

Duration	Malaria control initiative	Regulatory agencies involved	Nation affected	References
1940–1950	DDT spraying	World Health Organization (WHO)	Globally	[49, 50]
1955–1969	Global Malaria Eradication Programme (GMEP)	World Health Organization (WHO)	18 Nation eradication the disease by 1970	[51]
1965–1970	Research and Training in Tropical Diseases (TDR)	WHO with United Nations Development Programme and the World Bank	Globally	[52]
1975–1976	General health programs	UNICEF and other major collaborating agencies	Globally	[53]
1969–1978	Malaria eradication programmes	Several countries	Several countries	[54]
1978–1985	Malaria control strategy	WHO	Globally	[55]
1998–2000	Roll Back Malaria (RBM)	WHO, UNICEF, the United Nations Development Programme (UNDP), and the World Bank	Globally	[56]
2000–2004	Integrated vector management (IVM)	WHO	Globally	[57]
2008–2015	IVM and capacity building	WHO, UNEP, ICIPE, the USAID Environmental Health Project (EHP), The Hashimoto Initiative, and the Panel of Experts on Environmental Management (PEEM)	Globally	[58]
2016–2030	Global technical strategy	WHO	Globally	[25]

responded to the four years of obvious deterioration (1963–1967), nor had it adequately accounted for the 30 years of accumulated knowledge regarding the frequency of pandemic risk in the country [53]. The 22nd World Health Assembly, convened 14 years after the GMEP's founding, had to accept the existence of countries where eradication wasn't actually feasible in the medium term and that a control strategy was in fact the best path forward for future eradication within these regions [62].

When it became clear that eliminating malaria would not be accomplished through a short-term program, UNICEF and other significant cooperating organizations stopped funding malaria programs in lieu of general health initiatives. A strong La Nia in 1975–1976 contributed to this reduction in program funding, which led to major epidemics in a number of nations, most notably the Indian subcontinent and Turkey. The loss of highly skilled workers throughout the 1970s was another issue that became obvious. Campaigns tend to be less able to rethink their strategy as a result of all of these factors working together. Some, including Brazil and Indonesia, deliberately fostered both agriculture and mining to colonize their enormous first-growth forests, a process made more straightforward by the building of penetrating roads. These tactics led to prominent epidemics of malaria owing to mainstream malaria control's relative ineffectiveness. This stimulated a frenzied trade in all antimalarial medications, which helped develop drug resistance [53].

The aforementioned problems supported the idea that progress required the establishment of unique methods and approaches, and in an attempt to reinstate research findings significance for malaria control, WHO set up the Special Programme for Research and Training in Tropical Diseases (TDR) in conjunction with the United Nations Development Programme and the World Bank in the latter half of the 1970s. The TDR has made huge strides in the creation of new instruments as well as in laboratory and field research since its inception. However, most programs have not yet adopted the "problem-solving" strategy of field malarialogists in the first half of the twentieth century. Additionally, the divide between control and research – once referred to in India as "a curious rivalry between the malaria programme and outside research bodies" remains [52].

Several nations made attempts to incorporate marginally successful malaria eradication programs into disproportionately achieved basic health services during the decade 1969–1978. Certain of these developments resulted in worsening malaria scenarios and an end to the vertical strategy for malaria control [54].

The creation of a malaria control strategy was necessary for the 1978 primary healthcare strategy for the development of the health infrastructure [63]. The 38th World Health Assembly proposed an emergency overview of the malaria situation and current control activities for each endemic nation, both in regard to efficacy and viability in preserving the control level that might have been accomplished, through 1985. Regarding the Assembly's proposals, the Expert Committee examined the global epidemiological situation and the difficulties associated with implementing the 1978 control strategy in its eighteenth report [55].

In 1998, the WHO, UNICEF, United Nations Development Programme (UNDP), and World Bank launched the Roll Back Malaria (RBM) effort in response to the unacceptably high impact that malaria places on economies and public health. A neglected malaria control project that had been created after the abandonment of worldwide eradication objectives in 1969 was revived as a result of the effort, which refocused attention on malaria globally. Malaria eradication is not the goal of the new initiative, which is aimed at halving malaria-related mortality and morbidity by 2010, and it is emphasized that this can only be accomplished through strengthened local health systems [56].

It is defined as "a rational decision-making system for such effective the use vector control resources" and comprises of five essential components: First, use evidence when making decisions; second, use integrated strategies. Collaboration between the health sector and other sectors, advocacy, social mobilization, and legislation, and capacity building are the next four factors. In 2004, the WHO adopted IVM for the complete control of all illnesses transmitted by vectors. IVM creation and dissemination for countrywide malaria control regimes in Africa has raised significantly most recently, especially at a time when comprehensive malaria prevention initiatives are expanding with insecticide-treated nets (ITN) along with indoor residual spraying (IRS) coverage [57].

The WHO has advocated for the strengthening of IVM along with capacity building to be one of the key thrusts for action as an aspect of the worldwide effort to combat neglected tropical diseases over 2008–2015 [64]. The creation of strategic plans for the nation began with the initial phase of IVM adoption across 16 different countries. Assessments of needs for vector control and countrywide IVM accord workshops were carried out in nine different nations. The "Partnership for IVM in Africa," also known as "IVM Africa," originated in Harare in order to promote all of these campaigns. It originally included six organizations: WHO, UNEP, ICIPE, the USAID Environmental Health Project (EHP), The Hashimoto Initiative, and the Forum of Professionals on Environmental Management (PEEM) [58].

ITNs have been used for over 13 years on Boiko Island in Equatorial Guinea, and an IRS campaign was launched in 2004 [65]. Additionally, IVM is supporting vector control initiatives in 15 African nations under the President's Malaria Initiative of the US government, including the distribution and promotion of ITNs generally and IRS in targeted communities (6 million LLINs and 17 million people covered by IRS by the end of 2007). In some circumstances, IRS and ITNs are combined with larval source management. The focus shifts from personal protection to vector control as programs scale up from providing ITNs to vulnerable people to community-wide interventions, according to a technical review of Global Malaria Control and Elimination and the Global Malaria Business Plan. This necessitates not only a sizable financial investment in commodities and deployment, but the various contrived interventions also require an investment in human resources for planning, targeting, monitoring, and evaluating [66].

In Sub-Saharan Africa (SSA), the President's Malaria Initiative (PMI) was started in 2005 as a significant role in the prevention and treatment of malaria. By expanding malaria interventions, various specific to a nation analyses have demonstrated significant improvement in cutting deaths among children under five in PMI target countries. Findings show that PMI dramatically boosted malaria control intervention coverage and decreased under-five mortality in SSA [67].

Between 2005 and 2014, malaria scheme investments increased by more than 2.5 times, from US dollars 960 million to US dollars 2.5 billion, enabling an increase in treatment for malaria, safeguarding, and diagnostics efforts. In 2015, over 50% of the population in Sub-Saharan Africa awoke inside mosquito nets that contained chemical pesticides, up from 2% in 2000. A greater number of individuals can now get prompt and effective treatment as a result of the improved accessibility of fast diagnostic procedures and medication against malaria. The incidence and fatality rates of malaria have fallen by 37% and 60%, respectively, since 2000. In Sub-Saharan Africa, malaria therapies are considered to be responsible for 70% of the declines in case counts [68].

To address the ongoing and new problems, WHO created the Global Technical Strategy for Malaria 2016–2030 (GTS 2016–2030). The 68th World Health Assembly adopted the Global Technical Strategy for Malaria 2016–2030 in May 2015. It emphasizes the importance of increasing malaria control efforts and progressing toward elimination [69]. The development is projected to assist countries in minimizing and ultimately eradicating the harm caused by malaria, in addition to progressively achieving the Goals for Sustainable Development. The Worldwide Technical Agenda relies on three fundamental cornerstones, including two subsidiary aspects. The initial among the accompanying variables, leveraging innovation along with extending study findings, is also acknowledged as vital in the worldwide control and extinction of other disorders, including tuberculosis. The forecast recognizes that functional and execution studies are needed to make sure that existing treatments are being used successfully and effectively in various circumstances, as well as to make sure that, as new therapies become available, innovation is deployed effectively to have the greatest impact [70].

The first target, however, of reaching a 40% worldwide decline in deaths from malaria as well as cases per year through 2020, had not been met. Combating malaria financing has plateaued. Since 2015, malaria death and incidence rates have been constant. Additionally, there is projected to be 627,000 malaria fatalities and 241 million malaria cases in 2020, indicating a 12% and 7% rise over 2015. The disastrous failure of malaria control measures brought about by the COVID-19 epidemic could trigger a gloomy reaction. The Worldwide Fund, in collaboration with the RBM Coalition to Combat Malaria, including the United States President's Malaria Initiative, led the COVID-19 Response Mechanism, a vast effort to restructure resources and operations in order to avert the worst-case situation. In December 2021, the Gavi board approved the initial funding for the purchase and shipment of RTS,S (the world's initial efficacious malaria vaccine) to Sub-Saharan Africa. Afterwards in the year, the Global Fund will convene its seventh replenishment meeting. It is demanding a total of $18 billion in financing to assist malaria, HIV, and tuberculosis programs, as well as to build health systems and prepare for a pandemic. Malaria mortality will be minimized by 66% by 2026 if it is completely financed, based on the Global Fund's simulation, while cases per year would be dropped by 69%. The United Nations Sustainable Development Goal of eradicating malaria as an issue of public health by 2030 is at threat, as is the well-being of hundreds of millions of people worldwide annually [71].

20.5 Diagnosis and Treatment

The identification of malaria needs to be accurate and reliable in order to manage the disease successfully in any context; postponed therapy boosts the risk of fatalities or catastrophic neurological sequelae [72]. The signs of malaria (fever, a cold, migraines, pain in the body, nausea, and vomiting) and physical examination (fever, pallor) are vague and may have been a sign of an assortment of illnesses [73–76]. Microscopy, antigen identification, and polymerase chain reaction (PCR) are all techniques employed to detect malaria parasites in peripheral blood. Traditionally, malaria diagnosis relied on microscopic analysis of smears of blood for Plasmodium spp.; nevertheless, inventive diagnostic approaches are currently being used for recognizing malaria parasitemia throughout both chronic and non-endemic settings [77]. Malaria screening methods are classified as follows: (Figure 20.3).

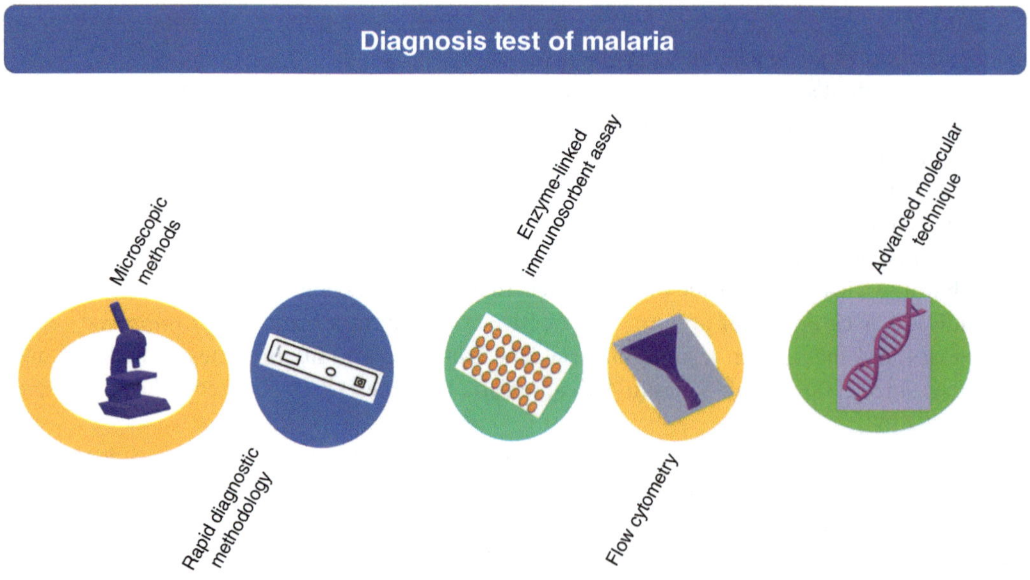

Figure 20.3 Various diagnosis test of malaria.

20.5.1 Microscopic Methods

Light microscopy was once thought to be the "gold standard" for diagnosing malaria [78]. Since 2010, the WHO has suggested that all worrisome instances of malaria be revealed in all clinical settings using both microscopy for parasitic infection measurements or Rapid diagnostic tests (RDT). Through a microscope, the parasite Plasmodium inside an erythrocyte may be noticed in the context of malaria [79].

Microscopy furthermore enables the determination of species of plasmodium and the quantitative determination of parasitic infestation infection density, particularly on thin smears. When *P. vivax* or *P. ovale* is identified, the appropriate treatment option is provided. Since microscopy can provide a quantitative tool for parasitemia diagnosis, it is the preferred method for treating patients with severe malaria. It also allows for the distinction of clinically significant asexual parasite stages from gametocytes, which contribute to ongoing parasite transmission but do not cause illness. For malaria microscopic examination, various staining techniques are available. Giemsa stain has classically been the favored staining technique for regular use in the profession and in research findings utilizing source microscopy [80].

Admittedly, the microscopic approach has numerous drawbacks as well. Under the microscope, the morphology of mature trophozoites, schizonts, and gametocytes from *P. knowlesi* along with *P. malariae*, in addition to the early trophozoite phase from *P. knowlesi* along with *P. falciparum*, could not be distinguished. These limitations can lead to incorrect diagnosis [81, 82]. To acquire an accurate light microscopy result, the blood smear procedure must be standardized, according to guidelines established by WHO, CDC, and another national regulation. A tiny volume of the blood (around 2–61) is being collected using a vein puncture as well as finger-tip blood sample [83].

Narrow and dense blood smear techniques can also be used to count parasites. The count is calculated by dividing the total amount of protozoa by the count of white blood cells. The comprehensive approach is based on WHO standards published in 1991 [84]. Microscopists ought to monitor each parasite species apparent under a microscope while doing parasite counts. Aside from that, they must distinguish between sexual as well as asexual , particularly while evaluating the reaction of schizonticide drugs, which have no effect on gametocytes [84].

20.5.2 Rapid Diagnostic Methodology (RDM) Adopted on Immunochromatography

Immunodiagnostic strategies that include immunochromatography-based RDM, enzyme-linked immunosorbent assay (ELISA), and flow cytometry are commonly used for assessing Plasmodium parasite infection. It is the most popular method for identifying particular antigens otherwise antibodies that correlate to the target infectious agent, and it has been used as an alternate test for the identification of malaria in the last decade when acceptable microscopic performance is inadequate as well as disappointing [80].

The main diagnosis of RDM depends on the reaction of dye-labeled antibodies with the antigen of interest in the blood, resulting in a visible band on the strip. RDM is a lateral flow immune-chromatographic test on a nitrocellulose strip that detects particular antigens in malaria cases ranging from one species (*P. falciparum*) to multi-species (*P. vivax*, malariae, and *P. ovale*). RDMs work by transferring a drop of blood combined with buffer through the test area to the stream. If the antigen is present in the sample, free dye-labeled Ab will bind to the antigen, and this complex will be bound to the bound Ab on the test line. The excess complexes are trapped and accumulate on the control line. The line color intensity may reflect the number of parasite antigens [80].

RDMs are essential for preventing malaria for an array of causes. (i) There should be an obvious advantage while the outcome of a positive parasitic infection becomes apparent in the clinical environment, as sensitivity ranges from 85% to 94.8% and specificity is typically greater than 95%; (ii) the RDMs can effectively monitor 50–100 parasites per l of the blood; (iii) a "cold chain" for the shipment and storage can be maintained; (iv) certified medical personnel can be offered, and (v) supported by local policy [85]. Despite major challenges, RDMs offer specific benefits over microscopes in remote or rural medical facilities. RDMs are simple, inexpensive, do not need specialized equipment, and provide swift findings [86–90].

20.5.3 Enzyme-Linked Immunosorbent Assay

PfHRP-2 can be identified in the bloodstream, the serum, and plasma of patients with malaria using the ELISA [91]. The immune-mediated reaction to an infection with Plasmodium has been assessed utilizing an antibody-based ELISA method. The majority of studies on human immunity toward malaria have focused on immunoglobulin G (IgG) functions. IgM, however, also plays a crucial part in immunity to malaria, according to recent research [92, 93]. Long-lasting antibodies termed merozoite-specific IgM play a vital role in the immune response to blood-stage parasites that assist in avoiding malaria. This IgM prevented merozoite infiltration of red blood cells in a complement-dependent mode in an ongoing cohort trial that was associated with a much lower incidence of clinical malaria [92].

20.5.4 Flow Cytometry

Van Vianen invented flow cytometry in 1993, becoming the first automated approach for identifying malaria parasites in blood samples [94].

In one of the studies, a flow cytometric assay (FCM) was developed in 2004 using reagents from the Japanese company Sysmex. The FCM used a notably quicker, overall simpler approach during the present study, delivering a level of sensitivity of 91.26%, a specificity of 86.28%, and an accuracy of 87.42% [95]. The tri-color technique (TCM), established by Malleret et al. [96], allows researchers to measure and categorize various malaria parasites in whole blood or in vitro-cultivated *P. falciparum* red blood cells. SYBR Green I can be employed instead of Hoechst staining, but with the limitation that pathogen development is not as readily apparent [96].

One of the well-known flow cytometry markers that are used to detect malaria infection is the malaria color hemozoin (Hz), which is created when intra-erythrocytic parasites of malaria devour host hemoglobin [97]. Leukocytes in the blood that carry Plasmodium species may be useful in predicting malaria [98]. The drawbacks include the need for skilled workers, the high price of equipment for testing, the labor needs, and the potential for false positives with other bacterial or viral infections. This method should therefore be utilized as a tool for screening for malaria [97].

20.5.5 Advanced Molecular Technique

Nucleic acid amplification technologies (NAATs), important molecular biology resources, ultimately arise as a result of the finding of DNA's helix. Due to their capacity to amplify incredibly little quantities of targeted DNA or RNA, NAATs are particularly sensitive [99].

NAATs are under the category of PCR, which is further divided into instantaneously multiplexes, including nested PCRs. They also belong within the classification of DNA isothermal amplification mediated by a loop (LAMP), including molecular-based point-of-care tests (POCT). One of the biggest and most important technological developments in molecular genetics in over a decade has been the establishment of PCR [100]. A polymerase chain reaction is the most often used genotyping technology in laboratories [101, 102] because of its stable thermal cycled strategy, specificity, quickness, and sensibility. PCR based assessments have been designed to address the drawbacks and shortcomings of traditional methods of detecting malaria [103].

Sequencing and combinatorial results from PCR may provide accurate identification as well as differentiation when severe morphological issues arise during efforts to identify pathogens at the species level. However, in low resource scenarios or at the moment of need, these methods are laborious, costly, and require an uninterrupted power supply. To solve this problem, isothermal DNA amplification techniques have been created, including clustered regularly interspaced short palindromic repeats (CRISPR), rolling circle amplification, LAMP, and recombinase polymerase amplification (RPA). RPA as well as CRISPR may be employed at room temperature because they have swift amplifying, powerful discrimination, and sensibility, as well as being consistent with multiplexing [104].

20.6 Prevention and Elimination of Malaria

The incidence of malaria decreased by 70% globally between 2010 and 2014, but progress toward additional reductions over the past five years has remained mostly stagnant [105]. The earlier decline of infections was explained by scaling up widespread interventions like the gratuitous distribution of long-lasting mosquito nets (LLINs) or nets treated with insecticides, recurring indoor residual spraying (IRS), swift treatment of recognized cases, and use of combination therapy based on artemisinin for the treatment of malaria caused by *P. falciparum* [106–109] (Figure 20.4). Although certain nations focus their approaches on avoiding malaria by promoting the administration of LLINs, IRS, larvicides, and chemoprophylaxis, along with malaria control initiatives that aim to reduce the disease burden to the point where it is no longer a public health concern, countries with fewer malaria cases aim for elimination to ensure sustained zero regional transmission of malaria as well in the population within a predetermined geography borders through a consolidated regional malaria control strategy. Community engagement (CE) is defined as "a process of working collaboratively with groups of people

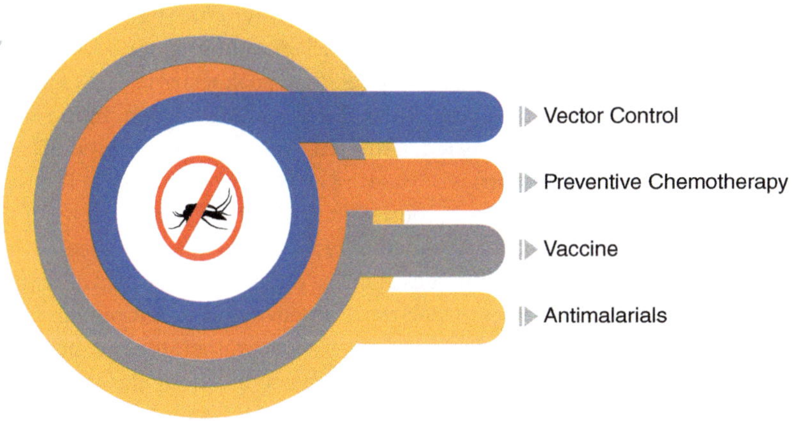

Figure 20.4 Prevention of malaria.

who are affiliated by geographic proximity, special interests, or similar situations" to address issues affecting their well-being [110]. According to the WHO International Malaria Policy 2016–2030, countries with low or middle incomes (LMICs) are adopting CE in their attempts to eradicate malaria by 2030 [111, 112].

In a range of national programs, CE has been used to co-design public health interventions and strategies for the prevention and control of malaria in a number of nations, including mass drug administration for malaria prevention in Myanmar and Laos [113, 114]; increased use of LLINs and promotion of early testing and treatment in Cambodia and Kenya [115]; and improved access to diagnosis and treatment in communities in Zambia [116]. Based on CE, a range of initiatives have been put into place for malaria prevention, control, and eradication.

In order to improve early detection and prompt treatment in rural areas with high rates of migration, these initiatives include the formation of community leadership groups made up of seniors, young people, and regional legislators, drama campaigns, and healthcare educational initiatives delivered in local languages in churches and schools, as well as collaborative malaria research led by community members [113–116].

Providing information, advising, making decisions together, working together, and independently advancing the interests of the community are just a few ways that CE may be included in public health initiatives. CE can be useful in addressing health inequities, particularly for disadvantaged communities that face structural, geographic, cultural, linguistic, and economic obstacles. A number of CE strategies that are most appropriate for the context and the target community have been implemented internationally to strengthen local ownership and offer year-round entry to cost-free medical care and testing in remote, hard-to-reach communities. For instance, Malawi's countrywide malaria campaign relied on community-based health animators (volunteers who offer peer education in Malawi) as peer influencers to raise awareness and promote healthier habits in nearby communities [117].

A laboratory test that has undergone quality assurance offers precise, trustworthy, and timely data that can help with clinical and public health decision-making. Instruments for molecular diagnostics, which are constantly improving, have a huge potential to contribute to the local and worldwide eradication of malaria. They assist in making an informed decision about the best medication and track its effectiveness. They polish data produced in a clinical environment, at the public health level, and at the research level [118].

Recurrence of malaria is a significant factor in maintaining the epidemiology of malaria. To overcome the existing obstacles to attempts to eradicate malaria, it is essential to comprehend the biology and epidemiology of malaria recurrence. Malaria-related morbidity and death rise when it occurs often. It keeps malaria transmission going by fostering gametocytogenesis. The efficacy of therapies is limited without identifying the precise Plasmodium species responsible for recurrence in certain locations. In order to eradicate malaria, disease-control strategies must be optimized to stop transmission and prevent relapse [119].

20.7 Vaccine for Malaria and Current Trends of Malaria Treatment

Malaria is a parasitic disease that may be treated and prevented, but it still affects billions of people globally and is a serious public health concern. Individuals who travel to areas with high malaria transmission, such as migrant workers, mobile populations, and tourists, are at significantly increased risk of acquiring malaria and becoming seriously ill [120].

These individuals include newborns, young children, pregnant women, individuals suffering from HIV/AIDS, and those with low immunity. Due to COVID-related disruptions, there were an additional 13 million cases of malaria and 63,000 malaria deaths during the two pandemic peak years of 2020 and 2021 [121]. The WHO African Region still carries a disproportionately large amount of the global malaria burden. In the Region in 2021, around 95% of all malaria cases and 96% of all malaria deaths occurred. Children under five were responsible for around 80% of all malaria-related deaths in the area. Four African countries – Nigeria (31.3%), the Democratic Republic of the Congo (12.6%), the United Republic of Tanzania (4.1%), and Nigeria – accounted for more than half of all malaria-related deaths worldwide [122].

Over the past 20 years, the worldwide burden of malaria has been dramatically reduced thanks to increased access to prevention strategies and techniques, such as efficient vector control and the use of preventative antimalarial drugs [123]. Efforts to control and eradicate malaria rely heavily on vector management, which is highly effective at reducing disease transmission and preventing infection. The two main therapies are nets sprayed with a pesticide and indoor residual spraying [124]. The development of pesticide resistance in Anopheles mosquitoes threatens the advancement of efforts to combat malaria worldwide. Preventive chemotherapy (MDA) is the term for the administration of medications, either alone or in combination. Malaria infections and the effects that occur are avoided. Regardless of whether the receivers are already infected, it asks for giving susceptible groups (typically newborns, children under 5, and pregnant women) a whole course of an antimalarial medicine at specified times during the period of peak malarial risk. MDA includes the following: perennial malaria chemotherapy (PMC), seasonal malaria chemotherapy (SMC), intermittent preventive treatment of malaria in pregnancy (IPTp) and school-aged children (IPTsc), post-discharge malaria chemotherapy (PDMC), and mass pharmaceutical administration. These low-risk, cost-effective measures are aimed at supplementing ongoing malaria prevention initiatives. According to WHO recommendations, starting in October 2021, children who reside in areas with a moderate to high *P. falciparum* malaria transmission rate should receive the RTS, S/AS01 malaria vaccine. There is evidence that the vaccine significantly reduces the risk of childhood malaria, including severe malaria, which can be fatal [125].

Despite progress throughout the first 15 years of this millennium, malaria control has stalled in recent years, with the resurgence and growing morbidity in several highly endemic nations made worse by service delays brought on by the COVID-19 pandemic. With 241 million cases and 627,000 predicted deaths from malaria in 2020, 77% of these deaths were reported in children under the age of five. Six nations – Nigeria, the Democratic Republic of the Congo, Uganda, Mozambique, Angola, and Burkina Faso – report 55% of all malaria cases worldwide, accounting for 90% of all cases and deaths from the disease that are recorded worldwide [126].

Malaria infections and deaths are still at unacceptable levels and are resurgent in many situations, despite recent breakthroughs that provide grounds for optimism. This contains the accomplishments of research investigating cutting-edge combinatorial medicines and novel vaccination candidates, as well as the effectiveness of the first malaria vaccine ever authorized.

Since the parasite was discovered in 1897, the need to create a malaria vaccine has been highlighted. Ronald Ross discovered the disease-carrying mosquitoes in 1897 [127]. The parasite can also only be transmitted by the female Anopheles mosquito. However, the pursuit of a malaria vaccine with great efficacy has given rise to a variety of novel strategies. Focus has turned to other control developments, including vaccines, as a result of the advent of resistant parasites and vectors. The creation of a potent vaccine can be a crucial component of malaria prevention plans. Regrettably, the extremely complex biology of malaria parasites, diverse and complex parasite genomes, immune evasion by the parasites, and the intricate architecture of the parasite infection cycle have all impeded the development of an effective vaccine for falciparum malaria [128]. The creation of malaria vaccines has advanced significantly. For the creation of each new vaccine, several elements should be considered, including the pathogen life cycle and epidemiology, immune control and evasion, antigen candidate and vaccine formulation, and preclinical and clinical findings [129]. The parasite's complex life cycle and the wide range of antigens present in each stage are two main barriers to vaccine development. The development of a malaria vaccine has been ongoing since the 1960s, with significant advancements made in the past 10 years. 6 October 2021 will be remembered as a significant day in the development of malaria vaccines because it marks the release of the WHO recommendation for the widespread use of the RTS, S/AS01 (RTS, S) malaria vaccine among children living in sub-Saharan Africa and other areas with moderate to high *P. falciparum* malaria transmission [130]. RTS, S, which is sold under the trade name Mosquirix, is the only malaria vaccine that is officially approved. One million kids who lived in places where malaria transmission was moderate to high have received the vaccination as of April 2022. With a fourth dose protecting for a further one to two years, it needs at least three doses in infants by the age of two. Around 30% fewer people with severe malaria end up in hospitals thanks to immunization. Several malaria vaccines are being researched. The most effective malaria vaccine is R21/Matrix-M, which has significantly higher antibody levels than the RTS, S vaccine and a preliminary efficacy rate of 77%. The WHO objective of a malaria vaccine with at least 75% efficacy has been met for the first time with this vaccine [131]. The Bill and Melinda Gates Foundation provided funding for the development of RTS, S by PATH Malaria Vaccine Initiative (MVI) and GlaxoSmithKline (GSK) [132]. It is a recombinant vaccine made from the pre-erythrocytic stage circumsporozoite protein (CSP) of *P. falciparum*. In addition to triggering a cellular response that allows the elimination of infected hepatocytes, the CSP antigen stimulates the development of antibodies that can stop the invasion of hepatocytes. The CSP vaccine's weak immunogenicity caused issues throughout the trial stage. To make a more effective and immunogenic vaccine, RTS, S combined the protein with a surface antigen from hepatitis B. The vaccine was the first malaria vaccine to be recommended for "wide use" in children by the WHO in October 2021. One million kids in Ghana, Kenya, and Malawi have at least one vaccination as of April 2022. UNICEF granted GSK a contract in August 2022 to buy 18 million doses of the RTS, S vaccine over three years. The vaccine is anticipated to be helpful in areas of more than 30 nations where malaria transmission is moderate to high [133].

Organizations like the Centers for Disease Control and Prevention (CDC) are actively working to control, monitor, and eradicate malaria. The primary goal of the CDC's work is to maintain public health, making it the leading data-driven, science-based service institution [134]. We have been using research for almost 70 years to help families, businesses, and communities prevent disease, maintain health, and fight against the disease. This has made it possible for kids to maintain their health so they can grow and study. In order to manage or eradicate malaria worldwide, the CDC conducts research ranging from basic research and development in the field and laboratory to strategic and applied research. Some of the CDC's primary goals include the following [135]:

- Improve the combination of existing malaria prevention measures.
- Create and incorporate fresh or updated interventions.
- Find opportunities to combine one's efforts with other initiatives.
- Conduct cutting-edge research and development in the lab and in the field, with a primary focus on the spread of the malaria parasite, fresh developments like drug resistance, and host immunological and pathological responses to malaria.
- Analyze and address potential risks to the control of malaria, such as pesticide and drug resistance.

To combat malaria, CDC has a long history of working with Ministries of Health and other partners. The CDC offers technical assistance in the creation of policies, the direction, and support of programs, scientific research, and the monitoring and evaluation of progress made toward the Roll Back Malaria objectives. The CDC also carries out strategically targeted research to be prepared to address the changes in malaria epidemiology that have emerged as a result of factors like the recent ramping up of malaria measures, climate change, and the migration of people. One way that the CDC is continuing to build on this foundation of strategically targeted research and program implementation is through the President's Malaria Initiative external icon, an ambitious interagency initiative created to address issues of antimalarial drug resistance and conduct drug quality surveillance in the Greater Mekong Subregion of Asia. The mission of the CDC's malaria research program is to advance our knowledge of the illness and produce more effective treatments. Research frequently involves both field and lab work and is carried out in partnership with other institutions. Field research sheds light on host reactions and transmission mechanisms. They frequently provide samples that, when examined further in labs in the US and overseas, yield useful information. The laboratories undertake more fundamental research, which can then be confirmed or built upon during field investigations (which are supplemented by insectaries and animal facilities). A WHO Collaborating Center for Malaria is housed in the CDC's research facilities on the disease [136].

20.8 Challenges for National Governments

To ensure that malaria is eradicated, effective malaria elimination strategies are essential. Understanding health behaviors regarding the adoption of an intervention and recognizing the obstacles to its implementation are necessary steps in the process of eliminating malaria. Although there has been significant progress against malaria and the death rate has decreased, there is still a need to identify innovative approaches to reach impacted populations with long-lasting and extended effective interventions to make sure they are used properly. Despite having successful malaria control strategies, the elimination of malaria in many nations fails.

There are several current challenges in malaria treatment, including:

1) Drug resistance: The creation of drug-resistant forms of the malaria parasite is one of the main obstacles to effective therapy for the disease. In many regions of the world, there has been reported resistance to conventional antimalarial

medications like chloroquine, sulfadoxine-pyrimethamine, and artemisinin, making treatment more challenging and expensive.

2) Limited access to effective treatments: In many regions of the world, notably in sub-Saharan Africa, where malaria is most prevalent, access to effective therapies for the disease remains a problem. This is partially caused by the price of some antimalarial medications, the scarcity of health facilities, and the lack of appropriately qualified healthcare professionals.
3) Misdiagnosis and underdiagnosis: Malaria can be difficult to diagnose, especially in areas where there is limited access to diagnostic tools such as microscopy or rapid diagnostic tests. Misdiagnosis and underdiagnosis can lead to delays in treatment, which can have serious consequences for patients.
4) Lack of new drugs: There aren't many new medications being developed for the treatment of malaria, despite the growing issue of drug resistance. This is partially brought on by the expensive process of developing new drugs and the dearth of financial incentives for pharmaceutical companies to make investments in cutting-edge medical procedures.
5) Vector control challenges: pesticide-treated bed nets and indoor residual spraying are two vector control strategies that are crucial in avoiding the spread of malaria, although they are hampered by pesticide resistance and uneven coverage in some regions.

Addressing these challenges will require a coordinated effort from governments, international organizations, and the private sector to ensure that effective treatments are available and accessible to all those who need them. As a result of better collaboration between Role Back Malaria partners, WHO, and more funding, the majority of endemic countries are on track to fulfill the malaria-specific Millennium Development Goal target of reducing malaria case incidence by 75%. In order to eliminate the scourge of the disease over the next 15 years and prevent its return, action, and investment to combat malaria 2016–2030 have been created and launched [137, 138].

20.9 Malaria Eradication Programme

After a protracted approval process that began in June 2013 and culminated with numerous consultations, the World Health Assembly endorsed the Global Technical Strategy for Malaria 2016–2030 in May 2015. Robert Newman, John Reeder, and Pedro Alonso, Directors of the Global Malaria Programme, were the principal architects of the approach, working closely with several colleagues and partners from throughout the globe.

This technology strategy provides a framework for the development of specialized programs that will speed the eradication of malaria. The foundation of both national and local malaria control programs should be the aforementioned framework. It outlines a precise and ambitious plan for malaria control and elimination for endemic countries and their international partners over the next 15 years. It highlights the need for full coverage of important anti-malarial therapies for all populations at risk as well as the significance of employing high-quality surveillance data to guide personalized measures in line with national or subnational goals. The strategy outlines the circumstances in which innovative thinking is required to accomplish its goals. It highlights the anticipated costs of implementing the strategy and provides an estimate of the costs related to research and development for cutting-edge new tools. WHO and the global malaria community both want to eradicate malaria. As a part of this vision, the plan defines global targets for 2030 that are both ambitious and attainable, as well as benchmarks for measuring performance in 2020 and 2025. The national or subnational targets that each country sets may be different from the global targets. Table 20.3 provides a list of the objectives, benchmarks, and targets.

Table 20.3 Technical strategy for malaria eradication by WHO.

Strategic goal in comparison to 2015	Milestone to be achieved by 2025	Milestone to be achieved by 2030
Global mortality rate	At least 75%	At least 90%
Global reduction in malaria case incidence	At least 75%	At least 90%
Eliminate malarial transmittance	At least 20 countries	At least 35 countries
Prevent re-establishment	Prevent re-establishment	Prevent re-establishment

WHO advises afflicted nations and the global malaria community to maximize the effectiveness of currently available life-saving instruments and strategies to hasten the process of elimination. To maximize response effectiveness and put an end to malaria fatalities that can be prevented, it is vital to embrace and broaden the implementation of all WHO-recommended initiatives while new and improved tools and methods are still being developed. The plan, which consists of three main pillars and two auxiliary components, directs international efforts to bring malaria under control. Below is a summary of them.

20.9.1 Pillar 1. Ensure that Everyone Has Access to Treatment, Diagnosis, and Prevention for Malaria

Morbidity and death can be considerably reduced by implementing the quality-assured vector control, chemoprevention, diagnostic testing, and therapy recommended by the WHO. The widespread availability of therapies in regions with moderate-to-high transmission should be a top priority for national malaria initiatives. WHO recommends adopting two sets of interventions in a complementary manner: (i) prevention initiatives based on vector control; and (ii) universal diagnosis and timely effective treatment of malaria in public and private health institutions as well as at the community level.

20.9.2 Pillar 2. Intensify Efforts to Eradicate the Disease and Make the World Malaria-Free

Specifically, in environments where transmission is limited, countries must step up their efforts to reduce the spread of new diseases within certain geographic boundaries. As part of a malaria surveillance and response program, targeting both parasites and vectors in clearly defined transmission foci will be necessary to achieve this goal, in addition to the basic interventions. In some situations, the use of prophylactic medications or other potential novel strategies to eliminate the infectious reservoir may be necessary to accomplish eradication once those strategies are advised by the WHO. To combat the spread of pesticide resistance and residual transmission, as well as to aim *P. vivax* hypnozoite reservoirs, novel methods must be created and put into practice.

20.9.3 Pillar 3. Establish Malaria Surveillance as a Focal Point of Treatment

To plan and carry out programs effectively, it is essential to strengthen malaria surveillance, which is also a key aspect of accelerating development. A robust health management and information system should be in place in all nations with endemic malaria and those that are susceptible to its re-establishment to help national malaria programs allocate resources to the populations that are most affected, identify coverage gaps, detect outbreaks, and assess the efficacy of interventions to guide changes in the programs' course. When the transmission is extremely low, monitoring should alert local authorities to any infections found, program coverage gaps, tool efficacy reductions, or outbreaks that may have occurred.

20.9.4 Supporting Element 1. Increasing Research and Utilizing Innovation

In order to promote these three pillars of malaria control, endemic countries and the global malaria community should boost their participation in basic, clinical, and implementation research. Progress will be considerably accelerated by successful innovation in the production of goods and the delivery of services. In order to develop more effective and creative vector control strategies, diagnostic and therapeutic medications, and other tools like vaccinations, basic research is required to better understand the parasites and their vectors. The key to maximizing impact and cost-effectiveness and allowing quick acceptance among populations at risk is execution research.

20.9.5 Supporting Element 2. Enhancing the Favorable Environment

Further development requires increased multisectoral cooperation, significant financial support, and strong political commitment. Maximizing national malaria responses also requires a general strengthening of health services and an improvement in the enabling environment. To reduce the burden of illnesses and the potential for parasite transmission to others, strong health systems – both public and private – are essential. They also enable the quickest possible adoption and application of new tools and processes. To improve health systems, such as laboratory services, mother and child health initiatives, and disease and entomological outbreak surveillance, the expansion of malaria interventions can, in turn, be used as a trampoline [139].

20.10 Conclusion

Plasmodium falciparum was the most prevalent species, with a total prevalence of 87 (21.1%), demonstrating that malaria is still a severe public health issue in the region. To both avoid the occurrence of the disease and control its attack, several programs have been developed. Preventive strategies include using insecticide-treated nets and curtains as part of vector control programs, and maintaining a clean environment. New attention – and innovative interventions – are unquestionably needed to achieve the goals set forth by the WHO "high burden to high impact" campaign to reduce malaria cases and fatalities in the nations most impacted by the disease. There are a number of reasons why the use of a vaccine for the treatment of malaria is now undergoing clinical trials and holds significant potential for antimalarial medication. Cautious optimism, including the success of trials evaluating potential vaccine candidates, the world's first malaria vaccine's licensure, and results from vaccination trials. However, there are still considerable research gaps, making it essential to prioritize, fund, and develop new malaria treatment, prevention, and vaccination alternatives.

References

1 https://www.britannica.com/science/malaria/Malaria-through-history Accessed on 21/11/2022.
2 Beard, M. (2015). *SPQR: A History of Ancient Rome*. Great Britain: Profile Books Ltd.
3 Carter, R. and Mendis, K. (2003). Evolutionary and historical aspects of the burden of malaria. *Clin. Microbiol. Rev.* 16: 173–1173.
4 Soren, D. and Soren, N. (1999). *A Roman Villa and a Late Roman Infant Cemetery: Excavation at Poggio Gramignano, Lugnano in Teverina*. Roma: L'Erma di Bretschneider.
5 Dagen, M. (2020). Chapter 1—History of malaria and its treatment. In: *Antimalarial Agents* (ed. G.L. Patrick), 1–48. Elsevier.
6 Cox, F.E. (2010). History of the discovery of the malaria parasites and their vectors. *Parasit. Vectors* 3 (5): 1–10.
7 Celli, A. (1933). *A History of Malaria in the Italian Campagna from Ancient Times*. London: Bale & Danielsson.
8 Stephens, J.W.W. (1937). *Blackwater Fever, A Historical Survey and Summary Made over a Century*. London: Hodder and Stoughton.
9 Scott, H.H. (1939). *A History of Tropical Medicine*, 1. London: Edward Arnold.
10 Russell, P.F. (1965). *Man's Mastery of Malaria*. London: Oxford University Press.
11 Foster, W.D. (1965). *A History of Parasitology*. Edinburgh: Livingstone.
12 Garnham, P.C.C. (1966). *Malaria Parasites and Other Haemosporidia*. Oxford: Blackwell Scientific Publications.
13 Garnham, P.C.C. (1971). *Progress in Parasitology*. London: Athlone Press.
14 Harrison, G. (1978). *Mosquitoes and Malaria: A History of the Hostilities Since 1880*. London: John Murray.
15 Bruce-Chwatt, L.J. (1988). History of malaria from prehistory to eradication. In: *Malaria: Principles and Practice of Malariology*, vol. vol. 1 (ed. W.H. Wernsdorfer and I. McGregor), 1–59. Edinburgh: Churchill Livingstone.
16 Desowitz, R. (1991). *The Malaria Capers*. New York: WW Norton.
17 McGregor, I. (1996). Malaria. In: *The Wellcome Trust Illustrated History of Tropical Diseases* (ed. F.E.G. Cox), 230–247. London: The Wellcome Trust.
18 Poser, C.M. and Bruyn, G.W. (1999). *An Illustrated History of Malaria*. New York: Parthenon.
19 Schlagenhauf, P. (2004). Malaria from pre-history to present. *Infect. Dis. Clin. N. Am.* 18: 189–205.
20 Laveran, A. (1880). Note sur un nouveau parasite trouve dans le sang de plusieurs malades atteint de fievre palustre. *Bull. Acad. Med.* 9: 1235–1236.
21 Garnham, P.C., Bird, R.G., Baker, J.R., and Bray, R.S. (1961). Electron microsope studies of motile stages of malaria parasites. II. The fine structure of the sporozoite of Laverania (Plasmodium) falcipara. *Trans. R. Soc. Trop. Med. Hyg.* 55: 98–102.
22 Nosten, F., Richard-Lenoble, D., and Danis, M. (2022). A brief history of malaria. *Presse Med.* 51 (3): 104130.
23 Brabin, B.J. (2014). Malaria's contribution to world war one—the unexpected adversary. *Malar. J.* 13: 497–501.
24 https://www.westernfrontassociation.com/world-war-i-articles/malaria-in-the-great-war Accessed on 21/11/2022.
25 Shortt, H.E. and Garnham, P.C.C. (1948). The pre-erythrocytic development of *Plasmodium cynomolgi* and *Plasmodium vivax*. *Trans. R. Soc. Trop. Med. Hyg.* 41: 785–789. https://doi.org/10.1016/S0035-9203(48)80006-4.
26 https://pharmeasy.in/blog/types-of-malaria-symptoms-causes-and-treatment/ Accessed on 21/11/2022.
27 https://redcliffelabs.com/myhealth/lab-test/blood-test/what-are-the-4-types-of-malaria Accessed on 21/11/2022.
28 https://www.news-medical.net/life-sciences/The-Malaria-Parasite-Life-Cycle.aspx Accessed on 21/11/2022.
29 https://www.vedantu.com/biology/plasmodium-life-cycle Accessed on 21/11/2022.

30 Herricks, J.R., Hotez, P.J., Wanga, V. et al. (2017). The global burden of disease study 2013: what does it mean for the NTDs? *PLoS Negl. Trop. Dis.* 11: e0005424.

31 World Health Organization (2019). *World Malaria Report*. Geneva, Switzerland: WHO https://www.who.int/malaria/publications/world-malaria-report-2019/en Accessed on 21/11/2022.

32 Al-Awadhi, M., Ahmad, S., and Iqbal, J. (2021). Current status and the epidemiology of malaria in the Middle East region and beyond. *Microorganisms* 9 (2): 338.

33 Carter, R. and Mendis, K.N. (2002). Evolutionary and historical aspects of the burden of malaria. *Clin. Microbiol. Rev.* 15 (4): 564–594.

34 Institute of Medicine (US) Committee on the Economics of Antimalarial Drugs (2004). 5, A brief history of malaria. In: *Saving Lives, Buying Time: Economics of Malaria Drugs in an Age of Resistance* (ed. K.J. Arrow, C. Panosian, and H. Gelband). Washington (DC): National Academies Press (US). PMID: 25009879.

35 https://www.who.int/news-room/fact-sheets/detail/malaria Accessed on 21/11/2022.

36 World Health Organization (2020). *World Malaria Report*. Geneva, Switzerland: WHO https://www.who.int/publications/i/item/9789240015791 Accessed on 21/11/2022.

37 World Health Organization. Health in 2015: From MDGs, Millennium Development Goals to SDGs, Sustainable Development Goals, 2015. https://apps.who.int/iris/bitstream/handle/10665/200009/9789241565110_eng.pdf;jsessionid=52F0BF417D Accessed on 21/11/2022.

38 World Health Organization. World Malaria Report 2019, 2019. https://apps.who.int/iris/rest/bitstreams/1262394/retrieve Accessed on 21/11/2022.

39 Liu, Q., Jing, W., Kang, L. et al. (2021). Trends of the global, regional and national incidence of malaria in 204 countries from 1990 to 2019 and implications for malaria prevention. *J. Travel Med.* 28 (5): 046–049.

40 World Health Organization. Global Technical Strategy for Malaria 2016–2030, 2015. https://apps.who.int/iris/bitstream/handle/10665/76712/9789241564991 Accessed on 21/11/2022.

41 https://www.who.int/news-room/feature-stories/detail/world-malaria-report-2019 Accessed on 25/11/2022.

42 https://www.who.int/docs/default-source/malaria/world-malaria-reports/world-malaria-report-2020-briefing-kit-eng.pdf Accessed on 25/11/2022.

43 www.nhp.gov.in/world-malaria-day-2022_ Accessed on 25/11/2022.

44 World Health Organisation (2021). World Malaria Report 2021.

45 Global Fund. India Funding Request Malaria 2020–22. www.globalfund.org/ (accessed 25 November 2022).

46 Dagen, M. (2020). History of malaria and its treatment. In: *Antimalarial Agents* (ed. G.L. Patrick), 1–48. Elsevier.

47 Malaria. https://www.who.int/news-room/fact-sheets/detail/malaria. Accessed 1 Mar 2020.

48 WHO (2019). *World Malaria Report 2019*. Geneva: World Health Organization https://apps.who.int/iris/handle/10665/330011. Accessed 1 Mar 2020.

49 Livadas, G. and WHO Expert Committee on Malaria (1952). *Is it Necessary to Continue Indefinitely DDT Residual Spraying Programmes? Relevant Observations made in Greece*. World Health Organization https://apps.who.int/iris/handle/10665/64195.

50 Pampana, E. (1969). *A Textbook of Malaria Eradication*, 2ee, 5–6. London Oxford University Press.

51 Meier, L., Casagrande, G., Abdulla, S., and Masanja, H. (2022). A brief history of selected malaria vaccine and medical interventions pursued by the Swiss Tropical and Public Health Institute and partners, 1943–2021. *Acta Trop.* 225: 106115.

52 Evaluation Committee (1985). *In-Depth Evaluation Report of Modified Plan of Operation under National Malaria Eradication Programme of India*. Delhi: National Malaria Eradication Programme. Government of India.

53 Nájera, J.A., González-Silva, M., and Alonso, P.L. (2011). Some lessons for the future from the Global Malaria Eradication Programme (1955–1969). *PLoS Med.* 8 (1): e1000412.

54 Najera, J.A. (1989). Malaria and the work of WHO. *Bull. World Health Organ.* 67 (3): 229.

55 World Health Organization. WHO Expert Committee on Malaria, Eighteenth Report. Technical Report Series No. 735.

56 Atta, H. and Zamani, G. (2008). The progress of Roll Back Malaria in the Eastern Mediterranean Region over the past decade. *East Mediterr. Health J.* 14: S82–S89.

57 Beier, J.C., Keating, J., Githure, J.I. et al. (2008). Integrated vector management for malaria control. *Malar. J.* 7: 1–16.

58 WHO (2003). *Vector Biology and Control Unit: Annual Report 2003*, 1–32. Harare, Zimbabwe: World Health Organization.

59 Perret-Gentil, A. (1945). L'observation des réfugiés malariens dans la section clinique et l e laboratoire de l'Institut Tropical Suisse. *Acta Trop.* 2: 97–121.

60 Meier L. Im Tropenfieber: das Schweizerische Tropeninstitut (STI) im Spannungsfeld zwischen ökonomischem Kalkül und humanitärer Tradition 1943–1961. Doctoral dissertation, Verlag nicht ermittelbar.

61 Packard, R.M. (2008). The Making of a Tropical Disease: A Short History of Malaria. *Emerg Infect Dis.* 14 (10): 1679.
62 World Health Organization (1973). *Executive Board. Handbook of Resolutions and Decisions of the World Health Assembly and the Executive Board*, vol. vol. 1, 1948–1972. World Health Organization.
63 Primary health care (1978). *Report of the International Conference on Primary Health Care*, 6–12. Alma-Ata: USSR, World Health Organization.
64 Barat, L.M., Palmer, N., Basu, S. et al. (2004). Do malaria control interventions reach the poor? A view through the equity lens. *Am. J. Trop. Med. Hyg.* 71: 174–178.
65 Pardo, G., Descalzo, M.A., Molina, L. et al. (2006). Impact of different strategies to control Plasmodium infection and anaemia on the island of Bioko (Equatorial Guinea). *Malar. J.* 5: 10.
66 WHO (2008). *Global Malaria Control and Elimination: Report of a Technical Review*. World Health Organization.
67 Ye, Y. and Duah, D. (2019). The President's malaria initiative contributed to reducing malaria burden in sub-Saharan Africa between 2004 and 2014: evidence from generalized estimating equation analysis. *PLoS One* 14 (5): e0217103.
68 Cibulskis, R.E., Alonso, P., Aponte, J. et al. (2016). Malaria: global progress 2000–2015 and future challenges. *Infect. Dis. Poverty* 5: 1–8.
69 World Health Organization (2015). *Global Technical Strategy For Malaria 2016–2030*, vol. vol. 11, 4. World Health Organization.
70 Ramsay, A., Olliaro, P., and Reeder, J.C. (2016). The need for operational research and capacity-building in support of the Global Technical Strategy for Malaria 2016–2030. *Malar. J.* 15 (1): 1–3.
71 Lancet, T. (2022). Malaria in 2022: a year of opportunity. *Lancet (London, England)* 399 (10335): 1573.
72 White, N.J. and Ho, M. (1992). The pathophysiology of malaria. *Adv. Parasitol.* 31: 83–173.
73 Kallander, K., Nsungwa-Sabiiti, J., and Peterson, S. (2004). Symptom overlap for malaria and pneumonia: policy implications for home management strategies. *Acta Trop.* 90 (2): 211–214.
74 O'Dempsey, T.J., McArdle, T.F., Laurence, B.E. et al. (1993). Overlap in the clinical features of pneumonia and malaria in African children. *Trans. R. Soc. Trop. Med. Hyg.* 87 (6): 662–665.
75 Reyburn, H., Mbatia, R., Drakeley, C. et al. (2004). Overdiagnosis of malaria in patients with severe febrile illness in Tanzania: a prospective study. *BMJ* 329 (7476): 1212.
76 Chandramohan, D., Jaffar, S., and Greenwood, B. (2002). Use of clinical algorithms for diagnosing malaria. *Tropical Med. Int. Health* 7 (1): 45–52.
77 Bronzan, R.N., McMorrow, M.L., and Patrick, K.S. (2008). Diagnosis of malaria. *Mol. Diagn. Ther.* 12 (5): 299–306.
78 Mouatcho, J.C. and Goldring, J.P.D. (2013). Malaria rapid diagnostic tests: challenges and prospects. *J. Med. Microbiol.* 62 (Pt 10): 1491–1505.
79 (2016). *Malaria Microscopy Quality Assurance Manual—Ver. 2*, 1–2. WHO Press.
80 Oyegoke, O.O., Maharaj, L., Akoniyon, O.P. et al. (2022). Malaria diagnostic methods with the elimination goal in view. *Parasitol. Res.* 121: 1867–1885.
81 Lee, K.S., Cox-Singh, J., and Singh, B. (2009). Morphological features and differential counts of Plasmodium knowlesi parasites in naturally acquired human infections. *Malar. J.* 8 (1): 1–10.
82 Singh, B., Sung, L.K., Matusop, A. et al. (2004). A large focus of naturally acquired Plasmodium knowlesi infections in human beings. *Lancet* 363 (9414): 1017–1024.
83 Kemenkes (2016). *Modul Peningkatan Kemampuan Teknis Mikroskopis Malaria*. Dirjen P2PTVZ.
84 WHO (1991). *Basic Malaria Microscopy, Part 1. Learner's Guide*. World Health Organization.
85 WHO The Role of RDTs in Malaria Control. 2015 https://www.who.int/malaria/areas/diagnosis/rapiddiagnostictests/role_in_malaria_control/en. Accessed July 30, 2020.
86 Faye, B., Nath-Chowdhury, M., Tine, R.C. et al. (2013). Accuracy of HRP2 RDT (Malaria Antigen P.f®) compared to microscopy and PCR for malaria diagnosis in Senegal. *Pathog. Glob. Health* 107 (5): 273–278.
87 Mbanefo, A. and Kumar, N. (2020). Evaluation of malaria diagnostic methods as a key for successful control and elimination programs. *Trop. Med. Infect. Dis.* 5 (2): 102.
88 Rezeki, S. and Pasaribu, A.P. (2018). The comparative accuracy of rapid diagnostic test with microscopy to diagnose malaria in subdistrict lima puluh batubara regency North Sumatera province. *IOP Conf. Ser. Earth. Environ. Sci.* 125 (1): 012020.
89 Wanja, E.W., Kuya, N., Moranga, C. et al. (2016). Field evaluation of diagnostic performance of malaria rapid diagnostic tests in western Kenya. *Malar. J.* 15 (1): 456.
90 Ling, X.X., Jin, J.J., Zhu, G.D. et al. (2019). Cost-effectiveness analysis of malaria rapid diagnostic tests: a systematic review. *Infect. Dis. Poverty* 8 (1): 104.

91 Kifude, C.M., Rajasekariah, H.G., Sullivan, D.J. et al. (2008). Enzyme-linked immunosorbent assay for detection of *Plasmodium falciparum* histidine-rich protein 2 in blood, plasma, and serum. *Clin. Vaccine Immunol.* 15 (6): 1012–1018.

92 Boyle, M.J., Chan, J.A., Handayuni, I. et al. (2019). IgM in human immunity to *Plasmodium falciparum* malaria. *Sci. Adv.* 5 (9): eaax4489.

93 Tomaras, G.D., Yates, N.L., Liu, P. et al. (2008). Initial B-cell responses to transmitted human immunodeficiency virus type 1: virion-binding immunoglobulin M (IgM) and IgG antibodies followed by plasma anti-gp41 antibodies with ineffective control of initial viremia. *J. Virol.* 82 (24): 12449–12463.

94 van Vianen, P.H., van Engen, A., Thaithong, S. et al. (1993). Flow cytometric screening of blood samples for malaria parasites. *Cytometry* 14: 276–280.

95 Wongchotigul, V., Suwanna, N., Krudsood, S. et al. (2004). The use of flow cytometry as a diagnostic test for malaria parasites. *Southeast Asian J. Trop. Med. Public Health* 35 (3): 552–559.

96 Malleret, B., Claser, C., Ong, A.S.M. et al. (2011). A rapid and robust tri-color flow cytometry assay for monitoring malaria parasite development. *Sci. Rep.* 1 (1): 118.

97 Tangpukdee, N., Duangdee, C., Wilairatana, P., and Krudsood, S. (2009). Malaria diagnosis: a brief review. *Korean J. Parasitol.* 47 (2): 93–102.

98 Hänscheid, T., Valadas, E., and Grobusch, M.P. (2000). Automated malaria diagnosis using pigment detection. *Parasitol. Today* 16 (12): 549–555.

99 Hadgu, A., Dendukuri, N., and Hilden, J. (2005). Evaluation of nucleic acid amplification tests in the absence of a perfect gold-standard test: a review of the statistical and epidemiologic issues. *Epidemiology* 16 (5): 604–612.

100 Francois, P., Tangomo, M., Hibbs, J. et al. (2011). Robustness of a loopmediated isothermal amplification reaction for diagnostic applications. *FEMS Immunol. Med. Microbiol.* 62 (1): 4.

101 Garibyan, L. and Avashia, N. (2013). Polymerase chain reaction. *J. Invest. Dermatol.* 133 (3): 1–4.

102 Liu, H.Y., Hopping, G.C., Vaidyanathan, U. et al. (2019). Polymerase chain reaction and its application in the diagnosis of infectious keratitis. *Med. Hypothesis Discov. Innov. Ophthalmol. J.* 8 (3): 152–155. https://pubmed.ncbi.nlm.nih.gov/31598517.

103 Anthony, C., Mahmud, R., Lau, Y.L. et al. (2013). Comparison of two nested PCR methods for the detection of human malaria. *Trop. Biomed.* 30 (3): 459–466.

104 Fitri, L.E., Widaningrum, T., Endharti, A.T. et al. (2022). Malaria diagnostic update: from conventional to advanced method. *J. Clin. Lab. Anal.* 36 (4): e24314.

105 Panda, S., Swaminathan, S., Hyder, K.A. et al. (2017). Drug resistance in malaria, tuberculosis, and HIV in South East Asia: biology, programme, and policy considerations. *BMJ* 358: j3545.

106 World Health Organization (2019). *World Malaria Report 2019*. Geneva: WHO.

107 World Health Organization (2018). *World Malaria Report 2018*. Geneva: WHO.

108 World Health Organization (2015). *WHA68 2: Global Technical Strategy and Targets for Malaria 2016–2030*. World Health Organization.

109 Bhatt, S., Weiss, D.J., Cameron, E. et al. (2015). The effect of malaria control on *Plasmodium falciparum* in Africa between 2000 and 2015. *Nature* 526: 207–211.

110 Control for Disease Control and Prevention (1997). *Principles of Community Engagement*. Atlanta, GA: CDC/ATSDR Committee on community engagement.

111 Epidemiology and Disease Control Division (2016). *Nepal Malaria Strategic Plan 2014–2025*. Department of Health Services.

112 World Health Organization (2015). *Global Technical Strategy and Targets for Malaria 2016–2030*. Geneva: WHO.

113 Adhikari, B., Pell, C., Phommasone, K. et al. (2017). Elements of effective community engagement: lessons from a targeted malaria elimination study in Lao PDR (Laos). *Glob. Health Action* 10: 1366136.

114 Kajeechiwa, L., Thwin, M.M., Nosten, S. et al. (2017). Community engagement for the rapid elimination of malaria: the case of Kayin state, Myanmar. *Wellcome Open Res.* 2: 59.

115 Lim, R., Tripura, R., Peto, T.J. et al. (2017). Drama as a community engagement strategy for malaria in rural Cambodia. *Wellcome Open Res.* 2: 95.

116 Gordon, A., Vander Meulen, R.J., and Maglior, A. (2019). The 2019 Isdell:Flowers cross border malaria initiative round table: community engagement in the context of malaria elimination. *Malar. J.* 18: 432.

117 Awasthi, K.R., Jancey, J., Clements, A.C., and Leavy, J.E. (2021). Community engagement approaches for malaria prevention, control and elimination: a scoping review protocol. *BMJ Open* 11 (10): e049812.

118 Alemayehu, A. (2023). Molecular diagnostic tools and malaria elimination: a review on solutions at hand, challenges ahead and breakthroughs needed. *Int. J. Clin. Exp. Med. Sci.* 9 (1): 7–20.

119 Alemayehu, A. Biology and epidemiology of malaria recurrence: implication for control and elimination. In: *Infectious Diseases Annual Volume 2022*, vol. vol. 12 (2) (ed. K. Garbacz, T. Jarzembowski, Y. Ran, et al.). IntechOpen.

120 World Health Organization (WHO), authors Geneva, Switzerland (2005). *Malaria and HIV Interactions and their Implications for Public Health Policy. Report of a Technical Consultation on Malaria and HIV Interactions and Public Health Policy Implications 2004.* World Health Organization.

121 Hogan, A.B., Hamlet, A., Watson, O.J. et al. (2020). The potential public health consequences of COVID-19 on malaria in Africa. *Nat. Med.* 26 (9): 1411–1416.

122 World Health Organization (WHO), authors Geneva, Switzerland (2021). *World Malaria Report 2021*. World Health Organization.

123 Hemingway, J., Shretta, R., Wells, T.N.C. et al. (2016). Tools and strategies for malaria control and elimination: what do we need to achieve a grand convergence in Malaria? *PLoS Biol.* 14 (3): 1–12.

124 Pluess, B., Tanser, F.C., Lengeler, C., and Sharp, B.L. (2010). Indoor residual spraying for preventing malaria. *Cochrane Database Syst. Rev.* 2010 (4): CD006657.

125 World Health Organization (WHO), authors Geneva, Switzerland (2021). *Preventive Chemotherapies*. Global Malaria Programme.

126 World Health Organization (WHO), authors Geneva, Switzerland (2021). *More Malaria Cases and Deaths in 2020 Linked to COVID-19 Disruptions*. World Health Organization.

127 Cox, F.E. (2010). History of the discovery of the malaria parasites and their vectors. *Parasit. Vectors* 3 (5): 1–3.

128 Pance, A. (2020). Diversify and conquer: the vaccine escapism of *Plasmodium falciparum*. *Microorganisms* 8 (11): 1748.

129 Brisse, M., Vrba, S.M., Kirk, N. et al. (2020). Emerging concepts and technologies in vaccine development. *Front. Immunol.* 11: 1–10.

130 Lozano, J.M., Parra, Z.R., Hernández-Martínez, S. et al. (2021). The search of a malaria vaccine: the time for modified immuno-potentiating probes. *Vaccine* 9 (2): 115.

131 Laurens, M.B. (2020). RTS,S/AS01 vaccine (Mosquirix™): an overview. *Hum. Vaccin. Immunother.* 16 (3): 480–489.

132 Bill Gates Announces $168 Million to Develop Next-Generation Malaria Vaccine. https://www.gatesfoundation.org/ideas/media-center/press-releases/2008/09/bill-gates-announces-168-million-to-develop-nextgeneration-malaria-vaccine

133 Millions More Children to Benefit from Malaria Vaccine as UNICEF Secures Supply. https://www.unicef.org/press-releases/millions-more-children-benefit-malaria-vaccine-unicef-secures-supply

134 World Health Organization (WHO) (2020). *Malaria Eradication: Benefits, Future Scenarios & Feasibility: A Report of the Strategic Advisory Group on Malaria Eradication*. World Health Organization https://apps.who.int/iris/handle/10665/331795.

135 Martens, P. and Hall, L. (2000). Malaria on the move: human population movement and malaria transmission. *Emerg. Infect. Dis.* 6 (2): 103–109.

136 Center for Disease Control and Prevention: Malaria. https://www.cdc.gov/parasites/malaria/index.html (accessed 3 December 2022).

137 Guyant, P., Corbel, V., Guérin, P.J. et al. (2015). Past and new challenges for malaria control and elimination: the role of operational research for innovation in designing interventions. *Malar. J.* 14: 279–280.

138 Shahandeh, K. and Basseri, H.R. (2019). Challenges and the path forward on malaria elimination intervention: a systematic review. *Iran. J. Public Health* 48 (6): 1004–1013.

139 World Health Organization (WHO) (2020). *Global Technical Strategy for Malaria 2016–2030*. World Health Organization https://apps.who.int/iris/handle/10665/176712 (accessed 25 November 2022).

21

Cryptosporidiosis: Recent Advances in Diagnostics and Management

Subhasundar Maji[1], Moitreyee Chattopadhyay[1], Debankini Dasgupta[2], Ananya Chanda[3], and Sandipan Dasgupta[1]

[1] *Department of Pharmaceutical Technology, Maulana Abul Kalam Azad University of Technology, Kolkata, West Bengal, India*
[2] *Department of Pharmacology, MGM College of Pharmacy, Patna, Bihar, India*
[3] *Department of Pharmaceutical Technology, School of Medical Sciences, ADAMAS University, Kolkata, West Bengal, India*

21.1 Introduction

Cryptosporidiosis is a highly prevalent parasitic disease caused by the insidious protozoan parasite Cryptosporidium. It has gained widespread notoriety as a major contributor to diarrheal illness, particularly in resource-limited regions. The parasitic transmission follows a fecal-oral route, predominantly occurring through the ingestion of water or food tainted with Cryptosporidium oocysts. Moreover, the parasite can be disseminated through encounters with infected humans or animals, amplifying its pervasiveness. In the light of the escalating prevalence of cryptosporidiosis, the accurate diagnosis and effective treatment of this debilitating condition have become arduous tasks, demanding meticulous attention and innovation [1].

The impact of Cryptosporidium extends beyond developing nations, as it has been recognized as a leading culprit behind waterborne disease outbreaks in the United States. Additionally, recreational water activities such as swimming pools and water parks have served as unsuspecting conduits for the transmission of this formidable parasite. Immunocompromised individuals, particularly those afflicted with HIV/AIDS, bear an alarming vulnerability to cryptosporidiosis. The parasite exhibits a relentless nature, manifesting in severe and persistent diarrhea that poses immense challenges in terms of therapeutic intervention. Recent strides in diagnostics and management have shed new light on the disease's intricate epidemiology, pathogenesis, and therapeutic avenues. Molecular diagnostic techniques like polymerase chain reaction (PCR) and next-generation sequencing (NGS) have revolutionized the detection and characterization of various Cryptosporidium species and subtypes, unraveling the enigmatic transmission dynamics that underlie this complex disease. Moreover, therapeutic breakthroughs, including the likes of nitazoxanide and paromomycin, have ushered in improved patient outcomes while simultaneously mitigating the risks associated with cryptosporidiosis [2].

Within the confines of this chapter, a comprehensive overview of recent advancements in the diagnosis and management of cryptosporidiosis is presented. The intricacies of disease epidemiology and modes of transmission are meticulously explored, elucidating the pivotal roles played by waterborne and foodborne outbreaks in perpetuating the burden of cryptosporidiosis. Furthermore, a critical appraisal of current diagnostic modalities is undertaken, with a keen emphasis on the nuanced merits and limitations inherent to each technique. Additionally, the burgeoning domain of molecular tools for Cryptosporidium detection and subtyping is surveyed, offering promising avenues for enhanced understanding and control of the disease [3].

Moreover, the available treatment options are comprehensively discussed, encompassing the notable efficacy and safety profiles of nitazoxanide, paromomycin, and other promising investigational therapies. Of particular concern are the unique challenges encountered in managing cryptosporidiosis among immunocompromised patients, including those afflicted with HIV/AIDS. In light of these challenges, prudent strategies for preventing and curtailing the outbreaks of this debilitating disease are expounded upon, aiming to arm healthcare professionals, researchers, public health officials, and policymakers with effective measures for combating cryptosporidiosis [4].

Rising Contagious Diseases: Basics, Management, and Treatments, First Edition. Edited by Seth Kwabena Amponsah, Ranjita Shegokar, and Yashwant V. Pathak.

In summary, this chapter serves as a compendious examination of the recent strides made in the diagnosis and management of cryptosporidiosis. The significance of implementing robust prevention and control measures to alleviate the burden imposed by this pernicious disease is emphasized throughout. Researchers, clinicians, public health professionals, and policymakers dedicated to the combat and management of parasitic infections will find this chapter to be an indispensable resource, guiding their endeavors toward the control and eradication of cryptosporidiosis.

21.2 Epidemiology

Cryptosporidiosis is an infectious disease caused by the protozoan parasite Cryptosporidium, which can infect humans and a variety of animals. The parasite is transmitted via the fecal-oral route, often through contaminated water or food, or through contact with infected individuals or animals. Once ingested, the parasite can cause gastroenteritis, leading to symptoms such as diarrhea, abdominal pain, and vomiting, particularly in immunocompromised individuals.

Globally, cryptosporidiosis is a significant cause of diarrheal disease, particularly in developing countries where access to clean water and sanitation is limited. According to World Health Organization (WHO), cryptosporidiosis is responsible for approximately 2–3% of all diarrheal disease cases worldwide, with an estimated 19–48 million cases annually [3].

Recent trends in cryptosporidiosis incidence and prevalence are difficult to determine due to variations in reporting and diagnosis methods, as well as differences in regional surveillance systems. However, several large outbreaks of cryptosporidiosis have been reported in recent years, often associated with contaminated water supplies or recreational water sources such as swimming pools. Notably, a large outbreak in Milwaukee, Wisconsin in 1993 affected over 400 000 individuals, highlighting the potential for waterborne transmission of the parasite [5, 6].

Advances in diagnostics and management of cryptosporidiosis have focused on improving the sensitivity and specificity of diagnostic tests, as well as the development of effective treatment options. Currently, the most commonly used diagnostic method is stool microscopy, which involves visualizing Cryptosporidium oocysts in fecal samples. However, newer molecular diagnostic methods such PCR and loop-mediated isothermal amplification (LAMP) offer improved sensitivity and faster turnaround times [7, 8].

Treatment of cryptosporidiosis is largely supportive, with rehydration and electrolyte replacement being the mainstay of management. However, several drugs, including nitazoxanide and paromomycin, are useful in lowering both the duration and intensity of symptoms, especially in immunocompetent people.

Overall, cryptosporidiosis remains an important public health concern, particularly in areas with inadequate sanitation and hygiene measures. Continued research and development of diagnostic and treatment options will be crucial in mitigating the impact of this parasitic disease.

21.3 What Is Cryptosporidium?

Cryptosporidium is a genus of protozoan parasites that causes a diarrheal disease known as cryptosporidiosis. The most common cause of diarrhea in young children is Cryptosporidium, especially those children who are malnourished or immunocompromised. Here is a brief description of the parasite, its lifecycle, transmission, and common symptoms–

- **Lifecycle:** The life cycle of Cryptosporidium involves several stages as follows–
  - **Stage 1:**
    The life cycle of Cryptosporidium starts when the parasite's eggs, called oocysts, are released into the environment through the feces of infected animals. These oocysts contaminate water sources and food items.
  - **Stage 2:**
    Human beings can become infected by Cryptosporidium by either inhaling or consuming water or food contaminated with the parasite's oocysts. Once the oocysts are ingested, they enter the human host.
  - **Stage 3:**
    Inside the human host, the oocysts go through a process called excystation. During this stage, the oocysts release sporozoites, which are an infective form of the parasite. The sporozoites invade and infect the epithelial cells, particularly those lining the gastrointestinal tract and respiratory tract. Within the epithelial cells, Cryptosporidium undergoes asexual reproduction through schizogony and merogony.

i) **Schizogony:** This is a form of asexual reproduction in which the parasite undergoes multiple fission, dividing into multiple daughter cells known as merozoites.
ii) **Merogony:** During merogony, the parasite replicates within its nucleus inside the host's cells. This process leads to the formation of male and female gametes (microgamonts and macrogamonts, respectively).
iii) **Gametogony:** The male and female gametes produced through merogony undergo fertilization, resulting in the formation of zygotes. These zygotes develop into oocysts within the infected human host.

- **Stage 4:**
  The mature oocysts are excreted from the human host through feces. Once outside the host, the oocysts become infective, allowing for the direct or indirect fecal-oral transmission of the parasite. Direct transmission can occur through contact with contaminated feces or surfaces, while indirect transmission can happen through the consumption of contaminated water or food.

  The life cycle of Cryptosporidium restarts when the infective oocysts are ingested by a new host, and the process begins again [9].

- **Transmission**
  Transmission of Cryptosporidium can be summarized as follows (also shown in Figure 21.1):
  - **Fecal-oral route:** Cryptosporidiosis is primarily transmitted through the fecal-oral route. This means that the Cryptosporidium parasite is spread when an individual ingests food or water that has been contaminated with the parasite. The parasite can be present in the feces of infected humans or animals, and if proper sanitation measures are not in place, the parasite can contaminate food or water sources.
  - **Contaminated food and water:** Consumption of contaminated food or water is a common mode of transmission. Cryptosporidium oocysts, which are the infective stage of the parasite, can survive for long periods in water sources, including lakes, rivers, and swimming pools, as well as in food and beverages. If these oocysts are ingested, they can cause infection in the digestive tract of the person consuming them.
  - **Person-to-person contact:** Cryptosporidium can also be transmitted through direct contact with infected individuals. This can occur through activities such as shaking hands, sharing personal items, or engaging in sexual contact. In settings where individuals are close to each other, such as hospitals, childcare centers, or households with infected individuals, there is an increased risk of transmission.
  - **Animal contact:** Contact with infected animals, particularly livestock such as cattle, sheep, and goats, can also lead to Cryptosporidium transmission. This is especially relevant in farm settings or areas where there is close interaction between humans and animals. Animals can shed the parasite in their feces, contaminating the environment and potentially infecting humans who come into contact with the contaminated areas or animals.

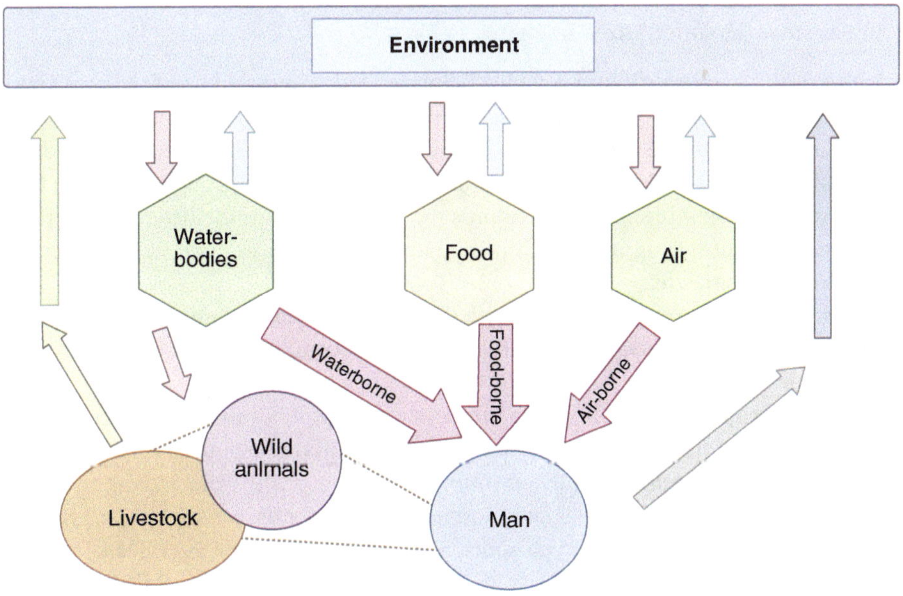

Figure 21.1 Transmission of Cryptosporidium in every medium.

- **Poor hygiene and sanitation:** Cryptosporidium infections are more prevalent in areas where hygiene and sanitation conditions are inadequate. Developing countries or regions affected by natural disasters may have limited access to clean water and proper sanitation facilities, increasing the risk of transmission. Lack of handwashing, improper disposal of feces, and inadequate water treatment systems contribute to the spread of the parasite in such settings.

Preventing and controlling the spread of cryptosporidiosis requires implementing public health initiatives focused on improving hygiene and sanitation practices. These include promoting proper handwashing with soap and clean water, ensuring access to safe drinking water, implementing effective sewage and wastewater treatment systems, and providing education on the importance of personal hygiene and sanitation [10].

- **Symptoms:**
  The symptoms of Cryptosporidium, or cryptosporidiosis, can vary in severity and duration. Here are the symptoms often associated with this parasitic infection:
  - **Watery diarrhea:** The most common symptom of cryptosporidiosis is watery diarrhea. This diarrhea may be persistent and can last for several weeks.
  - **Abdominal pain:** Cramping abdominal pain is frequently experienced alongside diarrhea. The pain may range from mild discomfort to severe cramps.
  - **Nausea and vomiting:** Many individuals infected with Cryptosporidium may experience nausea and vomiting, which can further contribute to dehydration.
  - **Fever:** Fever is a common symptom of cryptosporidiosis, especially if the infection is severe or prolonged. The body's elevated temperature is a response to the infection.
  - **Dehydration:** The combination of diarrhea, vomiting, and fever can lead to dehydration. Symptoms of dehydration include increased thirst, dry mouth, reduced urine output, fatigue, and dizziness.
  - **Weight loss:** Prolonged infection with Cryptosporidium can result in weight loss, particularly in individuals with compromised immune systems or those already suffering from malnutrition.

It is important to note that in healthy individuals, symptoms of cryptosporidiosis are typically self-limiting and resolve within a few weeks without specific treatment. However, people with weakened immune systems or underlying medical conditions may experience severe or chronic symptoms that necessitate medical intervention. Prompt diagnosis and appropriate treatment are crucial for managing cryptosporidiosis effectively and preventing complications [11].

21.4 Mechanism of Cryptosporidium Infection

Cryptosporidium is an intracellular parasite that infects the gastrointestinal tract of humans and animals, causing the disease known as cryptosporidiosis. The pathophysiology and mechanism of Cryptosporidium infection involve several complex steps (as summarized in Figure 21.2).

- **Ingestion:** Cryptosporidium is typically ingested via contaminated food or water as it is very contagious.
- **Attachment:** Once ingested, the oocysts move and survive in an acidic environment through the stomach and into the small intestine, where it attaches to the microvilli of the intestinal epithelial cells.
- **Invasion:** Once attached, the parasite invades the host cell through the formation of a parasitophorous vacuole, which is a protective membrane that surrounds the parasite within the host cell. Cryptosporidium can evade the host immune system by sequestering itself within the parasitophorous vacuole, which prevents the host cell from recognizing and attacking the parasite.
- **Replication:** Once inside the host cell, Cryptosporidium undergoes a process of asexual replication known as schizogony. During schizogony, the parasite undergoes several rounds of cell division, producing large numbers of daughter cells known as merozoites.
- **Immune response:** The host immune system responds to the presence of Cryptosporidium with a combination of innate and adaptive immune responses. Innate immune responses involve the activation of neutrophils, macrophages, and natural killer cells. Adaptive immune responses involve the activation of T and B lymphocytes. However, certain immune cells, such as CD4+ T cells and natural killer cells, can recognize and attack infected cells.
- **Inflammatory response:** Cryptosporidium infection triggers the release of inflammatory cytokines and chemokines, including interleukin-1β (IL-1β), IL-6, and tumor necrosis factor-α (TNF-α). The inflammatory response contributes to

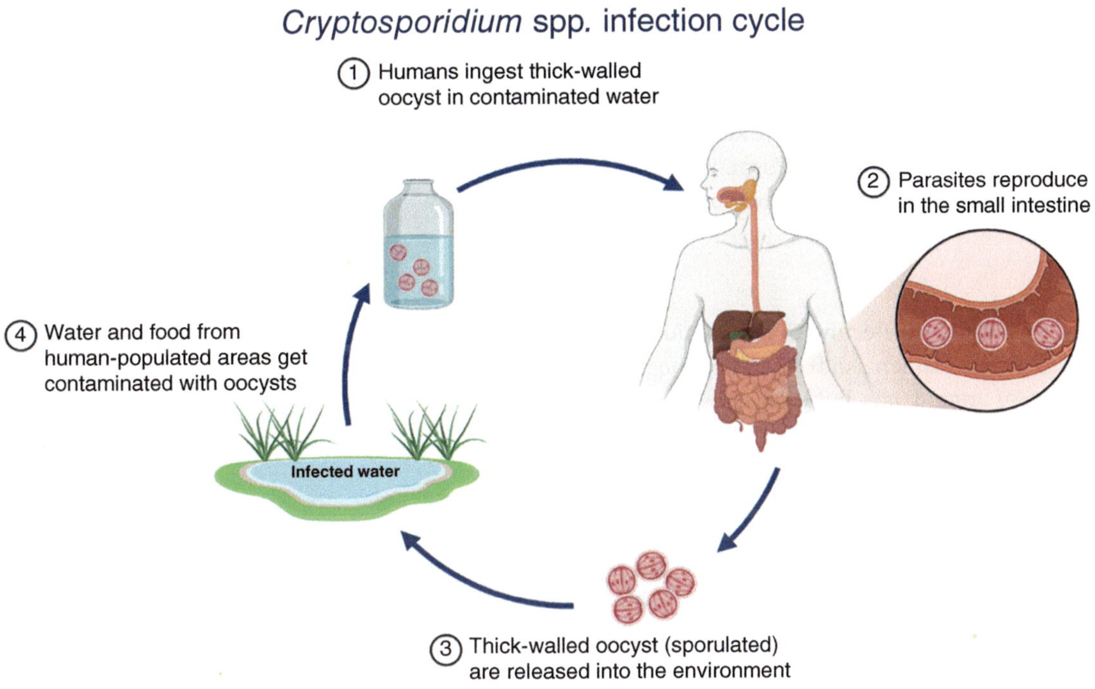

Figure 21.2 Life cycle of Cryptosporidium.

the pathogenesis of Cryptosporidium by damaging the intestinal epithelium and causing diarrhea, dehydration, and electrolyte imbalances. The inflammatory response can cause damage to the intestinal epithelium, leading to symptoms such as diarrhea, abdominal pain, and dehydration.

- **Shedding:** The shedding of Cryptosporidium oocysts in the feces of infected individuals is a critical step in the pathogenesis of Cryptosporidium, as it allows for transmission to other individuals and the environment.

21.5 Diagnostic Methods

Cryptosporidiosis is a parasitic infection caused by Cryptosporidium, which can affect both humans and animals. Accurate and timely diagnosis is essential for appropriate treatment and management of the disease. In recent years, several diagnostic methods have been developed and improved for detecting Cryptosporidium (Table 21.1).

Table 21.1 Diagnostic methods used to detect the cryptosporidium presence.

Method	Advantages	Disadvantages	Ref.
Acid-fast staining	Relatively simple and inexpensive	Low sensitivity and specificity	[12]
Immunofluorescence microscopy	More sensitive and specific than acid-fast staining	Requires specialized equipment and trained personnel	[13]
ELISA	High sensitivity and specificity	May produce false negatives due to low antigen concentrations in some samples	[14]
PCR	Can differentiate between different Cryptosporidium species	Requires specialized equipment and trained personnel. May be expensive	[15]
LAMP	High sensitivity and specificity. Can be performed with minimal equipment and training	Limited availability and standardization	[16]
Microscopy with infrared staining	High sensitivity and specificity. Can be performed with standard microscopy equipment	Limited availability and standardization	[17]

21.5.1 Acid-Fast Staining

Acid-fast staining is a well-established diagnostic method used to detect Cryptosporidium oocysts in stool samples. This method involves staining the stool sample with a red dye and then decolorizing it with acid-alcohol. Cryptosporidium oocysts are acid-fast, meaning they retain the red dye and can be visualized under a microscope. This method has a sensitivity of around 80–90% and is relatively easy to perform. However, acid-fast staining can produce false negatives, especially if the sample is heavily contaminated with other microorganisms.

21.5.2 Polymerase Chain Reaction

PCR is a highly sensitive method for detecting Cryptosporidium DNA in stool samples. This method involves amplifying specific DNA sequences using primers and detecting the amplified product using gel electrophoresis or real-time PCR. PCR has a sensitivity of around 95%, but it requires specialized equipment and expertise to perform.

21.5.3 Immunofluorescence Assay

Immunofluorescence assay (IFA) is a more sensitive method for detecting Cryptosporidium oocysts in stool samples. This method involves using fluorescent antibodies to bind to Cryptosporidium antigens, which can then be visualized under a fluorescent microscope. IFA has a sensitivity of around 95%, but it is more technically challenging and time-consuming than acid-fast staining.

21.5.4 Enzyme-Linked Immunosorbent Assay

Enzyme-linked immunosorbent assay (ELISA) is a widely used method for detecting Cryptosporidium antigens in stool samples. This method involves using specific antibodies to bind to Cryptosporidium antigens, which are then detected using an enzyme-linked colorimetric reaction. ELISA has a sensitivity of around 90%, but it can produce false positives if the sample is contaminated with other microorganisms or if the patient has recently been treated with antibiotics.

21.5.5 Loop-Mediated Isothermal Amplification

Loop-mediated isothermal amplification (LAMP) is a newer method for detecting Cryptosporidium DNA in stool samples. This method involves amplifying DNA sequences under isothermal conditions using a set of specific primers and detecting the amplified product using turbidity or fluorescent signals. LAMP has a sensitivity of around 90% and can be performed using a simple water bath or heating block, making it suitable for use in resource-limited settings.

21.5.6 Microscopy with Infrared Staining

Microscopy with infrared staining is a novel method for detecting Cryptosporidium oocysts in stool samples. This method involves staining the stool sample with a specific infrared dye that binds to Cryptosporidium oocysts, which can then be visualized under an infrared microscope. This method has a sensitivity of around 90%, but it requires specialized equipment and is not widely available.

21.6 Management

Traditional treatment options for cryptosporidiosis include hydration, antimicrobial therapy, and immune-based therapies. Additionally, some ayurvedic treatments have been suggested to alleviate the symptoms of cryptosporidiosis. Here is a more detailed explanation of each treatment option–

- **Hydration:**
  Oral rehydration therapy (ORT) is typically the first-line approach in treating mild to moderate dehydration associated with cryptosporidiosis. It involves the consumption of specially formulated oral rehydration solutions that contain a precise balance of salts, sugars, and water. These solutions help replenish the electrolytes and fluids lost during diarrhea,

promoting rehydration and restoring the body's vital functions. ORT is a simple and cost-effective method that can be easily administered at home, making it a preferred option in many cases [11].

- **Antimicrobial therapy:**
  In the battle against cryptosporidiosis, antimicrobial agents play a crucial role in controlling the parasite and curbing further spread of the infection. Among the available treatment options, nitazoxanide stands out as the only medication approved by the US Food and Drug Administration (FDA) for the treatment of cryptosporidiosis in individuals with intact immune systems. Though there are several drugs used in the treatment of cryptosporidiosis (Table 21.2). However, it is important to note that the effectiveness of nitazoxanide can vary, particularly in cases involving immunocompromised patients [18].
- **Immune-based therapies:**
  In cryptosporidiosis management, immune-based therapies have emerged as a potential avenue to enhance the host's immune response against the parasite. One such therapy is the utilization of Interferon-gamma (IFN-γ), which has been employed as an adjunctive treatment, particularly in immunocompromised individuals.

  Interferon-gamma is a naturally occurring cytokine that plays a pivotal role in regulating immune responses. By administering IFN-γ to patients with cryptosporidiosis, the aim is to bolster their immune system, enabling it to mount a more robust defense against the parasite. This therapy holds particular promise for immunocompromised patients, such as those living with HIV/AIDS or undergoing organ transplantation, who are at heightened risk of severe and chronic cryptosporidiosis infections [19].
- **Ayurvedic treatments:**
  Some ayurvedic treatments have been proposed as potential remedies for alleviating the symptoms of cryptosporidiosis. Among these, Kutaja (*Holarrhena antidysenterica*), Bilva (*Aegle marmelos*), and Amla (*Phyllanthus emblica*) have garnered attention for their purported therapeutic properties. Kutaja, known for its antidiarrheal properties, is believed to possess antimicrobial and anti-inflammatory effects that may aid in reducing the severity of diarrhea associated with Cryptosporidium infection [20]. Bilva, recognized for its digestive and antimicrobial properties, has been suggested as a potential remedy to alleviate gastrointestinal symptoms [21]. Amla, rich in antioxidants and known for its immune-enhancing effects, has been proposed as a supportive therapy to bolster the immune system and potentially aid in the management of cryptosporidiosis [22].

Recent advances in the management of cryptosporidiosis have paved the way for more effective strategies in the treatment of this challenging parasitic infection. These advancements encompass various aspects, from diagnosis to treatment, and have significantly enhanced our ability to address the impact of Cryptosporidium on public health.

Table 21.2 Drugs used for the cryptosporidium infection.

Drug class	Drug name	Route of administration	Common side effects
Antiprotozoal	Nitazoxanide	Oral	Abdominal pain, headache, nausea, urine discoloration
Aminoglycoside antibiotic	Paromomycin	Oral	Abdominal pain, headache, nausea
Macrolide antibiotic	Azithromycin Spiramycin Roxithromycin Erythromycin	Oral	Abdominal pain, headache, nausea, elevated liver transaminases, worsening diarrhea, skin rash
Antiprotozoal	Miltefosine	Oral	Nausea, diarrhea, increased serum creatinine, and vomiting
Somatostatin receptor agonist	Octreotide Vapreotide	IV	Abdominal discomfort, diarrhea, fever, cholecystitis, pancreatitis, discomfort at the point of the injection site
Antitubercular	Clofazimine	Oral	Abdominal pain, nausea, vomiting, malaise, anorexia, depression, weakness

- **New drug development:**
  Several new drugs, such as paromomycin inhalation powder, clofazimine, and halofuginone, are under development for the treatment of cryptosporidiosis. These drugs have shown promising results in preclinical studies and may be effective for both immunocompetent and immunocompromised patients [23].
- **Immunotherapy:**
  Immunotherapy is an emerging treatment strategy for cryptosporidiosis. Monoclonal antibodies (mAb), hen egg yolk antibodies (EYA), hyperimmune bovine colostrum (HBC), and hyperimmune bovine colostrum immunoglobulins (HBCIg), as well as oral administration of pooled human immunoglobulins for intravenous injection (Igiv), have shown efficacy against the parasite *in-vitro* and animal models. Clinical trials are underway to evaluate their safety and efficacy in humans [24].
- **Probiotics:**
  Probiotics have been suggested to alleviate the symptoms of cryptosporidiosis. *Lactobacillus rhamnosus* GG and Saccharomyces boulardii are two probiotics that have shown promise in clinical studies. They may help restore the gut microbiota and improve diarrhea symptoms [25, 26].
- **Other therapies:**
  Other therapies such as fecal microbiota transplantation, photodynamic therapy, and natural compounds, are also under investigation for the treatment of Cryptosporidiosis. They have shown some promise in preclinical studies but require further evaluation in clinical trials [27, 28].
- **Microbiome manipulation:**
  The gut microbiome plays a crucial role in the pathogenesis of cryptosporidiosis. Manipulating the gut microbiome through fecal microbiota transplantation (FMT) or dietary interventions, such as probiotics or prebiotics, may help in reducing the severity of the disease [29].

Although there is no specific drug approved for the treatment of cryptosporidiosis other than Nitrazoxanide, several drugs have been used off-label to manage the disease (Table 21.2). Here are some of the marketed drugs used for the treatment of cryptosporidiosis–

- **Nitazoxanide:**
  Nitazoxanide is an antiparasitic drug that has been approved by the US FDA for the treatment of diarrhea caused by Cryptosporidium and Giardia. Nitazoxanide works by interfering with the metabolism of the parasite, leading to its death. In both children and adults with cryptosporidiosis, it is beneficial in lowering the length and severity of diarrhea [30].
- **Paromomycin:**
  Paromomycin is an aminoglycoside antibiotic that has been used off-label for the treatment of cryptosporidiosis. It works by inhibiting protein synthesis in the parasite, leading to its death. Although paromomycin has been shown to reduce the length and severity of diarrhea in immunocompetent people, its effectiveness in those with immunocompromise is limited [31].
- **Azithromycin:**
  Azithromycin is a macrolide antibiotic that has been used off-label for the treatment of cryptosporidiosis. It works by inhibiting protein synthesis in the parasite, leading to its death. In immunocompetent people, azithromycin is useful in lowering the length and severity of diarrhea; it may also be used as a secondary therapy in immunocompromised patients [32].
- **Spiramycin:**
  Spiramycin is a macrolide antibiotic that has been used off-label for the treatment of cryptosporidiosis. It works by inhibiting protein synthesis in the parasite, leading to its death. Spiramycin has been shown to be successful at reducing both the duration and intensity of diarrhea in immunocompetent individuals and may be used as a second-line treatment in immunocompromised patients [33].
- **Roxithromycin:**
  Roxithromycin is a macrolide antibiotic that has been used off-label for the treatment of cryptosporidiosis. It works by inhibiting protein synthesis in the parasite, leading to its death. Roxithromycin has been shown to be successful at reducing both the duration and intensity of diarrhea in immunocompetent individuals and may be used as a second-line treatment in immunocompromised patients [34].
- **Erythromycin:**
  Erythromycin is a macrolide antibiotic that has been used off-label for the treatment of cryptosporidiosis. It works by inhibiting protein synthesis in the parasite, leading to its death. Erythromycin has been shown to be successful at reducing both the duration and intensity of diarrhea in immunocompetent individuals and may be used as a second-line treatment in immunocompromised patients [35].

- **Miltefosine:**
  Miltefosine is an antiprotozoal drug that has been used off-label for the treatment of cryptosporidiosis. It works by disrupting the cell membrane of the parasite, leading to its death. Miltefosine has been shown to be successful at reducing both the duration and severity of diarrhea in immunocompromised patients [36].
- **Octreotide and vapreotide:**
  These drugs are somatostatin analogs that have been used off-label for the treatment of cryptosporidiosis in immunocompromised patients. These drugs work by inhibiting the secretion of gastrointestinal hormones, leading to a reduction in intestinal motility and secretions. Octreotide and Vapreotide have been shown to improve diarrhea and reduce parasite shedding in patients with cryptosporidiosis. However, their use is limited by the need for subcutaneous injection and their potential side effects, such as nausea, vomiting, and abdominal pain [37, 38].
- **Clofazimine:**
  Clofazimine is an antimicrobial drug that has been used off-label for the treatment of cryptosporidiosis in immunocompromised patients. Clofazimine works by binding to DNA and interfering with its replication, leading to parasite death. Clofazimine has been shown to reduce the duration and severity of diarrhea caused by Cryptosporidium in immunocompromised patients. However, its use is limited by its potential side effects, such as skin discoloration and gastrointestinal disturbances [39].

21.7 Prevention and Control

Cryptosporidiosis is a waterborne and foodborne disease that can be prevented and controlled by various measures. Proper hygiene is one of the most important preventive measures against cryptosporidiosis. Individuals should wash their hands with soap and water before and after using the restroom and before eating. Avoiding consuming untreated water or food that may have been contaminated with Cryptosporidium is also crucial. People should avoid contact with fecal matter from infected animals or people. Surfaces that might have come into contact with contaminated feces should be cleaned. Some of the preventive measures for cryptosporidiosis are as follows (also summarized in Table 21.3)–

- **Water treatment:**
  Proper water treatment plays a crucial role in safeguarding the quality of drinking water, particularly in preventing the transmission of Cryptosporidium. Recent advancements in water filtration technologies have yielded promising results in effectively removing or killing Cryptosporidium from water sources. Membrane filtration and ultrafiltration techniques, for instance, have demonstrated high efficacy in eliminating Cryptosporidium oocysts from drinking water.

Table 21.3 Preventive measures to be taken for the cryptosporidium infection.

Measures	Description
Prevention	• Maintain proper hygiene by washing your hands with soap and water before and after using the restroom and before eating • Avoid consuming untreated water or food that may have been contaminated with Cryptosporidium • Avoid contact with fecal matter from infected animals or people • Surfaces that might have come into contact with contaminated faces should be cleaned
Treatment	• In healthy individuals, cryptosporidiosis usually resolves on its own within one to two weeks • Drink plenty of fluids to prevent dehydration • Antiparasitic medications, such as nitazoxanide or paromomycin, may be prescribed to speed up recovery • Immunocompromised individuals may require more aggressive treatment, including the use of antiretroviral drugs to treat HIV/AIDS or immunosuppressive drugs to manage organ transplant rejection
Follow-up-care	• Follow-up stool tests to confirm that the parasite has been eliminated • Immunocompromised individuals may require ongoing monitoring and treatment
Public health and interventions	• Outbreaks of cryptosporidiosis may require public health interventions, such as the closure of recreational • Water facilities or the distribution of clean water to affected communities • Proper sanitation and hygiene measures are critical to prevent the spread of Cryptosporidium

These processes involve the use of specialized membranes with fine pores that act as barriers, effectively trapping and removing the microscopic parasites.

Furthermore, advanced oxidation processes have emerged as an additional line of defense against Cryptosporidium contamination. Ultraviolet (UV) irradiation, which exposes the water to UV light, has shown remarkable success in inactivating Cryptosporidium. The UV light damages the genetic material of the parasites, rendering them unable to reproduce and causing their eventual demise. Ozone treatment is another effective method that involves the infusion of ozone gas into the water. Ozone has a powerful oxidizing effect, breaking down the structure of Cryptosporidium and rendering it non-viable.

- **Sanitation:**

One essential aspect of sanitation is the proper disposal of human and animal waste. Cryptosporidium can be present in feces, making it imperative to handle waste materials carefully. Appropriate containment and disposal methods, such as the use of sanitary toilets and septic systems, can prevent the release of Cryptosporidium into the environment, thus reducing the likelihood of contamination.

In addition to waste management, maintaining clean and disinfected food preparation areas and surfaces is crucial. Cryptosporidium can contaminate surfaces such as countertops, cutting boards, and utensils, posing a potential source of infection. Regularly cleaning these surfaces with soap and water, followed by disinfection using appropriate agents such as bleach, can effectively eliminate the parasite and prevent its transmission.

By emphasizing proper sanitation practices, individuals can significantly contribute to the prevention of Cryptosporidium transmission. These measures serve as a barrier against the spread of the parasite and its potential impact on human health.

- **Hygiene:**

Maintaining optimal personal hygiene practices plays a crucial role in preventing the transmission of Cryptosporidium, the parasitic culprit behind cryptosporidiosis. By adhering to simple yet effective measures, individuals can significantly reduce the risk of contracting and spreading the infection. One of the fundamental practices is thorough handwashing with soap and water, a practice that should be observed before and after using the restroom, before handling or consuming food, and after coming into contact with animals. These routine hygiene practices serve as barriers, effectively inhibiting the transmission of Cryptosporidium and safeguarding against the potentially debilitating consequences of cryptosporidiosis.

- **Vaccination:**

While there is currently no licensed vaccine available for cryptosporidiosis, recent advances in vaccine development have shown promise. Researchers are currently working on developing vaccines that target Cryptosporidium's unique life cycle, which could potentially provide long-lasting protection against the disease.

- **Early diagnosis and treatment:**

In addition to preventive measures, early diagnosis and treatment of cryptosporidiosis is critical. In healthy individuals, the disease usually resolves on its own within one to two weeks. However, antiparasitic medications, such as nitazoxanide or paromomycin, may be prescribed to speed up recovery. Immunocompromised individuals may require more aggressive treatment, including the use of antiretroviral drugs to treat HIV/AIDS or immunosuppressive drugs to manage organ transplant rejection.

A timely and accurate diagnosis followed by appropriate treatment is pivotal in minimizing the impact of cryptosporidiosis. It not only facilitates the resolution of symptoms and eradication of the parasite but also helps prevent potential complications and reduce the transmission of the disease to others.

- **Follow-up care:**

Follow-up care plays a crucial role in ensuring the complete eradication of the parasite and minimizing the risk of recurrence. After completing the prescribed treatment regimen, it is essential to undergo follow-up stool tests to confirm the elimination of the Cryptosporidium parasite. These tests help verify that the infection has been successfully cleared from the body.

Immunocompromised individuals, such as those with HIV/AIDS or other conditions that weaken the immune system, may require more intensive monitoring and treatment. Cryptosporidiosis can be particularly challenging to manage in these individuals due to the potential for severe and chronic diarrhea. Ongoing monitoring of their immune status and regular follow-up stool tests are crucial to detect any signs of a persistent or recurring infection.

In some cases, additional rounds of treatment may be necessary if the follow-up tests indicate the presence of the parasite. The healthcare provider will determine the appropriate course of action based on the individual's specific circumstances and response to treatment. It is important to promptly address any recurrent infections to prevent further complications and ensure the individual's overall well-being.

- **Public health and interventions:**
  In the case of outbreaks of cryptosporidiosis, public health interventions may be necessary. These interventions may include the closure of recreational water facilities or the distribution of clean water to affected communities. Proper sanitation and hygiene measures are critical to prevent the spread of Cryptosporidium.
- **Other prevention measures:**
  In addition to the advancements in diagnostics and treatment, it is essential to implement comprehensive prevention measures to effectively combat cryptosporidiosis. These preventive strategies encompass a range of practices that minimize the risk of exposure to Cryptosporidium and subsequent infection.

  One crucial preventive measure is to avoid direct contact with fecal matter from infected animals or individuals. Cryptosporidium can be shed in feces, making it important to practice proper hygiene, such as thorough handwashing after handling animals or caring for infected individuals, to prevent the transmission of the parasite.

  Furthermore, it is crucial to exercise caution when consuming water or food. Untreated water sources, such as streams, lakes, or wells, may contain Cryptosporidium oocysts. Hence, it is recommended to drink only treated or filtered water from reliable sources. Similarly, consuming raw or undercooked food, particularly fruits and vegetables that may have been washed with contaminated water, should be avoided. Ensuring proper food hygiene and thoroughly washing produce with safe water can help reduce the risk of infection.

  Proper management of swimming pools and recreational water facilities is another vital aspect of prevention. Cryptosporidium can survive in chlorinated water and is resistant to many common disinfectants. Therefore, maintaining adequate chlorine levels and regularly cleaning and disinfecting pools and water facilities are essential to minimize the risk of transmission. Additionally, individuals experiencing diarrhea or other gastrointestinal symptoms should refrain from swimming or using public water facilities to prevent the potential contamination of the water.

In the event of an outbreak of cryptosporidiosis, public health interventions may be necessary to control the spread of the disease. This may include the closure of recreational water facilities or the distribution of clean water to affected communities. Proper sanitation and hygiene measures are critical in preventing the spread of Cryptosporidium, and continued efforts to improve water treatment technologies and vaccine development will be important in controlling the disease in the future.

21.8 Future Perspectives

Despite significant advances in the diagnosis, treatment, and prevention of cryptosporidiosis, this disease remains a global health challenge. There is a need for continued research and development to identify new approaches for the prevention, diagnosis, and treatment of cryptosporidiosis.

- **Development of new therapeutic agents:**
  One potential direction for future research is the development of new therapeutic agents that specifically target Cryptosporidium. Currently, only a limited number of drugs are available for the treatment of cryptosporidiosis, and resistance to these drugs is a growing concern. New drugs that are more effective and have fewer side effects are needed to address this challenge.
- **Development of vaccines:**
  Another promising area of research is the development of vaccines for the prevention of cryptosporidiosis. Several vaccine candidates are currently in preclinical and clinical development, and recent advances in vaccine technology have shown promising results. A successful vaccine would have a significant impact on the prevention and control of cryptosporidiosis.
- **Improvement in water treatment and sanitation methods:**
  Improving water treatment and sanitation methods is crucial for the prevention of cryptosporidiosis. Advanced water filtration technologies, such as nanofiltration and reverse osmosis, may offer better removal of Cryptosporidium oocysts from water sources. Additionally, innovative approaches, such as the use of natural plant extracts or probiotics, may offer alternative methods for preventing and controlling cryptosporidiosis.

- **Improved understanding of Cryptosporidium biology and pathogenesis:**
  An improved understanding of the biology and pathogenesis of Cryptosporidium is needed. This includes identifying the genetic and molecular mechanisms that enable the parasite to evade the immune system and cause disease. Such knowledge can aid in the development of new therapeutic agents and vaccines.
- **Implementation of education and awareness campaigns:**
  Implementing education and awareness campaigns can improve public understanding of cryptosporidiosis and promote better hygiene and sanitation practices. This can help reduce the spread of the disease and its associated impact on public health.
- **Collaboration among stakeholders:**
  Collaboration between researchers, clinicians, public health officials, and policymakers is essential for the successful prevention and control of Cryptosporidiosis. Coordination among these groups can help identify emerging trends and facilitate the development and implementation of effective strategies for the prevention and control of this disease. By working together, stakeholders can pool their expertise and resources to tackle the challenges posed by cryptosporidiosis more effectively.

21.9 Conclusion

In conclusion, the chapter "Cryptosporidiosis: Recent Advances in Diagnostics and Management" provides a comprehensive overview of the progress made in understanding and combating cryptosporidiosis, a highly prevalent parasitic disease caused by the protozoan parasite Cryptosporidium. This chapter highlights the escalating global burden of cryptosporidiosis, particularly in resource-limited regions, and emphasizes the urgent need for accurate diagnostics and effective management strategies.

The chapter explores the intricate epidemiology of cryptosporidiosis, elucidating its transmission dynamics through the fecal-oral route, contaminated water, and food sources, person-to-person contact, and animal interactions. It underscores the significant impact of waterborne disease outbreaks, including those in developed nations, as well as the heightened vulnerability of immunocompromised individuals, such as those with HIV/AIDS, to the disease.

Advancements in diagnostic techniques are thoroughly examined, with a critical appraisal of their merits and limitations. Molecular diagnostic tools, including PCR and NGS, have revolutionized the detection and characterization of Cryptosporidium species and subtypes, offering valuable insights into disease transmission patterns. Additionally, traditional methods such as acid-fast staining, IFA, enzyme-linked immunosorbent assay (ELISA), LAMP, and microscopy with infrared staining are discussed, highlighting their respective roles in diagnosis.

Furthermore, the chapter delves into the available treatment options for cryptosporidiosis, focusing on the efficacy and safety profiles of drugs like nitazoxanide and paromomycin. The unique challenges faced in managing the disease among immunocompromised patients are also addressed, emphasizing the need for tailored therapeutic interventions.

The importance of implementing robust prevention and control measures is underscored throughout the chapter. It highlights the significance of improving hygiene and sanitation practices, access to safe drinking water, and effective sewage and wastewater treatment systems. The potential for future advancements, including the development of new therapeutic agents and vaccines, improvement in water treatment and sanitation methods, better understanding of Cryptosporidium biology and pathogenesis, and the implementation of education and awareness campaigns, is also explored.

References

1 Cacciò, S.M. and Chalmers, R.M. (2016). Human cryptosporidiosis in Europe. *Clin. Microbiol. Infect.* 22 (6): 471–480.
2 Brunkard, J.M., Ailes, E., Roberts, V.A. et al. (2011). Surveillance for waterborne disease outbreaks associated with drinking water – United States, 2007–2008. *MMWR Surveill. Summ.* 60 (12): 38–68. https://europepmc.org/article/MED/21937977.
3 Hunter, P.R. and Nichols, G. (2002). Epidemiology and clinical features of Cryptosporidium infection in immunocompromised patients. *Clin. Microbiol. Rev.* 15 (1): 145–154. https://pubmed.ncbi.nlm.nih.gov/11781272.
4 Checkley, W., White, A.C., Jaganath, D. et al. (2015). A review of the global burden, novel diagnostics, therapeutics, and vaccine targets for cryptosporidium. *Lancet Infect. Dis.* 15 (1): 85–94. https://pubmed.ncbi.nlm.nih.gov/25278220.

5 World Health Organization, Risk Assessment of Cryptosporidium in Drinking Water. https://www.who.int/publications/i/item/WHO-HSE-WSH-09.04 (accessed 27 April 2023).

6 Centers for Disease Control and Prevention, Parasites – Cryptosporidium (Also Known as "Crypto") CDC. https://www.cdc.gov/parasites/crypto/index.html (accessed 27 April 2023)

7 McHardy, I.H., Wu, M., Shimizu-Cohen, R. et al. (2014). Detection of intestinal protozoa in the clinical laboratory. *J. Clin. Microbiol.* 52 (3): 712–720. https://pubmed.ncbi.nlm.nih.gov/24197877.

8 Chalmers, R.M. and Katzer, F. (2013). Looking for Cryptosporidium: the application of advances in detection and diagnosis. *Trends Parasitol.* 29 (5): 237–251. https://pubmed.ncbi.nlm.nih.gov/23566713.

9 Pinto, D.J. and Vinayak, S. (2021). Cryptosporidium: host-parasite interactions and pathogenesis. *Curr. Clin. Microbiol. Rep.* 8 (2): 62.

10 Aldeyarbi, H.M., Abu El-Ezz, N.M.T., and Karanis, P. (2016). Cryptosporidium and cryptosporidiosis: the African perspective. *Environ. Sci. Pollut. Res.* 23 (14): 13811–13821.

11 Cleveland Clinic, Cryptosporidiosis (Crypto): Symptoms, Treatment & Prevention. https://my.clevelandclinic.org/health/diseases/21023-cryptosporidiosis (accessed 9 May 2023).

12 Elsafi, S.H., Al-Sheban, S.S., Al-Jubran, K.M. et al. (2014). Comparison of Kinyoun's acid-fast and immunofluorescent methods detected an unprecedented occurrence of Cryptosporidium in the eastern region of Saudi Arabia. *J. Taibah Univ. Med. Sci.* 9 (4): 263–267.

13 Stibbs, H.H. and Ongerth, J.E. (1986). Immunofluorescence detection of Cryptosporidium oocysts in fecal smears. *J. Clin. Microbiol.* 24 (4): 517–521. https://pubmed.ncbi.nlm.nih.gov/2429982.

14 Alhajj, M. and Farhana, A. (2023). *Enzyme Linked Immunosorbent Assay*. StatPearls https://www.ncbi.nlm.nih.gov/books/NBK555922 (accessed 9 May 2023).

15 Morgan, U.M. and Thompson, R.C.A. (1998). PCR detection of Cryptosporidium: the way forward? *Parasitol. Today* 14 (6): 241–245.

16 Becherer, L., Borst, N., Bakheit, M. et al. (2020). Loop-mediated isothermal amplification (LAMP) – review and classification of methods for sequence-specific detection. *Anal. Methods* 12 (6): 717–746. https://pubs.rsc.org/en/content/articlehtml/2020/ay/c9ay02246e.

17 Robinson, G. and Chalmers, R.M. (2020). Cryptosporidium diagnostic assays: microscopy. *Methods Mol. Biol.* 2052: 1–10. https://pubmed.ncbi.nlm.nih.gov/31452153.

18 Centers for Disease Control and Prevention, Cryptosporidiosis – Treatment. 2023

19 Ng, C.T., Fong, L.Y., and Abdullah, M.N.H. (2023). Interferon-gamma (IFN-γ): reviewing its mechanisms and signaling pathways on the regulation of endothelial barrier function. *Cytokine* 166: 156208.

20 Joshi, P., Maheshwari, V., Surana, S., and Mandan, S. (2009). The in vitro antidiarrhoeal activity of Holarrhena antidysenterica (bark) wall extracts. *Planta Med.* 75 (09): PD19. https://www.ayurtimes.com/holarrhena-antidysenterica-pubescens-kurchi-kutaj.

21 Deep Ayurveda, Bilva (*Aegle marmelos*) Herb Ayurvedic Overview. https://www.deepayurveda.com/bilva-aegle-marmelos-herb-ayurvedic-overview (accessed 9 May 2023).

22 Liu, X., Cui, C., Zhao, M. et al. (2008). Identification of phenolics in the fruit of emblica (Phyllanthus emblica L.) and their antioxidant activities. *Food Chem.* 109 (4): 909–915. https://www.ayurtimes.com/amla-indian-gooseberry-phyllanthus-emblica.

23 Aydogdu, U., Isik, N., Ekici, O.D. et al. (2018). Comparison of the effectiveness of halofuginone lactate and paromomycin in the treatment of calves naturally infected with Cryptosporidium parvum. *Acta Sci. Vet.* 46 (1): 9. https://www.researchgate.net/publication/325207407_Comparison_of_the_Effectiveness_of_Halofuginone_Lactate_and_Paromomycin_in_the_Treatment_of_Calves_Naturally_Infected_with_Cryptosporidium_parvum.

24 Crabb, J.H. (1998). Antibody-based immunotherapy of cryptosporidiosis. *Adv. Parasitol.* 40: 121–149. https://pubmed.ncbi.nlm.nih.gov/9554072.

25 Pickerd, N. and Tuthill, D. (2004). Resolution of cryptosporidiosis with probiotic treatment. *Postgrad. Med. J.* 80 (940): 112–113. https://pmj.bmj.com/content/80/940/112.

26 Moens, F., Duysburgh, C., van den Abbeele, P. et al. (2019). *Lactobacillus rhamnosus* GG and Saccharomyces cerevisiae boulardii exert synergistic antipathogenic activity in vitro against enterotoxigenic Escherichia coli. *Benefic. Microbes* 10 (8): 923–935. https://pubmed.ncbi.nlm.nih.gov/31965838.

27 Correia, J.H., Rodrigues, J.A., Pimenta, S. et al. (2021). Photodynamic therapy review: principles, photosensitizers, applications, and future directions. *Pharmaceutics* 13 (9): 1332. https://pubmed.ncbi.nlm.nih.gov/34575408.

28 Wang, J.W., Kuo, C.H., Kuo, F.C. et al. (2019). Fecal microbiota transplantation: review and update. *J. Formos. Med. Assoc.* 118 (Suppl 1): S23–S31. https://pubmed.ncbi.nlm.nih.gov/30181015.

29 Rasmussen, T.S., Koefoed, A.K., Jakobsen, R.R. et al. Bacteriophage-mediated manipulation of the gut microbiome – promises and presents limitations. *FEMS Microbiol. Rev.* 2020, 44 (4): 507–521. https://pubmed.ncbi.nlm.nih.gov/32495834.

30 Bloom, A.K. and Ryan, E.T. (2012). Nitazoxanide. In: *Hunter's Tropical Medicine and Emerging Infectious Disease*, 9th (ed. A.J. Magill, G.T. Strickland, J.H. Maguire, et al.), 1106–1107. Elsevier Health Sciences.

31 Hewitt, R.G., Yiannoutsos, C.T., Higgs, E.S. et al. (2000). Paromomycin: no more effective than placebo for treatment of cryptosporidiosis in patients with advanced human immunodeficiency virus infection. AIDS Clinical Trial Group. *Clin. Infect. Dis.* 31 (4): 1084–1092. https://pubmed.ncbi.nlm.nih.gov/11049793.

32 Kadappu, K.K., Nagaraja, M.V., Rao, P.V., and Shastry, B.A. (2002). Azithromycin as treatment for cryptosporidiosis in human immunodeficiency virus disease. *J. Postgrad. Med.* 48: 179–181. https://pubmed.ncbi.nlm.nih.gov/12432190.

33 Pilla, A.M., Rybak, M.J., and Chandrasekar, P.H. (1987). Spiramycin in the treatment of cryptosporidiosis. *Pharmacotherapy* 7 (5): 188–190. https://onlinelibrary.wiley.com/doi/full/10.1002/j.1875-9114.1987.tb04049.x.

34 Uip, D.E., Lima, A.L.L., Amato, V.S. et al. (1998). Roxithromycin treatment for diarrhoea caused by Cryptosporidium spp. in patients with AIDS. *J. Antimicrob. Chemother.* 41 (suppl_2): 93–97. https://academic.oup.com/jac/article/41/suppl_2/93/828099.

35 Connolly, G.M., Dryden, M.S., Shanson, D.C., and Gazzard, B.G. (1988). Cryptosporidial diarrhoea in AIDS and its treatment. *Gut* 29 (5): 593–597. https://gut.bmj.com/content/29/5/593.

36 Mahmood, M.N., Ramadan, F.N., Hassan, M.S. et al. (2016). Introducing miltefosine as an anti-cryptosporidial agent in immunocompromised mice. *J. Plant. Pathol. Microbiol.* 7 (5): 1–5. https://www.walshmedicalmedia.com/abstract/introducing-miltefosine-as-an-anticryptosporidial-agent-inimmunocompromised-mice-14111.html.

37 Diptyanusa, A. and Sari, I.P. (2021). Treatment of human intestinal cryptosporidiosis: a review of published clinical trials. *Int. J. Parasitol. Drugs Drug Resist.* 17: 128–138.

38 Vanathy, K., Parija, S., Mandal, J. et al. (2017). Cryptosporidiosis: a mini review. *Trop. Parasitol.* 7 (2): 72.

39 Iroh Tam, P., Arnold, S.L.M., Barrett, L.K. et al. (2021). Clofazimine for treatment of cryptosporidiosis in human immunodeficiency virus infected adults: an experimental medicine, randomized, double-blind, placebo-controlled phase 2a trial. *Clin. Infect. Dis.* 73 (2): 183–191. https://academic.oup.com/cid/article/73/2/183/5819176.

22

Leishmaniasis: Current Trends in Microbiology and Pharmacology

Ismaila Adams[1], Awo A. Kwapong[2], Eugene Boafo[1], Elizabeth Twum[3], and Seth K. Amponsah[1]

[1] *Department of Medical Pharmacology, University of Ghana Medical School, Accra, Ghana*
[2] *Department of Pharmaceutics and Microbiology, School of Pharmacy, University of Ghana, Accra, Ghana*
[3] *Department of Pharmaceutical Sciences, College of Pharmacy, University of Tennessee Health Science Center, Memphis, TN, United States*

22.1 Introduction

Leishmaniasis is a complex and widespread neglected tropical disease caused by protozoan parasites of the genus *Leishmania*. It affects millions of people worldwide, with diverse clinical manifestations, and significant public health implications [1]. The disease is transmitted through the bite of an infected female phlebotomine sandfly. Based on clinical presentation, the disease can be classified into cutaneous, mucocutaneous, and visceral forms [2].

The classification of *Leishmania* species and their clinical manifestations varies geographically, reflecting the diversity of the parasites and their vectors. Recent advances in the understanding *Leishmania* biology, including genomics and drug resistance mechanisms, have contributed to improved disease management and control [3]. These developments have paved the way for targeted control strategies that hold promise for reducing the burden of leishmaniasis [4].

22.1.1 Global Epidemiology and Burden of Disease

Leishmaniasis, caused by protozoan parasites of the genus *Leishmania*, remains a significant public health concern. This neglected tropical disease affects millions of people across various regions, particularly in tropical and subtropical areas that have limited resources and inadequate healthcare infrastructure. Leishmaniasis is endemic in more than 98 countries, with approximately 350 million people at risk of infection [4, 5]. The disease exhibits a wide range of distribution, with regional variations in prevalence, incidence, and transmission dynamics. Each year, there are approximately 1 million new cases of cutaneous leishmaniasis (CL) and up to 300,000 new cases of visceral leishmaniasis (VL) worldwide [6]. CL has a broad distribution that includes regions of the Americas, Africa, Asia, and the Mediterranean, while VL is predominantly found in parts of South Asia, East Africa, and South America, with the majority of cases concentrated in a few countries [7].

The burden of leishmaniasis extends beyond its direct health impacts. The disease exerts a substantial socioeconomic burden on affected individuals, communities, and healthcare systems. Impoverished populations living in rural or peri-urban areas with limited access to healthcare services are particularly vulnerable [8]. The debilitating nature of leishmaniasis, especially in severe forms like VL, can lead to prolonged illness, disability, and significant economic losses due to treatment costs, loss of productivity, and decreased quality of life [9]. Efforts to quantify the global burden of leishmaniasis have faced challenges due to underreporting, misdiagnosis, and limited surveillance systems in endemic regions. However, recent studies and initiatives have made significant strides in improving our understanding of the impact of the disease [8]. Development of standardized case definitions, improved diagnostic tools, and enhanced surveillance networks are enhancing the assessment of the epidemiology and burden of leishmaniasis [10].

Rising Contagious Diseases: Basics, Management, and Treatments, First Edition. Edited by Seth Kwabena Amponsah, Ranjita Shegokar, and Yashwant V. Pathak.

The distribution of leishmaniasis is closely linked to environmental factors, including climate, vegetation, and the presence of suitable vector species. Factors such as migration, urbanization, conflict, and changes in land use patterns can contribute to the spread and emergence of leishmaniasis [11]. To address the global burden of leishmaniasis, effective disease control strategies are needed, including improved access to healthcare services, vector control measures, early diagnosis, and appropriate treatment [10]. Additionally, community education and awareness programs can play an important role in prevention and control efforts. By implementing comprehensive approaches that consider the regional variations in epidemiology and burden, efforts can be made to reduce the impact of leishmaniasis [12].

22.1.2 Transmission and Vector Biology

Leishmaniasis, a parasitic disease caused by the genus *Leishmania*, is transmitted to humans through the bite of infected female phlebotomine sandflies. The transmission and vector biology of leishmaniasis play a crucial role in the epidemiology and control of the disease. Phlebotomine sandflies, belonging to the genera *Phlebotomus* in the Old World, and *Lutzomyia* in the New World, serve as the primary vectors for transmitting *Leishmania* parasites [13]. These small, blood-feeding insects have a worldwide distribution, with approximately 800 species identified to date. The life cycle of Leishmania parasites involves two main stages: the promastigote stage within the sandfly vector and the amastigote stage within the mammalian host [14], as shown in Figure 22.1a and b.

The transmission of leishmaniasis is influenced by various factors, including sandfly species, their distribution, behavior, and abundance. Different *Leishmania* species have distinct vector preferences, leading to variations in disease patterns [13]. The vectorial capacity of sandflies depends on factors such as their longevity, feeding preferences, competence as vectors, and ability to transmit parasites. Environmental factors, such as temperature, humidity, vegetation, and land use, also impact the abundance and distribution of sandfly populations [11]. Understanding the biology and behavior of sandflies, their interaction with *Leishmania* parasites, and the environmental factors influencing their distribution and abundance are critical for effective disease control. Vector control measures such as insecticide-treated bed nets, indoor residual spraying, and environmental management aim to reduce sandfly populations and interrupt disease transmission [15, 16].

Transmission dynamics can further be made complicated by the presence of reservoir hosts, which are infected mammals that serve as a source of *Leishmania* parasites for sandflies. Reservoir hosts can include wild animals, domestic animals, and sometimes humans. The interaction between sandflies, reservoir hosts, and infected humans contributes to the maintenance and spread of leishmaniasis within endemic areas [17]. Efforts to control leishmaniasis transmission include targeted surveillance and prevention strategies based on an understanding of vector biology. Additionally, research on sandfly behavior, host preferences, and environmental factors can inform the development of effective prevention and control strategies. A comprehensive understanding of the transmission and vector biology of leishmaniasis is essential for guiding surveillance efforts, implementing targeted interventions, and ultimately reducing the burden of the disease [18].

This chapter provides a comprehensive overview of the current trends in microbiology and pharmacology of leishmaniasis. It will delve into the classification of *Leishmania* species and their clinical manifestations, highlighting the molecular mechanisms underlying disease pathogenesis. Additionally, the chapter will discuss the challenges associated with accurate diagnosis of *Leishmania* species and available treatment options. Also, this chapter will explore recent developments in antileishmanial drug discovery, drug resistance, and vaccine development.

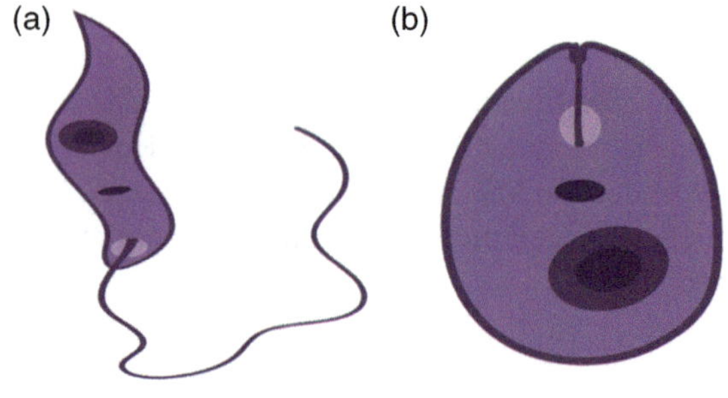

Figure 22.1 *Leishmania* parasite (a) promastigote stage, (b) amastigote stage.

22.2 Leishmania Species and Clinical Manifestations

22.2.1 Overview of *Leishmania* Species Causing Human Infection

Leishmaniasis is caused by various species of the *Leishmania* parasite, which can be classified into "Old World" and "New World" species. In the Old World, *Leishmania donovani* and *Leishmania infantum* are responsible for VL, while *Leishmania tropica* and *Leishmania major* cause CL [19]. In the New World, the *Leishmania braziliensis* complex, including *L. braziliensis*, *Leishmania guyanensis*, and *Leishmania panamensis*, is associated with CL and mucocutaneous leishmaniasis. Additionally, the *Leishmania mexicana* complex, including *L. mexicana*, causes localized CL primarily in Mexico, Central America, and the southern United States [20]. It is important to consider genetic diversity within each species, as it can influence disease manifestations and outcomes.

22.2.2 Cutaneous Leishmaniasis: Clinical Presentation and Pathogenesis

CL is characterized by localized skin lesions, and its clinical presentation and pathogenesis are influenced by factors such as the infecting *Leishmania* species and the host immune response. The initial manifestation of CL includes small papules or nodules at the site of the sandfly bite, which can progress into ulcers with raised borders and central ulcerated areas. The size, shape, and depth of the ulcers may vary, and multiple lesions can develop in certain cases [21].

The pathogenesis of CL involves a complex interaction between *Leishmania* parasites and the host immune response. After a sandfly bite, infective promastigotes are injected into the skin and engulfed by macrophages, where they transform into amastigotes and multiply. The destruction of infected cells by the multiplying amastigotes and the host's cellular immune response contribute to controlling the infection. Immune mediators, including cytokines and chemokines, are produced during pathogenesis, leading to tissue damage and the formation of characteristic skin lesions [22].

It is important to note that the clinical presentation and pathogenesis of CL can vary depending on geographic regions and specific *Leishmania* species. For example, certain species like *L. braziliensis* can cause mucocutaneous leishmaniasis, leading to severe and disfiguring lesions in the nose, mouth, and throat [23].

22.2.3 Visceral Leishmaniasis: Clinical Presentation and Pathogenesis

VL, also known as kala-azar, is the most severe form of leishmaniasis, characterized by systemic infection of internal organs. The clinical presentation of VL includes prolonged fever, fatigue, weight loss, hepatosplenomegaly, anemia, and a weakened immune system. If left untreated, VL can be fatal [24]. The pathogenesis of VL involves Leishmania parasites being transmitted to humans through sandfly bites. The parasites enter the bloodstream and are engulfed by macrophages, where they transform into replicative amastigotes. These amastigotes multiply and spread to various organs, including the liver, spleen, bone marrow, and lymph nodes [20]. The host immune response in VL is often compromised or dysregulated, allowing the parasites to evade and modulate the immune system, leading to persistent infection and disease progression. This dysregulated immune response contributes to the systemic involvement of multiple organs seen in VL [25].

In contrast to CL, which primarily affects the skin, VL exhibits distinct clinical features and a more severe course. The clinical presentation of VL includes systemic symptoms and enlargement of the liver and spleen, whereas CL presents with localized skin lesions. The pathogenesis of VL also differs from CL as the parasites spread through the bloodstream and infect multiple organs, whereas in CL, the parasites primarily reside in the skin. The immune response in VL is often impaired, while in CL, the immune response plays a significant role in controlling the infection. These differences highlight the distinct nature of VL compared to CL and emphasize the need for specific diagnostic and treatment approaches for each form of leishmaniasis [5, 26].

22.3 Microbiology of Leishmania

22.3.1 Life Cycle and Stages

Leishmania parasites, the causative agents of leishmaniasis, have a complex life cycle involving various stages and interactions between sandfly vectors and mammalian hosts (Figure 22.2).

Understanding the microbiology of *Leishmania*, including its life cycle and stages, is central to comprehending disease transmission, pathogenesis, and control strategies. The life cycle of *Leishmania* begins when an infected female sandfly belonging to the *Phlebotomus* genus in the Old World or the *Lutzomyia* genus in the New World takes a blood meal from a mammalian host, injecting promastigote forms of the parasite into the skin. Promastigotes are elongated, flagellated forms of *Leishmania* that reside in the midgut of the sandfly vector. Once inside the mammalian host, the promastigotes are phagocytized by macrophages. Inside the macrophages, the promastigotes differentiate into amastigotes, which are round-shaped, non-flagellated forms of the parasite. The amastigotes multiply and reside within specialized compartments called parasitophorous vacuoles within the host's cells.

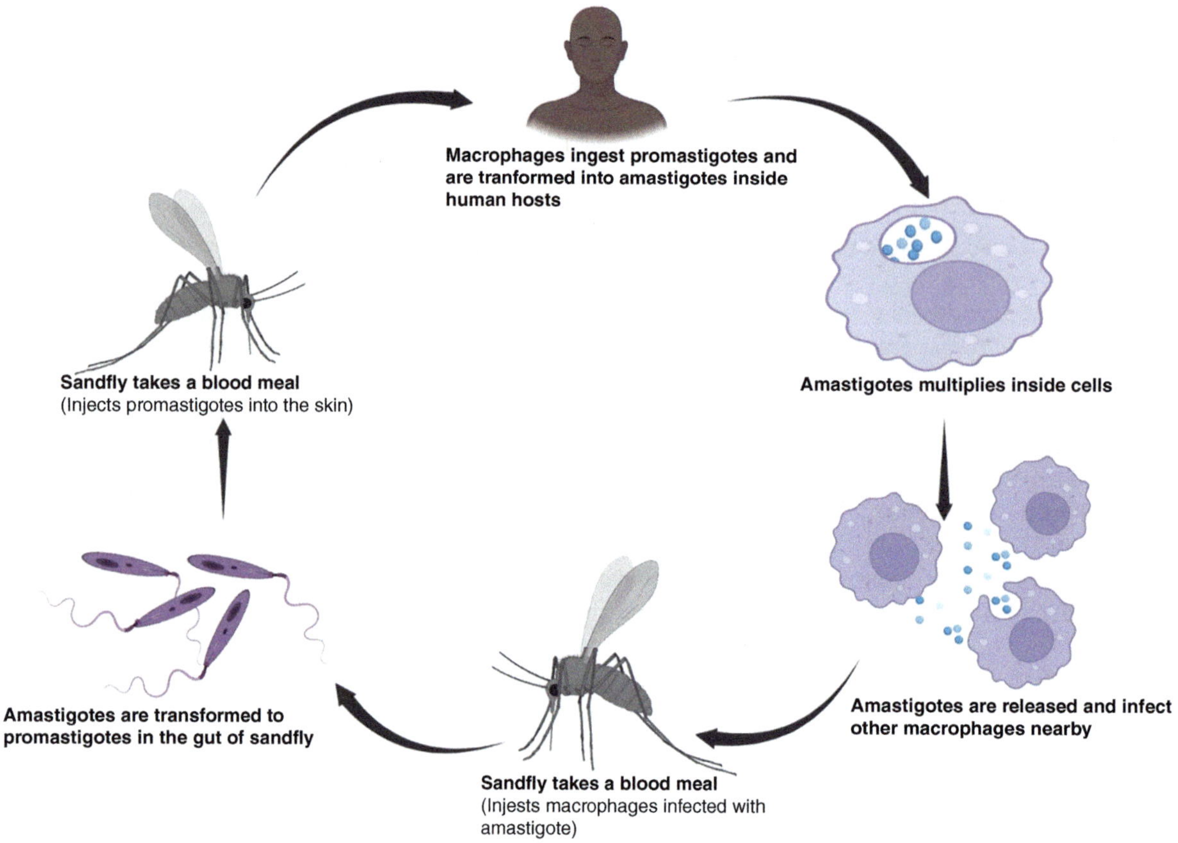

Figure 22.2 Life cycle of *Leishmania* parasite.

As the amastigotes multiply, they can cause damage to the infected cells, leading to clinical manifestations seen in leishmaniasis. The infected cells can rupture, releasing amastigotes into the surrounding tissues, where they can be taken up by other phagocytic cells or by sandflies during a blood meal. When an infected sandfly feeds on a mammalian host, it ingests the amastigotes along with the blood meal. The amastigotes differentiate into promastigotes within the midgut of the sandfly, attaching to the gut wall and replicating. The promastigotes then migrate to the proboscis of the sandfly, ready to be transmitted to another mammalian host during a subsequent blood meal.

The life cycle of *Leishmania* is completed when the promastigotes are injected into the skin of a mammalian host by an infected sandfly, perpetuating the cycle of infection [4, 27]. It is important to note that the life cycle and stages of *Leishmania* can vary depending on the *Leishmania* species and the specific sandfly vector involved. Different *Leishmania* species may exhibit variations in the duration of each stage and the preferred tissues or organs they infect within the mammalian host. Understanding the life cycle and stages of *Leishmania* is critical for multiple aspects of leishmaniasis research and control, including the development of diagnostic tools, drug discovery, and vaccine development.

22.3.2 Molecular Biology and Genomics of Leishmania

The study of molecular biology and genomics has significantly advanced our understanding of *Leishmania* parasites, revealing complex mechanisms of interaction between parasite and host. The availability of complete genome sequences has facilitated the identification of genes involved in various biological processes, such as metabolism, immune evasion, drug resistance, and virulence [28]. Genomic studies have highlighted a high level of genetic diversity within and between *Leishmania* species, influencing their virulence, drug susceptibility, and transmission dynamics. Advancements in molecular techniques, including polymerase chain reaction (PCR) and next-generation sequencing, have revolutionized the field [29]. These techniques have facilitated the development of new diagnostic methods and treatment strategies, including the discovery of novel drug targets. Furthermore, the exploration of the *Leishmania* genome has led to the identification of potential vaccine candidates. Continued research in this field promises to further improve the control and management of leishmaniasis and reduce the disease burden.

Comprehensive genomic studies have revealed a high degree of diversity among different *Leishmania* species and strains, which has implications for their virulence, drug resistance, and transmission dynamics [30]. Transcriptomic analyses have further illuminated the dynamic changes in gene expression that occur during different stages of the parasite life cycle and in response to environmental cues, providing insights into the key regulators and signaling pathways that enable survival and adaptation of the parasite [31, 32]. For instance, a study on *Leishmania chagasi* infection found unique gene signatures associated with different infection profiles, with symptomatic individuals showing greater changes in their transcriptome [33].

In addition, researchers have made strides in understanding the complex interactions between *Leishmania* parasites and the host immune system. They have identified several parasite-derived factors that modulate host immune responses, offering a clearer picture of the strategies these parasites use to evade the immune system [34]. A study on cryptic *L. donovani* infection revealed that ultraviolet B (UVB) exposure modifies skin immune-stroma cross-talk and promotes the recruitment of effector T cells, although the function of these cells may be limited by the distinct niches occupied by macrophages after UVB exposure [35]. Genomic and proteomic studies have also been instrumental in identifying potential drug targets and understanding the mechanisms of drug resistance in *Leishmania* [36]. Furthermore, the exploration of the *Leishmania* genome has accelerated the development of potential vaccines, with several promising candidates identified based on their ability to induce strong immune responses and provide protection in preclinical models [37]. The TriTrypDB resource, which integrates genomic and functional data from kinetoplastid parasites, including *Leishmania*, has been a valuable tool in these investigations [38].

22.3.3 Host–Parasite Interactions and Immune Response

The interactions between *Leishmania* parasites and their hosts, particularly the immune response, are crucial in determining the outcome of leishmaniasis. Recent research has shed light on the complex dynamics of these interactions. For instance, Oliveira et al. [39], explored the role of degrons, short peptide sequences that signal protein degradation, in the immune system of *Mus musculus* that are potential targets for *Leishmania* spp. proteases. Their findings suggest that degrons may play a role in the immune responses in leishmaniasis.

Zhang et al. [40], conducted a comprehensive immunometabolic profiling of Leishmania-infected macrophages (LIMs) and found a highly complex, mixed polarization phenotype and a unique bioenergetics signature. Their research revealed that *Leishmania* parasites actively inhibit maturation of not only infected but also bystander bone marrow-derived macrophages (BMDMs). This study highlighted the role of metabolism in the host–parasite interplay and disease establishment.

Margaroni et al. [41], investigated the transcriptional profile of dendritic cells (DCs) infected with *L. infantum*. Their findings suggested an important role of metabolism on DCs-Leishmania interplay and eventually disease establishment. They found that *Leishmania* parasites exploit macrophage phenotypic plasticity to establish chronic infections.

Lastly, Gonçalves et al. [42], addressed the role of inducing factors of the Th17 pathway in Leishmania-macrophage infection through computational modeling of gene regulatory networks (GRNs). Their study provided an integrative and dynamic view of Leishmania-macrophage interaction over time that extends beyond the analysis of single-gene expression.

These recent findings underscore the complex interplay between *Leishmania* parasites and the host immune system, and the role of metabolism and immune modulation in the pathogenesis of leishmaniasis.

22.4 Diagnosis of Leishmaniasis

22.4.1 Diagnostic Methods

The clinical diagnosis of leishmaniasis is a complex process that involves evaluation of various clinical signs and symptoms, epidemiological factors, and laboratory findings. Recent technological advances have enhanced our understanding and ability to diagnose this disease accurately. For instance, molecular techniques such as PCR have become increasingly utilized in the detection and identification of *Leishmania* DNA, offering high sensitivity and species-specific identification [43].

In addition to these molecular techniques, serological tests, including enzyme-linked immunosorbent assays (ELISAs) and immunochromatographic rapid diagnostic tests (RDTs), have been employed to detect antibodies against *Leishmania* parasites. These tests are valuable for screening purposes and can provide supportive evidence of infection, although they may not differentiate between current and past infections [44].

The differential diagnosis of leishmaniasis is crucial as it can mimic other diseases. For example, skin lesions in CL may resemble other skin infections, while VL can be mistaken for other febrile illnesses. Recent advances in imaging technologies have introduced promising modalities to aid differential diagnosis [43].

22.4.1.1 Microscopy

Microscopic examination of tissue smears or aspirates stained with Giemsa or Wright stains is a common technique used to visualize *Leishmania* parasites. The parasites appear as amastigotes (intracellular forms) within macrophages or as promastigotes (extracellular forms) in the case of culture-derived samples. Microscopy is relatively simple, cost-effective, and can provide rapid results. However, it requires trained personnel and may have limitations in terms of sensitivity, especially in cases with low parasite loads [45].

22.4.1.2 Culture

Culture techniques involve the growth of *Leishmania* parasites in specific culture media supplemented with appropriate supplements and growth factors. Culture allows for the expansion of parasites and the maintenance of their viability. Promastigote forms of *Leishmania* can be cultured from clinical samples, such as skin aspirates, bone marrow, or spleen aspirates. Culture-derived parasites can be used for further characterization, drug susceptibility testing, or other laboratory investigations. However, culture methods are time-consuming, may require specialized facilities, and have a risk of contamination [46].

22.4.1.3 Molecular Methods

The use of molecular techniques, such as PCR and its variants, has significantly improved the diagnosis of leishmaniasis. These methods aid in accurate detection and identification of *Leishmania* parasites, even in cases with low parasite loads. Other techniques, like deoxyribonucleic acid (DNA) sequencing and restriction fragment length polymorphism (RFLP) analysis, provide further insights into the genetic diversity and drug resistance profiles of *Leishmania* isolates [47]. Recent advancements include the development of point-of-care molecular diagnostics, which offer rapid results and are particularly useful in resource-limited settings. Overall, these molecular methods, in conjunction with traditional laboratory techniques like microscopy and culture, contribute to accurate diagnosis, surveillance, and research on leishmaniasis.

22.4.1.4 ELISA

ELISA employs *Leishmania*-specific antigens coated on a solid surface, such as a microplate. Patient serum is added, and if *Leishmania*-specific antibodies are present, they will bind to the antigens. Detection is achieved using enzyme-conjugated secondary antibodies, followed by a colorimetric reaction. ELISAs can provide quantitative or semi-quantitative results based on antibody titers [48].

22.4.1.5 Immunochromatographic RDT

RDTs are portable, and provide results within a short time frame. They consist of nitrocellulose strips containing *Leishmania*-specific antigens. The patient's serum is added to the strip, and if *Leishmania*-specific antibodies are present, they bind to the antigens, resulting in a visible line or color change [49].

22.4.1.6 Immunoblotting

Immunoblotting, also known as Western blotting, is a technique used to confirm serological results and assess antibody specificity. It involves separating *Leishmania* antigens using gel electrophoresis and transferring them onto a membrane. Patient serum is then incubated with the membrane, and if *Leishmania*-specific antibodies are present, they will bind to the corresponding antigens. Detection is achieved using enzyme-conjugated secondary antibodies [50].

22.4.1.7 Direct Agglutination Test (DAT)

The DAT is a serological assay used for both diagnosis and epidemiological studies. It measures the level of antibodies in patient serum that can agglutinate (clump together) *Leishmania* parasites. The test utilizes formalin-fixed promastigotes of

Leishmania as the antigen. Patient serum is serially diluted and mixed with the antigen. If *Leishmania*-specific antibodies are present, they will cause agglutination, which can be visually observed or quantified using a spectrophotometer [51].

22.4.1.8 Leishmanin Skin Test (LST)

The LST is an immunological assay used to assess an individual's delayed-type hypersensitivity (DTH) response to *Leishmania* antigens. It involves intradermal injection of a standardized *Leishmania* antigen, typically a crude extract or a purified antigen, into the forearm. After a period of time, the size of the induration (skin reaction) is measured. A positive reaction indicates a cellular immune response, suggesting exposure to *Leishmania* parasites [52].

22.5 Drug Therapy for Leishmaniasis

The management of leishmaniasis heavily relies on drug therapy. The choice of treatment is dependent on factors such as the form of the disease, the *Leishmania* species involved, and immune status of the patient. Pentavalent antimonials like sodium stibogluconate and meglumine antimoniate have been the primary treatment for many years, but drug resistance has emerged. Amphotericin B, available in different formulations, is often used as second-line treatment for severe or drug-resistant cases. Miltefosine, the only oral drug approved, is effective against various *Leishmania* species but its use is limited due to the development of resistance. Paromomycin, an aminoglycoside antibiotic, has shown efficacy against visceral and some forms of CL. Combination therapy is often employed to improve treatment outcomes and prevent drug resistance. The development of new drugs and treatment strategies remains an active area of research to improve treatment outcomes [53–55].

22.5.1 First-Line Drugs and Their Mechanism of Action

First-line drugs for the treatment of leishmaniasis vary depending on the specific clinical form and geographical region. Here are some commonly used first-line drugs and their mechanisms of action:

22.5.1.1 Sodium Stibogluconate (Pentostam) and Meglumine Antimoniate (Glucantime)

The pentavalent antimonials have been the traditional first-line drugs for leishmaniasis. They are administered through intravenous or intramuscular injections. The exact mechanism of action is not fully understood, but it is believed that these antimonials inhibit the enzymes involved in the parasite's energy metabolism, leading to its death. They may also induce oxidative stress and activate host immune responses against the parasite [56].

22.5.1.2 Amphotericin B

Amphotericin B is a polyene antifungal drug used as a first-line treatment for VL and some forms of CL. It can be administered as conventional amphotericin B deoxycholate or as lipid-based formulations, such as liposomal amphotericin B (AmBisome). Amphotericin B acts by binding to the ergosterol in the parasite's cell membrane, causing disruption and leakage of cellular components. This leads to the death of the parasite [57].

22.5.1.3 Miltefosine

Miltefosine is the only oral drug approved for the treatment of leishmaniasis and is effective against both cutaneous and visceral forms. It is administered as capsules or tablets. Miltefosine interferes with the parasite's cell membrane lipid metabolism, disrupting its integrity and leading to cell death. It can also modulate the host immune response against the parasite.

22.5.1.4 Paromomycin

Paromomycin is an aminoglycoside antibiotic used as a first-line treatment for VL. It can be administered intramuscularly or as a topical cream. Paromomycin inhibits protein synthesis in the parasite by binding to the 30S ribosomal subunit, interfering with translation, and causing the parasite's death [58].

22.5.2 Challenges and Limitations of Drug Therapy in Leishmaniasis

While drug therapy remains a cornerstone in the management of leishmaniasis, it is not without challenges and limitations. The emergence of drug resistance, particularly to pentavalent antimonials, is a significant concern, leading to treatment failure, and posing a threat to disease control [59]. The limited arsenal of drugs available for leishmaniasis treatment further exacerbates this issue, with some drugs, such as miltefosine, showing variable efficacy across different *Leishmania* species and geographical regions [60].

Toxicity and side effects associated with current treatments also pose challenges. For instance, amphotericin B, while highly effective, can cause serious nephrotoxic effects, and miltefosine is associated with gastrointestinal disturbances and teratogenicity [59]. These adverse effects necessitate careful monitoring and management to ensure patient safety and treatment success.

The prolonged duration of treatment, often several weeks or months, can have an impact on patient compliance and increase the risk of relapse [60]. Furthermore, access to appropriate and effective drugs can be limited in resource-constrained settings, where the disease is most prevalent. This issue is further complicated by the high cost of some drugs, such as liposomal amphotericin B [59]. Additionally, co-infections or comorbidities, such as HIV/AIDS, tuberculosis, or malaria, can complicate treatment [59].

The development of novel drug candidates targeting Leishmania's unique metabolic pathways [59], and the use of nanotechnology for drug delivery [60], offer promising avenues for improving leishmaniasis treatment. However, close collaboration between healthcare providers, researchers, policymakers, and affected communities is crucial to overcoming these challenges and improving the outcomes of leishmaniasis drug therapy.

22.6 Drug Resistance in *Leishmania*

Patterns of drug resistance in *Leishmania* can be intrinsic, acquired, cross-resistance, multidrug resistance, and geographically variable. Intrinsic resistance is due to inherent genetic variations or differences in drug target proteins. Acquired resistance occurs when initially susceptible parasites develop resistance following exposure to antileishmanial drugs. Cross-resistance refers to the development of resistance to one drug resulting in reduced susceptibility or resistance to other drugs within the same class or with similar mechanisms of action [61]. Multidrug resistance is characterized by resistance to multiple drugs with different mechanisms of action. Geographic variations in drug resistance patterns are due to factors such as local treatment practices, drug use, and parasite population genetics.

Molecular markers and surveillance of drug resistance in *Leishmania* parasites are critical for monitoring and managing the emergence and spread of drug resistance. Genotypic markers such as single nucleotide polymorphisms (SNPs), copy number variations (CNVs), and gene expression profiling can provide insights into the genetic changes associated with resistance [60]. Phenotypic markers, such as *in vitro* drug sensitivity assays can help identify resistance patterns. Surveillance methods include field monitoring, molecular surveillance, global collaboration and databases [62].

A case study reported successful treatment of resistant VL with a repeated trial of Liposomal Amphotericin B, highlighting the challenges of managing drug resistance in clinical practice [63]. Another study emphasized the importance of understanding the causes of variable drug efficacy related to parasite susceptibility, host immunity, and drug pharmacokinetics to delay the emergence of resistance to existing and new drugs. Here is an overview of molecular markers and surveillance methods used for detecting and tracking drug resistance in Leishmania.

22.6.1 Genotypic Markers

22.6.1.1 Single Nucleotide Polymorphisms (SNPs)

SNPs are variations in a single nucleotide within the DNA sequence. Genotyping SNPs in genes associated with drug resistance can help identify specific mutations or polymorphisms linked to resistance. Comparative genomics studies and genome-wide association studies (GWAS) can identify SNPs associated with drug resistance phenotypes [64].

22.6.1.2 Copy Number Variations (CNVs)

CNVs refer to variations in the number of copies of a specific genomic region. Amplifications or deletions of genes associated with drug targets or resistance mechanisms can contribute to drug resistance. Techniques such as quantitative PCR or next-generation sequencing can detect CNVs and their association with drug resistance [65].

22.6.1.3 Gene Expression Profiling

Changes in gene expression profiles can provide insights into the molecular mechanisms underlying drug resistance. Transcriptomic approaches, such as microarray analysis or RNA sequencing, can identify differentially expressed genes associated with resistance phenotypes [66].

22.6.2 Phenotypic Markers

22.6.2.1 *In Vitro* Drug Sensitivity Assays

These assays involve testing the sensitivity of *Leishmania* isolates to various antileishmanial drugs in the laboratory. The determination of the minimum inhibitory concentration (MIC) or the IC50 value (the concentration of drug that inhibits 50% of parasite growth) provides a measure of drug susceptibility. Comparative analysis of MICs between susceptible and resistant isolates can help identify resistance patterns [67].

22.6.3 Surveillance Methods

22.6.3.1 Field Monitoring

Surveillance of drug resistance in the field involves collecting clinical samples from patients before and after treatment. These samples are then assessed for parasite response to treatment and drug susceptibility using phenotypic or genotypic assays. Monitoring treatment outcomes and evaluating changes in drug susceptibility over time can help detect emerging resistance [68].

22.6.3.2 Molecular Surveillance

Molecular surveillance involves the collection and analysis of *Leishmania* isolates from various geographic regions. Genotypic markers associated with resistance are identified through molecular techniques such as PCR, sequencing, or hybridization-based assays. Comparative analysis of genetic markers across different populations or time periods can provide insights into the prevalence and distribution of resistant strains [69].

22.6.3.3 Global Collaboration and Databases

International collaboration and the establishment of databases are crucial for sharing data on drug resistance in *Leishmania*. This allows for the pooling of information from different regions, facilitating the identification of global resistance trends, hotspots, and the potential cross-border spread of resistant strains [68, 69].

By combining genotypic and phenotypic markers with surveillance methods, molecular surveillance provides a comprehensive understanding of drug resistance in Leishmania. This knowledge can inform treatment guidelines, aid in the development of new drugs, and contribute to the implementation of effective control strategies to combat drug resistance and ensure successful leishmaniasis management.

22.7 Strategies to Overcome Drug Resistance in Leishmania

Addressing drug resistance in *Leishmania* parasites necessitates a comprehensive approach that integrates both preventive and therapeutic strategies. The following strategies can be employed to counteract drug resistance in leishmaniasis:

22.7.1 Combination Therapy

This strategy involves the concurrent use of two or more drugs with distinct mechanisms of action, aiming to enhance treatment efficacy and minimize the emergence of resistance [70]. By targeting multiple pathways within the parasite, combination therapy reduces the likelihood of resistance developing against all the drugs simultaneously. The selection of suitable drug combinations should consider their synergistic effects, non-overlapping resistance mechanisms, and minimal overlapping toxicities.

22.7.2 Optimized Treatment Regimens

The rational use of existing drugs is crucial to prevent the emergence of drug resistance. This includes ensuring accurate diagnosis, appropriate drug selection based on species and geographic considerations, and adherence to standardized treatment regimens [59]. Treatment duration, dosage, and frequency should be optimized to ensure complete parasite clearance and minimize the risk of relapse. Additionally, efforts should be made to improve patient compliance and provide adequate follow-up care.

22.7.3 Development of New Drugs

The development of novel antileishmanial drugs with different mechanisms of action is essential to overcome drug resistance [60]. Targeting unique pathways or proteins within the parasite can circumvent existing resistance mechanisms. Investment in research and development for new drugs, including the exploration of alternative compounds, repurposing of existing drugs, and identification of novel drug targets, is crucial to expand the treatment armamentarium.

22.7.4 Drug Combination Screening

High-throughput screening of large compound libraries can identify new drug combinations that exhibit synergistic effects against *Leishmania* parasites [70]. By systematically testing combinations of existing drugs, new drug candidates, or drug repurposing, potential synergistic interactions can be identified. This approach may lead to the discovery of novel combination therapies effective against drug-resistant strains.

22.7.5 Drug Delivery Optimization

Improving the delivery of antileishmanial drugs can enhance treatment outcomes. Strategies such as liposomal formulations, nanoparticle-based drug delivery systems, or topical formulations can improve drug efficacy, reduce toxicity, and enhance drug accumulation at the site of infection [71]. These approaches can improve drug penetration, bioavailability, and target specificity, thereby optimizing therapeutic outcomes.

22.7.6 Molecular Surveillance and Monitoring

Vigilant surveillance of drug resistance is crucial for detecting and tracking resistance patterns. Molecular surveillance, including genotypic and phenotypic characterization of resistant strains, can help identify emerging resistance and guide treatment decisions [72]. Surveillance programs should be established to monitor treatment outcomes, assess drug efficacy, and evaluate changes in drug susceptibility over time.

22.7.7 Combination of Drug Therapy with Immunomodulation

Immune modulation strategies aim to enhance the host immune response against the parasite, which can synergize with drug therapy. Immunomodulatory agents, such as immunostimulants or cytokines, can be used in combination with antileishmanial drugs to improve treatment outcomes [73]. These approaches aim to augment the host immune response, thereby aiding in parasite clearance and reducing the likelihood of drug resistance.

22.7.8 Education and Awareness

Public education and awareness campaigns are crucial to promote responsible use of antileishmanial drugs. Proper understanding of the importance of complete treatment courses, adherence to prescribed regimens, and the consequences of incomplete or inappropriate drug use can help mitigate the development and spread of drug resistance [74].

22.8 New Developments in Antileishmanial Drugs

Target-based drug discovery approaches involve identifying and designing compounds that specifically target essential proteins or metabolic pathways within the *Leishmania* parasite. These approaches aim to enhance drug efficacy, reduce toxicity, and overcome drug resistance. Here are some new developments in target-based drug discovery for antileishmanial drugs, supported by recent research:

22.8.1 Enzymes and Metabolic Pathways

22.8.1.1 Folate Metabolism

Folate metabolism plays a crucial role in the growth and survival of *Leishmania* parasites. Enzymes involved in this pathway, such as dihydrofolate reductase-thymidylate synthase (DHFR-TS), have been targeted for drug development. Inhibitors of DHFR-TS, such as methotrexate analogs or trimethoprim-sulfamethoxazole combinations, have shown promise as antileishmanial agents [75].

22.8.1.2 Sterol Biosynthesis

Sterol biosynthesis is essential for maintaining the integrity of the parasite's cell membrane. Enzymes involved in this pathway, including sterol 14α-demethylase (CYP51), have been targeted. Azole compounds, such as posaconazole and ravuconazole, inhibit CYP51 and have shown efficacy against leishmaniasis [76].

22.8.1.3 Proteases

Leishmania parasites utilize various proteases for survival, immune evasion, and host cell invasion. Proteases such as cysteine proteases (e.g., cathepsins) and serine proteases (e.g., metalloproteases) are potential targets. Inhibitors targeting these proteases have demonstrated antileishmanial activity and can disrupt vital cellular processes [77, 78].

22.8.2 Cell Signaling Pathways

Leishmania parasites rely on specific signaling pathways for growth, differentiation, and adaptation to the host environment. Targeting key enzymes or receptors involved in these pathways can disrupt parasite survival. For example, inhibitors of protein kinases, such as mitogen-activated protein kinases (MAPKs) or protein kinase B (PKB/AKT), have shown potential as antileishmanial agents [79].

22.8.3 Transporters

Leishmania parasites utilize transporters for the uptake of nutrients and drugs and for the efflux of toxic compounds. Targeting these transporters can enhance drug efficacy and reduce drug resistance. For instance, inhibitors of the P-glycoprotein-like ABC transporters involved in drug efflux have been explored to overcome multidrug resistance in *Leishmania* [80].

22.8.4 DNA Topoisomerases

DNA topoisomerases are essential enzymes involved in DNA replication, transcription, and repair. Inhibitors targeting these enzymes, such as topoisomerase I and II, have shown antileishmanial activity. Compounds that selectively inhibit *Leishmania* topoisomerases can disrupt DNA replication and induce parasite death [81].

22.8.5 High-Throughput Screening and Repurposing

High-throughput screening of compound libraries, including natural product extracts or synthetic chemical libraries, allows for the identification of compounds with activity against Leishmania. Screening approaches can target specific pathways or proteins of interest, enabling the discovery of novel compounds or the repurposing of existing drugs with antileishmanial activity [82].

22.8.6 Nanotechnology and Drug Delivery Systems

Advancements in nanotechnology have enabled the development of novel drug delivery systems for improved efficacy and targeted drug delivery. Nanoparticles, liposomes, and polymeric carriers can enhance drug stability, bioavailability, and cellular uptake, thereby improving the delivery of antileishmanial drugs to the parasite.

These target-based drug discovery approaches offer promising avenues for the development of novel antileishmanial drugs. By specifically targeting essential proteins or pathways within the *Leishmania* parasite, these approaches aim to increase treatment efficacy, reduce toxicity, and overcome drug resistance. Continued research, collaboration between academia and pharmaceutical industries, and investment in drug discovery efforts are crucial to further advance these developments and improve the therapeutic options for leishmaniasis.

22.9 Repurposing Existing Drugs

Repurposing existing drugs for the treatment of leishmaniasis has gained significant attention in recent years. This approach involves exploring drugs that are already approved for other medical conditions and assessing their efficacy against Leishmania parasites. Repurposing offers several advantages, including reduced time and costs compared to developing new drugs from scratch. Here are some aspects of repurposing existing drugs for leishmaniasis:

22.9.1 Drug Screening

Drug screening involves systematically testing existing drugs or libraries of compounds for their activity against *Leishmania* parasites. High-throughput screening methods, including phenotypic screening or target-based assays, can help identify drugs with potential antileishmanial activity. Screening libraries of approved drugs or compounds that have undergone extensive safety testing allows for a faster transition to clinical trials [82].

22.9.2 Mechanism of Action

Repurposing efforts focus on understanding the mechanisms of action of existing drugs and determining whether they can be effective against *Leishmania* parasites. Some drugs may have direct antiparasitic activity by targeting essential parasite proteins or metabolic pathways. Others may exert their effect indirectly by modulating the host immune response or interfering with parasite–host interactions.

22.9.3 Synergistic Combinations

Repurposed drugs can also be used in combination with existing antileishmanial drugs to enhance treatment efficacy and overcome drug resistance. Synergistic combinations can improve parasite clearance and reduce the likelihood of resistance development. Combinatorial screening approaches can identify effective drug combinations with enhanced antileishmanial activity [83].

22.9.4 Safety and Pharmacokinetics

Repurposed drugs have the advantage of already having established safety profiles in humans. Safety data from their approved indications can guide the use of these drugs in leishmaniasis patients. However, it is essential to evaluate the pharmacokinetics, drug interactions, and potential toxicities specific to the treatment of leishmaniasis to ensure optimal dosing and minimize adverse effects [84].

22.9.5 Clinical Trials

Repurposed drugs that show promising activity in preclinical studies can advance to clinical trials for further evaluation. Phase II and III trials assess the efficacy, safety, and optimal dosing regimens of repurposed drugs in leishmaniasis patients. These trials provide critical data on treatment outcomes, side effects, and potential drug interactions in the target population.

22.9.6 Accessibility and Affordability

Repurposing existing drugs can potentially make antileishmanial treatments more accessible and affordable. As these drugs are already approved and available in the market, they may be more readily accessible in resource-limited settings where leishmaniasis is endemic. Repurposing can also reduce the cost and time associated with developing new drugs, making them more affordable for patients and healthcare systems [85].

22.10 Vector Control and Leishmaniasis Prevention

Leishmaniasis prevention and control can be broadly categorized into environmental and personal protection measures, and public health interventions.

22.10.1 Environmental and Personal Protection Measures

Environmental control measures aim to reduce the breeding and resting sites of sandflies, the vectors of *Leishmania* parasites. These measures include insecticide spraying, the use of bed nets and screens, and improving housing conditions to minimize sandfly entry. Personal protection measures focus on reducing contact with sandflies. These include wearing long-sleeved clothing, applying insect repellents to exposed skin, and avoiding outdoor activities during peak sandfly activity periods [86].

22.10.2 Public Health Interventions

Public health interventions involve active surveillance and case detection to identify and diagnose individuals with leishmaniasis. This helps in understanding disease burden, monitoring disease trends, and implementing appropriate control measures. Public health interventions ensure access to diagnosis, treatment, and care for individuals with leishmaniasis. This includes providing effective antileishmanial drugs, managing complications, and supporting patient compliance. Public health programs implement vector control measures to reduce sandfly populations. This includes indoor and outdoor residual spraying of insecticides, environmental modifications, and community education on vector control practices [86].

22.10.3 Future Perspectives and Challenges

With ongoing research, there is a growing understanding of the intricate biology of *Leishmania* parasites. Advances in genomics, proteomics, and other omics technologies have provided insights into the parasite's molecular mechanisms, virulence factors, and drug resistance mechanisms. Continued research on leishmaniasis vaccines aims to develop highly effective and safe vaccines that provide long-lasting protection against multiple *Leishmania* species. Further exploration of host–parasite interactions is crucial for understanding the immune response against *Leishmania* parasites. Advancing our understanding of the ecology, behavior, and distribution of sandfly vectors will help refine vector control strategies and predict disease outbreaks.

Drug resistance in *Leishmania* parasites remains a significant challenge. Continuous monitoring of drug resistance patterns, the development of molecular markers for early detection, and the exploration of novel drug targets and combination therapies are essential to overcome this obstacle. Ensuring access to accurate diagnosis and effective treatment for all individuals affected by leishmaniasis is crucial. Integrated control programs that combine vector control, case management, health education, and community engagement need to be strengthened.

22.11 Conclusion

Leishmaniasis continues to pose a significant global health issue, especially in endemic areas. This publication has illuminated various facets of leishmaniasis, encompassing its microbiology, clinical presentations, pharmacotherapy, and vaccine development. Progress in comprehending the biology of *Leishmania* parasites and host–parasite interactions has enriched our understanding of the disease. Encouraging research avenues, such as the development of vaccines and targeted drug therapies, hold promise for enhancing the prevention, diagnosis, and treatment of leishmaniasis. The ongoing work on effective vaccines, including whole-parasite vaccines, subunit vaccines, and viral vector-based vaccines, demonstrates considerable potential for offering enduring protection against this parasitic infection.

Nevertheless, numerous challenges remain in the control and management of leishmaniasis. Issues such as drug resistance, restricted access to diagnosis and treatment, and the necessity for integrated control programs demand sustained focus and collaborative efforts from researchers, healthcare providers, policymakers, and communities. Tackling these challenges and capitalizing on the advances in understanding *Leishmania* biology and promising research directions will be vital to alleviate the impact of leishmaniasis. Through the implementation of comprehensive control strategies, raising awareness, and investing in research and healthcare infrastructure, we can aim to decrease the incidence, morbidity, and mortality associated with leishmaniasis. By uniting our efforts, we can work toward the prevention, control, and eventual eradication of this neglected tropical disease.

References

1 WHO Expert Committee on the Control of the Leishmaniases & World Health Organization. (2010). Control of the Leishmaniases: Report of a Meeting of the WHO Expert Committee on the Control of Leishmaniases, Geneva, 22–26 March 2010. Control de Las Leishmaniasis: Informe de Una Reunión Del Comité de Expertos de La OMS Sobre El Control de Las Leishmaniasis, Ginebra, 22 a 26 de Marzo de 2010. https://apps.who.int/iris/handle/10665/44412

2 Hong, A., Zampieri, R.A., Shaw, J.J. et al. (2020). One health approach to leishmaniases: understanding the disease dynamics through diagnostic tools. *Pathogens* 9 (10): 809. https://doi.org/10.3390/pathogens9100809.

3 de Menezes, J.P.B., Guedes, C.E.S., de Petersen, A.L.O.A. et al. (2015). Advances in development of new treatment for leishmaniasis. *Biomed. Res. Int.* 2015: 815023. https://doi.org/10.1155/2015/815023.

4 Morgado, F.N., Conceição-Silva, F., Pimentel, M.I.F., and Porrozzi, R. (2023). Advancement in leishmaniasis diagnosis and therapeutics. *Trop. Med. Infect. Dis.* 8 (5): 5. https://doi.org/10.3390/tropicalmed8050270.

5 Torres-Guerrero, E., Quintanilla-Cedillo, M.R., Ruiz-Esmenjaud, J., and Arenas, R. (2017). Leishmaniasis: a review. *F1000Research* 6: 750. https://doi.org/10.12688/f1000research.11120.1.

6 WHO. (2023). Leishmaniasis. https://www.who.int/news-room/fact-sheets/detail/leishmaniasis

7 Ghatee, M.A., Taylor, W.R., and Karamian, M. (2020). The geographical distribution of cutaneous leishmaniasis causative agents in Iran and its neighboring countries, a review. *Front. Public Health* 8: 11. https://www.frontiersin.org/articles/10.3389/fpubh.2020.00011.

8 Okwor, I. and Uzonna, J. (2016). Social and economic burden of human leishmaniasis. *Am. J. Trop. Med. Hyg.* 94 (3): 489–493. https://doi.org/10.4269/ajtmh.15-0408.

9 Grifferty, G., Shirley, H., McGloin, J. et al. (2021). Vulnerabilities to and the socioeconomic and psychosocial impacts of the leishmaniases: a review. *Res. Rep. Trop. Med.* 12: 135–151. https://doi.org/10.2147/RRTM.S278138.

10 Alvar, J., den Boer, M., and Dagne, D.A. (2021). Towards the elimination of visceral leishmaniasis as a public health problem in East Africa: reflections on an enhanced control strategy and a call for action. *Lancet Glob. Health* 9 (12): e1763–e1769. https://doi.org/10.1016/S2214-109X(21)00392-2.

11 Abdullah, A.Y.M., Dewan, A., Shogib, M.R.I. et al. (2017). Environmental factors associated with the distribution of visceral leishmaniasis in endemic areas of Bangladesh: modeling the ecological niche. *Trop. Med. Health* 45 (1): 13. https://doi.org/10.1186/s41182-017-0054-9.

12 Boelaert, M. and Consortium, T.N. (2016). Clinical research on neglected tropical diseases: challenges and solutions. *PLoS Negl. Trop. Dis.* 10 (11): e0004853. https://doi.org/10.1371/journal.pntd.0004853.

13 Cecílio, P., Cordeiro-da-Silva, A., and Oliveira, F. (2022). Sand flies: basic information on the vectors of leishmaniasis and their interactions with leishmania parasites. *Commun. Biol.* 5: 305. https://doi.org/10.1038/s42003-022-03240-z.

14 Centers for Disease Control and Prevention. (2020, February 18). CDC—Leishmaniasis—Biology. https://www.cdc.gov/parasites/leishmaniasis/biology.html

15 Faber, C., Montenegro Quiñonez, C., Horstick, O. et al. (2022). Indoor residual spraying for the control of visceral leishmaniasis: a systematic review. *PLoS Negl. Trop. Dis.* 16 (5): e0010391. https://doi.org/10.1371/journal.pntd.0010391.

16 Picado, A., Dash, A.P., Bhattacharya, S., and Boelaert, M. (2012). Vector control interventions for visceral leishmaniasis elimination initiative in South Asia, 2005–2010. *Indian J. Med. Res.* 136 (1): 22–31.

17 Mozafari, O., Sofizadeh, A., and Shoraka, H.R. (2020). Distribution of leishmania infection in humans, animal reservoir hosts and sandflies in Golestan Province, Northeastern Iran: a systematic review and meta-analysis. *Iran. J. Public Health* 49 (12): 2308–2319. https://doi.org/10.18502/ijph.v49i12.4813.

18 Stockdale, L. and Newton, R. (2013). A review of preventative methods against human leishmaniasis infection. *PLoS Negl. Trop. Dis.* 7 (6): e2278. https://doi.org/10.1371/journal.pntd.0002278.

19 Akhoundi, M., Kuhls, K., Cannet, A. et al. (2016). A historical overview of the classification, evolution, and dispersion of leishmania parasites and sandflies. *PLoS Negl. Trop. Dis.* 10 (3): e0004349. https://doi.org/10.1371/journal.pntd.0004349.

20 Steverding, D. (2017). The history of leishmaniasis. *Parasit. Vectors* 10: 82. https://doi.org/10.1186/s13071-017-2028-5.

21 Khosravi, A., Sharifi, I., Fekri, A. et al. (2017). Clinical features of anthroponotic cutaneous leishmaniasis in a major focus, southeastern Iran, 1994–2014. *Iran. J. Parasitol.* 12 (4): 544–553.

22 Liu, D. and Uzonna, J.E. (2012). The early interaction of leishmania with macrophages and dendritic cells and its influence on the host immune response. *Front. Cell. Infect. Microbiol.* 2: 83. https://doi.org/10.3389/fcimb.2012.00083.

23 Mann, S., Frasca, K., Scherrer, S. et al. (2021). A review of leishmaniasis: current knowledge and future directions. *Curr. Trop. Med. Rep.* 8 (2): 121–132. https://doi.org/10.1007/s40475-021-00232-7.

24 Centers for Disease Control and Prevention (2023, June 13). CDC—Leishmaniasis—Resources for Health Professionals.https://www.cdc.gov/parasites/leishmaniasis/health_professionals/index.html

25 Faleiro, R.J., Kumar, R., Hafner, L.M., and Engwerda, C.R. (2014). Immune regulation during chronic visceral leishmaniasis. *PLoS Negl. Trop. Dis.* 8 (7): e2914. https://doi.org/10.1371/journal.pntd.0002914.

26 Yasmin, H., Adhikary, A., Al-Ahdal, M.N. et al. (2022). Host–pathogen interaction in leishmaniasis: immune response and vaccination strategies. *Immuno* 2 (1): 1. https://doi.org/10.3390/immuno2010015.

27 Bates, P.A. (2007). Transmission of leishmania metacyclic promastigotes by phlebotomine sand flies. *Int. J. Parasitol.* 37 (10–3): 1097–1106. https://doi.org/10.1016/j.ijpara.2007.04.003.

28 Samarasinghe, S.R., Samaranayake, N., Kariyawasam, U.L. et al. (2018). Genomic insights into virulence mechanisms of *Leishmania donovani*: evidence from an atypical strain. *BMC Genomics* 19: 843. https://doi.org/10.1186/s12864-018-5271-z.

29 Mohammadiha, A., Dalimi, A., Mohebali, M. et al. (2018). Molecular identification and phylogenetic classification of *Leishmania* spp. isolated from human cutaneous leishmaniasis in Iran: a cross-sectional study. Iran. *J. Parasitol.* 13 (3): 351–361.

30 Zheng, Z., Chen, J., Ma, G. et al. (2020). Integrative genomic, proteomic and phenotypic studies of *Leishmania donovani* strains revealed genetic features associated with virulence and antimony-resistance. *Parasit. Vectors* 13 (1): 510. https://doi.org/10.1186/s13071-020-04397-4.

31 Lee, H.J., Georgiadou, A., Otto, T.D. et al. (2018). Transcriptomic studies of malaria: a paradigm for investigation of systemic host-pathogen interactions. *Microbiol. Mol. Biol. Rev.* 82 (2): e00071–e00017. https://doi.org/10.1128/MMBR.00071-17.

32 Yang, K.Y., Chen, Y., Zhang, Z. et al. (2016). Transcriptome analysis of different developmental stages of amphioxus reveals dynamic changes of distinct classes of genes during development. *Sci. Rep.* 6 (1): 1. https://doi.org/10.1038/srep23195.

33 da Matta, V.L.R., Gonçalves, A.N., Gomes, C.M.C. et al. (2023). Gene signatures of symptomatic and asymptomatic clinical-immunological profiles of human infection by *Leishmania* (L.) *chagasi* in Amazonian Brazil. *Microorganisms* 11 (3): 3. https://doi.org/10.3390/microorganisms11030653.

34 Gabriel, Á., Valério-Bolas, A., Palma-Marques, J. et al. (2019). Cutaneous leishmaniasis: the complexity of host's effective immune response against a polymorphic parasitic disease. *J Immunol Res* 2019: 2603730. https://doi.org/10.1155/2019/2603730.

35 Almeida, F.S., Vanderley, S.E.R., Comberlang, F.C. et al. (2023). Leishmaniasis: immune cells crosstalk in macrophage polarization. *Trop. Med. Infect. Dis.* 8: 276. https://doi.org/10.3390/tropicalmed8050276.

36 Rabaan, A.A., Bakhrebah, M.A., Mohapatra, R.K. et al. (2022). Omics approaches in drug development against leishmaniasis: current scenario and future prospects. *Pathogens* 12 (1): 39. https://doi.org/10.3390/pathogens12010039.

37 Srivastava, S., Shankar, P., Mishra, J., and Singh, S. (2016). Possibilities and challenges for developing a successful vaccine for leishmaniasis. *Parasit. Vectors* 9 (1): 277. https://doi.org/10.1186/s13071-016-1553-y.

38 Shanmugasundram, A., Starns, D., Böhme, U. et al. (2023). TriTrypDB: an integrated functional genomics resource for kinetoplastida. *PLoS Negl. Trop. Dis.* 17 (1): e0011058. https://doi.org/10.1371/journal.pntd.0011058.

39 Oliveira, A.S., Aredes-Riguetti, L.M., Pereira, B.A.S. et al. (2023). Degron pathways and leishmaniasis: debating potential roles of Leishmania spp. proteases activity on guiding hosts immune response and their relevance to the development of vaccines. *Vaccine* 11 (6): 1015. https://doi.org/10.3390/vaccines1106015.

40 Ty, M.C., Loke, P., Alberola, J. et al. (2019). Immuno-metabolic profile of human macrophages after Leishmania and Trypanosoma cruzi infection. *PLoS One* 14 (12): e0225588. https://doi.org/10.1371/journal.pone.0225588.

41 Margaroni, M., Agallou, M., Vasilakaki, A. et al. (2022). Transcriptional profiling of *Leishmania infantum* infected dendritic cells: insights into the role of immunometabolism in host-parasite interaction. *Microorganisms* 10 (7): 1271. https://doi.org/10.3390/microorganisms10071271.

42 Gonçalves, L.O., Pulido, A.F.V., Mathias, F.A.S. et al. (2022). Expression profile of genes related to the Th17 pathway in macrophages infected by *Leishmania major* and *Leishmania amazonensis*: the use of gene regulatory networks in modeling this pathway. *Front. Cell. Infect. Microbiol.* 12: https://www.frontiersin.org/articles/10.3389/fcimb.2022.826523.

43 Gow, I., Smith, N.C., Stark, D., and Ellis, J. (2022). Laboratory diagnostics for human Leishmania infections: a polymerase chain reaction-focussed review of detection and identification methods. *Parasit. Vectors* 15: 412. Retrieved July 16, 2023, https://www.ncbi.nlm.nih.gov/pmc/articles/PMC9636697.

44 Mahdavi, R., Shams-Eldin, H., Witt, S. et al. (2023). Development of a novel enzyme-linked immunosorbent assay and lateral flow test system for improved serodiagnosis of visceral Leishmaniasis in different areas of endemicity. *Microbiol. Spectr.* 11 (3): e04338–e04322. https://doi.org/10.1128/spectrum.04338-22.

45 Abd El-Salam, N.M., Ayaz, S., and Ullah, R. (2014). PCR and microscopic identification of isolated *Leishmania tropica* from clinical samples of cutaneous Leishmaniasis in human population of Kohat region in Khyber Pakhtunkhwa. *Biomed. Res. Int.* 2014: 861831. https://doi.org/10.1155/2014/861831.

46 Castelli, G., Oliveri, E., Valenza, V. et al. (2023). Cultivation of protozoa parasites in vitro: growth potential in conventional culture media versus RPMI-PY medium. *Vet. Sci.* 10 (4): 4. https://doi.org/10.3390/vetsci10040252.

47 Thakur, S., Joshi, J., and Kaur, S. (2020). Leishmaniasis diagnosis: an update on the use of parasitological, immunological and molecular methods. *J. Parasit. Dis.* 44 (2): 253–272. https://doi.org/10.1007/s12639-020-01212-w.

48 Jamal, F., Shivam, P., Kumari, S. et al. (2017). Identification of *Leishmania donovani* antigen in circulating immune complexes of visceral leishmaniasis subjects for diagnosis. *PLoS One* 12 (8): e0182474. https://doi.org/10.1371/journal.pone.0182474.

49 Ejazi, S.A., Choudhury, S.T., Bhattacharyya, A. et al. (2021). Development and clinical evaluation of serum and urine-based lateral flow tests for diagnosis of human visceral Leishmaniasis. *Microorganisms* 9 (7): 1369. https://doi.org/10.3390/microorganisms9071369.

50 Heidari, S., Gharechahi, J., Mohebali, M. et al. (2019). Western blot analysis of *Leishmania infantum* antigens in sera of patients with visceral Leishmaniasis. *Iran. J. Parasitol.* 14 (1): 10–19.

51 Mohebali, M., Keshavarz, H., Shirmohammad, S. et al. (2020). The diagnostic accuracy of direct agglutination test for serodiagnosis of human visceral leishmaniasis: a systematic review with meta-analysis. *BMC Infect. Dis.* 20 (1): 946. https://doi.org/10.1186/s12879-020-05558-7.

52 Carstens-Kass, J., Paulini, K., Lypaczewski, P., and Matlashewski, G. (2021). A review of the leishmanin skin test: a neglected test for a neglected disease. *PLoS Negl. Trop. Dis.* 15 (7): e0009531. https://doi.org/10.1371/journal.pntd.0009531.

53 Cavassin, F.B., Baú-Carneiro, J.L., Vilas-Boas, R.R., and Queiroz-Telles, F. (2021). Sixty years of amphotericin B: an overview of the main antifungal agent used to treat invasive fungal infections. *Infect. Dis. Ther.* 10 (1): 115–147. https://doi.org/10.1007/s40121-020-00382-7.

54 Sundar, S. and Olliaro, P.L. (2007). Miltefosine in the treatment of leishmaniasis: clinical evidence for informed clinical risk management. *Ther. Clin. Risk Manag.* 3 (5): 733–740.

55 Wiwanitkit, V. (2012). Interest in paromomycin for the treatment of visceral leishmaniasis (kala-azar). *Ther. Clin. Risk Manag.* 8: 323–328. https://doi.org/10.2147/TCRM.S30139.

56 Haldar, A.K., Sen, P., and Roy, S. (2011). Use of antimony in the treatment of leishmaniasis: current status and future directions. *Mol. Biol. Int.* 2011: 571242. https://doi.org/10.4061/2011/571242.

57 Mesa-Arango, A.C., Scorzoni, L., and Zaragoza, O. (2012). It only takes one to do many jobs: amphotericin B as antifungal and immunomodulatory drug. *Front. Microbiol.* 3: 286. https://doi.org/10.3389/fmicb.2012.00286.

58 Fernández, M.M., Malchiodi, E.L., and Algranati, I.D. (2011). Differential effects of paromomycin on ribosomes of *Leishmania mexicana* and mammalian cells. *Antimicrob. Agents Chemother.* 55 (1): 86–93. https://doi.org/10.1128/AAC.00506-10.

59 Singh, O.P. and Sundar, S. (2022). Visceral leishmaniasis elimination in India: progress and the road ahead. *Expert Rev. Anti-Infect. Ther.* 20 (11): 1381–1388. https://doi.org/10.1080/14787210.2022.2126352.

60 Ponte-Sucre, A., Gamarro, F., Dujardin, J.-C. et al. (2017). Drug resistance and treatment failure in leishmaniasis: a 21st century challenge. *PLoS Negl. Trop. Dis.* 11 (12): e0006052. https://doi.org/10.1371/journal.pntd.0006052.

61 Chakravarty, J. and Sundar, S. (2010). Drug resistance in leishmaniasis. *J. Global Infect. Dis.* 2 (2): 167–176. https://doi.org/10.4103/0974-777X.62887.

62 Ndiaye, Y.D., Hartl, D.L., McGregor, D. et al. (2021). Genetic surveillance for monitoring the impact of drug use on *Plasmodium falciparum* populations. *Int. J. Parasitol. Drugs Drug Resist.* 17: 12–22. https://doi.org/10.1016/j.ijpddr.2021.07.004.

63 Singh, O.P., Singh, B., Chakravarty, J., and Sundar, S. (2016). Current challenges in treatment options for visceral leishmaniasis in India: a public health perspective. *Infect. Dis. Poverty* 5: 19. https://doi.org/10.1186/s40249-016-0112-2.

64 Malkki, M. and Petersdorf, E.W. (2012). Genotyping of single nucleotide polymorphisms by 5′ nuclease allelic discrimination. *Methods Mol. Biol.* 882: 173–182. https://doi.org/10.1007/978-1-61779-842-9_10.

65 Pös, O., Radvanszky, J., Buglyó, G. et al. (2021). DNA copy number variation: main characteristics, evolutionary significance, and pathological aspects. *Biomed. J.* 44 (5): 548–559. https://doi.org/10.1016/j.bj.2021.02.003.

66 Singh, K.P., Miaskowski, C., Dhruva, A.A. et al. (2018). Mechanisms and measurement of changes in gene expression. *Biol. Res. Nurs.* 20 (4): 369–382. https://doi.org/10.1177/1099800418772161.

67 Ginouvès, M., Simon, S., Nacher, M. et al. (2017). In vitro sensitivity of cutaneous leishmania promastigote isolates circulating in French Guiana to a set of drugs. *Am. J. Trop. Med. Hyg.* 96 (5): 1143–1150. https://doi.org/10.4269/ajtmh.16-0373.

68 Tzani, M., Barrasa, A., Vakali, A. et al. (2021). Surveillance data for human leishmaniasis indicate the need for a sustainable action plan for its management and control, Greece, 2004 to 2018. *Eurosurveillance* 26 (18): 2000159. https://doi.org/10.2807/1560-7917.ES.2021.26.18.2000159.

69 der Auwera, G.V., Davidsson, L., Buffet, P. et al. (2022). Surveillance of leishmaniasis cases from 15 European Centres, 2014 to 2019: a retrospective analysis. *Eurosurveillance* 27 (4): 2002028. https://doi.org/10.2807/1560-7917.ES.2022.27.4.2002028.

70 Alves, F., Bilbe, G., Blesson, S. et al. (2018). Recent development of visceral leishmaniasis treatments: successes, pitfalls, and perspectives. *Clin. Microbiol. Rev.* 31 (4): e00048–e00018. https://doi.org/10.1128/CMR.00048-18.

71 Patra, J.K., Das, G., Fraceto, L.F. et al. (2018). Nano based drug delivery systems: recent developments and future prospects. *J. Nanobiotechnol.* 16 (1): 71. https://doi.org/10.1186/s12951-018-0392-8.

72 Morel, C.M., de Kraker, M.E.A., and Harbarth, S. (2021). Surveillance of resistance to new antibiotics in an era of limited treatment options. *Front. Med.* 8: 652638. https://doi.org/10.3389/fmed.2021.652638.

73 Adriaensen, W., Dorlo, T.P.C., Vanham, G. et al. (2018). Immunomodulatory therapy of visceral leishmaniasis in human immunodeficiency virus-coinfected patients. *Front. Immunol.* 8: 1943. https://doi.org/10.3389/fimmu.2017.01943.

74 Alvar, J., Vélez, I.D., Bern, C. et al. (2012). Leishmaniasis worldwide and global estimates of its incidence. *PLoS One* 7 (5): e35671. https://doi.org/10.1371/journal.pone.0035671.

75 Vickers, T.J. and Beverley, S.M. (2011). Folate metabolic pathways in leishmania. *Essays Biochem.* 51: 63–80. https://doi.org/10.1042/bse0510063.

76 Choi, J.Y., Podust, L.M., and Roush, W.R. (2014). Drug strategies targeting CYP51 in neglected tropical diseases. *Chem. Rev.* 114 (22): 11242–11271. https://doi.org/10.1021/cr5003134.

77 Gupta, G., Oghumu, S., and Satoskar, A.R. (2013). Mechanisms of immune evasion in leishmaniasis. *Adv. Appl. Microbiol.* 82: 155–184. https://doi.org/10.1016/B978-0-12-407679-2.00005-3.

78 Siklos, M., BenAissa, M., and Thatcher, G.R.J. (2015). Cysteine proteases as therapeutic targets: does selectivity matter? A systematic review of calpain and cathepsin inhibitors. *Acta Pharm. Sin. B* 5 (6): 506–519. https://doi.org/10.1016/j.apsb.2015.08.001.

79 Ochoa, R., Ortega-Pajares, A., Castello, F.A. et al. (2021). Identification of potential kinase inhibitors within the PI3K/AKT pathway of Leishmania species. *Biomolecules* 11: 1037. Retrieved July 16, 2023, https://www.ncbi.nlm.nih.gov/pmc/articles/PMC8301987.

80 Waghray, D. and Zhang, Q. (2018). Inhibit or evade multidrug resistance P-glycoprotein in cancer treatment. *J. Med. Chem.* 61 (12): 5108–5121. https://doi.org/10.1021/acs.jmedchem.7b01457.

81 Yang, G., Choi, G., and No, J.H. (2016). Antileishmanial mechanism of diamidines involves targeting kinetoplasts. *Antimicrob. Agents Chemother.* 60 (11): 6828–6836. https://doi.org/10.1128/AAC.01129-16.

82 Annang, F., Pérez-Moreno, G., García-Hernández, R. et al. (2015). High-throughput screening platform for natural product–based drug discovery against 3 neglected tropical diseases: human African trypanosomiasis, leishmaniasis, and chagas disease. *SLAS Discov.* 20 (1): 82–91. https://doi.org/10.1177/1087057114555846.

83 Nettey, H., Allotey-Babington, G.L., Somuah, I. et al. (2017). Assessment of formulated amodiaquine microparticles in *Leishmania donovani* infected rats. *J. Microencapsul.* 34 (1): 21–28. https://doi.org/10.1080/02652048.2017.1280094.

84 Allotey-Babington, G.L., Amponsah, S.K., Nettey, T. et al. (2020). Quinine sulphate microparticles as treatment for leishmaniasis. *J. Trop. Med.* 2020: 5278518. https://doi.org/10.1155/2020/5278518.

85 Charlton, R.L., Rossi-Bergmann, B., Denny, P.W., and Steel, P.G. (2018). Repurposing as a strategy for the discovery of new anti-leishmanials: the-state-of-the-art. *Parasitology* 145 (2): 219–236. https://doi.org/10.1017/S0031182017000993.

86 Centers for Disease Control and Prevention (2020, February 19). CDC—Leishmaniasis—Prevention & Control. https://www.cdc.gov/parasites/leishmaniasis/prevent.html

23

Recent Trends in Toxoplasmosis Diagnosis and Management

Mehul Chorawala[1], Nirjari Kothari[1], Aayushi Shah[1], Aanshi Pandya[1], Ishika Shah[1], and Bhupendra G. Prajapati[2]

[1] *Department of Pharmacology and Pharmacy Practice, L. M. College of Pharmacy, Gujarat University, Ahmedabad, Gujarat, India*
[2] *Department of Pharmaceutics and Pharmaceutical Technology, Shree S. K. Patel College of Pharmaceutical Education and Research, Ganpat University, Mehsana, Gujarat, India*

23.1 Introduction

Toxoplasmosis is one of the most significant zoonotic diseases worldwide which has a very cumbersome epidemiology and a large number of manifestations [1]. The causative agent for this zoonotic disorder is a protozoan parasite, *Toxoplasma gondii (TG)*. It is an obligate single-cell eukaryotic parasite from the subgroup Alveolata, phylum Apicomplexa, class Conoidasida, Order: Eucoccidiorida, and Family: Sarcocystidae. Although it is the only species from the genus Toxoplasma, it is closely related to *Neospora caninum* and *Hammondia hammondi* [2]. Toxoplasmosis in humans is primarily acquired through three mechanisms or ways or means: (i) consuming tissue cysts found in raw or undercooked meat; (ii) ingesting oocysts found in contaminated food, water, fruits, and vegetables, and (iii) possible way of mother-to-fetus transmission during pregnancy, if the mother becomes infected [3].

Toxoplasmosis caused by the parasite *T. gondii* can be difficult to diagnose because the symptoms can vary greatly, and often mimic other diseases. Additionally, the parasite can persist in the body for years without causing any symptoms, making it difficult to determine whether a person is in the acute or chronic phase of infection. Serological tests, such as ELISA or Western blot, can detect antibodies to *T. gondii* in the blood, but they cannot differentiate between an acute and chronic infection. Polymerase chain reaction (PCR)-based tests can detect the parasite's DNA in blood or tissues, but they are not widely available and can be expensive. There is still a need for more specific and reliable diagnostic tools for toxoplasmosis that can differentiate between the acute and chronic phases of infection and accurately diagnose the disease in all populations [4, 5].

Treatment of toxoplasmosis has traditionally been based on two main approaches: **chemotherapy and immunotherapy** [6]. Chemotherapy targets the replicating forms of *T. gondii* (tachyzoites) with drugs such as sulfadiazine, pyrimethamine, and clindamycin. These drugs have been used for decades and have been shown to be effective against acute infections. However, they do have some limitations, such as toxicity, the need for long-term treatment, and the development of resistance by the parasite [7]. Immunotherapy, on the other hand, aims to enhance the host's immune response against the parasite by stimulating the production of antibodies or by activating T cells. This approach has been used with some success in animal models, but its efficacy in humans has been limited. New strategies to develop anti-Toxoplasma drugs have been focused on identifying new targets in the parasite's life cycle, such as enzymes involved in energy production, protein synthesis, and DNA replication. Other approaches include the use of small-molecule inhibitors, the development of vaccines, and the use of natural compounds such as essential oils and plant extracts [8].

Additionally, the use of combination therapy has been proposed as a way to enhance the efficacy of current treatments. This involves the use of two or more drugs that target different stages of the parasite's life cycle, or that work through different mechanisms. Overall, the development of new, effective, and well-tolerated drugs for the management of toxoplasmosis remains a major challenge. However, ongoing research in this field has the potential to revolutionize the current approach to treatment and improve outcomes for patients with this infection.

Rising Contagious Diseases: Basics, Management, and Treatments, First Edition. Edited by Seth Kwabena Amponsah, Ranjita Shegokar, and Yashwant V. Pathak.

In this review, we will address the current advancements in the serological diagnosis and treatments available for human toxoplasmosis. In serodiagnosis part, we mainly focus on the benefits of using novel serological testing and offer a glimpse into the potential future of these techniques. The use of novel serodiagnostic techniques has become a prevalent trend in recent times and presents various advantages over conventional antigens such as improved specificity, accuracy, and sensitivity. These methods have the potential to overcome some of the limitations of traditional serological tests and be more cost-effective and accessible in the future, particularly in areas with limited resources. In the context of therapeutic armamentarium, we will discuss current therapy for pregnant women and newborns as well as for reactivation episodes in immuno-compromised patients, and explore the various strategies to develop novel drugs or regimens against toxoplasmosis (anti-Toxoplasmic agents). Therefore, it is of great importance to identify novel potent candidates that would be well-tolerated in pregnant women, newborns, and immuno-compromised patients.

23.1.1 Epidemiology

Every year, WHO estimates about 1 million cases of foodborne toxoplasmosis in the European region [9]. In the United States, it is estimated that 11% of the population aged six years and older have been infected with Toxoplasma. It has been observed that more than 60% of the population throughout the globe has been infected with the parasite. The prevalence and incidence are high in regions with hot and humid climates and in regions at lower altitudes [10]. In 2009, a study carried out in the United States identified eating uncooked seafood like oysters, clams, and mussels are prominent risk factors associated with *T. gondii* infection. Several risk factors previously associated with toxoplasmosis include the consumption of raw meat, unpasteurized goat milk, drinking untreated water, cats as pets, any job that requires contact with dirt, soil, or other material that could contain cat feces, and consumption of unwashed fruits and veggies [11, 12]. A systemic review and meta-analysis carried out provided evidence-based findings on the seroprevalence of *T. gondii* and the prevalence of active infections in patients suffering from HIV. The data of the review suggested a high global burden of *T. gondii* among patients suffering from HIV, with the median global seroprevalence of 44.22% for *T. gondii* and the prevalence for active *T. gondii* infection recorded was 3.24% by IgM and 26.22% by molecular methods [13]. The factors responsible for higher susceptibility to *T. gondii* in HIV patients are impaired IL-2 and IFN-γ production, and the activity of cytotoxic T cells irrespective of CD4+ cell count [14]. A case–control study concluded high seroprevalence of *T. gondii* infection in patients who undergo hemodialysis when compared to healthy volunteers. As the patients are immunocompromised toxoplasmosis may lead to severe complications with poor prognosis [15]. Another study carried out in western Romania concluded a high prevalence of *T. gondii* in pregnant women, and this infection is associated with older age, working with meat, having pets, low level of education, higher gravidity, and histories of spontaneous abortions [16]. The ultrastructure of *T. gondii* is depicted in Figure 23.1.

This structure depicts ultrastructure of tachyzoite stage. As displayed in the schematic diagram, tachyzoites as other stages of *Toxoplasma gondii* are very highly polarized cells. They contain several organelles that are specialized in the secretion of virulence factors. This is nucleated structure having conoid, rhoptry, and micronemes at apical region. Dense granules are also present with vacuolar and endosomal vesicles and apicoplast. Other organelles present are golgi body, Endoplasmic reticulum, mitochondria and a posterior pole can also be seen.

23.1.2 Life Cycle and Pathophysiology

The life cycle of TG is much more complex as it consists of more than one specific infective form and has several pathways for transmission [17]. The parasite *T. gondii* goes through sexual and asexual reproduction in the context of a heterogenic life cycle involving the Felidae family member as a definitive host and any warm-blooded vertebrate as an intermediate host. The life forms of *T. gondii* can be categorized into the ones that can be clonally replicated and those that can be generated as a result of gamete combination [18]. The forms that are asexually or clonally replicated can be distinctly identified as tachyzoites and bradyzoites. These forms are the ones that develop in intermediate warm-blooded hosts [19]. Tachyzoites are an extremely blooming form responsible for acute infection, reactivation, and vertical transmission. On the other end, bradyzoites are responsible for chronic infection as they are latent, even though active metabolically, slow growing encysted form [20]. The bradyzoite is linked with persistence, immune evasion, and refractory to the presently available therapeutic regimens [21, 22]. Both these forms follow endodyogeny, a cell division scheme that consists of a round of DNA replication with the aid of semiclosed nuclear mitosis [23, 24]. The cell division for tachyzoites is rapid as it generates two daughter cells in six to eight hours, while the replication for bradyzoites is slower [19]. When bradyzoite gains access to the feline

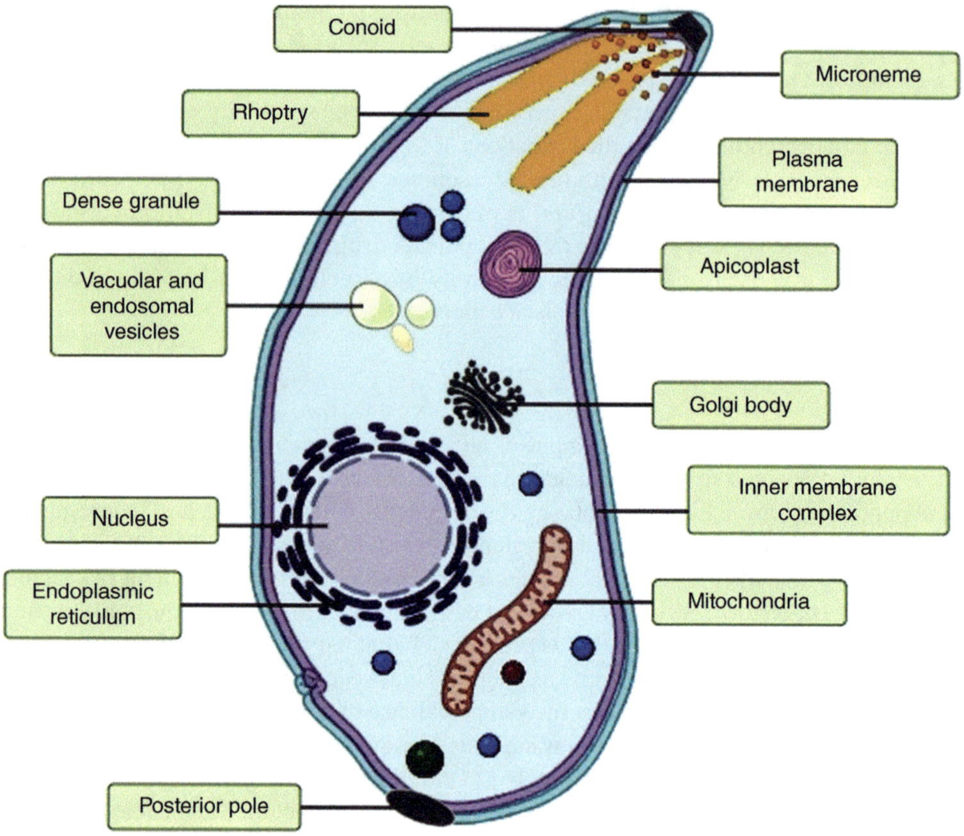

Figure 23.1 Ultrastructure of *Toxoplasma gondii.*

gastrointestinal tract, it either develops into a tachyzoite or a merozoite [25]. The sexual replication happens only in the feline host that is the definitive host for *T. gondii*. The parasite in cat gut cells develops into male and female gametocytes leading to sexual reproduction. The cat then ultimately shed the oocytes that contain four haploid sporozoites [26]. The parasite is transmitted when one is exposed to cat feces or eats undercooked meat. It can also vertically transmit from the infected mother to the fetus [27]. The lifecycle of parasite is depicted in Figure 23.2.

The schematic representation depicts the life cycle of *T. gondii* in both definitive and intermediate hosts. Feline being the definitive host, the sexual replication is carried out in it and the formation of zygote takes place. The oocyst then in the form of fecal matter enters the external environment and converts to sporulated oocyst. This can be ingested into the human host either by contaminated food or by contaminated water. Another route for transmission is via consumption of raw or uncooked meat. The asexual replication into tachyzoite, bradyzoite, or merozoite takes place in both the hosts.

Invasion in the host cells followed by rapid asexual replication is the crucial step of the parasite life cycle. This allows the expansion of population in the definitive and intermediate hosts, through the merozoites and tachyzoite developmental stages. This invasion procedure and successive steps of asexual replication and egress from the host cell are collectively referred to as the lytic cycle [28]. The mechanisms of host cell invasions include PV formation that consists of a distinctive replicative niche that provides defense from the host tissue and access to the nutrient source [29]. This invasion and establishment of *T. gondii* are attainable due to exocytosis of three specialized secretory organelles viz., micronemes, rhoptries, and dense granules. The micronemes are rod-like organelles clustered at the apical region of the parasite containing a large assemblage of proteins (MICs) that are vital for the invasion process. Some MICs play a crucial role in the exocytosis of rhoptries [30]. Rhoptries are club-shaped organelles that constitute proteins localized to discrete sub-compartments. The proteins known as ROPs present in the bullous part are responsible for the sabotage of host cell functions, while another set of proteins named as RONs whose function is specifically linked with host cell invasion is found in the neck region of the parasite [31]. Both micronemes and rhoptries are involved in the secretion of factors that will account for the moving junctions (MJ), a structure secreted by *T. gondii* into the host plasma membrane to adhere itself firmly before

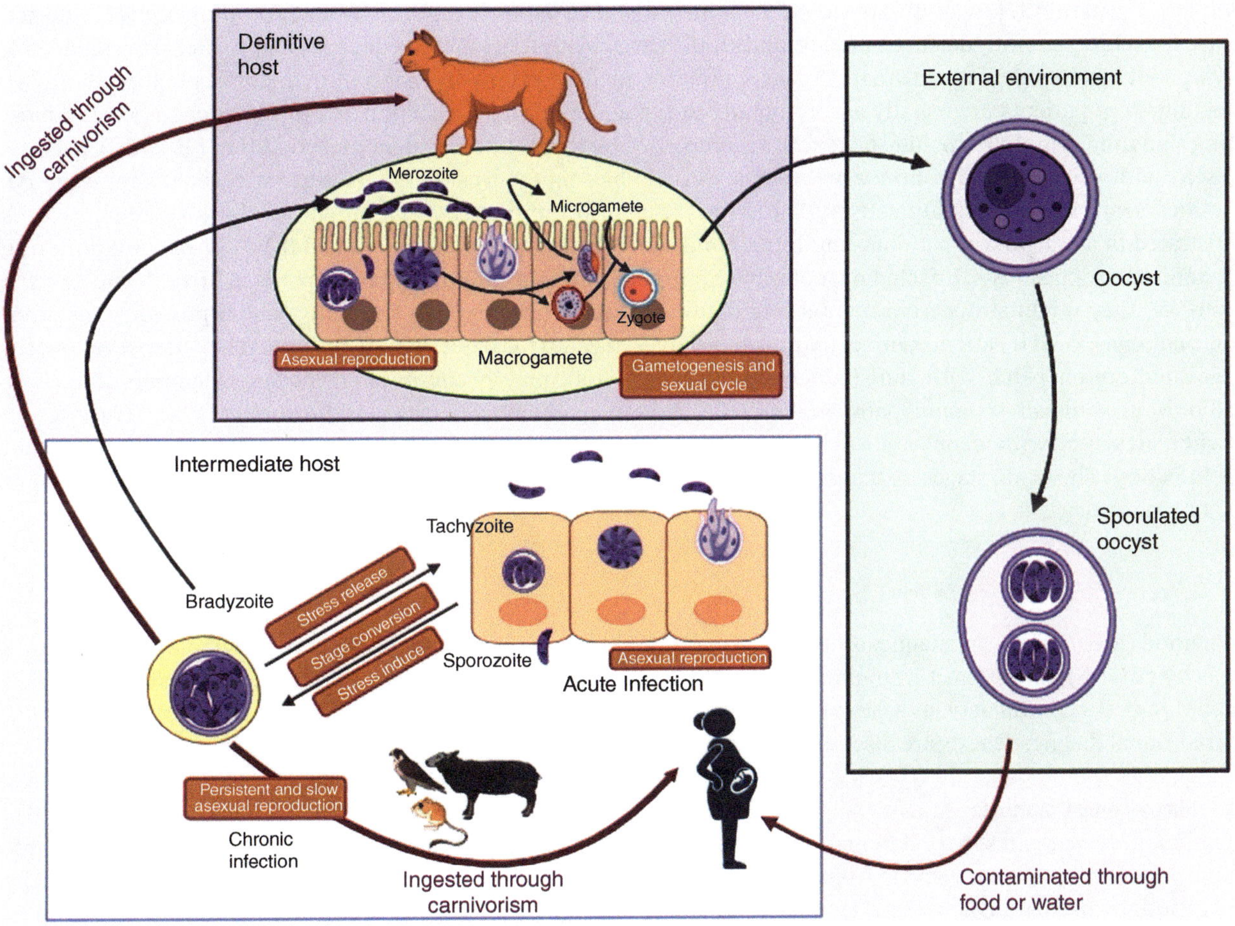

Figure 23.2 Lifecycle of *T. gondii* and its transmission.

entering [32]. Once these tachyzoites are inside the host cells, they secrete factors that modify host cells in a way that facilitates replication [33]. These factors include ROPs and dense granule effectors (GRAs). Some of these GRAs are responsible for the genesis of an intravacuolar membrane network as these modifications will aid to scavenge essential nutrients from the host for intracellular growth and development of parasites [34]. As mentioned above tachyzoites replicate asexually through endodyogeny. After successive rounds of replication, they egress out from PV and host cell which causes destruction [35]. Once the tachyzoites are in the extracellular environment, they need to invade a new host cell for initiating a new lytic cycle to continue the duplication process [30]. Bradyzoites, a latent and persistent form found in the intermediate host, also reproduce asexually. The replication takes place by a combination of two processes viz., endodyogeny and endopolygeny. Endopolygeny is a process by which more than two daughter cells form within the mother cell [20, 30]. After asexual multiplication, the sexual cycle commences with gametogenesis wherein the formation of micro- and macrogametes take place which will later develop into gametes. The rate of macrogamete fertilization is not known, but *T. gondii* manages to achieve the maximum number of oocysts, with hundreds of millions of them being shed potentially by a feline [18].

23.1.3 Clinical Presentation

It has been observed that only 10–20% of toxoplasmosis cases in adults are symptomatic. Patients suffering from congenital toxoplasmosis may manifest as mild or severe neonatal disease. They present with a wide variety of manifestations during the perinatal period [36]. Clinical manifestation of toxoplasmosis depends upon the age and immune condition of the patient [37]. A study carried out demonstrated that patients suffering from *T. gondii* infection reported headaches, muscle aches, fatigue, fever, sweats, and chills. Other symptoms include decreased appetite, arthralgia, dark urine, lymphadenopathy, and sore throat. Ocular symptoms include eye pain, vision blurring, and photophobia [38]. Immunocompetent patients are usually asymptomatic in the acute phase of infection but in the later stages, the most presenting signs and symptoms include headache accompanied by fever and altered mental condition [39]. Also, patients may present with seizures, and common neurologic deficits like motor weakness and slurred speech. Cranial nerve abnormalities, visual field defects, and sensory disturbances may also be present in these individuals [40]. Retinochoroiditis is demonstrated in about 15% of patients, and intracranial calcifications are observed in about 10% of patients suffering from congenital toxoplasmosis. Infected neonates have jaundice, anemia, and thrombocytopenia at birth. If the affected babies survive, they might suffer from mental retardation, seizures, hearing loss, spasticity, vision problems, or other neurologic deficits [36, 41]. Patients with ocular toxoplasmosis develop focal necrotizing retinitis. They have a yellowish-white, elevated cotton patch with indistinct margins. The lesions may occur in small clusters. Symptoms include impaired or blurred vision, scotoma, pain, photophobia, red eyes, metamorphopsia, and floaters [42]. An HIV-infected patient when presented with toxoplasmosis, the main clinical manifestation is encephalitis. This takes place in the latent phase of infection. The acute stages of acquired toxoplasmosis in HIV-infected patients demonstrate the involvement of multiple organs [43].

23.1.4 Current Diagnostic Tests for Toxoplasmosis

Several methods are used for the diagnosis of *T. gondii* infection, but the gold standard tests are enzyme-linked immunosorbent assay (ELISA) and indirect immunofluorescence assay (IFA). These tests detect toxoplasma-specific antibodies; IgG and IgM [44]. For central nervous system (CNS) toxoplasmosis, MRI is the best initial screening radiological procedure. Certain traditional diagnostic tests are discussed hereafter in brief.

23.1.4.1 Microscopic Diagnosis

The detection of the parasites in fecal matter, water, environmental, and tissue samples has been done through a simple microscopic examination. The oocysts from these samples can be stained to separate them from the host cells. The stains traditionally used are Giemsa and Hematoxylin and Eosin (HE). Periodic acid Schiff (PAS) stains the amylopectin granules in bradyzoites. These methods are time-consuming and require skilled labor and hence are difficult for daily use [4, 45].

23.1.4.2 Serological Assays

In most cases, *T. gondii* infection shows no clinical signs and symptoms and hence the diagnosis mainly relies upon serological analysis. Various serological tests are available that includes dye test (DT), modified agglutination test (MAT), enzyme-linked immunosorbent assays (ELISA), indirect fluorescent antibody test (IFAT), and indirect hemagglutination assays (IHA). These assays are employed to detect different classes of antibodies or antigens.

DT has been considered a gold standard to diagnose *T. gondii* infection in humans, but it requires live parasites which is a major drawback of this test. The test is hazardous and requires a high degree of technical expertise limiting its use [46].

In the MAT test, formalin-fixed tachyzoites are used as samples. It detects IgG antibodies but may produce a fairly high toll of negative results in active infection [47]. When acetone is used in place of formalin, this test can detect IgG antibodies for acute infections which is very useful in diagnosing toxoplasmosis in patients suffering from AIDS, and acute glandular toxoplasmosis [48].

Indirect ELISA and sandwich ELISA are developed to detect the presence of *T. gondii*-specific antibodies and antigens. The indirect ELISA detects anti-*T. gondii* IgG, IgM, and IgA antibodies and not antigens. The sandwich ELISA detects both antigens and antibodies. Improvising the format of ELISA can help us in detecting *T. gondii*-specific IgM, IgG, and IgA antibodies, and circulating antigens [4]. Another modified ELISA assay technique is dot-ELISA which detects both antigens and antibodies specific to the parasite infection [49]. ISAGA detects IgM antibodies and is easier and simpler to perform than IgM-ELISA.

IFAT is used to detect both IgG and IgM antibodies and is widely employed to diagnose toxoplasmosis infection in both humans and animals [50, 51]. The IHA test detects IgG antibodies but detection takes place later than DT, and hence the diagnosis of acute and congenital infections can be missed [52]. This test is rapid and simple and thus is recommended for mass screening during epidemiologic surveys.

23.1.4.3 Imaging Techniques

Computed tomography, MRI, ultrasonography, and nuclear imaging facilitate the diagnosis of toxoplasmosis although they are not specific. These methods are also used to monitor the therapeutic effects [53]. Encephalitis may occur in immunocompetent patients, and hence CT and MRI aid to detect and locate the lesions. For congenital toxoplasmosis, physicians recommend the use of the US for prenatal diagnosis [4].

23.1.4.4 Molecular Methods

These methods are used in addition to the conventional diagnosis methods to aid in the diagnosis of toxoplasmosis infection. They play the role of confirmational tests to the findings of traditional diagnostic tests. Several methods are employed out of which conventional PCR have the main purpose of species detection [54–56].

23.2 Recent Advances in Diagnosis

The currently used traditional techniques for diagnosis have certain limitations, and hence advancements are carried out to obtain more precise results. Diagnosing toxoplasmosis, caused by the parasite *T. gondii*, remains a challenge as the symptoms can be varied and often mimic other diseases, and determining whether a person is in the acute or chronic phase of infection can be difficult as the parasite can persist without causing symptoms for years. The current diagnostic tools, such as serological tests (ELISA or Western blot) which detect antibodies to *T. gondii* in the blood, or PCR-based tests that detect the parasite's DNA in blood or tissues, are not always specific or widely available. Therefore, there is a need for more effective diagnostic methods that can differentiate between the acute and chronic phases of infection and accurately diagnose the disease in all populations.

Trials for several new techniques are carried out to detect the toxoplasmosis infection. Advances and modifications in traditional serological and molecular test have been developed to detect the infection with much more precision. Different novel diagnostic techniques are mentioned hereafter. Certain serologic tests carried out to detect the antibodies and antigens include immunochromatographic test (ICT) [57], piezoelectric immune agglutination assay (PIA) [58], Western Blot (WB), Avidity testing, and Latex agglutination test (LAT) [4].

23.2.1 Novel Enhanced Dot Blot Immunoassay That Uses Colorimetric Bioassay for *T. gondii* Detection

This test is a sensitive and simple method developed for naked-eye differentiation of acute and chronic stages of *T. gondii* infection which is based on Au-NPs immunoconjugates. The principle on which this method works is the formation of sandwich layer of acute anti-TLA antibodies (IgM) or chronic anti-TLA antibodies (IgG) between the NC membrane-surface bound TLA and anti-IgM or anti-IgG Au-NP immunoconjugates, respectively [59]. Although this study carried out is the first point of care for detecting *T. gondii* infection, there are reports on rapid diagnosis of other infections with the aid of Au-NP-based immunoassays [60, 61]. In this study, to synthesize Au-NPs, low molecular weight chitosan were used as reducing and caping reagent according to the literature available [62]. Another study used ZnO-NPs covered by chitosan for detecting anti-*T. gondii* IgG Abs [63]. In this present method, the specific anti-toxoplasmosis antibodies, IgG and IgM, react with the Chi-Au-NPs-Ab conjugates that results in the formation of a Chi-Au-NPsAb complex [64]. Although the developed method rapidly detects the presence of acute and chronic infection, improvement is needed by tuning the NC membrane type to differentiate between the sages of infection.

23.2.2 A Fluorescent Immunosensor with Chitosan-ZnO-Nanoparticles

A study is conducted to analyze the sensitivity of microfluidic LIF immunosensor with chitosan-ZnO-nanoparticles for the quantitative detection of anti-*T. gondii* IgG-specific antibodies. The principle of novel LIF immunosensor is based on the incorporation of CH-ZnO-NPs into the microfluidic channel for sensitive quantification of IgG antibodies. The results of the study demonstrated improved sensitivity when integrated with CH-ZnO-NPs. The proposed immunosensor was seen to be superior for antibody detection when compared with traditional assays. Also, this developed method turned out to be more reliable, specific, and highly reproducible with a low detection limit [63].

23.2.3 YKL-40 as a Novel Diagnostic Biomarker in Toxoplasmosis

The YKL-40, also known as chitinase-3-like protein 1, is a glycoprotein that is secreted by a number of cell types in different patterns which is associated with various pathological reactions namely inflammation, fibrosis, and tissue remodeling and it is also a disease-specific biomarker for neuroinflammation [65, 66]. One study was conducted to check if YKL-40 increases in toxoplasmosis or not [67]. Previously, the studies have indicated that YKL-40 plays a critical role in inflammatory reactions, remodeling and repair, and tissue injuries [68–70]. Also, it has been noted that certain cells that secrete YKL-40 engage in fighting the parasite and eventually producing an immune response [66, 71–73]. Upon final evaluation, which involved comparing ROC curves and AUC values between individuals with acute toxoplasmosis and healthy controls, it was revealed that YKL-40 serves as an excellent biomarker. This outcome was consistent when examining patients with chronic toxoplasmosis as well. Thus, this study designates YKL-40 as a sophisticated biomarker capable of distinguishing between infections in the acute and chronic phases. The same results were obtained for chronic toxoplasmic patients. From this study YKL-40 was rendered as a sophisticated biomarker which can detect that whether the infection is in acute phase or chronic [67].

23.2.4 Advances in Serological Methods Based on Recombinant Antigen of *T. gondii*

As mentioned aforehand that ELISA is a gold standard for diagnosis of toxoplasmosis, instead of TLA recombinant antigens as an alternative approach to improve the sensitivity of the tests [74]. In recent times, numerous recombinant antigens are expressed by the method of gene cloning and expression techniques and their potential for detecting IgG and IgM antibodies for *T. gondii* infection has been assessed [53]. These antigens are including surface antigens SAG1 (P30), SAG2 (P22), SAG3 (P43), and P35; dense granule antigens GRA1 (P24), GRA2 (P28), GRA4, GRA5, GRA6 (P32), and GRA7 (P29; microneme antigens MIC2, MIC3, MIC4, and MIC5; matrix antigens MAG1; and rhoptry antigensROP1 (P66) and ROP2 (P54) [4, 75–78]. It has been demonstrated that mixture of GRA7, GRA8, ROP1 and SAG2A, GRA2, GRA4, ROP2, GRA8, and GRA7 recombinant antigens have the potential to detect IgM and IgG, respectively [53]. Another alternative approach is to use chimeric or synthetic peptide antigen to elevate the sensitivity and specificity of the test.

23.2.5 Advances in Serological Methods Based on Chimeric Antigens and Multiepitope Peptides of *T. gondii*

A growing number of studies are carried out that demonstrate the potentials of chimeric or peptide-based antigens when used to detect the antibodies specific for *T. gondii* infection. Certain advantages are associated with this approach which includes high sensitivity, low contamination with proteins of the organisms, reducing the costs of production, and increasing the probability of discriminating different stages of toxoplasmosis [79–82]. In recent years different types of innovative tools are developed for predicting specific epitopes and their localization on the parasite. These tools include peptide microarray analysis by bioinformatic methods, epitope mapping, phage display of cDNA libraries, and reactivity with monoclonal antibodies [76, 83, 84]. A study evaluating diagnostic utilization of five chimeric antigens in compression with three recombinant antigen mixtures demonstrated that chimeric ones are more reactive compared to the recombinant one. The study also elucidated that chimeric antigens composed of SAG2–GRA1–ROP1L are more sensitive to detect the *T. gondii* infection [75]. Another study revealed that the above-mentioned team of chimeric antigens are highly specific and sensitive to IgG chemiluminescence assays (CLIA) [85].

23.2.6 Advanced Molecular Technique

Advances in the molecular methods during recent decades induce a great revolution in diagnosing toxoplasmosis infection. These techniques include use of real-time PCR, nested PCR, and loop-mediated isothermal amplification (LAMP) that detects the parasite DNA in the biological sample [86]. Multiplex PCR is another technique in which simultaneously multiple targets can be analyzed from a single sample. LAMP is a one-step modified PCR technique that is rapid, specific, efficient, and highly sensitive and also there is no need for the isolation of the parasite [87]. The other methods focus on genotyping and it includes microsatellite analysis, multilocus sequence typing, high-resolution melting (HRM) analysis, random amplified polymorphic DNA-PCR (RAPD-PCR), and PCR-RFLP.

Although it is known that serological diagnostic test are the gold standards but modifying them and combining with other novel approaches results in an accurate and definitive diagnosis of *T. gondii* infection.

23.3 Current Management of Toxoplasmosis

Toxoplasmosis is a parasitic infection caused by the parasite *T. gondii*, and current chemotherapy options for this infection are limited. The main target of anti-Toxoplasma drugs is the folate pathway, which is involved in DNA synthesis, and drugs such as pyrimethamine (PYR) and trimethoprim (TMP) act on the dihydrofolate reductase (DHFR) and dihydropteroate synthetase (DHPS) enzymes, respectively. However, these drugs are unable to distinguish between the parasite's enzymes and the human host's enzymes, and therefore, they need to be combined with sulfonamides to block DHPS. Timeline for conventional management is depicted in Figure 23.3. Therapeutic strategies for the management of toxoplasmosis in pregnant women and newborns as well as immuno-compromised patients will be discussed in the below section. Currently, numerous clinical trials are ongoing and have shown beneficial effects in patients suffering from congenital toxoplasmosis as well as shown to improve immunity in immunocompromised patients (Table 23.1).

23.3.1 Congenital Toxoplasmosis (CT)

Approximately 40% pregnancies vertically transmit the parasite in which the mothers have primary exposure during the gestational period. 90% of infected mothers are devoid of symptoms at the onset of infection and 50% of expectant mothers who give birth deny to recollect having any array of symptoms or possible exposure to the parasite [12]. Mothers report to experience flu-like symptoms such as fever, malaise, and cervical lymphadenopathy [92]. Moreover, transmission of parasite occurs rarely when the exposure has occurred prior to conception unless the reactivation of *T. gondii* is aided due to suppression of immunity [93]. The anti-*T. gondii* treatment can be initiated at two specific time points of pathogenesis: (i) prenatal stage and (ii) postnatal stage [94]. Treating former stage aims at preventing the materno-fetal transmission of parasite (MFTP) and declining the fetal damage while postnatal treatment is administered with a purpose to prevent long-term sequelae and alleviate the pathological burden on the infected neonate [95].

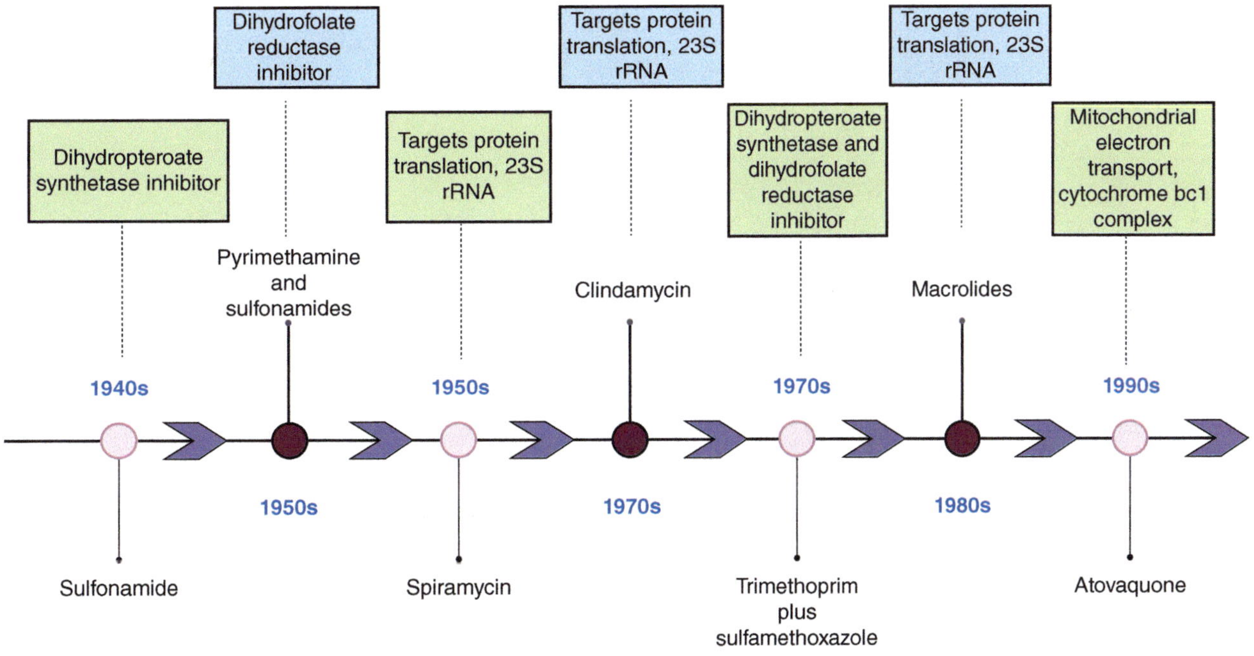

Figure 23.3 Timeline of conventional therapy.

Table 23.1 Ongoing clinical trials pertaining to *Toxoplasmosis gondii* infection.

Sr no.	NCT number	Study type and allocation	Status	Phase	Title	Condition	Patient no.	Intervention	Outcomes	Reference
1	NCT04341155	Interventional and randomized	Recruiting	II	Dexamethasone for Cerebral Toxoplasmosis (De-Tox)	Cerebral toxoplasmosis	138	Dexamethasone Comparator arm: Placebo	1° outcome: Mortality rate 2° outcome: AEs, change in consciousness, neurologic response, cognitive function, and neuroradiological response	[88]
2	NCT00004317	Interventional and randomization	Recruiting	IV	Pyrimethamine, Sulfadiazine, and Leucovorin in Treating Patients With Congenital Toxoplasmosis	Toxoplasmosis	600	Leucovorin calcium, Pyrimethamine, Spiramycin, Sulfadiazine	1° outcome: persistent motor abnormality, vision, hearing, new chorioretinal lesion, IO less than 70	[89]
3	NCT02936921	Observational and cohort	Recruiting	–	Lyon Cohort of Maternal and Congenital Toxoplasma Infections	Congenital toxoplasmosis	4030	–	1° outcome: Long-term ophthalmological outcome of congenital infection	[90]
4	NCT04825600	Interventional	Not yet recruiting	N/A	Diagnosis of Toxoplasma Gondii Infection by Exploration of Cellular Immunity (TOXCELL) (TOXCELL)	Toxoplasmosis Toxoplasma Infections Toxoplasmosis, Congenital Toxoplasmosis Recurrent	60	Diagnostic Test: anti-Toxoplasma gondii IgG and IgM dosage Diagnostic Test: cellular test	1° outcome: Implement a cell test with lymphocytic stimulation by toxoplasmic antigen and screening for *T. gondii* infection to assess cellular immunity against *T. gondii* 2° outcome: Compare cell diagnosis with serological diagnosis and assess whether or not cellular and humoral immunity against *T. gondii*	[91]

23.3.1.1 Prenatal Treatment

Pregnant women are not exposed to Pyrimethamine (PYR) in the first trimester due to its terogenic effects and spiramycin (SPI) is preferably administered which is a potent macrolide antibiotic [96]. SPI concentrates in placenta which is one of the ideal characteristics required for an anti-*T. gondii* drug to prevent MFTP [97]. Safety parameters of SPI make it a comfortable and reliable treatment alternative until amniocentesis begins [6]. On the contrary, the applicability of SPI reduces in an already established fetal infection as the drug does not potentially cross the placental barrier to provide fetal protection [98]. 16th gestational week onward pyrimethamine-sulfadiazine (PYR-SDZ) combination can be prescribed while it must be absolutely avoided during first 14 weeks. SDZ as a monotherapy may be prescribed during the first trimester followed by late addition of PYR after the crucial stage of fetal growth has been surpassed [99]. This practice must only be implemented if the benefits of the treatment outweigh the possibilities of risk to the fetus.

PYR and SDZ acts synergistically inhibit DNA synthesis in *T. gondii* tachyzoites, but their actions also extend to interrupt DNA synthesis in tissues such as bone marrow and epithelium possessing high metabolic activity [100]. And to prevent this influence on such tissues while administering the PYR-SDZ treatment, folinic acid supplementation must be practiced which bypasses the adverse events [101]. Folic acid supplements are inefficacious and the infected women must be advised to drink minimum 2 l of fluid per 24 hours along with consuming a citrus-based diet to alkalinize the urine [97, 102]. A caution must be considered if the pregnant woman is G6PD deficient and the regimen must be avoided to eradicate the risk of anemia. Moreover, monitoring of patient is essential every 1–2 weeks through a diagnosis of blood cell count (BCC). If neutropenia is evident (neutrophil count $<1500/mm^3$), cease the PYR-SDZ treatment and continue intake of folic acid [103]. Follow this course by a new BCC test after one week and obtain opinion of the working group before reinitiating the regimen.

23.3.1.2 Postnatal Treatment

After retrieval of a confirmed diagnosis of congenital infection, the patient is subjected to the postnatal treatment to treat the clinical manifestations at birth and prevent the progression of infection and its relapse rates. Postnatal treatment also offers prophylactic protection against the retinal disease occurring as a result of toxoplasmosis progression. Chicago Collaborative Treatment Trial (CCTT) is a landmark study that favored implementation of PYR-SDZ regimen in postnatal treatment of CT [104]. This result was supported by an encouraging outcome of PYR-SDZ treatment given for a year in 120 infected neonates followed up during 1981 and 2004. Among severely infected neonates at birth, 80% presented with normal unaffected motor function, 64% did not progress to retinal involvement, and none had sensory or neural hearing impairment [99]. Before initiating the PYR-SDZ regimen in neonates immediately after birth, a prenatal diagnosis or neonatal screening such as serology, ophthalmological examination, and CNS imaging is beneficial. A minimum of three-month course has been proposed in asymptomatic newborns at the risk of developing retinochoroiditis due to the presence of cerebral calcifications [99, 105].

The range of adverse events accompanying this treatment led to an exploration of alternatives including PYR-clindamycin (PYR-CLD), PYR-azithromycin, cotrimoxazole – a combination therapy of trimethoprim with sulfamethoxazole (TMP-SMX) and atovaquone [106]. Other than lack of compliance due to adverse events, treatment failure can be justified due to possible resistance issues met during the treatment of toxoplasmic encephalitis (TE), chorioretinitis, and CT [107].

23.3.2 Immunocompromised Patients

Immunocompromised patients with chronic infection develop toxoplasmosis in the form of a severe opportunistic infection which results due to the reactivation of the latent dormant cysts [108]. TE is a highly prevalent clinical manifestation observed in HIV-positive patients, but this prevalence number has now been altered due to an increasing number of immunocompromised HIV-negative patients [109]. In non-HIV patients the outcome is poorer and the pathogen disseminates more often rather than remaining confined in CNS. Life-threatening and severely debilitating disease form is observed in bone marrow and hematopoietic stem cell transplant (HSCT) patients [110]. The mortality incidences range from 38% to 67% in such cases [99]. This raises a need of anti-*T. gondii* treatment to be precisely effective and capable of surpassing the blood–brain barrier. Common drugs are prescribed for both HIV and non-HIV patients which suggests a lack of wide therapeutic options for treating toxoplasmosis. This concludes into an urgent need of introducing newer drug molecules before the species adapt to show resistance to the entire arsenal of currently prescribed drugs. Multiple clinical trials have been conducted to compare different PYR-SDZ dosage regimens ranging from 50 mg to 100 mg/day and/or compared to PYR/

CLD or TMP-SMX in HIV-positive patients with toxoplasmosis [111]. No significant differences were observed among these treatment arms except that PYR-SDZ had most efficiency [112]. Although PYR-SDZ was capable of reducing the relapse rates the most, it resulted in more adverse events than PYR-CLD or TMP-SMX. For the induction phase, a 200-mg loading dose of PYR on day 1 has been recommended by Infectious Disease Society of America (IDSA). This must be followed by a dosage cycle of 75 mg/day in patients weighing >60 kg or 50 mg/day in patients weighing <60 kg [113]. SDZ can be combined with the former PYR treatment in the dose of 1000 mg/day if patients weigh ≤60 kg and 1500 mg BID in patients weighing ≥60 kg. Simultaneous administration of folinic acid in the dose of 10–25 mg/day is essential [111]. Furthermore, if the alternative regimen has to be considered, then PYR-CLD is prescribed in the dose of 600–900 mg four times a day along with folinic acid supplementation [111, 114]. 5 mg/kg BID TMP can be recommended combined with PYR or SDZ in addition to folinic acid. Another alternative regimen includes Azithromycin macrolide 900–1200 mg plus PYR and folinic acid [115]. Monotherapy of CLD is unfavorable due to the adverse events and possibility of higher relapse rates. Maintenance therapy must not be initiated until the induction therapy has been taken for a minimum of six weeks until clinical and radiographic improvement is evident. Maintenance therapy must be continued until CD4+ cell count remains over 200 cells/mm^3 [116, 117].

23.4 Need for Novel Treatment

The chemotherapeutic armamentarium battling toxoplasmosis is limited mainly to targeting the folate pathway involved in the synthesis of DNA utilizing dihydrofolate reductase (DHFR) and dihydropteroate synthetase (DHPS) enzymes [99]. PYR and TMP are two major drugs used in the management of acute toxoplasmosis. The applicability of these drug molecules is limited as a cause of their incapability in distinguishing between the parasite and human DHFR enzyme [118]. The monotherapy prescriptions of these drugs fail to contribute optimal benefits which demands combination therapy with sulfonamides which target DHPS [119]. This therapeutic benefit came with a potential risk of myelotoxicity along with a spectrum of life-threatening events which necessitated discontinuation of this therapy or more frequently, affected medical adherence and patient compliance [120]. This is a significant setback in therapeutic success as immunocompromised patients and congenitally infected neonates require prolonged courses of anti-Toxoplasma drugs. Moreover, drugs in current practice are incapable of erasing the *T. gondii* cysts existing in the quiescent form unless the immune system has the potential to obstruct their reactivation into tachyzoites [121]. Poor adherence, possibility of adverse effects, and drug resistance are some of the major issues for therapeutic failure in eradicating *T. gondii*. Taking this note, the development of a resistant *T. gondii* strain had led to the failure of the prolonged therapy of PYR regimen [122]. Furthermore, in Brazil, SDZ-resistant *T. gondii* strain was also isolated in congenitally infected neonates [123]. Formerly mentioned instances prove the possibility of development of *T. gondii* resistant strain; however, the spread of this strain is limited because of slow multiplication and transmission route of this disease [99]. This raises a necessity of exploring new drug moieties that would eliminate both tachyzoites and cyst forms of pathogen while being a safe therapy option for pregnant women and neonates. The parameters attributed to the ideal treatment include an acceptable bioavailable formulation able to concentrate in placenta and fetal compartment. The drug must cross the blood–brain barrier and diffuse in CNS and eye compartment to rule out the cysts [124]. However, development of such a drug that checklists all these requirements has presented an ample load of problems. The novel treatments must be nontoxic and aim to treat chronic toxoplasmosis alongside preventing the reactivation of *T. gondii* species. A shorter duration of treatment and well tolerability of drug would improve medical adherence and optimize the therapeutic efficacy. Several preclinical studies have been conducted to evaluate the efficacy of various novel and repurposed compounds in treating toxoplasmosis. Some of these compounds have demonstrated a prolonged survival rate in infected animals in murine models of acute as well as chronic toxoplasmosis. Furthermore, they have also shown the ability to reduce parasite loads in brain and muscle tissues in models of chronic toxoplasmosis (Table 23.2).

23.5 Novel or Repurposed Therapeutic Molecules as Anti-Toxoplasma Activity

The foundation of new candidate molecules is based on their efficiency to aim for bradyzoite along with tachyzoite stages. Continuous efforts are made to develop promising anti-Toxoplasmosis drug entities, but the achievement of an ideal drug with formerly mentioned characteristics seems a distant goal [135]. Both in vivo and in vitro preclinical studies of some

Table 23.2 Preclinical trials for toxoplasmosis infection.

Sr no.	Drug	Animal	Strain	Type of infection	Main result	Effectivity	Reference
1	Compound 32 (Tg CDPK1 inhibitor)	Female CF-1 CBA/J	RH ME49	Acute, chronic	Reduces infection in spleen and brain. Also, reduction in brain cyst was observed	Effective	[125]
2	Resveratrol and sulfamethoxazole-trimethoprim	Male Swiss webster	VEG	Chronic	Reduction in cysts in brain was observed	Coadministration more effective	[126]
3	Propranolol	BALB/c	RH	Acute, chronic	Decrease in parasite load observed in brain, eye, and spleen tissue	Effective	[127]
4	Dextran sulfate	Pigs	RH	Acute	High dose of the drug caused reversible hepatocellular degeneration of liver	Effective	[128]
5	Nitrofurantoin	Female-ICR mice	RH	Acute	Biological parameters and hematology indicators reduced significantly	Effective	[129]
6	Fluphenazine and thioridazine	BALB/c	Tehran strain	Chronic	Drugs when given in higher dose reduce the number of cysts	Effective, not significant	[130]
7	Pyrimethamine (PYR) and sulfadiazine (SDZ) combined with levamisole and echinacea	BALB/c	RH	30 days after treatment	Survival rate: 33.3–89.9%	Effective	[131]
8	Aripiprazole	BALB/c	Tehran strain	Chronic	No significant results were obtained	No effective	[132]
9	Propranolol	BALB/c	RH	Acute	Propranolol combined with pyrimethamine is more effective	Effective	[133]
10	Ciprofloxacin derivatives	Female Swiss mice	RH	Acute	Survival of mouse increased significantly	Effective	[134]

compounds returned affirmative results with a low IC50 value, improved survival rates in murine species affected with acute toxoplasmosis, high selectivity index and declined parasite load in muscles and CNS in chronic toxoplasmosis models [136].

Development of an advanced drug arsenal for toxoplasmosis can be undertaken through two major approaches: (i) finding molecules with increasing target specificity and (ii) repurposing of the available drugs or promising candidates directed against other pathogens [99]. Some of the known or suggested mechanisms of action and targets of reviewed drugs and compounds are presented in Table 23.3.

23.5.1 Searching for Compounds with Effects on Specific Parasitic Targets

Ideally, the pathways targeted by the drugs must not superimpose with the pathways present in the host to avoid the occurrence of adverse events. Drugs acting on the important cascades such as gliding motility, parasite differentiation process, host cell invasion, and egress may show promising results. Moreover, several investigations function to improve drug resistance complications by improving current therapies working on folate or electron transport pathway. One of the examples is of a PYR analog named dihydrotriazine JPC-2067-B87 with a potent activity against *T. gondii* in vitro with a low 1C50 value [147]. The drug was tested in a prodrug formulated form JPC-2056 given orally in murine models suffering from acute toxoplasmosis. On the contrary, its impact on parasitic cyst forms is not precisely elucidated and demands more investigation [99, 148].

Table 23.3 Overview of novel and repurposed drugs against toxoplasmosis.

Sr no.	Drug	Mechanism of action	Target	Activity	References
1	Triclosan	Fatty acid synthesis II pathway	Enoyl-acetyl carrier protein reductase	Effect was observed during in-vivo study on both tachyzoites and bradyzoites	[137, 138]
2	Bumped kinase inhibitor 1294	Ca-dependent pathway	Calcium-dependent protein kinase 1	The activity was seen in in-vitro studies and in-vivo tachyzoites	[139, 140]
3	MMV689480 (buparvaquone)	–	–	Showed effects in in-vitro trials and effects were seen in in-vivo tachyzoites for which vertical transmission model was used	[140, 141]
4	Miltefosine	Anticancer	Enzymes involved in phospholipid metabolism	The effect is seen in in-vitro studies as well as both in-vivo tachyzoites and bradyzoites, but the effect is poor on tachyzoites	[142, 143]
5	Compounds with anti-inflammatory and anticancer properties	Anti-inflammatory and anticancer	Uncertain	The efficacy was observed in in-vitro trials	[144]
6	Guanabenz	Translational control	TgIF2α phosphorilation	The activity was observed in in-vitro trial as well as in both in-vivo tachyzoites and bradyzoites	[145]
7	Tanshinone IIA	Anticancer	Uncertain	The effectivity was seen when an in-vitro study was conducted	[146]

23.5.2 Identifying Promising Compounds or Repurposing Potential Drugs Active Against Other Pathogens

The ultrastructure of *T. gondii* species has an apicoplast organelle where metabolic pathways take place. Targeting these pathways through drugs may alleviate the influence of *T. gondii* [149]. One new promising marker is enoyl-acetyl carrier protein reductase (ENR) of the type II fatty acid synthesis (FAS II) cascade. *T. gondii* scavenges the reservoir of fatty acids while those produced in apicoplast are necessary for organelle biogenesis [99, 149]. Triclosan, a drug targeting and inhibiting ENR, is an antibacterial drug that diminished the burden and reduced the viability of cysts and tachyzoites [137]. Furthermore, a drug named Thiolactomycin is also studied to interfere with the same pathway. Thiolactomycin obstructs the elongation of fatty acids by inhibiting β-ketoacyl-acyl carrier protein synthase [150]. Eight analogs of this drug were tested among which only one showed deficient activity. It was derived that the length of the compound side group had also an impact over drug interaction with its target. Two analogs with longer alkyl chains in vitro setting had the least IC50s and were capable of altering morphology of *T. gondii* [99].

The isoprenoid pathway of apicoplast plays a crucial role in facilitating parasite survival as it is associated with a bifunctional enzyme farnesyl disphosphate/geranyl diphosphate synthase which is of importance in synthesis of sterols and polyisoprenoid compounds [151]. A range of newly synthesized bisphosphonates inhibit this enzyme and cause an interference in isoprenoid pathway, ultimately affecting the survival of this parasite. These compounds exhibited low toxicity and remarkable efficacy against *T. gondii* both in vitro and in vivo [152].

The gliding motility and host invasion machinery of *T. gondii* is regulated by a calcium-dependent protein kinase 1 (TgCDPK1) which is different than kinases present in humans [153]. Investigations of pyrazolo-pyrimidine (PP)-based compounds inhibiting TgCDPK1 showed promising result in terms of efficacy against *T. gondii* [148]. A bumped kinase inhibitor BKI 1294 met obstacles in successful therapeutic implementation because of a risk of cardiotoxicity occurring due to inhibition of human Ether-à-go-go-Related Gene (hERG) ion channel [154]. Further modifications in BKI 1294 were made to retain its anti-*T. gondii* potential and eliminate its toxicity on hERG which resulted in a drug molecule capable of reducing the parasite burden in murine model affected with acute toxoplasmosis [148]. Another benefit obtained through such a modification was an ability to control brain cyst formation through the penetration of CNS when administered five weeks postinfection [155].

One of the fundamental events of parasite life cycle is of interconversion between bradyzoite and tachyzoite stages [22]. Blocking the differentiation into bradyzoites aids in treating the acute infection while blocking differentiation into tachyzoites helps in improving clinical manifestations of chronically infected immunocompromised patients, especially those at the higher risk of reactivation [1]. Histone acetylase (HAT) and histone deacetylase (HDAC) regulate histone acetylation levels and control gene expression during this interconversion process [156]. *T. gondii* HDAC3 (TgHDAC3) inhibitors disrupt the balance of histone acetylation throughout the genome and so, these drugs can be implemented in the therapy regimen. Cyclopeptide FR235222 is a TgHDAC3 inhibitor which interrupts the *T. gondii* intracellular growth and induces conversion of tachyzoites into bradyzoites in vitro. The drug also approaches the bradyzoites present within the ex vivo cysts and hinders their conversion into tachyzoites [121]. Anti-*T. gondii* efficacy of FR235222 and its analogs is yet to be evaluated in vivo.

Inhibition of Phosphodiesterase-4 (PDE4) also interferes in the interconversion of bradyzoite and tachyzoite through an alteration in cAMP signaling pathways and suppression of Th1 cytokines which are pro-inflammatory in nature [99, 157]. Rolipram is a drug that works through this mechanism and is reported to reduce the brain cyst load by 74% [149]. Despite offering positive results, rolipram usage is limited because of severe nausea and vomiting associated with it [99].

To consider an example of repurposed drug, Guanabenz, FDA-approved antihypertensive drug, dephosphorylates *T. gondii* eukaryotic initiation factor 2 (TgeIF2) without targeting the host eIF2 [145]. Guanabenz targets both tachyzoite and bradyzoite parasitic forms and provides protection against acute toxoplasmosis. In addition, it also surpasses the blood–brain barrier to eliminate around 69% of brain cysts in mice affected with chronic toxoplasmosis [158].

Beyond these investigations, morpholinos were explored as a novel approach in identifying and characterizing new *T. gondii* biomarkers [159]. Efforts were made to facilitate entry into cells through modification of morpholinos with covalently attached chemical groups. This internalization process further decreases expression of specific *T. gondii* enzymes [160].

Artemisinin and its derivatives namely artemisone and artemiside showed increased survival rates and reduced the burden of CNS cysts in mice positive to be infected with type II toxoplasma strain tachyzoites [148]. Mortality reported upon drug discontinuation indicated the specificity of artemisinin on tachyzoites and no action on bradyzoites [161].

An in vitro screening of open access Medicines for Malaria Venture (MMV) Malaria Box consisting of 400 blood-stage-active anti-Plasmodium compounds yielded seven potential anti-*T. gondii* drugs out of which a piperazineacetamide MMV007791 was the most effective [162]. These compounds differed from conventional anti-*T. gondii* drugs in having novel chemical scaffolds with different pyrimidine moiety than other drugs prescribed in toxoplasmosis [99]. An identical screening procedure was performed for a Pathogen box comprising 400 compounds targeting a variety of distinct pathogens and 15 of them had anti-*T. gondii* activity [141]. Eight out of fifteen had selective and favorable in vitro effects on tachyzoite growth. A focus was set upon early preclinical results while screening the Malaria and Pathogen boxes [163]. Furthermore, the antimalarial drugs used in current practice and late preclinical development were also assessed concluding a limited potential of these drugs against *T. gondii* in vitro. Antimalarial drugs falling in drug class 4-amino and 8-aminoquinolines had little or modest anti-*T. gondii* activity while previous literatures reported promising results observed with newly synthesized quinoline compounds and one 4-aminoquinoline [99, 164].

More investigational campaigns were conducted by screening another compound library provided by the Drug Discovery Initiative (University of Tokyo, Japan) and two in vitro effective inhibitors were obtained which were devoid of host toxicity [99]. One of those compounds was Tanshinone IIA and the other was an antihistamine Hydroxyzine. Tanshinone IIA also works to arrest the tumor cell growth [165]. Surprisingly, both of these compounds inhibited the growth of intermediately differentiated bradyzoites which is a characteristic rarely observed in currently prescribed drugs [146].

In addition, a large library of natural products along with FDA-approved drugs was screened in which parasite growth inhibitor drugs were majorly possessing anticancer and anti-inflammatory properties [144]. The evidence was not surprising as it was based on the fact that multiple drugs actively fighting toxoplasmosis are also prescribed in cancer and immunotherapy [146]. Miltefosine, a drug with a wide consideration for the treatment of visceral leishmaniasis, failed to control acute toxoplasmosis infection. On the contrary, administration of Miltefosine 60 days after the murine infection led to a decrement in brain cyst quantity and size [166]. Moreover, an anticancer drug named Tetraoxane also prolonged survival rates significantly in infected mice [167].

All of these studies draw the toxoplasmosis treatment arsenal toward an expansion with some of them to hold more anti-*T. gondii* potential. However, the results obtained in vitro must be supported by investigations performed in vivo in order to strengthen the observations and accelerate the implementation on novel drugs in current practice.

23.5.3 Immunotherapeutic Arsenal Against Toxoplasmosis

The clinical manifestations are absent or the patient is usually asymptomatic in an acute setting of toxoplasmosis [168]. The resident immune system is responsible for such a presentation owing to its immune control resulting in toxoplasma encystation [169]. On the contrary, partial subversion of the host immunity allows the parasite to persist for a prolonged time. This coexistence of parasite and host depends on a tightly controlled balance over replication of pathogenic parasite [170]. An instance of any immune defect, especially those involving T-cell machinery, occurring mostly as a secondary to HIV infection or immunosuppressive therapies, allows the parasite to be reactivated. This further initiates the involvement of CNS leading to severe encephalitis or disseminated infection. The therapies mentioned formerly are incapable of fighting this course of infection and due to this, the infected host is always at the risk of relapses [171]. Considering this aspect of toxoplasmosis infection, there has been a necessity raised to develop a beneficial and promising approach which allows immunocompromised patients to have an adequate immune response [99].

The immune response formed against *T. gondii* results from complex cascades of both innate and adaptive immunity [172]. Protection against acute infection of toxoplasmosis is provided by innate immune components while long-term defensive capacity is formed by an adaptive immune response [173]. Cellular effectors like IFNγ-producing T lymphocytes control the replication and spread of *T. gondii* along with maintenance of latency. The CD8+ T cells and NK cells lyse the infected cells and present the parasite to other immunologic machinery to render them harmless [174]. Immunoglobulins, macrophages, dendritic cells along with complement system function to recognize and lyse the presented parasite [174]. Pro-inflammatory response generated by Th1 cells is important during both acute and chronic infection. Cytokines such as IL-12, IFNγ, TNFα along with IL-7 and IL-15 contribute in pro-inflammatory response and diminish the parasite burden in acute infection [99, 175]. Further, they also exert immune pressure to maintain parasite encystment in chronic form of toxoplasmosis. Conversely, IL-10, an immunosuppressive cytokine, impairs the immunity which eases the reactivation of latent *T. gondii* cysts in CNS by contributing in parasite intracerebral penetration [176].

Observation of such a course of IL-10 led to the development of a strategy involving anti-IL-10 antibodies tested in mice model of chronic encephalitis. The intracerebral load of parasite depleted due to enhanced IFNγ and TNFα function. Additionally, an increased recruitment of CD4+ and CD8+ cells was noted in CNS. However, the mice with IL-10 deficiency subjected to anti-IL-10 antibodies quickly succumbed to necrotizing hepatitis due to the response formed by CD4+ T cells due to an overproduction in IFNγ, TNFα, and IL-12 [99, 177].

Another mechanism studied for developing immunotherapy strategies included the pathway of CD8+ T cell functionality [99]. Deficient CD8+ T cell activity has a forehand in reactivation of parasites. Such a deficient or exhausted function of CD8+ T cells is related to the inhibitory lymphocyte receptor Programmed Death-1 (PD-1) on the surface of T cells which causes apoptotic cascade activation leading to apoptosis and progressive attrition of CD8+ T cell activity [178]. The dysfunctional CD8+ T cells can be rescued by blocking the PD-1/PDL-1 cascade, but this benefit accompanies a risk of autoimmunity. Chronically infected mice models showed a significant improvement in the form of lower count of parasite-infected leukocytes in CNS and systemic circulation and prolonged survival compared to untreated mice [176].

Additionally, cellular-based therapies ushered and the evaluation of adoptive transfer of immune CD8+ cells was undertaken. This therapy alternative proved to temporarily restrict the breakdown of cysts and provided a transient protection against reactivation of *T. gondii* but failed to rescue the exhaustive machinery of CD8+ T cells for a long term [179]. The underlying reason for this was the failure of donor cells to survive for a prolonged time. Application of this therapy now remains limited to the high-risk immunocompromised patients requiring a prevention therapy against toxoplasma reactivation [6].

Although many advances have been made in vaccination strategies against *T. gondii*, the therapeutic potency of immunization therapies is hampered in immunocompromised patients because of deficient cellular immunity [180]. Thus, an interest was raised to explore the passive immunization options that provide direct protection against reactivation. Specific recombinant antibodies directed toward Toxoplasma block the parasite spreading at early stage, restrict parasite from attaching and invading host cells, and obstruct the replication process of intracellular tachyzoites. Among others, antibodies working on SAG1 antigen show a promising possibility for drug development as SAG1 antigen is abundantly expressed on the parasite surface. Fab antibody fragments to SAG1 increased survival by 50% after lethal challenge [6, 181]. The underlying mechanism was a direct blockage of cellular invasion, but such therapies can be considered for adjuvant or prophylactic management only when the course of parasite invasion is slowed down [6].

On the other hand, additional supplementation or inhibition of certain specific ILs such as anti-IL-10 in the form of adjunctive therapy can drive the conventional treatments toward delivering a highly potent anti-*T. gondii* effect [1]. The entire immunotherapy system has been successfully explored in the preclinical models; only the results have to resonate with the clinical models which can only be possible through conduction of more and more investigations.

23.6 Conclusion and Outlook

The diagnosis of *T. gondii* infection remains a challenge, and the current serological methods have limitations in terms of accuracy and reliability. The lack of standardization in the production of *T. gondii* whole-cell lysates used in these tests contributes to these limitations. As a result, there is a need to consider alternative diagnostic options that may provide more accurate and reliable results. This could involve the development of new serological tests or the use of molecular techniques, such as amplification of nucleic acid (PCR-based tests) and recombinant protein-based method, which have shown promise in the diagnosis of *T. gondii* infections. The identification of more accurate and reliable diagnostic methods will be important in the control and management of toxoplasmosis. The development of new diagnostic methods, such as the Mn-PCR-RFLP method, has greatly improved our ability to genetically characterize *T. gondii*. This method provides high-resolution results, making it possible to differentiate between different strains of the parasite. The use of serotyping methods based on polymorphic polypeptides also has potential to become the preferred method for typing *T. gondii* in both humans and animals. These new methods offer the promise of more accurate and reliable diagnosis of *T. gondii* infections, which is crucial for the effective control and management of toxoplasmosis. The integration of molecular and bioinformatic technologies is becoming increasingly important in the study of *T. gondii* and the development of new diagnostic methods for toxoplasmosis. The increased use of genomic, transcriptomic, and proteomic technologies, along with the development of multilocus genotyping methods, provides a wealth of information about the genetic makeup of *T. gondii*. This information can be used to design new diagnostic methods that are more accurate, reliable, and efficient. The use of these technologies will be crucial in the ongoing efforts to control and manage toxoplasmosis and in the development of new therapeutic strategies for the treatment of this disease.

There are several different approaches that have been shown to be effective in controlling toxoplasma infection in animal and human, but much more work is needed to determine if these approaches are safe and effective in humans. Clinical trials in humans are necessary to validate these strategies and to determine the best approach for the control of toxoplasmosis in immunocompromised patients. Additionally, the field of toxoplasmosis research may offer valuable insights into immunomodulation interventions for the control of this condition. There is a growing body of evidence that suggests that the modulation of the host immune response can be a powerful tool in the fight against infectious diseases, and this approach has been studied extensively in the context of toxoplasmosis. By building on this work, researchers may be able to develop new strategies for controlling Toxoplasmosis that are safe and effective for use in humans.

The poor response to conventional treatment for toxoplasmosis has led to the exploration of alternative therapies that could improve treatment outcomes. Therefore, enhancing the host immune response is a promising therapeutic approach for the treatment of toxoplasmosis. Boosting the immune system could help to prevent reactivation of the infection in immunocompromised patients, potentially with fewer adverse effects than the current standard treatment with cotrimoxazole. It may also be useful in treating severe reactivation episodes in combination with conventional drugs. However, further research is needed in this field to fully understand the potential benefits and limitations of this approach, and to develop safe and effective treatments that can be used in clinical practice. The integration of various diagnostic and therapeutic technologies, as well as a better understanding of the host–parasite interactions, will be important in the ongoing efforts to control and manage toxoplasmosis.

Acknowledgment

The authors are grateful to **Prof. Gaurang B. Shah**, Department of Pharmacology, L. M. College of Pharmacy, Ahmedabad, Gujarat, India for kind support and guidance in manuscript preparation. The authors also extend their appreciation to L. M. College of Pharmacy, Ahmedabad, India for providing continuous library resources support throughout literature survey and data collection.

References

1 Smith, N.C., Goulart, C., Hayward, J.A. et al. (2021). Control of human toxoplasmosis. *Int. J. Parasitol.* 51 (2–3): 95–121.

2 *Toxoplasma gondii* [Available from: https://www.biologyonline.com/dictionary/toxoplasma-gondii.

3 Moncada, P.A. and Montoya, J.G. (2012). Toxoplasmosis in the fetus and newborn: an update on prevalence, diagnosis and treatment. *Expert Rev. Anti-Infect. Ther.* 10 (7): 815–828.

4 Liu, Q., Wang, Z.-D., Huang, S.-Y., and Zhu, X.-Q. (2015). Diagnosis of toxoplasmosis and typing of *Toxoplasma gondii*. *Parasit. Vectors* 8 (1): 292.

5 Calderaro, A., Peruzzi, S., Piccolo, G. et al. (2009). Laboratory diagnosis of *Toxoplasma gondii* infection. *Int. J. Med. Sci.* 6 (3): 135–136.

6 Dunay, I.R., Gajurel, K., Dhakal, R. et al. (2018). Treatment of toxoplasmosis: historical perspective, animal models, and current clinical practice. *Clin. Microbiol. Rev.* 31 (4): e00057–e00017.

7 Masur, H. (2017). Management of opportunistic infections associated with human immunodeficiency virus infection. *Infection* 4: 5.

8 Robert, M.G., Brenier-Pinchart, M.-P., Garnaud, C. et al. (2021). Molecular diagnosis of toxoplasmosis: recent advances and a look to the future. *Expert Rev. Anti-Infect. Ther.* 19 (12): 1529–1542.

9 WHO Estimates of the Global Burden of Foodborne Diseases: Toxoplasmosis Fact Sheet 2015 [Available from: https://www.euro.who.int/__data/assets/pdf_file/0011/294599/Factsheet-Toxoplasmosis-en.pdf.

10 Toxoplamosis: Epidemiology and Risk Factors [Available from: https://www.cdc.gov/parasites/toxoplasmosis/epi.html#:~:text=Toxoplasmosis%20is%20caused%20by%20the,have%20been%20infected%20with%20Toxoplasma.

11 Jones, J.L., Dargelas, V., Roberts, J. et al. (2009). Risk factors for *Toxoplasma gondii* infection in the United States. *Clin. Infect. Dis.* 49 (6): 878–884.

12 Kaye, A. (2011). Toxoplasmosis: diagnosis, treatment, and prevention in congenitally exposed infants. *J. Pediatr. Health Care* 25 (6): 355–364.

13 Safarpour, H., Cevik, M., Zarean, M. et al. (2020). Global status of *Toxoplasma gondii* infection and associated risk factors in people living with HIV. *AIDS* 34 (3): 469–474.

14 Basavaraju, A. (2016). Toxoplasmosis in HIV infection: an overview. *Trop. Parasitol.* 6 (2): 129.

15 Soltani, S., Kahvaz, M.S., Soltani, S. et al. (2020). Seroprevalence and associated risk factors of *Toxoplasma gondii* infection in patients undergoing hemodialysis and healthy group. *BMC Res. Notes* 13 (1): 1–5.

16 Olariu, T.R., Ursoniu, S., Hotea, I. et al. (2020). Seroprevalence and risk factors of *Toxoplasma gondii* infection in pregnant women from Western Romania. *Vector Borne Zoonotic Dis.* 20 (10): 763–767.

17 Attias, M., Teixeira, D.E., Benchimol, M. et al. (2020). The life-cycle of *Toxoplasma gondii* reviewed using animations. *Parasit. Vectors* 13: 1–13.

18 Tomasina, R. and Francia, M.E. (2020). The structural and molecular underpinnings of gametogenesis in *Toxoplasma gondii*. *Front. Cell. Infect. Microbiol.* 10: 608291.

19 Weiss, L.M. and Kim, K. (2000). The development and biology of bradyzoites of Toxoplasma gondii. *Front Biosci.* 5: D391–D405.

20 Watts, E., Zhao, Y., Dhara, A. et al. (2015). Novel approaches reveal that *Toxoplasma gondii* bradyzoites within tissue cysts are dynamic and replicating entities in vivo. *MBio* 6 (5): e01155–e01115.

21 Sinai, A.P., Watts, E.A., Dhara, A. et al. (2016). Reexamining chronic *Toxoplasma gondii* infection: surprising activity for a "dormant" parasite. *Curr. Clin. Microbiol. Rep.* 3: 175–185.

22 Cerutti, A., Blanchard, N., and Besteiro, S. (2020). The bradyzoite: a key developmental stage for the persistence and pathogenesis of toxoplasmosis. *Pathogens* 9 (3): 234.

23 White, M.W. and Suvorova, E.S. (2018). Apicomplexa cell cycles: something old, borrowed, lost, and new. *Trends Parasitol.* 34 (9): 759–771.

24 Gubbels, M.-J., Keroack, C.D., Dangoudoubiyam, S. et al. (2020). Fussing about fission: defining variety among mainstream and exotic apicomplexan cell division modes. *Front. Cell. Infect. Microbiol.* 10: 269.

25 Calero-Berna, R. and Gennari, S. (2019). Clinical toxoplasmosis in dogs and cats: an update. *Front. Vet. Sci.* 6: 54.

26 Kochanowsky, J.A. and Koshy, A.A. (2018). *Toxoplasma gondii*. *Curr. Biol.* 28 (14): R770–R771.

27 Toxoplasmosis [Available from: https://www.cell.com/current-biology/pdf/S0960-9822(18)30671-7.pdf.

28 Blader, I.J., Coleman, B.I., Chen, C.-T., and Gubbels, M.-J. (2015). Lytic cycle of *Toxoplasma gondii*: 15 years later. *Annu. Rev. Microbiol.* 69: 463–485.

29 Clough, B. and Frickel, E.-M. (2017). The toxoplasma parasitophorous vacuole: an evolving host–parasite frontier. *Trends Parasitol.* 33 (6): 473–488.

30 Sanchez, S.G. and Besteiro, S. (2021). The pathogenicity and virulence of *Toxoplasma gondii*. *Virulence* 12 (1): 3095–3114.

31 Ben Chaabene, R., Lentini, G., and Soldati-Favre, D. (2021). Biogenesis and discharge of the rhoptries: key organelles for entry and hijack of host cells by the Apicomplexa. *Mol. Microbiol.* 115 (3): 453–465.

32 Besteiro, S., Dubremetz, J.F., and Lebrun, M. (2011). The moving junction of apicomplexan parasites: a key structure for invasion. *Cell. Microbiol.* 13 (6): 797–805.

33 Coppens, I. and Romano, J.D. (2018). Hostile intruder: toxoplasma holds host organelles captive. *PLoS Pathog.* 14 (3): e1006893.

34 Gold, D.A., Kaplan, A.D., Lis, A. et al. (2015). The toxoplasma dense granule proteins GRA17 and GRA23 mediate the movement of small molecules between the host and the parasitophorous vacuole. *Cell Host Microbe* 17 (5): 642–652.

35 Caldas, L.A., Attias, M., and de Souza, W. (2018). A structural analysis of the natural egress of *Toxoplasma gondii*. *Microbes Infect.* 20 (1): 57–62.

36 Toxoplasmosis Clinical Presentation [Available from: https://emedicine.medscape.com/article/229969-clinical.

37 Naqi, R., Azeemuddin, M., and Ahsan, H. (2010). Cerebral toxoplasmosis in a patient with acquired immunodeficiency syndrome. *J. Pak. Med. Assoc.* 60 (4): 316.

38 Schumacher, A.C., Elbadawi, L.I., DeSalvo, T. et al. (2020). Toxoplasmosis outbreak associated with *Toxoplasma gondii*-contaminated venison—high attack rate, unusual clinical presentation, and atypical genotype. *Clin. Infect. Dis.* 72 (9): 1557–1565.

39 Basit, K.A., Nasir, S., Vohra, E., and Shazlee, M.K. (2018). Toxoplasmosis in an immunocompetent patient. *Pak. J. Med. Sci.* 34 (6): 1579.

40 Ramachandran, R., Radhan, P., Anand, R. et al. (2014). CNS toxoplasmosis in an immunocompetent individual. *Radiol. Case Rep.* 9 (1): 908.

41 Salviz, M., Montoya, J.G., Nadol, J.B., and Santos, F. (2013). Otopathology in congenital toxoplasmosis. *Otol. Neurotol.* 34 (6): 1165.

42 Fabiani, S., Caroselli, C., Menchini, M. et al. (2022). Ocular toxoplasmosis, an overview focusing on clinical aspects. *Acta Trop.* 225: 106180.

43 Weiss, L.M. and Dubey, J.P. (2009). Toxoplasmosis: a history of clinical observations. *Int. J. Parasitol.* 39 (8): 895–901.

44 Ahmadpour, E., Daryani, A., Sharif, M. et al. (2014). Toxoplasmosis in immunocompromised patients in Iran: a systematic review and meta-analysis. *J. Infect. Dev. Ctries.* 8 (12): 1503–1510.

45 da Silva, P.C., Shiraishi, C.S., da Silva, A.V. et al. (2010). *Toxoplasma gondii*: a morphometric analysis of the wall and epithelial cells of pigs intestine. *Exp. Parasitol.* 125 (4): 380–383.

46 Udonsom, R., Buddhirongawatr, R., and Sukthana, Y. (2010). Is Sabin-Feldman dye test using *T. gondii* tachyzoites from animal inoculation still the best method for detecting *Toxoplasma gondii* antibodies? *Southeast Asian J. Trop. Med. Public Health* 41 (5): 1059.

47 Macrì, G., Sala, M., Linder, A.M. et al. (2009). Comparison of indirect fluorescent antibody test and modified agglutination test for detecting *Toxoplasma gondii* immunoglobulin G antibodies in dog and cat. *Parasitol. Res.* 105: 35–40.

48 Montoya, J.G., Berry, A., Rosso, F., and Remington, J.S. (2007). The differential agglutination test as a diagnostic aid in cases of toxoplasmic lymphadenitis. *J. Clin. Microbiol.* 45 (5): 1463–1468.

49 Pour Azami, S.J., Keshavarz, H., Rezaian, M. et al. (2011). Rapid detection of *Toxoplasma gondii* antigen in experimentally infected mice by Dot-ELISA. *Iran. J. Parasitol.* 6 (1): 28.

50 Sucilathangam, G., Palaniappan, N., Sreekumar, C., and Anna, T. (2010). IgG – indirect fluorescent antibody technique to detect seroprevalence of *Toxoplasma gondii* in immunocompetent and immunodeficient patients in southern districts of Tamil Nadu. *Indian J. Med. Microbiol.* 28 (4): 354–357.

51 Rodrigues, I.M.X., Castro, A.M., Gomes, M.B.F. et al. (2009). Congenital toxoplasmosis: evaluation of serological methods for the detection of anti-*Toxoplasma gondii* IgM and IgA antibodies. *Mem. Inst. Oswaldo Cruz* 104: 434–440.

52 Dubey, J.P. (2016). *Toxoplasmosis of Animals and Humans*. CRC Press.

53 Rostami, A., Karanis, P., and Fallahi, S. (2018). Advances in serological, imaging techniques and molecular diagnosis of *Toxoplasma gondii* infection. *Infection* 46: 303–315.

54 Su, C., Shwab, E., Zhou, P. et al. (2010). Moving towards an integrated approach to molecular detection and identification of *Toxoplasma gondii*. *Parasitol. Res.* 137 (1): 1–11.

55 Wang, Y., Wang, G., Zhang, D. et al. (2013). Detection of acute toxoplasmosis in pigs using loop-mediated isothermal amplification and quantitative PCR. *Korean J. Parasitol.* 51 (5): 573.

56 Yu, H., Huang, B., Zhuo, X. et al. (2013). Evaluation of a real-time PCR assay based on the single-copy SAG1 gene for the detection of *Toxoplasma gondii*. *Vet. Parasitol.* 197 (3–4): 670–673.

57 Terkawi, M.A., Kameyama, K., Rasul, N.H. et al. (2013). Development of an immunochromatographic assay based on dense granule protein 7 for serological detection of *Toxoplasma gondii* infection. *Clin. Vaccine Immunol.* 20 (4): 596–601.

58 Cabral-Miranda, G., de Jesus, J., Oliveira, P.S. et al. (2014). Detection of parasite antigens in *Leishmania infantum*-infected spleen tissue by monoclonal antibody-, piezoelectric-based immunosensors. *J. Parasitol.* 100 (1): 73–78.

59 Safarpour, H., Pourhassan-Moghaddam, M., Spotin, A. et al. (2021). A novel enhanced dot blot immunoassay using colorimetric biosensor for detection of *Toxoplasma gondii* infection. *Comp. Immunol. Microbiol. Infect. Dis.* 79: 101708.

60 Jazayeri, M.H., Aghaie, T., Avan, A. et al. (2018). Colorimetric detection based on gold nano particles (GNPs): an easy, fast, inexpensive, low-cost and short time method in detection of analytes (protein, DNA, and ion). *Sens. Bio-Sens. Res.* 20: 1–8.

61 Safarpour, H., Majdi, H., Masjedi, A. et al. (2021). Development of optical biosensor using protein A-conjugated chitosan–gold nanoparticles for diagnosis of cystic echinococcosis. *Biosensors* 11 (5): 134.

62 de Lima, C.A., da Silva, P.S., and Spinelli, A. (2014). Chitosan-stabilized silver nanoparticles for voltammetric detection of nitrocompounds. *Sensors Actuators B Chem.* 196: 39–45.

63 Medawar-Aguilar, V., Jofre, C.F., Fernández-Baldo, M.A. et al. (2019). Serological diagnosis of toxoplasmosis disease using a fluorescent immunosensor with chitosan-ZnO-nanoparticles. *Anal. Biochem.* 564–565: 116–122.

64 Aly, I., Taher, E.E., El Nain, G. et al. (2018). Advantages of bioconjugated silica-coated nanoparticles as an innovative diagnosis for human toxoplasmosis. *Acta Trop.* 177: 19–24.

65 Oudah, N.A.R., Al-Teea, K.S.C., and Mohammed, A.A. (2019). Using the level of serum YKL-40 as an indicator to the pathogenesis of allergic asthma and helminths infection. *Indian J. Public Health Res. Dev.* 10 (10): 2898.

66 Zhao, T., Su, Z., Li, Y. et al. (2020). Chitinase-3 like-protein-1 function and its role in diseases. *Signal Transduct. Target. Ther.* 5 (1): 201.

67 Shahad, F.A.A.L.-Z., Mohammed, A.A., and Jasim, G.A. (2022). YKL-40 as a novel diagnostic biomarker in toxoplasmosis. *J. Popul. Ther. Clin. Pharmacol.* 29 (02): e61–e70.

68 Majewski, S., Szewczyk, K., Jerczyńska, H. et al. (2022). Longitudinal and comparative measures of serum chitotriosidase and YKL-40 in patients with idiopathic pulmonary fibrosis. *Front. Immunol.* 13: 760776.

69 Gandhi, P., Khare, R., Vasudev Gulwani, H., and Kaur, S. (2018). Circulatory YKL-40 & NLR: underestimated prognostic indicators in diffuse glioma. *Int. J. Mol. Cell. Med.* 7 (2): 111.

70 Coriati, A., Bouvet, G.F., Massé, C. et al. (2021). YKL-40 as a clinical biomarker in adult patients with CF: implications of a CHI3L1 single nucleotide polymorphism in disease severity. *J. Cyst. Fibros.* 20 (6): e93–e99.

71 Salomon, J., Matusiak, Ł., Nowicka-Suszko, D., and Szepietowski, J.C. (2017). Chitinase-3-like protein 1 (YKL-40) reflects the severity of symptoms in atopic dermatitis. *J Immunol Res* 2017: 5746031.

72 Jankowska-Konsur, A., Łyko, M., Rubas, K. et al. (2022). Chitinase-3-like protein 1 (YKL-40): a new biomarker of inflammation in pyoderma gangrenosum. *Acta Derm. Venereol.* 102: 978. adv00646-adv.

73 Kwon, Y., Kim, J.H., Ha, E.K. et al. (2020). Serum YKL-40 levels are associated with the atherogenic index of plasma in children. *Mediat. Inflamm.* 2020: 8713908.

74 Zhang, K., Lin, G., Han, Y., and Li, J. (2016). Serological diagnosis of toxoplasmosis and standardization. *Clin. Chim. Acta* 461: 83–89.

75 Ferra, B., Holec-Gąsior, L., and Kur, J. (2015). A new *Toxoplasma gondii* chimeric antigen containing fragments of SAG2, GRA1, and ROP1 proteins—impact of immunodominant sequences size on its diagnostic usefulness. *Parasitol. Res.* 114: 3291–3299.

76 Holec-Gąsior, L. (2013). *Toxoplasma gondii* recombinant antigens as tools for serodiagnosis of human toxoplasmosis: current status of studies. *Clin. Vaccine Immunol.* 20 (9): 1343–1351.

77 Arab-Mazar, Z., Javadi Mamaghani, A., Fallahi, S. et al. (ed.) (2021). Immunodiagnosis and molecular validation of *Toxoplasma gondii*-recombinant dense granular (GRA) 5 protein for the detection of toxoplasmosis in hemodialysis patients. *Semin. Dial.* 34: 332–337. Wiley Online Library.

78 Selseleh, M., Keshavarz, H., Mohebali, M. et al. (2012). Production and evaluation of *Toxoplasma gondii* recombinant GRA7 for serodiagnosis of human infections. *Korean J. Parasitol.* 50 (3): 233–238.

79 Dai, J., Jiang, M., Wang, Y. et al. (2012). Evaluation of a recombinant multiepitope peptide for serodiagnosis of *Toxoplasma gondii* infection. *Clin. Vaccine Immunol.* 19 (3): 338–342.

80 Dai, J.-f., Jiang, M., Qu, L.-l. et al. (2013). *Toxoplasma gondii*: enzyme-linked immunosorbent assay based on a recombinant multi-epitope peptide for distinguishing recent from past infection in human sera. *Exp. Parasitol.* 133 (1): 95–100.

81 Holec-Gąsior, L., Ferra, B., Drapała, D. et al. (2012). A new MIC1-MAG1 recombinant chimeric antigen can be used instead of the *Toxoplasma gondii* lysate antigen in serodiagnosis of human toxoplasmosis. *Clin. Vaccine Immunol.* 19 (1): 57–63.

82 Holec-Gąsior, L., Ferra, B., and Drapała, D. (2012). MIC1-MAG1-SAG1 chimeric protein, a most effective antigen for detection of human toxoplasmosis. *Clin. Vaccine Immunol.* 19 (12): 1977–1979.

83 Maksimov, P., Zerweck, J., Maksimov, A. et al. (2012). Peptide microarray analysis of in silico-predicted epitopes for serological diagnosis of *Toxoplasma gondii* infection in humans. *Clin. Vaccine Immunol.* 19 (6): 865–874.

84 Cardona, N., de-la-Torre, A., Siachoque, H. et al. (2009). *Toxoplasma gondii*: P30 peptides recognition pattern in human toxoplasmosis. *Exp. Parasitol.* 123 (2): 199–202.

85 Holec-Gąsior, L., Ferra, B., Czechowska, J. et al. (2016). A novel chemiluminescent immunoassay for detection of *Toxoplasma gondii* IgG in human sera. *Diagn. Microbiol. Infect. Dis.* 85 (4): 422–425.

86 Fallahi, S., Tabaei, S.J.S., Pournia, Y. et al. (2014). Comparison of loop-mediated isothermal amplification (LAMP) and nested-PCR assay targeting the RE and B1 gene for detection of *Toxoplasma gondii* in blood samples of children with leukaemia. *Diagn. Microbiol. Infect. Dis.* 79 (3): 347–354.

87 Gallas-Lindemann, C., Sotiriadou, I., Mahmoodi, M.R., and Karanis, P. (2013). Detection of *Toxoplasma gondii* oocysts in different water resources by loop mediated isothermal amplification (LAMP). *Acta Trop.* 125 (2): 231–236.

88 Dexamethasone for Cerebral Toxoplasmosis (De-Tox) [Available from: https://clinicaltrials.gov/ct2/show/NCT04341155?recrs=abdf&cond=Toxoplasmosis&phase=0123&draw=2&rank=1.

89 Pyrimethamine, Sulfadiazine, and Leucovorin in Treating Patients with Congenital Toxoplasmosis [Available from: https://clinicaltrials.gov/ct2/show/NCT00004317?recrs=abdf&cond=Toxoplasmosis&phase=0123&draw=2&rank=2.

90 Lyon Cohort of Maternal and Congenital Toxoplasma Infections [Available from: https://clinicaltrials.gov/ct2/show/NCT02936921?recrs=abd&cond=Toxoplasmosis%2C+Congenital&draw=2&rank=1.

91 Diagnosis of *Toxoplasma Gondii* Infection by Exploration of Cellular Immunity (TOXCELL) (TOXCELL) [Available from: https://clinicaltrials.gov/ct2/show/NCT04825600?recrs=abd&cond=Toxoplasmosis%2C+Congenital&draw=2&rank=3.

92 Feleke, D.G., Gebreweld, A., and Zewde, G. (2019). Toxoplasmosis in pregnant women and HIV/AIDS patients in Ethiopia: a systematic review and meta-analysis. *J. Parasitol. Res.* 2019: 4670397.

93 Al-Malki, E.S. (2021). Toxoplasmosis: stages of the protozoan life cycle and risk assessment in humans and animals for an enhanced awareness and an improved socio-economic status. *Saudi J. Biol. Sci.* 28 (1): 962–969.

94 Robert-Gangneux, F. (2014). It is not only the cat that did it: how to prevent and treat congenital toxoplasmosis. *J. Infect.* 68: S125–S133.

95 Ahmed, M., Sood, A., and Gupta, J. (2020). Toxoplasmosis in pregnancy. *Eur. J. Obstet. Gynecol. Reprod. Biol.* 255: 44–50.

96 Leke, A., Dolk, H., Loane, M. et al. (2020). *Macrolide Antibiotic Exposure in the First Trimester of Pregnancy and Risk of Congenital Anomaly: A European Case-Control Study*. Authorea Preprints.

97 Peyron, F., L'ollivier, C., Mandelbrot, L. et al. (2019). Maternal and congenital toxoplasmosis: diagnosis and treatment recommendations of a French multidisciplinary working group. *Pathogens* 8 (1): 24.

98 Bakhraibah, A.O. (2022). Warm-blooded animal toxoplasmosis and its connection to COVID-19: a review. *Microb. Biosyst.* 7 (1): 18–26.

99 Konstantinovic, N., Guegan, H., Stäjner, T. et al. (2019). Treatment of toxoplasmosis: current options and future perspectives. *Food Waterborne Parasitol.* 15: e00036.

100 Hopper, A.T., Brockman, A., Wise, A. et al. (2019). Discovery of selective *Toxoplasma gondii* dihydrofolate reductase inhibitors for the treatment of toxoplasmosis. *J. Med. Chem.* 62 (3): 1562–1576.

101 Lapinskas, P.J. and Ben-Harari, R.R. (2019). Perspective on current and emerging drugs in the treatment of acute and chronic toxoplasmosis. *Postgrad. Med.* 131 (8): 589–596.

102 Rahman, F., Birowo, P., Widyahening, I.S., and Rasyid, N. (2017). Effect of citrus-based products on urine profile: a systematic review and meta-analysis. *F1000Res.* 6: 220.

103 Picone, O., Fuchs, F., Benoist, G. et al. (2020). Toxoplasmosis screening during pregnancy in France: opinion of an expert panel for the CNGOF. *J. Gynecol. Obstet. Hum. Reprod.* 49 (7): 101814.

104 Maldonado, Y.A., Read, J.S., Byington, C.L. et al. (2017). Diagnosis, treatment, and prevention of congenital toxoplasmosis in the United States. *Pediatrics* 139 (2): e20163860.

105 Kieffer, F., Wallon, M., Garcia, P. et al. (2008). Risk factors for retinochoroiditis during the first 2 years of life in infants with treated congenital toxoplasmosis. *Pediatr. Infect. Dis. J.* 27 (1): 27–32.

106 Feliciano-Alfonso, J.E., Muñoz-Ortiz, J., Marín-Noriega, M.A. et al. (2021). Safety and efficacy of different antibiotic regimens in patients with ocular toxoplasmosis: systematic review and meta-analysis. *Syst. Rev.* 10: 1–14.

107 Garweg, J.G. and Pleyer, U. (2021). Treatment strategy in human ocular toxoplasmosis: why antibiotics have failed. *J. Clin. Med.* 10 (5): 1090.

108 Wang, Z.-D., Liu, H.-H., Ma, Z.-X. et al. (2017). *Toxoplasma gondii* infection in immunocompromised patients: a systematic review and meta-analysis. *Front. Microbiol.* 8: 389.

109 Vidal, J.E. (2019). HIV-related cerebral toxoplasmosis revisited: current concepts and controversies of an old disease. *J. Int. Assoc. Provid. AIDS Care* 18: 2325958219867315.

110 Cetinkaya, F.I., Mumcu, N., Unuvar, G.K. et al. (2022). Toxoplasmic encephalitis after allogeneic hematopoietic stem cell transplantation. *North. Clin. Istanb.* 9 (3): 282.

111 Dian, S., Ganiem, A.R., and Ekawardhani, S. (2023). Cerebral toxoplasmosis in HIV-infected patients: a review. *Pathog. Glob. Health* 117 (1): 1–10.

112 Foroutan, M., Ghaffarifar, F., Sharifi, Z. et al. (2018). Bioinformatics analysis of ROP8 protein to improve vaccine design against *Toxoplasma gondii*. *Infect. Genet. Evol.* 62: 193–204.

113 Kaplan, J.E., Benson, C., Holmes, K.K. et al. (2009). Guidelines for prevention and treatment of opportunistic infections in HIV-infected adults and adolescents: recommendations from CDC, the National Institutes of Health, and the HIV Medicine Association of the Infectious Diseases Society of America. *MMWR Recomm. Rep.* 58 (RR-4): 1–207; quiz CE1-4.

114 Schmidt, D.R., Hogh, B., Andersen, O. et al. (2006). Treatment of infants with congenital toxoplasmosis: tolerability and plasma concentrations of sulfadiazine and pyrimethamine. *Eur. J. Pediatr.* 165: 19–25.

115 La Hoz, R.M., Morris, M.I., and Transplantation IDCoPotASo (2019). Tissue and blood protozoa including toxoplasmosis, Chagas disease, leishmaniasis, Babesia, Acanthamoeba, Balamuthia, and Naegleria in solid organ transplant recipients—Guidelines from the American Society of Transplantation Infectious Diseases Community of Practice. *Clin. Transpl.* 33 (9): e13546.

116 Hosoda, T., Mikita, K., Ito, M. et al. (2021). Cerebral toxoplasmosis with multiple hemorrhage lesions in an HIV infected patient: a case report and literature review. *Parasitol. Int.* 81: 102280.

117 Trasia, R.F. (2021). Current treatment and primary prophylaxis for toxoplasmosis. *J. Pharm. Care Sci.* 1 (2): 132–135.

118 Montazeri, M., Mehrzadi, S., Sharif, M. et al. (2018). Drug resistance in *Toxoplasma gondii*. *Front. Microbiol.* 9: 2587.

119 Griffith, E.C., Wallace, M.J., Wu, Y. et al. (2018). The structural and functional basis for recurring sulfa drug resistance mutations in *Staphylococcus aureus* dihydropteroate synthase. *Front. Microbiol.* 9: 1369.

120 King, L.D., Sia, H., and Anoopkumar-Dukie, S. (2021). Trimethoprim-sulfamethoxazole as *Pneumocystis jiroveci* pneumonia prevention in patients undergoing methotrexate therapy for hematological malignancies: a review of the literature. *Hematol. Oncol.* 39: 170–175.

121 Silva, M., Teixeira, C., Gomes, P., and Borges, M. (2021). Promising drug targets and compounds with anti-*Toxoplasma gondii* activity. *Microorganisms* 9 (9): 1960.

122 Villena, I., Aubert, D., Leroux, B. et al. (1998). Pyrimethamine-sulfadoxine treatment of congenital toxoplasmosis: follow-up of 78 cases between 1980 and 1997. Reims Toxoplasmosis Group. *Scand. J. Infect. Dis.* 30 (3): 295–300.

123 Silva, L.A., Reis-Cunha, J.L., Bartholomeu, D.C., and Vítor, R.W.A. (2017). Genetic polymorphisms and phenotypic profiles of sulfadiazine-resistant and sensitive *Toxoplasma gondii* isolates obtained from newborns with congenital toxoplasmosis in Minas Gerais, Brazil. *PLoS One* 12 (1): e0170689.

124 Wohlfert, E.A., Blader, I.J., and Wilson, E.H. (2017). Brains and brawn: toxoplasma infections of the central nervous system and skeletal muscle. *Trends Parasitol.* 33 (7): 519–531.

125 Vidadala, R.S.R., Rivas, K.L., Ojo, K.K. et al. (2016). Development of an orally available and central nervous system (CNS) penetrant *Toxoplasma gondii* calcium-dependent protein kinase 1 (Tg CDPK1) inhibitor with minimal human ether-a-go-go-related gene (hERG) activity for the treatment of toxoplasmosis. *J. Med. Chem.* 59 (13): 6531–6546.

126 Bottari, N.B., Baldissera, M.D., Tonin, A.A. et al. (2016). Synergistic effects of resveratrol (free and inclusion complex) and sulfamethoxazole-trimetropim treatment on pathology, oxidant/antioxidant status and behavior of mice infected with *Toxoplasma gondii*. *Microb. Pathog.* 95: 166–174.

127 Montazeri, M., Ebrahimzadeh, M.A., Ahmadpour, E. et al. (2016). Evaluation of propranolol effect on experimental acute and chronic toxoplasmosis using quantitative PCR. *Antimicrob. Agents Chemother.* 60 (12): 7128–7133.

128 Kato, K., Murata, Y., Horiuchi, N. et al. (2016). Dextran sulfate inhibits acute *Toxoplama gondii* infection in pigs. *Parasit. Vectors* 9 (1): 134.

129 Yeo, S.-J., Jin, C., Kim, S., and Park, H. (2016). In vitro and in vivo effects of nitrofurantoin on experimental toxoplasmosis. *Korean J. Parasitol.* 54 (2): 155.

130 Saraei, M., Ghaderi, Y., Mosavi, T. et al. (2016). The effect of fluphenazine and thioridazine on *Toxoplasma gondii* in vivo. *Iran. J. Parasitol.* 11 (2): 226.

131 Köksal, Z.Ş., Yanik, K., Bilgin, K. et al. (2016). In vivo efficacy of drugs against *Toxoplasma gondii* combined with immunomodulators. *Jpn. J. Infect. Dis.* 69 (2): 113–117.

132 Saraei, M., Samadzadeh, N., Khoeini, J. et al. (2015). In vivo anti-toxoplasma activity of aripiprazole. *Iran. J. Basic Med. Sci.* 18 (9): 938.

133 Montazeri, M., Daryani, A., Ebrahimzadeh, M. et al. (2015). Effect of propranolol alone and in combination with pyrimethamine on acute murine toxoplasmosis. *Jundishapur J. Microbiol.* 8 (9): e22572.

134 Martins-Duarte, E.S., Dubar, F., Lawton, P. et al. (2015). Ciprofloxacin derivatives affect parasite cell division and increase the survival of mice infected with *Toxoplasma gondii*. *PLoS One* 10 (5): e0125705.

135 Escotte-Binet, S., Huguenin, A., Aubert, D. et al. (2018). Metallopeptidases of *Toxoplasma gondii*: in silico identification and gene expression. *Parasit. Vectors* 25: 26.

136 McFarland, M.M., Zach, S.J., Wang, X. et al. (2016). Review of experimental compounds demonstrating anti-toxoplasma activity. *Antimicrob. Agents Chemother.* 60 (12): 7017–7034.

137 El-Zawawy, L.A., El-Said, D., Mossallam, S.F. et al. (2015). Triclosan and triclosan-loaded liposomal nanoparticles in the treatment of acute experimental toxoplasmosis. *Exp. Parasitol.* 149: 54–64.

138 El-Zawawy, L.A., El-Said, D., Mossallam, S.F. et al. (2015). Preventive prospective of triclosan and triclosan-liposomal nanoparticles against experimental infection with a cystogenic ME49 strain of *Toxoplasma gondii*. *Acta Trop.* 141: 103–111.

139 Doggett, J.S., Ojo, K.K., Fan, E. et al. (2014). Bumped kinase inhibitor 1294 treats established *Toxoplasma gondii* infection. *Antimicrob. Agents Chemother.* 58 (6): 3547–3549.

140 Müller, J., Aguado-Martínez, A., Ortega-Mora, L.-M. et al. (2017). Development of a murine vertical transmission model for *Toxoplasma gondii* oocyst infection and studies on the efficacy of bumped kinase inhibitor (BKI)-1294 and the naphthoquinone buparvaquone against congenital toxoplasmosis. *J. Antimicrob. Chemother.* 72 (8): 2334–2341.

141 Spalenka, J., Escotte-Binet, S., Bakiri, A. et al. (2018). Discovery of new inhibitors of *Toxoplasma gondii* via the pathogen box. *Antimicrob. Agents Chemother.* 62 (2): e01640–e01617.

142 Ni Nyoman, A.D. and Lüder, C.G. (2013). Apoptosis-like cell death pathways in the unicellular parasite *Toxoplasma gondii* following treatment with apoptosis inducers and chemotherapeutic agents: a proof-of-concept study. *Apoptosis* 18: 664–680.

143 Eissa, M.M., Barakat, A.M., Amer, E.I., and Younis, L.K. (2015). Could miltefosine be used as a therapy for toxoplasmosis? *Exp. Parasitol.* 157: 12–22.

144 Adeyemi, O.S., Sugi, T., Han, Y., and Kato, K. (2018). Screening of chemical compound libraries identified new anti-*Toxoplasma gondii* agents. *Parasitol. Res.* 117: 355–363.

145 Benmerzouga, I., Checkley, L.A., Ferdig, M.T. et al. (2015). Guanabenz repurposed as an antiparasitic with activity against acute and latent toxoplasmosis. *Antimicrob. Agents Chemother.* 59 (11): 6939–6945.

146 Murata, Y., Sugi, T., Weiss, L.M., and Kato, K. (2017). Identification of compounds that suppress *Toxoplasma gondii* tachyzoites and bradyzoites. *PLoS One* 12 (6): e0178203.

147 Trotsko, N., Bekier, A., Paneth, A. et al. (2019). Synthesis and in vitro anti-*Toxoplasma gondii* activity of novel thiazolidin-4-one derivatives. *Molecules* 24 (17): 3029.

148 Alday, P.H. and Doggett, J.S. (2017). Drugs in development for toxoplasmosis: advances, challenges, and current status. *Drug Des. Dev. Ther.* 11: 273–293.

149 Montazeri, M., Sharif, M., Sarvi, S. et al. (2017). A systematic review of in vitro and in vivo activities of anti-toxoplasma drugs and compounds (2006–2016). *Front. Microbiol.* 8: 25.

150 Cheon, D., Lee, W.C., Lee, Y. et al. (2019). Structural basis of branched-chain fatty acid synthesis by Propionibacterium acnes β-ketoacyl acyl carrier protein synthase. *Biochem. Biophys. Res. Commun.* 509 (1): 322–328.

151 Okada, M., Rajaram, K., Swift, R.P. et al. (2022). Critical role for isoprenoids in apicoplast biogenesis by malaria parasites. *elife* 11: e73208.

152 Hajj, R.E., Tawk, L., Itani, S. et al. (2021). Toxoplasmosis: current and emerging parasite druggable targets. *Microorganisms* 9 (12): 2531.

153 Foroutan, M. and Ghaffarifar, F. (2018). Calcium-dependent protein kinases are potential targets for *Toxoplasma gondii* vaccine. *Clin. Exp. Vaccine Res.* 7 (1): 24–36.

154 Sánchez-Sánchez, R., Ferre, I., Re, M. et al. (2019). Treatment with bumped kinase inhibitor 1294 is safe and leads to significant protection against abortion and vertical transmission in sheep experimentally infected with *Toxoplasma gondii* during pregnancy. *Antimicrob. Agents Chemother.* 63 (7): e02527–e02518.

155 Rutaganira, F.U., Barks, J., Dhason, M.S. et al. (2017). Inhibition of calcium dependent protein kinase 1 (CDPK1) by pyrazolopyrimidine analogs decreases establishment and reoccurrence of central nervous system disease by *Toxoplasma gondii*. *J. Med. Chem.* 60 (24): 9976–9989.

156 Ma, H., Li, L., Gai, Y. et al. (2021). Histone acetyltransferases and deacetylases are required for virulence, conidiation, DNA damage repair, and multiple stresses resistance of *Alternaria alternata*. *Front. Microbiol.* 12: 783633.

157 Haidar, M., Ramdani, G., Kennedy, E., and Langsley, G. (2017). PKA and apicomplexan parasite diseases. *Horm. Metab. Res.* 49 (04): 296–300.

158 Martynowicz, J., Doggett, J.S., and Sullivan, W.J. Jr. (2020). Efficacy of guanabenz combination therapy against chronic toxoplasmosis across multiple mouse strains. *Antimicrob. Agents Chemother.* 64 (9): e00539–e00520.

159 McPhillie, M., Zhou, Y., El Bissati, K. et al. (2016). New paradigms for understanding and step changes in treating active and chronic, persistent apicomplexan infections. *Sci. Rep.* 6 (1): 29179.

160 Lykins, J.D., Filippova, E.V., Halavaty, A.S. et al. (2018). CSGID solves structures and identifies phenotypes for five enzymes in *Toxoplasma gondii*. *Front. Cell. Infect. Microbiol.* 8: 352.

161 McPhillie, M.J., Zhou, Y., Hickman, M.R. et al. (2020). Potent tetrahydroquinolone eliminates apicomplexan parasites. *Front. Cell. Infect. Microbiol.* 10: 203.

162 Subramanian, G., Belekar, M.A., Shukla, A. et al. (2019). Erratum for Subramanian et al., "Targeted phenotypic screening in Plasmodium falciparum and *Toxoplasma gondii* reveals novel modes of action of medicines for malaria venture malaria box molecules". *Msphere* 4 (2): e00159–e00119.

163 Radke, J.B., Burrows, J.N., Goldberg, D.E., and Sibley, L.D. (2018). Evaluation of current and emerging antimalarial medicines for inhibition of *Toxoplasma gondii* growth in vitro. *ACS Infect. Dis.* 4 (8): 1264–1274.

164 Secrieru, A., Costa, I.C., O'Neill, P.M., and Cristiano, M.L. (2020). Antimalarial agents as therapeutic tools against toxoplasmosis—a short bridge between two distant illnesses. *Molecules* 25 (7): 1574.

165 Montazeri, M., Mehrzadi, S., Sharif, M. et al. (2018). Activities of anti-toxoplasma drugs and compounds against tissue cysts in the last three decades (1987 to 2017), a systematic review. *Parasitol. Res.* 117: 3045–3057.

166 Antczak, M., Dzitko, K., and Długońska, H. (2016). Human toxoplasmosis–searching for novel chemotherapeutics. *Biomed. Pharmacother.* 82: 677–684.

167 Wang, Y.-L., Jin, L.-L., Cheng, X. et al. (2022). Synthesis and evaluation of in vitro and in vivo anti-*Toxoplasma gondii* activity of tetraoxane-substituted ursolic acid derivatives. *Nat. Prod. Res.* 14: 1–9.

168 Kota, A.S. and Shabbir, N. (2022). *Congenital Toxoplasmosis*. Treasure Island (FL): StatPearls Publishing.

169 Zhao, X.-Y. and Ewald, S.E. (2020). The molecular biology and immune control of chronic *Toxoplasma gondii* infection. *J. Clin. Invest.* 130 (7): 3370–3380.

170 Ghartey-Kwansah, G., Adu-Nti, F., Aboagye, B. et al. (2020). Autophagy in the control and pathogenesis of parasitic infections. *Cell Biosci.* 10: 1–11.

171 Waymack, J.R. and Sundareshan, V. (2023). *Acquired Immune Deficiency Syndrome*. StatPearls Publishing.

172 Sasai, M. and Yamamoto, M. (2019). Innate, adaptive, and cell-autonomous immunity against *Toxoplasma gondii* infection. *Exp. Mol. Med.* 51 (12): 1–10.

173 Khan, I.A., Ouellette, C., Chen, K., and Moretto, M. (2019). Toxoplasma: immunity and pathogenesis. *Curr. Clin. Microbiol. Rep.* 6: 44–50.

174 Ivanova, D.L., Krempels, R., Denton, S.L. et al. (2020). NK cells negatively regulate CD8 T cells to promote immune exhaustion and chronic *Toxoplasma gondii* infection. *Front. Cell. Infect. Microbiol.* 10: 313.

175 Maspi, N., Abdoli, A., and Ghaffarifar, F. (2016). Pro-and anti-inflammatory cytokines in cutaneous leishmaniasis: a review. *Pathog. Glob. Health* 110 (6): 247–260.

176 Schlüter, D. and Barragan, A. (2019). Advances and challenges in understanding cerebral toxoplasmosis. *Front. Immunol.* 10: 242.

177 Savarin, C. and Bergmann, C.C. (2018). Fine tuning the cytokine storm by IFN and IL-10 following neurotropic coronavirus encephalomyelitis. *Front. Immunol.* 9: 3022.

178 Bhadra, R., Gigley, J., and Khan, I. (2012). PD-1-mediated attrition of polyfunctional memory CD8(+) T cells in chronic toxoplasma infection. *J. Infect. Dis.* 206: 125–134.

179 Khan, I.A., Hwang, S., and Moretto, M. (2019). *Toxoplasma gondii*: CD8 T cells cry for CD4 help. *Front. Cell. Infect. Microbiol.* 9: 136.

180 Chu, K.-B. and Quan, F.-S.J. (2021). Advances in *Toxoplasma gondii* vaccines: current strategies and challenges for vaccine development. *Vaccine* 9 (5): 413.

181 Wang, Y. and Yin, H. (2014). Research progress on surface antigen 1 (SAG1) of *Toxoplasma gondii*. *Parasit. Vectors* 7: 1–9.

24

Recent Trends in Neurocysticerosis Diagnosis and Management

Sachin P. Bhatt[1], *Avani N. Joshi*[1], *and Bhupendra G. Prajapati*[2]

[1] *Gyanmanjari Pharmacy College Gujarat Technological University, Bhavnagar, Gujarat, India*
[2] *Department of Pharmaceutics and Pharmaceutical Technology, Shree S K Patel College of Pharmaceutical Education and Research, Ganpat University, Mehsana, Gujarat, India*

Abbreviations

AED	Antiepileptic medications
AEDs	Antiepileptic drugs
AHD	Anthelminthic medicines
ALB	Albendazole
CD8	Cluster of differentiation 8
CNS	Central nervous system
CSF	Cerebrospinal fluid
CSF	Cerebral spinal fluid
CT	Computed tomography
CT	Computed tomography scan
DWI	Diffusion-weighted imaging
EEG	Electroencephalographic
EIA	Enzyme immunoassay
EITB	Enzyme linked immunoelectrotransfer blot
ELISA	Enzyme-linked immunosorbent assay
ETIBI	Enzyme-linked immunoelectrotransfer blot test
FLAIR	Fluid-attenuated inversion recovery
IgG	Immuno globulin G
MRI	Magnetic resonance imaging
NCC	Neurocysticercosis
PZQ	Praziquantel
VP	Ventriculoperitoneal

24.1 Introduction

The most prevalent parasite condition affecting the human CNS is NCC [1]. NCC, caused by the larval form of the cestode, *Taenia solium*, is commonly referred to as the "pork tapeworm" [2]. *T. solium* is a multi-host parasite that travels from the final or accidental transitional pig host to the human victim. Its zoonotic spread cycle is intricate [3]. Humans may become infected with *T. solium*, which is brought on by consuming infected water or food, as well as insufficiently cooked or raw pork, and this infection can lead to taeniasis (caused by the mature tapeworm living in the small

Rising Contagious Diseases: Basics, Management, and Treatments, First Edition. Edited by Seth Kwabena Amponsah, Ranjita Shegokar, and Yashwant V. Pathak.

intestine) and/or cysticercosis or NCC. When *T. solium* live cysts are ingested by humans in pork, the cysticercal grows into tapeworms in the human intestine and eliminates eggs that can then be seen in human stool. The helminthic infection of NCC is the primary cause of acquired epilepsy worldwide. About 1.7–3 million people globally suffer from epilepsy as a result of NCC [4]. As results, and the unreliability of various tests used to detect anti-cysticercal blood immunoglobulins or antigens The variety of clinical manifestations of NCC, the absence of precision of the majority of brain scans findings, and the reliability of different investigations used to detect anti-cysticercal antibodies or antigens in blood are all factors that make this parasite disease difficult to diagnose [5]. The clinical manifestations of NCC are influenced by the quantity, distribution, bulk, and phase of growth of the invading parasite larvae in addition to the immune/erythrogenic response from the host [4]. More importantly, extra parenchymal NCC infections live in the ventricles or subarachnoid spaces and are associated with intracranial hypertension, which is a major contributor to mortality and morbidity. Intraparenchymal NCC infections, on the opposite hand, are less abrasive and are connected to headaches, seizures, and tremors [6].

24.1.1 Life Cycle, Biology, and Transmission

Taenia solium's life cycle is cestode has two hosts – humans and pigs (Figure 24.1). Pork and humans might act as secondary hosts for the tapeworm's larval development, but humans are the tapeworm's only true adult host (cysticercus). The mature tapeworm, which is attached to the human intestinal wall, excretes a few gravid of these creatures on a regular basis along with feces. Each proglottid releases thousands of infectious eggs into the atmosphere [8]. Pigs have had access to human feces in certain regions that contain *T. solium* ova with poor swine husbandry and feces disposal. While inside the pig's digestive system, eggs release oncospheres that pass through the intestinal wall and enter the circulation before being transported to develop tissues for metacestodes, which then mature into larvae (cysticercus). The digestive enzymes release cysticerci when people eat contaminated, improperly cooked pork. These cysticercievaginate stick to the gut wall, and these creatures begin to proliferate until the adult tapeworm is formed [9]. The eggs in the host's intestinal structure become uncoated, exposing the oncosphere-like larvae found within. These enter the intestinal wall and travel via the bloodstream to different physical zones, such as the brain, eyes, skin, and skeletal muscles, in which they are lodged [10]. Oncospheres within these tissues separate to form metacestodes, which progress through several developmental stages before establishing as cysticerci. The initial form is referred to as the "viable or vesicular stage," because it is at this point that each oncosphere creates a membrane that creates a vesicle with clear fluid encircling the parasite head, or scolex.

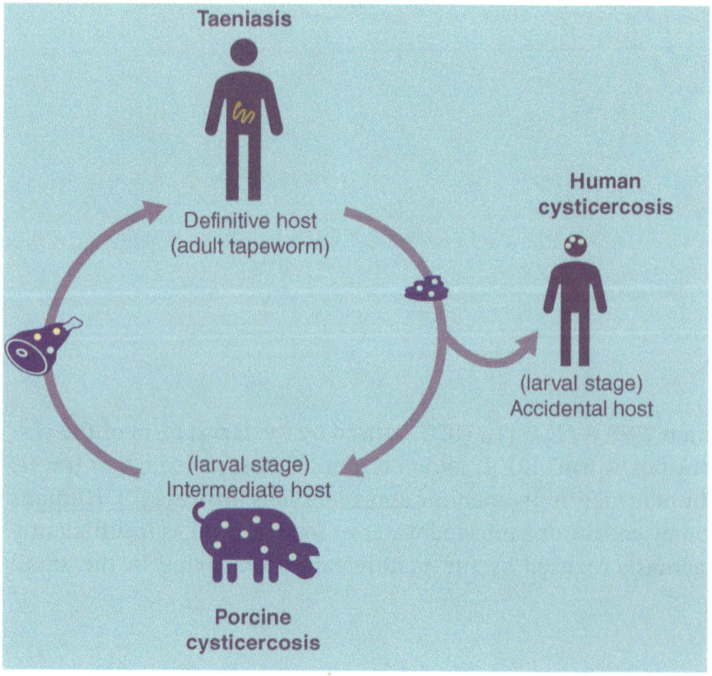

Figure 24.1 Lifecycle of *Taenia solium*. *Source:* Ref. [7] / with permission of Elsevier.

Parasites can be found in either their adult stage or the stage of larval development in the human intestines (taeniasis), brain (NCC), or muscles (cysticercosis). Similar to other helminth species, mature solium strain has a very high reproductive and ovulation rate, which increases the chance of the life cycle to continue and the species to survive. Due to all of these variables, this parasite vector is the main aim for inhibition measures [11].

24.1.2 Etiopathogenesis

The intricate life cycle of *T. solium* involves two hosts. Pigs and humans can both act as additional homes for the tapeworm cysticercus' larvae, but only humans are the tapeworm's true definitive home. During the typical cycle of transmission, the mature *T. solium* resides in the small intestine of humans where its granular stage connects it to the intestinal wall. Ultimately, a mineralized nodule representing the parasite remnants is seen [12]. The development of a fluid consisting of lymphocytes, eosinophils, multinucleated giant cells, filaments of collagen, and hyalinized parasitical walls, as a result of meningeal cysticerci, typically results in a severe inflammatory response in the subarachnoid space and aberrant leptomeningeal swelling. The chiasm of the optic nerve, cranial nerves that come from the brainstem, small penetrating arteries that emerge via the circle of Willis, and other structures could be harmed by this inflammation as it spreads throughout the body. As shown in Figure 24.2, the second scenario could lead to the vessel's lumen becoming blocked, which would then lead to the creation of a brain infarct [13]. Moreover, it has been proposed that patients with NCC experience cellular immunological dysfunction as a result of elevated subpopulations of CD8 T-lymphocytes, impaired lymphocyte proliferation, and aberrant cytokine concentrations. The documented correlation between immunodeficiency-related illnesses and the development of gliomas and neurocysticercosis has led to the hypothesis that this suppressed cellular immunity may be to blame [14].

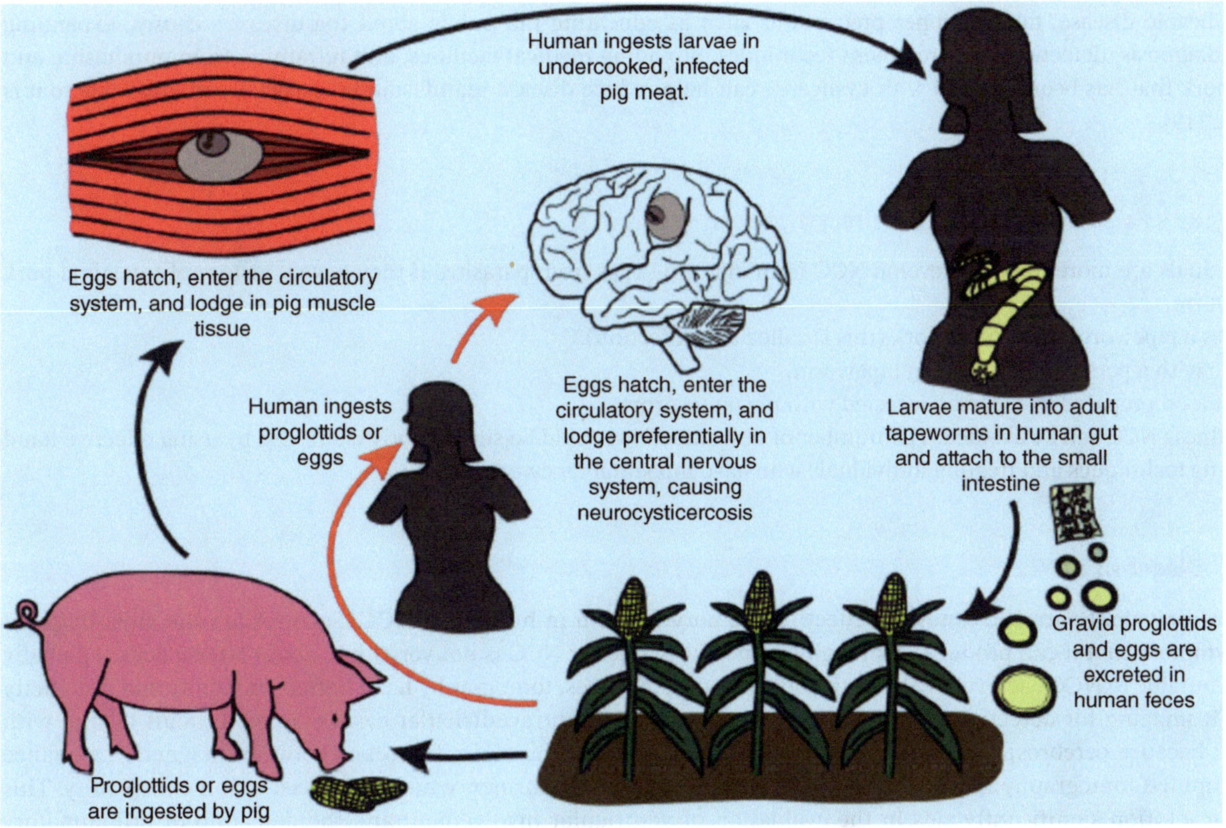

Figure 24.2 Schematic representation illustrating *Taenia solium* and its effect on human which leads to NCC. *Source:* Ref. [12] / with permission of Cambridge University Press.

24.1.3 Symptoms and Characteristics of Cysticerci

The vesicular wall and the scolex are the two basic components of cysticerci. Viable cysticerci feature a transparent cylindrical mucus, a transparent membrane, as well as a martyr invaginated scolex that resembles *T. solium* in mature state. This stage, often known as the "vesicular stage," appears after the majority of cysticerci have entered the brain and spinal cord. Given that cysticerci can remain in this condition for decades or due to a host immune reaction, they may start a degenerative process that results in the parasite's death and transformation into a mineralized nodule (the "calcified stage"). Cysticerci frequently passes through the "colloidal" and "granular" stages of involution on their way from the vesicular to the calcified state. Assuming that some parasites could be eliminated right away after invading the nervous system of the body (even in their early metacestodes phase) and progress to NCC, the nervous-system form of cysticercosis is frequently associated with nonspecific symptoms that may include diarrhea, abdominal pain, constipation, altered appetite, and autonomic signs that include seizures, headache, and increased intrusiveness, it appears that not all parasites follow this orderly process of degeneration [4, 5, 15]. Focal neurological impairments and long-lasting neurological consequences like epilepsy and hydrocephalus are possible [16]. Due to the death of rigorous population-based investigations, the incidence of the illness in the circumstances of India is not properly described. India has a high prevalence of all risk factors and taeniasis-transmission indicators [17].

24.1.4 Epidemiology

As systematic population-based research is rare in the majority of the nation, it is challenging to determine the true disease burden in India. Yet, new CT and MRI findings indicate that the incidence of NCC in India is startlingly high, particularly among the poorer, meat-eating sectors of the population who either do not receive intervention or have an intervention gap of more than 90% [18]. Kerala has a very low epidemic rate, which is ascribed to the state's high literacy rate, exclusively vegetarian population, and strict sanitation laws. Because a majority of the people in Jammu and Kashmir are Muslim and do not eat pork for religious reasons, the prevalence is also low there. Since cysticercosis is a treatable, curable, preventable, and eradicable disease, taking proper precautions such as educating the public about the disease's causes, expanding disease diagnosis, detection, and treatment techniques, providing medical facilities, and refraining from purchasing and eating pork that has been infected with cysticerci, can help reduce disease manifestation, especially in areas where it is endemic [19].

24.1.5 Risk Factors for Acquiring Neurocysticercosis

- Individuals are more likely to develop NCC from ingesting eggs from parasites if they have a history of intestinal pork tapeworm.
- Possess a tapeworm invasion in pork (this is called autoinfection).
- Reside with a person who has a pig tapeworm.
- Ingest food prepared by a person infected with pig tapeworms.

  The illness NCC can be avoided. The number of new infections could be significantly decreased by using effective hand washing techniques and treating individuals who have intestinal tapeworms [20].

24.1.6 Diagnosis of NCC

The most prevalent parasitic condition affecting the nerve system in humans is NCC. Seizures are the most frequent brain symptom that it can produce. Although the mortality rate for NCC is not very high, about 50 000 deaths globally are attributable to NCC each year [21]. In the majority of NCC cases, tomography has a better sensitivity and specificity than MR imaging for detecting hardened granulomas. However, intraventricular cysts may be difficult to find with imaging because cerebrospinal and cyst fluid attenuate similarly. The main difference between magnetic resonance and computed tomography is the greater contrast resolution of the former, which enhances lesion conspicuity. This higher resolution significantly aids in the evaluation of ventricular involvement and the detection of inflammatory changes [22].

Diagnosis mainly involves–:

24.1.7 Neuroimaging

24.1.7.1 Parenchymal NCC

The cerebral hemispheres are usually affected by NCC, which typically shows up in the parenchyma of the brain. The cerebellum, brainstem, and basal ganglia may be involved too. Most likely as a result of the larvae being deposited in the terminal small vessels of these areas, the lesions are frequently located at the gray-white matter intersection. However, the majority of recent parenchymal disease is still treated as a unique form of NCC in therapeutic and radiologic studies [23].

24.1.7.2 Subarachnoid NCC

The three subarachnoid cysticercosis varieties are as follows – Cysticerci in the cerebral convexities' gyri are present in some individuals. Active parenchymal NCC is comparable in appearance, clinical presentation, and pathogenesis to this condition. Giant cysticerci are cysticerci that can grow to be several centimeters in diameter when discovered in fissures, particularly the Sylvian fissure. Similar to parenchymal cysts, isolated cysticerci in the fissures have an outlook, and many symptoms are brought on by inflammation of the parenchyma. However, the cysts occasionally grow to the point where they alter the midline. Cysticerci in the parenchyma or basilar cisterns frequently accompanies giant cysticerci [24].

24.1.7.3 Ventricular NCC

In comparison to the parenchymatous form, 15–54% of individuals have intraventricular and subarachnoid (cisternal) forms of NCC, which have more aggressive clinical feature presentation. The choroid plexus is how the oncosphere enters the ventricular chamber. 38% of patients arrive with a fast clinical decline. The intraventricular type progresses quickly clinically and necessitates immediate treatment. Most people experience it in isolation, but in 24% of cases parenchymatous disease may be connected [25].

24.1.7.4 Spinal cord NCC

Magnetic resonance scanning is the diagnostic test of choice for evaluating cerebrospinal NCC because it provides noninvasive multiplanar images of a significant part of the spinal cord, the cauda equina, and any potential intraspinal pathological entity. Because the contents are frequently isointense to CSF while the tumor itself cannot be removed, the cyst wall is seen on weighted MR scans. These lesions can best be seen on magnetic resonance imaging using weighted sequences; the cyst's signal intensity increases, probably due to a greater protein content. Additionally, pericystic edema in the spinal cord tissue can be seen using -weighted MR imaging. On MR images, the real scolex can occasionally be seen as a mural nodule [10, 24]. Small subarachnoid spinal NCC tumors can be found using myelography and post myelography CT [26].

24.2 Immunologic Diagnosis

24.2.1 Neuroimaging

Neuroimaging continues to be the gold standard in the diagnosis of cysticercosis despite new developments in serodiagnosis. Degenerating parenchymal lesions can be identified on images from CT and MRI as nodular or ring enhancing spots. However, MRI is more accurate for scolex demonstration (which increases specificity of diagnosis) and for detecting extra parenchymal cysticerci (e.g. in the ventricles or in the cisterns). As cysticercosis becomes more prevalent in endemic areas, MRI is expected to play a bigger part in the diagnosis of the disease. It was briefly addressed in a recent case study how cysticercosis cysts and brain abscesses can be distinguished using diffusion-weighted imaging (DWI). It can be difficult to locate cysticerci within the ventricular system or in the subarachnoid regions using neuroimaging because the cyst fluid in a viable cysticercosis lesion is usually isointense with CSF [27]. Since they only react with serum from individuals with NCC. The ability to diagnose NCC earlier may result from the identification of immune complexes or circulating parasite proteins in CSF samples [28]. NCC neuroimaging features– (i) In a vesicular cyst in a contrast-enhanced CT image without scolex (ii) Vesicular tumor in weighted MRI without scolex (iii) In gadolinium-enhanced weighted MRI, ring-enhancing tumors represent colloidal cysticerci. (iv) Brain calcifications in the parenchyma on a non-enhanced CT image [29].

24.2.2 Muscle and Cranium X-Rays

These could occasionally display calcifications, which are frequently cigar shaped. Because the yield is so low, X-rays should only be used as a supplement to other diagnostic methods and not as a confirmatory test for NCC. In the juvenile age range, cranial sutures may separate. A head CT scan is necessary to identify NCC. Calcifications, which are readily detected by CT scans, are present in about 50% of cases of NCC. A single or several round tumors that resemble cysts can have an eccentric nodule that stands in for the scolex. When an inflammatory response or granuloma formation begins, rings are enhanced. Various kinds of NCC that are encephalitic may have significant edema as well as perilesional edema. However, ventricular and cisternal forms of NCC are much more difficult to identify using a CT scan. MRI is the most effective modality for pinpointing the lesion and determining the parasite's evolutionary state. It is very effective at detecting cysts in the eyes, bones, base of the brain, brain stem, and cerebellum. It can also detect very slight perilesional edema. On a T-1 weighted MRI, the scolex shows up as a mural nodule with strong signal strength. On a T-1 weighted MRI, a degenerating cyst has cloudy fluid that shows up as a high signal intensity lesion. In the granulo-nodular stage, there is varying perilesional edema and ring enhancement on gadolinium injection. On a CT image, calcified cysts can best be seen [25]. To allow consistent diagnosis of human NCC, experts developed recommendations for diagnosis and divided them up into three distinct groups– Absolute, MRI, and clinical/exposure are the first three. Absolute requirements include the parasite being clearly visible through imaging of recognizable cysts harboringscolex, sub-retinal cysts, pathology examinations, or sub-retinal cysts' presence. Patients which meet these requirements can get a precise diagnostic, prognosis, and treatment plan. Despite the fact that histological confirmation is the "gold standard" for diagnosing NCC, it is not the most suggested method because it requires invasive surgical operations. The three categories of neuroimaging criteria are (i) main, (ii) confirmative, and (iii) minor. The presence of enhancing lesions, calcifications, and cystic lesions without a distinctive scolex are important factors. Confirmatory criteria include migrating ventricular cysts seen on consecutive neuroimages, the disappearance of cysts after anti-parasitic therapy. Leptomeningeal enhancement and hydrocephalus are minor neuroimaging factors. Antigens from the metacestode in conjunction with epidemiological factors like proof of *T. solium* infection in the home or residing in endemic areas [30]. The determination of vague medical symptoms, such as seizures, which frequently exhibit distinctive results on brain CT or MRI scans, as well as the utilization of specific serological tests, are required for the diagnosis of cysticercosis of the CNS. Although objective clinical, imaging, immunological, and epidemiological data-based diagnostic criteria have been suggested, they are not typically applied in regions where the disease is endemic. The identification of *T. solium* seropositivity requires serology. However, there is a significant lacuna between a positive serology and the evaluation of an NCC diagnosis. A positive *T. solium* serology may persist years after the infection has ended and is not always indicative of a neurological location [21].

24.2.3 Immunologic Diagnosis

ELISA (enzyme-linked immunosorbent assay) or an immunoblot are commonly used for the initial antibody identification in the laboratory diagnosis of cysticercosis. Antibody testing for *T. solium* is a popular technique for diagnosing infections, but it has many drawbacks because it may be difficult to identify a single cyst carrier. Western blots and ELISA are the two main commercially accessible tests. The "gold standard" test for finding particular antibodies against *T. solium* is a western blot [31]. Three factors were initially chosen as "absolute," as they are the only ones that can be used to diagnose NCC with 100% confidentiality using just one criterion– The parasite can be seen histologically after the biopsy of a brain or spinal cord injury. The appearance of parasitic membranes or microscopic visualization of all neuroimaging forms of cysticerci presentation. Fundoscopic inspection enables direct subretinal parasite visualization. Patients accompanied by subretinaltumors are believed to have NCC because the retina is regarded as a component of the CNS. These cysts typically have a central dark spot that stands in for the scolex and are yellowish in appearance [32]. When looking for specific antibodies against *T. solium* larvae to diagnose NCC, EIA standardized with a crude extract, and the ETIBI, human serum and blood were used to compare with dvconrotein immune system antibodies. In EIA, saliva had a marginally higher sensitivity (82.1%) than blood (74.1%). Serum was significantly more responsive to EITB (100%) than saliva (70.4%). On the other hand, using saliva and EIA together is a good combination for diagnosis, particularly epidemiology [33]. ELISA was used to compare brain CT results in cases of suspected NCC with cysticercus-specific IgG antibody levels and to determine ELISA's clinical relevance. Brain CT and ELISA tests were performed on 123 individuals with neurologic conditions. In confirmed instances, ELISA's sensitivity and specificity were 94% and 91%, respectively. Patients were found to be positive by ELISA when they had multiple cysts without enhancement on a CT scan, which were believed to represent viable cysticerci. The ELISA test was also positive for those with enhancing cysts, weakly defined low densities, or enhancing nodules, which indicated cysticerci on degeneration. By ELISA, those with numerous spotty calcifications, which signifies cysticerci in the dormant stage, were found to be negative. The above-mentioned mixed CT findings, which show the patient's different stages of infection, were tested positive by ELISA. Others, believed to be additional inflammatory granulomas, had only enhancing nodules with encircling edema but tested negative by ELISA. Patients can be positively identified as having ventricular cysticercosis if they have only hydrocephalus on a CT scan and are ELISA positive. The findings suggest that ELISA is helpful as an additional diagnostic tool, particularly in patients with ambiguous NCC CT findings [34].

24.2.4 Prognosis

In general, if sufficient symptomatic therapy is given, parenchymal NCC appears to have a benign prognosis. Antiparasitic drugs should be used to remove cysts to reduce the chance of further seizure relapses and the development of illness. On the other hand, extra parenchymal NCC is not innocuous. Both subarachnoid NCC and intraventricular cysts have the potential to advance to the point of killing the patient, causing encephalopathy that is hindered with unexpected mortality. In the past, it was thought that deadly cases of subarachnoid NCC accounted for 20–50% of all cases. Despite being under the most favorable managerial circumstances, this kind of NCC still entails substantial risks, even though these rates are probably much lower now [35].

24.2.5 Diagnostic Criteria and Degrees of Diagnostic Certainty for Neurocysticercosis

The diagnosis of NCC is made using a variety of data, including epidemiology, symptoms, signs, serology, brain imaging, and pathology, if available. Although some findings are therapeutic, it is not always the case. The mixture of data that supports the diagnosis is so unusual in other disease states that it is virtually certain that NCC is the cause. Although useful, published diagnostic criteria have not been systematically validated [36]. In CSF cisterns, cystic lesions frequently cover large areas, displace nearby structures, and have a multilobulated appearance. In individuals with subarachnoid NCC, magnetic resonance imaging is a helpful benign diagnostic method to demonstrate regional cerebral artery narrowing or blockage [37]. Especially for arachnoiditis, extra parenchymal locations are primarily linked to inflammatory changes in the CSF. In individuals with NCC, electroencephalographic (EEG) findings have little bearing on symptoms and CT lesions. In one research, 30–50% of patients with NCC-related seizures had abnormal EEGs. Only 15–30% of patients have been found to have a favorable correlation between CT lesions and localizing or lateralizing EEG abnormalities. Similar to this, there is a 7–20% correlation between the sort of seizure and EEG abnormalities since residual calcifications from cysticercotic lesions presumably have low epileptogenicity. The enzyme EITB assay, which employs LLGP to identify serum *T. solium* antibodies, is the serological test with the finest body of research. People who have more than one positive serology if they have more than one active cyst or subarachnoid illness at the time of testing because EITB sensitivity is around 98% for patients with two or more live parasites in the nervous system. Heterologous viruses and EITB do not interact. When a patient's serology results are negative, other diagnoses should be looked into. (90% versus 100%), the sensitivity of antibody detection by EITB appears to be marginally lower in CSF. Due to EITB's poor sensitivity (50–60%) in patients with a single intracranial cysticercus, the condition cannot be ruled out by a negative test result [1]. Anticysticercal antibody detection in blood or cerebral spinal fluid (CSF) is a crucial diagnostic component. Their use may be limited in developing nations with high infection rates due to their high expense and limited availability [4]. Clinical and neuroimaging NCC manifestations occasionally lack specificity, and immunological studies have a low positive predictive value. As a result, identifying NCC is still challenging. Furthermore, because the densities of the cyst fluid and CSF are so comparable, cyst visualization in MRI scans is unlikely, making extra parenchymal NCC diagnosis more challenging than parenchymal NCC. Since there have been fewer NCC cases in China over the past few years, a regular exam no longer includes an immunodiagnostic test. Furthermore, the use of a constant state imaging technique is uncommon. A diagnosis of NCC may be difficult if medical and common neuroimaging indications lack specificity, which precludes physicians from performing the immunodiagnostic test [38].

Numerous illnesses are present in NCC. A diagnosis, which requires clinical suspicion, is commonly made by combining scans, symptoms, and serologic tests. Early management of NCC symptoms, such as seizure control, cerebral edema management, and intracranial pressure decrease, should be the main focus of care. Instead of active parasites, NCC symptoms are a result of the host's reaction. The benefit of albendazole in reducing lesions on imaging and in reducing generalized seizures when viable cysticerci are present has been shown in two placebo-controlled studies. Diagnosis of cysticercosis can be difficult. The majority of cases requires a combination of neuroimaging findings, exposure history, and serologic confirmation, even though the parasite is rarely visible immediately in the eye [39].

24.2.6 Limitation and Prospects for Improvement in Neurocysticercosis

Images are frequently inconclusive, and distinctive imaging results could be imitated by different circumstances. The most effective immunodiagnostic tests are not widely accessible, and each method has limitations. To standardize immunodiagnostic internationally, simpler, more accurate tests based on synthetic antigens or molecular diagnosis are required. Making an accurate and timely diagnosis of NCC is difficult for a variety of reasons, in addition to the restricted availability of diagnostic tools. Patients might not always have access to sophisticated diagnostic methods or limited access to medical care in endemic areas. Additionally, attending physicians who are unfamiliar with NCC may become perplexed by the wide variation in the lengthy pre-patent era and the present symptoms [40].

24.3 Management Approaches by Neurocysticercosis Form

24.3.1 Management of Neurocysticerosis

Depending on the NCC variety, the course of treatment should be adjusted. An NCC patient's doctor must always keep in mind that treatment involves a mix of symptomatic and antiparasitic treatments, such as the use of painkillers, AED, cysticidal medications, operative removal of lesions, and the installation of ventricle shunts [41]. NCC treatment is debatable. Most types of NCC have few natural history studies, so until relatively recently, the benefits of treatments were not founded upon studies that were well-designed. Despite this insufficient supporting data, clinical findings show that different forms or types of NCC require different treatment strategies. The anatomical location of cysts, their stage of evolution, the number of cysts, the level of associated inflammation, the size, and the severity of symptoms are all factors that are taken into account when choosing treatments. Treatment choices are frequently made more difficult by the existence of multiple cysts in various anatomical locations and at various stages of evolution. There are various kinds of treatments used, just like with many complex diseases that develop over time. Although the majority of studies have concentrated on treatment with the cysticidal drugs, albendazole and praziquantel, other treatments are also used and are crucial. These include the use of antiepileptic drugs (AEDs), surgical procedures, general supportive measures, and the use of corticosteroids and other immunosuppressive/anti-inflammatory drugs to regulate harmful and possibly harmful host inflammatory responses. Despite being frequently used to manage signs and symptoms, corticosteroids have not been thoroughly researched. However, long-term treatment is frequently necessary in complicated NCC, and side effects are frequent, severe, and even life-threatening. Short-term use is linked with transient side effects that are typically tolerated [42].

Consumers with NCC can choose from a variety of therapy options, such as–

1) Cysticants, which destroy larvae.
2) Corticosteroids, which lessens or stops an inflammatory response.
3) Anti-seizure medication, which is used to prevent or reduce the occurrence and severity of seizures.
4) Operation for the removal of the tumor or placement of a cerebrospinal device. This option was chosen after analysis of the available data and the publication of consensus treatment recommendations for NCC [43].

24.3.2 Role of Antiparasitic Drugs

Since the previous 30 years, antihelminthic drugs have been prescribed but their therapeutic efficacy is debatable. Praziquantel (PZQ) was used for the first time in cases of infection in Mexico, and ALB were utilized for the first time in China [44, 45]. ALB and PZQ are only helpful for live lesions, according to research. When compared to the use of AHD medications alone, ALB and PZQ together were discovered to be more effective at addressing AHD [46]. The results of the trial, which evaluated the combined use of AHD and individual ALB treatment in children with focal seizures, were not very significant [47]. According to a different study, patients who receive ALB treatment alone recover at a rate of 36%, whereas patients who receive ALB and PZQ treatment together recover at a rate of 64% [48]. The study, which was based

Table 24.1 Intervention of antiparasitic regimen for NCC.

Sr. no	Types of NCC	Intervention	Reference
1.	Parenchymal NCC	Between 60% and 85% of parenchymal brain cysticerci are killed by the antiparasitic drugs praziquantel and albendazole, and most studies indicate that albendazole administration results in a higher cost reduction. However, the majority of these investigations have been unstructured, casual imaging research. The majority of studies assessed praziquantel at a dose of 50 mg/kg/day for two wks, but there have also been studies reporting a single-day regimen	[52]
2.	Extra parenchymal NCC	The treatment of subarachnoid disease has not been the subject of any regulated studies. A study of 33 patients with giant cysticerci in the Sylvian fissure treated with albendazole (15 mg/kg/day for four wks) found only one death from aplastic anemia at 59 mo, with patients needing multiple courses of therapy. In a series of patients treated with only CSF diversion, 50% died at a median follow-up of 8 yrs and 11 mo. Therefore, a single run of treatment for patients with subarachnoid disease is probably insufficient, and some patients may need long-term therapy (months)	[53]

on treatment recommendations for parenchymal cysts, discovered that ALB therapy, is presumably effective at lowering the frequency of long-term seizures and the quantity of radiologically detectable cysts in both adults and children, whether corticosteroids are used or not [49]. Praziquantel is less effective against extra parenchymal forms and cannot reach the CSF as well as albendazole. Albendazole is also more commonly prescribed. These medications may make high blood pressure with cysticercosis meningitis worse in some individuals. In patients with NCC, other medications such metrifonate and flubendazole have been tested, but their usage has been discontinued due to unfavorable side effects or a lack of efficacy. Ivermectin may be beneficial in some NCC patients, according to one study [50]. A benzimidazole molecule called oxyfendazole has been shown to be effective in treating swine cysticercosis; if it can be used safely in people, it could replace the cysticidal medications currently in use [51]. As indicated in Table 24.1, even though positive outcomes are seen in numerous clinical studies, the use of cysticidal medications has come under scrutiny, which has caused confusion and led to poor management choices. For instance, it has been proposed that while cysticidals kill the cysts, they do not alter the disease's clinical trajectory [54].

24.3.3 Complications of Anti-Parasitic Drugs

Cysticidal medications can kill cysts, but the inflammatory response that follows may be harmful, manifesting either acutely as swelling in the brain or chronically as a gliotic scar [4]. Some medical professionals argue against using cysticidal drugs and suggest symptomatic therapy (AED) and/or surgery to remove the cysts in order to prevent complications from a rise in pressure inside the skull, seizures, or epileptic scar. While mannitol can lower intracranial pressure, AED medications usually control seizures in individuals with calcified cysticerci adequately. When combined with cysticidal medications, steroids can reduce edema brought on by the lesions and the inflammatory reaction linked to the removal of viable cysts [3].

24.3.4 Role of Corticosteroids

Steroid administration is crucial in controlling the acute inflammatory response that follows the destruction of viable cysts that is triggered by NCC-related inflammation in the CNS. Dexamethasone is used most frequently, at doses that vary from 4.5 to 12 mg daily. When lasting steroid therapy is needed, prednisone at 1 mg/kg/day may be used rather than dexamethasone. For chronic cysticercosis arachnoiditis or encephalitis, where up to 32 mg of dexamethasone per day is required to reduce the brain edema accompanying this condition, corticosteroids are frequently used to reduce neurological symptoms resulting from the death of the parasite. Severe intracranial hypertension brought on by NCC is another condition for which mannitol is prescribed at dosages of 2 g/kg/day. It is suggested to taper gradually because abrupt steroid discontinuation may cause inflammation to return. Although there aren't any clear indications and there aren't many instances, methotrexate has been used successfully as a steroid substitute [55].

24.3.5 Role of Anti-Epileptic's Drugs

Seizure activity is more common in patients with declining cysts and rare in patients with only viable, noninflamed cysts; seizures are the most distinctive symptom of parenchymal brain cysticercosis, and they are possible in any phase of the parasite, from viable noninflamed cysts, deteriorating cysts, to calcified lesions [43]. It is unclear when and if an epileptogenic circuit forms around a cysticerci, but patients frequently experience recurrent seizures of the same kind, sometimes years after the parasitic larvae have died and become calcified. One frequently referred to clinical symptom of parenchymal NCC is seizures. Anti-epileptic drug therapy may be necessary in instances with obvious symptoms and multiple viable cysts. Many times, drugs like phenobarbitone and carbamazepine are effective at managing seizures associated with NCC [56]. Up to 50% of patients getting anti-epileptic therapies for NCC experienced relapses in seizure control following drug withdrawal, according to one study. This finding suggests that the therapy was having an inhibitory impact on seizures that were sparked by the presence of cysts. Once more, the use of these medications has generated dispute with certain research indicating no change in seizure result [57].

24.3.6 Surgical Management for Neurocysticercosis

A consensus statement on the treatment of NCC was released by a group of international cysticercosis experts. For NCC, they evaluated the following surgical management options–

1) Extra parenchymal NCC
   a) Ventricular tumors
   b) Racemose tumors are the cause of hydrocephalus
   c) NCC-induced ependymitis-induced hydrocephalus
2) Intramedullary and extramedullary spinal cysticercosis [42]

Excision may be necessary due to a solid impact, blocked hydrocephalus, a fourth ventricular cyst, or an unclear diagnostic. Ventricular tumors are surgically removed using endoscopy or microsurgery. When ventricular cysts are the cause of acute hydrocephalus, a ventriculostomy must be performed immediately to lower the elevated cerebral pressure. The cyst must then be removed. If there is concomitant ependymitis, a ventriculoperitoneal shunt must be implanted. According to reports, shunt failure rates can reach 30–70%. Patients might therefore need numerous shunt revisions, which have a high mortality risk. It has been documented how to remove fourth-ventricular cysts through a posterior fossa craniotomy and how to remove lateral or third-ventricular cysts through a supratentorial open or stereotactic craniotomy. Endoscopic methods, however, are now more frequently used as a form of therapy. The neuroendoscopic method has the benefit of keeping ventricle navigation straightforward and allowing the removal of numerous cysts with few complications following the procedure. Intraventricular bleeding is one of the neuroendoscopic approach's possible drawbacks, and this procedure may be dangerous if there are ependymitis and dense adhesions present [58]. In order to alleviate pressure and remove cysts, surgical intervention is typically urgently necessary in intraventricular NCC. The standard course of treatment includes the installation of an emergency shunt and the removal of the cyst through an open craniotomy to enable for antihelmintic therapy, which is typically ineffective. The surgical procedure of preference, neuroendoscopic cyst resection with ventriculostomy, and placement of an external ventricular drain or permanent shunt, may be curative without the use of an anthelmintic and may result in fewer complications and higher cure rates. Although it is still required in patients with severe disorder and ependymitis, ventriculostomy performed at the time of the operation can frequently relieve hydrocephalus without the need for outer drainage, which poses a risk of microbial infection [59]. Surgical NCC resection may significantly enhance the patient's quality of life and boost PFS. Because MRI frequently misses some intraventricular NCC, examination of all ventricles and basal cisterns is essential. A stiff endoscope, however, struggles to access the intracranial CSF system's more difficult-to-reach regions. In the bilateral occipital and temporal horns of the lateral ventricle, the third and fourth ventricles, and the basal cistern, the use of a flexible endoscope can lessen traction on the brain and make it simpler and shielded to find and reset NCC. Therefore, using both stiff and flexible endoscopes simultaneously can increase the efficiency of ventriculoscopic incision. There are some restrictions on the bendable endoscope though. First, compared to rigid endoscopes, their imaging resolution is weaker. Additionally, because of its narrow working channel, some tools cannot be used. For instance, managing intraoperative bleeding is challenging because the channel can only be used for low-power monopolar coagulation [60].

The signs that spinal neurocysticercosis should be surgically treated include extremely intense indicators, symptom irritation, failure of medical therapy, and acute worsening of neurological signs while undertaking therapy. The aim of the surgery is to detach as much of the injury as feasible without causing neurological deterioration. Arachnoiditis can occasionally cause cystic lesions to exhibit significant adhesion to the tissues around them. This severe arachnoiditis adhesion may obstruct the flow of CSF, necessitating a shunt or duroplasty to reform the CSF. Cystic migration may take place, and the lesion seen on an MRI or CT scan may not match the lesion experienced during operation. Therefore, if there is a significant gap in time between the MRI or CT and the surgery, another imaging procedure should be explored [61]. If a ventricular cyst is adherent or surgery is not a possibility, high-dose corticosteroids with anthelminthics for one to three months or longer may be explored. The ongoing inflammation that results in swelling and hydrocephalus may be suppressed by taking anti-inflammatory medications like corticosteroids or methotrexate on a regular basis. Medical treatment is available for lateral ventricle non-obstructing cysts, but entrapment disorders can occasionally occur. It is unknown whether a second inflammatory response to leftover cysts, membranes, or antigens causes hydrocephalus to develop later. A comparatively large number of patients with fourth ventricular lodged cysts were successfully treated, suggesting that medical therapy, which occasionally required a shunt to control hydrocephalus, could be successfully used. However, the majority of experts advise endoscopic or surgical excision because even patients who have previously undergone successful treatment for cysts in the fourth ventricle occasionally need additional shunts and the long-term benefit is unclear [62]. This approach has several advantages, including the immediate removal of the obstruction, reducing inflammation through elimination of the source of antigen, and a lower chance of developing blocked and communicating hydrocephalus due to elimination of the cause of inflammation [63]. In the preponderance of cases of spinal NCC, surgery is necessary. Acute neurological deterioration or the existence of severe, progressing symptoms are indications for surgical action. In most accounts on surgical management of racemose cysticercosis, the cysts are removed using a microsurgical technique. Because of arachnoid scarring, the excision of extramedullary tumors is frequently challenging [64].

24.3.7 Issues and Challenges for Surgery of Neurocysticerosis

With antiparasitic treatment, cysticercosis of the retina or vitreous may get worse. Therefore, before beginning antiparasitic treatment, a fundoscopic examination should be done (ideally an indirect fundoscopy with pupil dilation). Prolonged treatment with steroids can also raise the risks of TB and strongyloidiasis, which are more prevalent in groups at risk for NCC. Since tapeworm carriers can also contract NCC, some people have suggested testing patients and their close acquaintances for tapeworms. The yield of screening is incredibly poor overall. However, certain patients should be screened, for example cases that were acquired outside of endemic regions. Only a small portion of illnesses are found by stool microscopy. Although more sensitive, copro-antigen tests and serologic tests for tapeworm-stage antigens are at present only accessible in labs for research [40].

24.4 Conclusion

NCC is a significant problem in global health that is still not fully known. While researching NCC in people has yielded valuable insights into the condition, there are inherent barriers that can only be surmounted by using further improved animal models. On the disease mechanisms in NCC, model systems using *T. solium*, *T. crassiceps*, and *M. corti* in mice, rats, pigs, and even rhesus monkeys provided invaluable data. But animal models of NCC that cause convulsions and epilepsy are urgently required. Basic practices like using toilets and sanitary facilities and sticking to demanding food hygiene regulations must be improved. To stop the spread of disease on a national scale, pig farmers and breeders ought to promote the vaccination of pigs. To take the place of the expensive radiological examination in emerging nations, low-cost diagnostic serological tests are required. Although CSF is a rich reservoir of antigen and the antibody-based ELISA test uses specific antigen, it is risky to isolate the antigen from the human body and requires lumbar puncture, which is unpopular with patients. Considering MRI tests, patients from developing nations cannot purchase them, nor are advanced technologies easily accessible in far-flung locations. Antihelminthic pharmaceuticals ALB and PZQ are very effective in treating patients with parenchymal cysts, but they function poorly in cases of intraparenchymal cysts. More time is required to create novel medications that are more efficient in treating extra parenchymal NCC. Another problem with using corticosteroids as anti-inflammatory therapies in NCC is that they have other side effects from long-term use, so anti-inflammatory medications with few adverse reactions are required for NCC therapy.

References

1 Del Brutto, O.H. (2012). Neurocysticercosis: a review. *Sci. World J.* 2012: 159821.
2 Fogang, Y.F., Camara, M., and Ndiyaye, M.M. (2012). Cerebral neurocysticercosis mimicking or comorbid with episodic migraine? *BMC Neurol.* 14: 138.
3 Gripper, L.B. and Welburn, S.C. (2017). Neurocysticercosis infection and disease – a review. *Acta Trop.* 166: 218–224.
4 Garcia, H.H., Nash, T.E., and Del Brutto, O.H. (2014). Clinical symptoms, diagnosis, and treatment of neurocysticercosis. *Lancet Neurol.* 13: 1202–1215.
5 Leon, A., Saito, E.K., Mehta, B. et al. (2015). Calcified parenchymal central nervous system cysticercosis and clinical outcomes in epilepsy. *Epilepsy Behav.* 43: 77–80.
6 Rajshekhar, V., Joshi, D.D., Doanh, N.Q. et al. (2003). *Taenia solium* taeniosis/cysticercosis in Asia: epidemiology, impact and issues. *Acta Trop.* 87: 53–60.
7 (2021). *Guideline for Preventive Chemotherapy for the Control of Taenia solium Taeniasis*. Washington (DC): Pan American Health Organization.
8 Del Brutto, O.H. (2014). Clinical management of neurocysticercosis. *Expert. Rev. Neurother.* 14 (4): 389–396.
9 Garcia, H.H., Gonzalez, A.E., and Gilman, R.H. (2020). *Taenia solium* cysticercosis and its impact in neurological disease. *Clin. Microbiol. Rev.* 33: e00085–e00119.
10 Nakao, M., Lavikainen, A., Yanagida, T., and Ito, A. (2013). Phylogenetic systematics of the genus Echinococcus (Cestoda: Taeniidae). *Int. J. Parasitol.* 43: 1017–1029.
11 Coral-Almeida, M., Gabriel, S., Abatih, E.N. et al. (2015). *Taenia solium* human cysticercosis: a systematic review of sero-epidemological data from endemic zones around the world. *PLoS Negl. Trop. Dis.* 9: 1–20.
12 de Lange, A., Mahanty, S., and Raimondo, J.V. (2019). Model systems for investigating disease processes in neurocysticercosis. *Parasitology* 146 (5): 553–562.
13 Elliott, E.J. and Landaker, E.J. (2017). Worsening migraine due to neurocysticercosis. *Cleve. Clin. J. Med.* 84: 196–198.
14 Fogang, Y.F., Camara, M., Diop, A.G., and Ndiaye, M.M. (2014). Cerebral neurocysticercosis mimicking or comorbid with episodic migraine? *BMC Neurol.* 1 (14): 138.
15 Carabin, H., Ndimubanzi, P.C., Budke, C.M. et al. (2011). Clinical manifestations associated with neurocysticercosis: a systematic review. *PLoS Negl. Trop. Dis.* 5: e1152.
16 Carpio, A. (2002). La neurocisticercosisen el Ecuador: unnuevoparadigmaepidemiológico y clínico de la cisticercosis cerebral. In: *El Cóndor, la Serpiente y el Colibrí. La OPS/OMS y la SaludPúblicaen el Ecuador del Siglo XX* (ed. R. Fierro, C. Hermida, E. Granda, et al.), 283–288. Quito: OPS/OMS.
17 Prasad, K.N., Prasad, A., Verma, A. et al. (2008). Human cysticercosis and Indian scenario: a review. *J. Biosci.* 33: 571–580.
18 Ahmad, R., Khan, T., Ahmad, B. et al. (2017). Neurocysticercosis: a review on status in India, management, and current therapeutic interventions. *Parasitol. Res.* 116: 21–33.
19 Prasad, K.N., Prasad, A., Gupta, R.K. et al. (2007). Prevalence and associated risk factors of *T. solium* taeniasis in a rural pig farming community of north India. *Trans. R. Soc. Trop. Med. Hyg.* 101: 1241–1247.
20 Herrick, J.A., Maharathi, B., Kim, J.S. et al. (2018). Inflammation is a key risk factor for persistent seizures in neurocysticercosis. *Ann. Clin. Transl. Neurol.* 5 (5): 630–639.
21 Malik, A.A., Ashraf, U., and Abdali, N. (2014). Role of immunological markers in the diagnosis, treatment and follow-up of human neurocysticercosis. *J. Neuroinfect. Dis.* 5 (156): 2.
22 Kimura-Hayama, E.T., Higuera, J.A., Corona-Cedillo, R. et al. (2010). Neurocysticercosis: radiologic-pathologic correlation. *Radiographics* 30 (6): 1705–1719.
23 Del Brutto, O.H. (2014). Neurocysticercosis. *Handb. Clin. Neurol.* 121: 1445–1459.
24 Zhao, J.L., Lerner, A., Shu, Z. et al. (2015). Imaging spectrum of neurocysticercosis. *Radiol. Infect. Dis.* 1 (2): 94–102.
25 White, A.C. Jr. (2000). Neurocysticercosis: updates on epidemiology, pathogenesis, diagnosis, and management. *Annu. Rev. Med.* 51 (1): 187–206.
26 Sinha, S. and Sharma, B.S. (2009). Neurocysticercosis: a review of current status and management. *J. Clin. Neurosci.* 16 (7): 867–876.
27 Alsina, G.A., Johnson, J.P., McBride, D.Q. et al. (2002). Spinal neurocysticercosis. *Neurosurg. Focus.* 12 (6): 1–7.
28 Del Brutto, O.H. (2022). Human neurocysticercosis: an overview. *Pathogens* 11 (10): 1212.
29 Yancey, L.S., Diaz-Marchan, P.J., and White, A.C. (2005). Cysticercosis: recent advances in diagnosis and management of neurocysticercosis. *Curr. Infect. Dis. Rep.* 7 (1): 39–47.

30 Flisser, A. (1988). Neurocysticercosis in Mexico. *Parasitol. Today* 4 (5): 131–137.

31 Gadea, N.A., Matamoros, G., and Rueda, M.M. (2018). Recent advances in the diagnosis of neurocysticercosis. *Curr. Treat. Options Infect. Dis.* 10: 410–420.

32 Butala, C., Brook, T.M., Majekodunmi, A.O., and Welburn, S.C. (2021). Neurocysticercosis: current perspectives on diagnosis and management. *Front. Vet. Sci.* 8: 615703.

33 Del Brutto, O.H. (2012). Diagnostic criteria for neurocysticercosis, revisited. *Pathog. Glob. Health* 106 (5): 299–304.

34 Feldman, M., Plancarte, A., Sandoval, M. et al. (1990). Comparison of two assays (EIA and EITB) and two samples (saliva and serum) for the diagnosis of neurocysticercosis. *Trans. R. Soc. Trop. Med. Hyg.* 84 (4): 559–562.

35 Garcia, H.H. (2018). Neurocysticercosis. *Neurol. Clin.* 36 (4): 851–864.

36 Del Brutto, O.H., Rajshekhar, V., White, A.C. Jr. et al. (2001). Proposed diagnostic criteria for neurocysticercosis. *Neurology* 57 (2): 177–183.

37 Garg, R.K. (2004). Diagnostic criteria for neurocysticercosis. *Neurol. India* 52 (2): 171.

38 Prasad, A., Prasad, K.N., Yadav, A. et al. (2008). Lymphocyte transformation test: a new method for diagnosis of neurocysticercosis. *Diagn. Microbiol. Infect. Dis.* 61 (2): 198–202.

39 Fei, X., Li, C., Zhang, Y. et al. (2020). Next-generation sequencing of cerebrospinal fluid for the diagnosis of neurocysticercosis. *Clin. Neurol. Neurosurg.* 193: 105752.

40 Webb, C.M. and White, A.C. (2016). Update on the diagnosis and management of neurocysticercosis. *Curr. Infect. Dis. Rep.* 18: 1–8.

41 Guzman, C., Garcia, H.H., and Cysticercosis Working Group in Peru (2021). Current diagnostic criteria for neurocysticercosis. *Res. Rep. Trop. Med.* 12: 197–203.

42 Garcia, H.H. and Del Brutto, O.H. (2005). Neurocysticercosis: updated concepts about an old disease. *Lancet Neurol.* 4 (10): 653–661.

43 Nash, T.E., Singh, G., White, A.C. et al. (2006). Treatment of neurocysticercosis: current status and future research needs. *Neurology* 67 (7): 1120–1127.

44 Xiao, Z.X., Zhao, C.Y., and Liu, L.P. (1986). Albendazole treatment in cerebral cysticercosis. *Chin. J. Integr. Med.* 25: 100–102.

45 Wu, W., Jia, F., Wang, W. et al. (2013). Antiparasitic treatment of cerebral cysticercosis: lessons and experiences from China. *Parasitol. Res.* 112: 2879–2890.

46 Romo, M.L., Wyka, K., Carpio, A. et al. (2015). The effect of albendazole treatment on seizure outcomes in patients with symptomatic neurocysticercosis. *Trans. R. Soc. Trop. Med. Hyg.* 109: 738–746.

47 Kaur, S., Singhi, P., Singhi, S. et al. (2009). Combination therapy with albendazole and praziquantel versus albendazole alone in children with seizures and single lesion neurocysticercosis: a randomized, placebo -controlled double blind trial. *Pediatr. Infect. Dis.* 28: 403–406.

48 Baird, R.A., Wiebe, S., Zunt, J.R. et al. (2013). Evidence-based guideline: treatment of parenchymal neurocysticercosis: report of the guideline development Subcommittee of the American Academy of Neurology. *Neurology* 80: 1424–1429.

49 Diazgranados-Sanchez, J.A., Barrios-Arrazola, G., Costa, J.L. et al. (2008). Ivermectinacomoalternativaterape'uticaenneurocisticercosisresistente al tratamientofarmacolo'gicoconvencional. *Rev. Neurol.* 46 (11): 671–674.

50 Singhi, P. and Suthar, R. (2015). Neurocysticercosis. *Indian J. Paediatr.* 82: 166–171.

51 Salinas, R., Counsell, C., Prasad, K. et al. (1999). Treating neurocysticercosis medically: a systematic review of randomized, controlled trials. *Tropical Med. Int. Health* 4 (11): 713–718.

52 Coyle, C.M. and Tanowitz, H.B. (2009). Diagnosis and treatment of neurocysticercosis. *Interdiscip. Perspect. Infect. Dis.* 2009: 180742.

53 Kelesidis, T. and Tsiodras, S. (2012). Extraparenchymal neurocysticercosis in the United States. *Am J Med Sci* 344 (1): 79–82.

54 Garcia, H.H., Evans, C.A.W., Nash, T.E. et al. (2002). Current consensus guidelines for treatment of neurocysticercosis. *Clin. Microbiol. Rev.* 15: 747–756.

55 Fogang, Y.F., Savadogo, A.A., Camara, M. et al. (2015). Mangingneurocysticercosis: challenges and solutions. *Int. J. Gen. Med.* 8: 333–344.

56 Escobar, A. and Weidenheim, K.M. (2002). The pathology of neurocysticercosis. In: *Taenia solium Cysticercosis. From Basic to Clinical Science* (ed. G. Singh and S. Prabhakar), 289–305. Oxon, UK: CAB International.

57 Del Brutto, O.H. (2008). Stroke and vasculitis in patients with cysticercosis. In: *Uncommon Causes of Stroke* (ed. L.R. Caplan), 53–58. New York, NY: Cambridge University Press.

58 Rajshekhar, V. (2010). Surgical management of neurocysticercosis. *Int. J. Surg.* 8 (2): 100–104.

59 Mahale, R.R., Mehta, A., and Rangasetty, S. (2015). Extraparenchymal (racemose) neurocysticercosis and its multitude manifestations: a comprehensive review. *J. Clin. Neurol.* 11 (3): 203–211.

60 Jensen, T.O. and Post, J.J. (2016). Intraventricular neurocysticercosis: presentation, diagnosis and management. *Asian Pac J Trop Med* 9 (8): 815–818.

61 Zhenye, L., Chuzhong, L., Xuyi, Z. et al. (2017). Ventriculoscopic approach for intraventricular neurocysticercosis: a single neurosurgical center's experience. *World Neurosurg.* 107: 853–859.

62 Han, S.B., Kwon, H.J., Choi, S.W. et al. (2014). Lumbar intraduralneurocysticercosis: a case report. *Korean J. Spine* 11 (3): 205.

63 Nash, T.E. and Garcia, H.H. (2011). Diagnosis and treatment of neurocysticercosis. *Nat. Rev. Neurol.* 7 (10): 584–594.

64 Yacoub, H.A., Goldstein, I., El-Ghanem, M. et al. (2017). Spinal racemose cysticercosis: case report and review. *Hosp. Pract.* 45 (3): 99–103.

25

Trichinosis: History and Current Trends

Moitreyee Chattopadhyay[1], *Sandipan Dasgupta*[1], *Ananya Chanda*[2], *Subhasundar Maji*[1], *and Shwetlana Bandyopadhyay*[3]

[1] *Department of Pharmaceutical Technology, Maulana Abul Kalam Azad University of Technology, Kolkata, West Bengal, India*
[2] *Department of Pharmaceutical Technology, School of Medical Sciences, ADAMAS University, Kolkata, West Bengal, India*
[3] *MGM College of Pharmacy, Chiraura, Patna, India*

25.1 Background

Trichinosis, the other name for trichinellosis, is a disease related to the helminthes *Trichinella* species, the causative agent. The genus *Trichinella* is a nematode, and commonly spiral worms. In the world, this disease has kept people worried about its major outbreak. The human health is not only the major issue of concern for the disease, but its effect has also influenced the pork production industry and the food safety status. The helminthes as reported over the years have infected pigs at a much higher rate. Pork is the meat consumed mostly, so the risk of infection among humans has increased. To improve food quality and maintain a quality food chain, the aim has been to promote safe and zoonotic infection-free veterinary products for consumption [1]. There was a worldwide effort to eliminate the infection of *Trichinella* species. The outbreaks of Trichinosis reflected that not only the meat from pigs was infected with the *Trichinella* species, but the meat of wild boar and horses was also infected. After the outbreaks steps were taken to control the contamination but later on the modernization, increased population and economic and political changes the huge production of meat and consumption increased the rate of contamination [2, 3].

The spiral worm, initially called *Trichina spiralis*, was identified as a causative organism responsible for Trichinellosis in 1835 by two scientists, Sir Richard Owen and Sir James Paget. They observed a mass of roundworms present along the diaphragm lining of a corpse. Later in 1860, Friedrich Zenker provided strong evidence regarding the nematode transmission from animals to humans [4].

Though Trichinellosis is a disease prevalent worldwide over centuries, identifying the organism and the outbreak in different parts of the world has made people aware of its related condition. The disease is due to consuming raw meat unknowingly the contamination caused by *Trichina spiralis*. The symptoms of the disease are minor to major conditions that can be life-threatening. The sudden increase in this disease needs to be controlled with food production and consumption quality.

25.2 Historical Reports of Trichinosis Infection

The time from the existence of mankind on the earth has used natural sources such as food. Initially, man ate fruits until they came to know about hunting. When man started hunting, they consumed animals firstly as raw meat, and later, after the invention of fire, cooked food came into existence. So, food contamination with other organisms was expected over a long period. But lack of evidence from early humans could not establish whether mankind suffered from Trichinellosis.

Rising Contagious Diseases: Basics, Management, and Treatments, First Edition. Edited by Seth Kwabena Amponsah, Ranjita Shegokar, and Yashwant V. Pathak.

In the later periods of different civilizations, the ingestion of meat came into documentation. The consumption of pork is not a dish of the modern world, but various observations and evidence support the fact that pork was a common food from ancient times. The history of trichinosis has very ancient historical roots which date back before Christ. This was proven when the autopsy was conducted on an Egyptian mummy about 3200 years old. The researchers were surprised to find the cyst of the nematode in the skeletal muscles. According to a report from 1977, the researchers reported that the cause of the nematode infection in that type may be the fact "Nakht's [mummy's] diet probably contained little meat, yet his association with a funerary temple may have given him access to more meat than the average working-class Egyptian enjoyed." Another condition of a soldier was reported from 404 BCE who had symptoms of "severe dysentery and acute pains in all parts of the body, anguish and great depression of both mind and body," while fighting against Sicily under Himilco in 396 BC. The clinical manifestations of the soldier were similar to trichinellosis [5, 6]. Apart from this, in biblical times there were restrictions that impeded on pork consumption. The writing from the New American Bible stated ". . . and the pig, which indeed has hoofs and is cloven-footed, but does not chew the cud and is therefore unclean for you. Their flesh you shall not eat, and their dead bodies you shall not touch." The symptoms diagnosed after pork consumption in that period were similar to the present-day trichinellosis [6]. In ancient times, the Jews prevented pork consumption as they often faced the dangerous and risky situations of ingesting infected animals. The disease caused due to pork eating remained through centuries. The massive risk arising from such food consumption was a concern to society's authorities in previous times. Many religions imposed pork-based restrictions to combat such a situation [7, 8].

With the passage of time, at different times in history, there were reports of symptoms similar to the Trichinosis infection. The names were different for each type of symptom in different countries. But during those years, the spiral worm was still not identified. Therefore, people could not identify the disease as Trichinosis at that time. Especially massive outbreaks were observed in England and Ireland, where it was named "English fever." Apart from these, European countries like Spain, France, Germany, and Italy faced diseases with symptoms similar to Trichinellosis. The common symptoms in all the cases were fever with inflammation, pain, loss of strength, weakness, rigorous headache, and a troublesome stomach. These conditions were named differently, like "sweating sickness," "dandy fever," "the sweating plague," or "military pestilence." These conditions were observed mainly in the fifteenth and sixteenth centuries in Europe. But later, with a wide search, it was found that this type of condition was also present in the American continent, as Glazier reported in 1881. Furthermore, there are reports of such infectious fever on Saint Thomas Island in the West Indies in the early nineteenth century. So, it was well established that Trichinosis infections were common and affected a huge population at different ages [6]. In all these circumstances, it is evident that pork consumption was the leading cause of such infections. Pork was the common meat for festive dishes, and pigs were the common animals at those ages used for domestic purposes. The disease came to the limelight after the sudden discovery, and the cause became clear in that state. But later reports were also available from South America, Argentina [9].

25.3 The Discovery of Trichinosis

The Trichinosis infection was troubling men over the years. As observed through the different historical periods, the disease flourished in humans after eating meat on various occasions. Though in ancient times and till the early medieval period, people did not pay much attention to the cause of the disease. However, alarming situations from the eighteenth century made people gain interest in discovering the causative organism responsible for the disease. When the discovery of the pathogenic parasite started, it was seen that the entire nineteenth century was the era for the deep search for various aspects of *T. spiralis*. It was not only the species that were studied but the pathogenesis, life cycle, structure, and mode of transmission were all studied thoroughly.

When the history of identifying the scientist involved with the discovery of the worm causing Trichinosis is considered, then the vigorous attempt of a group of scientists comes to light of discovery. When the roadmap of the discovery is followed, it is seen that it is difficult to identify the first person who was wholly involved in the discovery of the organism. It was not the effort of a single scientist, but many scientists got keenly involved in discovering the new disease Trichinosis. As rightly described by Thomas Spencer Cobbold, English helminthologist, the discovery was not only restricted to the organism involved in causing the disease but every scientist involved with all aspects of Trichinosis should be given the same weightage. According to him, the worm, the structure of the worm, the pathological appearance, and every criterion were important [6]. Friedrich Tiedemann, the anatomist in Germany, observed the first indication of infected tissue in 1822. He never reported the cysts he observed but only indicated them as "little bodies" [10].

According to the publication of *The Lancet*, James Paget was given the achievement of discovering the causative worm. Paget was then a young student at St. Bartholomew's Hospital, London, dissecting a cadaver of a middle-aged Italian who died of Tuberculosis. During that condition, he found white spots on the diaphragm, which were animal particulate. This discovery took place on 2 February 1835. The discovery was declared on 6 February of that year to the medical student society of St. Bartholomew's Hospital, Abernethian Society [11]. There was a dispute regarding the discovery of the parasite involved with Trichinosis. Though Paget was the first person to identify it, he wanted to withdraw the credit of the discovery and gave the entire credit to his mentor parasitologist, Richard Owen. Richard Owen was the first scientist to give an official declaration to the Zoological Society of London. He identified it as the "flesh worm." Though Paget wanted Richard Owen to receive credit for the discovery, the latter gave lesser importance to Paget's effort in identifying the helminth. History says that there were writings from Paget where he was satisfied with the credit Owen achieved and never regretted the neglected nature of Richard Owen. According to James Paget, the representation done by his mentor was more perfect, which he could not have done so perfectly. Thomas Spencer Cobbold supported the work of Paget and, thereby, the first man to discover the nematode was James Paget [12].

There lies a story in the naming of the nematode helminth for Trichinosis. After discovering the nematode, Richard Owen and James Paget named the worm *Trichina spiralis*. It was in the year 1835. But it was seen that the genus *Trichina* already existed and was assigned to an insect in 1830. Ultimately, French helminthologist Louis-Joseph Alcide Railliet, in the year 1895, renamed the parasitic nematode as *Trichinella spiralis* [13, 14]. The research works of Joseph Leidy, the renowned parasitologist and paleontologist from the University of Pennsylvania, are related to the anatomy of various surviving and extinct organisms. He was the first scientist to recognize the transmission of *Trichina spiralis* from uncooked pork to humans in 1846. Moreover, he noted that insects and protozoa were responsible for escalating the disease among the public [15, 16]. After this discovery, interest aroused regarding the transmission process of the infection from animal to human. There are reports of scientists who researched different animals. The disease was so common and infected a massive population that European scientists from different places were urged to study it. Later, three scientists in chronological order studied the pathology and presence of infectious worms in various animals. Ernst Herbst 1850 found that the parasite resides in the muscles after entering the body. Rudolph Leuckart, in 1857, continued his research work with mice and found that the nematode larvae accommodated themselves in the small intestine of the experimental animals. A similar study was done with the dogs by Rudolph Virchow in 1859 and found the presence of larvae in the small intestine [8, 17]. As Cobbold mentioned, the research was spread to a wider area immediately after the nematode discovery controversy. The first accurate experimental work regarding the *Trichinellosis* species, its pathology, the way the transmission of the disease occurs from animal to human, and the infections related to the disease was elaborated by Friedrich Zenker around the year 1860 [8, 17]. These works on Trichinosis cleared various doubts and solved many problems related to the disease and people became aware of such disease prevailed while consuming half-cooked or raw meat. Further, clinical symptoms that were the results of Trichinellosis were studied and reports from Baltimore on eosinophilia were given by Thomas R. Brown in 1897. His study was related to the subjects in the John Hopkins Hospital where he analyzed the samples of blood and muscle of the infected subjects. It was evident in that condition the eosinophil and neutrophil count increased which can be treated as a measuring parameter suffering from Trichinosis. Such a finding was effective for diagnostic purposes [18]. The discovery that took place around 200 years ago was a milestone success in the field of medicine and research. The search for newer trends is still under experimental process as the disease is related to public health.

The historical data on the entire discovery of the nematode and corresponding data is tabulated as the outstanding milestone in Table 25.1.

25.4 Major Outbreaks of Trichinosis

The broad inner and deep knowledge of Trichinellosis developed in the nineteenth century as reported earlier. Before this knowledge about the disease, it has been found that multiple historical events related to the disease came to light. As of the documents available, various instances of the disease were found in the ancient world, but based on the symptoms, they were supposed to be Trichinellosis. But the report of a vast outbreak was not found in any literature. Specific incidents came into existence, and the correlated cultural and societal impacts and food habits of different civilizations and religions from ancient times to the medieval world indicated that the consumption of food from animals was restricted for better public health. In ancient times, hog meat was restricted among various cultural backgrounds for various reasons, including contaminated food. If seen in the Mediterranean region, pigs were not much in the diet in

Table 25.1 Trichinosis milestones – The significant discovery of Trichinellosis in the Nineteenth Century.

Trichinosis milestones		
Year	**Scientist**	**Discovery**
1822	Friedrich Tiedemann (Germany)	Little bodies
	Richard Owen – mentor of James Paget	Declared about the worm in the Zoological Society of London
1835	**James Paget (London)**	**Credit for the discovery of the worm**
1846	Joseph Leidy (parasitologist and paleontologist, University of Pennsylvania)	Transmission of *Trichina spiralis from uncooked pork to humans*
1850	Ernst Herbst	Reported parasite resides in the muscles after its entry into the body
1857	Rudolph Leuckart	Reported nematode larvae accommodated in the small intestine
1859	Rudolph Virchow	
1860	Friedrich Zenker	Pathology and transmission from animal to human
1895	Louis-Joseph Alcide Railliet (French helminthologist)	Named *Trichinella spiralis*
1897	Thomas R. Brown	Clinical symptoms of Trichinellosis

ancient times. The keeping of pigs was costly at that time. Moreover, the Israelites, the people of Babylon, and later Islam prohibited the use of hog meat. All these conditions are the reports from 7000 BC till the medieval period [19]. But pork is a common diet of people in many parts of the world. It is not only pork but meats from other animals that are included in the diet; these also cause the disease Trichinellosis. Significant outbreaks of the disease have occurred all over the world. In the modern world, according to the Centers for Disease Control and Prevention 2019 report, food-borne disease affects approximately 10,000 cases every year. The disease outbreak in the world is independent of the country's socioeconomic development. The disease can be of both types: acute and chronic. Depending upon the load of larval concentration within the body's muscles, the disease's condition is evaluated. In this part, we will check on the significant parts of the world that have suffered from infection.

25.4.1 Germany

Suppose a detailed report is studied regarding the cases of Trichinellosis. In that case, it will be observed that the country has had many episodes of this epidemic disease since the nineteenth century until precautions were taken in the early twentieth century to protect public health. The period from 1861 to 1890 showed frequent epidemics of trichinellosis with deaths. When the incidence of Trichinellosis increased in the last years of the nineteenth century, then trichinoscopy was introduced. That reduced the cases of Trichinellosis in the country and the stringent laws framed made it possible to keep the infection under control till World War II. After World War II, the illegal use of pork, and wild boar were the reason for the increase in cases. Prussia and Saxony were the kingdoms with maximum endemicity during World War I. The trichinoscopy when carried out in Prussia, Trichinella spiralis existed in the meat. The people of Prussia and Saxony usually took uncooked food for their diet. Post World War II disease spread to the rest of the country as the use of uncooked pork started among people. It was found that there were huge differences in the process of pig farming all over the country. The places where the law and legal procedures were maintained had a lesser risk of the disease. Though the pig is the reservoir of this infection, other animals like the wild boar and dogs are noted. Between 1904 and 1912, a report of trichinae infected 126 among 55,749 dogs was found. But in the present day, it is better that dogs are not used as a source of meat. But wild boars are used which were found to be infected with the nematode. But the consumption of wild boars has also reduced in Germany. Not only are the consumed animals infected with *Trichinellosis spiralis,* but animals like red foxes and badgers were also found to have the larvae of the *Trichinellosis spiralis*. The most important observation that can be made is that the life cycle of the worm is very much active and can easily be transmitted to humans. Though recently, in the twenty-first century, the number of infected animals has dramatically reduced and the disease is under control [20].

25.4.2 Poland

As in the case of Poland, it also suffered from Trichinellosis infection in the nineteenth century. It was a mystery; no solution was available to free the public from the infection. Initially, it was the pork that was responsible for infectious human health. The infections at times were so severe that they led to the death of many of the population. Still, epidemiological data got reduced by the end of the twentieth century. Pork was most commonly used in food until in the 1990s, it was replaced by wild boars. The cases of Trichinellosis observed were more due to the meat of wild boar rather than pork in the 1990s. The causative organisms that infected the wild were *Trichinella spiralis* and *Trichinella britovi*. The outbreaks of 2005 and in the later years were mainly due to wild boar consumption. Though the incidents reduced over the years they never got eradicated. In 2020, Trichinellosis was spread through the sausage prepared from wild boar meat. Proper awareness of the hygienic condition must be maintained for meat use [21, 22].

25.4.3 Russia

The wave of Trichinellosis outbreaks is estimated in Russia from the late twentieth century to the early years of the twenty-first century. It will be seen that Trichinellosis infection among the human race is quite common though the number of such cases has reduced over the years. Between the years 1998 and 2002, the incidence of the disease got reduced to 35.8% from 80%, as found during the years before 1997, but the main reason for the infection was the uncooked or poorly cooked or preserved pork. Due to its geographical location and severe climatic conditions, year-round meat consumption is widespread in Russia. Russia extends from Asia to Europe. The diet among people also changes accordingly. The result of this has reflected that the European part is infected mainly due to the use of pork. In contrast, the infection of Trichinellosis in Asia is from wild animals like bears, badgers, and wild dogs. The natural hosts of Trichinosis are usually not infected, but the consumer gets the infection due to the high larval load in the tissues. In most of the outbreaks in Russia, children are infected the most. The cause of the attacks was a need for more vigilance and proper control over the preservation of the meat [23, 24].

25.4.4 Outbreaks in Other European Countries

Trichinosis infection was reported in many countries of Europe. The European Union identified four significant Trichinella species causing the disease in humans and animals. Of these four species *Trichinella britovi* is the leading causing agent for the infectious outbreaks post World War II in Italy. But the number of incidences reduced over time until the outbreak of *Trichinella pseudospiralis* was reported in 2015 from Italy after beef consumption. There were reports of *Trichinella pseudospiralis* from Denmark and Great Britain, apart from Italy [25]. Reports of Trichinellosis infection were also found in France and Serbia in the twenty-first century. The cause of the disease was the pork and the meat of wild boars and horses. According to the European Union, in most cases, infections were observed among tourists unaware of the source of the meat they consume [26]. In Europe, the epidemiologic patterns were segregated into four zones. Firstly, the countries like Bosnia, Bulgaria, Croatia, Romania, Lithuania, and Latvia, where cases were low, and comprise the Eastern part. Secondly, Slovakia and Slovenia region, where the risk of the disease is high. Thirdly, in those countries with large outbreaks but because of large populations, the ratio becomes less. In another case, the number of cases of the disease is less. Fourthly, those countries with very few incidences. The categorical division depended on the incidences from the late twentieth to early twenty-first century [27].

25.4.5 Africa

Though most people in Africa included animal meat in their diet, until the 1950s and 1960s, outbreak report of Trichinellosis was usually not observed. The Kikuyu tribe in Kenya was infected with helminthiasis through wild pigs. But data of African origin predicted the disease among the Christian community rather than Muslims. In one such outbreak, Muslims were detected with the disease because beef used in their food was illegally contaminated with the infected meat. Otherwise, the report of an epidemic of Trichinellosis is not found in the history of Africa. The meat of wild boars used is often the cause of the disease. The area which generally shows the presence of the disease is along the coastline of the Mediterranean region and an area with Christian migrants from other countries. The soldiers and policemen in Ethiopia who are Christians report most of the Trichinellosis cases [28].

25.4.6 Asia and Pacific Region

There were multiple reports of Trichinellosis outbreaks in the South East Asian and Pacific regions. The countries like Laos, Vietnam, Cambodia, and Thailand have frequent outbreaks of the disease post World War II. These countries are involved with pig farming. The quality and hygienic conditions remain questionable in these regions. South East Asian countries, by tradition and culinary habits, consume raw and uncooked meat, which has become the leading cause of the disease. Reports from Myanmar and the Philippines were also found. The condition is more prominent among males than females, though there are infection data among the children. The leading cause for such diversity in infection was that men tend to take more raw meat in their diet and during hunting than females who avoid taking such uncooked food. Between 2001 and 2021, there were multiple outbreaks of infections in different parts of the South East Asia [29]. The first case of Trichinellosis was reported in 1969 in New Zealand. But proper Government regulations have controlled the spread of infection in New Zealand in the later years. Japan also has few reports of the disease in the country. Though there are few cases, the cause was raw and uncooked brown bear meat [30].

The first official report of an infection outbreak was noted in 1964 in Tibet. Later, till 2020 there were multiple cases of infection in the Peoples's Republic of China. Severe symptoms and even death were observed in all cases reported from 2009 to 2020. The common cause of the infectious outbreak in China was consuming raw and uncooked food. Though in recent times, proper surveillance and hygienic conditions from government authorities have been set in the country. Traditional pig rearing has been improved, and the commercialization of pork with advanced technologies has been set to reduce the infection rate in China [31].

Common disease investigations revealed that encapsulated and nonencapsulated species were involved in the infections. *Trichinella spiralis, Trichinella britovi*, and *Trichinella nativa* are the capsulated species, whereas *Trichinella pseudospiralis* and *Trichinella papuae* are non-capsulated in nature. A new species *Trichinella* T9 was identified in wild animals like bears, raccoons, and red foxes in Japan, but reports in human samples were not found [30, 32].

25.4.7 North and South America

T. nativa and *Trichinella* T6 are the two common species responsible for the disease in Canada from walruses (*Odobenus rosmarus*) and bears. There is a report of Trichinellosis in Ontario, Canada from the dried meat of a Black bear. The causative organism was *T. nativa*. The symptoms in that condition were cases of thrombosis. The same organism, Trichinella T6, was responsible for the infection that spread in Ohio, USA. It was found that the bear meat was imported from Canada. Another wild animal responsible for spreading the disease in Canada and Northern America is the adult male Wolverine. These animals are found to be susceptible to many strains of *Trichinella*. Actually, the main cause for the spread of the disease from this wild animal is that the remains of the dead animals, after collecting fur, are not cleaned from the area. The extreme winter climate allows the larva of *Trichinella* to survive. This has made disease transmission in humans very easy [33].

If records are analyzed, it is seen that the cases of *Trichinella* infection in the United States have reduced post-1990. The proper vigilance and control over pig farming and meat processing have been controlled strictly in the United States of America, considering human health. In South America, most countries have controlled the outbreak of the disease, but Argentina and Chile still have a large number of cases. There are outbreaks in Mexico, Bolivia, Brazil, and Ecuador where it is mainly sporadic and the number of incidences is low. The common cause is unhygienic food and pig husbandry. The two species *T. spiralis* and *T. patagoniensis* are most common in South America. Another cause for the increase in cases in Argentina is the entry of a huge number of migrants from Europe [27, 34].

The outbreak of the disease is prevalent in the entire world irrespective of geographical location and climatic conditions. The various species of *Trichinella* are responsible for spreading the disease among humans through various wild animals. Though the number of incidences has reduced over the years strict constant monitoring is required for the control of the disease. The disease intensity of the world for the year 2022 is in Figure 25.1.

25.5 Trichinosis Episodes in Animals

Currently, pork is the most susceptible source for spreading infection from animal to animal and from animal to human. But the report of disease in animals was first found in badgers in 1850 by scientist Herbst. But now, it is seen that multiple wild animals are the reservoir for the infection. The common animals are listed below:

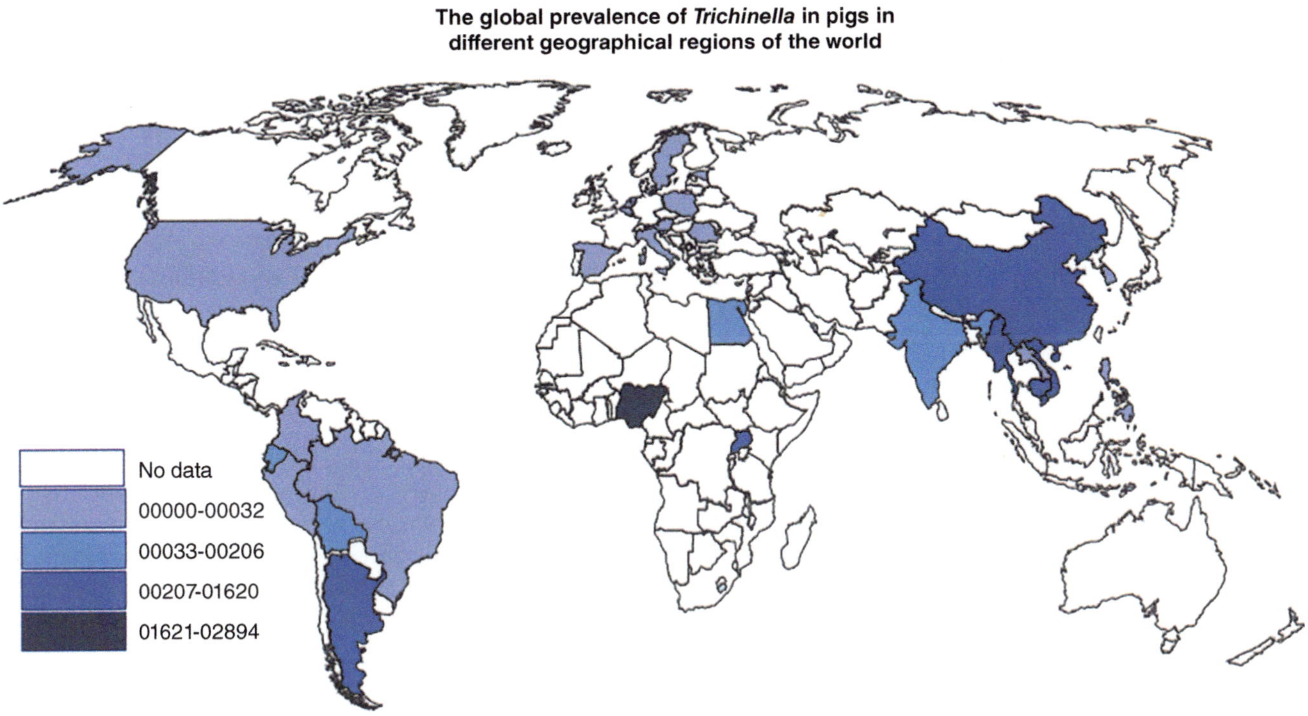

Figure 25.1 The global prevalence of *Trichinella* in pigs in different geographical regions of the world based on the included studies (as on 2022).

25.5.1 Pigs

Near the end of the nineteenth century, pigs were found to be infected with Trichinosis in many European countries. Most of the cases were from Germany, but infected pigs were also found in Western European countries. Before the twentieth century pigs in the United States of America and Latin American countries were manifested with the infection. To date, the pig is the most crucial carrier of Trichinosis, and most cases are due to pigs as pork is highly consumed meat worldwide.

25.5.2 Horses

Horse meat was also responsible for trichinellosis, confirmed in 1975. Approximately, data of around 3000 cases were reported in France and Italy. The incidence of contaminated meat consumption led to the death of five people. The meat was imported from North America (Canada, United States of America, and Mexico) and Eastern Europe [35].

25.5.3 Wild Boars

The intake of wild boars have drastically increased since the twentieth century with the increase in population. Wild boars are the available animals over large geographical areas, so wild boars came in contact with domestic animals and humans. The farming and hunting of animals by people have made it easy for contamination. As the demand for pork increased, to meet up the demand, pork was mixed with contaminated wild boars. Wild boar meat is used as food for pigs and other domestic carnivorous animals. Wild boar is the second largest animal used as food. *T. spiralis*, *T. britovi*, and *T. papuae* are common worms that infect humans through wild boars [36].

25.5.4 Dog

Dogs are the common source of meat among humans especially in many parts of Europe and in Asian countries like Korea, Thailand, and China. These animals usually become infected with the nematode because of their scavenger nature.

25.5.5 Bear

These animals are common among the people of the Arctic region, Alaska, and North America. The different species of bears like black bears, grizzly bears, and polar bears are also sources of meat. Many outbreaks of Trichinellosis have been reported due to uncooked bear, especially among hunters [37].

25.5.6 Marine Mammals

Marine mammals of the Arctic zone are infected with the Trichinella species. These include walruses, bearded and ringed seals, and infection in Beluga whales was also reported. In most cases, it was assumed that carcasses of infected animals were consumed by these mammals from which the disease got transferred [38].

25.5.7 Other Animals

Trichinellosis infection is common for carnivorous animals as they consume carcasses. There are many carnivorous animals that are consumed rarely by different races of people. Badgers, jackals, foxes, and warthogs are a few uncommon animals consumed. There are causes for the transfer of Trichinellosis infection in carnivorous animals, but Trichinellosis was found in the meat of herbivorous animals. It is supposed to be that the meat of herbivorous animals may have been contaminated with infected pork. That may be the major cause of infected beef, sheep, and goat [39–41]. Cases were found where infection of Trichinellosis got transferred from reptiles like turtles, pythons, lizards, and crocodiles. But all these are carnivorous reptiles. Therefore, the infection is common for the reptiles and eventually, these are consumed in different sections of the world.

The genesis of transmission of the disease from animals to humans is because of uncooked or half-cooked meat and consumed processed meat.

25.6 The Nematode: *Trichinella* sp.

When *Trichinella* nematode was initially discovered, it was found that it was monospecific and did not have any genotypes. *Trichinella spiralis* was thought to be the only species responsible for the disease in humans from infected meat. The main source of infection of Trichinellosis in humans was pig meat, and the pig was infected by *T. spiralis*. But with time, when meat consumption increased from various carnivorous animals, it was found that the symptoms and causative organisms were different from *T. spiralis*. Later in the 1960s, the search for newer species of *Trichinella* intensified. There is a capsule present in *T. spiralis*. When the search continued, it was found that there were capsules present in some of the species and there was a lack of capsules in many. According to the geographical location, different species have been identified to date. According to various phylogenetic studies, the encapsulated species of *Trichinella* are *T. nelsoni, T. nativa, T britovi, T. murrelli, T. patagoniensis, T. chanchalensis*, T6, T8, and T9. Similarly, the non-encapsulated species found are *T. pseudospiralis, T. papuae*, and *T. zimbabwensis* [42]. A brief details of the various *Trichinella*sp. is predicted in Table 25.2.

Let us have a discussion over the various species of *Trichinella*:

Trichinella spiralis: This species is widely distributed globally and present almost in all synanthropic animals of which, wild pigs and domestic pigs are most common. This species is responsible for most human infections as pork is the most consumed meat. Apart from pigs this species also infect rats, horses, dogs, cats which are in close contact with the human population.

Trichinella nativa and T6: *Trichinella nativa* survive mostly in the cold frigid zones of the entire world including North America, Asia, and Northeastern Europe. But T6 is found only in the frigid zones of Canada and North America. These two species are usually present in both terrestrial and marine animals. The most significant characteristic of these two species is that they survive in a dormant state in freezing conditions for over five years, as a result of which, when consumed, infections easily get transmitted.

Trichinella britovi: Apart from *T. spiralis,* this species is the most common causative agent for infectious disease. This species is usually spread in Europe, Africa, and Asia. Domestic pigs are infected with *T. britovi* mainly after feeding on grazing land and on carcasses.

Table 25.2 *Trichinella* Species, Host, Pathogenicity level in humans, region of distribution, Larval form, and Cycle.

Species	Region	Host	Pathogenicity level in humans	Larval form	Cycle
T. native	Asia, Arctic and subarctic, Europe, America	Dogs, wild carnivores	High	Sylvatic	Encapsulated
T. spiralis	Worldwide	Pigs, wild boar, rats, carnivores	High	Sylvatic, domestic	Encapsulated
T. pseudospiralis	Australia, New Zealand, Thailand	Birds and mammals	High	Sylvatic	Non-encapsulated
T. nelsoni	Eastern-Southern Africa	Carnivores	Low	Sylvatic	Encapsulated
T. papuae	Thailand, Papua New Guinea	Mammals and reptiles	Moderate	Sylvatic, domestic	Non-encapsulated
T. patagoniensis	Argentina	Carnivores	Unknown	Sylvatic	Encapsulated
T. zimbabwensis	South Africa, Mozambique, Ethiopia, Zimbabwe	Mammals and reptiles	Unknown	Sylvatic	Non-encapsulated
T. murrelli	Canada and USA	Carnivores	Moderate	Sylvatic	Encapsulated
T. britovi	Middle east countries, Europe, Asia	Jackal, wild boar, dog	High	Sylvatic	Encapsulated

Trichinella murrelli: These organisms are absent in pork but are present in sylvatic carnivores. These are found in the states of the United States and Canada. The most common cause for the spread of this species is the consumption of raw meat by the hunters of the region.

Trichinella nelsoni: The Eastern African countries have widespread of this species. The warthogs and wild pigs are the hosts for this species. Kenya and Tanzania report the maximum number of cases of this infection.

Types T8 and T9: T8 is similar to *T. britovi*, whereas T9 has resemblance with *T. murrelli*. These are rare species found in certain special cases, but frequent reports or infections in humans are still not found.

Trichinella pseudospiralis: These species reside in both mammals and birds, but are higher in animals. Its distribution is found in the Arctic and Australia and recent reports are from South East Asia. The species infect both the domestic and sylvatic animals.

Trichinella papuae: This species is common in mammals and reptiles and human infection was also reported especially in areas where reptile consumption is widespread.

Trichinella zimbabwensis: The reptiles of the African country are usually infected with this species of *Trichinella*. Though animal infection reports are there, but reports from a human are absent.

The infection through *Trichinella* species is very common. As Trichinella infects a vast number of animals, it is prevalent that newer species or strains may develop over time [43].

25.7 Life Cycle of Trichinosis

Trichinellosis is caused by nematodes named the genus *Trichinella*. Undercooked pork or other contaminated meat causes trichinellosis disease. *Trichinella spiralis* species commonly causes this condition that affects humans due to improperly prepared raw meat (pork). Mammals like horses, goats and carnivores can be infection sources [44]. The infection reaches the human after residing in the body of animals. The animals usually show no clinical manifestations as humans do.

Trichinella spiralis (*T. spiralis*) infection poses a significant threat to the host's immune system. The parasite's whole life cycle is completed in a single host. As a result, elements from each stage of the parasite life cycle – adult, newborn larvae (NBL), and muscle larvae – impact the host organism. *T. spiralis* produces an environment that supports the host organism's existence and, by extension, the parasite's survival to complete its life cycle [43]. This is accomplished by systems that inhibit or modify the host's immune reactions to both their own antigens and those of other parties who might jeopardize the host.

The two generations of *Trichinella* sp. life cycle occur in the same host. Continuous studies on the life cycle of all *Trichinella* sp. reported that these nematodes execute their life cycle in multiple animals, including mammals, birds, reptiles, and even insects. But the manifestations of the infection are observed in humans. After consuming the pregnant mother worm, NBL travels directly into the host's lymphatic and blood vessels. The worms reside in intestinal mucosa of host. This enables them to penetrate preference locations (highly oxygenated muscles) by being delivered there. It is likely that NBL uses its stylet to access the striated muscle cells. Enzyme-based penetration mechanisms have been theorized but have not yet been proven [45].

When newborn larvae NBL was administered directly into muscles during experimental infections, the penetration started to happen as soon as 10 minutes after injection. NBL develops to infective muscle-stage larvae within such muscle nurse cells without molting. Within about 15 days, this maturation comes to an end. Parasite larvae can live for a very long time in muscle nurse cells (up to 40 years in humans) [46]. This is due to the calcification of the encapsulation that occurs in the nurse cells and larva of the nematode. This process is activated because when the larva is in the host cell the immune response becomes active. The *Trichinella* species or genotype thereby gets calcified. It remains in this hypo biotic stage until it is consumed by a new host [47].

After ingesting, the parasite's larvae are released in the new host during gastric digestion. From the gastric region, the larvae pass to the duodenum of the host, where they find the intestinal mucosa is the best place to reside. Here the larvae undergo four molts and quickly convert to adult form within just two days. Within a week (5–7 days), postinfection mating occurs, new NBL generations start to be born from the females. Within a few weeks, an intestinally immune-mediated host response is established, immune effector mechanisms affect the viability of the female parasites, leading to a continual expulsion of adult worms [48].

The etiopathogenesis occurs in two stages with respect to the life cycle of larvae.

Intestinal phase: In the initial phase, the larvae are released in the stomach from the infected undercooked pork during the gastrointestinal phase when due to the actions of hydrochloric acid and proteolytic enzyme (pepsin). Larvae then permeate to the small intestine. After becoming adults, larvae mate [49]. The gastrointestinal phase of *Trichinella* worms is accomplished with the production of larvae. After the ingestion of meat infected with L1-*T. spiralis*, the release of L1-*T. spiralis* occurs in the stomach. These are then migrated to the small intestine and mature into female and male adult worms. Reproduction of adult worms of *T. spiralis* to produce new-born larvae [50].

Muscle phase: This is also known as the systemic phase where larvae enter the lymphatic system before moving into the blood circulation and reaching skeletal muscles, the heart, and the brain NBL *T. spiralis* invades the skeletal muscle cells to develop into the infective stage of *T. spiralis* forming the complex nurse cell (NC) and L1-*T. spiralis*. This stage can result in myocarditis and encephalitis as well as systemic symptoms such as myositis, fevers, myalgias, and periorbital edema [51].

The well diagrammatic view of *Trichinella* sp. is represented in Figure 25.2.

25.7.1 Symptoms

Above 400 trichinellosis cases per year were reported to the Centers for Disease Control and Prevention (CDC) in the 1940s [52]. According to estimates, trichinellosis affects approximately 10,000 people annually. Nine different *Trichinella* species and 12 genotypes of *Trichinella* are now recognized. While *Trichinella nelson*, *Trichinella pseudospiralis*, *Trichinella murelli*, and *Trichinella nativa* are all causative microorganisms of human beings. The most prevalent type of *Trichinella spiralis* mostly infects human beings. Possible symptoms range from myalgia and fever to more serious ones like myocarditis and encephalitis [53].

The earliest signs of symptoms of trichinellosis are digestive problems due to pork, which is generally ingested raw or in an undercooked form can spread this disease. The incubation period is one to six weeks. The degree of human illness is influenced by the quantity of ingested larvae. They frequently manifest themselves two to seven days after eating raw or undercooked meat. Along with abdominal pain, the symptoms include nausea, vomiting, and diarrhea. A common symptom is muscle pain, in the chest, belly, and face. Fever often starts at two weeks and peaks at four weeks, with readings as high as 40 or 41 °C in extreme cases.

Diarrhea and other gastrointestinal problems may persist during the parenteral period. There may also be coughing, and dyspnea. Dyspnoea is mostly brought on by parasite invasion and the ensuing inflammation of the diaphragm and other breathing muscles. Infarction and bronchopneumonia may also be present [54]. It was noted that breathing was difficult during the first few days of albendazole treatment in a *T. spiralis* outbreak, most likely as a result of toxic parasite by-products being released [55].

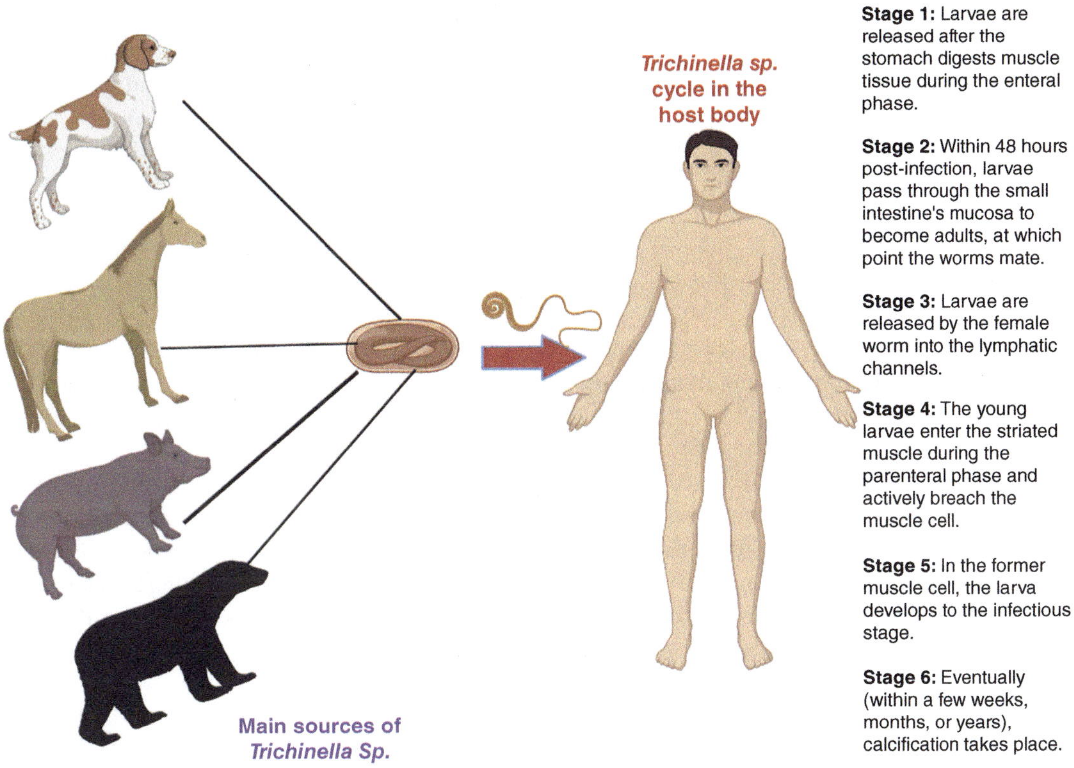

Figure 25.2 Life cycle of *Trichinella* sp.

According to some reports, 10–24% of cases involve neurological symptoms, which are more prevalent in severe infections [56]. Meningoencephalitis symptoms were the most often detected clinical signs in 55 individuals with neuro trichinosis (96%), followed by localized paralysis or paresis (73%), delirium (71%), and psychosis [57]. Trichinellosis patients frequently experience headaches, which are exacerbated by head movements [58].

Ocular symptoms, such as blurred vision, conjunctival hemorrhages, and ocular pain can be seen. In almost 2000 trichinellosis cases evaluated, periorbital edema occurred in 17–100% of patients [59].

The symptoms of trichinellosis onsets for two weeks and last up to eight weeks after consuming undercooked meat. Other signs and symptoms include chills, fever, weakness, facial edema, myalgias, fatigue, and persistent diarrhea. Conjunctivitis and subconjunctival hemorrhages affect most patients [52]. Furthermore, retinal hemorrhages and subungual splinter hemorrhages can happen. Urticaria, petechiae, and splinter hemorrhage can occur. Also present are the less common symptoms of dysphagia, rash, headache, coughing, and dyspnea. Other serious side effects include arrhythmias, secondary bacterial pneumonia, meningitis, encephalitis, hematuria, and respiratory myositis [60].

Other clinical symptoms are facial edema, muscle soreness, swelling, weakness, and fever are all symptoms of trichinellosis syndrome; conjunctivitis, headaches, anorexia, and urticaria occur less commonly [61]. During the parenteral phase, muscles, notably the heart, are primarily impacted [62].

25.7.2 Diagnosis

The first diagnosis is often made using clinical signs and symptoms. Serology can be used to confirm a diagnosis; on rare occasions, a muscle biopsy can be required. Through blood count, leukocytosis and eosinophilia can be identified which are related to the finding of the number of worms causing the infection [63].

The parasite that results in the breakdown of skeletal muscle, aldolase, aminotransferases, creatine kinase, and lactate dehydrogenase might all be elevated. Moreover, individuals may develop hypokalemia, and increased blood IgE hypoalbuminemia [64].

A Computed Tomography scan is to find out neurological impairment. Premature ventricular contractions (PVCs), conduction block, PR interval prolongation, and atrial arrhythmias are possible. There are several serological testing options, including Enzyme-linked immunosorbent assay (ELISA), immunofluorescence (indirect IF), and latex agglutination assays. Results of serologic tests can be confirmed by Western blot analysis. Often serology is not accurate, especially in the early stages of an illness during the first three weeks or more. Other helminth infections and autoimmune diseases may cause a false positive serologic diagnosis [65].

Another diagnostic method is a muscle biopsy. Accuracy is high if the biopsy is performed four weeks after the infection. A negative response can be observed if carried out at an early stage of the disease [66].

ELISA is the method that is most frequently employed. In order to find anti-Trichinella antibodies in serum samples, cryostat sections of infected rodent muscle, frozen sections of free muscle larvae, or formalin-fixed entire larvae are used. The genotype of the host and antigens exposure in the immediate environment and food can alter the result of the ELISA [67].

Immunoblotting has been found to be another promising technique for excluding false-positive serological data or verifying positive instances. It shows more accurate results than traditional ELISA [68]. Due to the time consumption, this method may be employed as confirming rather than screening tests. The diagnosis is etiologically supported by the direct identification of larvae in the muscle stage through adequate evaluation of muscle samples [67].

Anti-Trichinella antibodies present in serum can be used for the diagnosis of the disease. After a primary Trichinella larvae infection, the human patient's seroconversion time determines whether antibodies can be detected, and there are differences between the various immunoglobulin classes. In the acute stage of infection, the levels of antibody IgE levels get elevated. The allergic reactions like rash or edema, observed at various stages of infection, are due to different classes of antibodies in other infectious conditions. However, in the absence of this antibody class, trichinellosis cannot be ruled out. IgE is, therefore, not thought to have diagnostic value for standard diagnosis [55].

25.7.3 Management and Treatment with Medicines

Mild infections with symptoms are treated with antipyretics and anti-inflammatory medications. And systemic infections are treated with corticosteroids and antiparasitic drugs. Mebendazole (200–400 mg) can be given three times daily for the first three days and 400–500 mg three times daily for the following 10 days. Another alternative is to take 500 mg of albendazole orally two times a day for 10–14 days. In adverse conditions, corticosteroids (prednisone) can be given for a total of 10–14 days (30–60 mg) daily [65].

According to the World Health Organization, they recommended not administering mebendazole or albendazole to children or pregnant women (first trimester). Pregnant women are prescribed antihelminthic drugs until after their first trimester, according to the World Health Organization (mebendazole, albendazole, pyrantel, or levamisole) can be used [69].

Antiparasitic and steroidal anti-inflammatory medications are utilized in the pharmacotherapy of trichinellosis. Benzimidazoles, particularly albendazole, and mebendazole, are being utilized as an antiparasitic treatment for Trichinellosis because they are effective against the parasite [65]. Additionally, various antiparasitic medications including ivermectin, nitazoxanide, quinfamide, and flubendazole have been evaluated with encouraging outcomes. Since they kill adult parasites, these medications are the most effective treatments when the disease is just getting started. A recent study revealed that thiabendazole was a potent and curable medicine since its efficacy was almost effective to eliminate intestinal worms, even though albendazole is better tolerated [70].

The most common medication used to treat the signs and symptoms of the inflammatory response brought on by *T. spiralis* infection is glucocorticoids (GC). These drugs have strong anti-inflammatory action that controls transcriptional pathways in a variety of biological situations, including development, homeostasis, metabolism, and inflammation [71]. Glucocorticoids inhibit inflammation by two ways:

a) promote the expression of numerous genes that code for proteins that have anti-inflammatory properties, including IL-10, the IL-1 antagonist receptor, and leukocyte-inhibitory secretory protein [72];
b) Pro-inflammatory genes are inhibited from being expressed by protein–protein interactions between transcription factors like NF-B and activating protein (AP)-1, which also control the expression of pro-inflammatory cytokines and enzymes like cyclooxygenase (COX)-2, inducible phospholipase A2 (cPLA2), and inflammatory receptors [73].

Immunodominant antigens have been used as immunotherapy for Trichinellosis. These antigens stimulate the host's immune system, which lowers the intestinal parasite burden and reduces the fertility of adult female worms [74]. Several studies have demonstrated that *T. spiralis* total soluble (TS) immunotherapy provides protection in mouse models because a decline in muscle parasite burden was estimated [75]. Antigens with an M.W. range of 14–97 kDa were found in a study based on a pig model infected with *T. spiralis*, to which immunotherapy with *T. spiralis* TS antigen was applied [42]. Comparatively to the control group (absence of immunotherapy), immunotherapy with *T. spiralis*, TS antigen elicited a primary immune response, with a decrease in parasite burden as well as damage in the muscle phase of the infection [76].

25.8 Control of Trichinella Infection in Humans

Trichinellosis infection is caused due to ingestion of undercooked food by humans. In the last 100 years when infection was most during pre-World War II, the Governments of different countries developed multiple rules, laws, and policies to restrict the spread of infection. Despite stringent regulations, there are cases of numerous outbreaks in other parts of the world. The International Commission on Trichinellosis (ICT) has developed specific strategies to control the *Trichinella* larvae in the meat and, thereby, control the spread of the disease.

If we see the policies, we will find that the control of *Trichinella* larvae has been implemented after harvesting the meat. The source of meat consumed by humans is carnivorous animals. The initial checkpoint for the infection is testing raw meat for the larva. In many cases, there are traditional habits of raw meat consumption. So, testing raw meat is an essential criterion to control the infection. Trichinoscope method is the direct method for the analysis of raw meat. The serological method is restricted to detection in raw meat and according to ICT, it is recommended for study during outbreaks.

ICT has recognized three treatment methods that can be adopted to remove the larvae from the infected meat. The three methods are

a) **Heating or cooking**: Controlled temperature with time is the most acceptable and trusted method used to remove the *Trichinella* larvae from infected raw meat. There are various combinations of temperature and time used for treating the meat.
b) **Freezing**: It is the method for preparing larval-free meat from domestic pigs. Actually, the larva of *Trichinella spiralis* is the common nematode for domestic pigs. The larva of this species can easily be removed by freezing, but the species that survive at a very low temperature over years cannot be made infection free by this method.
c) **Irradiation**: The irradiation method is implemented for packed meat items and has been found safe for intake by humans.

It has been seen that during the processing of meat, treatment of meat is very much essential. The treatment, also known as curing meat, does not follow any particular protocol presently. But based on the chemistry of meat the temperature and time-controlled treatment is the most common and effective method for the removal of *Trichinella* larvae from the various sources of meat obtained. It has been seen that the microwave is not an appropriate or suitable method for removing larvae from meat as uniform heating does not occur in such cooking.

The other way of controlling Trichinellosis infection among humans after consumption is by increasing knowledge about the risk of contamination on intake of carnivorous meat is to be increased. Consumers need to make aware of the conditions of ingesting raw meat. They are needed to be made aware of the various methods of processing the meat before consumption. The food processing industries are to make aware of the treatment of meat and various methods adopted to remove the larva.

It is also necessary to make people aware of the risk of consuming raw and uncooked meat. The hunters are to be educated regarding the intake of raw meat from carnivorous animals. Another way of checking the contamination is by maintaining hygiene in the areas of animal farming. In many cases, the carcasses are given to carnivorous animals, which should be infection free so that transmission of the disease is controlled.

The main effort is to reduce the risk of infection among people as Trichinellosis is a very common infectious disease and is spread all over the world. All governing authorities have tried to develop policies in many ways to control the spread of the disease among animals and further to humans.

25.9 Current Trend in Trichinella Epidemiology

The nematode *Trichinella* sp. usually are found in the wild carnivorous animals. When humans ingest infected predatory animals, the transmission of the disease occurs with increased larval load in humans. But in the contemporary world, global warming has brought about climate changes and as a result, the world's vegetation is getting altered. As a result of which, the changes in animal diversity became prominent. Many of the wildlife species to which the nematode entered and moved to the human cycle are getting extinct. Therefore, there may be chances that *Trichinella* sp. get reduced. The increased temperature due to global warming has increased drought, glacier melting, and forest fire. All these factors have brought a change in the wildlife population. Moreover, this increased dryness and heat in the environment can lead to the death of larvae of the nematode. The environmental changes are also due to increased population, and demand for increased urbanization has changed the world's ecosystem. The reduction in humidity and increased temperature is due to the pollution caused by increased population with increased demand and misuse of resources. Moreover, the lifestyle of people has changed with time. People have shifted to more organized vegetative farming and hunting has reduced. These are also the causes for the loss of forest life, and the survival of larvae has become a risk. With urbanization, wild animals have changed their habitat. Wild animals have changed their food habits from carnivores to urbanized diets. This has resulted in a lack of deposition of the *Trichinella* in the animals. The survival of nematode has been at risk.

Apart from the environment, the nematode infection pattern has altered due to globalization, and changes will be observed in the coming years. Human behavior and lifestyle, especially in rural areas, have made this species a part of the household due to improper unhygienic farming of the wild carnivores. Farmers easily imported the infected pigs from different countries and infection moved from one corner of the world to the other due to lack of knowledge about the risk of infection due to *Trichinella* sp. Another cause of the spread of *Trichinella* infection in society is feeding household animals with infected animal products. The migration of animals from one continental part to another is another reason for the spread of infection from one region to the other. Especially animals like the raccoon, raccoon dog, and mink, which are the reservoir of *Trichinella* infection were imported from America to Europe for various farming industries. Therefore, those animals increased in population in European countries and the risk of Trichinellosis has increased in Europe. On the other hand, for the native wild animals of Europe (wolverines, lynxes, and wolves), the reservoirs of *Trichinella* sp. are increasing. But though there are chances of increased *Trichinella,* presently the animals are changing their food habits [36].

The outbreaks observed in Trichinellosis cases from the late-twentieth century till date were sporadic or very few cases in parts of the European or American countries. In most of the issues, it was observed that the infection outbreak was due to the illegal contamination of the meat from various sources to meat consumed on different occasions. The meat of horses and wild boar that was consumed was contaminated. The export of meat in the present day is under supervision and contamination free. The pigs that are mostly consumed are farmed with controlled monitoring. In the future, there will be a reduced risk of *Trichinella* infection. The reason is that the world environment has become conscious of the enormous CO_2 concentration in the environment. Meat production increases the concentration of CO_2 in the atmosphere. So, to reduce the concentration of the pollutant gas, meat production through different sources will reduce in the near future. So, if meat farming reduces, the chances of *Trichinella* infection may drop with control over meat production [55].

The environmental changes have dropped *Trichinella* infection among the animals in the cold region. The survival of *Trichinella* infection is most in temperatures below −20 °C and they are characterized as subnivium. High humid condition with extremely low temperature favors the growth of the larvae. In recent years, transmission of the disease among wild animals has reduced. The causes may be a decrease in the snowy layers in the mountains of Europe or it may be that decreased hunting has reduced the dead matter to get contaminated or it may be that the food pattern has changed among people [42].

25.10 Conclusion

The disease Trichinellosis is rare but common. The historical evidences proved that cause for the disease remained the same though the causative organism has changed and are found different. As *Trichinella* sp. is common among the carnivorous animals, though data of the infection is also reported among the herbivores, it is necessary to maintain hygiene and proper cooking of food before intake. Control measures are to be adopted to reduce the risk of transmission from one organism to the other. It was till the middle of twentieth century that huge contamination of the parasite among food was observed. Prompt actions to maintain quality of food and preserve the human health and hygiene, strict regulations were incorporated in the society. It is very much necessary to control the spread of infection, as with passage of time the chances of development of new species will happen. The significance of this strain is, they survive extreme cold condition and therefore the infection is common among the frigid and temperate zone.

The prominent way of controlling Trichinellosis is by cooking the meat at optimum temperature and maintaining the optimum time. Hygienic conditions, disposal of carcasses, and maintaining scientific methods of meat treatment according to guidelines can control Trichinellosis in the animal kingdom.

References

1 Wilson, N.O., Hall, R.L., Montgomery, S.P., and Jones, J.L. (2015). Trichinellosis surveillance—United States, 2008–2012. *Morb. Mortal. Wkly. Rep. Surveill. Summ.* 64 (1): 1–8.

2 Blaga, R., Durand, B., Antoniu, S. et al. (2007). A dramatic increase in the incidence of human trichinellosis in Romania over the past 25 years: impact of political changes and regional food habits. *Am. J. Trop. Med. Hyg.* 76 (5): 983–986.

3 Boireau, P., Vallee, I., Roman, T. et al. (2000). *Trichinella* in horses: a low frequency infection with high human risk. *Vet. Parasitol.* 93 (3–4): 309–320.

4 Cox, F.E. (2002). History of human parasitology. *Clin. Microbiol. Rev.* 15 (4): 595–612.

5 De Boni, U., Lenczner, M.M., and Scott, J.W. (1977). Autopsy of an Egyptian mummy. 6. *Trichinella spiralis* cyst. *Can. Med. Assoc. J.* 117 (5): 472.

6 Neghina, R., Moldovan, R., Marincu, I. et al. (2012). The roots of evil: the amazing history of trichinellosis and *Trichinella* parasites. *Parasitol. Res.* 110: 503–508.

7 Neghina, R. (2010). Trichinellosis, a Romanian never-ending story. An overview of traditions, culinary customs, and public health conditions. *Foodborne Pathogens Dis.* 7 (9): 999–1003.

8 Kiple, K.F. (2003). *The Cambridge Historical Dictionary of Disease*. Cambridge University.

9 Despommier, D.D. (1983). Biology. In: *Trichinella and Trichinosis* (ed. W.C. Campbell), 75–151. New York: Plenum Press.

10 Zarlenga, D., Thompson, P., and Pozio, E. (2020). *Trichinella* species and genotypes. *Res. Vet. Sci.* 133: 289–296.

11 Nelson, G.S. (1988). More than a hundred years of parasitic zoonoses: with special reference to trichinosis and hydatid disease. *J. Comp. Pathol.* 98 (2): 135–153.

12 Morley, N.J. (2017). Richard Owen: an overlooked parasitologist. *Arch. Nat. Hist.* 44 (1): 19–30.

13 Mahajan, M. (2021). *Trichinella spiralis*. *Emerg. Infect. Dis.* 27 (12): 3155.

14 Esch, G. (2007). *Parasites and Infectious Disease: Discovery by Serendipity and Otherwise*. Cambridge University Press.

15 Wilson, L.G. (1999). *Joseph Leidy: The Last Man Who Knew Everything*, 715–717. Yale University Press.

16 Kean, B.H., Mott, K.E., and Russell, A.J. (1978). *Tropical Medicine and Parasitology. Classic Investigations. Vols I and II*. Cornell University Press.

17 Cox, F.E. (2002). History of human parasitology. *Clin. Microbiol. Rev.* 15 (4): 595–612.

18 Brown, T.R. (1898). Studies on trichinosis, with especial reference to the increase of the eosinophilic cells in the blood and muscle, the origin of these cells and their diagnostic importance. *J. Exp. Med.* 3 (3): 315.

19 Harris, M. (1991). *Cannibals and Kings: Origins of Cultures*. Vintage.

20 Wagner, J.A., Schnell, M., and Frank, W. (1988). Zur Vorkommen von *Trichinella* beim einheimischen wild. *Berl. Munch. Tierarztl. Wochenschr.* 101: 413–416.

21 Golab, E. and Sadkowska-Todys, M. (2006). Epidemiologia wlosnicy w Polsce dawniej i dzis. *Ann. Parasitol.* 52 (3): 181–187.

22 Różycki, M., Korpysa-Dzirba, W., Bełcik, A. et al. (2022). Analysis of a Trichinellosis outbreak in Poland after consumption of sausage made of wild boar meat. *J. Clin. Med.* 11 (3): 485.

23 Lipilkina, T., Popov, I., Lipilkin, P., and Ermakov, A. (2021). Trichinosis in the Rostov oblast. *IOP Conf. Ser. Earth Environ. Sci.* 937 (2): 022013. IOP Publishing.

24 Ozeretskovskaya, N.N., Mikhailova, L.G., Sabgaida, T.P., and Dovgalev, A.S. (2005). New trends and clinical patterns of human trichinellosis in Russia at the beginning of the XXI century. *Vet. Parasitol.* 132 (1–2): 167–171.

25 Gómez-Morales, M.A., Mazzarello, G., Bondi, E. et al. (2020). Second outbreak of *Trichinella pseudospiralis* in Europe: clinical patterns, epidemiological investigation and identification of the etiological agent based on the western blot patterns of the patients' serum. *Zoonoses Public Health* 68 (1): 29–37.

26 Barruet, R., Devez, A., Dupouy-Camet, J. et al. (2020). A common source for a trichinellosis outbreak reported in France and Serbia in 2017. *Euro Surveill.* 25 (24): 1900527.

27 Murrell, K.D. and Pozio, E. (2011). Worldwide occurrence and impact of human trichinellosis, 1986–2009. *Emerg. Infect. Dis.* 17 (12): 2194.

28 Forrester, A.T., Nelson, G.S., and Sander, G. (1961). The first record of an outbreak of trichinosis in Africa south of the Sahara. *Trans. R. Soc. Trop. Med. Hyg.* 55 (6): 503–513.

29 Yera, H., Bory, S., Khieu, V., and Caron, Y. (2022). Human trichinellosis in Southeast Asia, 2001–2021. *Food Waterborne Parasitol.* 28: e00171.

30 Tada, K., Suzuki, H., Sato, Y. et al. (2018). Outbreak of *Trichinella* T9 infections associated with consumption of bear meat, Japan. *Emerg. Infect. Dis.* 24 (8): 1532.

31 Zhang, X.Z., Wang, Z.Q., and Cui, J. (2022). Epidemiology of trichinellosis in the People's Republic of China during 2009–2020. *Acta Trop.* 229, 106388.

32 Takahashi, Y., Mingyuan, L., and Waikagul, J. (2000). Epidemiology of trichinellosis in Asia and the Pacific Rim. *Vet. Parasitol.* 93 (3–4): 227–239.

33 Sharma, R., Harms, N.J., Kukka, P.M. et al. (2021). High prevalence, intensity, and genetic diversity of *Trichinella* spp. in wolverine (Gulo gulo) from Yukon, Canada. *Parasit. Vectors* 14 (1): 1–9.

34 Landaeta-Aqueveque, C., Ayala, S., Poblete-Toledo, D., and Canals, M. (2021). Temporal and geographic analysis of trichinellosis incidence in Chile with risk assessment. *Parasit. Vectors* 14 (1): 1–6.

35 Dupouy-Camet, J., Soule, C.L., and Ancelle, T. (1994). Recent news on trichinellosis: another outbreak due to horsemeat consumption in France in 1993. *Parasite* 1 (2): 99–103.

36 Manlick, P.J. and Pauli, J.N. (2020). Human disturbance increases trophic niche overlap in terrestrial carnivore communities. *Proc. Natl. Acad. Sci.* 117 (43): 26842–26848.

37 Furnas, B.J., Goldstein, B.R., and Figura, P.J. (2022). Intermediate fire severity diversity promotes richness of forest carnivores in California. *Divers. Distrib.* 28 (3): 493–505.

38 Forbes, L.B. (2000). The occurrence and ecology of *Trichinella* in marine mammals. *Vet. Parasitol.* 93 (3–4): 321–334.

39 Theodoropoulos, G., Kapel, C.M., Webster, P. et al. (2000). Infectivity, predilection sites, and freeze tolerance of *Trichinella* spp. in experimentally infected sheep. *Parasitol. Res.* 86: 401–405.

40 Reina, D., Munoz-Ojeda, M.C., Serrano, F. et al. (1996). Experimental trichinellosis in goats. *Vet. Parasitol.* 62 (1–2): 125–132.

41 Smith, H.J., Snowdon, K.E., Finley, G.G., and Laflamme, L.F. (1990). Pathogenesis and serodiagnosis of experimental Trichinella spiralis spiralis and Trichinella spiralis nativa infections in cattle. *Can. J. Vet. Res.* 54 (3): 355.

42 Bruschi, F., Dupouy-Camet, J., Kociecka, W. et al. (2002). Opinion on the diagnosis and treatment of human trichinellosis. *Expert Opin. Pharmacother.* 3 (8): 1117–1130.

43 Gottstein, B., Pozio, E., and Nöckler, K. (2009). Epidemiology, diagnosis, treatment, and control of trichinellosis. *Clin. Microbiol. Rev.* 22 (1): 127–145.

44 Blaxter, M.L., De Ley, P., Garey, J.R. et al. (1998). A molecular evolutionary framework for the phylum Nematoda. *Nature* 392 (6671): 71–75. https://doi.org/10.1038/32160. PMID: 9510248.

45 Campbell, W.C. (1988). Trichinosis revisited — another look at modes of transmission. *Parasitol. Today* 4 (3): 83–86.

46 Fröscher, W., Gullotta, F., Saathoff, M., and Tackmann, W. (1988). Chronic trichinosis. *Eur. Neurol.* 28 (4): 221–226.

47 Born, E.W. and Henriksen, S.A. (1990). Prevalence of *Trichinella* sp. in polar bears (Ursus maritimus) from northeastern Greenland. *Polar Res.* 8 (2): 313–315.

48 Pozio, E. and Zarlenga, D.S. (2005). Recent advances on the taxonomy, systematics and epidemiology of *Trichinella*. *Int. J. Parasitol.* 35 (11–12): 1191–1204.

49 Rostami, A., Gamble, H.R., Dupouy-Camet, J. et al. (2017). Meat sources of infection for outbreaks of human trichinellosis. *Food Microbiol.* 64: 65–71.

50 Pozio, E., Paterlini, F., Pedarra, C. et al. (1999). Predilection sites of *Trichinella spiralis* larvae in naturally infected horses. *J. Helminthol.* 73 (3): 233–237.

51 Murrell, K.D. and Pozio, E. (2000). Trichinellosis: the zoonosis that won't go quietly. *Int. J. Parasitol.* 30 (12–13): 1339–1349. https://doi.org/10.1016/s0020-7519(00)00132-6.

52 Capó, V. and Despommier, D.D. (1996). Clinical aspects of infection with *Trichinella* spp. *Clin. Microbiol. Rev.* 9 (1): 47–54. https://doi.org/10.1128/CMR.9.1.47.

53 MacLean, J.D., Poirier, L., Gyorkos, T.W. et al. (1992). Epidemiologic and serologic definition of primary and secondary trichinosis in the Arctic. *J. Infect. Dis.* 165 (5): 908–912. https://doi.org/10.1093/infdis/165.5.908.

54 Neghina, R., Neghina, A.M., and Marincu, I. (2011). Reviews on trichinellosis (III): cardiovascular involvement. *Foodborne Pathog. Dis.* 8 (8): 853–860. https://doi.org/10.1089/fpd.2010.0815.

55 Bruschi, F. and Murrell, K.D. (2002). New aspects of human trichinellosis: the impact of new Trichinella species. *Postgrad. Med. J.* 78 (915): 15–22. https://doi.org/10.1136/pmj.78.915.15.

56 Capo, V.A., Despommier, D.D., and Polvere, R.I. (1998). *Trichinella spiralis*: vascular endothelial growth factor is up-regulated within the nurse cell during the early phase of its formation. *J. Parasitol.* 84 (2): 209.

57 Muñoz-Carrillo, J.L., Muñoz-López, J.L., Muñoz-Escobedo, J.J. et al. (2017). Therapeutic effects of resiniferatoxin related with immunological responses for intestinal inflammation in trichinellosis. *Korean J. Parasitol.* 55 (6): 587–599.

58 Kramer, M.D. and Aita, J.F. (1972). Trichinosis with central nervous system involvement: a case report and review of the literature. *Neurology* 22 (5): 485–485.

59 Gvozdenovic, E., Mitrović, N., Korac, M. et al. (2014). Analysis of clinical picture of trichinellosis in patients hospitalised in clinic for infectious diseases, Belgrade, Serbia. *Int. J. Infect. Dis.* 21: 153.

60 Stewart, G.L. (1989). Biological and immunological characteristics of *Trichinella pseudospiralis*. *Parasitol. Today* 5 (11): 344–349.

61 Jongwutiwes, S., Chantachum, N., Kraivichian, P. et al. (1998). First outbreak of human trichinellosis caused by *Trichinella pseudospiralis*. *Clin. Infect. Dis.* 26 (1): 111–115.

62 MacLean, J.D., Viallet, J., Law, C., and Staudt, M. (1989). Trichinosis in the Canadian Arctic: report of five outbreaks and a new clinical syndrome. *J Infect Dis* 160 (3): 513–520.

63 Wang, Z.Q., Shi, Y.L., Liu, R.D. et al. (2017). New insights on serodiagnosis of trichinellosis during window period: early diagnostic antigens from *Trichinella spiralis* intestinal worms. *Infect. Dis. Poverty* 6 (1).

64 Thawornkuno, C., Nogrado, K., Adisakwattana, P. et al. (2022). Identification and profiling of *Trichinella spiralis* circulating antigens and proteins in sera of mice with trichinellosis. *PLoS One* 17 (3): e0265013. https://pubmed.ncbi.nlm.nih.gov/35271623.

65 Watanabe, N., Bruschi, F., and Korenaga, M. (2005). IgE: a question of protective immunity in *Trichinella spiralis* infection. *Trends Parasitol.* 21 (4): 175–178.

66 Mahannop, P., Setasuban, P., Morakote, N. et al. (1995). Immunodiagnosis of human trichinellosis and identification of specific antigen for *Trichinella spiralis*. *Int. J. Parasitol.* 25 (1): 87–94.

67 Bruschi, F., Tassi, C., and Pozio, E. (1990). Parasite-specific antibody response in *Trichinella* Sp. 3 human infection: a one year follow-up. *Am. J. Trop. Med. Hyg.* 43 (2): 186–193.

68 Engvall, E. and Ljungström, I. (2009). Detection of human antibodies to *Trichinella spiralis* by enzyme-linked immunosorbent assay, ELISA. *Acta Pathol. Microbiol. Scand. Sect. C Immunol.* 83C (3): 231–237.

69 Taher, E.E., Méabed, E.M.H., El Akkad, D.M.H. et al. (2017). Modified dot-ELISA for diagnosis of human trichinellosis. *Exp. Parasitol.* 177: 40–46. https://doi.org/10.1016/j.exppara.2017.04.002.

70 Watt, G., Saisorn, S., Jongsakul, K. et al. (2000). Blinded, placebo-controlled trial of antiparasitic drugs for trichinosis myositis. *J Infect Dis* 182 (1): 371–374. https://doi.org/10.1086/315645.

71 Gyorkos, T.W., Larocque, R., Casapia, M., and Gotuzzo, E. (2006). Lack of risk of adverse birth outcomes after deworming in pregnant women. *Pediatr. Infect. Dis. J.* 25 (9): 791–794. https://doi.org/10.1097/01.inf.0000234068.25760.97.

72 Fouresti, V., Bougnoux, M.E., Ancelle, T. et al. (1988). Randomized trial of albendazole versus tiabendazole plus flubendazole during an outbreak of human trichinellosis. *Parasitol. Res.* 75 (1): 36–41.

73 Kocięcka, W. (2000). Trichinellosis: human disease, diagnosis and treatment. *Vet. Parasitol.* 93 (3–4): 365–383.

74 Chen, C.W., Lee, S.T., Wu, W.T. et al. (2003). Signal transduction for inhibition of inducible nitric oxide synthase and cyclooxygenase-2 induction by capsaicin and related analogs in macrophages. *Br. J. Pharmacol.* 140 (6): 1077–1087.

75 Etewa, S.E., Fathy, G.M., Abdel-Rahman, S.A. et al. (2018). The impact of anthelminthic therapeutics on serological and tissues apoptotic changes induced by experimental trichinosis. *J. Parasit. Dis.* 42 (2): 232–242.

76 *Trichinella spiralis*. https://www.cabidigitallibrary.org/doi/abs/10.1079/9780851994239.0103. https://doi.org/10.1079/9780851994239.0103 CABI Compendium (2022). CABI Compendium.

26

Schistosomiasis: Recent Clinical Reports and Management

Emmanuel K. Ofori[1] *and Akua O. Forson*[2]

[1] *Department of Chemical Pathology, University of Ghana Medical School, Accra, Ghana*
[2] *Department of Medical Laboratory Sciences, School of Biomedical and Allied Health Sciences, University of Ghana, Accra, Ghana*

26.1 Introduction

Schistosomiasis, sometimes referred to as Bilharzia, is a parasitic disease caused by blood flukes (trematode worms) of the genus Schistosoma [1, 2]. It is estimated that at least 251 million individuals worldwide are afflicted with schistosomiasis, and nearly 90% of fatalities occur in tropical and sub-tropical African regions [3–5]. Humans can become infected with schistosomes when larval forms (cercariae) of the parasite released by freshwater snails penetrate the skin during contact with infested water. There are two major forms of schistosomiasis: intestinal and urogenital – caused by five main species of blood fluke [6, 7]. Urogenital schistosomiasis is caused by *Schistosoma haematobium* and intestinal schistosomiasis by *S. guineensis*, *S. intercalatum*, *S. mansoni*, *S. japonicum*, and *S. mekongi* [1, 8].

The life cycle of schistosomes requires a stage of asexual reproduction or development within an intermediate host, which is typically a snail (*Biomphalaria* spp. snails for *S. mansoni*, *Bulinus* spp. snails for *S. haematobium*) [9]. The intermediate host snail releases the free-swimming, infectious cercariae stage of the schistosome. Cercariae can infect humans if they penetrate the skin [10]. Every day, the females can lay anywhere from 100–1000 of eggs. Eggs can be passed out of the body through urine or feces, where they can remain viable for up to seven days. The miracidium is released from the egg upon coming into touch with water. It is guided in its search for the intermediate host, which are snails that live in freshwater, by both light and chemical cues. Once inside the snail, the miracidia begin to reproduce asexually, first as sporocysts and then, later, as cercarial larvae. The larvae of cercariae are characterized by a split tail and the presence of embryonic suckers. After four to six weeks of infection, the cercariae begin to exit the snail and swim around in the water for up to 72 hours in an attempt to find the skin of a definite host that is acceptable for them. When cercariae encounter a suitable host, they burrow through the skin, travel through the bloodstream via the lungs to the liver, and eventually develop into juvenile worms known as schistosomula. Schistosomula develops in the portal vein for about four to six weeks, mates, and then moves to their destination in the mesenteric or perivenous capillaries, where the cycle begins again. Adult schistosomes have a white or grayish body with two terminal suckers, a complicated tegument, a blind digestive tract, and reproductive organs. They can grow to a length of 7–20 mm [2, 11]. Schistosomes can live anywhere from three to five years as adults, but some have been known to live as long as 30 [4, 12].

Three stages of illness progression can be identified in patients: the early, acute infection stage; the latter, chronic infection stage; and the established, active infection stage [13]. Acute schistosomiasis manifests itself in travelers to schistosomiasis-endemic areas with myalgia, abdominal pain in the right upper quadrant, diarrhea (with or without blood), lethargy, malaise, fever, and, in the case of *S. haematobium* infection, hematuria. People who have just returned from a region where schistosomiasis is endemic are at risk for developing an acute case (blood in urine). People in poor rural regions who have had infections for a long period are disproportionately affected by established active and late chronic diseases. Because different species of Schistosoma nest in different anatomical places, chronic and active infections can cause inflammation and organ obstruction due to immunopathological reactions to schistosome eggs trapped in host tissues [13, 14].

Rising Contagious Diseases: Basics, Management, and Treatments, First Edition. Edited by Seth Kwabena Amponsah, Ranjita Shegokar, and Yashwant V. Pathak.

In established active and late chronic infections, inflammation and obstruction occur from the host's immune response to schistosome eggs that have been stuck in tissues [1]. The liver and spleen are also targets of the advanced illness (hepato-splenic schistosomiasis). Late, persistent infection with *S. haematobium*, which lays its eggs in the pelvic venous plexus, leads to urogenital schistosomiasis. Lesions of the bladder wall are the most common manifestation of urogenital schistosomiasis [15, 16]. High-intensity infections (infections with a significant worm burden), particularly infections caused by *S. mansoni* and *S. japonicum*, are associated with a high risk of morbidity [17, 18]. Malnutrition, exercise intolerance (the inability to complete a physical exercise at what would be considered the regularly expected level or duration), bloody diarrhea, chronic pain, and anemia have all been associated with schistosomiasis.

26.1.1 Acute Schistosomiasis

Once a cercaria can enter a host and mature into a schistosomula, the infection progresses to the acute stage, which is characterized by symptoms [19]. Cercarial dermatitis or Swimmer's itch develops a few hours after contact with waterbodies carrying infective cercariae, with the development of an itchy maculopapular rash, limited to areas of the body immersed in water [20]. Acute infection established active infection, and late chronic infections are the three broad stages of schistosome infection progression that overlap and are impacted by the length of time an individual has been infected with schistosomes. The clinical signs and symptoms, in addition to the rates of egg excretion in stool or urine, serve to differentiate between the stages of infection.

Acute schistosomiasis appears in naive individuals one week to three months following their first encounter with *Schistosoma* spp. [19]. Acute schistosomiasis manifests with sudden onset of flu-like symptoms, including high body temperature, muscle aches, fatigue, a dry cough, and gastrointestinal distress. In the days, weeks, or months after exposure, most people experience mild to moderate symptoms. These mild and transient symptoms can be misdiagnosed as malaria, salmonellosis, or acute hepatitis, all of which also have an early onset and are associated with a fever [17]. Acute schistosomiasis often has a benign clinical course, with patients recovering without treatment within 2–10 weeks. However, in some cases, the disease can have a more severe clinical course, causing patients to lose weight, have difficulty breathing, have diarrhea, and develop hepatomegaly [17, 21, 22].

Diagnosing acute schistosomiasis requires a different battery of tests than those used to detect latent infections or advanced stages of the disease. In endemic areas, feces and urine are typically examined under the microscope for schistosome eggs [23]. The reason for this is that by the time an infection is discovered in a person living in these areas, it is usually either at an advanced stage or already active. These microscopic methods often lack the sensitivity required to detect acute infections, which are characterized by low intensity and are most common in newly arrived immigrants or returning travelers from a schistosomiasis-endemic location [4].

26.1.2 Established-Active-Infection

Established-active-infection occurs when full-grown adult worms and robust egg production are present in a *Schistosoma* infected person. During this phase, the existing adult worms in the blood arteries do not create any local irritation or symptoms [13, 24]. This finding can be explained by several mechanisms, including the ability of adult worms to regenerate their surface tegument (the outer protective layer) from somatic stem cells and the binding of host antigens to the surface tegument, which masks the worm antigens on the surface from the immune system of the host [24]. The ability of mature worms to repair their surface tegument using somatic stem cells is another possible explanation [25].

The majority of symptoms and organ-related diseases associated with patent infections are caused by inflammatory reactions to parasite eggs. Antigenic glycoproteins are actively secreted by schistosome eggs. Their job is to help the eggs move from the blood vessel lumen (where they are placed) to the intestinal or urinary bladder lumen (thereby promoting transmission) [26]. When an inflammatory reaction is triggered, there is an accumulation of inflammatory cells such as eosinophils, neutrophils, lymphocytes, and macrophages in granulomas when eggs are present [27]. In addition to triggering an immune response, soluble egg antigens can cause granulomas to develop around entrapped eggs. Eggs stuck in the tissue can cause granulomas to form [28]. Inflammation and granulomatous reactions caused by eggs are the root cause of organ-specific clinical symptoms. High lymphocyte proliferative responses elicited by soluble egg antigens at this stage of the infection gradually decrease as the illness advances and becomes chronic [29].

26.1.3 Late Chronic Infection

Most people who live in schistosomiasis-endemic areas and are constantly exposed to the parasite have steadily decreasing worm burdens beginning in their early teenage years and continuing into adulthood [1]. This happens because the body builds up some resistance to future infections and the number of long-established worms as a result of previous infections naturally decreases with time [13]. The decrease in worm population leads to a concomitant decrease in the number of eggs being shed or laid in the affected person. Due to immune downregulation, newly generated granulomas are smaller in size and produce less inflammation than previously formed granulomas, and the previously formed granulomas gradually vanish when the eggs contained inside them are destroyed [1, 29]. After this, the granulomas transform into scar tissue, also known as fibrous tissue. This is what ultimately alleviates the condition's symptoms [30]. The severity of symptoms often corresponds with the intensity of the infection, and organ-specific morbidity might occur during established active and late chronic stages due to the deposition of parasite eggs and the production of fibrosis. Damage to organs is species-specific because adult worms of different *Schistosoma* species get localized and continue to lay eggs in separate preferred regions. Furthermore, adult worms or eggs can have disastrous effects on their host ecosystem if they locate or swept to unconventional places (e.g., liver, brain, or spinal cord) [1, 31].

26.1.4 Immunology and Host–Parasite Interactions

Schistosomes constitute a particularly challenging case of host–parasite interaction due to the possibility of the simultaneous presence of numerous stages of the parasite's life cycle within the host's body [1, 9]. The life cycle of a parasite consists of several phases, such as tissue-penetrating cercariae, migratory schistosomula, fully developed adult worms, and entrapped parasite eggs. Resistance to reinfection in schistosomiasis has been established in part by immune responses against cercariae and maybe other parasite stages, while pathogenesis is caused by responses against egg antigens [29, 32].

26.1.5 Intestinal Schistosomiasis

The intestinal schistosomiasis-causing adult worms of the species *S. mansoni* and *S. japonicum* reside in the venous plexuses of the colon and lower ileum and the portal system of the liver of their host [33]. Microulcerations, superficial bleeding, and even pseudo-polyposis have been linked to intestinal schistosomiasis because of the inflammation of mucosal granulomatous tissue [34]. Intestinal inflammation is triggered by the transit of parasite eggs or their capture by intestinal tissues. During the healing phase of recurring cycles of ulceration, pseudopolyps emerge as protruding masses of inflammatory tissue. Schistosomiasis of the digestive tract has been related to several symptoms, including abdominal pain and discomfort, loss of appetite, and occasionally bloody diarrhea [9, 35].

26.1.6 Hepato-Splenic Schistosomiasis

Some of the eggs of *S. japonicum* or *S. mansoni* are carried in the bloodstream to the small portal branches of the liver through the portal vein, where they become entrapped in the presinusoidal periportal tissues [22, 34]. When granulomas grow around the eggs, it can lead to an enlarged spleen and liver, especially the left hepatic lobe, and in cases of severe, long-standing active infection often referred to as hepato-splenic schistosomiasis [33, 36]. Hepato-splenomegaly is a common complication of severe childhood diseases and with continued exposure to *Plasmodium* spp. infection, the parasite responsible for malaria can exacerbate the condition [37, 38]. The severity of hepato-splenomegaly decreases over time due to a decline in worm infestation, egg secretion, and egg deposit. Symmer's pipestem fibrosis (also known as severe periportal fibrosis) can develop in some patients after a granulomatous response to eggs [39].

26.1.7 Urogenital Schistosomiasis

The pelvic venous plexus is home to adult *S. haematobium* worms [33]. The transmission of urogenital schistosomiasis is intimately linked to the passage of parasite eggs through the urinary bladder wall and the subsequent deposition of eggs in the tissue of the urinary bladder and genital organs, where they cause symptoms and pathological alterations [9, 33]. Sexually mature *S. haematobium* worms can be found in the pelvic venous plexus. Clusters of viable eggs are found in the urogenital tissues of an individual with a chronic active infection, and these eggs are surrounded by significant inflammatory reactions and robust tissue eosinophilia [27]. Bleeding often accompanies the passage of egg clusters through the epithelial layer and into the lumen of the bladder. Eggs can cause severe tissue inflammation, which in turn can cause tumors and pseudopolyps in the bladder [40, 41]. Inflammation and granuloma formation around the opening of the ureter in the bladder can restrict the passage of urine, leading to swelling of a kidney due to an accumulation of urine [42]. Even if there is no longer an active infection with eggs in the urine, it is still possible to develop bladder cancer after years of exposure to infections and inflammation of the urinary tract [43]. During a cystoscopic examination of the bladder mucosa, late-stage chronic infections are distinguished by an accumulation of calcified tissue eggs that have died. These tissue eggs appear as yellow sandy patches on the bladder's surface. This occurs when the eggs become calcified during late chronic infections, which then blocks egg excretion. The presence of highly prominent cervical lesions is diagnostic of an *S. haematobium* infection [40, 44]. Active-stage lesions, which have severe tissue inflammation and the presence of viable eggs, are included in this category. Analysis of sperm samples from men with genital schistosomiasis reveals an increase in the percentage of leukocytes and the concentration of inflammatory cytokines [45]. Indeed, male and female *S. haematobium* infection symptoms may contribute to HIV transmission [46].

26.2 Parasitology Diagnosis

To detect and confirm an active infection, it is necessary to look for eggs in the stool, urine, or rectal biopsy samples [47]. The use of microscopy is useful for this purpose. Parasitological diagnosis of schistosomiasis is often made by filtering a standard volume of stool or urine and then counting all of the eggs that are present in that volume under a microscope in areas where the disease is endemic [48]. The sensitivity of this microscopic analyses methods is low since egg excretion may vary significantly at different times and days from one person to another. This variability necessitates collecting a large number of biological samples over several days or doing multiple analyses on the same specimen. Further, despite the severity of the clinical indications, egg excretion may become very low or absent if the infection has progressed to a late chronic stage [14]. Since infection rates are often quite low in people who have traveled or been exposed in the past, this diagnostic approach is rarely sensitive enough; thus, additional methods, such as concentration and sedimentation protocols, are required [49]. Rectal snip samples and tissue biopsies have a high detection rate for schistosomes, especially *S. haematobium*, and can be used to diagnose infections in people who have traveled from schistosome-endemic areas [50].

26.3 Biomarkers: Screening, Detection of Worm Antigens and Specific Antibodies

Schistosomiasis can negatively impact a person's quality of life, but assessing this impact can be challenging due to the disease's complex life cycle. Direct parasitological methods (microscopy), although easy to use and inexpensive, are less sensitive [23, 51]. Due to schistosome size and structure, their characteristic lateral or terminal spine, and the living miracidium (in fresh samples) that has movable cilia and pulsating excretory cells, the eggs are simple to see and identify using microscopy. However, the direct wet slide approach is not very sensitive for egg detection, although concentration methods can be employed; however, the concentration methods can miss eggs in schistosomiasis [52]. Urine samples should be collected about noon or after vigorous exercise, and the urine should be concentrated using sedimentation, centrifugation, or filtering. A predetermined volume of urine, which is typically 10 ml, is poured through a filter made of paper or nitrocellulose so that the eggs can be counted immediately under a microscope. This procedure is performed so that a quantitative estimate of the severity of the infection may be obtained. After that, the intensity can be stated as eggs per 10 ml.

One of the most common and easy-to-use screening techniques for *S. haematobium*-related microhaematuria is the use of a reagent strip test [53]. A urine sample is far more easily obtained than a stool or blood sample, so it can be tested right away, which is a major logistical plus. Schistosomes consume hemoglobin in their guts when they feed on red blood cells [54]. Following hemoglobin digestion in the schistosome gut, peptides and heme and glycoproteins are released into the circulation of infected individuals [55]. Circulating anodic (CCAs) and cathodic (CAAs) antigens are two types of proteoglycans that are produced in the body [56]. They can be found in blood and urine by using an enzyme-linked immunosorbent test (ELISA) or a monoclonal-antibody-based lateral flow assay. Immunodiagnosis of schistosomiasis has employed the detection of CAA in serum and CCA in urine as an indication of a full-fledged infection with real, living worms [55]. Studies of schistosomiasis epidemiology have benefited greatly from the use of these assays for circulating antigens;

however, these tests may not be the best option in areas with a low worm burden or among travelers, who may only have a few worms [57], [58].

A point-of-care (POC) test for detecting *S. mansoni* CCA in urine has been developed by Rapid Medical Diagnostics of Pretoria, South Africa [59]. Communities with *S. mansoni* infections are being prescreened for antigen initiatives using this assay. Because it is more sensitive than traditional parasitological methods, PCR-based detection of parasite DNA in feces or urine is becoming increasingly popular in high-resource settings [60, 61]. It has been shown that DNA from *Schistosoma* spp. may be detected in vaginal lavage and cerebrospinal fluid samples for the diagnosis of urogenital schistosomiasis and neuro schistosomiasis, respectively [51]. Although serological techniques for the detection of anti-schistosome-specific antibodies can be very helpful, they cannot tell the difference between active infection and past exposure in schistosomiasis-endemic regions; hence, many patients in these areas remain seropositive for years after therapy despite being cured [62]. To successfully achieve morbidity reduction and eventually disease elimination, accurate detection of *Schistosoma* spp. infections in both snail and mammalian hosts would be essential. This is especially true in locations where comprehensive control measures have successfully reduced the prevalence of schistosomiasis to very low levels. As a result, the development of new diagnostic assays that are both more cost-effective and user-friendly in the field should be prioritized in any future applied schistosomiasis research that takes place.

26.4 Radiology

Ultrasonography can be utilized to see the pathological alterations associated with active and late-stage chronic schistosomiasis. Scientists have been using portable ultrasonography equipment to study population-wide shifts in organ-specific diseases for over 20 years [63]. Ultrasonography is an essential instrument for monitoring the direct effect of treatments on the severity of schistosomiasis. Additionally, there are well-established methodologies available for the classification and grading of pathological lesions that can be applied in conjunction with ultrasonography. Radiological procedures such as intravenous pyelography, a radiographic examination in which contrast material is administered, can assist in the identification of kidney injury. These treatments allow for the visualization of renal, ureteral, and bladder disease. Cystoscopy and endoscopy are diagnostic procedures that are used to evaluate the digestive system and the urinary bladder, respectively, for signs of disease or dysfunction. Cystoscopy examines the urinary bladder. Endoscopy examines the digestive tract [64]. Both magnetic resonance imaging (MRI) and computed tomography (CT) are considered to be the gold standards when it comes to making a diagnosis of neuro schistosomiasis [65].

26.5 Management

Treatment for schistosomiasis involves eliminating the parasites responsible for the sickness (or at least relieving the symptoms) and attending to any complications that may arise as a result of the infection. Schistosomiasis is controlled by giving specific therapy to those who contract the disease in an endemic area or who bring it back with them after visiting or migrating elsewhere [19]. Curative therapy is given in the form of the anti-schistosomal drug praziquantel; however, numerous doses of the drug are required before the infection can be treated (i.e., until no eggs are detected upon microscopic examination of feces or urine after therapy). Global community-based schistosomiasis control programmes with a major focus on mass drug administration (MDA) can drastically reduce morbidity [66, 67].

26.5.1 Praziquantel

Praziquantel, an acylated quinoline-pyrazine derivative, has been the gold standard for treating schistosomiasis for more than three decades [68]. Although the exact mechanism of action of oral praziquantel against adult worms of all *Schistosoma* spp. has not yet been discovered, it is safe and effective [16, 69]. Worldwide, praziquantel is the backbone of MDA-focused community-based schistosomiasis control strategies aiming to reduce morbidity [70]. Praziquantel is safe and effective for use in pregnant women after the first trimester and in children older than four years old [71]. Human testing of this potential anti-schistosomal medicine started in 1978, and by 1988, Bayer had released it to the public under the trade name Biltricide. The WHO recommended dose of praziquantel (40 mg/kg) is supported by evidence from meta-analyses that cured 94.7% of *S. japonicum* infections, 77.1% of *S. haematobium* infections, 76.7% of *S. mansoni* infections, and 63.5% of combination *S. haematobium* and *S. mansoni* infections [72, 73]. Praziquantel is well absorbed, and its hepatic clearance is also high (that is, over 80% of the drug is metabolized in the liver to inactive metabolites that are eliminated by the kidneys in the form of urine).

Also useful in combating helminth infections transmitted through the soil is the medicine albendazole. In places like Africa and Asia where both *Schistosoma* species and soil-transmitted helminths are co-endemic, people often carry both types of worms simultaneously. In these areas, the WHO advises a combined preventative medication of praziquantel and albendazole, with dosing schedules determined by the community prevalence of the illnesses [5, 74]. Praziquantel medication has been linked to several undesirable side effects, and these include dizziness, headache, abdominal pain, temporary nausea, pruritus (itch), and rash [75]. These adverse reactions are reported to be linked to the drug's interaction with antigens released from dead or dying worms (ref). Schistosomiasis patients who also have cysticercosis (cysts in the central nervous system, skin, eyes, and muscle), an infection caused by *Taenia solium* (pork tapeworm), are at increased risk of developing permanent eye lesions and seizures due to the severe inflammatory reactions brought on by the dying *T. solium* worms [76]. Very tiny quantities of praziquantel in its active form are released in breast milk; therefore, it is not expected that there will be any negative effects on the kid. However, it is advised that mothers do not breastfeed on the day of their praziquantel treatment and for the subsequent 72 hours [17].

26.5.2 Limitations of Praziquantel

There have also been isolated cases of *S. mansoni* and *S. haematobium* infections that show little to no response to praziquantel even in areas where the drug has been widely prescribed [2, 77]. Furthermore, some experimental evidence in mice revealed that schistosome parasites in mice that were repeatedly treated *in vivo* with praziquantel developed into adults that were less sensitive to the drug, and that this effect appears to be compounded with each succeeding generation of worms [78]. In addition, schistosome worms that have developed a resistance to drugs have higher than normal amounts of a protein called P-glycoprotein [79]. Individuals with schistosomiasis who have been treated with many regimens of medication over a long period have shown no deterioration in the drug's effectiveness in some studies conducted in varied endemic settings [80]. The emergence of resistance to praziquantel cannot be ruled out, as MDA is being expanded, there is also growing worry about the dangers of over-reliance on a single medication [81]. The potential for drug resistance and the ineffectiveness of praziquantel further emphasize the need for ongoing study into the development of novel compounds and the enhancement of praziquantel formulations.

26.5.3 Artemisinin and its Derivatives

Artemisinin, the active component of the medicinal herb *Artemisia annua*, is the basis for the antimalarial drugs artesunate and artemether; however, these drugs only have half-lives of one to two hours, and migrating schistosomula are resistant to them [82]. Artemether has been demonstrated to kill schistosomes by reacting with haemin, an iron-containing molecule [83]. However, it is not advised to employ artemether and artemisinin-based combinations for schistosomiasis chemoprophylaxis in malaria-endemic areas due to the risk of selecting artemisinin-resistant *Plasmodium* spp. Parasites. Artemether and other artemisinins, in conjunction with praziquantel, have been proposed as a viable therapy to increase cure rates (CR) in areas of the world where schistosomiasis transmission is ongoing [84]. Further, a meta-analysis demonstrated that compared to praziquantel monotherapy, the CR achieved by combination therapy was twice as high in human patients [85]. However, species of the parasite [80], parasite stage [86], and infection intensity can influence the treatment outcomes with praziquantel. Studies conducted in western Kenya have revealed children with *S. mansoni* infection responded better to praziquantel monotherapy than to another combination therapy consisting of artemether with sulfalene plus pyrimethamine [87]. Adults infected with *S. mansoni* also responded better to praziquantel monotherapy [88].

26.5.4 Adjuvants

Corticosteroids, such as prednisone, are commonly used for the treatment of acute schistosomiasis usually within two months of cercarial exposure [89]. Praziquantel can be combined with corticosteroids and anticonvulsants to treat patients with neuro-schistosomiasis. Whenever neuro-schistosomiasis is diagnosed and treated promptly, the prognosis is usually good. The central nervous system inflammation brought on by schistosome egg granulomas can be reduced or avoided altogether with the help of corticosteroids. In addition, if an acute allergic reaction does occur, corticosteroids can help lessen its impact. Anticonvulsants are effective against the seizures brought on by neuro schistosomiasis, but long-term use of these drugs is usually not necessary [17, 89]. People who have been diagnosed with intestinal schistosomiasis have a risk of developing the more severe form of the disease known as hepato-splenic schistosomiasis, which can affect both the liver and the spleen [90, 91]. This condition can cause esophageal varices and gastrointestinal bleeding even if liver function is preserved [92]. Portal hypertension and splenomegaly are symptoms of this condition.

26.5.5 New Anti-Schistosomal Drugs

There is an essential need to identify an alternative or complementary treatment, to supplement or compete with praziquantel. This is because the potential benefits of praziquantel are not fully realized, as well as the possibility that resistance will develop. As an alternate form of treatment, new chemical compounds or repurposed pharmaceuticals are required, and the hunt for their development should continue. The annotated genome sequences of common Schistosoma spp. that infect humans have just been made publically available. This has rekindled the quest for new therapeutic targets that are effective against schistosomes [93]. Additionally, pharmacological screening of anticancer drugs and other medicines with repurposing potential for use in treating schistosomiasis has been conducted *in vitro* and *in vivo* [94, 95]. Today, drug development initiatives for schistosomiasis utilize molecular modeling, drug repurposing research, structure-based drug design, and target-focused chemical screening. Metabolic rate-limiting processes in schistosomes can now be predicted thanks to the biochemical mapping of metabolic pathways made feasible by these methods. This, in turn, can help find compounds that are existing drug targets or have the potential to become new therapeutic targets in other organisms. Recent advances in genome and transcriptomics sequencing have allowed the identification of essential molecular components of the apoptosis signaling pathway in schistosomes, one of the most critical signaling mechanisms in these parasites [96]. Many drugs, especially those used to treat cancer, have been created specifically to induce or hasten apoptosis. Several of these compounds are now being considered as potentially beneficial anti-schistosomal therapy [97, 98].

26.5.6 Vaccines

The elimination of schistosomiasis will require additional interventions including the development of schistosomiasis vaccines and multisectoral approaches. There is presently no schistosomiasis vaccine available for widespread use. Numerous putative, protective antigens that could be candidates for vaccines have been identified through a procedure that entails identifying protective antigens, selecting and cloning the most promising among them, and then ensuring that large-scale manufacture is feasible [99]. The National Institute of Allergy and Infectious Diseases (NIAID) in the United States and the Bill & Melinda Gates Foundation have bolstered funding for the development of schistosomiasis vaccines and their evaluation in public health programmes.

26.6 Conclusion

There is a growing interest in the field of parasitic helminth infection. Schistosomiasis is an extremely dangerous threat to public health, and it is more prevalent in economically disadvantaged regions. This condition can have an effect on virtually every organ in the body, and there is a diverse range of clinical symptoms. Schistosomiasis can show its symptoms in a broad range of different ways, depending on the stage the disease is in and the particular clinical manifestation that the patient has. Schistosomiasis can go untreated for a very long time, which is when secondary clinical indications will begin to appear. The apoptotic signaling channel in schistosomes is one of the most significant signaling systems in these parasites, and recent achievements in research may enable us to make advances in the discovery of alternative medicines to the usage of praziquantel and possibly succeed in the development of effective vaccines.

References

1 Colley, D.G. et al. (2014). Human schistosomiasis. *The Lancet* 383 (9936): 2253–2264.
2 Gryseels, B. et al. (2006). Human schistosomiasis. *The Lancet* 368 (9541): 1106–1118.
3 Carbonell, C. et al. (2021). Clinical Spectrum of schistosomiasis: an update. *J. Clin. Med.* 10 (23): 5521.
4 Aula, O.P. et al. (2021). Schistosomiasis with a focus on Africa. *Trop. Med. Infect. Dis.* 6 (3): 109.

5 World Health Organization, Schistosomiasis (Bilharzia). 2022: https://www.who.int/health-topics/schistosomiasis#tab=tab_1.
6 Deol, A.K. et al. (2019). Schistosomiasis—assessing progress toward the 2020 and 2025 global goals. *N. Engl. J. Med.* 381 (26): 2519–2528.
7 World Health Organization (2022). *WHO Guideline on Control and Elimination of Human Schistosomiasis*. World Health Organization.
8 Chuah, C. et al. (2019). Schistosomiasis in Malaysia: a review. *Acta Trop.* 190: 137–143.
9 Nelwan, M.L. (2019). Schistosomiasis: life cycle, diagnosis, and control. *Curr. Ther. Res. Clin. Exp* 91: 5–9.
10 McKerrow, J. and Salter, J. (2002). Invasion of skin by Schistosoma cercariae. *Trends Parasitol.* 18 (5): 193–195.
11 Adenowo, A.F. et al. (2015). Impact of human schistosomiasis in sub-Saharan Africa. *Braz. J. Infect. Dis.* 19: 196–205.
12 Verjee, M.A. (2019). Schistosomiasis: still a cause of significant morbidity and mortality. *Res. Rep. Trop. Med.* 10: 153.
13 McManus, D.P., Bergquist, R., Cai, P. et al. (2020). Schistosomiasis—from immunopathology to vaccines. *Sem. Immunopathol.* 42: 355–371.
14 Clerinx, J. and Van Gompel, A. (2011). Schistosomiasis in travellers and migrants. *Travel Med. Infect. Dis.* 9 (1): 6–24.
15 Botelho, M.C., Figueiredo, J., and Alves, H. (2015). Bladder cancer and urinary schistosomiasis in Angola. *J. Nephrol. Res.* 1 (1): 22.
16 Bamgbola, O.F. (2014). Urinary schistosomiasis. *Pediatr. Nephrol.* 29 (11): 2113–2120.
17 Gray, D.J. et al. (2011). Diagnosis and management of schistosomiasis. *BMJ* 342: d2651.
18 Relman, D.A. and Choffnes, E.R. (ed.) (2011). *The Causes and Impacts of Neglected Tropical and Zoonotic Diseases: Opportunities for integrated intervention strategies*, 323–341. National Academies Press.
19 Inobaya, M.T. et al. (2014). Prevention and control of schistosomiasis: a current perspective. *Res. Rep. Trop. Med.* 5: 65.
20 Tracz, E.S. et al. (2019). Outbreak of Swimmer's itch in Denmark. *Acta Derm. Venereol.* 99 (12): 1116–1120.
21 Rollinson, D., Knopp, S., Levitz, S. et al. (2013). Time to set the agenda for schistosomiasis elimination. *Acta Trop. 128* (2): 423–440.
22 Nelwan, M.L. (2019). Schistosomiasis: life cycle, diagnosis, and control. *Curr. Ther. Res. 91*: 5–9.
23 Ajibola, O. et al. (2018). Tools for detection of schistosomiasis in resource limited settings. *Med. Sci.* 6 (2): 39.
24 Nepal, P. et al. (2019). Multisystem imaging review of human schistosomiasis: characteristic imaging findings. *Clin. Imaging* 54: 163–171.
25 Salawu, O.T. and Odaibo, A.B. (2014). Maternal schistosomiasis: a growing concern in sub-Saharan Africa. *Pathog. Glob. Health* 108 (6): 263–270.
26 Schwartz, C. and Fallon, P.G. (2018). Schistosoma "eggs-iting" the host: granuloma formation and egg excretion. *Front. Immunol.* 9: 2492.
27 Costain, A.H., MacDonald, A.S., and Smits, H.H. (2018). Schistosome egg migration: mechanisms, pathogenesis and host immune responses. *Front. Immunol.* 9: 3042.
28 Hams, E., Aviello, G., and Fallon, P.G. (2013). The schistosoma granuloma: friend or foe? *Front. Immunol.* 4: 89.
29 Colley, D. and Secor, W. (2014). Immunology of human schistosomiasis. *Parasite Immunol.* 36 (8): 347–357.
30 Wynn, T.A. et al. (2004). Immunopathogenesis of schistosomiasis. *Immunol. Rev.* 201 (1): 156–167.
31 Barsoum, R.S., Esmat, G., and El-Baz, T. (2013). Human schistosomiasis: clinical perspective. *J. Adv. Res.* 4 (5): 433–444.
32 Hagan, P. (1992). Reinfection, exposure and immunity in human schistosomiasis. *Parasitol. Today* 8 (1): 12–16.
33 Nation, C.S. et al. (2020). Schistosome migration in the definitive host. *PLoS Negl.Trop. Dis.* 14 (4): e0007951.
34 Elbaz, T. and Esmat, G. (2013). Hepatic and intestinal schistosomiasis: review. *J. Adv. Res.* 4 (5): 445–452.
35 Elbaz, T. and Esmat, G. (2013). Hepatic and intestinal schistosomiasis. *J. Adv. Res.* 4 (5): 445–452.
36 Shaker, Y., Samy, N., and Ashour, E. (2014). Hepatobiliary schistosomiasis. *J. Clin. Transl. Hepatol.* 2 (3): 212.
37 Semenya, A.A. et al. (2012). Schistosoma mansoni infection impairs antimalaria treatment and immune responses of rhesus macaques infected with mosquito-borne Plasmodium coatneyi. *Infect. Immun.* 80 (11): 3821–3827.
38 Anchang-Kimbi, J.K. et al. (2017). Coinfection with Schistosoma haematobium and Plasmodium falciparum and anaemia severity among pregnant women in Munyenge, Mount Cameroon area: a cross-sectional study. *J. Parasitol. Res.* 2017: 6173465.
39 Wilson, S. et al. (2010). Health implications of chronic hepatosplenomegaly in Kenyan school-aged children chronically exposed to malarial infections and Schistosoma mansoni. *Trans. R. Soc. Trop. Med. Hyg.* 104 (2): 110–116.
40 Cozzi, D. et al. (2020). Ultrasound findings in urogenital schistosomiasis: a pictorial essay. *J. Ultrasound* 23 (2): 195–205.
41 Ferrara, D. et al. (2018). Sonographic early findings in a case of bladder schistosomiasis. *J. Ultrasound* 21 (3): 259–263.
42 Beer, A. and Rummeny, E.J. (2016). Kidney, urinary tract, and bladder. In: *Diagnostic and Interventional Radiology* (ed. T.J. Vogl, W. Reith, and E. Rummeny), 875–915. Berlin, Heidelberg: Springer.
43 Saginala, K. et al. (2020). Epidemiology of bladder cancer. *Med. Sci.* 8 (1): 15.

44 Rafferty, H. et al. (2021). Association between cervical dysplasia and female genital schistosomiasis diagnosed by genital PCR in Zambian women. *BMC Infect. Dis.* 21 (1): 691.

45 Chohan, K.R., Kling, C.A., and Byler, T.K. (2021). Schistosoma haematobium ova in human semen: a case report. *F&S Rep.* 2 (1): 126–128.

46 Wall, K.M. et al. (2018). Schistosomiasis is associated with incident HIV transmission and death in Zambia. *PLoS Negl.Trop. Dis.* 12 (12): e0006902.

47 Wichmann, D. et al. (2009). Diagnosing schistosomiasis by detection of cell-free parasite DNA in human plasma. *PLoS Negl. Trop. Dis.* 3 (4): e422.

48 Utzinger, J. et al. (2015). New diagnostic tools in schistosomiasis. *Clin. Microbiol. Infect.* 21 (6): 529–542.

49 Ngwese, M.M. et al. (2020). Diagnostic techniques of soil-transmitted helminths: impact on control measures. *Trop. Med. Infect. Dis.* 5 (2): 93.

50 Tzanetou, K. et al. (2010). Intestinal schistosomiasis caused by both Schistosoma intercalatum and Schistosoma mansoni. *Travel Med. Infect. Dis.* 8 (3): 184–189.

51 Weerakoon, K.G. et al. (2015). Advances in the diagnosis of human schistosomiasis. *Clin. Microbiol. Rev.* 28 (4): 939–967.

52 Alemu, M., Zigta, E., and Derbie, A. (2018). Under diagnosis of intestinal schistosomiasis in a referral hospital, North Ethiopia. *BMC Res. Notes* 11 (1): 1–5.

53 Mazigo, H.D. et al. (2021). Primary health care facilities capacity gaps regarding diagnosis, treatment and knowledge of schistosomiasis among healthcare workers in North-Western Tanzania: a call to strengthen the horizontal system. *BMC Health Serv. Res.* 21 (1): 1–9.

54 Skelly, P.J. et al. (2014). Schistosome feeding and regurgitation. *PLoS Pathog.* 10 (8): e1004246.

55 Oliveira, M.F. et al. (2005). Structural and morphological characterization of hemozoin produced by Schistosoma mansoni and Rhodnius prolixus. *FEBS Lett.* 579 (27): 6010–6016.

56 van Lieshout, E.A. (1996). *Detection of the Circulating Antigens CAA and CCA in Human Schistosoma Infections: Immunodiagnostic and Epidemiological Applications*. Leiden University.

57 Corstjens, P.L. et al. (2020). Circulating anodic antigen (CAA): a highly sensitive diagnostic biomarker to detect active Schistosoma infections—improvement and use during SCORE. *Am. J. Trop. Med. Hyg.* 103 (1 Suppl): 50.

58 Hawkins, K.R. et al. (2016). Diagnostic tests to support late-stage control programs for schistosomiasis and soil-transmitted helminthiases. *PLoS Negl.Trop. Dis.* 10 (12): e0004985.

59 Colley, D.G. et al. (2020). Evaluation, validation, and recognition of the point-of-care circulating cathodic antigen, urine-based assay for mapping Schistosoma mansoni infections. *Am. J. Trop. Med. Hyg.* 103 (1 Suppl): 42.

60 Guegan, H. et al. (2019). Real-time PCR for diagnosis of imported schistosomiasis. *PLoS Negl.Trop. Dis.* 13 (9): e0007711.

61 Fuss, A., Mazigo, H.D., and Mueller, A. (2021). Detection of Schistosoma mansoni DNA using polymerase chain reaction from serum and dried blood spot card samples of an adult population in North-Western Tanzania. *Infect. Dis. Poverty* 10 (1): 15.

62 Hinz, R. et al. (2017). Serological approaches for the diagnosis of schistosomiasis–a review. *Mol. Cell. Probes* 31: 2–21.

63 Cimini, A. et al. (2021). Medical imaging in the diagnosis of schistosomiasis: a review. *Pathogens* 10 (8): 1058.

64 Akiyama, Y. et al. (2020). Interstitial cystitis/bladder pain syndrome: the evolving landscape, animal models and future perspectives. *Int. J. Urol.* 27 (6): 491–503.

65 Kamal, I. et al. (2020). Diagnosis of neuroschistosomiasis: how to overcome this challenge. *Libyan J. Med. Sci.* 4 (1): 35.

66 Burnim, M., Ivy, J.A., and King, C.H. (2017). Systematic review of community-based, school-based, and combined delivery modes for reaching school-aged children in mass drug administration programs for schistosomiasis. *PLoS Negl.Trop. Dis.* 11 (10): e0006043.

67 Chong, N.S. et al. (2021). Modelling the ability of mass drug administration to interrupt soil-transmitted helminth transmission: community-based deworming in Kenya as a case study. *PLoS Negl.Trop. Dis.* 15 (8): e0009625.

68 Siqueira, L.D.P. et al. (2017). Schistosomiasis: drugs used and treatment strategies. *Acta Trop.* 176: 179–187.

69 Fukushige, M. et al. (2021). Efficacy of praziquantel has been maintained over four decades (from 1977 to 2018): a systematic review and meta-analysis of factors influence its efficacy. *PLoS Negl.Trop. Dis.* 15 (3): e0009189.

70 Toor, J. et al. (2018). Are we on our way to achieving the 2020 goals for schistosomiasis morbidity control using current World Health Organization guidelines? *Clin. Infect. Dis.* 66 (suppl_4): S245–S252.

71 Stothard, J.R. et al. (2013). Schistosomiasis in African infants and preschool children: let them now be treated! *Trends Parasitol.* 29 (4): 197–205.

72 Zwang, J. and Olliaro, P.L. (2014). Clinical efficacy and tolerability of praziquantel for intestinal and urinary schistosomiasis—a meta-analysis of comparative and non-comparative clinical trials. *PLoS Negl.Trop. Dis.* 8 (11): e3286.

73 Hailegebriel, T., Nibret, E., and Munshea, A. (2021). Efficacy of praziquantel for the treatment of human schistosomiasis in Ethiopia: a systematic review and meta-analysis. *J. Trop. Med.* 2021: 2625255.

74 Kabatende, J. et al. (2022). Safety of praziquantel and albendazole coadministration for the control and elimination of schistosomiasis and soil-transmitted helminths among children in Rwanda: an active surveillance study. *Drug Saf.* 45 (8): 909–922.

75 Onakpoya, I.J. (2019). Antihelminthic drugs. In: *Side Effects of Drugs Annual* (ed. S.D. Ray), 339–349. Elsevier.

76 Wang, W., Dai, J.R., Li, H.J. et al. (2012, 2012). The sensitivity of Schistosoma japonicum to praziquantel: a field evaluation in areas with low endemicity of China. *Am. J. Trop. Med. Hygiene 86* (5): 834–836.

77 Wang, W. et al. (2010). Is there reduced susceptibility to praziquantel in Schistosoma japonicum? Evidence from China. *Parasitology* 137 (13): 1905–1912.

78 Melman, S.D. et al. (2009). Reduced susceptibility to praziquantel among naturally occurring Kenyan isolates of Schistosoma mansoni. *PLoS Negl.Trop. Dis.* 3 (8): e504.

79 James, C.E., Hudson, A.L., and Davey, M.W. (2009). Drug resistance mechanisms in helminths: is it survival of the fittest? *Trends Parasitol.* 25 (7): 328–335.

80 King, C.H. et al. (2011). Utility of repeated praziquantel dosing in the treatment of schistosomiasis in high-risk communities in Africa: a systematic review. *PLoS Negl.Trop. Dis.* 5 (9): e1321.

81 Doenhoff, M.J., Cioli, D., and Utzinger, J. (2008). Praziquantel: mechanisms of action, resistance and new derivatives for schistosomiasis. *Curr. Opin. Infect. Dis.* 21 (6): 659–667.

82 Lam, N.S. et al. (2018). Artemisinin and its derivatives in treating helminthic infections beyond schistosomiasis. *Pharmacol. Res.* 133: 77–100.

83 Xiao, S. et al. (2000). Effect of artemether against Schistosoma haematobium in experimentally infected hamsters. *Int. J. Parasitol.* 30 (9): 1001–1006.

84 Mnkugwe, R.H. et al. (2020). Efficacy and safety of praziquantel and dihydroartemisinin piperaquine combination for treatment and control of intestinal schistosomiasis: a randomized, non-inferiority clinical trial. *Int. J. Parasitol.* 14 (9): e0008619.

85 Moser, W., Schindler, C., and Keiser, J. (2019). Drug combinations against soil-transmitted helminth infections. *Adv. Parasitol.* 103: 91–115.

86 Coles, G.C. and Kinoti, G.K. (1997). Defining resistance in Schistosoma. *Parasitol. Today* 13 (4): 157–158.

87 Obonyo, C.O., Muok, E.M., and Mwinzi, P.N. (2010). Efficacy of artesunate with sulfalene plus pyrimethamine versus praziquantel for treatment of Schistosoma mansoni in Kenyan children: an open-label randomised controlled trial. *Lancet Infect. Dis.* 10 (9): 603–611.

88 Bergquist, R., Utzinger, J., and Keiser, J. (2017). Controlling schistosomiasis with praziquantel: how much longer without a viable alternative? *Infect. Dis. Poverty* 6 (1): 1–10.

89 Ross, A.G. et al. (2012). Neuroschistosomiasis. *J. Neurol.* 259 (1): 22–32.

90 Papamatheakis, D.G. et al. (2014). Schistosomiasis-associated pulmonary hypertension. *Pulm Circ.* 4 (4): 596–611.

91 Abdela, S.G. et al. (2020). Hepatosplenic schistosomiasis, the ignored morbidity: experience from a referral hospital in Ethiopia. *Trans. R. Soc. Trop. Med. Hyg.* 115 (1): 57–62.

92 Opio, C.K. et al. (2021). Hepatic schistosomiasis, upper gastrointestinal bleeding, and health related quality of life measurements from the Albert Nile Basin. *J. Patient-Rep. Outcomes* 5 (1): 1–13.

93 Swain, M.T. et al. (2011). Schistosoma comparative genomics: integrating genome structure, parasite biology and anthelmintic discovery. *Trends Parasitol.* 27 (12): 555–564.

94 Cowan, N. and Keiser, J. (2015). Repurposing of anticancer drugs: in vitro and in vivo activities against Schistosoma mansoni. *Parasit. Vectors* 8 (1): 1–9.

95 Gouveia, M.J. et al. (2018). Drug repurposing for schistosomiasis: combinations of drugs or biomolecules. *Pharmaceuticals* 11 (1): 15.

96 Liu, Z. et al. (2022). Pathology and molecular mechanisms of Schistosoma japonicum-associated liver fibrosis. *Front. Cell. Infect. Microbiol.* 12: 1035765.

97 Lee, E.F. and Fairlie, W.D. (2015). Repurposing apoptosis-inducing cancer drugs to treat schistosomiasis. *Future Med. Chem.* 7 (6): 707–711.

98 Lee, E.F. et al. (2014). Apoptosis in schistosomes: toward novel targets for the treatment of schistosomiasis. *Trends Parasitol.* 30 (2): 75–84.

99 Tebeje, B.M. et al. (2016). Schistosomiasis vaccines: where do we stand? *Parasit. Vectors* 9 (1): 1–15.

27

Granulomatous Amebic Encephalitis: Evolutionary Dynamics, Advances in Diagnostics and Therapeutic Interventions

Lucy Mohapatra[1], *Alok S. Tripathi*[2], *Bhupendra G. Prajapati*[3], *Alka*[1], *Deepak Mishra*[1], *Mohammad Yasir*[1], *and Rahul K. Maurya*[1]

[1] *Amity Institute of Pharmacy, Lucknow, Amity University Uttar Pradesh, Noida, Uttar Pradesh, India*
[2] *Department of Pharmacology, ERA College of Pharmacy, ERA University, Lucknow, Uttar Pradesh, India*
[3] *Department of Pharmaceutics and Pharmaceutical Technology, Shree S K Patel College of Pharmaceutical Education and Research, Ganpat University, Mehsana, Gujarat, India*

27.1 Introduction

GAE is a type of infection involving the brain and related systems that can range from being acute to becoming chronic. It is caused by amoebae including *Acanthamoeba species, Balamuthia mandrillaris*, and *Naegleria fowleri* [1]. Acanthamoebic encephalitis carried on by this infection is more prevalent in immunocompromised people [2]. GAE can result in symptoms such as headaches, meningism, fever, changes in mental state, visual abnormalities, ataxia, hemiparesis, and seizures [3, 4]. Although the precise incubation period is unclear, it is most presumably weeks or months. An estimated duration of a week to two months may pass before the sickness manifests itself in about 39 days on an average [5, 6]. Approximately, 97–98% of GAE patients die from it every year. Susceptibility testing needs to be done if a culture is positive [7]. Multiple antimicrobial substances can make *Acanthamoeba amoebae* vulnerable. Depending on their point of entrance, these amoebae either spread via the lungs, sinuses, or skin via a hematogenous route. A skin lesion or lung infection that emerged a few months prior served as a predictor to the impending CNS infection in a number of these cases. Over 95% of instances of GAE caused by *B. mandrillaris* has resulted in death, which is a very high fatality rate [8]. There is an improved prognosis for these patients since early diagnosis may be possible in around half of the cases where the patient survived from GAE caused by the species *B. mandrillaris* [9]. Typically, miltefosine and pentamidine isethionate are included in multidrug regimens used to treat patients with GAE caused by *B. mandrillaris*. Particularly in cases of CNS balamuthiasis, very few cases of survival have been observed [10].

Successful patient outcomes depend heavily on early diagnosis and aggressive treatment. Other CNS pathogens, such as viruses, bacteria, and fungi, have symptoms that are similar to those of GAE. Because of this, diagnosing GAE can be difficult and demands considerable skill. CT scan or MRI scans of the brain may unfold tumors or lesions that are suggestive of CNS abnormalities. Even though they do not confirm GAE, the cerebrospinal fluid (CSF) results are useful for identifying CNS involvement [11, 12]. Pathogenic protozoans known as *Acanthamoeba species* are prevalent and can be found in soil, air, and water. Both the trophozoite (active) and cyst (dormant) forms of *Acanthamoeba* exists [13]. The cyst can sustain and exist for years in various of unfavorable conditions such as freezing waters, chlorinated swimming pools, hot springs and they can also evolve into trophozoites in the presence of a suitable environment. The pathophysiology of GAE is believed to include unknown pathways. The stimulation of pro-inflammatory signals, invasion inside the blood brain barrier (BBB) and connective tissue, and neuronal injury resulting in brain dysfunction, however, are complicated physiological characteristics related to CNS that are supposed to get developed [14]. The lower respiratory tract is one of the points of entry, which might result in amoeba invasion of the intravascular space and subsequent spread of hematogenous. An amoeba can penetrate the body through skin openings, where they can then spread hematogenous to the brain and other organs. Furthermore, olfactory neuroepithelium offers a different entrance into the CNS [7].

Rising Contagious Diseases: Basics, Management, and Treatments, First Edition. Edited by Seth Kwabena Amponsah, Ranjita Shegokar, and Yashwant V. Pathak.

The cornerstone of successful therapy is to detect the condition of the disease very early. The majority of instances are discovered in advanced stages, which results in significant mortality in these cases. Anti-tubercular drugs like amphotericin B, rifampin, flucytosine along with the macrolide antibiotics such as azithromycin and amikacin, as well as pentamidine, cotrimoxazole, propamidine isethionate, azoles like fluconazole, itraconazole, and voriconazole are some of the more often used antibiotics [15]. Miltefosine, an anticancer medication, as well as the antipsychiatry medications – phenothiazines and thioridazine – have all recently entered the market and have shown efficacy in treating GAE infections. However, the latter two had to be discontinued during therapy due to significant toxicity. GAE infections have often been successfully treated using a cocktail of almost four to five different antibiotics [6].

27.2 Epidemiology for Granulomatous Amebic Encephalitis

Particularly in patients with impaired immune systems, *Acanthamoeba* can induce an uncommon but lethal brain infection [16]. Approximately 97–98% of GAE instances lead to death [17]. Since the past three decades, more than 70 cases of patients with GAE were brought on because of major *Acanthamoeba spp*, as per the documentation of PubMed. The majority of patients were from the United States, but there were also cases from Austria, China, Turkey, Italy, Thailand, Japan, Spain, Sweden, Canada, Saudi Arabia, Mexico, Taiwan, Venezuela, and the United Kingdom [18]. A better prognosis occured in these people due to the potential of early identification because half of the total number of the surviving cases of GAE occurred due to *B. mandrillaris* and acquired a pre-existing lesion on the skin. The majority of the time, a space-occupying lesion is seen on radiological examination as a hypodense/ring-enhancing lesion [19, 20]. Furthermore, it is suspected that several instances were not really recognized or have been incorrectly labeled as pathogenic bacteria, adenoviruses, or fungi infections as a result of serious medical conditions and clinical issues [21, 22]. Although GAE is document for those individual who have compatible immune system, immunocompromised people seem to experience it more frequently. Around 12 of the 19 patients (63%) who recovered from cerebral *Acanthamoeba spp.* infection were immunocompetent. All patients got combination therapy for the treatment, expect for one patient. Immunosuppressive conditions like AIDS, hematological malignancies, organ transplantation, steroid use, or autoimmune disorder like lupus erythematosus, type 1 diabetes, addiction to alcoholism, cirrhosis of the liver, pregnancy, starvation, surgical trauma, fires wounds, and radiation therapy are additional immunosuppressive treatments which are risk factors for GAE [23]. Around 97–98% of GAE cases result in death [16]. Since 1990–2020, 75 cases of patients with GAE brought on by *Acanthamoeba species* have been documented in PubMed [18]. But it is assumed that many instances were missed or not properly detected as bacteria, different forms of virus, or some other fungal infection because of scarcity of particular symptoms and diagnostic challenges [24]. The majority of individuals with GAE are immunocompromised, but cases of immunocompetent patients with GAE have also been documented [25].

27.3 Pathogenesis of Granulomatous Amebic Encephalitis

Acanthamoeba spp.-caused GAE often affects those who are seriously ill and immunosuppressed patients, such as those with human immunodeficiency virus/acquired immunodeficiency syndrome, and individuals who got a prescription of antibiotics or chemotherapeutic drugs [14, 26]. However, the other major parasites involved in the disease prognosis are *B. mandrillaris* and *N. fowleri*. Figure 27.1 depicts how these major causative agents leads to the precipitation of GAE. Here is a brief outline of the basic pathogenesis of GAE arising as a result of different parasite infection (Figure 27.2).

27.3.1 *Acanthamoeba Species* Intricated GAE

Acanthamoeba, an amoeba, is a member of the family *Acanthamoebidae* and causes GAE, which is a life-threatening condition of *amebic keratitis* (AK) and CNS [27, 28]. One of the most common protozoa to be discovered in the environment is *Acanthamoeba* spp. [29]. They are found in dust, air, and water, as well as other pollutants in yeast and bacteria among other places [30, 31]. They are present all throughout the planet.

Amoeba can get into the body through skin openings and spread hematogenous to the brain and other organs [27]. Mannose-binding protein (MBP) is located over the surface of *Acanthamoeba* and binds to mannose-containing glycoproteins on the exterior of the epithelial cells, and the outermost cell is destroyed [32]. The release of serine proteases by *Acanthamoeba* induces the breakdown of collagen, fibronectin, IgG, IgA, and elastin enhancing the porosity of the basement membrane [33]. Iron and copper-zinc superoxide dismutase are the two-superoxide dismutase that have been found

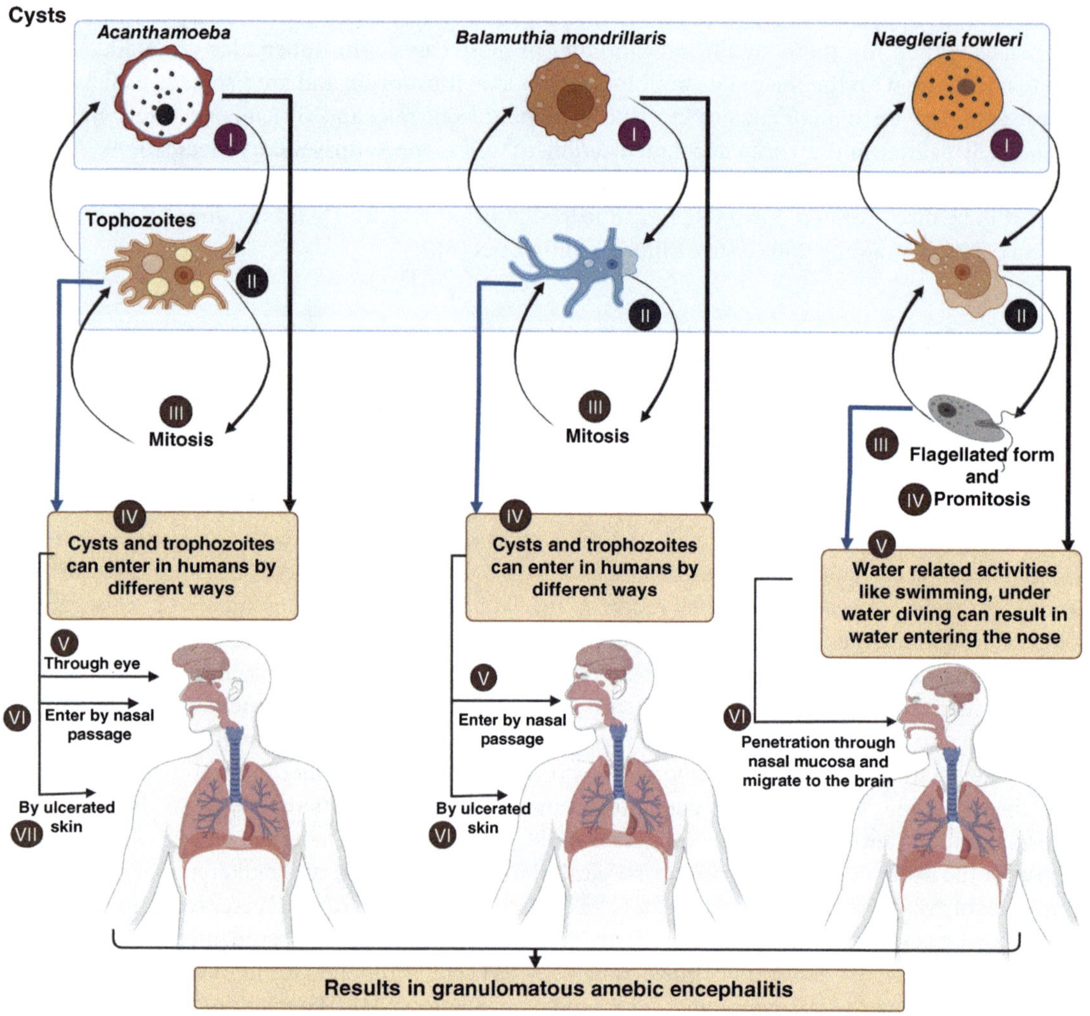

Figure 27.1 Mechanisms involved in pathogenesis of GAE caused due to the infection of *Acanthamoeba spp.*, *B. Mandrillaris*, and *N. fowleri*. Amoeba can get into the body through skin openings and spread hematogenous to the brain and other organs in case of *Acanthamoeba* species, whereas the cyst and trophozoite are the two natural forms of *B. mandrillaris*, which can infect people through open skin wounds or through the respiratory system. In case of *N. fowleri's* infectious, GAE originates when the form enters through the eyes, nose, respiratory system, or through skin ulcers. It starts to intrude on the surface of the frontal lobe. In the *acanthamoeba* and *B. mondrillaris*, trophozoites enter the body through different ways after the completion of mitosis, but for *N. fowleri's* infection, initially the trophozoites are converted into flagellated form before they enter the body, and then promitosis occurs and mainly invade through water bodies during any activities such as swimming via nasal route and penetrate the nasal mucosa and then enters the human brain to cause GAE.

in *Acanthamoeba* to facilitate antioxidant defense mechanisms [6]. *Acanthamoeba* activates the plasminogen which then catalyzes the host plasminogen into plasmin. Plasmin then causes a matrix of metalloproteinases, which activates and leads to the degradation of the basement membranes [34]. *Acanthamoeba* causes cell cycle arrest in microvascular endothelial cells of the human brain by preventing pRb phosphorylation [35]. *Acanthamoeba's* plasma membrane stores a lot of Ca^{2+} ions. An elevation in intracellular calcium brought on by the *Acanthamoeba*, injecting it through its acanthopodia, results in the death of the host cell [36].

27.3.2 *Balamuthia mandrillaris* Intricated GAE

A free-living amoeba called *B. mandrillaris* can be found in freshwater and soil. This was discovered in a pregnant monkey's brain [37]. Cyst and trophozoite are the two natural forms of *B. mandrillaris*, which can infect people through open skin wounds or through the respiratory system [38]. They have one massive nucleus and are bigger in size as compared to *Acanthamoeba* trophozoites.

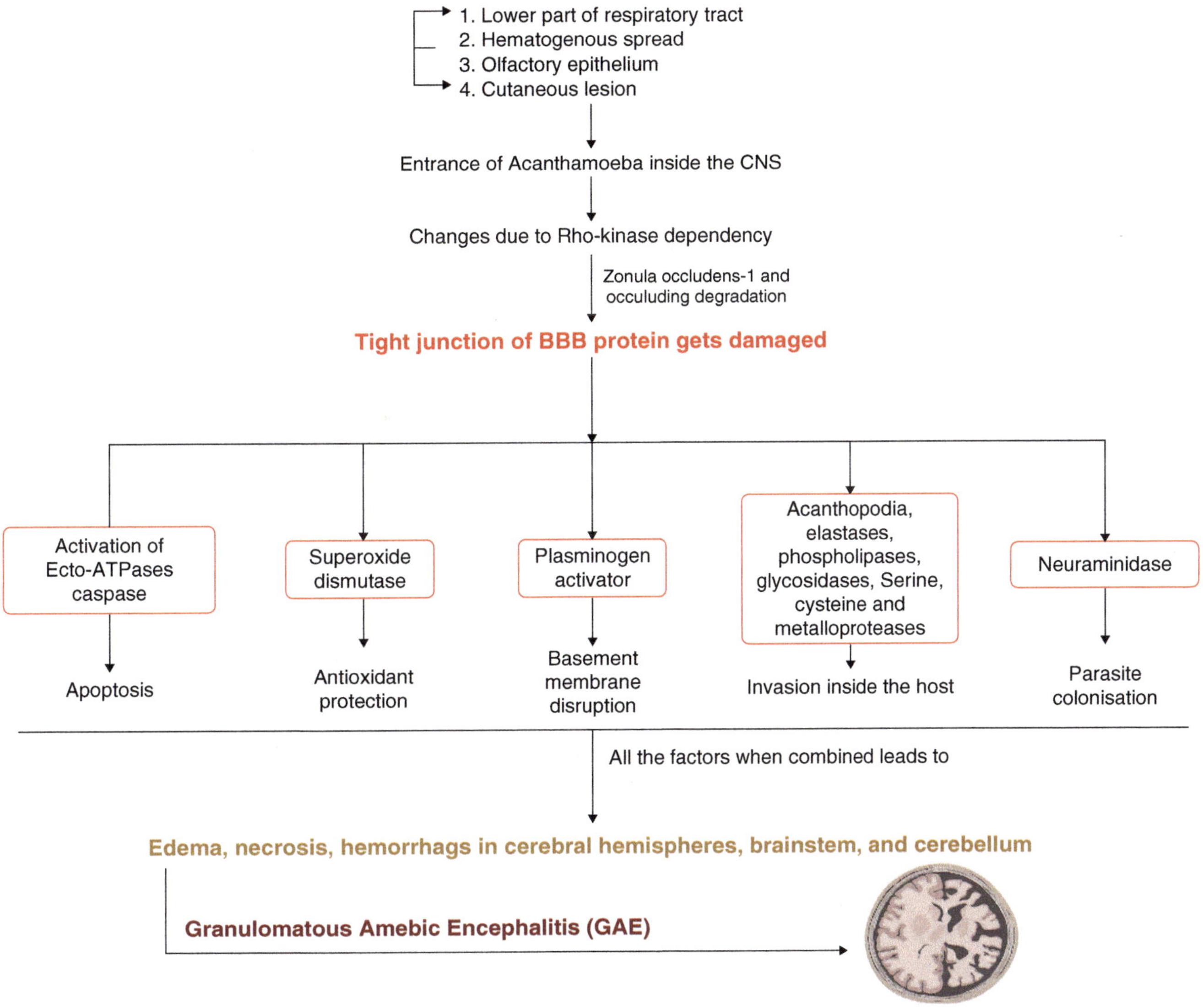

Figure 27.2 Showing the pathogenesis and combined factors leading to granulomatous amebic encephalitis. IgG, IgA, fibronectin which is a the most essential component of extracellular matrix, type I, III, IV collagen, elastin along with the tight junction proteins ZO-1 and occludin, and Hb, as well as the release of serine proteases by *Acanthamoeba*, all cause breakdown, increasing the permeability of the basement membrane. The activation of caspase-3 and ecto-ATPases are further elements that could play a role in *Acanthamoeba* pathogenesis. In the colonization of the parasite, which results in changes to glycolipids linked to meningoencephalitis, *Acanthamoeba's* neuraminidase activities may be important for the pathogenesis of GAE.

The human body can be infected with *B. mandrillaris* via a wound, nasal cavity, and organ transplant. The activation of ecto-ATPases cleave ATP is generated by activated T cells, and the amoeba successfully evades host immunological responses in blood [39]. The most affected organ through *Balamuthia* is the brain and skin. By modulating the expression of adhesion molecules, IL-6 makes BBB more permeable, aiding the amoeba in breaking through it and thus starting the early inflammatory response [40]. After the amoeba penetrates across the BBB, it efficiently kills the host cells by inducing apoptosis and phagocytosis. The trophozoites are abundant and seem to be both smooth and polymorphic in shape. Both types could bind to several host cells simultaneously before starting to swallow and attack the target cells. *Balamuthia* also generates metalloproteinases, including phospholipase D and phospholipase A2 that could help in the breakdown and migration via living tissues [7, 41].

27.3.3 *Naegleria fowleri* Intricated GAE

Naegleria fowleri, a thermophilic amoeba that is a protozoan environmental parasite, may live without a specific host and can withstand temperatures of up to 45 °C and survives at temps over 30 °C. They are found mostly in freshwater environments like lakes and rivers. GAE and primary meningoencephalitis (PAM) are two separate clinical syndromes that can be brought on by this free-living amoeba [42].

GAE originates when *N. fowleri's* infectious form enters through the eyes, nose, or respiratory system, or through skin ulcers. It starts to intrude on the surface of the frontal lobe. Clinical signs of olfactory epithelium involvement and invasion of the brain tissue includes respiratory illness and changes in the sensation of smell [43]. When *N. fowleri* is introduced into tissue culture cells, it has a cytopathic effect that kills the cell monolayer (CPE). The cells of tissue culture are "nibbled" at by the amoeba's sucker-like appendages, called amebostomes. The formation and release of phospholipase along with neuraminidase, the formation of holes in target cell membranes, and vigorous phagocytic activity are potential virulence factors present in *N. fowleri* trophozoites resulting in lysis and brain tissue necrosis [44].

27.3.4 Immune Responses and GAE

In the early stages of communication, immune response processes are crucial to prioritize defense since it takes time for a specific immune system response to be activated. Toll-like receptors (TLRs) are receptors that function as the first line of defense against bacteria or other microorganisms. They can recognize by invading pathogens and endogenous particles which are released by dead cells [45]. TLR's primary function is the stimulation of the cellular activity of the system of defense [46]. The experimental studies demonstrate that TLR2 ligands are leftovers of damaged cell breakdown [47], whereas the TLR4 ligands are thermal attack proteins that are present on the surface of the *Acanthamoeba* species [48]. Species of TLR2 and TLR4 may have a role in the pathogenesis of GAE [49]. TLR activation results in the release of cytokines which enhance the differentiation of Th1 cells and suppress the differentiation of Th2 cells, which also primarily release IL [4, 5, 13, 50].

27.3.4.1 Tissue Inhibitors of MMPs (TIMPs) and Matrix Metalloproteinases (MMPs)

MMPs are amino acids and are crucial in controlling a variety of activities, including cell migration and differentiation, control growth action, and the onset of swelling. Additionally, inter-cell communication, cell survival, and the initiation of apoptosis are all affected by MMPs [51]. A higher number of MMPs are seen in brain lesions, including sclerosis, neuropathy, or stroke; this elevated level of enzyme is brought on by an increase in leukocyte synthesis, but also by an increase in synthesis in endothelial cells and brain neurons [52]. It has been hypothesized that inhibiting MMPs would delay the early invasion process since metalloproteinases are secreted during parasite invasion into the brain [53]. *Acanthamoeba spp.* are known to trigger a few virulence that enables the organism to get through the host's cellular defense. *Acanthamoeba spp.* manufactures CDM while adhering to infected cells, and according to Lorenzo-Morales et al., these proteolytic enzymes cause the breakdown of cell walls leading to the death of infected cells [54]. *Acanthamoeba spp.* of IC mice, at the start of the disease, showed increased MMP-9 action, but *Acanthamoeba spp.* of IS mice showed decreased MMP-9 rates relative to an uninfected IS animal [55]. The researchers also assert that MMPs may be an important treatment to stop the overall opening of BBB in amebic brain infections that have been caused [56].

27.3.4.2 Neurotrophins: Neurotrophin-4 (NT-4) and Brain-Derived Neurotrophic Factor (BDNF)

The family of neurotrophin includes NT-4 and BDNF. From neural stem cells, BDNF both promotes and regulates neurogenesis [57]. When mice first became infected with *Acanthamoeba spp.*, the hippocampus region and frontal lobe of the mice were found to have the highest levels of BDNF (8 dpi). Increased BDNF levels were associated with neurological symptoms in *Acanthamoeba spp.*-infected both IC and IS animals. According to the study, increasing levels of BDNF in the brain may have neuroprotective benefits, and decreased levels of BDNF may be linked to neuronal degradation that is frequently seen in the cognitive form of acanth amoebiasis [58]. When it comes to the growth of the nervous system, neurotrophin-4 is crucial [59]. On the eighth day following *Acanthamoeba spp.* Infection, the brain of IC and IS animals displayed the highest NT-4 activity. The amount of NT-4 activity reduced with time in infected immunocompetent mice (Figure 27.3).

27.3.5 Histopathological Changes

The researchers discovered that mice infected with *Acanthamoeba* spp. Died 4 days after receiving amoebas intranasally. In histological preparations, mucosal ulceration of the nose was seen, and *amoebas* were found in the brain's frontal

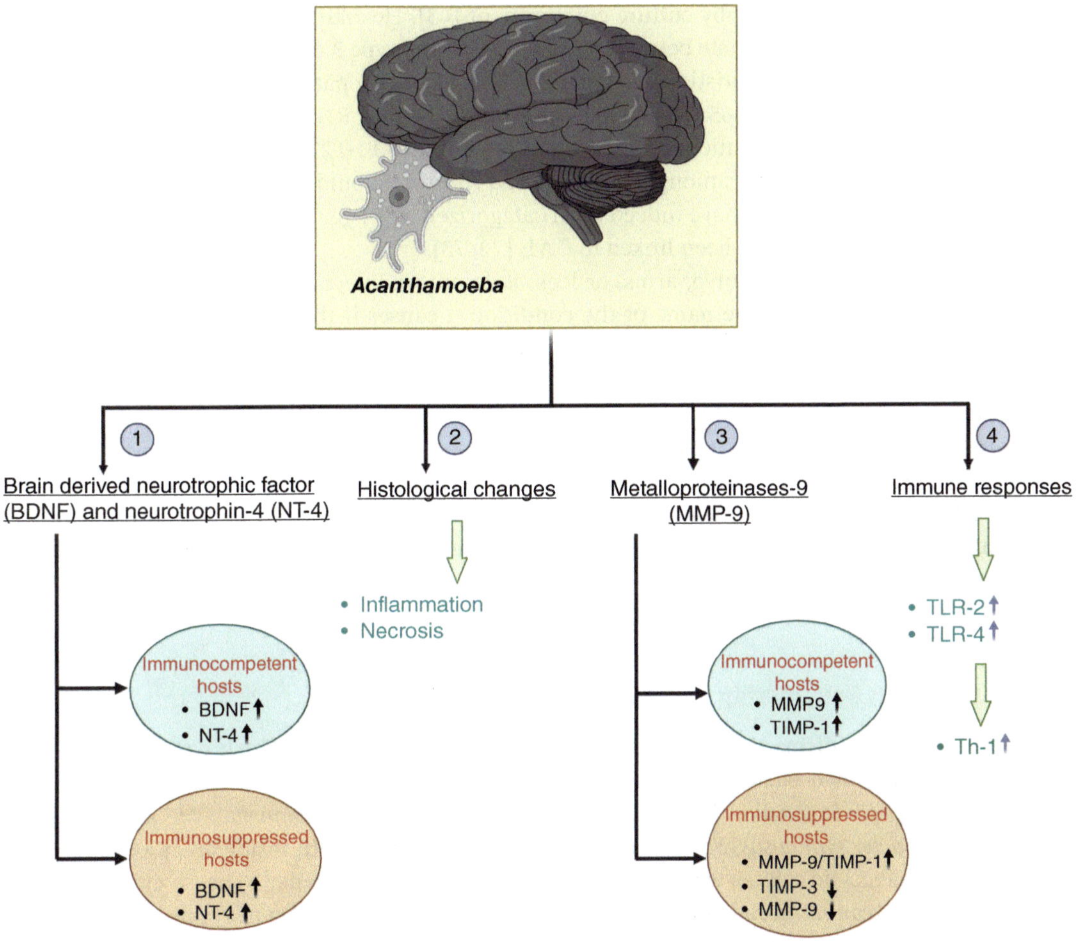

Figure 27.3 Defense mechanism of the host against several species of Acanthamoeba infection of the brain. (1) NT-4 and BDNF activity get modulated by the infection of *Acanthamoeba spp.* in both IC and IS hosts; (2) The amoebas caused necrotic and inflammatory lesions in the structural components of the brain; (3) In IC hosts, anthamoebiasis increases the activity of MMP-9 and the TIMP-1, but in immunosuppressed hosts, it decreases MMP-9 and TIMP-3 (IS). The elevation in the level of the MMP-9/TIMP-1 ratio was also seen in immunocompromised hosts; (4) The toll-like receptor (TLR) expression increase are influenced by an infection of the *Acanthamoeba species*, particularly TLR-2 and 4. Additionally, here, the lymphatic cells in brain of an *Acanthamoeba species* were antigen sensitive. Th1 cytokine-producing cells predominate mostly in infected mice.

lobe [60]. According to Culbertson et al. findings in a different study, *Acanthamoeba spp.* can cause catastrophic alterations in the brain, frequently coupled with abnormalities in the lungs, by damaging olfactory receptors and the mucosal layer of the nose [60]. Newly found strains of *Acanthamoeba spp.* that caused granulomas to grow in the brain, primarily in the olfactory region, were identified in 1966 by Culbertson et al. [61]. The presence of cysts and trophozoites of *Acanthamoeba spp.* were found in the brain of GAE patients, according to Martinez. Additionally, it was shown that only *Acanthamoeba spp.* trophozoites appeared in the brains of amoeba-infected animals [62]. Patients suffering from *Acanthamoeba spp.* infection had multiple necrotic and hemorrhagic regions throughout their brains, as seen by histopathological investigations. Other findings in the perivascular areas were selective build-up of trophozoites and cysts of *Acanthamoeba spp.* A study of the patient's CSF found lymphocyte-predominant pleocytosis, high protein levels, and low blood sugar levels [43].

27.4 Evolutionary Dynamics Involved in Granulomatous Amebic Encephalitis

27.4.1 Evolution Regarding Genotypes of Pathogens in Granulomatous Amebic Encephalitis

According to their morphology and cyst morphology, *Acanthamoeba species* has been categorized into three separate categories (I–III), but this classification was considered unsatisfactory in consideration of the observed variations in cyst structure among identical isolates and strain induced by culture environment [63]. Hexokinase, esterase, and acid phosphatase were used by Maura et al. in 1992 to discriminate between type 1, type 2, and type 3 of *Acanthamoeba species* [64]. Subsequently, the variations in physiological characteristics and antigen patterns among pathogenic and non-pathogenic *Acanthamoeba* were utilized to differentiate them [65, 66]. Depending on ribosomal RNA (rRNA) gene sequencing, *Acanthamoeba* has been divided into 21 different evolutionary lineages, or haplotypes (T1-T21), due to the use of advanced molecular techniques [67–70]. The sequence diversity among *Acanthamoeba variants* should be at minimum 5% [71]. The majority of *Acanthamoeba strains* isolated from GAE have indeed been categorized as T4 genotypes [72–74]. Genotypes of Acanthamoeba like T (1, 2, 5, 10, 12 and 18) have also been linked to GAE [72, 73].

Beginning with a skin wound on the face, chest, torso, arms, or legs, Balamuthia infection symptoms might develop depending on the genotype of the species. GAE is the name of the condition it causes if the infection affects the brain. Because the symptoms of Balamuthia GAE are not unique to that condition, diagnosis might be challenging. The illness may first seem moderate, but it can worsen over a period of weeks to months. 90% of people who have the disease die from it due to lack of diagnosis in the time between the infection and the onset of symptoms. After initial contact with the amoeba, it can take weeks or months before Balamuthia GAE symptoms appear.

There are three types of *N. fowleri*. The protozoa really gets converted into cyst, which are not active metabolically, characterized as a spherical shape with a diameter of 7–12 m, a broad width endocyst, a narrow width ectocyst, and some voids assisted by mucoid, when the atmosphere has become too unfavorable [74]. Because of the cyst's remarkable resistance and ability to survive a harsh physiological condition such as temperatures as cold as 2–4 °C, such type of amoeba can withstand the chilly winter months and multiply during the summer [75].

27.4.2 *Acanthamoeba*-Endosymbionts Relationship

Endosymbionts, a diverse set of microorganisms that *Acanthamoeba* is known to host, and which can have a variety of effects, are well-known. Many well-known bacterial endosymbionts of *Acanthamoeba* include *Betaproteobacterium, Amoebophilus asiaticus, A. asiaticus, C. legionella, M. pseudomonas Chlamydophila, Legionella, M. pseudomonas, Burkholderia, Candidatus, Paracaedibacter symbiosus, Salmonella, and Vibrio* [76, 77]. Growth inside an amoeba may aid bacterial adaptation to mammalian phagocytic cells as well as bacterial development of resistance to harsh environmental conditions [6]. *Legionella species* prefer *Acanthamoeba* as their environment-specific host [78]. Gomes et al. (2018) demonstrated in a recent report that yet another strain of *Legionella feeleii*, passing across Acanthamoeba, resulted in enhanced production of virulent protein effectors (SdhA, SidK, and LegK2), which exacerbated the pathogenicity of this bacteria [79]. Another aquatic bacterium that causes severe diarrheal disorders in humans is *Vibrio cholerae* [80]. The enhanced viability of *V. cholerae* in the vicinity of *Acanthamoeba* has indeed been established by several authors in the literature [81–83]. Thus, there is a greater potential for the association involving Acanthamoeba and waterborne pathogens to impair quality of water and spread diseases among the vulnerable human population. *Campylobacter jejuni*, like other food-borne pathogens, is also found to survive inside the targeted amoeba. Survival of *C. jejuni* in Acanthamoeba intracellularly has been firmly hypothesized by a number of writers [84–86]. But more research examining the molecular basis of the *Campylobacter-Acanthamoeba* association would assist us in realizing the possibility of infection among susceptible populations. Researchers that are actively working in this domain have identified a number of new amoebic endosymbionts. Furthermore, an amoebic endosymbiont known as *Paenibacillus species.* was identified [87]. *Parachlamydia* and *Paracaedibacte* are the two endosymbionts which are differentiated by their morphology on the diagnostic strain on which *Acanthamoeba hatchetti* were discovered to be present [88]. This interaction between *Acanthamoeba* and its endosymbionts may encourage evolutionary mechanistic pathways and the growth of pathogenic characterization [89]. Acanthamoeba plays a carrier and a vector together for a variety of infectious pathogens, imposing a significant harm to population health. This underlines the need for extensive research studies to determine the mechanism and consequences of the Acanthamoeba-endosymbiont connection in the upcoming times.

27.4.3 Evolutionary Dynamics Related to Lesion Inside Host

GAE typically manifests as one or more lesions that take up space in the various areas of the brain such as basal ganglia, cerebellum, hemisphere of cerebrum along with brain stem. There have already been studies of imaging results and numerous augmented patterns of GAE, notably gyriform [90], nodular [91, 92], ring, or partial ring, [93, 94] and heterogeneity [95] amplification. The varying enhanced patterns have been speculated to represent distinct lesions growth stages [96], despite a lack of any convincing data to demonstrate it. This notion has been provided concrete support in this article by the changes in increased patterns in lesions caused during the condition of GAE. The hypothetical cyst of GAE is angiotropic in the early phases and has a tendency to gather around along with the trophozoites and crosses through the blood vessels. The BBB is harmed as a result, leading to nodular augmentation. Occasionally, it damages the leptomeninges and brain surface [97]. As the condition intensifies, scar tissue and swelling will develop both on vertical and horizontal edges of the nodule amplification, providing the region a ring-enhanced aspect. A heterogeneous augmentation would eventually evolve as the necrosis and inflammation within the GAE lesion expands as a consequence of the lesion's progression and the clustering of adjacent daughter lesions. At every stage of the lesion's progression, there are a diversity of diagnoses [97]. At the outset, it is important to distinguish between the periventricular interface which has many follicular extension, and vasculitis caused by other granuloma-toxic encephalopathy, pathogenic or non-infectious causes, neurodegenerative lesions, and malignancies. However, at the round-enhanced stage, round shaped neoplasm, various granulomatous encephalitis, and demyelination illness must be thought about. Just at the nonoverlapping increased phase, additional merged granulomatous encephalopathy, dissemination, and cancer neurodegenerative disease should also be taken into consideration [98]. Eventually, GAE can commence as a lesion with a nodular augmentation or a gyriform enlargement. Ring expansion could be seen if the lesion gets worse and produces more tissue injury and swelling. Due to the additional tissue destruction and edema, and the accumulation of nearby necrosis spots, the uneven augmentation may finally become visible. A subacute to chronic patient history and the cluster of imaging abnormalities may serve as an indicator that this condition may be prevalent. Recognizing the successive alterations also support detection and therapy of the condition [99].

27.5 Diagnosis for Granulomatous Amebic Encephalitis

27.5.1 Traditional Method for Diagnosis

The forecast for GAE, an unusual condition, is quite lackluster. An effective solution for the sufferer allows for quick detection and pharmacological intervention. GAE symptoms are like other CNS pathogens such as viruses, bacteria, and other microbes. This confuses the diagnostic parameters of GAE and demands for a particularly significant degree of suspicion and expertise [6].

The traditional method for GAE detection and diagnosis commonly include microscopy and CSF culture.

27.5.1.1 Microscopy

In GAE patients, CSF frequently depicts normal levels of glucose, but protein levels may be raised up. Typically, lymphocytic pleocytosis is observed. One or more diverse, quasi, spatially, hypodense injuries may be evident inside the striatum, frontal lobes, subcortical white tissue, brainstem, and cerebellar region, which may indicate an intracerebral hematoma or brain tumor and brain abscess [100, 101]. To prevent amoeba from deformation or rupture, 2–3 ml of CSF is centrifuged at 250 g for approximately 10 minutes before being used for microscopic examination of amoebae inside CSF [102]. The trophozoites might be observed during fluid investigation. A generic comparison can be made using these characteristics– A small portion of the fluid is combined with 1 ml of distilled water after which it is studied after an hour for checking the presence of *N. fowleri*, which are quite tiny and measure between 10 and 25 mm [103]. *Balamuthia* trophozoites are bigger in size, of about 60 mm, have extensive branching along with false pseudopodes, travel slowly, and reject the enflagellation test whereas trophozoites of *Acanthamoeba spp.* are 15–40 m in size, with prickly projections called acanthopodia, and reject the enflagellation test [76].

27.5.1.2 Cell Culture

Amoeba causing GAE cultivated by three different ways, xenic media (contaminated with bacteria), and axenic media (Media specifically for the isolation of parasite) were used to culture the tissue [104]. Xenic culture is carried out when bacteria are present in a non- or low-nutrient agar. First, the most common bacterial isolates used in xenic cultures are nonmucoid colonies of *E. coli* or other forms of bacterial colony, that are supposed to be injected inside the culture media and then cultured at 37 °C. Plates should be observed for up to seven days [105]. A low-nutrient, non-nutrient agar was chosen to prevent bacterial proliferation [66, 106].

The Axenic method is carried out on a culture medium that has been supplemented with antibiotics like aminoglycosides and penicillin. Antifungals should be exempted, especially Amp-B and some other anti-fungal like fluconazole, as they are able to hinder the growth or recovery of these amoeba. The different types of examples of axenic media used frequently for the culture of *Acanthamoeba* include- Shukla's M-11, Byer's DGM-21A, and others, and BM-3 like medium are employed to check *B. mandrillaris*. Although the process is straightforward and quick to carry out, it is uncertain how reliable the test will be and how much time it will take [18, 107].

Tissue culture has been developed for the efficient recovery and cultivation of these amoeba. In tissue culture, the cell line that is used is human pulmonary tissue, a rat glioblastoma cells, a monkey hepato-cell line, and human brain microvascular endothelial cell line [108, 109].

27.5.2 Advances in Diagnosis of Granulomatous Amoebic Encephalitis

Recent advanced techniques have been employed for the diagnosis of GAE. The different modern techniques used for the purpose of diagnosis includes – molecular method, neuroimaging, serology, and histopathology.

27.5.2.1 Molecular Method of Diagnosis

Acanthamoeba, Balamuthia, and *N. fowleri* infections can now be diagnosed using a real-time multiplex PCR [110]. Using amoebae in cultivation, precise detection limits for many assays have been validated. A detection limit of roughly 1–2 amoebas/specimen was demonstrated by the nucleus Ssu-rRNA and Rnase-P genes which are focused primarily in PCR. PCR has been reported that the detection limit for *B. mandrillaris* mitochondrial 18S rDNA is 0.2 amoeba/sample because many replicas can remain inside only one body of amoeba's mitochondrial DNA [111]. Different investigations describing a Taqman real-time PCR Assay encoding the sequence analysis of rRNA gene can be employed to detect *N. fowleri*, Acanthamoeba species, and *B. mandrillaris* [112].

27.5.2.2 Neuroimaging

MRI and CT scans can be employed to take brain images [101, 113]. The cerebellar membrane, pons, subcortical white tissue, striatum, or cortex are typically affected by one or more enhancing lesions with a minor mass effect [42, 114]. The thalamus along with the brain stem, diencephalon, and posterior fossa components are frequently affected by multifocal lesions with a ring-like pattern of enhancement and T2 hyperintensity or heterogeneous according to MR imaging results [115, 116]. Some authors believe that intralesional bleeding is a crucial diagnostic indicator and that it can be proven due to the presence of necrotizing angiitis seen in severe GAE cases [117].

27.5.2.3 Serology

It is a non-invasive process and facilitates the early detection of instances, so the detection of antibodies in GAE cases is dependable and preferred. The detection of antibodies has been done by using indirect immunofluorescence (IIF), ELISA, and flow cytometry [118]. Immunohistochemistry (IHC) and molecular methods validated GAE instances with high antibody titers (1 : 128–1 : 1024) [119]. IIF involves fixing amoebic antigenic extracts onto a slide in aliquots, adding sequential donor serum aliquot, and then adding their antibody labeled with fluorescein isothiocyanate to visualize fluorescence release. There was not a single instance of *Acanthamoeba* or *Balamuthia* infections interacting with one another [120]. To distinguish between healthy and ill people, higher titter cut-offs should be used, according to the results of fluorescence-associated cell sorting. For the early detection of these GAE instances, western blot and silvery stain testing is highly recommended [121, 122].

27.5.2.4 Histopathology

While frozen or paraffin-installed brain portions are explored histologically, the hit areas display tissue damage and internal bleeding, fibrinoid chronic granulomatous, thrombus formation, polymorphonuclear, peri-vascular rotational granulation tissue, and frothy monocytes, lymphoplasmacytic inflammation infiltrate, and poly-nucleated gigantic cells. In the vascular wall and in inflammatory and non-inflammatory zones, trophozoites and cysts are often dispersed [123]. Grocott's GMS-187, trichrome, calcofluor white, hematoxylin and eosin, and other stains have all been employed to identify 186 cysts and trophozoites [124].

27.6 Advances in Therapeutic Interventions for Granulomatous Amebic Encephalitis

27.6.1 Treatment Therapy Used for GAE

Acanthamoeba species infections of the brain and lungs do not currently have a treatment that is consistently effective. To establish synergistic effects and optimal therapeutic benefits, the majority of documented instances used multiple medication combinations. Due to the inability of the majority of medicines to get through the various body barriers and crossing the tissue of the brain or lung at high enough doses to eliminate the amoeba, there are indeed few medical cases showing successful treatment outcomes. Anti-fungal drugs such as Amp-B, itraconazole, fluconazole, ketoconazole, and other antibiotic drugs like paromomycin, azithromycin, pentamidine, sulfadiazine, cotrimoxazole, rifampicin, along with pyrimethamine, and 5-fc are the FDA-approved medications in Acanthamoeba-related species infections [124]. The IDSA recommended either ketoconazole+ rifampicin+ trimethoprim-sulfamethoxazole, or sulfadiazine+ fluconazole+ pyrimethamine for Acanthamoeba CNS infection [125]. Miltefosine, an antitumor and drug used in leishmaniasis, is now accessible for therapy for CNS infection, according to CDC [126]. Miltefosine was administered in six cases in combination as shown in Table 27.1. In immunocompetent patients, miltefosine and oral voriconazole were found to reduce serological titters and brain lesions in the individuals suffering from GAE as shown by Webster et al. [134].

As new therapeutical applicants with possible impact on Acanthamoeba, Elsheikha et al. [135] addressed alpha-blockers like amlodipine, prochlorperazine, loperamide, and the cumulative effects of antipsychotics like chlorpromazine along with antibiotic like rokitamycin. Procyclidine, digoxin, corifungin, tigecycline, heterocyclic alkyl-phosphocholines which is a structural analogue of miltefosine along with chloroquine.

The best treatment for GAE has yet to be found. There are nine survivors of *B. mandrillaris* caused encephalitis [117, 129, 136–138], and in the majority of cases, medication combinations comprising fluconazole, flucytosine, pentamidine, amphotericin B, sulfadiazine, clarithromycin, albendazole, and miltefosine were employed (Table 27.2). A case of a brain abscess complication managed with a combination of antibacterial medicines and surgical excision was recently reported [141]. In this example, clarithromycin, pentamidine, albendazole, and flucytosine were added to fluconazole in the initial step of therapy [19, 141]. Although pentamidine and clarithromycin were withdrawn due to adverse effects, the combination of the remaining treatments allowed the patient to achieve recovery [139]. We assumed a diagnosis of fungal meningitis early on and gave fluconazole, which may have reduced amoeba proliferation, resulting in recovery. The discovery of novel antimicrobial medications is essential, as is proving the efficacy and safety of existing antimicrobials when taken continuously for long periods of time, particularly when used in combination [140].

Table 27.1 Treatment regimen for GAE caused due to Acanthamoeba species.

Sr no.	Information about subject	Sample taken from	Therapeutic intervention	Reference
1.	Immunocompetent	Cerebrospinal fluid	Fluconazole, rifampicin, cotrimoxazole (two months)	[126]
2.	Acute lymphoblastic leukemia	Cerebrospinal fluid	Trimethoprim, sulfamethoxazole, fluconazole, pentamidine, and miltefosine	[127]
3.	Recipient of transplanted liver	Damaged tissue of brain	Rifampicin, co-trimoxazole (for three months)	[128]
4.	Immunocompetent	Serum sample, brain biopsy	Voriconazole, miltefosine	[129]
5.	Immunocompetent	Cerebrospinal fluid	Fluconazole, rifampicin, metronidazole	[130]
6.	Immunocompetent	Brain tissue biopsy	Trimethoprim, sulfamethoxazole, fluconazole and miltefosine	[115]
7.	HIV patient	Cerebrospinal fluid	Trimethoprim, sulfamethoxazole, flucytosine, fluconazole, miltefosine	[112]
8.	AIDS patient	Biopsy of the brain	Surgical excision of lesions followed up by fluconazole, sulfadiazine	[131]
9.	Chronic malnutrition child	Cerebrospinal fluid	–	[132]
10.	Malnourished child	Cerebrospinal fluid	Cotrimoxazole, rifampicin, ketoconazole	[133]

Table 27.2 Treatment regimen for GAE caused due to *Balamuthia mandrillaris*.

Sr no.	Author and year	Risk factor	Means of diagnosis	Effective treatment regimen	Outcomes	Reference
1.	Martinez et al. (2010)	Immunocompetent	Skin biopsy, immunofluorescence, PCR (biopsy)	Albendazole, fluconazole, meltefosone (7.5 months)	Skin and CNS lesions healed	[139]
2.	Deetz et al. (2003)	Immunocompetent	Brain biopsy, immunofluorescence	Flucytosine, fluconazole, thioridazine, clarithromycin (1.9 years), maintenance: fluconazole, clarithromycin	Moderate performance	[113]
3.	Deetz et al. (2003)	Immunocompetent	Brain biopsy, immunofluorescence	Clarithromycin (two years), flucytosine, fluconazole, sulfadiazine (five years), maintenance: fluconazole, sulfadiazine	Enable to walk independently, to communicate well	[113]
4.	Doyle et al. (2011)	Immunocompetent	Skin/brain biopsy, PCR (biopsy)	Flucytosine, azithromycin, itraconazole, sulfadiazine (seven months), surgical excision	Recovered completely	[140]
5.	Jung et al. (2004)	Immunocompetent	Brain biopsy, immunofluorescence	Pentamidine, fluconazole, sulfadiazine, clarithromycin (unreported period of regimen)	Recovered completely	[141]
6.	Orozco et al. (2011)	Kidney recipient	PCR (CSF), culture	Flucytosine, azithromycin, miltefosine, extra ventricular drainage (EVD)	Severely disabled require the care by near ones	[8]
7.	Seas et al. (2006)	Not reported	Not reported	Albendazole, itraconazole (14 months)	Mild left hemiparsis	[139]
8.	Present case	Immunocompetent	Microscopy of CSF	Albendazole, flucytosine (four months), maintenance: fluconazole, albendazole (>17 months)	Gait disturbance, delusions	[138]

27.6.2 Advances in the Treatment for Granulomatous Amebic Encephalitis

Traditional therapy has still not been validated to be very efficacious for the therapeutic intervention of GAE, so some of the newer approaches have been tried for the same, but proper treatment regimen is yet to be discovered. Some newer approaches that can be employed for combating GAE are shown below.

27.6.2.1 Gold-Conjugated Curcumin

Curcumin has antioxidant activities and inhibits lipid peroxidation in the rat brain at low concentrations. Curcumin at 170 nmol per micromolar concentration also inhibited superoxide and hydroxyl radicals, confirming its antioxidant properties. Curcumin exhibits action against inflammation in both chronic and acute inflammation models [142, 143].

27.6.2.2 Alkylphosphocholines (APC)

In rats, APC-1 can permeate the BBB, and a tissue dispersal study revealed that there can be up to 170 nmol g of APC-1 in the cortex. It is the same as 170 M in fluid, which is too excessive for Acanthamoeba bodies to sustain. APC-1 could also maintain strong cysticidal capabilities, destroying more than 99% of cyst and inhibiting the excystation of the remaining parts [144].

27.6.2.3 New Combination Approach

Amlodipine, loperamide, pro-chlorperazine, a combination of antipsychotics like chlorpromazine along with antibiotic rokitamycin, procyclidine, digoxin, corifungin, tigecycline, heterocyclic APC, which is a structural analogue of miltefosine, and chloroquine were presented and discussed by Elsheikha et al. [125] as potential anti-Acanthamoebic candidates.

27.7 Conclusion

Acanthamoeba is a protozoan that is gradually being reported in both normal and unsuitable conditions. There is a high probability that it could infect the susceptible human population to and cause deadly diseases like GAE. Currently, there are more and more GAE incidences being reported. *Acanthamoeba, B. mandrillaris, and N. fowleri* are some of the organisms responsible for GAE. The pathogenesis and factors determining the development of GAE are best understood when understanding the molecular and biochemical pathways known to date, which are participating in the response to *Acanthamoeba spp.* infections in the hosts. This chapter discussed in detail the etiology, associated clinical symptoms, diagnostics method, and treatment of GAE in individuals brought on by *Acanthamoeba*. *Acanthamoeba* is a serious hazard to global civilization since it can infect both immunocompetent and immunocompromised individuals, increasing mortality and morbidity in both populations. The proper therapy of infected patients would benefit from early diagnosis of the *Acanthamoeba* condition. Successful therapy depends on increased awareness, significant anticipation based on clinical markers, early diagnosis, and a strong therapeutic approach using a combination of such medications.

References

1 Bloch, K.C. and Schuster, F.L. (2005). Inability to make a premortem diagnosis of Acanthamoeba species infection in a patient with fatal granulomatous amebic encephalitis. *J. Clin. Microbiol.* 43 (6): 3003–3006.

2 Pemán, J., Jarque, I., Frasquet, J. et al. (2008). Unexpected postmortem diagnosis of Acanthamoeba meningoencephalitis following allogeneic peripheral blood stem cell transplantation. *Am. J. Transplant.* 8 (7): 1562–1566.

3 Lass, A., Guerrero, M., Li, X. et al. (2017). Detection of Acanthamoeba spp. in water samples collected from natural water reservoirs, sewages, and pharmaceutical factory drains using LAMP and PCR in China. *Sci. Total Environ.* 584: 489–494.

4 Juarez, M.M., Tartara, L.I., Cid, A.G. et al. (2018). Acanthamoeba in the eye, can the parasite hide even more? Latest developments on the disease. *Cont. Lens Anterior Eye* 41 (3): 245–251.

5 Sissons, J., Alsam, S., Goldsworthy, G. et al. (2006). Identification and properties of proteases from an Acanthamoeba isolate capable of producing granulomatous encephalitis. *BMC Microbiol.* 6 (1): 1–8.

6 Marciano-Cabral, F. and Cabral, G. (2003). Acanthamoeba spp. as agents of disease in humans. *Clin. Microbiol. Rev.* 16 (2): 273–307.

7 Schuster, F.L. and Visvesvara, G.S. (2004). Free-living amoebae as opportunistic and non-opportunistic pathogens of humans and animals. *Int. J. Parasitol.* 34 (9): 1001–1027.

8 Orozco, L., Hanigan, W., Khan, M. et al. (2011). Neurosurgical intervention in the diagnosis and treatment of *Balamuthia mandrillaris* encephalitis: report of 3 cases. *J. Neurosurg.* 115 (3): 636–640.

9 Schuster, F.L. and Visvesvara, G.S. (2004). Opportunistic amoebae: challenges in prophylaxis and treatment. *Drug Resist. Updat.* 7 (1): 41–51.

10 Mungroo, M.R., Khan, N.A., and Siddiqui, R. (2019). *Naegleria fowleri*: diagnosis, treatment options and pathogenesis. *Expert Opin. Orphan Drugs* 7 (2): 67–80.

11 Heggie, T.W. (2010). Swimming with death: *Naegleria fowleri* infections in recreational waters. *Travel Med. Infect. Dis.* 8 (4): 201–206.

12 Sau, K., Mambula, S.S., Latz, E. et al. (2003). The antifungal drug amphotericin B promotes inflammatory cytokine release by a toll-like receptor-and CD14-dependent mechanism. *J. Biol. Chem.* 278 (39): 37561–37568.

13 Pettit, D.A., Williamson, J., Cabral, G.A., and Marciano-Cabral, F. (1996). In vitro destruction of nerve cell cultures by Acanthamoeba spp.: a transmission and scanning electron microscopy study. *J. Parasitol.* 82: 769–777.

14 Martinez, A.J. and Visvesvara, G.S. (1997). Free-living, amphizoic and opportunistic amebas. *Brain Pathol.* 7 (1): 583–598.

15 Siddiqui, R. and Khan, N.A. (2008). Balamuthia amoebic encephalitis: an emerging disease with fatal consequences. *Microb. Pathog.* 44 (2): 89–97.

16 Zamora, A., Henderson, H., and Swiatlo, E. (2014). Acanthamoeba encephalitis: a case report and review of therapy. *Surg. Neurol. Int.* 5: 68.

17 Kot, K., Lanocha-Arendarczyk, N.A., and Kosik-Bogacka, D.I. (2018). Amoebas from the genus Acanthamoeba and their pathogenic properties. *Ann. Parasitol.* 64 (4): 299–308.

18 Kalra, S.K., Sharma, P., Shyam, K. et al. (2020). Acanthamoeba and its pathogenic role in granulomatous amebic encephalitis. *Exp. Parasitol.* 208: 107788.

19 Centers for Disease Control and Prevention (CDC) (2008). Balamuthia amebic encephalitis – California, 1999–2007. *MMWR Morb. Mortal. Wkly Rep.* 57 (28): 768–771.

20 Blake, D.P. and Tomley, F.M. (2014). Securing poultry production from the ever-present Eimeria challenge. *Trends Parasitol.* 30 (1): 12–19.

21 Yagi, S., Schuster, F.L., and Visvesvara, G.S. (2008). Demonstration of Balamuthia and Acanthamoeba mitochondrial DNA in sectioned archival brain and other tissues by the polymerase chain reaction. *Parasitol. Res.* 102 (2): 211–217.

22 Reddy, R., Vijayasaradhi, M., Uppin, M.S. et al. (2011). Acanthamoeba meningoencephalitis in an immunocompetent patient: an autopsy case report. *Neuropathology* 31 (2): 183–187.

23 Trabelsi, H., Dendana, F., Sellami, A. et al. (2012). Pathogenic free-living amoebae: epidemiology and clinical review. *Pathol. Biol.* 60 (6): 399–405.

24 Martinez, A.J. (1991). Infection of the central nervous system due to Acanthamoeba. *Rev. Infect. Dis.* 13 (Supplement_5): S399–S402.

25 Das, S., Gunasekaran, K., Ajjampur, S.S. et al. (2020). Acanthamoeba encephalitis in immunocompetent hosts: a report of two cases. *J. Family Med. Prim. Care* 9 (2): 1240.

26 Martinez, A.J. (2019). *Free-Living Amebas: Natural History, Prevention, Diagnosis, Pathology, and Treatment of Disease.* CRC Press.

27 Culbertson, C.G. With the Technical Assistance of, Overton, W.M., and Reveal, M.A.(1961). Pathogenic Acanthamoeba (Hartmanella). *Am. J. Clin. Pathol.* 35 (3): 195–202.

28 Jones, D.B., Visvesvara, G.S., and Robinson, N.M. (1975). Acanthamoeba polyphaga keratitis and Acanthamoeba uveitis associated with fatal meningoencephalitis. *Trans. Ophthalmol. Soc. U. K.* 95 (2): 221–232.

29 Mergeryan, H. (1991). The prevalence of Acanthamoeba in the human environment. *Rev. Infect. Dis.* 13 (Supplement_5): S390–S391.

30 Casemore, D.P. (1977). Free-living amoebae in home dialysis unit. *Lancet* 310 (8047): 1078.

31 Neilson, J.B., Ivey, M.H., and Bulmer, G.S. (1978). Cryptococcus neoformans: pseudohyphal forms surviving culture with Acanthamoeba polyphaga. *Infect. Immun.* 20 (1): 262–266.

32 Clarke, D.W. and Niederkorn, J.Y. (2006). The pathophysiology of Acanthamoeba keratitis. *Trends Parasitol.* 22 (4): 175–180.

33 Khan, N.A. and Siddiqui, R. (2009). Acanthamoeba affects the integrity of human brain microvascular endothelial cells and degrades the tight junction proteins. *Int. J. Parasitol.* 39 (14): 1611–1616.

34 Lösch, F. (1875). Massenhafte entwickelung von amöben im dickdarm. *Arch. Pathol. Anat. Physiol. Klin. Med.* 65 (2): 196–211.

35 Khan, N.A. (2006). Acanthamoeba: biology and increasing importance in human health. *FEMS Microbiol. Rev.* 30 (4): 564–595.
36 Khan, N.A. *Acanthamoeba: Biology and Pathogenesis*. Caister Acad.
37 Visvesvara, G.S., Schuster, F.L., and Martinez, A.J. (1993). *Balamuthia mandrillaris*, NG, N. Sp., agent of amebic meningoencephalitis in humans and other animals. *J. Eukaryot. Microbiol.* 40 (4): 504–514.
38 The Editors (2001). *Balamuthia mandrillaris* infection. *J. Med. Microbiol.* 50 (3): 205–207.
39 Matin, A., Siddiqui, R., Jayasekera, S., and Khan, N.A. (2008). Increasing importance of *Balamuthia mandrillaris*. *Clin. Microbiol. Rev.* 21 (3): 435–448.
40 Matin, A., Stins, M., Kim, K.S., and Khan, N.A. (2006). *Balamuthia mandrillaris* exhibits metalloprotease activities. *FEMS Immunol. Med. Microbiol.* 47 (1): 83–91.
41 Haider, R. (2007). Colourimertic determination of phospholipase activities in *Balamuthia mandrillaris*. *Am. J. Biochem. Biotechnol.* 3: 171–179.
42 Marciano-Cabral, F. (1988). Biology of Naegleria spp. *Microbiol. Rev.* 52 (1): 114–133.
43 Visvesvara, G.S., Moura, H., and Schuster, F.L. (2007). Pathogenic and opportunistic free-living amoebae: Acanthamoeba spp., *Balamuthia mandrillaris*, *Naegleria fowleri*, and *Sappinia diploidea*. *FEMS Immunol. Med. Microbiol.* 50 (1): 1–26.
44 Kawasaki, T. and Kawai, T. Toll-like receptor signalling pathways. *Front. Immunol.* 5: 461.
45 Ford, A.L., Goodsall, A.L., Hickey, W.F., and Sedgwick, J.D. (1995). Normal adult ramified microglia separated from other central nervous system macrophages by flow cytometric sorting. Phenotypic differences defined and direct ex vivo antigen presentation to myelin basic protein-reactive CD4+ T cells compared. *J. Immunol.* 154 (9): 4309–4321.
46 Sauter, B., Albert, M.L., Francisco, L. et al. (2000). Consequences of cell death: exposure to necrotic tumor cells, but not primary tissue cells or apoptotic cells, induces the maturation of immunostimulatory dendritic cells. *J. Exp. Med.* 191 (3): 423–434.
47 Vabulas, R.M., Ahmad-Nejad, P., Ghose, S. et al. (2002). HSP70 as endogenous stimulus of the toll/interleukin-1 receptor signal pathway. *J. Biol. Chem.* 277 (17): 15107–15112.
48 Wojtkowiak-Giera, A., Derda, M., Kolasa-Wołosiuk, A. et al. (2016). Toll-like receptors in the brain of mice following infection with Acanthamoeba spp. *Parasitol. Res.* 115 (11): 4335–4344.
49 Robinson, D.S. and O'Garra, A. (2002). Further checkpoints in Th1 development. *Immunity* 16 (6): 755–758.
50 Parks, W.C., Wilson, C.L., and López-Boado, Y.S. (2004). Matrix metalloproteinases as modulators of inflammation and innate immunity. *Nat. Rev. Immunol.* 4 (8): 617–629.
51 Raeeszadeh-Sarmazdeh, M. and Do LD, H.B.G. (2020). Metalloproteinases and their inhibitors: potential for the development of new therapeutics. *Cells* 9 (5): 1313. https://doi.org/10.3390/cells9051313.
52 Bruschi, F. and Pinto, B. (2013). The significance of matrix metalloproteinases in parasitic infections involving the central nervous system. *Pathogens* 2 (1): 105–129.
53 Khezri, A., Fallah, E., Mostafazadeh, M. et al. (2016). Molecular and morphometric characterization of Acanthamoeba spp. from different water sources of Northwest Iran as a neglected focus, co-bordered with the country of Iraq. *Jundishapur J. Microbiol.* 9 (11): e38481.
54 Lam, C., Jamerson, M., Cabral, G. et al. (2017). Expression of matrix metalloproteinases in *Naegleria fowleri* and their role in invasion of the central nervous system. *Microbiology* 163 (10): 1436–1444.
55 Łanocha-Arendarczyk, N., Baranowska-Bosiacka, I., Gutowska, I. et al. (2018). The activity of matrix metalloproteinases (MMP-2, MMP-9) and their tissue inhibitors (TIMP-1, TIMP-3) in the cerebral cortex and hippocampus in experimental acanthamoebiasis. *Int. J. Mol. Sci.* 19 (12): 4128.
56 Zigova, T., Pencea, V., Wiegand, S.J., and Luskin, M.B. (1998). Intraventricular administration of BDNF increases the number of newly generated neurons in the adult olfactory bulb. *Mol. Cell. Neurosci.* 11 (4): 234–245.
57 Kot, K., Łanocha-Arendarczyk, N., and Kosik-Bogacka, D. (2021). Immunopathogenicity of Acanthamoeba spp. in the brain and lungs. *Int. J. Mol. Sci.* 22 (3): 1261.
58 Ibáñez, C.F. (1996). Neurotrophin-4: the odd one out in the neurotrophin family. *Neurochem. Res.* 21 (7): 787–793.
59 Culberson, C.G., Smith, J.W., and Minner, J.R. (1958). Acanthamoeba: observations on animal pathogenicty. *Science* 127: 1506–1509.
60 Culbertson, C.G., Ensminger, P.W., and Overton, W.M. (1966). Hartmannella (acanthamoeba). Experimental chronic, granulomatous brain infections produced by new isolates of low virulence. *Am. J. Clin. Pathol.* 46 (3): 305–314.
61 Martinez, A.J. (1980). Is *Acanthamoeba* encephalitis an opportunistic infection? *Neurology* 30: 567–574. https://doi.org/10.1212/wnl.30.6.567.

62 Lanocha, N., Kosik-Bogacka, D., Maciejewska, A. et al. (2009). The occurrence Acanthamoeba (free living amoeba) in environmental and respiratory samples in Poland. *Acta Protozool.* 48 (3): 271.

63 De Jonckheere, J.F. (1983). Isoenzyme and total protein analysis by agarose isoelectric focusing, and taxonomy of the genus Acanthamoeba. *J. Protozool.* 30 (4): 701–706.

64 Moura, H., Wallace, S., and Visvesvara, G.S. (1992). Acanthamoeba healyi n. sp. and the isoenzyme and immunoblot profiles of Acanthamoeba spp., groups 1 and 3. *J. Protozool.* 39 (5): 573–583.

65 Walochnik, J., Sommer, K., Obwaller, A. et al. (2004). Characterisation and differentiation of pathogenic and non-pathogenic Acanthamoeba strains by their protein and antigen profiles. *Parasitol. Res.* 92 (4): 289–298.

66 Khan, N.A., Jarroll, E.L., and Paget, T.A. (2002). Molecular and physiological differentiation between pathogenic and nonpathogenic Acanthamoeba. *Curr. Microbiol.* 45 (3): 197–202.

67 Fuerst, P.A. (2014). Insights from the DNA databases: approaches to the phylogenetic structure of Acanthamoeba. *Exp. Parasitol.* 145: S39–S45.

68 Corsaro, D., Köhsler, M., Montalbano Di Filippo, M. et al. (2017). Update on Acanthamoeba jacobsi genotype T15, including full-length 18S rDNA molecular phylogeny. *Parasitol. Res.* 116 (4): 1273–1284.

69 Corsaro, D., Walochnik, J., Köhsler, M., and Rott, M.B. (2015). Acanthamoeba misidentification and multiple labels: redefining genotypes T16, T19, and T20 and proposal for Acanthamoeba micheli sp. nov.(genotype T19). *Parasitol. Res.* 114 (7): 2481–2490.

70 Nuprasert, W., Putaporntip, C., Pariyakanok, L., and Jongwutiwes, S. (2010). Identification of a novel T17 genotype of Acanthamoeba from environmental isolates and T10 genotype causing keratitis in Thailand. *J. Clin. Microbiol.* 48 (12): 4636–4640.

71 Gast, R.J., Ledee, D.R., Fuerst, P.A., and Byers, T.J. (1996). Subgenus systematics of Acanthamoeba: four nuclear 18S rDNA sequence types. *J. Eukaryot. Microbiol.* 43 (6): 498–504.

72 Rao, K.R., Kauser, F., Srinivas, S. et al. (2005). Analysis of genomic downsizing on the basis of region-of-difference polymorphism profiling of *Mycobacterium tuberculosis* patient isolates reveals geographic partitioning. *J. Clin. Microbiol.* 43 (12): 5978–5982.

73 Ledee, D.R., Booton, G.C., Awwad, M.H. et al. (2003). Advantages of using mitochondrial 16S rDNA sequences to classify clinical isolates of Acanthamoeba. *Invest. Ophthalmol. Vis. Sci.* 44 (3): 1142–1149.

74 Qvarnstrom, Y., Nerad, T.A., and Visvesvara, G.S. (2013). Characterization of a new pathogenic Acanthamoeba species, *A. byersi* n. sp., isolated from a human with fatal amoebic encephalitis. *J. Eukaryot. Microbiol.* 60 (6): 626–633.

75 Garcia, L.S. and Procop, G.W. (2016). Diagnostic medical parasitology. In: *Manual of Commercial Methods in Clinical Microbiology: International Edition* (ed. A.L. Truant), 284–308. Wiley Blackwell.

76 Gillespie, S.H. and Pearson, R.D. (ed.) (2001). *Principles and Practice of Clinical Parasitology*. Toronto, ON: Wiley.

77 Turankar, R.P., Lavania, M., Darlong, J. et al. (2019). Survival of *Mycobacterium leprae* and association with Acanthamoeba from environmental samples in the inhabitant areas of active leprosy cases: a cross sectional study from endemic pockets of Purulia, West Bengal. *Infect. Genet. Evol.* 72: 199–204.

78 Barker, J. and Brown, M.R. (1994). Trojan horses of the microbial world: protozoa and the survival of bacterial pathogens in the environment. *Microbiology* 140 (6): 1253–1259.

79 Chaabna, Z., Forey, F., Reyrolle, M. et al. (2013). Molecular diversity and high virulence of *Legionella pneumophila* strains isolated from biofilms developed within a warm spring of a thermal spa. *BMC Microbiol.* 13 (1): 17.

80 Verelst, F., Kuylen, E., and Beutels, P. (2020). Indications for healthcare surge capacity in European countries facing an exponential increase in coronavirus disease (COVID-19) cases, March 2020. *Euro Surveill.* 25 (13): 2000323.

81 Khodr, A., Kay, E., Gomez-Valero, L. et al. (2016). Molecular epidemiology, phylogeny and evolution of Legionella. *Infect. Genet. Evol.* 43: 108–122.

82 Gomes, T.S., Gjiknuri, J., Magnet, A. et al. (2018). The influence of acanthamoeba–legionella interaction in the virulence of two different legionella species. *Front. Microbiol.* 9: 2962.

83 Morens, D.M., Folkers, G.K., and Fauci, A.S. (2004). The challenge of emerging and re-emerging infectious diseases. *Nature* 430 (6996): 242–249.

84 Abd, H., Saeed, A., Weintraub, A. et al. (2007). Vibrio cholerae O1 strains are facultative intracellular bacteria, able to survive and multiply symbiotically inside the aquatic free-living amoeba *Acanthamoeba castellanii*. *FEMS Microbiol. Ecol.* 60 (1): 33–39.

85 Sandström, G., Saeed, A., and Abd, H. (2010). *Acanthamoeba polyphaga* is a possible host for *Vibrio cholerae* in aquatic environments. *Exp. Parasitol.* 126 (1): 65–68.

86 Shanan, S., Abd, H., Hedenström, I. et al. (2011). Detection of *Vibrio cholerae* and Acanthamoeba species from same natural water samples collected from different cholera endemic areas in Sudan. *BMC. Res. Notes* 4 (1): 1–4.

87 Baré, J., Sabbe, K., Huws, S. et al. (2010). Influence of temperature, oxygen and bacterial strain identity on the association of *Campylobacter jejuni* with *Acanthamoeba castellanii*. *FEMS Microbiol. Ecol.* 74 (2): 371–381.

88 Olofsson, J., Axelsson-Olsson, D., Brudin, L. et al. (2013). *Campylobacter jejuni* actively invades the amoeba *Acanthamoeba polyphaga* and survives within non digestive vacuoles. *PLoS One* 8 (11): e78873.

89 Snelling, W.J., Stern, N.J., Lowery, C.J. et al. (2008). Colonization of broilers by *Campylobacter jejuni* internalized within *Acanthamoeba castellanii*. *Arch. Microbiol.* 189 (2): 175–179.

90 Maschio, V.J., Corção, G., Bücker, F. et al. (2015). Identification of paenibacillus as a symbiont in acanthamoeba. *Curr. Microbiol.* 71 (3): 415–420.

91 Müller, A., Walochnik, J., Wagner, M., and Schmitz-Esser, S. (2016). A clinical Acanthamoeba isolate harboring two distinct bacterial endosymbionts. *Eur. J. Protistol.* 56: 21–25.

92 Wessel, H.B. (1980). Granulomatous amebic encephalitis (GAE) with prolonged clinical course CT scan findings, diagnosis by brain biopsy, and effect of treatment. *Neurology* 30: 442.

93 Kidney, D.D. and Kim, S.H. (1998). CNS infections with free-living amebas: neuroimaging findings. *AJR Am. J. Roentgenol.* 171 (3): 809–812.

94 Lowichik, A., Rollins, N., Delgado, R. et al. (1995). Leptomyxid amebic meningoencephalitis mimicking brain stem glioma. *Am. J. Neuroradiol.* 16 (4): 926–929.

95 Zagardo, M.T., Castellani, R.J., Zoarski, G.H., and Bauserman, S.C. (1997). Granulomatous amebic encephalitis caused by leptomyxid amebae in an HIV-infected patient. *Am. J. Neuroradiol.* 18 (5): 903–908.

96 Gardner, H.A., Martinez, A.J., Visvesvara, G.S., and Sotrel, A. (1991). Granulomatous amebic encephalitis in an AIDS patient. *Neurology* 41 (12): 1993.

97 Matson, D.O., Rouah, E., Lee, R.T. et al. (1988). Acanthameba meningoencephalitis masquerading as neurocysticercosis. *Pediatr. Infect. Dis. J.* 7 (2): 121–124.

98 Falcone, S., Quencer, R.M., and Post, M.J. (1994). Magnetic resonance imaging of unusual intracranial infections. *Top. Magn. Reson. Imaging* 6 (1): 41–52.

99 Deol, I., Robledo, L., Meza, A. et al. (2000). Encephalitis due to a free-living amoeba (*Balamuthia mandrillaris*): case report with literature review. *Surg. Neurol.* 53 (6): 611–616.

100 Parija, S.C., Dinoop, K.P., and Venugopal, H. (2015). Management of granulomatous amebic encephalitis: laboratory diagnosis and treatment. *Trop. Parasitol.* 5 (1): 23.

101 Pearl, M., Visvesvara, G.S., Martínez, A. et al. (1990). Naegleria and Acanthamoeba infections. *Rev. Infect. Dis.* 12: 490–513.

102 Khan, N.A. (2009). Acanthamoeba biology and pathogenesis. University of Nottingham, UK. Caister Academic Press. *Acta Parasitol.* 54: 1230–2821.

103 Garcia, L.S. (2006). *Diagnostic Medical Parasitology*. American Society for Microbiology Press.

104 Khan, N.A. and Paget, T.A. (2002). Molecular tools for speciation and epidemiological studies of Acanthamoeba. *Curr. Microbiol.* 44 (6): 444–449.

105 Weekers, P.H., Bodelier, P.L., Wijen, J.P., and Vogels, G.D. (1993). Effects of grazing by the free-living soil amoebae *Acanthamoeba castellanii*, *Acanthamoeba polyphaga*, and *Hartmannella vermiformis* on various bacteria. *Appl. Environ. Microbiol.* 59 (7): 2317–2319.

106 Khan, N.A. (2003). Pathogenesis of acanthamoeba infections. *Microb. Pathog.* 34 (6): 277–285.

107 Schuster, F.L. (2002). Cultivation of pathogenic and opportunistic free-living amebas. *Clin. Microbiol. Rev.* 15 (3): 342–354.

108 Walochnik, J., Scheikl, U., and Haller-Schober, E.M. (2015). Twenty years of acanthamoeba diagnostics in Austria. *J. Eukaryot. Microbiol.* 62 (1): 3–11.

109 Yagi, S., Booton, G.C., Visvesvara, G.S., and Schuster, F.L. (2005). Detection of Balamuthia mitochondrial 16S rRNA gene DNA in clinical specimens by PCR. *J. Clin. Microbiol.* 43 (7): 3192–3197.

110 Qvarnstrom, Y., Visvesvara, G.S., Sriram, R., and da Silva, A.J. (2006). Multiplex real-time PCR assay for simultaneous detection of Acanthamoeba spp., *Balamuthia mandrillaris*, and *Naegleria fowleri*. *J. Clin. Microbiol.* 44 (10): 3589–3595.

111 Dowell, J.D., Mukherjee, S., Raghavan, P., and Rehm, P.K. (2015). AIDS presenting as granulomatous amebic encephalitis: PET and MR imaging correlation. *J. Neuroimaging* 25 (6): 1047–1049.

112 El Sahly, H., Udayamurthy, M., Parkerson, G., and Hasbun, R. (2017). Survival of an AIDS patient after infection with Acanthamoeba sp. of the central nervous system. *Infection* 45 (5): 715–718.

113 Deetz, T.R., Sawyer, M.H., Billman, G. et al. (2003). Successful treatment of Balamuthia amoebic encephalitis: presentation of 2 cases. *Clin. Infect. Dis.* 37 (10): 1304–1312.

114 Schumacher, D.J., Tien, R.D., and Lane, K. (1995). Neuroimaging findings in rare amebic infections of the central nervous system. *Am. J. Neuroradiol.* 16 (4): 930–935.

115 Modica, S., Miracco, C., Cusi, M.G. et al. (2018). Non-granulomatous cerebellar infection by Acanthamoeba spp. in an immunocompetent host. *Infection* 46 (6): 885–889.

116 Kiderlen, A.F., Radam, E., and Tata, P.S. (2009). Assessment of *Balamuthia mandrillaris*-specific serum antibody concentrations by flow cytometry. *Parasitol. Res.* 104 (3): 663–670.

117 Schuster, F.L., Yagi, S., Gavali, S. et al. (2009). Under the radar: balamuthia amebic encephalitis. *Clin. Infect. Dis.* 48 (7): 879–887.

118 Schuster, F.L., Yagi, S., Wilkins, P.P. et al. (2008). *Balamuthia mandrillaris*, agent of amebic encephalitis: detection of serum antibodies and antigenic similarity of isolates by enzyme immunoassay. *J. Eukaryot. Microbiol.* 55 (4): 313–320.

119 Tavares, M., Correia da Costa, J.M., Carpenter, S.S. et al. (2006). Diagnosis of first case of Balamuthia amoebic encephalitis in Portugal by immunofluorescence and PCR. *J. Clin. Microbiol.* 44 (7): 2660–2663.

120 Schuster, F.L., Honarmand, S., Visvesvara, G.S., and Glaser, C.A. (2006). Detection of antibodies against free-living amoebae *Balamuthia mandrillaris* and Acanthamoeba species in a population of patients with encephalitis. *Clin. Infect. Dis.* 42 (9): 1260–1265.

121 Huang, Z.H., Ferrante, A., and Carter, R.F. (1999). Serum antibodies to *Balamuthia mandrillaris*, a free-living amoeba recently demonstrated to cause granulomatous amoebic encephalitis. *J. Infect. Dis.* 179 (5): 1305–1308.

122 Kucerova, Z., Sriram, R., Wilkins, P.P., and Visvesvara, G.S. (2011). Identification of antigenic targets for immunodetection of *Balamuthia mandrillaris* infection. *Clin. Vaccine Immunol.* 18 (8): 1297–1301.

123 Visvesvara, G.S. (2013). Infections with free-living amebae. *Handb. Clin. Neurol.* 114: 153–168.

124 Siddiqui, R., Jarroll, E.L., and Khan, N.A. (2009). *Balamuthia mandrillaris*: staining properties of cysts and trophozoites and the effect of 2, 6-dichlorobenzonitrile and calcofluor white on encystment. *J. Eukaryot. Microbiol.* 56 (2): 136–141.

125 Elsheikha, H.M., Siddiqui, R., and Khan, N.A. (2020). Drug discovery against acanthamoeba infections: present knowledge and unmet needs. *Pathogens* 9 (5): 405.

126 Lackner, P., Beer, R., Broessner, G. et al. (2010). Acute granulomatous Acanthamoeba encephalitis in an immunocompetent patient. *Neurocrit. Care.* 12 (1): 91–94.

127 Maritschnegg, P., Sovinz, P., Lackner, H. et al. (2011). Granulomatous amebic encephalitis in a child with acute lymphoblastic leukemia successfully treated with multimodal antimicrobial therapy and hyperbaric oxygen. *J. Clin. Microbiol.* 49 (1): 446–448.

128 Fung, K.T., Dhillon, A.P., McLaughlin, J.E. et al. (2008). Cure of Acanthamoeba cerebral abscess in a liver transplant patient. *Liver Transpl.* 14 (3): 308–312.

129 Webster, D., Umar, I., Kolyvas, G. et al. (2012). Case report: treatment of granulomatous amoebic encephalitis with voriconazole and miltefosine in an immunocompetent soldier. *Am. J. Trop. Med. Hyg.* 87 (4): 715.

130 Petry, F., Torzewski, M., Bohl, J. et al. (2006). Early diagnosis of Acanthamoeba infection during routine cytological examination of cerebrospinal fluid. *J. Clin. Microbiol.* 44 (5): 1903–1904.

131 Seijo Martinez, M., Gonzalez-Mediero, G., Santiago, P. et al. (2000). Granulomatous amebic encephalitis in a patient with AIDS: isolation of Acanthamoeba sp. group II from brain tissue and successful treatment with sulfadiazine and fluconazole. *J. Clin. Microbiol.* 38 (10): 3892–3895.

132 Megha, K., Sehgal, R., and Khurana, S. (2018). Genotyping of Acanthamoeba spp. isolated from patients with granulomatous amoebic encephalitis. *Indian J. Med. Res.* 148 (4): 456.

133 Khurana, S., Mewara, A., Verma, S., and Totadri, S.K. (2012). Central nervous system infection with Acanthamoeba in a malnourished child. *BMJ Case Rep.* 2012: bcr2012007449.

134 Tunkel, A.R., Glaser, C.A., Bloch, K.C. et al. (2008). The management of encephalitis: clinical practice guidelines by the Infectious Diseases Society of America. *Clin. Infect. Dis.* 47: 303–327.

135 Cope, J.R. (2013). Investigational drug available directly from CDC for the treatment of infections with free-living amebae. *MMWR Morb. Mortal. Wkly Rep.* 62 (33): 666.

136 Shirabe, T., Monobe, Y., and Visvesvara, A.G. (2002). An autopsy case of amebic meningoencephalitis. The first Japanese case caused by *Balamuthia mandrillaris*. *Neuropathology* 22 (3): 213–217.

137 Bando, Y., Takahashi, T., Uehara, H. et al. (2012). Autopsy case of amebic granulomatous meningoencephalitis caused by *Balamuthia mandrillaris* in Japan. *Pathol. Int.* 62 (6): 418–423.

138 Kato, H., Mitake, S., Yuasa, H. et al. (2013). Successful treatment of granulomatous amoebic encephalitis with combination antimicrobial therapy. *Intern. Med.* 52 (17): 1977–1981.

139 Martínez, D.Y., Seas, C., Bravo, F. et al. (2010). Successful treatment of *Balamuthia mandrillaris* amoebic infection with extensive neurological and cutaneous involvement. *Clin. Infect. Dis.* 51 (2): e7–e11.

140 Doyle, J.S., Campbell, E., Fuller, A. et al. (2011). *Balamuthia mandrillaris* brain abscess successfully treated with complete surgical excision and prolonged combination antimicrobial therapy: case report. *J. Neurosurg.* 114 (2): 458–462.

141 Jung, S., Schelper, R.L., Visvesvara, G.S., and Chang, H.T. (2004). *Balamuthia mandrillaris* meningoencephalitis in an immunocompetent patient: an unusual clinical course and a favorable outcome. *Arch. Pathol. Lab. Med.* 128 (4): 466–468.

142 Sharma, O.P. (1976). Antioxidant activity of curcumin and related compounds. *Biochem. Pharmacol.* 25 (15): 1811–1812.

143 Ruby, A.J., Kuttan, G., Babu, K.D. et al. (1995). Anti-tumour and antioxidant activity of natural curcuminoids. *Cancer Lett.* 94 (1): 79–83.

144 Kötting, J., Berger, M.R., Unger, C., and Eibl, H. (1992). Alkylphosphocholines: influence of structural variation on biodistribution at antineoplastically active concentrations. *Cancer Chemother. Pharmacol.* 30 (2): 105–112.

28

Epidemiology and Current Treatment Trends in "*Thelaziasis*"

Rihana B. Patnool[1], Badrud D. Mohammad[2], Vishwas H. Nagendra[1], Bhupendra G. Prajapati[3], Muhasina K. Muhamadkazim[4], and Sivasankaran Ponnusankar[1]

[1] Department of Pharmacy Practice, JSS College of Pharmacy, JSS Academy of Higher Education & Research, Ooty, Tamil Nadu, India
[2] Department of Pharmaceutical Chemistry, GRT Institute of Pharmaceutical Education and Research, Tirutanni, Tamil Nadu, India
[3] Department of Pharmaceutics and Pharmaceutical Technology, Shree S K Patel College of Pharmaceutical Education and Research, Ganpat University, Mehsana, Gujarat, India
[4] Department of Pharmacognosy, JSS College of Pharmacy, JSS Academy of Higher Education & Research, Ooty, The Nilgiris, Tamil Nadu, India

28.1 Introduction

Thelaziasis is an arthropod-borne illness of the eyes instigated by *Thelazia* species, a nematode worm spread to humans and carnivores by *drosophilid* flies. A group of parasitic roundworms known as *Thelazia* infects the eyes of many domesticated animals, particularly cattle, goats, sheep, dogs, cats, and other farm animals. A lot of other animals and birds are also impacted. The tissues around the eyes, including the eyelids, lacrimal ducts, and glands, are home to *thelazia* worms [1]. Eyeworms are another name for Thelazia nematodes. It is generally recognized as Oriental Eye Worm because of its limited dissemination in South Asian nations and Russia. It is frequently underreported and not given the therapeutic attention it deserves [2, 3]. Figure 28.1 shows small, pale, motile Thelazia worms wriggled over bulbar conjunctiva of an eye infected with thelaziasis [4].

The first reported case of thelaziasis in animals was in a dog in India in the early 1900s. The dog had been experiencing eye irritation and discharge, and upon examination, veterinarians discovered small, thread-like worms in its conjunctival sac. These worms were identified as *Thelazia* species and were removed from the dog's eyes through a surgical procedure. The first reported human case of thelaziasis dates back to 1910, when a patient in Sri Lanka was diagnosed with the disease. The patient, a 56-year-old man, presented with complaints of eye irritation and redness. Upon examination, doctors discovered multiple small worms, measuring between 6 and 8 mm in length, in the patient's conjunctival sacs. These worms were later identified as *Thelazia* species, and the patient was successfully treated with surgical removal of the worms. Since then, several cases have been reported worldwide, with the majority occurring in Asia and Europe. Thelaziasis has become a growing concern in recent years due to its increasing prevalence and potential to cause permanent damage to the eyes [5].

Although Thelazia infestations in both animal and human hosts might go undetected, they commonly result in conjunctivitis, watery eyes (epiphora), corneal ulcers, or corneal opacity. Individuals who have been infected have also described a "foreign body sensation," or having some irritation in their eyes. *Thelaziasis callipaeda* has been linked to ocular infections in carnivores in France. Whereas in Italy, both dogs and cats are susceptible to this virus. Additionally, thelaziasis instances in imported carnivores have been documented in Switzerland, the Netherlands, and Germany. Human thelaziasis case reports have risen in numerous Asian regions, where it mostly affects the old and young, and it has become widespread in rural places with minimal socioeconomic standards. No human cases have been reported, despite an increase in accounts of thelaziasis infection in carnivores in various European nations. Thus, the majority of doctors and ophthalmologists are unaware of this eyeworm infection [6].

We wanted to make the scientific community more aware of the potential health risks associated with this emerging nematode and the necessity of incorporating this parasitic infection into the differential identification of ocular infections.

Rising Contagious Diseases: Basics, Management, and Treatments, First Edition. Edited by Seth Kwabena Amponsah, Ranjita Shegokar, and Yashwant V. Pathak.

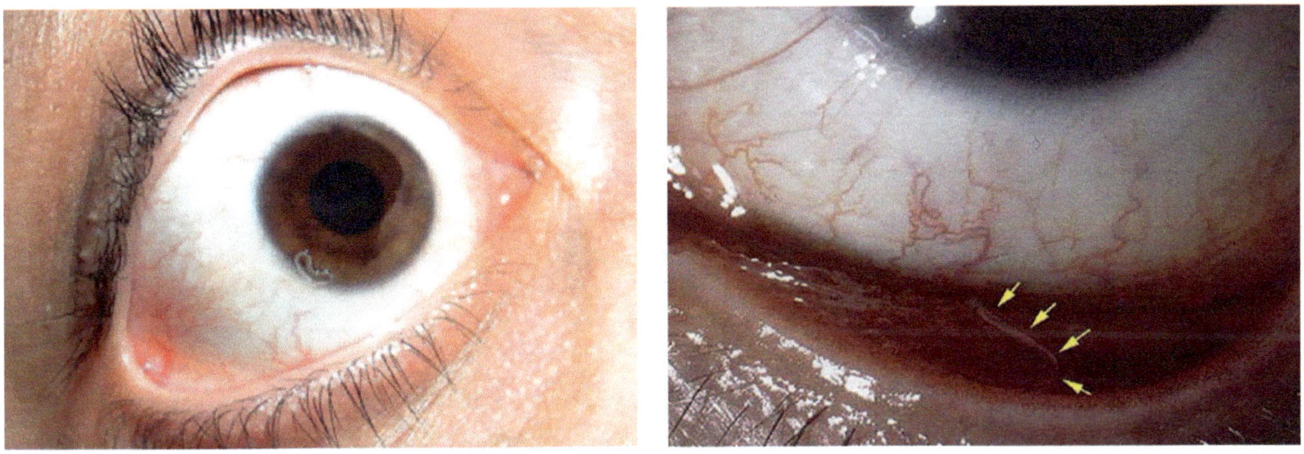

Figure 28.1 A slit-lamp examination eye infected with thelaziasis [4]. *Source:* Naem et al. [4] under CC BY version 4.0.

Thelaziasis, its causative organisms, hosts and vectors, modes of transmission, geographic distribution, clinical manifestations, diagnosis, treatment, preventive measures, and recent advancements in disease control are therefore the primary topics of this chapter.

28.2 Causative Agents

Thelazia species are nematode parasites that affect humans and carnivores alike and cause thelaziasis, an arthropod-borne illness of the eye and adnexa that is spread by drosophilid flies. It is widely referred to as the "Oriental Eye Worm" due to its limited spread, which is mostly to South Asian nations and Russia. In general, *T. callipaeda* is a human parasite, and *Thelazia spirurid* nematodes are predominantly veterinary parasites; however, they can infrequently infect humans. Most zoonotic infections are caused by *T. callipaeda*, also known as the Oriental eye worm. Less frequent causal agents include *Thelazia californiensis*, also regarded as the California eye worm, and *Thelazia gulosa*, better known as the Cattle eyeworm. Nematode worms belonging to the genus Thelazia parasitize the eyes and related areas of several avian and mammal organisms, including humans, which serve as definitive hosts. The infection caused by the nematode, *Thelazia* species is called as "thelaziasis," and they are sometimes dubbed "eyeworms" (rarely spelled as "thelaziosis"). Well-matured forms of this organism are often reported in the third eyelid, tear ducts, tear glands, or eyelids (nictitating membrane). On rare occasions, they are discovered inside the eyeball itself, either in the vitreous cavity or below the conjunctiva (the skin membrane that protects the white area of the eye). There are several species of *Thelazia* that infect the eyes of humans as well as several other mammals, namely cats, dogs, cattle, and horses. The various *Thelazia* species that cause infection are listed below [7]. Table 28.1 elaborates the articularity of the geographical invasion of *Thelazia nematodes* and the respective year of reporting.

28.2.1 *Thelazia anolabiata*

Thelazia anolabiata: studies found that *T. anolabiata* commonly causes thelaziasis in birds like the Andean (*Rupicola peruviana*) and many other Brazilian birds in the South American region.

28.2.2 *Thelazia bubalis*

Thelazia bubalis: mainly found in Indian water buffalo. There is not much data available on this species in the literature.

28.2.3 *Thelazia callipaeda*

T. callipaeda: it is also well-known as the Oriental eye worm, which is a species of *Thelazia that* is responsible for the majority of thelaziasis (or eyeworm infection) in people, canines, and cats. In 1910, it was first found in the eyes of a dog

Table 28.1 Invasion of thelaziasis infection globally.

Species name	The geographical location of reported cases	Year of first reported
Thelazia bubalis	India	1952
Thelazia californiensis	USA and Brazil	1930
Thelazia callipaeda	Europe, Asia, and India	1910
Thelazia ershowi	Former Union of Soviet Socialist Republics	1931
Thelazia gulosa	Australia, former USSR, Europe, and North America	1910
Thelazia lacrymalis	North and South America, Europe, North Africa, and Asia	1831
Thelazia leesei	Europe, Africa, India, and Central Asia	1910
Thelazia rhodesii	Europe, Zambia, Afghanistan, Asia UK, Ghana, Japan, and North America	1828
Thelazia skrjabini	Europe, former USSR, India, Australia, and North America	1928

in China. Over 280 human instances have been recorded in the medical literature thus far. *T. callipaeda's* definitive hosts include the raccoon dog, the wolf (*Canis lupus*), the European rabbit, the red fox (*Vulpes fulva*), and, in addition to people, dogs and cats. India, China, Germany, Indonesia, Taiwan, France, Italy, Korea, the Netherlands, Thailand, Switzerland, Croatia, Japan, and Russia have all discovered this species.

T. callipaeda invasion was much more common in domestic and wild cats, dogs, and foxes. Furthermore, there is no indication of a host sex propensity, but prior research has given inconsistent results in this regard. In terms of cats, it is known that all of the hosts in the Motta et al. research had outdoor access and male, which might be risk elements for *T. callipaeda* infection, perhaps related to provincial and tracking behavior, but also due to a larger wandering area.

The environment of the nations where *T. callipaeda* has been found ranges from temperate in the Russian Federation to hot (Indonesia) and semitropical (Japan). Cats, dogs, and foxes were evaluated in research that was conducted in two different regions of Italy (northwestern Italy and southern Italy). In southern Italy, *T. callipaeda* was linked to a high frequency, with about 60% in dogs and 49.3% in foxes [8].

28.2.4 *Thelazia californiensis*

T. californiensis: also known as the *California eye worm*; the name was given as *Thelazia for Dr. Thelaz; californiensis* for the place from which the cases were collected. *Thelazia* spp. was discovered in a dog's eye that was gathered in the Los Angeles, California, region, according to Allerton (1929). Dr. Ackert named the worms in this paper as *T. callipaeda*. Price named the worm *T. californiensis* in 1930 after identifying it as a new species. The majority of complaints have come from dogs. These worms, which may be seen on the conjunctiva, behind the lids, and on the nictitating membrane, are parasites of the eye's orbits. So far, about 260 instances of *T. callipaeda* infection in humans have been reported. This species is majorly distributed in Europe and Asia regions [9].

28.2.5 *Thelazia erschowi*

T. erschowi: was found first by Oserskaja in the year 1931 in Pig (*Sus domesticus*) and is majorly distributed in post-Soviet states. There are not many reported cases of this species.

28.2.6 *Thelazia gulosa*

T. gulosa: is an ovoviviparous parasite, also known as the cattle eyeworm, that affects the conjunctiva of the primary host and belongs to the genus Thelazia, family Thelaziidae, order Spirurida, and phylum Nematoda. *T. gulosa* is found in all the northern U.S. states as well as southern Canada in North America. *T. gulosa* is a parasite that mostly affects cattle and is extensively dispersed over Europe, North America, Australia, and Central Asia. The most frequent clinical signs are increased lacrimation, follicular enlargement of the conjunctiva, moderate conjunctival irritation, and a feeling of a foreign mass. Sometimes the worms move over the eye's surface, damaging the cornea and leading to opacity and blindness [3]. Because the conjunctiva is the only area where the Thelazia infection is present, and because removing the worms with a cotton swab or forceps is adequate, typically, when the parasites are eliminated, the clinical symptoms go away rapidly [10].

28.2.7 *Thelazia lacrymalis*

Thelazia lacrymalis: is the most prevalent eyeworm in horses; it was first discovered in Ontario, Canada, in 1970 and later spread to several other USA states. In comparison to the rest of Europe, the UK recorded the first horse case in the 1930s. Thelaziasis in horses caused by *T. lacrymalis* has just been documented in France. Clinical manifestations of *T. lacrymalis* might range from mild conjunctivitis, photophobia, lacrimation, and, less frequently, ulcerative keratitis and blindness, to no visible lesions. Cattle (*Bos taurus*) and Horses (*Equus caballus*) are the species' definitive hosts, while face flies (*Musca autumnalis*) and *Musca osiris* serve as intermediate hosts. South and North America, the Middle East, Asian countries, and Europe are the main distribution regions for this specialty [11].

28.2.8 *Thelazia leesei*

Thelazia leesei: is a rarely seen eye worm mainly observed in Bactrian camel (*Camelus bactrianus*) and Dromedary (*Camelus dromedarius*). Flies like *Musca lucidulus* act as intermediate hosts for this species. This species was mainly found in post-Soviet states and in India. There is not much data available about this species.

28.2.9 *Thelazia rhodesii*

Thelazia rhodesii: *T. rhodesii* is mainly found in bovids, buffalo, cattle, zebu, horses, bison, and sometimes domestic sheep and goats. Face flies, namely *Musca sorbens, M. larvipara, and M. autumnalis* are the recommended intermediate hosts for this species. *T. rhodesii* is more harmful than a few other species's cuticle because its coarse and prominent cuticle serration is just faintly ragged. Frequently, *T. rhodesii* is discovered in the conjunctival sac. Conjunctivitis that ranges in severity from mild to severe and eyelid inflammation (red, irritating, and itchy) are typical symptoms. With *T. rhodesii* infections in cattle, keratitis (corneal inflammation), including opacity, ulceration, perforation, and persistent fibrosis, may also manifest in severe cases. Cattle infected with *T. rhodesii* are found in Europe, North America, and Asia [12].

28.2.10 *Thelazia skrjabini*

Thelazia skrjabini: it was distinguished from other *Thelazia* species known to affect cattle based on the following factors: In comparison to other *Thelazia* spp. from cattle, *T. skrjabini* has a tiny, shallow buccal capsule, short, subequal spicules, and more pairs of preanal papillae. Dr. J. Ralph Lichtenfels, Animal Parasitology Institute, USDA, ARS, Beltsville, Maryland, USA, verified *T. skrjabini's* identity. The USDA National Parasite Collection received voucher specimens that were placed with the following entry number: USNM: 79414. Cattle (*B. taurus*) and yak (*Bos grunniens*) are the reported definitive hosts for this species, whereas face flies, namely *Musca vitripennis*, *M. autumnalis*, and *Musca amica*, are the intermediate hosts. Geographically, these species are primarily fund in North America and Europe [13].

28.3 Morphology

All of the worms were white, small, and threadlike up close. All of the male worms that were found in the first case ranged in length from 9.58 to 10.1 mm and in width from 0.25 to 0.41 mm. However, the male worms found in the second case ranged in length from 11 to 11.2 mm and were around 0.3 mm wide. All the worms' bodies were covered in transverse cuticular striations. There were no lips or teeth-like features in the buccal cavity, which was rectangular. The length of the lengthy spicule, which extended from the cloaca, ranged from 0.85 to 1.31 mm. Due to the posterior end's spiral coiling, not all of the preanal papillae were visible [13]. Figure 28.2 represents the scanning electron microscopic image of adult *Thelazia* [14].

Figure 28.2 Microscopic image of adult *Thelazia* species [14]. *Source:* Lebon et al. [15] / EDP Sciences / CC BY 4.0.

The mature worms have a milky white hue and are tiny and slender. Males may grow to a maximum length of 12 mm, and females can reach 20 mm. The frontal edge of the buccal cavity is overturned and split into six swags; four of these seem to be filled by a tiny refractile papilliform organ. The mouth lacks lips. There are four submedian and two lateral cephalic papillae. Transverse striae may be present throughout the body. A blunt tail, which is a lacking caudal alae, was observed in male species. Several pre-cloacal papillae, of which three or four pairs are post-cloacal, and one of which is unpaired and situated in front of the anal orifice. The spicules differ in size. The female's hind end is bluntly rounded. Table 28.2 clearly illustrates the morphological features of various *Thelazia* species. The detailed morphological features of the reported species are mentioned below.

28.3.1 Morphology of *Thelazia gulosa*

The mouth lacked lips, and the buccal entrance was spherical, measuring 33–47 m in diameter and encircled by four cephalic papillae and two amphids. On each side, there were two lateral cervical papillae. The female's body was slender, pale, and attenuated at both ends, measuring 8.0–11.5 mm long and 350–460 mm broad at its widest point. The vulva was found between 460 and 610 m from the cephalic region. The anal orifice was positioned 70–120 μm from the posterior end of the body. Each lateral side of the female's tail has one button-like subterminal phasmid. The males were 7.5–8.0 mm in length and 300–370 μm broad at the maximal width. Its tail appeared blunt without caudal alae and curled ventrally [10]. Figure 28.3 illustrates the posterior ends of *T. gulosa*, which clearly show cephalic papillae (CP) and the two different spicula (A) [11].

28.3.2 Morphology of *Thelazia californiensis*

In the case of *T. californiensis,* the males have spicules that are different lengths – of a maximum 1.5–1.7 mm and a minimum 0.15–0.19 mm in length, and the overall length is amount 12–13 mm. The females measure 15–17 mm in length. Whereas the female vulva is about 0.8–1 mm in length from the frontal end and situated behind the esophagus intestinal junction. The uterine eggs are 52 mm by 29 mm in size. In contrast to *T. callipaeda*, this species' vulva is located forward to the point where the intestine and esophagus converge. The tail end of a female *T. californiensis* consists of phasmids and a projecting anal orifice, as well as the anterior end, which has a tiny, thin buccal cavity and noticeable, broadly spaced grooves [12].

28.3.3 Morphology of *Thelazia rhodesii*

This species has comparable cephalic areas among male and female types. The mouth lacked lips and was divided into two swags, beginning with a 10.0–14.5 μm wide tiny capsule at the oral aperture (Figure 28.4) [4]. There were eight submedian tiny, conoidal cephalic papillae and two adjacent amphidal openings around the mouth. Additionally, two horizontal cervical papillae, one on each side, were discovered 350–384 μm from the frontal end. Females measured 12.5–20.5 mm in length and 300–500 μm in width. On the front end of the body, the vulva was positioned in the esophageal area. The shape of the cuticle surrounding the vulva was distinct. The anal orifice was 80–84 μm away from the tail end. Figure 28.4 shows the scanning electron micrograph of the frontal end of the male *T. rhodesi*, displaying a mouth with its frontal edge pointed back and cut into two festoons (arrowheads), cephalic papillae (stars), and amphidal apertures, and the posterior end showing an annal pore [14].

28.3.4 Morphology of *Thelazia skrjabini*

The mouth was orbicular with no lips, an anterior border that was flipped over, and six grooves beginning with a tiny capsule 11.0–13.5 m wide at the buccal aperture. Two rings of cephalic papillae were found around the mouth. Six papillae

Table 28.2 Morphological characteristics of various *Thelazia* species.

			Morphology of				
Species name	**Length (cm)**	**Width (mm)**	**Buccal cavity**	**Vulval opening**	**Position opening**	**Midsection ridges**	**Tail**
T. bubalis	0.65–0.78	0.25–0.30	DNA	DNA	DNA	DNA	DNA
T. californiensis	1.20–1.88	0.33–0.42	Cup-like deep	No unique characteristics	Bottom	Widely spread, rounded, and relatively conspicuous	Protruded anus from tip of the tail, without unilateral tapering
T. callipaeda	1.07–1.85	0.29–0.51	Cup-like wide shaped	A little flap at the opening	Front	Serrated, quite noticeable, and closely spaced	Protruded anus from tip of the tail, without unilateral tapering
T. ershowi	0.55–0.87	0.18–0.21	DNA	DNA	DNA	DNA	DNA
T. gulosa	0.80–1.15	0.35–0.46	Cup-like deep	No unique characteristics	At same level	Rounded, conspicuous, and widely distributed	Two postanal phasmids, unilateral postanal taper, no protruding anus
T. lacrymalis	1.05–1.25	0.28–0.29	Cup-like deep	No unique characteristics	Bottom	Rounded, unobtrusive, and closely spaced	Two postanal phasmids, a unilateral postanal taper, and an absence of anus protrusion
T. leesei	1.40–2.10	0.40–0.41	DNA	DNA	DNA	DNA	DNA
T. rhodesi	1.25–2.05	0.30–0.50	Cup-like deep	Distinct cuticular patterns and protruding	Bottom	Widely spread, conspicuous, and serrated	Postnatal unilateral tapering with a plasmid protrudes from tip of the tail
T. skrjabini	1.10–1.90	0.18–0.38	Indistinct and short	Obtrudes	Bottom	Widely spread, hardly discernible, and rounded	Two postanal phasmids extend from the anus, and it is spherical without unilateral tapering

DNA, data not available.

Figure 28.3 SEM image of *Thelazia gulosa*'s posterior ends [11]. *Source:* Zhang et al. [11] under CC BY version 4.0.

were found in the inner circle, while four submedian cephalic papillae were found in the outer circle. Two amphids were found on both lateral surfaces of the cranium. There were two lateral cervical papillae, one on each side. Fine, barely perceptible transverse striations ran across the cuticle. Females measured 11–19 mm in length and 178–378 mm in width. The vulva protruded from the cephalic area and was positioned 410–710 m away. The anal orifice was 65–90 m from the tail's tip, and two phasmids were found at the tip. Figure 28.5 illustrates the anterior and posterior ends of *T. skrjabini*, showing the cephalic region, cervical papilla (CP), anal pore, and plasmid [15].

28.3.5 Morphology of *Thelazia lacrymalis*

The female's body was slender, pale, elongated at both ends, 12.5 mm long, and up to 279 μm wide, according to a SEM investigation. The buccal aperture was spherical, without lips, and encircled by two amphids, four cephalic papillae, and four cephalic papillae. On each side, there have been two lateral cervical papillae. The vulva was 493 μm away from the front end of the body on the ventral side.

28.4 Mode of Transmission

Only secretophagous flies may transmit *Thelazia* species, which are larviparous. During the day, these flies swarm around the eyes of animals and people, landing on the eyes and dispersing the infectious larvae onto the conjunctiva of the host. Theliasis can only spread in this way, according to current knowledge. When feeding on the lacrimal secretions of afflicted animals, these white nematodes are spread by non-biting dipteran muscid flies or drosophila fruit flies (intermediary hosts). During the day, these flies fly around the eyes of animals and people, settling on the eyes and spreading infective larvae onto the host conjunctiva. That is the only way Thelazia may be transmitted. The worm is spread by secretophagous flies that feed on the lacrimal discharges of infected animals and/or people that contain Thelazia first-stage larvae. These

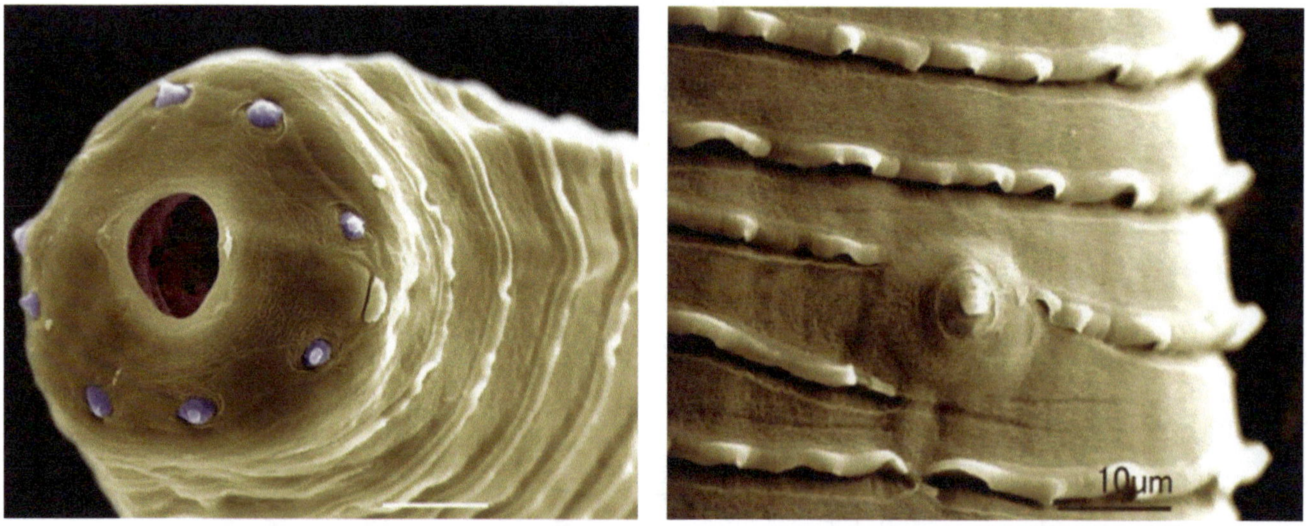

Figure 28.4 Microscopic image of *Thelazia rhodesii* anterior and posterior ends [4]. *Source:* Naem et al. [4] under CC BY version 3.0.

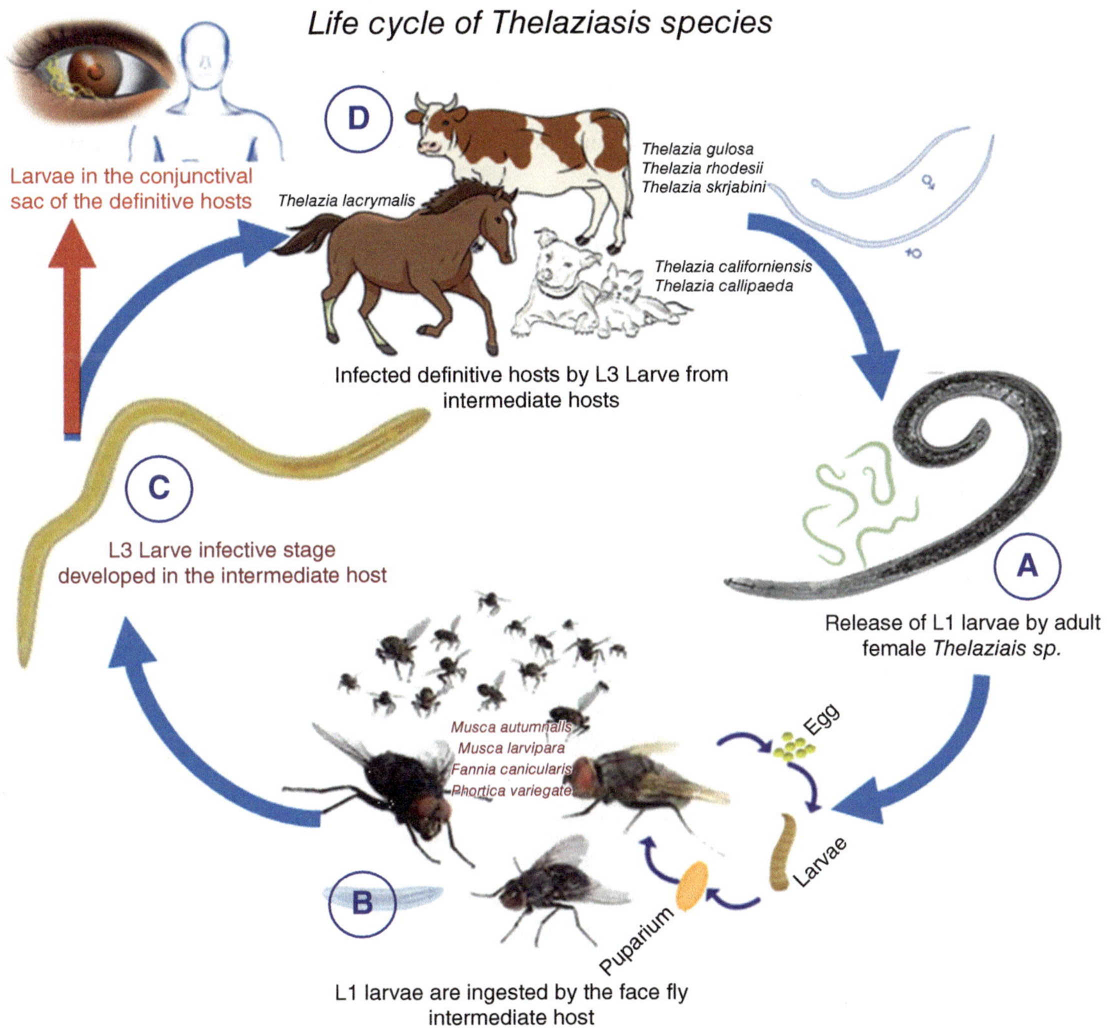

Figure 28.5 Transmission cycle of Thelaziasis species.

larvae mature into third-stage larvae in the fly, which are then released back into the orbit. The illness primarily affects the elderly and kids in rural settings where living standards and hygiene are deplorable. Humans are prone to ocular thelaziasis due to cattle husbandry, interaction with stray dogs, hilly terrain, and the rainy season [16].

It is assumed that the larva enters the body through the skin or by use of contaminated drinking water. The adult worms are creamy-white, filaria-like nematodes that can be seen in the conjunctival sac or on the cornea under a slit light. Mechanical extraction of the nematodes from the conjunctival sac is part of the treatment and can be done with forceps or a sterile saline wash.

28.5 Hosts and Vectors

Hosts:
The definitive hosts for Thelazia are typically mammals, including humans, dogs, cats, horses, cattle, sheep, and goats. Some species of Thelazia can also infect birds.

Vectors:
Thelazia are transmitted by various species of flies and other biting insects, which act as intermediate hosts for the parasite. The most common vectors are different species of flies, including *Musca autumnalis* (the face fly), *Musca domestica*

Table 28.3 Details of *Thelazia* species invading hosts and year of detection.

Species name	Intermediate hosts (reported)	Primary definitive hosts (reported)
Thelazia anolabiata	Not known	Andean cock-of-the-rock (*Rupicola peruviana*), Brazilian birds
Thelazia bubalis	Not repos	Water buffalo
Thelazia californiensis	House flies *Fannia benjamini* *Fannia canicularis*	Human (*Homo sapiens*), American black bear (*Ursus americanus*), Mule deer (*Odocoileus hemionus*), Domestic sheep (*Ovis aries*), Dogs (*Canis familiaris*), Coyote (*Canis latrans*), Cats (*Felis catus*)
Thelazia callipaeda	Fruit flies (*Phortica variegate*) – Europe, and *Phortica okadai* in China	Over 260 reported humans, Dogs (*C. familiaris*), Cats (*F. catus*), Gray wolf (*Canis lupus*), Raccoon dog (*Nyctereutes procyonoides*), red fox (*Vulpes vulpes*), European rabbit (*Oryctolagus cuniculus*)
Thelazia ershowi	Not known	Pig (*Sus domesticus*)
Thelazia gulosa	Face fly (*Musca autumnalis*) – North America and Europe *Musca larvipara* – Ukraine, *Musca vitripennis* – Crimea, *Musca amica* – Far East	Yak (*Bos grunniens*), Cattles (*Bos taurus*) and Rarely in human (*H. sapiens*)
Thelazia lacrymalis	Face flies *M. autumnalis*, *Musca osiris*	Horses (*Equus caballus*) and Cattles (*B. taurus*)
Thelazia leesei	Flies (*Musca lucidulus*)	Dromedary (*Camelus dromedarius*) and Bactrian camel (*Camelus bactrianus*)
Thelazia rhodesii	*M. autumnalis* *Musca larvipara* *Musca sorbens*	Goat (*Capra hircus*) Bovids, cattles (*B. taurus*) Domestic sheep (*O. aries*) Zebu (*Bos indicus*) Water buffalo (*Bubalus bubalis*) Bison (*Bison bonasus*) Horse (*E. caballus*) Dromedary (*C. dromedarius*)
Thelazia skrjabini	Face flies *M. autumnalis*, *Musca vitripennis*, and *Musca amica*	Cattle (*B. taurus*) Yak (*B. grunniens*) Water buffalo (*B. bubalis*)

(the house fly), and *Haematobia irritans* (the horn fly). These flies feed on the ocular secretions of infected animals and transmit the parasite to other animals by regurgitating the parasite onto their eyes. It is important to note that not all species of Thelazia are transmitted by the same vector. For example, *T. callipaeda* is primarily transmitted by a type of fruit fly known as *Phortica variegata*, which is commonly found in parts of Europe and Asia.

The major definitive hosts for thelaziasis are both wild and domestic canids. Natural infections in felids, mustelids, and lagomorphs have also been discovered. *T. californiensis* lesions have been observed in a variety of animals, most notably canids (wild and domestic), but also felids, bears, cervids, sheep, and jackrabbits (*Lepus californicus*). Several of these might possibly be species misidentifications. *T. gulosa* is a parasite that lives on cattle and other big ruminants. Table 28.3 provides detailed information on the Primary definitive hosts and intermediate hosts for the *Thelazia* species.

Drosophilid flies, which depend on lacrimal secretions (*lacrimophagous*), are intermediate hosts or vectors for *Thelazia* spp., and relationships are parasite-specific. For instance, the predominant intermediate hosts for *T. callipaeda* are *P. variegata* and *Phortica okadai*. *T. californiensis* is spread by *Fannia species*; including the smaller house fly (*Fannia canicularis*) and the canyon fly (*Fannia benjamini*). The face fly, *M. autumnalis*, appears to be the main vector for *T. gulosa* [17].

28.5.1 *Thelazia's* Relationship to Their Specific Hosts

Thelazia eyeworm biology and epidemiology are largely determined by the existence, biology, and seasonality of their vectors, as well as by the vulnerability of their final hosts. Bovine thelaziosis is the most frequently reported animal thelaziosis, and it has received extensive research in North America and Europe, likely as a result of cows' susceptibility to Thelazia and the likelihood that free-grazing animals will be exposed to fly vectors of the eyeworm more frequently than other animals (such as carnivores).

Thelazia has a long lifespan (more than a year), is unlikely to elicit a protective immunological response, and can reinfect an animal at any time during its lifetime, all of which contribute to the greater incidence and severity of infection in adult cattle and dogs as opposed to young cattle and dogs. Thelazia parasitism in cattle has a pattern with two peaks, one in early summer (adults that overwinter) and one in late summer, despite the year-round existence of adult parasites [18].

28.5.2 *Thelazia's* Relationship to Their Definitive Hosts

The difficulty in extracting larvae from vectors due to the low incidence and mean intensity of infected flies has restricted the number of studies on the distribution of *Thelazia* in intermediate hosts in recent years. *Thelazia* vector research has been conducted on species affecting cows in North America, Slovakia, and the ex-USSR by dissecting experimental and infected flies. Several species, including *M. autumnalis* and *Musca larvipara* (commonly known as face flies), have been implicated in the spread of eyeworms. Due to the difficulty of dissecting flies and the need to compare larval eyeworms retrieved from laboratory-infected flies in order to identify a species, the detection of Thelazia 3rd stage larvae (L3) in the intermediate hosts is particularly difficult. Although *T. callipaeda* thelaziosis poses a threat to public health in Asian nations, little is known about this parasitic worm's natural vectors and their role in the intermediate host's life cycle. There is some evidence to suggest that *T. callipaeda* may be spread in the Asia-Pacific area by *Musca domestica* (Diptera, Muscidae), *Amiota variegata*, and *Amiota okadai* (Diptera, Drosophilidae) [19].

28.6 Life Cycle of *Thelazia* Species

Adult eyeworms can be found in the lachrymal glands of animals, on their lids and nictitating membranes, in their conjunctival sacs, nasolacrimal ducts, and excretory ducts. The very first larvae (L1) that are secreted in the lacrimal gland secretions of the host organisms by adult females, after mating with the adult males, are transmitted by scatophagous non-biting flies that feed on animal lachrymal secretions. Based on the *Thelazia* species, the larvae travel through developmental phases, first from L1 to the pathogenic 3rd larvae (L3), while staying encapsulated in various locations on the body of the vector. The L3s appear in the labella of infected flies, feed on animal lachrymal secretions, and mature into the adult stage in the eye cavity [20].

Eye illness is caused by the mature and larvae stages of the telaziasis species and manifests as a variety of symptoms, including lacrimation, conjunctivitis, epiphora, ocular discharge, corneal opacity, or keratitis. While obvious indications have been more commonly reported in gravid females, asymptomatic, subclinical thelaziosis is more common when only male worms infect animals, highlighting the crucial part larvae play in the etiology of conjunctivitis. Since flies feed on lachrymal secretions that include L1s, the adjacent serration of the *T. cuticle* mechanically harms the conjunctival and ocular epithelium and induces ophthalmic secretion, which is important for worm transmission [21].

The growth proceeds up to the contagious L3 phase in the flies. The infectious larva moves closer to the fly's head. The matured worm afflicts a different host when a fly finds food near its eyes; the larva enters a new host animal. When the larva penetrates the ophthalmic area of the host, it matures into an adult worm. The growth process takes around a month. *T. callipaeda* contaminates a variety of wild species, including martens, wolves, and foxes, which can serve as reservoirs for parasitic incursions in the wild. Several factors influence the life cycle pathways, including the host's age and immunological state, as well as the infection route.

Figure 28.5 clearly elucidates the entire life cycle of Thelaziasis species in both definitive hosts as well as the intermediate hosts. Adults live in the primary host's conjunctival sac, where the females produce the first-phase (L1) larvae unsheathed in a shielded membrane. The face fly intermediate host consumes L1 larvae while consuming tears and lachrymal discharges. L1 larvae get exsheathed in the gastrointestinal system of the intermediate host and infiltrate numerous host tissues, including the hemocoel, fat body, testis, and egg follicles, where they mature in capsules. The enclosed larvae molt twice before becoming infective L3 larvae. Fully mature L3 larvae emerge from the capsules and travel to the fly's mouthparts, where they remain until the fly feeds on the tears of the final host. During the summer, first-stage larvae (L1) have been discovered in the lachrymal discharges of infected dogs. This is likely because the larvae are ready to be consumed by flies feeding around the eyes. But because L4s could be descended from L3s released by flies in the spring and/or late summer to early fall, the discovery of *T. callipaeda* fourth-stage larvae (L4) during these times is consistent with the existence of intermediate hosts during these times and with nematode generations that overlap.

28.7 Molecular Acumens of Thelaziasis Species

Extensive research on several target genes [mitochondrial DNA (mtDNA) and ribosomal DNA (rDNA)] during the last 20 years has offered methods for detecting parasitic nematodes and studying their systematics, epidemiology, and biology regardless of developmental stage. The ribosomal internal transcribed spacer 1 (ITS1) sequences of *Thelaziasis callipaeda, Thelaziasis gulosa, Thelaziasis lacrymalis, Thelaziasis rhodesi,* and *Thelaziasis skrjabini* have been categorized in order to advance our understanding of the rDNA in Spirurida (about which little is known compared to the Strongylida and Ascaridida Orders). Furthermore, these findings will aid in the taxonomic and genetic identification of all these worms in their primary and intermediate hosts.

One of the most intriguing aspects of the Thelazia ITS1 sequence is that its length varies significantly across species, unlike other nonspirurid parasitic worms of the same genus. The inter-specific size polymorphism, in addition to the high AT content and the inclusion of microsatellites inside the ITS1 sequences, distinguishes spirurids. The presence of icrosatellites causes intraspecific variation in the length of the ITSs in eyeworms and other Spirurida nematodes (e.g. *Mansonella ozzardi*, *Onchocerca volvulus*, and *Habronema* spp.). The ITS1 polymorphism was found to be critical for the simultaneous identification of bovine *Thelazia* species using a PCR-RFLP test [18].

28.8 Geographical Distribution

Thelaziasis is found in many parts of the world, including Asia, Europe, North and South America, and Africa. In Asia, the infection is most commonly reported in rural areas, particularly in China, Japan, Korea, Thailand, and the Philippines. In Europe, the infection has been reported in countries such as Italy, Spain, Greece, and Russia. In North and South America, the infection has been reported in the United States, Canada, Mexico, Brazil, and Argentina. In Africa, the infection has been reported in countries such as Ethiopia, Sudan, and Nigeria. The prevalence of thelaziasis varies depending on the region and the species of Thelazia involved. In some areas, the infection is more common in certain animal populations, such as dogs or cattle, while in other areas, the infection is more prevalent in humans. The distribution and prevalence of thelaziasis are also affected by factors such as climate, geography, and the presence of suitable intermediate hosts.

As public knowledge of the disease grows, there are more reports of thelaziasis infections. The infection, also known as the "oriental eye worm," is particularly prevalent in Asian nations and can cause mild to severe symptoms of lacrimation, epiphora, conjunctivitis, and occasionally corneal ulcers. Dogs, cats, and foxes are hosts to Thelazia parasites, which reside in the conjunctiva and lachrimal sac. Human infections are sporadic and zoonotic. When flies that feed on lacrimal secretions eat the larvae that adults generate and transmit via their tears, the larvae encapsulate and grow into infective stages in the arthropod. Infectious larvae move to the fly's mouthparts while feeding on lacrimal secretions in order to enter a new host.

Most human infections occur in rural locations with low socioeconomic standards. China is the nation with the highest prevalence of infections, with an exponential rise in cases over the previous 20 years. No gender preference was seen in a study of 179 patients conducted in China; infections predominantly affected children under the age of 6, were unilateral in 88% of cases, and 4 individuals experienced anterior chamber invasion. From Japan, Korea, India, Russia, Taiwan, and Thailand, further instances have been recorded. Studies in Europe demonstrate that human infections as well as transmission occur in France and Italy. In California, *T. californiensis* has infected people. *T. callipaeda* is found in southern Europe (Italy, Switzerland, Portugal, France, and Spain), the former Soviet Union, and the Far East. The *T. californiensis* nematode is indigenous to northern America and is infectious to dogs. *Thelazia* species are both zoonotic [22].

28.9 Clinical Presentations of Thelaziasis

Clinical symptoms of thelaziasis include continued lacrimation of the eye, eye irritation, and a feeling of a dust particle in varying degrees. Conjunctivitis, epiphora, edema, photophobia, corneal ulcers, and may appear in more severe infections. The symptoms of Thelazia may vary based on the seriousness and duration of the infection. Here are some of the common clinical presentations of thelaziasis:

- **Conjunctivitis:** One of the most common symptoms of thelaziasis is conjunctivitis, which is characterized by redness, inflammation, and discharge from the eyes.
- **Eye irritation:** Thelazia worms may cause irritation and discomfort in the eyes, resulting in itching, burning, and a foreign body sensation.
- **Epiphora:** Epiphora, which is excessive tearing of the eyes, may occur as a result of Thelazia infection.
- **Visual impairment:** In severe cases of thelaziasis, the parasites may obstruct the visual pathway, resulting in visual impairment or even blindness.
- **Corneal ulcers:** In rare cases, the Thelazia worms may cause corneal ulcers, which can lead to permanent scarring and vision loss.
- **Eyelid edema:** Thelazia worms may cause swelling of the eyelids, which might cause difficulty in opening and closing the eyes.

28.10 Diagnosis of Thelaziasis

Adult worms of thelaziasis species are mechanically extracted using forceps and saline flushing. Because worms may conceal in places like tear ducts, medical attention is also advised. The most often used macrocyclic lactones are imidaclopride and moxidectin, selamectin, or milbemycin oxime. Traveling canines and migrating wild animals may transmit Thelazia because various fly species can function as vectors. In endemic locations, some dogs receive continuous prophylactic therapy with macrocyclic lactones.

These worms can infect the conjunctival sac and tear ducts of humans and other animals, causing irritation, discomfort, and, in some cases, more severe symptoms. Diagnosis of thelaziasis is typically made through a combination of clinical presentation and laboratory testing. Symptoms of the condition may include:

- Eye pain or discomfort
- Itching or irritation of the eyes
- Redness and swelling of the conjunctiva
- Sensation of a foreign body in the eye
- Excessive tearing
- In severe cases, corneal ulcers or vision loss

In order to confirm the diagnosis, a physician may perform a physical exam and collect a sample of the discharge from infected eyes. This sample can be examined under a microscope to identify the presence of Thelazia worms. In some cases, blood tests or imaging tests may also be ordered to rule out other potential causes of eye symptoms. Prompt diagnosis and treatment of thelaziasis is important to prevent the infection from causing more serious complications, so it is important to seek medical attention if anyone is experiencing any symptoms of the condition.

28.11 Treatment Options for Thelaziasis

The treatment of thelaziasis involves the removal of the worms from the affected area, usually with the help of a medical professional. Here are some common treatments for thelaziasis:

- **Manual removal:** If the worms are visible, a healthcare professional may be able to remove them manually with the help of forceps or a cotton swab.
- **Antimicrobials:** Antiparasitic eye drops containing ivermectin or albendazole can be used to kill the worms and help flush them out of the eye. Treatment for thelaziasis typically involves the use of anthelmintic drugs to kill the parasites.

Ivermectin is a commonly used drug for the treatment of thelaziasis. It is an anthelmintic medication that works by paralyzing the parasites, thereby causing them to release their grip on the host's tissues and ultimately killing them. In some cases, surgical removal of the worms may also be necessary, especially in cases where the infection has caused significant damage to the eye or surrounding tissues.

The dosage and duration of treatment with ivermectin may vary based on the seriousness of the infection and the individual's reaction to the medication. It is important to follow the prescribed treatment regimen and to consult with a healthcare professional if there are any concerns or if symptoms persist after treatment. Levamisole, injected subcutaneously or intramuscularly, is also beneficial in the treatment of thelaziasis in horses and cattle. In human cases, topical therapy with thiabendazole or cocaine has also been observed to destroy the worms.

- **Systemic medication:** In instances where the infection has spread beyond the eyes, oral medications such as ivermectin or albendazole may be prescribed to eliminate the parasites.
- **Surgery:** In rare cases where the infection has caused significant damage to the eye or surrounding tissues, surgical intervention may be necessary.

Preventative measures such as avoiding contact with infected animals, wearing protective eyewear in high-risk areas, and maintaining good hygiene can also help to reduce the risk of thelaziasis infection.

28.12 Precautionary Measures

The following are some preventive measures that can help reduce the risk of contracting thelaziasis:

- **Personal hygiene:** Maintain great individual hygiene by cleaning hands regularly, preventing touching the eyes with dirty hands, and keeping the face clean.
- **Eye protection:** Wear goggles or other forms of protective eyewear when working outdoors, especially in areas where the disease is prevalent.
- **Insect control:** Use insect repellents, mosquito nets, or other methods to keep insects away from your eyes.
- **Animal control:** Avoid contact with domestic animals, especially those that are infected with thelaziasis. If you have pets, keep them clean and take them for regular check-ups with a veterinarian.
- **Environmental sanitation:** Keep the surroundings clean and free of debris to discourage the breeding of flies and other insects that transmit thelazia.
- **Education:** Educate the public, especially those who are at high risk, such as farmers, about the causes, symptoms, and prevention of the disease.

28.13 Preventive Measures for Thelaziasis

While treatment is available, there are some precautions and contraindications to be aware of when dealing with this disease.

Precautions:

- **Good hygiene:** Practicing good personal hygiene, including washing hands frequently, is important to prevent the spread of the disease.
- **Avoid rubbing eyes:** Avoid rubbing the infected eye, as this can cause further damage and increase the risk of secondary infections.
- **Eye protection:** Use protective eyewear, such as goggles, to avoid contact with insects that can transmit the disease.
- **Avoid contact with infected animals:** Avoid contact with animals, especially those that are known to be infected with thelaziasis.

Contraindications: There is data available pertaining to the contraindication with respect to thelaziasis but to avoid contraindication, it is better to follow the below mentioned points.

- **Avoid self-treatment:** Do not attempt to treat the disease at home without medical advice, as this can lead to complications and worsen the condition.

- **Avoid certain medications:** Certain medications, such as topical corticosteroids, may exacerbate the condition and should be avoided.
- **Pregnancy and breast-feeding:** There is limited information about the use of medications to treat thelaziasis during pregnancy and breastfeeding. It is essential to consult a doctor before taking any medication.
- **Severe cases:** In severe cases of thelaziasis, surgery may be necessary. However, surgery is a last resort and should only be performed by an experienced ophthalmologist.

28.14 Recent Trends in Thelaziasis Control

Recent trends in the control of the disease include the following:

- **Vector control:** Vector control is an effective way of controlling the spread of *Thelazia* species. The use of insecticides and other chemicals can help to kill the vectors and prevent the spread of the disease.
- **Mass treatment:** Mass treatment of infected animals can also help control the disease. The use of deworming drugs and other anthelmintic drugs can be used to eliminate the parasites from the infected animals.
- **Public awareness:** Public awareness campaigns can help reduce the spread of thelaziasis. Education on the importance of personal hygiene and sanitation can be emphasized in areas where the disease is prevalent.
- **Surveillance:** Surveillance of thelaziasis is important for effective control. Surveillance can help detect outbreaks of the disease and provide timely treatment to prevent the spread of the disease.
- **Research:** Research into the ecology and biology of *Thelazia* species can help identify new control strategies. This includes the development of new drugs and vaccines to prevent the disease.

Overall, the control of thelaziasis requires a multidisciplinary approach that involves the collaboration of veterinarians, public health officials, and the community to prevent the spread of the disease.

28.15 Conclusion

Thelaziasis is a rare but important parasitic infection that can cause discomfort and complications if left untreated. Prevention measures such as good hygiene and avoiding contact with infected animals can help reduce the risk of contracting the disease. Early diagnosis and prompt treatment are essential to prevent complications and promote recovery. While the disease can be successfully treated with medication and the removal of the eyeworms, it is important to seek medical attention if you suspect that you have thelaziasis or have any symptoms. Education about thelaziasis and its prevention is crucial in reducing the risk of transmission and raising awareness about the disease. With proper prevention and treatment, thelaziasis can be effectively managed, and individuals can enjoy a healthy and active life.

Because of their interaction with definitive and intermediate hosts, *Thelazia* spp. are perhaps one of the most unusual parasitic nematode taxa. *Thelazia* are endoparasitic nematodes that live in the anterior eye chamber of the definitive host and are therefore exposed to the air and the outside world, exactly like ectoparasites. Because of the eyeworm's habitat, *thelaziosis* is the only nematode illness that can be treated topically, with medications directly instilled into the eyes. *Thelazia*, along with a few other Spirurida Order worms (*Habronema* spp., *Stephanofilaria* spp., *and Parafilaria* spp.), are the only nematodes transmitted by secretophagous flies.

Accurate molecular detection of *Thelazia* spp. is critical for managing eyeworm infection in definitive hosts, especially because various species can cross-infect the same animals and different species are susceptible to different anthelminthic medicines. *Thelazia* epidemiology, prevalence in flies, and the involvement of different insect species as intermediary hosts may all be studied using molecular identification of larval eyeworms in their vectors. The use of molecular techniques to research eyeworms overcomes several limitations.

References

1 Otranto, D., Cantacessi, C., Mallia, E., and Lia, R.P. (2007). First report of *Thelazia callipaeda* (Spirurida, Thelaziidae) in wolves (Canis lupus) in Italy. *J. Wildl. Dis.* 43 (3): 508–511.
2 Otranto, D. and Dutto, M. (2008). Human thelaziasis, Europe. *Emerg. Infect. Dis.* 14 (4): 647.

3 Tripathi, A. and Bhalla, S. (2020). Wriggly creatures coming out of eye: case report on human ocular thelaziasis. *Trop. J. Ophthalmol. Otolaryngol.* 5 (8): 250–253.

4 Naem, S. (2011). Thelazia species and conjunctivitis. In: *Conjunctivitis – A Complex and Multifaceted Disorder* (ed. Z. Pelikan), 201–232. InTech doi:10.5772/28335.

5 Krishnachary, P.S., Shankarappa, V.G., Rajarathnam, R., and Shanthappa, M. (2014). Human ocular thelaziasis in Karnataka. *Indian J. Ophthalmol.* 62 (7): 822–824. https://doi.org/10.4103/0301-4738.138183. PMID: 25116781; PMCID: PMC4152658.

6 Sah, R., Khadka, S., Adhikari, M. et al. (2018). Human thelaziasis: emerging ocular pathogen in Nepal. *Open Forum Infect. Dis.* 5 (10): ofy237.

7 Schwartz, A.B., Lejeune, M., Verocai, G.G. et al. (2021). Autochthonous *Thelazia callipaeda* infection in dog, New York, USA, 2020. *Emerg. Infect. Dis.* 27 (7): 1923.

8 Naem, S. (2011). Thelazia species and conjunctivitis. In: *Conjuctivitis – A Complex and Multifaceted Disorder*, vol. 23 (ed. Z. Pelikan), 201–232. Croatia, Europe: Intech Open Science/Open Minds.

9 Price, E.W. (1930). A new nematode parasitic in the eyes of dogs in the United States. *J. Parasitol.* 17: 112–113.

10 Otranto, D., Tarsitano, E., Traversa, D. et al. (2003). Molecular epidemiological survey on the vectors of *Thelazia gulosa*, *Thelazia rhodesi* and *Thelazia skrjabini* (Spirurida: Thelaziidae). *Parasitology* 127 (4): 365–373.

11 Zhang, X., Shi, Y.L., Wang, Z.Q. et al. (2017). Morphological and mitochondrial genomic characterization of eyeworms (*Thelazia callipaeda*) from clinical cases in central China. *Front. Microbiol.* 8: 1335.

12 Bradbury, R.S., Breen, K.V., Bonura, E.M. et al. (2018). Case report: conjunctival infestation with *Thelazia gulosa*: a novel agent of human thelaziasis in the United States. *Am. J. Trop. Med. Hyg.* 98 (4): 1171.

13 Priyadarshini, D.R. (2022). Thelaziasis. In: *Textbook of Parasitic Zoonoses* (ed. S.C. Parija and A. Chaudhury), 557–564. Singapore: Springer Nature Singapore.

14 Naem, S. (2007). Morphological differentiation among three Thelazia species (Nematoda: Thelaziidae) by scanning electron microscopy. *Parasitol. Res.* 101: 145–151.

15 Lebon, W., Guillot, J., Álvarez, M.J. et al. (2019). Prevention of canine ocular thelaziosis (*Thelazia callipaeda*) with a combination of milbemycin oxime and afoxolaner (Nexgard Spectra®) in endemic areas in France and Spain. *Parasite* 26: 1.

16 Otranto, D. and Dantas-Torres, F. (2015). Transmission of the eyeworm *Thelazia callipaeda*: between fantasy and reality. *Parasit. Vectors* 8: 273.

17 Klesov, M.D. (1950). Contribution to the question of the biology of two nematodes of the same genus Thelazia Bosc 1819, parasites of the eyes of cattle. *Dokl. Akad. Nauk SSSR* 75: 591–594.

18 Otranto, D. and Traversa, D. (2005). Thelazia eyeworm: an original endo- and ecto-parasitic nematode. *Trends Parasitol.* 21 (1): 1–4. https://doi.org/10.1016/j.pt.2004.10.008.

19 Geden, C.J. and Stoffolano, J.G. (1981). Geographic range and temporal patterns of parasitization of *Musca autumnalis* De Geer by Thelazia spp. in Massachusetts with observation on *Musca domestica* as an unsuitable host. *J. Med. Entomol.* 18: 449–456.

20 Ikeme, M.M. (1967). Keratoconjunctivitis in cattle in the plateau area of northern Nigeria. A study of *Thelazia rhodesi* as a possible aetiological agent. *Bull. Epizoot. Dis. Afr.* 15: 363–367.

21 Smeal, M.G. (1968). Observations on the occurrence of Thelazia or eyeworm infection of cattle in Northern New South Wales. *Aust. Vet. J.* 44: 516–521.

22 Saari, S., Näreaho, A., and Nikander, S. (2019). Chapter 5 – Nematoda (roundworms). In: *Canine Parasites and Parasitic Diseases*, 83–149. Cambridge, MA: Academic Press.

29

Trypanosomiasis: Current Trends in Microbiology and Pharmacology

Verner N. Orish

Department of Microbiology and Immunology, School of Medicine, University of Health and Allied Sciences, Ho, Volta region Ghana

29.1 Introduction

Trypanosomiasis is a vector-borne parasitic disease caused by a hemoflagellate protozoan called *Trypanosoma brucei* [1, 2]. Trypanosomiasis affects both humans and animals from the bite of the tsetse flies, causing a huge toll on economic growth as a result of the morbidity and mortality of both humans and domestic animals [1]. The disease is widely distributed in sub-Saharan Africa, largely due to spread of the tsetse fly vector [1]. Trypanosomiasis caused by *T. brucei is aptly* referred to as African trypanosomiasis because of its predominance in sub-Saharan Africa [1], clearly differentiated from the American trypanosomiasis (also called Chagas disease, Chapter 17) caused by another species of Trypanosoma called *Trypanosoma cruzi,* which is transmitted by a totally different bug from the tsetse fly, called the triatomine bug or assassin bug [3].

African trypanosomiasis affects both humans and animals, including domestic livestock as well as wild animals [1, 4]. The tsetse fly vector is responsible for transmission in both humans and animals, but the Trypanosoma species that causes animal infection is somewhat different from the *T. brucei* species responsible for human infection [1, 4]. African trypanosomiasis in animals is also called Animal Trypanosomiasis of African Origin (ATAO) [4]. Nagana disease is another name given to ATAO, a name that means powerless or useless in the local dialect and is derived from the behavior of infected livestock [1, 4]. ATAO is caused by a variety of species of trypanosomes, including *Trypanosoma congolense*, *Trypanosoma vivax, Trypanosoma simiae*, *Trypanosoma godfreyi*, and others [1, 4]. Of these species, *T. vivax* is the most prevalent; however, the most pathogenic is *T. congolense* causing severe disease mostly in domestic animals compared to wild animals [4, 5]. *T. brucei* subspecies – *T. brucei (T.b.) gambiense* and *T. brucei (T.b.) rhodesiense*, are responsible for Human African Trypanosomiasis (HAT) [4].

HAT is the name given to African trypanosomiasis found in humans [6, 7]. It is also called sleeping sickness due to the severe somnolence that is a significant feature at the late stage of the disease [8]. It is caused by the *T.b. gambiense* and *T.b. rhodesiense*, with *T.b. rhodesiense* involved in zoonotic transmission in humans [7]. The disease is prevalent in sub-Saharan Africa, with *T.b. gambiense* responsible for HAT in west Africa and *T.b. rhodesiense* in east Africa [6, 7]. HAT caused by *T.b. rhodesiense* is more severe compared to that caused *T.b. gambiense* [2, 6, 7]. Millions of people in sub-Saharan Africa are at risk, especially in rural areas [7].

This chapter is focused on HAT, with highlights on the history and epidemiology, vector and transmission. It will also discuss the life cycle, pathogenesis as well as clinical manifestations, diagnosis, and current trends in treatment. This chapter will conclude on the control and intervention of HAT, the successes recorded, as well as the challenges of control programs.

Rising Contagious Diseases: Basics, Management, and Treatments, First Edition. Edited by Seth Kwabena Amponsah, Ranjita Shegokar, and Yashwant V. Pathak.

29.2 History

Ancient history is abundant with written documents and folklore tales of African trypanosomiasis as it impacted human civilization in ancient times [1]. From prehistoric times to early modern civilization, the impact of the parasite has been felt by humans, both wild and domestic, as well as farm animals. From ancient Egypt to central and west Africa, documentary evidence has described the withering sleeping disease among animals and humans [1].

Dissecting the evolutionary history of the parasitic organism that causes HAT, the trypanosome is a Salivarian trypanosome that evolved from a much older form (Stercorarian or non-Salivarian) of the parasite that infects non-mammalian hosts, such as fish, amphibians, and reptiles [1, 9]. This separation took place several million years ago (300 million years), even before the appearance of the vector Tsetse fly [1, 9]. The appearance of the Tsetse fly much later down the line (35 million years), ensured the effective transmission of trypanosomes to mammals stating from wild animals [1, 9]. This ancient evolutionary relationship between Salivarian trypanosome and wild animals elucidates the reason why wild animals are more tolerant to trypanosomes compared to domestic animals, whose encounter with trypanosome and tsetse fly happened very much later [1].

Ancient human ancestors, the hominids encountered the oldest of trypanosomes that infects humans including *T. brucei* [10]. This probably would have been responsible for the human resistance to all Trypanosomes except *T.b. gambiense* and *T.b. rhodesiense* [1, 10]. This probably suggests that *T.b. gambiense* and *T.b. rhodesiense* are subsequent evolutionary derivatives of *T. brucei* [1, 10]. Constant exposure to the vector has explained the evolutionary resistance in the hominids and modern humans [10]. Arboreal primates are more susceptible to trypanosomes because they live in trees far above the reach of Tsetse fly, which cannot fly beyond three meters high [1, 10]. On the other hand, baboons are the only primates that are resistant to all forms of trypanosomes because they are the only terrestrial animal historically known to have had a constant exposure to tsetse fly and trypanosomes [10]. Again, in humans, the constant exposure to *T.b. gambiense* compared to occasional exposure of *T.b. rhodesiense* from wild animals makes the former more tolerated among humans than the latter [1].

Historical evidence suggests that Tsetse fly had a wider distribution involving the whole of Africa and larger portion of North America due to the typical tropical vegetation and climate prevalent in these areas several millions of years ago [10, 11]. However, as the years went by, climatic changes occurred, such as the historical Pleistocene glaciation which took place in North America, there was a total decimation of the tsetse fly population [10]. Climatic changes also occurred in Africa restricting the tsetse fly population to regions in sub-Saharan Africa [1, 10, 11]. This is very much evident in Egypt, which at present times has no case of trypanosomiasis because tsetse fly population is virtually nonexistent. This situation was totally different from ancient Egypt with historical evidence of lush green tropical forest extending to the river Nile encouraging the presence of tsetse fly and transmission of trypanosomiasis [1, 11]. Ancient practices among herders of livestock farmers in ancient Egypt such as keeping their domestic animals in proximity with wild animals with the intension of creating trypanotolerant domestic animals, further lay credence to the presence of trypanosomiasis in ancient Egypt [1]. Ancient writing from Egypt also reports a cattle disease that looks like the Nagana disease of Eastern Africa [1].

Writings from Arabian scholars who encountered African in the Middle Ages showed evidence of the presence of trypanosomiasis in western and eastern Africa [1]. Historical writings from a highly acclaimed Arabian scholar reported an affliction closely resembling trypanosomiasis affecting animals and humans in certain village in the Ghana/Wagadu heartland of west Africa [1, 12]. Another Arabian scholar reported a tale of the death of the king of Mali in the fourteenth century, from a common disease in the locality that had uncanny similarities with African trypanosomiasis [1].

African trypanosomiasis made its impact inslave trade of the early modern times [1, 13]. Many slaves taken from Africa died from the disease, prompting the slave traders tasking their doctors to find a solution to the problem [1, 13]. This led to the first medical report of African trypanosomiasis by Dr. Atkin in 1734, who accurately described the central nervous system manifestation of the late phase of the disease [1, 13]. Early clinical manifestations of HAT were reported in 1803 by Dr. Thomas Winterbottom, a physician who accurately described the Winterbottom sign, pathognomonic clinical feature of swollen lymph node at the posterior triangle (back) of the neck of infected patients [1, 13]. This feature was used by Arabian slave traders in the Middle Ages as they avoid slaves with this swelling in their neck [1].

In 1852, the link between the vector and African trypanosomiasis was made by a missionary named David Livingstone, who described how his cows died after being bitten by Tsetse flies [1, 13]. In 1895, a Scottish microbiologist called David Bruce discovered Trypanosome parasite as the cause of African trypanosomiasis in cattle, while Dr. Robert Michael Forde was the first person to isolate the parasite in a human in 1901 and a year later in 1902 Joseph Everett Dutton accurately named the organism *Trypanosoma gambiense* [1, 13]. David Bruce in 1909 described the life cycle of the parasite within the vector [1, 13]. In 1910, *T. brucei rhodesiense* was discovered by two parasitologists John William Watson Stephens and Harold Benjamin Fantham [1, 13].

29.3 Parasite

29.3.1 Classification of Trypanosomes

Trypanosome is a protozoan parasite with a flagellum (Trypanosomatida) and similar to Leishmania, it has a kinetoplast (kinetoplastae) [14]. Trypanosome is a protozoan parasite whose existence dates back over 300 million years ago [1]. It includes several species that infect a wide range of animals including mammals, fish, amphibians, reptiles, and others (Figure 29.1) [1, 9, 15]. Some 300 million years ago even before the emergence of Tsetse fly, these trypanosomes that infect mammals, further evolved into Salivarian and Stercorarian trypanosomes (Figure 29.1) [1, 9]. Salivarian trypanosomes are those that reside in the upper gastrointestinal tract of the tsetse fly vectors and transmitted in the saliva during a blood meal [16]. The Stercorarian trypanosomes reside in the distal gastrointestinal tract of the insect and are transmitted in feces during defecation, and through a broken skin, the parasite can enter the body [15]. The grouping or classification of these parasites especially the ones that are not responsible for African trypanosomiasis- ATAO and HAT- is a topic of heated debate among evolutionary biologists [17]. One of the reasons for this is, while their mode of transmission through feces or through blood meal can classify them into either Stercorarian or Salivarian trypanosomes respectively, molecular typing have seen similarities among species in different classes [17]. However, Figure 29.1 is a simplified classification with emphasis of the trypanosomes that cause African trypanosomiasis especially HAT. The trypanosomes that cause African trypanosomiasis are exclusively classified as Salivarian trypanosomes [1, 9, 15]. They include trypanosomes that infects animals and those that infects human [1, 9, 15]. Salivarian trypanosomes that infect animals include, *Trypanosoma evansi, Trypanosoma equiperdum, T. brucei* and sub specie *T.b. brucei, T. congolense, T. vivax* and *T. godfreyi* (Figure 29.1) [1, 15]. These trypanosomes are exclusively parasites of animals as reports have shown infections in variety of animals [18–20]. However, it is important to note that human infections with these trypanosomes of animal infections, have been reported in endemic areas [21, 22]. Salivarian trypanosomes that cause HAT include *T.b. gambiense* and *T.b. rhodesiense*, subspecies of *T. brucei*.

29.3.2 *Trypanosoma brucei* Specie

Trypanosoma brucei is one of the early species of the Trypanosoma genus, one of the earliest known Salivarian trypanosomes [1, 10]. It is the species that causes HAT, or sleeping sickness, through its sub species, *T.b. gambiense* and *T.b. rhodesiense*. There is a third sub species of *T. brucei* called *T. brucei* (T.b.) brucei that is responsible for animal trypanosomiasis [1].

T.b. brucei is more likely the oldest sub species of the 3, with earlier and prolong exposure with human ancestors compared to the latter, *T.b. gambiense* and *T.b. rhodesiense* (Table 29.1). This is probably the reason why humans are resistant

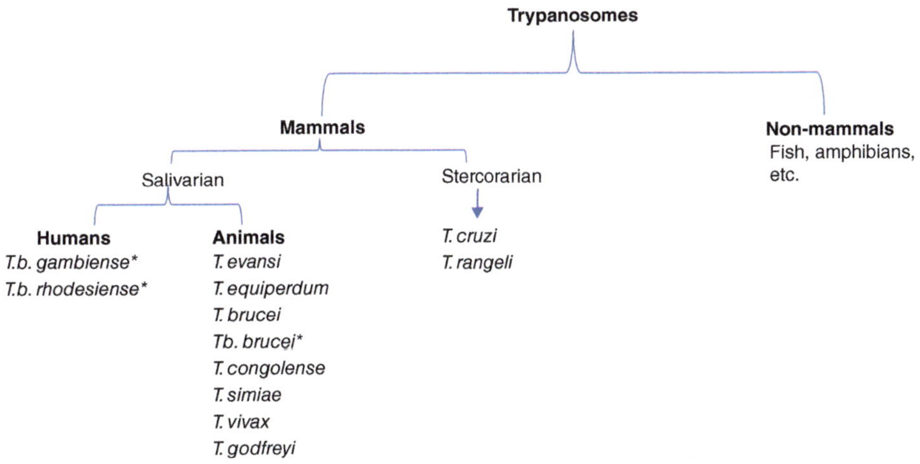

Figure 29.1 Classification of Trypanosomes. * Sub species.

Table 29.1 Taxonomy of *Trypanosoma brucei*.

Domain	Eukaryota
Kingdom	Protista
Phylum	Euglenozoa
Class	Kinetoplastae
Order	Trypanosomatida
Genus	Trypanosoma
Sub genus	Trypanozoon
Species	*Trypanosoma brucei*
Sub species	*Trypanosoma brucei gambiense* *Trypanosoma brucei rhodesiense*

Source: Adapted from Romera-Meza et al. [23].

to *T.b. brucei* but *susceptible to T.b. gambiense* and *T.b. rhodesiense* [10]. *Both T.b. gambiense* and *T.b. rhodesiense* are predominantly found on the African continent [23, 24]. *T.b.* gambiense is found in about 24 countries in western and central Africa, and it is responsible for more than 95% of all cases of HAT [1, 24]. *T. b. rhodesiense*, on the other hand, causes HAT in East Africa [15]. While *T.b. gambiense* causes anthroponotic infection, *T.b. rhodesiense* is seen in infected animals; thus, it is known for zoonotic infection, causing a far more severe disease in humans [1, 24]. Before infecting humans, these trypanosomes live within the Tsetse fly, mainly located in sub-Saharan Africa and that is the reason why Trypanosoma parasite and HAT is found in this region [1, 10, 15].

29.3.3 Tsetse Fly – Genus Glossina

29.3.3.1 Classification

Tsetse fly is a biting insect that belongs to the order Diptera from the super family Hippoboscoidea and the genus glossina [25]. The over 30 species of tsetse flies have been divided into three major Sub-genera, namely, Austenina, Nemorhina, and Glossina sensu stricto (Table 29.2) [24, 25]. The Subgenus Nemorhina has the following major species of Tsetse flies namely, *Glossina palpalis, Glossina tachinoides*, and *Glossina fuscipes* [24, 25]. *G. palpalis* is the most prominent of all the species in the Nemorhina Subgenus; hence the subgenus is also aptly called the palpalis group (Figure 29.3) [24, 25]. The Subgenus *Glossina* sensu stricto has *Glossina morsitans, Glossina pallidipes, Glossina longipalpis, Glossina swynnertoni*, and *Glossina austeni* as species [24, 25]. Subgenus Glossina sensu stricto is also called the morsitans group because *G. morsitans* is the most prominent tsetse fly under this subgenus [24, 25]. *Glossina fusca, Glossina tabaniformis, Glossina medicorum, Glossina longipennis*, and *Glossina brevipalpis* species belong to the Austenina subgenus, also known as the fusca group, because *G. fusca* is the most prominent tsetse fly specie in the group (Table 29.2) [24, 25].

Table 29.2 Classification of tsetse fly.

Genus		Glossina	
Sub-genus	Nemorhina	Glossina sensu stricto	Austenina
Group	Palpalis	Morsitan	Fusca
Specie	*Glossina palpalis, G. tachinoides, G. fuscipes*	*Glossina morsitans, G. pallidipes, G longipalpis, Gswynnertoni, G. austeni*	*Glossina fusca, G.tabaniformis, G. medicorum, G. longipennis, G. brevipalpis*
Trypanosomes	*T.b. gambiense*	*T.b. rhodesiense*	*T.b. gambiense /T.b. rhodesiense*
Location in Africa	West Africa	East Africa	West Africa
Habitat	Riverine/Lacustrine/ Agriculture site	Savanna wood land, open country	Forest (rain, swam and mangrove)

29.3.3.2 Reproduction

Unlike many other insect vectors, the tsetse fly has restricted fecundity, ensured by a single viviparous reproduction with a relatively prolong time (over 32 days) of maturation from larva to adult tsetse fly [25, 26]. Tsetse fly uses Adenotrophic Viviparity to reproduce, a unique style of reproduction involving the maturation of a larva within the reproductive organ of the mother tsetse fly [25–27]. A mature adult female tsetse fly moves its egg from the ovary to its uterus, where it gets fertilized by the sperm stored in the spermatheca, an organ of the reproductive system close to the oviduct of the fly [27, 28].

The stored sperm is gotten from a single sexual encounter with a male fly, despite mating multiple times in its life time [27, 28]. The fertilized egg lasts in the uterus for five days, after which it hatches and releases the larva [27, 28]. The larva remains in the uterus and is nourished by a unique milk gland in the fly [27–29]. This milk gland provides nourishment for the larva in the form of fat and protein, and it also contains a symbiont bacterium, which ensures that adult females developed from the larva will maintain the ability to reproduce in the future [27–29]. Six days after receiving nourishment from the milk gland, the larva is ready to exit the uterus [27, 28]. However, before the egress of the larva from the uterus, the mother tsetse fly most actively selects a humane soil habitat suitable to accommodate the fragile larva [27, 28]. Once on the soil, the larva tunnels into it and subsequently forms a carapace to ensure safety, becoming the pupal stage of the developing tsetse fly (Figure 29.2) [27, 28]. After about three weeks in the soil, the pupa matures into an adult tsetse fly and is ready to begin the cycle again if it is a female tsetse fly (Figure 29.2) [27, 28]. This newly formed fly is often referred to as a teneral adult tsetse fly [24].

29.3.3.3 Feeding Habit

Tsetse fly is hematophagous, and this is true for both male and female tsetse fly [24, 25]. After the newly formed tsetse fly comes up from the soil, they must be fed to survive [24, 25]. Usually, they feed on blood from any mammals available at the time. The fly feeds during the day, and it is attracted to humans or animals with a bright color, strong smell of urine, feces, or body secretion, shape, form, and size [24, 30]. They bite more in the lower parts of the body, like the legs and feet, so as to avoid being killed by body movement [24]. It was recently discovered that at times when blood is not available, the adult tsetse fly feeds on non-blood food sources such as sugars and water [31]. Solano and colleagues (2015) reported sugar and water residue in the stomach of both laboratory tsetse fly and those found in nature [31]. During blood meal, tsetse fly get infected if the mammalian host is infected with trypanosomes, and the risk of acquisition of infection is higher in teneral fly with their first blood meal compared to older ones that have had previous blood meals [24]. However, taken together,

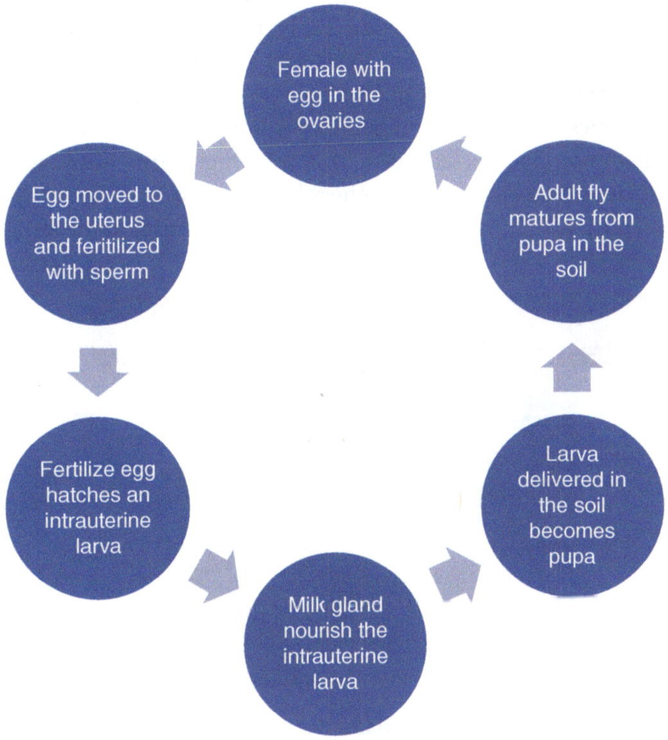

Figure 29.2 Reproductive cycle of tsetse fly.

the tsetse fly has a lower infection rate compared to other insect vectors because many of the trypanosomes ingested during a blood meal do not survive in the gut of the fly [24, 25]. This is due to the resilient innate immune system of the gut of the tsetse fly [25]. This is obviously a significant impediment to the tsetse fly transmission of trypanosomes; however, their longevity and persistent infectivity throughout their relatively longer life span still make transmission of trypanosomiasis a significant problem in endemic areas [24, 25].

29.3.3.4 Survival Strategies

Relatively, the population of tsetse fly in sub-Saharan Africa is smaller compared to other insect vectors [24, 25]. This restriction in the size of the tsetse fly population is chiefly due to the long reproduction time coupled with a single egg carriage of the adult female tsetse fly [24, 32]. In addition to the reproductive constraints, the tsetse fly has to deal with negative physical (harsh climate) and biotic (predators) threats to its existence, which are replete in sub-Saharan Africa [32]. For these reasons, the tsetse fly are conservative flyers, resting most times and only flying short distances and short periods of time (less than an hour in males and much shorter in females), and they do these (resting and flying) below a height of 3 m from the ground [10, 24, 32]. However, passive migration by animals and vehicles can cause the spread of tsetse fly over a wider area [24].

29.4 Epidemiology of Human African Trypanosomiasis (HAT)

29.4.1 Transmission Dynamics

Ultimately, it is the contact of humans with infected tsetse fly that causes HAT, as the distribution of the fly in sub-Saharan Africa is the major driver of HAT in the region [24, 33]. However, before the tsetse fly can have the potential to infect, it must first pick up the trypanosome, either from an infected human or animal [24]. In west Africa, there exists the epidemiologic triad of *T.b. gambiense*, tsetse fly, and humans (Figure 29.3), which result in a milder chronic and endemic HAT [24, 33]. Thus, infection spreads from human to human via the bite of the tsetse fly [24]. Although *T.b. gambiense* is not known to be zoonotic, studies have reported animals with *T.b. gambiense*, which might contribute significantly to the burden of HAT and impede effective control of HAT caused by *T.b. gambiense* [24, 34]. For *T.b. rhodesiense,* the transmission cycle is usually between animals with the help of tsetse fly, with human infection being an occasional occurrence (Figure 29.3) [24]. *T.b. rhodesiense* is picked up by the fly from animals, both wild and domestic, before it transmits it to a human, resulting in a more acute and severe disease [24].

Transmission of HAT is unevenly distributed in endemic areas of sub-Saharan Africa, with no report of HAT in some areas even with suitable habitat and climate for the tsetse fly [33]. Transmission is confined to discrete, multiple areas referred to as foci [33]. In all, there are over 300 foci distributed in about 36 countries in sub-Saharan Africa where trypanosomes, tsetse fly, animals and/or humans interact, usually in remote rural areas [24, 33]. In these foci, the risk of humans encountering infected tsetse fly is quite high, either through simple human activities like farming, fishing, hunting, cattle grazing in deep forests, or others [24]. For *T.b. gambiense*, the risk of transmission increases with human activities in the vicinity of the tsetse fly habitat, such as farming, hunting, fishing, swimming, and washing in rivers and streams [24]. In the case of *T.b. rhodesiense*, coming into contact with wildlife is one of the major factors that increase the risk of transmission, especially in foci where wildlife is the main reservoir, as it inevitably exposes humans to the tsetse flies that feed on these wild animals [24]. These activities include hunting, livestock grazing deep in the forest, as in the case of herders, living close to game reserves, and others [24, 35]. *T.b. rhodesiense* foci where domestic animals are the reservoir, livestock farming activities pose a risk of infection, especially for the family involved and those who live close to these farms or market places where the domestic animals are sold [24, 36].

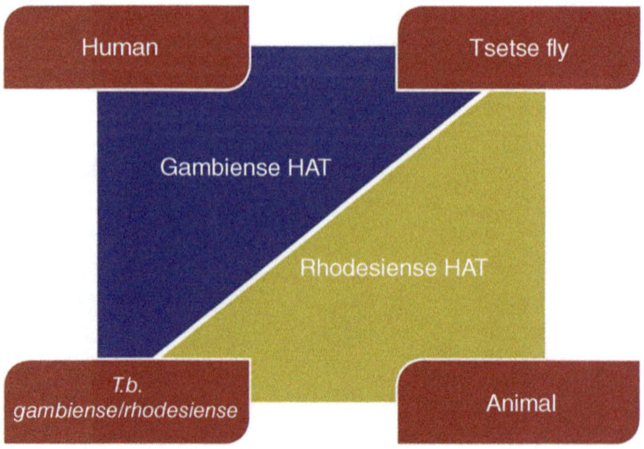

Figure 29.3 Transmission triads.

Like for almost all vector-borne diseases, climatic factors, especially the ongoing climate changes, also affect HAT transmission [37]. Unfavorable climatic changes, such as increased heat, can disrupt tsetse fly habitats, causing them to migrate to areas of conducive temperature and humidity, probably encroaching into human settlement [24, 37]. Conversely, the areas where the tsetse fly migrated away because of the increase in temperature will notice a low population of the fly and hence a reduced prevalence of HAT [38]. In the rainy season, the tsetse fly lives longer and thus procreates more; there is more pupal survival; and it flies higher and longer distances. All these translate to increased transmission of HAT during the rainy season [24].

29.4.2 Burden of HAT in Endemic Areas

In sub-Saharan Africa, over 70 million people are at risk of HAT with 10 to over 20,000 infections yearly, predominantly caused by *T.b. gambiense* [24, 33]. Before 2010, HAT was responsible for loss of 1.7 million Disability-adjusted life years (DALYs), but from 2016 it dropped significantly, an obvious dividend of the intensified trypanosomiasis control program initiated two decades ago [39]. HAT is considered a neglected tropical disease, and it contributes significantly to mortality and morbidity caused by this group of diseases [39, 40]. Mortality is almost 100% for those left untreated with severe clinical manifestations of the disease, and mortality is often seen with HAT caused by *T.b. rhodesiense* [24]. Asymptomatic and oligosymptomatic infections followed by resolution without treatment have been reported, albeit very rare, especially with *T.b. gambiense* [24]. Those who survive the illness usually suffer loss of motor, sensory, and mental function, which can result in stigmatization, low self-esteem, and some antisocial behaviors [24].

29.4.3 Impact on Children and Women

Children are usually less affected than adults because the majority of human activities that increase the risk of the acquisition of *T.b. gambiense and T.b. rhodesiense* such as hunting, farming, fishing, mining activities, and others are carried out by older adults [24]. However, recreational activities such as swimming in streams and rivers can be undertaken by children, especially in riparian areas [24]. Infected school-aged children are known to have poor academic performance, culminating in a high rate of school dropout due to poor concentration and amnesia [24, 40]. Children from affected parents also suffer various social deprivations due to disease,death of parents, divorce, and domestic abuse [40].

Women are generally not involved in the high-risk activities that predispose them to bite of tsetse fly, thus the women are less likely to get HAT [40]. However, it is opined that the passive case detection of HAT impedes the detection of women with HAT [40]. Women in deprived areas where HAT is common are a vulnerable group because they hardly seek care because of financial constraints and other social factors [40]. The prevalence of women with HAT increased when active case detection was employed [40]. Central nervous system dysfunction and musculoskeletal deformity among women who have recovered from HAT cause disability, resulting in a loss of self-esteem [40].

29.4.4 Burden of HAT in Non-Endemic Areas

Cases of HAT have been reported in non-endemic areas [6, 41]. These are by no means cases of active transmission of tsetse fly in these non-endemic areas, but rather imported cases of migrants or travelers from endemic areas [6, 41]. Non-endemic countries such as the United States, South Africa, China, and some countries in Europe have reported cases of HAT, albeit with a decreasing incidence of cases, with 108 cases in the first decade of the twenty-first century compared to 15 cases in the second decade [6]. Majority of these imported cases are *T.b. rhodesiense* HAT traced back to eastern and southern Africa [6]. Imported HAT in non-endemic areas does not show the predominance of *T.b. gambiense* HAT, suggesting probably that people who imported HAT visited the eastern part of Africa where *T.b. rhodesiense* HAT is predominant [6]. With increasing global travel and several activities of industrialized nations in many endemic countries in Africa, imported HAT will increase, posing a challenge in the diagnosis and management of a relatively rare condition in these non-endemic areas [6, 41].

29.5 Life Cycle of Trypanosoma

T.b. gambiense and *T.b. rhodesiense* are morphologically indistinguishable, and the same is true with their heteroxenous cycles in humans and tsetse fly [24]. However, *T.b. rhodesiense* can have a life cycle involving animals and tsetse, with the capacity to infect humans sometimes [24]. Tsetse fly is recognized as the definitive host because of the genetic exchange

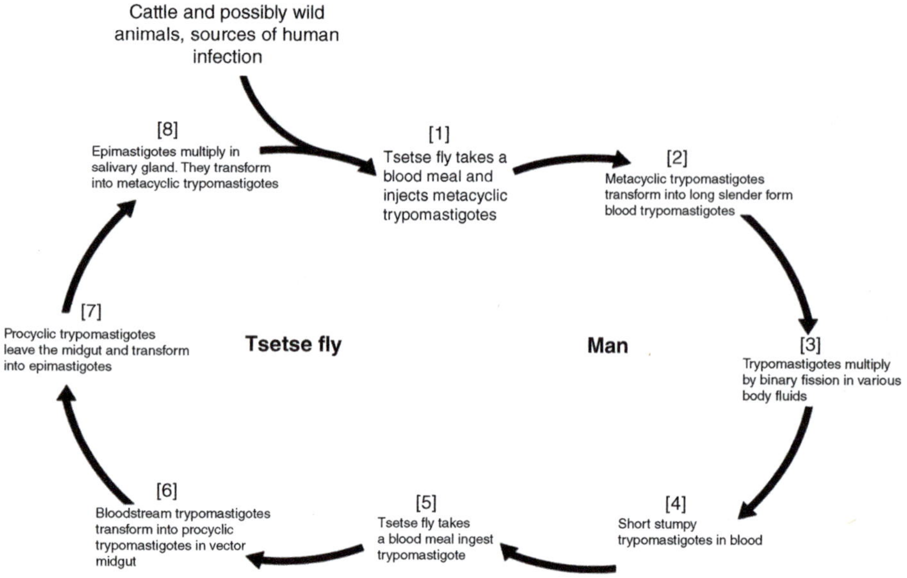

Figure 29.4 Life cycle *T.b. gambiense* and *rhodesiense.*

that takes place during the complex life cycle of the parasite within it (Figure 29.4) [42]. The life cycle starts during the blood meal of the fly, releasing metacyclic trypomastigotes into the dermis, where they transit into lymphatics, and then finally into the blood stream of the human host [24, 43]. In the blood stream, trypomastigotes assume a long, slender form and are aptly referred to as blood trypomastigotes [24]. They remain extra cellular, dividing via binary division and traveling the blood to many tissues in the human host [24, 43]. Blood trypomastigotes transform into short stumpy forms and are inadvertently picked up by Tsetse fly during a blood meal. Inside the gut of the fly, the blood trypomastigotes transform into polycyclic trypomastigotes, which then leave the gut, transforming into epimastigotes [24, 43]. The epimastigotes continue their multiplication in the salivary gland to form metacyclic trypomastigotes, which is the infectious form of the parasite [24, 43].

29.6 Other Mechanisms of Transmission

It is an incontrovertible fact that HAT is a vector-borne disease with the tsetse fly as the sole transmission agent. However, some cases of HAT have been reported in humans that have no link to the bite of the tsetse fly.

29.6.1 Congenital Transmission

This kind of transmission came to light when newborns and young children born in non-endemic areas came from mothers who had lived in endemic areas [24]. Newborns diagnosed with HAT establish the fact that *T.b gambiense* and *T.b. rhodesiense* have joined the league of parasites that are transplacentally transmitted from mother to fetus [44]. Cases reported in the literature have buttressed the point that congenital HAT might be a common occurrence [45–49], but a low index of suspicion and poor diagnostic capacity in endemic areas might be impeding accurate diagnosis [48, 49]. Moreso, misdiagnosis might also be an issue because the symptoms of congenital HAT are similar to the many infectious diseases threatening the lives of newborns and young children in endemic areas [48, 49]. Furthermore, congenital HAT might present with atypical symptoms compared to HAT in adults, adding another confusion to the diagnosis even in endemic areas [47].

29.6.2 Sexual Transmission

Cases of congenital HAT heightened the suspicion of sexually transmitted HAT, as congenital HAT can result from a successful sexually transmitted HAT in a woman [50, 51]. Sexually transmitted HAT is usually considered in an infected woman or in a mother of a child with HAT who has never visited an endemic area, and there is no other possible exposure other than having a sexual partner who was infected with *T.b. gambiense* or *T.b. rhodesiense* [51]. Most infections have only reported women getting infected by their infected male partners, not vice versa. Though there are no reports in the literature of trypomastigote seen in the semen of people with HAT, animal experiments have reported trypanosomes seen in semen but not in vaginal secretions [52].

29.6.3 Extremely Rare Transmission Route

Although it has been asserted that transmission by the blood transfusion and organ transplant is possible, as in the case of some blood-borne parasites, reports of acquisition of HAT through these routes in literature are extremely scarce or probably nonexistent at the moment [24].

Needle prick injuries as part of laboratory accidental exposure have been reported in literature [53].

29.7 Pathogenesis of HAT

29.7.1 Trypomastigotes Induced Inflammation in the Skin

There are two major stages of the disease process in HAT, which involve both the hemolymphatic system and the central nervous system [33]. However, before these stages of the disease are initiated by the entry of trypomastigotes into the blood stream, there is an inflammatory response induced by the trypomastigote initially deposited in the dermis [33, 54]. This inflammatory response is predominantly infiltrates of B lymphocytes and plasma cells (lymphoplasmacytic infiltrates) together with histiocytes, fibroblasts, and endothelia cells surrounding actively dividing metacyclic trypomastigotes [54]. Other innate immune cells, such as neutrophils and natural killer cells, have also been reported to contribute to the immune response induced by the trypomastigote in the dermis [55]. These trypomastigotes in the dermis have proclivity for the collagen bundles and fibroblasts in the reticular portion of the dermis [54]. These inflammatory cells, coupled with the actively dividing trypomastigotes, form a nodule at the bite site on the skin, popularly known as the trypanosomal chancre [54]. The inflammatory response succeeds in destroying almost half of the population of the metacyclic, and the surviving ones become resistant to destruction [54]. More recently, it has been discovered that the resistant trypomastigotes in the dermis serve as reservoirs for the infection of tsetse fly after biting infected persons, especially asymptomatic individuals with chronic HAT caused by *T.b. gambiense* [54, 56, 57]. However, for eventual symptomatic individuals, the skin trypomastigotes traverse the dermis, gain entry into draining lymph nodes through the afferent lymphatics, and enter the blood stream [55].

29.7.2 Host-Trypomastigotes Immune Interactions in the Blood and Lymphatics

The metacyclic trypomastigotes in the blood stream transform into blood trypomastigotes and remain entirely extracellular. The blood trypomastigotes are assailed by both innate and acquired immune responses in the blood [33, 58]. The immediate innate immune response capable of destroying the blood trypomastigote is the apolipoprotein-A1 component of High-density lipoprotein (HDL) circulating in the blood [33, 59]. This trypanosome lytic factor is the reason why humans are resistant to all other forms of Trypanosoma because they destroy their trypomastigotes as soon as they enter the blood species except *T.b. gambiense* and *T.b. rhodesiense* stream [21]. Blood trypomastigote of *T.b. gambiense* and *T.b. rhodesiense* has a surface coat containing Variable Surface Glycoprotein (VSG) responsible for evading the immune attack by trypanosome lytic factor (apolipoprotein-A1) [33, 59]. The VSG also resists destruction from other innate immune responses such as the complement alternative pathway [58]. VSG stimulates a large amount of immunoglobin M, which mounts a somewhat successful response, killing some the blood trypomastigote [33]. Other blood trypomastigotes evade the antibody killing by creating new surface VSG, which is new to the circulating antibodies [33]. This allows for uninhibited asexual division of the blood trypomastigotes. Eventually, new antibodies will be produced to target the new VSG, but the trypomastigotes will alter the VSG again to evade killing, and the cycle will keep repeating itself [33]. These trypomastigotes, with their ever-changing VSG, induce massive cytokine production through the production of autoantibodies, and together they are responsible for the intense tissue damage that is characteristic of HAT [33, 58].

Blood trypomastigote have been reportedly seen in the lymph nodes, although there are arguments about whether they are blood trypomastigote or trypomastigotes that solely reside in lymph nodes [54]. The lymph nodes experience proliferation of macrophages and plasma cells induced by the complex systemic host-trypomastigotes immune interactions [54].

29.7.3 Central Nervous System Injury

The blood trypomastigotes eventually enter the brain, heralding the CNS phase of HAT. Time taken for the blood trypomastigotes to gain access to the central nervous system, specifically the brain, will depend on the load or density of the circulating blood trypomastigote [60]. Contrary to the belief that the trypomastigote only crosses the blood brain barrier to gain access to the brain [61], it is now believed they also enter through the blood cerebrospinal fluid barrier to the choroid plexus and subsequently into the sub arachnoid space [60, 62]. Studies conducted using animal models suggest that there is a fluctuating number of blood trypomastigotes in the brain, which is regulated by the amount in the blood [62]. Pro-inflammatory cytokines such as interferon gamma (INF-y) and tumor necrosis factor (TNF) are responsible for facilitating the crossing of the blood cerebrospinal fluid barrier and the initiation of inflammatory injury in the brain [63]. The inflammatory brain injury is seen more in the region of the white matter of the brain, characterized by lymphocyte and plasma cell infiltration with meningoencephalitis [63].

29.8 Clinical Manifestations

The symptoms and signs of HAT are generally divided into the early-stage and the late-stage clinical features. However, before the early-stage symptoms, there was a unique clinical feature at the bite site known as the trypanosomal chancre.

29.8.1 Trypanosomal Chancre

This is a characteristic painful lesion at the bite site on the skin seen between 2 and 15 days after the bite of the tsetse fly [33, 54, 64]. It is a typical inflammatory lesion with redness, swelling, pain, and warmth. It is rarely seen in persons with HAT caused *T.b. gambiense* but is typically seen in HAT caused by *T.b. rhodesiense* [33]. However, among indigenes with *T.b. rhodesiense,* trypanosomal chancre is still not very common but is very common among travelers to endemic areas [54, 65]. There is a resolution of the chancre within a few weeks of onset [54, 65].

29.8.2 Early Stage

This stage is noted for non-specific symptoms generated by the interactions between the immune system and blood trypomastigote in the blood and lymphatics. It is noted with febrile symptoms, weight loss, lymphadenopathy, Winterbottom sign, splenomegaly, edema, tachycardia, arrythmias, pruritus, and dermatosis [24, 33]. In *T.b. gambiense,* this stage begins months to a few years after an infective bite and is more chronic, lasting months to years with mild to moderate recurrent symptoms. The onset of symptoms for *T.b. rhodesiense* is about one to three weeks after an infection, and it produces more acute and severe symptoms and rapid progressions to the late stage, which is indicative of CNS involvement [24].

29.8.3 Late Stage

This stage is heralded by the manifestation of neurologic features, and the earliest of these is Kerandel's sign characterized by deep hyperaesthesia accompanied by delayed sensations of pain [33, 64, 66]. The typical sleep disturbance that gave HAT the notorious name of sleeping sickness manifests at this stage. It is characterized by the disturbance of the normal circadian rhythm, with initial sleepiness during the day and insomnia at night [8]. However, as the disease progresses, the individual becomes somnolent all through the day as well as at night [8]. There are a plethora of other neurologic symptoms at this stage, such as changes in personality, confusion, distorted speech, uncoordinated movement, muscle weakness, dementia, and others [33, 64, 66]. More of these symptoms are commonly seen in *T.b. gambiense* than *T.b. rhodesiense* probably because the latter progresses rapidly to death, thereby not allowing time to manifest all these symptoms [66].

29.9 Diagnosis

Clinical diagnosis of trypanosomiasis is extremely difficult, as symptoms are usually nonspecific (especially for the early-stages) and can easily be categorized as symptoms of other endemic diseases, coupled with a low index of suspicion among clinicians and health care providers. There is a three-steps approach employed in the diagnosis of HAT in endemic areas, and this includes screening of suspected individuals using serology, confirmation of disease, and staging of disease using microscopy [33, 67].

29.9.1 Serology

This is not used as a definitive diagnosis but as a screening tool for *T.b.* gambiense, but it is unreliable as a screening tool for *T.b. rhodesiense* [67]. Serology screening test detects antibodies against antigens on VSG of the blood trypomastigotes. Screening for *T.b. gambiense* has commonly been carried out using the card agglutination test for trypanosomiasis (CATT), a screening tool heavily relied upon in many HAT control and intervention programs in endemic areas [33]. In CATT, the antigens are already prepared as dry antigens fixed on the card, which will react with the available specific antibodies in the blood samples being tested [67]. CATT is cheap and has a cost effect, but its inherent sensitivity issues make it only a screening tool; therefore the result should be confirmed by a definitive laboratory investigation. SD Bioline HAT RDT, manufactured in South Korea, is a more sensitive and cost-effective screening tool packaged as a rapid diagnostic test (RDT) has been introduced for community screening of HAT in endemic areas [68]. Definitive diagnosis is very important because HAT treatment should only be commenced on someone with confirmed disease owing to the toxicity of the drugs used in treatment [64].

29.9.2 Microscopy

Visualizing the trypomastigotes in body fluids such as blood, lymphatics, aspirates from the lymph node and chancre, cerebrospinal fluid (CSF), and bone marrow of an infected human is very definitive [33, 64]. Microscopy with blood samples is not very sensitive, but concentration techniques such as the buffy coat method using blood centrifugation have been employed [33, 64]. Visualizing the parasite in the blood is much more likely with *T.b. rhodesiense* because of the high parasitemia compared to the low and periodic parasitemia seen in *T.b. gambiense*, where lymph node aspirate is preferred instead of blood [64]. The easier detection of *T.b. rhodesiense* using microscopy is a relief because there is no serology screening for it but only a reliance on clinical features [33]. Establishing the late stage or CNS involvement of HAT is very critical, as the treatment modalities are totally different from the early-stage. To do this, parasites and/or white blood cells of more than 5 cells per microliter of CSF must be detected in the CSF of patients with microscopic evidence of early-stage HAT [33, 64].

29.9.3 Molecular Diagnosis

Polymerase chain reaction has been reported to have close to 99% and 98% sensitivity and specificity, respectively [64]. When compared to CATT and microscopy, it had the best sensitivity [69]. The only drawback of this method is that it is expensive and logistically difficult to carry in the field [64].

29.10 Treatment

One of the negative impacts of the categorization of HAT as a neglected tropical disease was the lack of interest and investment by various stakeholders in the production of new, more effective treatment [33, 64]. Anti-trypanosomal treatment evolved from extremely toxic agents such as Dyes and Arsenicals, which were both toxic to animals and humans [70]. Although there has been some improvement with newly synthesized forms from earlier toxic forms and improvement in the treatment regimen in the past 10 years compared to the past 50 years, there are still grappling issues of drug toxicities and drug failures [64]. Despite these issues, these drugs are still in use in endemic areas.

29.10.1 Pentamidine Isethionate

This drug, also known as pentamidine, is a derivative of aromatic diamidines. It was first used in the treatment of trypanosomiasis in 1937 and was registered as a drug in the treatment of HAT in 1940 [64, 70]. Its trypanocidal effect comes from its ability to disrupt complex mitochondrial function after accumulation within the cytoplasm of the trypanosome [71, 72]. It is used only in the treatment of the early-stage of *T.b. gambiense* HAT, as it is not effective in the treatment of both stages of *T.b. rhodesiense* HAT [33, 64, 71, 72]. Pentamidine is highly protonated, making it unsuitable for oral route administration; thus, it is given parenterally [72]. There are side effects reported with the use of pentamidine, such as nephrotoxicity, hypotension, and hypoglycemia as well as severe Stephen Johnson syndrome, pancreatic, and hepatic disruption of function have been reported [33, 64, 71]. Taken together, pentamidine is the most tolerated of the drugs used in the treatment of HAT when given at the required dosage of 4 mg/kg daily for seven days, and resistance has not been reported despite many years of use [71].

29.10.2 Suramin

Suramin, also known as Germanin, is a derivative of naphthalene urea, a colorless dye that has some trypanocidal effects [70]. Suramin was the most potent compound discovered after a series of failed experiments to get an effective compound able to treat HAT from the dye trypan red [70, 72]. It was discovered in 1917 but put into use in 1924 and since then it has been in use for the treatment of HAT [70, 72]. Suramin inhibits the glycolysis of the blood trypomastigotes by disrupting the enzymes that catalyze the pathway in the parasite [71, 72]. It is effective in the treatment of early-stages of *T.b. rhodesiense* [33, 64, 71, 72]. It is highly water-soluble, so it is given intravenously [72] with a test dose of 4–5 kg/kg/day and weekly dose 20 mg/kg for five weeks [33]. Side effects can be serious such as anaphylactic shock, renal failure, severe diarrhea, and others like hemolytic anemia, severe dermatoses, nausea, and others [33, 64, 71, 72]. Like pentamidine, resistance to suramin is not common, probably because suramin blocks the activities of several enzymes involved in the glycolysis of blood trypomastigotes [71].

29.10.3 Melarsoprol

Also known as Mel B, it is a derivative of an arsenical compound effective in killing Trypanosoma, discovered after series of experiments in 1946 and introduced as a treatment for HAT [70–72]. Melarsoprol disrupts many metabolic pathways inside the Trypanosoma, especially redox reactions and the glycolytic pathway [71, 72]. It is effective in the treatment of late stage for both *T.b. gambiense* and *T.b. rhodesiense* [33, 64, 71, 72]. Though it is not soluble in water, Melarsoprol is given intravenously dissolved in an irritant propylene glycol [71, 72]. It is given at 2.2 mg/kg for 10 days [33]. It is a highly toxic drug with the side effect of severe post-treatment encephalopathy, which occurs in about 10% of those on treatment, and almost half of these die from this side effect [33, 64, 71, 72]. Other side effects include inflammation and hardening of the veins, peripheral neuropathy, hepatic toxicity, dermatoses, and others [33, 64, 71, 72]. Due to high failure rates and very disturbing side effects, especially post treatment encephalitis, the drug has been replaced by a more effective and less toxic late stage drug called eflornithine [33, 64, 71, 72].

29.10.4 Eflornithine

This drug was synthesized much later (1970) compared to the other agents used in the treatment of HAT, and treatment for HAT was commenced in 1981 [66, 70]. It is a fluorinated analogue of an amino acid capable of irreversibly inhibiting ornithine decarboxylase (ODC), which catalyzes the polyamine biosynthetic pathway that produces polyamine, a compound that helps in cell division and growth [71, 72]. It was for this reason that it was used to treat cancer but failed because mammalian cells have fluctuating levels (high turnover) of ODC [70, 71]. However, it was more effective in interrupting the polyamine biosynthetic pathway of the parasite because of a more stable (low turnover) ODC [70–72]. It is used in the treatment of the late stage of *T.b. gambiense* but not for *T.b. rhodesiense* because of the high turnover of the ODC enzyme in the later stage [70–72]. It is given at 400 mg/kg/day in a laborious, slow infusion daily for seven days [33, 71, 72]. It has minimal side effects and is better tolerated than Melarsoprol [71, 72].

29.10.5 Nifurtimox

This drug has been around since 1960s, used in the treatment of Chagas disease and its anti-trypanosomal activity was later discovered. It was first used in the treatment of HAT in 1971 [66, 70, 71]. It is also called Lampit and is a derivative of 5-nitrofuran produced by the German pharmaceutical company Bayer [70, 72]. It is the only oral agent used in HAT, often in combination with Eflornithine, in the treatment of the late stage of HAT caused by either *T.b. gambiense* and *T.b. rhodesiense* [33, 64, 66, 71, 72]. The mechanism of action is the production of superoxide and other free radicals that cause oxidative stress damage to the trypomastigotes-damaging cell membranes and DNA [71, 72]. It has been found to interrupt trypanothione metabolism, which is a potent antioxidant [66]. It is given at 15 mg/kg in three divided doses a day for seven days in combination with intravenous Eflornithine (NECT) an effective combination in the treatment of late stage HAT [33, 64, 66, 71, 72]. It has a minimal side effect of gastrointestinal disturbance and some minor neurologic disturbances [66].

29.11 Control and Prevention

History has proven that neglect of HAT results in a resurgence of the disease, as evident by three epidemics that Africa experienced in the early 1900s, early 1920s, and 1990s, killing millions in the process [1, 73]. Control efforts mounted at those times obviously interrupted these epidemics, but challenges faced, ranging from political, war, environmental, and poor socioeconomic issues, squelched the progress made in the control of HAT [1, 73]. In the beginning of the twenty-first century, stakeholders intensified efforts, with several of them, such as WHO, organization of African unity (OAU), and others, initiating multifaceted control and prevention programs in partnership with pharmaceutical companies willing to provide safe and effective anti-trypanosomal agents for free [1, 73, 74]. These efforts resulted in a steady decline of cases and deaths from the beginning of this millennium [1, 73, 74] and as of 2015, there were less than 5000 cases reported [74, 75]. Stakeholders were enlivened by the progress made, and the goal of control and prevention went beyond the focus on reducing the burden of HAT in rural West African populations where HAT foci are replete to the elimination of HAT caused by *T.b. gambiense* [75]. To achieve this in 2012, WHO set a 2020 target for the elimination of HAT as public health problem [74, 75]. Although HAT was not eliminated at the set date, there was an impressive reduction in cases during the period. Therefore, the effort was aptly extended to 2030 with the goal of successful elimination of *T.b. gambiense* HAT [75]. *T.b. rhodesiense* HAT is not the focus because of the impossibility to eliminate it due to the animal reservoir components of its transmission and the apparent difficulty in identifying all the domestic and wild animals infected [1, 73–75].

Three major areas have been the focus of control and prevention of HAT all these years, and these include vector control, identifying cases and reservoirs of HAT, and adequate treatment of the infected.

29.11.1 Vector Control

Vector control has successful historical precedence, as it was employed in the earlier epidemic of the twentieth century [1, 73]. This strategy focuses on reducing the population of tsetse fly since total elimination of the vectors is impossible. Aerial sprays, sterile insect techniques, and some others have been used [74, 76]. Destroying the natural habitat of the tsetse fly like bush burning was employed in the past but was abandoned because of environmental unfriendly issues [76]. Alone, vector control might not do much, but in combination with other control strategies such as case identification and treatment, it produces the desired result of a reduction in the burden of HAT [74, 76].

29.11.2 Identifying and Treatment of Cases

Active case identification involving mass screening of high-risk populations using RDTs tools, followed by parasitological confirmation, has proven to be very beneficial in the control of HAT [73, 74]. It does not only ascertain the true burden of the disease in the population but also confirm those that need treatment. Adequate treatment is recommended for all cases identified during community screening in active case detection and those who come to the hospital with symptoms and signs suggested of HAT in passive case detection [1, 74]. The combination therapy of oral Nifurtimox and parenteral Eflornithine is an effective combination used in the treatment of HAT (NECT) [74].

29.11.3 Other Control Methods

Preventing bites is also very important and for this reason, it is recommended to wear long sleeves and trousers to cover the lower extremities [77]. These clothes should preferably be of neutral or dull colors, as tsetse fly are attracted to bright colors [77]. Use of insect repellents is also helpful but might not totally prevent bites [77]. Avoid bushes, especially during the day, and stay inside the car if you are visiting a game reserve or national park [77]. Genetic manipulation of the bacteria symbiont that live in the gut of the fly, aiding the infectivity of the fly can further reduce the infectivity of the Tsetse fly. This is another means of vector control that has been employed [74, 78].

29.12 Challenges of Control and Prevention

One of the major challenges to the control of *T.b. gambiense* and *T.b. rhodesiense*, like all neglected tropical diseases, is a lack of knowledge of HAT and the true burden of the disease in the human population [73, 79]. There is a lack of awareness among people and among healthcare workers in many sub-Saharan African countries, which can result in the missed diagnosis of HAT because of the indistinguishable symptoms from other diseases in endemic areas [73]. This lack of awareness among healthcare workers leads to a low index of suspicion, allowing cases to go undiagnosed in healthcare settings [73]. The case fatality rate of the disease is almost 100% if left untreated.

Conflict and war have been linked with epidemics of HAT in sub-Saharan Africa [24, 80]. Many countries with wars, conflict, and political tension see a surge of infection in epidemic proportions, especially in the 1970s during the post-independence era [24, 80]. These conflicts and unrest disrupted control and prevention programs, worsened poor health infrastructure and health systems, and exposed the fleeing population to bites of the tsetse fly [24, 75, 80].

It is now considered that the tsetse fly population might not be the only explanation for the endemicity or periodic epidemic of the disease in Africa [75]. This arose from findings showing several tsetse flies examined in the field had no trypanosomes in them, coupled with the inherently poor infectivity of the vector [75]. This opened up the possibility that asymptomatic infections and probably another mode of transmission might be contributing to the transition of HAT in sub-Saharan Africa [75]. Asymptomatic infections, especially with HAT caused by *T.b. gambiense*, pose a serious challenge to the control and prevention of HAT [75]. This is because these asymptomatic individuals might be negative on screening but are potent in driving transmission since parasites can be picked up by the tsetse fly after a blood meal. These asymptomatic individuals can survive with *T.b. gambiense* for a long period of time, and during this time they can trigger an epidemic of HAT [75, 81]. Sexual and vertical transmissions are potent modes of transmission that do not require tsetse fly bites. It is believed that these might be significantly contributing to cases of HAT in endemic areas, and logically asymptomatic individuals might be the major drivers of this mode of transmission [75]. There is an argument concerning the route of infection if the newborn is diagnosed with HAT after five days of life, for it is considered to be due to the bite of a tsetse fly [48, 75]. However, the idea that children less than two years of age are already in the late stages of HAT suggests that it is likely to be congenital HAT [75, 82].

29.13 Conclusion

HAT is a neglected tropical disease that causes significant morbidity and mortality among those infected. Despite the decreasing prevalence of HAT, there is a worrying possibility that an epidemic is just around the corner owing to the animal component of the transmission dynamic, asymptomatic individuals, as well as the tsetse fly independent transmission such as sexual and vertical transmission. Poor knowledge and awareness among the endemic population and healthcare workers might be contributing to the spread of HAT in sub-Saharan Africa. Wars and conflicts that are not uncommon in endemic areas contribute significantly to the burden of HAT.

Periodic epidemiological studies should be carried out in high-risk populations, backed up with campaigns to keep the population alert as well as healthcare workers. With the ongoing unrest in some parts of Africa, it is possible that these conflict zones might experience a resurgence of HAT cases. Thus, control measures should be intensified in these areas as soon as wars and conflicts are resolved. Control measures should target asymptomatic individuals, especially women of reproductive age. Improving diagnostic intervention by making available sensitive diagnostic tools in rural areas that will be easy to use with rapid, accurate results. Creating a safer and more effective treatment that can be easily administered orally to those who are infected albeit asymptomatic. Improving all these might bring an end to *T.b. gambiense* HAT by the year 2030.

References

1 Steverding, D. (2008). The history of African trypanosomiasis. *Parasit. Vectors* 1 (1): 3.

2 Álvarez-Rodríguez, A., Jin, B.K., Radwanska, M., and Magez, S. Recent progress in diagnosis and treatment of Human African Trypanosomiasis has made the elimination of this disease a realistic target by 2030. *Front. Med.* 9: 1037094. Available from: https://www.frontiersin.org/articles/10.3389/fmed.2022.1037094/full.

3 Hemmige, V., Tanowitz, H., and Sethi, A. (2012). *Trypanosoma cruzi* infection: a review with emphasis on cutaneous manifestations. *Int. J. Dermatol.* 51 (5): 501–508.
4 Desquesnes, M., Gonzatti, M., Sazmand, A. et al. (2022). A review on the diagnosis of animal trypanosomoses. *Parasit. Vectors* 15 (1): 64.
5 Bengaly, Z., Sidibe, I., Ganaba, R. et al. (2002). Comparative pathogenicity of three genetically distinct types of *Trypanosoma congolense* in cattle: clinical observations and haematological changes. *Vet. Parasitol.* 108 (1): 1–19.
6 Gao, J.M., Qian, Z.Y., Hide, G. et al. (2020). Human African trypanosomiasis: the current situation in endemic regions and the risks for non-endemic regions from imported cases. *Parasitology* 147 (9): 922–931.
7 Fevre, E.M., Wissmann, B.V., Welburn, S.C., and Lutumba, P. (2008). The burden of human African trypanosomiasis. *PLoS Negl. Trop. Dis.* 2 (12): e333.
8 Rijo-Ferreira, F., Carvalho, T., Afonso, C. et al. (2018). Sleeping sickness is a circadian disorder. *Nat. Commun.* 9 (1): 62.
9 Haag, J. (1998). The molecular phylogeny of trypanosomes: evidence for an early divergence of the Salivaria. *Mol. Biochem. Parasitol.* 91 (1): 37–49.
10 Lambrecht, F.L. (1985). Trypanosomes and hominid evolution. *Bioscience* 35 (10): 640–646.
11 Jolly, D., Prentice, I.C., Bonnefille, R. et al. (1998). Biome reconstruction from pollen and plant macrofossil data for Africa and the Arabian peninsula at 0 and 6000 years. *J. Biogeogr.* 25 (6): 1007–1027.
12 Kea, R.A. (2004). Expansions and contractions: world-historical change and the Western Sudan world-system (1200/1000 B.C. 1200/1250 A.D.). *J. World-Syst. Res.* 10: 723–816.
13 Cox, F.E.G. (2004). History of sleeping sickness (African trypanosomiasis). *Infect. Dis. Clin. N. Am.* 18 (2): 231–245.
14 Tibayrenc, M. and Ayala, F.J. (1999). Evolutionary genetics of Trypanosoma and Leishmania. *Microbes Infect.* 1 (6): 465–472.
15 Radwanska, M., Vereecke, N., Deleeuw, V. et al. (2018). Salivarian trypanosomosis: a review of parasites involved, their global distribution and their interaction with the innate and adaptive mammalian host immune system. *Front. Immunol.* 9: 2253.
16 Cayla, M., Rojas, F., Silvester, E. et al. (2019). African trypanosomes. *Parasit. Vectors* 12 (1): 190.
17 Stevens, J. and Gibson, W. (1999). The evolution of Salivarian trypanosomes. *Mem. Inst. Oswaldo Cruz* 94 (2): 225–228.
18 Nimpaye, H., Njiokou, F., Njine, T. et al. (2011). *Trypanosoma vivax*, *T. congolense* "forest type" and *T. simiae*: prevalence in domestic animals of sleeping sickness foci of Cameroon. *Parasite* 18 (2): 171–179.
19 Nkinin, S.W., Njiokou, F., Penchenier, L. et al. (2002). Characterization of *Trypanosoma brucei* s.l. subspecies by isoenzymes in domestic pigs from the Fontem sleeping sickness focus of Cameroon. *Acta Trop.* 81 (3): 225–232.
20 Solano, P., Michel, J.F., Lefrançois, T. et al. (1999). Polymerase chain reaction as a diagnosis tool for detecting trypanosomes in naturally infected cattle in Burkina Faso. *Vet. Parasitol.* 86 (2): 95–103.
21 Truc, P., Büscher, P., Cuny, G. et al. (2013). Atypical human infections by animal trypanosomes. *PLoS Negl. Trop. Dis.* 7 (9): e2256.
22 Truc, P., Gibson, W., and Herder, S. (2007). Genetic characterization of *Trypanosoma evansi* isolated from a patient in India. *Infect. Genet. Evol.* 7 (2): 305–307.
23 Romero-Meza, G. and Mugnier, M.R. (2020). *Trypanosoma brucei*. *Trends Parasitol.* 36 (6): 571–572.
24 Simarro, P., Franco, J., Diarra, A., and Jannin, J. (2014). Epidemiology of human African trypanosomiasis. *Clin. Epidemiol.* 6: 257.
25 Krafsur, E. (2009). Tsetse flies: genetics, evolution, and role as vectors. *Infect. Genet. Evol.* 9 (1): 124–141.
26 Gooding, R.H. and Krafsur, E.S. (2005). Tsetse genetics: contributions to biology, systematics, and control of tsetse flies. *Annu. Rev. Entomol.* 50 (1): 101–123.
27 Benoit, J.B., Attardo, G.M., Baumann, A.A. et al. (2015). Adenotrophic viviparity in tsetse flies: potential for population control and as an insect model for lactation. *Annu. Rev. Entomol.* 60 (1): 351–371.
28 Attardo, G.M., Guz, N., Strickler-Dinglasan, P., and Aksoy, S. (2006). Molecular aspects of viviparous reproductive biology of the tsetse fly (*Glossina morsitans morsitans*): regulation of yolk and milk gland protein synthesis. *J. Insect Physiol.* 52 (11–12): 1128–1136.
29 Attardo, G.M., Lohs, C., Heddi, A. et al. (2008). Analysis of milk gland structure and function in *Glossina morsitans*: milk protein production, symbiont populations and fecundity. *J. Insect Physiol.* 54 (8): 1236–1242.
30 Tirados, I., Esterhuizen, J., Rayaisse, J.B. et al. (2011). How do tsetse recognise their hosts? The role of shape in the responses of tsetse (*Glossina fuscipes* and G. palpalis) to artificial hosts. *PLoS Negl. Trop. Dis.* 5 (8): e1226.
31 Solano, P., Salou, E., Rayaisse, J.B. et al. (2015). Do tsetse flies only feed on blood? *Infect. Genet. Evol.* 36: 184–189.
32 Mebourou, E.K., Bernáth, B., Schenker, D., and Guerin, P.M. (2018). Vision and genesis of survival strategies in tsetse flies: a laboratory study. *J. Insect Physiol.* 107: 212–223.
33 Malvy, D. and Chappuis, F. (2011). Sleeping sickness. *Clin. Microbiol. Infect.* 17 (7): 986–995.

34 Cordon-Obras, C., Berzosa, P., Ndong-Mabale, N. et al. (2009). *Trypanosoma brucei gambiense* in domestic livestock of Kogo and Mbini foci (Equatorial Guinea). *Tropical Med. Int. Health* 14 (5): 535–541.

35 Kinung'hi, S., Malele, I., Kibona, S. et al. (2009). Knowledge, attitudes and practices on tsetse and sleeping sickness among communities living in and around Serengeti National Park, Tanzania. *Tanzan. J. Health Res.* 8 (3): 168–172. Available from: http://www.ajol.info/index.php/thrb/article/view/45115.

36 Zoller, T., Fèvre, E.M., Welburn, S.C. et al. (2008). Analysis of risk factors for *T. brucei rhodesiense* sleeping sickness within villages in south-east Uganda. *BMC Infect. Dis.* 8 (1): 88.

37 Moore, S., Shrestha, S., Tomlinson, K.W., and Vuong, H. (2012). Predicting the effect of climate change on African trypanosomiasis: integrating epidemiology with parasite and vector biology. *J. R. Soc. Interface* 9 (70): 817–830.

38 Lord, J.S., Hargrove, J.W., Torr, S.J., and Vale, G.A. (2018). Climate change and African trypanosomiasis vector populations in Zimbabwe's Zambezi Valley: a mathematical modelling study. *PLoS Med.* 15 (10): e1002675.

39 Sutherland, C.S., Stone, C.M., Steinmann, P. et al. (2017). Seeing beyond 2020: an economic evaluation of contemporary and emerging strategies for elimination of *Trypanosoma brucei gambiense*. *Lancet Glob. Health* 5 (1): e69–e79.

40 Mwiinde, A.M., Simuunza, M., Namangala, B. et al. (2017). Estimating the economic and social consequences for patients diagnosed with human African trypanosomiasis in Muchinga, Lusaka and Eastern Provinces of Zambia (2004–2014). *Infect. Dis. Poverty* 6 (1): 150. Available from: http://idpjournal.biomedcentral.com/articles/10.1186/s40249-017-0363-6.

41 Simarro, P.P., Franco, J.R., Cecchi, G. et al. (2012). Human African trypanosomiasis in non-endemic countries (2000–2010). *J. Travel Med.* 19 (1): 44–53.

42 Urwyler, S., Vassella, E., Abbeele, J.V.D. et al. (2005). Expression of procyclin mRNAs during cyclical transmission of *Trypanosoma brucei*. *PLoS Pathog.* 1 (3): e22.

43 Magez, S., Pinto Torres, J.E., Oh, S., and Radwanska, M. (2021). Salivarian trypanosomes have adopted intricate host-pathogen interaction mechanisms that ensure survival in plain sight of the adaptive immune system. *Pathogens* 10 (6): 679.

44 Reynolds, S.J., Bollinger, R.C., and Quinn, T.C. (2008). Parasitic diseases during pregnancy. *Glob. Libr. Womens Med.* https://doi.org/10.3843/GLOWM.10188.

45 De Kyvon, M.A.L.C., Maakaroun-Vermesse, Z., Lanotte, P. et al. (2016). Congenital trypanosomiasis in child born in France to African mother. *Emerg. Infect. Dis.* 22 (5): 935–937.

46 Ngoma, M.S. (2004). Congenital trypanosomiasis. *J. Trop. Pediatr.* 50 (6): 377–378.

47 Olowe, S.A. (1975). A case of congenital tryi'anosomiasis in Lagos. *Trans. R. Soc. Trop. Med. Hyg.* 69 (1): 57–59.

48 Lindner, A.K. and Priotto, G. (2010). The unknown risk of vertical transmission in sleeping sickness – a literature review. *PLoS Negl. Trop. Dis.* 4 (12): e783.

49 Howie, S., Guy, M., Fleming, L. et al. (2006). A Gambian infant with fever and an unexpected blood film. *PLoS Med.* 3 (9): e355.

50 Rocha, G., Martins, A., Gama, G. et al. (2004). Possible cases of sexual and congenital transmission of sleeping sickness. *Lancet* 363 (9404): 247.

51 Burnaz, D. and Tătar, S. (2019). GP53 Elusive cause for a recurrent fever in a 2-year-old boy. In: *Abstracts*. BMJ Publishing Group Ltd and Royal College of Paediatrics and Child Health pp. A51.1–A51. [Internet]. [cited 2023 May 23]. Available from: https://adc.bmj.com/lookup/doi/10.1136/archdischild-2019-epa.119.

52 Sereno, D., Akhoundi, M., Sayehmri, K. et al. (2020). Noninvasive biological samples to detect and diagnose infections due to trypanosomatidae parasites: a systematic review and meta-analysis. *Int. J. Mol. Sci.* 21 (5): 1684.

53 Herwaldt, B.L. (2001). Laboratory-acquired parasitic infections from accidental exposures. *Clin. Microbiol. Rev.* 14 (4): 659–688.

54 Crilly, N.P. and Mugnier, M.R. (2021). Thinking outside the blood: perspectives on tissue-resident *Trypanosoma brucei*. *PLoS Pathog.* 17 (9): e1009866.

55 Alfituri, O.A., Quintana, J.F., MacLeod, A. et al. (2020). To the skin and beyond: the immune response to African trypanosomes as they enter and exit the vertebrate host. *Front. Immunol.* 11: 1250.

56 Camara, M., Soumah, A.M., Ilboudo, H. et al. (2021). Extravascular dermal trypanosomes in suspected and confirmed cases of *gambiense* human African trypanosomiasis. *Clin. Infect. Dis.* 73 (1): 12–20.

57 Capewell, P., Cren-Travaillé, C., Marchesi, F. et al. (2016). The skin is a significant but overlooked anatomical reservoir for vector-borne African trypanosomes. *elife* 5: e17716.

58 Vincendeau, P. and Bouteille, B. (2006). Immunology and immunopathology of African trypanosomiasis. *An. Acad. Bras. Cienc.* 78 (4): 645–665.

59 Samanovic, M., Molina-Portela, M.P., Chessler, A.D.C. et al. (2009). Trypanosome lytic factor, an antimicrobial high-density lipoprotein, ameliorates Leishmania infection. *PLoS Pathog.* 5 (1): e1000276.

60 Mogk, S., Meiwes, A., Boßelmann, C.M. et al. (2014). The lane to the brain: how African trypanosomes invade the CNS. *Trends Parasitol.* 30 (10): 470–477.

61 Masocha, W. and Kristensson, K. (2012). Passage of parasites across the blood-brain barrier. *Virulence* 3 (2): 202–212.

62 Mogk, S., Meiwes, A., Shtopel, S. et al. (2014). Cyclical appearance of African trypanosomes in the cerebrospinal fluid: new insights in how trypanosomes enter the CNS. *PLoS One* 9 (3): e91372.

63 Rodgers, J., Steiner, I., and Kennedy, P.G.E. (2019). Generation of neuroinflammation in human African trypanosomiasis. *Neurol. Neuroimmunol. Neuroinflamm.* 6 (6): e610.

64 Kennedy, P.G. (2013). Clinical features, diagnosis, and treatment of human African trypanosomiasis (sleeping sickness). *Lancet Neurol.* 12 (2): 186–194.

65 Urech, K., Neumayr, A., and Blum, J. (2011). Sleeping sickness in travelers – do they really sleep? *PLoS Negl. Trop. Dis.* 5 (11): e1358.

66 Kennedy, P.G.E. (2004). Human African trypanosomiasis of the CNS: current issues and challenges. *J. Clin. Invest.* 113 (4): 496–504.

67 Biéler, S., Waltenberger, H., Barrett, M.P. et al. (2016). Evaluation of antigens for development of a serological test for human African trypanosomiasis. *PLoS One* 11 (12): e0168074.

68 Bessell, P.R., Lumbala, C., Lutumba, P. et al. (2018). Cost-effectiveness of using a rapid diagnostic test to screen for human African trypanosomiasis in the Democratic Republic of the Congo. *PLoS One* 13 (9): e0204335.

69 Habeeba, S., Khan, R.A., Zackaria, H. et al. (2022). Comparison of microscopy, card agglutination test for *Trypanosoma evansi*, and real-time PCR in the diagnosis of trypanosomosis in dromedary camels of the Abu Dhabi Emirate, UAE. *J. Vet. Res.* 66 (1): 125–129.

70 Steverding, D. (2010). The development of drugs for treatment of sleeping sickness: a historical review. *Parasit. Vectors* 3 (1): 15.

71 Babokhov, P., Sanyaolu, A.O., Oyibo, W.A. et al. (2013). A current analysis of chemotherapy strategies for the treatment of human African trypanosomiasis. *Pathog. Glob. Health* 107 (5): 242–252.

72 Fairlamb, A.H. (2003). Chemotherapy of human African trypanosomiasis: current and future prospects. *Trends Parasitol.* 19 (11): 488–494.

73 Franco, J.R., Simarro, P.P., Diarra, A. et al. (2014). The journey towards elimination of gambiense human African trypanosomiasis: not far, nor easy. *Parasitology* 141 (6): 748–760.

74 Aksoy, S., Buscher, P., Lehane, M. et al. (2017). Human African trypanosomiasis control: achievements and challenges. *PLoS Negl. Trop. Dis.* 11 (4): e0005454.

75 Welburn, S.C., Molyneux, D.H., and Maudlin, I. (2016). Beyond tsetse – implications for research and control of human African trypanosomiasis epidemics. *Trends Parasitol.* 32 (3): 230–241.

76 Solano, P., Torr, S.J., and Lehane, M.J. (2013). Is vector control needed to eliminate gambiense human African trypanosomiasis? *Front. Cell. Infect. Microbiol.* 3: 33. Available from: http://journal.frontiersin.org/article/10.3389/fcimb.2013.00033/abstract.

77 Sinha, A., Grace, C., Alston, W.K. et al. (1999). African trypanosomiasis in two travelers from the United States. *Clin. Infect. Dis.* 29 (4): 840–844.

78 Aksoy, S., Weiss, B., and Attardo, G. (2008). Paratransgenesis applied for control of tsetse transmitted sleeping sickness. In: *Transgenesis and the Management of Vector-Borne Disease*, Advances in Experimental Medicine and Biology, vol. vol. 627 (ed. S. Aksoy), 35–48. New York, NY: Springer New York Available from: http://link.springer.com/10.1007/978-0-387-78225-6_3.

79 Elenga, V.A., Lissom, A., Elion, D.O.A. et al. (2022). Risk factors and prevalence of human African trypanosomiasis in individuals living in remote areas of the republic of Congo. *BMC Public Health* 22 (1): 2322.

80 Berrang-Ford, L., Lundine, J., and Breau, S. (2011). Conflict and human African trypanosomiasis. *Soc. Sci. Med.* 72 (3): 398–407.

81 Sudarshi, D., Lawrence, S., Pickrell, W.O. et al. (2014). Human African trypanosomiasis presenting at least 29 years after infection – what can this teach us about the pathogenesis and control of this neglected tropical disease? *PLoS Negl. Trop. Dis.* 8 (12): e3349.

82 Eperon, G., Schmid, C., Loutan, L., and Chappuis, F. (2007). Clinical presentation and treatment outcome of sleeping sickness in Sudanese pre-school children. *Acta Trop.* 101 (1): 31–39.

30

Emerging and Re-Emerging Infectious Diseases of the Decade: Current Challenges and Future Directions

Seth K. Amponsah[1], Ranjita Shegokar[2], and Yashwant V. Pathak[3]

[1] Department of Medical Pharmacology, University of Ghana Medical School, Accra, Ghana
[2] Capnomed GmbH, Tübingen, Germany
[3] USF Health Taneja College of Pharmacy, University of South Florida, Tampa, FL, United States

30.1 Introduction

Emerging infectious diseases (EIDs) are infections that have appeared for the first time in a population, or infectious diseases that have previously been reported but have currently increased in incidence or geographic range [1]. EIDs can be new, hitherto indistinct infectious diseases, and already existing diseases with new features. The new features may be that the disease is occurring in a new population (for example, the disease may appear among the youth when previously it only affected the elderly), new location, new clinical outcomes (for example resistance to available treatments), or a rapid increase in disease spread. Re-emerging infectious diseases (REIDs) can be described as the reappearance of a disease that was once endemic and eventually eradicated or controlled but has now become a public health problem [2, 3]. Among the various pathogens, influenza viruses have been found to constantly emerge and re-emerge.

The morbidity and mortality associated with these emerging and re-emerging infectious diseases cannot be overemphasized. About 15 million (>25%) annual deaths worldwide (estimated to be 57 million) are known to be of infectious disease origin [4]. Furthermore, infectious disease related morbidity and mortality has been found to occur more in resource-poor settings [5], and predominantly in children [4]. In resource-rich countries, infectious disease appears to disproportionately affect the disadvantaged minorities [6].

Over the last few decades, momentous events related to infectious diseases (with epidemic potential) have occurred. Some of these include the cholera outbreak in Yemen (2015–2018), Ebola virus disease (EVD) outbreak in West Africa (2013–2016), the SARS pandemic (2002–2004), Nipah virus in India and South Asia (2017–2018), Zika virus outbreak in the Americas and Southeast Asia (2016–2018), and Lassa fever (LF) in Nigeria (2018) [7]. Furthermore, in the last three years the world had to deal with the novel coronavirus disease 2019 (COVID-19), which has caused major disease conditions and deaths across the world. SARS coronavirus 2 (SARS-CoV-2), etiological agent of COVID-19, was first identified in China in December 2019. COVID-19 spread to many countries; eventually becoming a global pandemic [8, 9]. Currently, a number of these EIDs and REIDs are global threats and require cooperation between regions and nations [10].

Most EIDs and REIDs have emerged from animals (zoonotic origin) and crossed the species barrier [11]. Furthermore, a majority of these diseases originate from wildlife, while a few others have come from pets and intensive animal farming [12–14]. It is noteworthy, however, that not all EIDs and REIDs have emerged from animals. These infectious diseases can be transmitted to humans from animals through direct contact, fomites, droplets, food, water, or vectors [15], as shown in Figure 30.1.

Multi-drug resistant (MDR) organisms, such as *Candida auris* and vancomycin-resistant *Staphylococcus aureus* exist and are deemed non-zoonotic. These infections occur as a result of antibiotic misuse or overuse. Regardless of the origin of infection, a pathogen must be introduced into a susceptible population, and have the ability to spread and cause disease [16].

Rising Contagious Diseases: Basics, Management, and Treatments, First Edition. Edited by Seth Kwabena Amponsah, Ranjita Shegokar, and Yashwant V. Pathak.

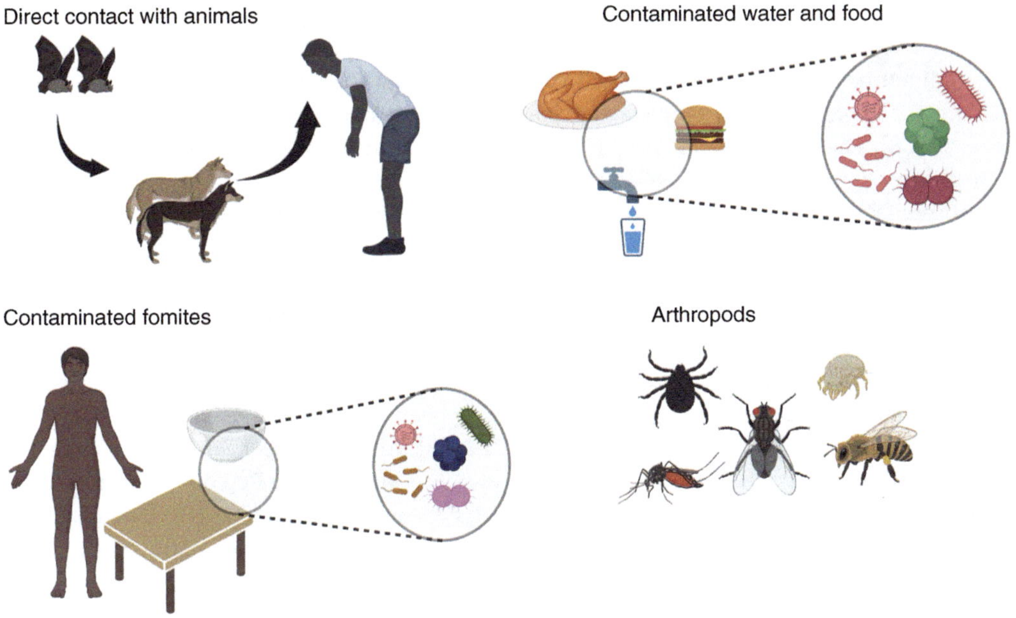

Figure 30.1 Infectious disease transmission to humans.

30.2 Factors Leading to the Emergence and Re-Emergence of Infectious Diseases

A number of factors are known to promote the emergence or re-emergence of infectious diseases. These factors enable pathogens to reach and adopt to susceptible hosts, evolve in ecological niches, and spread among these hosts. Some of these factors include international travel, urbanization, migration, changing ecosystems, overcrowding, and rapid population growth. Other factors include irrational antimicrobial use and the development of resistance in humans and livestock, microbial adaptation and change, and breaches in sanitation and vaccination programs [17–19]. It is, therefore, obvious that the emergence and re-emergence of infectious diseases are closely linked to the activities of humans.

Furthermore, a large number of infectious diseases emerge when causative agents in animals are spread to humans (also known as zoonoses). With the rise in human population (and into new geographical zones), the chances of humans coming into close contact with animal species that are possible carriers of infectious agents is high. By this increase in contact between man and animals, which is further compounded by increase in human mobility and density, one can conclude that these factors are a serious threat to global health [20].

Climate change is gradually becoming a significant factor in the emergence of infectious diseases. As the climate of the earth gets warmer, habitats are usually altered, and diseases can spread into new geographic spheres. For instance, high temperatures usually allow mosquitoes – and the diseases they transmit – to expand into regions where they hitherto were not found [20]. Most recently, surges in malaria infections have been recorded following floods in Pakistan and cyclones in Mozambique [21]. This was likely due to the resulting large pools of standing water which attracts mosquitoes. Indigenes of highlands in Ethiopia and Kenya are now known to succumb to malaria infections. This is due to the shift in low temperatures that once made the area less habitable for mosquitoes [21].

Over the years, re-emergence of infectious diseases has occurred as a result of antimicrobial resistance. Microorganisms can evolve with time and develop resistance to drugs that were once effective in pathogen treatment [20]. Antimicrobial resistance has been shown to be the consequence of adaptation and genetic changes in microorganisms that facilitates infection of human cells and the spread of diseases. These microorganisms can become resistant to multiple antimicrobials within a short period of time [22–26] via mechanisms where there is exchange or incorporation of new genetic material into their genomes [27]. Antimicrobials that are used as additives in plant agriculture (vegetables and fruits) and animal rearing, more predominantly food-producing animals, can lead to an increase in antimicrobial resistance [28]. Also, resistant genes are spread from animals to humans through feeding, through environmental spread, or direct contact with farmed animals [29–32].

Furthermore, the inappropriate use of antimicrobial medications, particularly in resource-poor settings, mostly due to poor state of health facilities, easy access to "prescription-only" drugs, and self-medication can contribute to antimicrobial resistance [33–35]. Up to 50% of all antibiotics that are prescribed are misused in these resource-poor settings [36, 37]. Discontinuation of antibiotic treatment by patients is also perceived to play a major role in the development of virulent and resistant strains of microorganisms [38]. These aforementioned factors contribute significantly to genetic selection pressure for the emergence of multi-drug resistant microorganisms in the community.

Limited vaccine coverage is another factor that can cause a disease to re-emerge. When an effective and safe vaccine is introduced and a sizeable number of people decide not to get vaccinated, there is the possibility that a previously eradicated infectious disease may re-emerge. This has been observed with the measles vaccine. Measles was eliminated from the United States of America (USA) in 2000 and from the Western Hemisphere in 2016. However, the disease has re-emerged due to an increase in the number of people opting not to take the measles vaccine. In various parts of the world, there exists anti-vaccine movements founded largely on a questionable study that reported a link between vaccine against measles and autism [20]. Due to the general decline in vaccine coverage, measles cases have been on the high this decade, with more than 1000 cases of measles reported in the USA in the first six months of 2019 [20]. Also, this has been the case with the COVID-19 vaccine, and new cases of the disease continue to be recorded. In a speech made by the Regional Director of the World Health Organization (WHO), Africa, Dr. Matshidiso Moeti, he stated that "The end of the COVID-19 pandemic is within sight, but as long as Africa lags behind the rest of the world in reaching widespread protection, there is a dangerous gap which the virus can exploit to come roaring back." To improve the coverage of vaccination in Africa, WHO has set about a number of courses of action that include giving support to countries to appraise the readiness for vaccination campaigns at district and regional levels, keeping track of vaccinations among primary groups of concern, carrying out high-level advocacy to increase vaccine uptake, and helping countries integrate COVID-19 vaccines into other planned mass vaccination operations [39].

30.3 Current Challenges of Emerging and Re-Emerging Infectious Diseases

It has been more than 20 years since the genomic era, however, there is still paucity to a complete understanding of infectious diseases. Indeed, this is as a result of a number of challenges, a few enumerated in this chapter. A detailed molecular understanding of the dynamics of" "host-pathogen" interaction is missing for many viral, bacterial, parasitic, and zoonotic pathogens. Furthermore, details of the origin and route of infection, transmission, latency period, and host immunity for many infectious diseases are sketchy. The genetic complication of individual host, adaptive prospects of pathogens, and spatiotemporal disease dynamics often pose challenges to drug development [40]. This is especially true for polymicrobial infections. For example, it is not very clear how an infection by a single microorganism affects susceptibility to infections by other microorganisms. There is paucity of data to explain when and how the host immune surveillance capacity becomes saturated during co-infections. Details of how septic hyper-inflammation affects the host rather than the pathogen is also sketchy [40]. Many viral, bacterial, and fungal pathogens often drive lethal septic shocks due to progressive dysregulation of host immunity. The complex interplay of environmental factors, epigenetic and genetic predispositions on pathogen susceptibility is also enigmatic. Additionally, the lack of species-specific and rapid clinical diagnosis of several pathogens is yet another clinical challenge [40].

Concerted interdisciplinary research into infectious diseases is still at its infancy. Validation of basic research into clinical patient settings is also relevant. It is important to ascertain common denominators in host immunity to various pathogens [41]. There is the need for better understanding on how host immune surveillance drives the co-evolution of pathogens with their hosts [42], particularly in the case of adaptable and zoonotic pathogens. Besides immune defense, other factors may drive the evolution of pathogens, some of which include chemical diversity of drugs with adverse impact on immunity. Furthermore, environmental parameters such as global warming and climate change can affect infectious disease susceptibility [43]. Such shifts may have contributed to emergence of pathogens from the environment or wildlife to humans; an example is influenza from wild birds to poultry. Indeed, change in climate is considered the main trigger for the transmission of the thermotolerant fungus *Candida auris* [44] from environmental niches to birds, and then to becoming commensals in humans [45].

Furthermore, tissue microbiomes have recently emerged as key players in infectious diseases and cancers. There is convincing evidence that show that dysregulated microbiota contributes to and cause many diseases [46]. However, it remains a conundrum how microbiomes which circumscribe several bacterial and fungal species in a metabolic equilibrium, can alter predisposition to infections, cancer, and autoimmune diseases. A better understanding of the impact of microbiomes on metabolism and well-being of humans remains relevant. It will require interdisciplinary and translational efforts of scientists in many fields to unravel association between immunity-pathogen-microbiota interactions and diseases [47, 48].

Drug discovery with the aim of developing novel anti-infective agents has always been challenging and outrageously costly; usually only affordable in high-income countries. For several years, drugs used in the treatment of infections have been plagued with multi-drug resistance (MDR). Pharma companies were unwilling to develop costly antimicrobials or vaccines, also because most people affected would be unable to bear the costs of such treatments. The first malaria vaccine marked a milestone in the treatment of this major diseases that affect man [49]. Pan-anti-infective drugs for groups of pathogens have become less likely, given the continuous threat of MDR. Furthermore, the process from basic research to drug approval for each new drug often takes about 15 years with estimated costs of about 2 billion USD/Euros. However, these timelines are unpractical when new infections or pandemics emerge and when there appears to be no safe drug treatment available. Again, COVID-19 has thought us the need for drug repurposing [50].

Validated drug safety with the use of antimicrobials is another important factor which needs approval by health authorities. For example, geriatrics sometimes use multiple drugs on regular basis, yet predicting adverse reactions and drug–drug interactions have remained a clinical challenge [51]. This makes the implementation of fast-track drug discovery and use in the population challenging. Furthermore, short-cuts in drug development process can be risky, and drug safety evaluation should be uncompromised. The drug discovery process has to follow a translational, sometimes a rocky road that requires trial and error; often with the use of animal models [52]. Animal models are, and will remain, indispensable in the drug discovery process, ensuring appropriate translation of lead drugs into clinical use. Moreover, pharma industries may have to embrace new strategies in developing anti-infective drugs. For example, drugs that target host immunity can be beneficial during infections, especially for pathogens that drive fatal septic immune responses [53]. Modulating or targeting the immune system of the host rather than killing pathogens may also reduce the risk of MDR microorganisms [54].

Finally, a better understanding of the harmful or advantageous roles of host immunity in disease outcome will be relevant in precision or personalized medicine. Research into the human genome has achieved a lot of positive strides and promoted personalized medicine strategies for years. However, it is still evident that there are many challenges and pitfalls [55]. While some cancer treatments based on patient data turned out successful [50], the same for infectious diseases remain scarce. During SARS-CoV-2 infection, symptoms (or the lack thereof in some cohorts) is primarily determined by tolerance to the virus by the host, often making therapies and clinical outcome a gamble at best [56]. Indeed, most disease pathogenesis seem to follow the laws of game theory, whereby the looser always pays [57]. At the end, it may simply be the fitness (or tolerance) of the host, or the fitness of the pathogen to resist the defense of the host, that determines clinical outcome [58].

30.4 Strategies/Possible Interventions

30.4.1 Strengthening Surveillance and Rapid Response Mechanisms

Active and passive surveillance systems with prompt reporting and analysis of data are important in early detection of emerging and re-emerging infectious diseases. However, many developing countries do not have functional surveillance systems. Over the last few years, a lot of countries have developed national plans of action for critical epidemiological outbreaks such as influenza, viral hemorrhagic fevers, among others [59]. In India, the Integrated Disease Surveillance Project (IDSP) was established in 2004 in 101 districts [60]. Since then, it has been expanded to cover other districts in the country. This means each district now has a surveillance unit and a rapid response team to manage disease outbreak [60].

Indeed, surveillance is the cornerstone for the control of emerging and re-emerging infections. The implementation of global surveillance programs for the rapid detection of disease-causing microorganisms is of principal importance. The "One Health" concept that is been promoted in many nations aims to improve human health [61]. Efforts are to be made to reduce the environmental effects from land use, urbanization, and climate change that may increase the risk of pathogen spillover from animals to humans [62, 63]. Multidisciplinary efforts could have positive impact, especially in developing countries, as they are the most afflicted by zoonoses [64]. Also, integrative methodologies to improve surveillance programs for the prevention and control of emerging and re-emerging infections in countries remain relevant.

30.4.2 Conforming to International Health Regulations

In 2005, the International Health Regulations (IHR) bill was passed by 194 member countries. The purpose of this bill was to ensure the inhibition of the spread of infectious disease across borders, with restricted interference to international traffic and trade. In order to do so, IHR bill required every country to evaluate their surveillance and response potential, and make sure that primary capacities were implemented by 2012 [65]. Countries were only required to notify WHO of yellow fever, cholera, and plague outbreaks under the IHR. After the emergence of SARS, it became evident that the rules required an update considering the increase in international travel and trade. A revised version was thus developed and approved by the World Health Assembly. The purpose and scope of the new regulations were not limited to any specific disease or manner of transmission, but rather was aimed at addressing illness or medical conditions, irrespective of origin or source, that could be harmful to humans [65].

30.4.3 Building Capacity in Epidemiology

Through various short- and long-term field epidemiology training programs (FETP), capacities at the national and local levels can be improved. In India, the National Centre for Disease Control (NCDC) has been conducting short term (three month) FETPs from 1963. Also from 1996, a three-month regional FETP for WHO South-East Asia region is organized each year for senior and middle level health professionals [66]. A second, two-year FETP was started as a degree-program, granting Master of Public Health in Field Epidemiology [67]. The National Institute of Epidemiology in Chennai has been offering two years FETP, which is an in-service training program in applied epidemiology. The training has participants that include public health leaders and health authorities in India [68]. Similarly, a new mentorship driven two-year in-service training program modeled on Epidemic Intelligence Service (EIS) is run by CDC, Atlanta. This training program is aimed at creating a team of well-trained field epidemiologists at national, state, and district levels [69].

30.4.4 Strengthening of Laboratory and Networks

For an effective and prompt response to emerging and re-emerging infectious diseases, global healthcare systems must have laboratories which have capacity to detect, identify, and transmit results of newly identified pathogens. It is therefore important that all countries in the various WHO regions are strengthened and given the capacity to rapidly detect, report, and confirm outbreaks of emerging and epidemic-prone infectious diseases [70]. To be able to offer timely and effective responses to these emerging infectious diseases, as well as being able to mitigate their negative impact, there is the need to have the capacity to do high-quality risk assessment of these public health emergencies. The role of laboratories in this regard can therefore not be underestimated as these facilities aid rapid detection, risk assessment, and continuous monitoring of any outbreak. These laboratories must not work in isolation but in consonance with other equally functional and well-equipped facilities forming a network of laboratories. Such a laboratory network is important as surveillance and outbreak management depend on the rapid detection and identification of the causative pathogen(s) [70, 71]. For these laboratories to be fit for purpose, they must maintain a capacity to meet surge testing demands during epidemics and pandemics and not just their ability to do routine diagnostic tests. Another element of importance is that there must be well-established, well-defined, and readily accessible referral network systems between individual laboratories as this will help to provide advanced, rapid, and well-coordinated diagnostic services in record time [71, 72]. This can be within countries, sub-regions, and even between continents.

Also, these laboratories must be highly specialized and possess the appropriate biosafety levels as well as the ability to accurately detect these high-risk pathogens. After the passage of the IHR bill in 2005, all member states party to the agreement are required to establish laboratories with diagnostic capacity for diseases which are considered to be of high priority and to implement comprehensive biosafety and biosecurity legislation [73]. This must include standards for the safe storage, use, transport, and disposal of high-risk pathogens, as well as improving the capacity of regulatory bodies to keep in check all these individual laboratories. Emphasis must be placed on regular training and simulation exercises for all laboratory staff in order to keep them abreast with current and relevant knowledge of diseases of global public health importance [72, 73].

30.4.5 Research and Development

Research is the bedrock on which knowledge is generated and utilized in the effective management of diseases. Scientific knowledge is often needed to guide and drive public policy in that it helps in the development of evidenced-based policies such as rational use of antimicrobial in human populations, horticulture, and animal husbandry. Scientific research plays an important role during outbreaks, especially in the identification of the etiological agent, development of diagnostic tools, case management modules, and preventive strategies [73, 74]. Knowledge ought to be produced through research and simplified, analyzed, and transferred to ameliorate practices for prevention and control of emerging infectious diseases. Most research is incremental and builds on current and existing knowledge in order to attain new discoveries. According to the WHO, for research to be useful in the combat of emerging infectious diseases, it must be focused on the understanding of the following thematic areas– environmental and host factors which facilitate emergence, and transmission of these diseases. Research must also take a critical look at the evolution of pathogens responsible for these infections and how these changes culminate in changes in their infectivity, virulence, and transmissibility. Other factors of research importance should include the development of new diagnostic tools which will aid in the rapid and accurate diagnosis of pathogens during and after outbreaks. Lastly, research should be intensified to identify the role of climatic variability in the emergence of new and extra virulent microbes [70, 74].

30.4.6 Information Sharing and Partnerships

The recent outbreak of SARS-CoV-2 has highlighted the need for international collaborations, as well as timely and accurate dissemination of information. This outbreak showed that, sharing information during an outbreak remains a cornerstone to effective control. Successful pandemic response depends on countries rapidly communicating accurate information, including pathogen identification, incidence, transmission patterns, and mortality to the international community [75–78]. When this is done, it affords other countries or health organizations the opportunity to implement measures that allow for immediate and effective control of the outbreak. Effective communication has become essential now more than ever in the era of "fake news" or misinformation [77].

One of the key components of the IHR, which is the overarching legal agreements in international health, stipulates the need for timely, precise, and coordinated information sharing during public health emergencies. It requires governments to keep WHO updated on all proceedings within the confines of their states that may be regarded as a public health emergency of international concern (PHEIC) [78]. Depending on the intelligence a country presents, the WHO can proclaim a PHEIC and issue guidelines to control travel and trade [78].

The devastating effects of delaying information sharing has been quite telling on the world in many instances, the most recent being the case of China during the COVID-19 outbreak. Certain countries like the United States and Australia reprimanded China for their non-transparency and delay in sharing information [79] and a deliberate plan to underestimate or report the number of COVID-19 cases recorded in the country [80].

30.5 Future Directions

The fight against emerging and re-emerging infectious diseases is one that involves or requires continues planning, innovation, and collaborations. It requires considerable financial commitment from all governments of the world, including the private sector collaborations. These areas must be considered as we forge ahead in our bid to overcome or better respond to the ever-growing threat of emerging infectious diseases–

- Research, surveillance, and control of antimicrobial resistance.
- Research into the role of zoonoses in the emergence of new pathogens and the evolution of the same utilizing the One-health approach. To meet future demands, new research programs must assume a holistic and interdisciplinary view of emerging and re-emerging infectious diseases, so as to break barriers between human health funding and other research funding.
- International efforts at minimizing the disparities/gaps in health-research infrastructure between the developed and third world countries. This must include empowering and encouraging investment in research and digital innovation in health especially in Africa and Southeast Asia.
- Development of international guidelines on data sharing (epidemiological and microbiological) in addition to other materials such as access to sample, and other frameworks for quick initiation of phase 3 clinical trials during outbreaks of the magnitude of a pandemic. This is relevant especially for vaccines and novel drugs/medicines as was demonstrated during the COVID-19 pandemic [81].

30.6 Conclusion

In the last two decades, the world has witnessed outbreaks of at least three different diseases, each of which could be classified either as an epidemic or pandemic, the latest being COVID-19. These emerging infectious diseases have caused considerable socio-economic hardships to many countries, as well as mortality estimated in millions. The nature of these outbreaks and the characteristics of the causative pathogens only reinforce the notion that emerging and re-emerging infectious diseases (caused by newer microorganisms) will continue to plague man. Available data over the years have shown that these emerging and re-emerging infectious diseases can cause significant morbidity and mortality, as well as affect other facets of society including the economy. It is important to note that the COVID-19 pandemic exposed many lapses in global healthcare systems and also highlighted disparities between healthcare infrastructure and provision systems between high-income and low-middle-income countries. The lessons learnt provide many opportunities for research collaborations, strengthening of global health systems, as the need to bridge several gaps identified in health systems worldwide. Going forward, a collective approach by the international community to the fight these emerging and re-emerging infectious diseases must be the target. While it may be uncertain which new infectious diseases will emerge or re-emerge in the future, we must all be prepared and be a step ahead of these microorganisms. The aforementioned can be achieved by creating and sustaining robust and vigilant public health systems. Environmental and socio-economic disparities which increase the world's vulnerability to these infectious diseases must be tackled effectively through international commitment, strengthened public health infrastructure, skilled and competent human resources development, intersectoral collaboration, and intercountry cooperation.

References

1 National Institute for Allergy and Infectious Diseases. (2018). NIAID Emerging Infectious Diseases/Pathogens.
2 Dikid, T., Jain, S.K., Sharma, A. et al. (2013). Emerging & re-emerging infections in India: an overview. *Indian J. Med. Res.* 138: 19–31.
3 Tabish, S.A. (2009). Recent trends in emerging infectious diseases. *Int. J. Health Sci.* 3: V–VIII.
4 World Health Organization (2004). *The World Health Report.* Genève: World Health Organization.
5 Guerrant, R.L. and Blackwood, B.L. (1999). Threats to global health and survival: the growing crises of tropical infectious diseases—an "unfinished" agenda. *Clin. Infect. Dis.* 28: 966–986.
6 Butler, J.C. et al. (2001). Emerging infectious diseases among indigenous peoples. *Emerg. Infect. Dis.* 7: 554–555.
7 Zumla, A. and Hui, D.S.C. (2019). Emerging and reemerging infectious diseases: global overview. *Infect. Dis. Clin. N. Am.* 33: xiii–xix.
8 Lin, C.Y., Su, S.B., and Chen, K.T. (2022). An overview of gastrointestinal diseases in patients with COVID-19—epidemiology and pathophysiology: a narrative review. *Medicine* 101: e30297.
9 World Health Organization. (2022). COVID-19 Weekly Epidemiological Update. http://www.WHO.int/emergencies/disease/novel-cononavirus-2019
10 Muraille, E. (2021). Promoting health in a globalized world requires adopting a *One Health* perspective. *Glob. Health Promot.* 28: 3–5.
11 World Health Organization (2014). *A Brief Guide to Emerging Infectious Diseases and Zoonosis*. Geneva, Switzerland: World Health Organization.
12 Wolfe, N.D., Dunavan, C.P., and Diamond, J. (2007). Origins of major human infectious diseases. *Nature* 447: 279–283.
13 Jones, K.E., Patel, N.G., Levy, M.A. et al. (2008). Global trends in emerging infectious diseases. *Nature* 451: 990–993.
14 Stephens, P.R., Gottdenker, N., Schatz, A.M. et al. (2021). Characteristics of the 100 largest modern zoonotic disease outbreaks. *Philos. Trans. R. Soc. B* 376: 20200535.
15 Rahman, M.T., Sobur, M.A., Islam, M.S. et al. (2020). Zoonotic diseases: etiology, impact, and control. *Microorganisms* 8: 1405.
16 McArthur, D.B. (2019). Emerging infectious diseases. *Nurs. Clin. N. Am.* 54: 297–311.
17 Mazonakis, N., Markaki, I., and Spernovasilis, N. (2021). How human behavior will lead to the next pandemic. *Infect. Control Hosp. Epidemiol.* 43: 1–2.
18 Spernovasilis, N., Markaki, I., Papadakis, M. et al. (2021). Epidemics and pandemics: is human overpopulation the elephant in the room? *Ethics Med. Public Health* 19: 100728.

19 Poulakou, G. and Plachouras, D. (2016). Planet's population on the move, infections on the rise. *Intensive Care Med.* 42: 2055–2058.

20 Baylor College of Medicine (2023). *Emerging Infectious Diseases.* Baylor College of Medicine Retrieved January 24, 2023. https://www.bcm.edu/departments/molecular-virology-and-microbiology/emerging-infections-and-biodefense/emerging-infectious-diseases.

21 de Kretser, L. (2023). *Davos 2023: Climate Change Leads to More Malaria, Tuberculosis up in a Recession*. Medscape, http://www.medscape.com/viewarticle/986997.

22 Centers for Disease Control and Prevention (CDC) (2014). Notes from the field: New Delhi metallo-beta-lactamase-producing *Escherichia coli* associated with endoscopic retrograde cholangiopancreatography – Illinois, 2013. *MMWR Morb. Mortal. Wkly Rep.* 62: 1051.

23 Centers for Disease Control and Prevention (CDC) (2013). Notes from the field: hospital outbreak of carbapenem-resistant Klebsiella pneumoniae producing New Delhi metallobeta-lactamase—Denver, Colorado, 2012. *MMWR Morb. Mortal. Wkly Rep.* 62: 108.

24 Centers for Disease Control and Prevention (CDC) (2011). Carbapenem-resistant *Klebsiella pneumoniae* associated with a long-term care facility—West Virginia, 2009–2011. *MMWR Morb. Mortal. Wkly Rep.* 60: 1418–1420.

25 Doi, Y. and Paterson, D.L. (2015). Carbapenemase-producing Enterobacteriaceae. *Semin. Respir. Crit. Care Med.* 36: 74–84.

26 McGann, P., Hang, J., Clifford, R.J. et al. (2012). Complete sequence of a novel 178-kilobase plasmid carrying bla(NDM-1) in a *Providencia stuartii* strain isolated in Afghanistan. *Antimicrob. Agents Chemother.* 56: 1673–1679.

27 Ochman, H. and Moran, N.A. (2001). Genes lost and genes found: evolution of bacterial pathogenesis and symbiosis. *Science* 292: 1096–1099.

28 Vidaver, A.K. (2002). Uses of antimicrobials in plant agriculture. *Clin. Infect. Dis.* 34: S107–S110.

29 Aarestrup, F. (2012). Sustainable farming: get pigs off antibiotics. *Nature* 486: 465–466.

30 Addis, Z., Kebede, N., Worku, Z. et al. (2011). Prevalence and antimicrobial resistance of *Salmonella* isolated from lactating cows and in contact humans in dairy farms of Addis Ababa: a cross sectional study. *BMC Infect. Dis.* 11: 222.

31 Doyle, M.E. (2015). Multidrug-resistant pathogens in the food supply. *Foodborne Pathog. Dis.* 12: 261–279.

32 Wall, B., Mateus, A., Marshall, L. et al. (2016). *Drivers, Dynamics and Epidemiology of Antimicrobial Resistance in Animal Production*, vol. 68, 109441. Rome: Food and Agriculture Organization.

33 Laxminarayan, R., Duse, A., Wattal, C. et al. (2013). Antibiotic resistance-the need for global solutions. *Lancet Infect. Dis.* 13: 1057–1098.

34 World Health Organization. (2019). Assessing Non-prescription and Inappropriate Use of Antibiotics: Report on Survey.

35 Okeke, I.N., Laxminarayan, R., Bhutta, Z.A. et al. (2005). Antimicrobial resistance in developing countries. Part I: recent trends and current status. *Lancet Infect. Dis.* 5: 481–493.

36 Fridkin, S., Baggs, J., Fagan, R. et al. (2014). Vital signs: improving antibiotic use among hospitalized patients. *MMWR Morb. Mortal. Wkly Rep.* 63: 194–200.

37 Marston, H.D., Dixon, D.M., Knisely, J.M. et al. (2016). Antimicrobial resistance. *JAMA* 316: 1193–1204.

38 Sabin, N.S., Calliope, A.S., Simpson, S.V. et al. (2020). Implications of human activities for (re)emerging infectious diseases, including COVID-19. *J. Physiol. Anthropol.* 39: 29.

39 World Health Organisation African Region. (2022). COVID-19 Vaccination Roll Out Stagnates in Africa.

40 Cloeckaert, A. and Kuchler, K. (2020). Grand challenges in infectious diseases: are we prepared for worst-case scenarios? *Front. Microbiol.* 11: 613383.

41 Netea, M.G., Dominguez-Andres, J., Barreiro, L.B. et al. (2020). Defining trained immunity and its role in health and disease. *Nat. Rev. Immunol.* 20: 375–438.

42 Woolhouse, M.E., Webster, J.P., Domingo, E. et al. (2002). Biological and biomedical implications of the co-evolution of pathogens and their hosts. *Nat. Genet.* 32: 569–577.

43 Rocklov, J. and Dubrow, R. (2020). Climate change: an enduring challenge for vector-borne disease prevention and control. *Nat. Immunol.* 21: 479–483.

44 Lockhart, S.R., Etienne, K.A., Vallabhaneni, S. et al. (2017). Simultaneous emergence of multidrug-resistant *Candida auris* on 3 continents confirmed by whole-genome sequencing and epidemiological analyses. *Clin. Infect. Dis.* 64: 134–140.

45 Casadevall, A., Kontoyiannis, D.P., and Robert, V. (2019). On the emergence of *Candida auris*: climate change, azoles, swamps, and birds. *MBio* 10: 01397.

46 Fiers, W.D., Gao, I.H., and Iliev, I.D. (2019). Gut mycobiota under scrutiny: fungal symbionts or environmental transients? *Curr. Opin. Microbiol.* 50: 79–86.

47 Plichta, D.R., Graham, D.B., Subramanian, S., and Xavier, R.J. (2019). Therapeutic opportunities in inflammatory bowel disease: mechanistic dissection of host- microbiome relationships. *Cell* 178: 1041–1056.

48 Walter, J., Armet, A.M., Finlay, B.B., and Shanahan, F. (2020). Establishing or exaggerating causality for the gut microbiome: lessons from human microbiota-associated rodents. *Cell* 180: 221–232.

49 The Lancet Infectious Diseases (2019). Malaria vaccination: a major milestone. *Lancet Infect. Dis.* 19: 559.

50 Singh, T.U., Parida, S., Lingaraju, M.C. et al. (2020). Drug repurposing approach to fight COVID-19. *Pharmacol. Rep.* 72: 1–30.

51 Weber, E.W., Maus, M.V., and Mackall, C.L. (2020). The emerging landscape of immune cell therapies. *Cell* 181: 46–62.

52 Le Magnen, C., Dutta, A., and Abate-Shen, C. (2016). Optimizing mouse models for precision cancer prevention. *Nat. Rev. Cancer* 16: 187–196.

53 Cecconi, M., Evans, L., Levy, M., and Rhodes, A. (2018). Sepsis and septic shock. *Lancet* 392: 75–87.

54 De Oliveira, D.M.P., Forde, B.M., Kidd, T.J. et al. (2020). Antimicrobial resistance in ESKAPE pathogens. *Clin. Microbiol. Rev.* 33: 00181.

55 Willis, J.C. and Lord, G.M. (2015). Immune biomarkers: the promises and pitfalls of personalized medicine. *Nat. Rev. Immunol.* 15: 323–329.

56 Singer, M. (2018). Personalizing sepsis care. *Crit. Care Clin.* 34: 153–160.

57 Chang, S.L., Piraveenan, M., Pattison, P., and Prokopenko, M. (2020). Game theoretic modelling of infectious disease dynamics and intervention methods: a review. *J. Biol. Dyn.* 14: 57–89.

58 Medzhitov, R., Schneider, D.S., and Soares, M.P. (2012). Disease tolerance as a defense strategy. *Science* 335: 936–941.

59 Löscher, T. and Prüfer-Krämer, L. (2009). Emerging and re-emerging infectious diseases. *Mod. Infect. Dis. Epidemiol.* 28: 39–67.

60 Integrated Disease Surveillance Project. (2012). http://www.idsp.nic.in. Accessed 26 April 2012.

61 One Health Commission [OHC] (2020). One Health. https://www.onehealthcommission.org

62 Okello, A.L., Gibbs, E.P., Vandersmissen, A., and Welburn, S.C. (2011). One Health and the neglected zoonoses: turning rhetoric into reality. *Vet. Rec.* 169: 281–285.

63 Rabozzi, G., Bonizzi, L., Crespi, E. et al. (2012). Emerging zoonoses: the "one health approach". *Saf. Health Work* 3: 77–83.

64 Bidaisee, S. and Macpherson, C.N. (2014). Zoonoses and one health: a review of the literature. *J. Parasitol. Res.* 2014: 874345.

65 World Health Organization (WHO) (2005). *International Health Regulations 2005*. Geneva; Switzerland: World Health Organization.

66 Dhariwal, A.K., Jain S.K. (2010). Field epidemiology training programme in India: ten years of experience. In: *Proceedings of South East Asia Regional Conference on Epidemiology*, 285–288. New Delhi: World Health Organizations, Regional Office for South-East Asia.

67 Sharma, K. and Zodpey, S. (2011). Public health education in India: need and demand paradox. *Indian J. Community Med.* 36: 178–181.

68 Murhekar M.V., Manickam P., Bhatnagar T., Kaur P., Ramachandran V., Ramakrishnan R., et al. (2010). Ten years of field epidemiology training at the National Institute of Epidemiology, Chennai, India, 2001–2010. In: *Proceedings of South East Asia Regional Conference on Epidemiology* 289–292. New Delhi: World Health Organizations, Regional Office for South-East Asia.

69 Government of India, Ministry of Health and Family Welfare (MOHFW) (2012). *Compendium of Operational Guidelines for Epidemic Intelligence Service (EIS) like Training Programme in India*. New Delhi, India: MOHFW.

70 World Health Organization (2005). *Combating Emerging Infectious Diseases in the South-East Asia Region (No. SEA-CD-139)*. WHO Regional Office for South-East Asia.

71 Nsubuga, P., White, M.E., Thacker, S.B. et al. (2006). Public health surveillance: a tool for targeting and monitoring interventions. In: *Disease Control Priorities in Developing Countries*, 2ee (ed. D.T. Jamison, J.G. Breman, and A.R. Measham), 997–1015. Washington (DC): The International Bank for Reconstruction and Development / The World Bank. Available from: https://www.ncbi.nlm.nih.gov/books/NBK11770/.

72 Respiratory Virus Infections Working Group (2020). Nosocomial infection surveillance: Canadian public health laboratory network best practices for COVID-19. *Can. Commun. Dis. Rep.* 46 (5): 112.

73 Balajee, S.A., Arthur, R., and Mounts, A.W. (2016). Global health security: building capacities for early event detection, epidemiologic workforce, and laboratory response. *Health Secur.* 14 (6): 424–432.

74 Meng, Z., Guo, S., Zhou, Y. et al. (2021). Applications of laboratory findings in the prevention, diagnosis, treatment, and monitoring of COVID-19. *Signal Transduct. Target. Ther.* 6 (1): 316.

75 Kilpatrick, A.M. and Randolph, S.E. (2012). Drivers, dynamics, and control of emerging vector-borne zoonotic diseases. *Lancet* 380 (9857): 1946–1955.

76 Lencucha, R. and Bandara, S. (2021). Trust, risk, and the challenge of information sharing during a health emergency. *Glob. Health* 17 (1): 1–7.

77 United Nations (2020). *UN Tackles 'infodemic' of Misinformation and Cybercrime in COVID-19 Crisis*. United Nations.

78 Dute, J. (2005). World Health Organization – revision of the International Health Regulations. *Eur. J. Health Law* 12 (3): 269–271.

79 Gostin, L.O., DeBartolo, M.C., and Katz, R. (2017). The global health law trilogy: towards a safer, healthier, and fairer world. *Lancet* 390 (10105): 1918–1926.

80 Hancock, T. and Xueqiao, W. (2020). *How China's Slow Response Aided Coronavirus Outbreak*. Financial Times.

81 Hill-Cawthorne, G.A. and Sorrell, T.C. (2016). Future directions for public health research in emerging infectious diseases. *Public Health Res. Pract.* 26: e2651655.

Index

Note: Page numbers followed by *f* refer to figures and *t* refer to tables.

Rising Contagious Diseases: Basics, Management, and Treatments, First Edition. Edited by Seth Kwabena Amponsah, Ranjita Shegokar, and Yashwant V. Pathak.

b

d

e

o

p

t